CLIMATE CHANGE AND CULTURAL DYNAMICS

A Global Perspective on Mid-Holocene Transitions

The front cover image shows Mounds G, H, and I at Caral, a Mid-Holocene ceremonial center in Peru. (Photograph by Daniel H. Sandweiss.)

CLIMATE CHANGE AND CULTURAL DYNAMICS

A Global Perspective on Mid-Holocene Transitions

Edited by

David G. Anderson
Department of Anthropology, University of Tennessee
Knoxville, TN 37996, USA

Kirk A. Maasch
Climate Change Institute and Department of Earth Sciences, University of Maine
Orono, ME 04469, USA

Daniel H. Sandweiss
Department of Anthropology, Climate Change Institute, and Graduate School, University of Maine, Orono, ME 04469, USA

ELSEVIER

Amsterdam – Boston – Heidelberg – London – New York – Oxford
Paris – San Diego – San Francisco – Singapore – Sydney – Tokyo
Academic Press is an imprint of Elsevier

ACADEMIC PRESS

Academic Press is an imprint of Elsevier
84 Theobald's Road, London WCIX 8RR, UK
Radarweg 29, PO Box 211, 1000 AE Amsterdam, The Netherlands
Linacre House, Jordan Hill, Oxford OX2 8DP, UK
30 Corporate Drive, Suite 400, Burlington, MA 01803, USA
525 B Street, Suite 1900, San Diego, CA 92101-4495, USA

First edition 2007

ISBN: 978-0-12-088390-5

Printed and bound in USA

07 08 09 10 11 10 9 8 7 6 5 4 3 2 1

DEDICATION

Dedicated to the memory of Thor Heyerdahl (1914–2002), and all those exploring relationships between climate and culture, perhaps the greatest challenge facing our species in the twenty-first century

Contents

List of Contributors

Heather Almquist Department of Geography, University of Montana, Missoula, MT 59812, USA

Atholl Anderson Research School of Pacific and Asian Studies, Australian National University, ACT 0200, Australia

David G. Anderson Department of Anthropology, University of Tennessee, Knoxville, TN 37996-0720, USA

C. Fred T. Andrus Department of Geological Sciences, University of Alabama, USA

Kevin G. Cannariato Department of Earth Sciences, University of Southern California, Los Angeles, CA 90089, USA

Brendan J. Culleton Department of Anthropology, University of Oregon, Eugene, OR 97403-1218, USA

Ann Dieffenbacher-Krall Climate Change Institute, University of Maine, Orono, ME 04469, USA

Jon M. Erlandson Department of Anthropology, University of Oregon, Eugene, OR 97403-1218 USA

Michael Gagan Research School of Earth Sciences, Australian National University, ACT 0200, Australia

Martin Grosjean NCCR Climate and Institute of Geography, University of Bern, 9a Erlachstrasse, 3012 Bern, Switzerland

Wibjörn Karlén Geografiska Annaler Ser. A, Institutionen för Kulturgeografi, Box 513, 751 20 Uppsala, Sweden

Douglas J. Kennett Department of Anthropology, University of Oregon, Eugene, OR 97403-1218, USA

James P. Kennett Department of Earth Science and Marine Science Institute, University of California, Santa Barbara, CA 93106, USA

Lars Larsson University of Lund, Department of Archaeology and Ancient History, Sandgatan 1, Lund, S-223 50, Sweden

Tracey L.-D. Lu Department of Anthropology, The Chinese University of Hong Kong, Shatin, N.T., Hong Kong

Konstantin A. Lutaenko Institute of Marine Biology, Far East Branch of the Russian Academy of Sciences, Vladivostok 690041, Russia

Kirk A. Maasch Department of Earth Sciences, Climate Change Institute, University of Maine, Orono, ME 04469, USA

Paul A. Mayewski Department of Earth Sciences, Climate Change Institute, University of Maine, Orono, ME 04469, USA

Betty J. Meggers Smithsonian Institution, PO Box 37012, NMNH MRC-112, Washington, DC 20013, USA

Sarah E. Metcalfe School of Geography, University of Nottingham, Nottingham, NG7 2RD, UK

Yuri A. Mikishin Coastal Research Center, Far East State University, Vladivostok 690000, Russia

Madonna L. Moss Department of Anthropology, University of Oregon, Eugene, OR 97402-1218, USA

Lautaro Núñez Instituto de Investigaciones Arqueológicas y Museo, Universidad Católica del Norte, San Pedro de Atacama, Chile

Dorothy M. Peteet NASA/Goddard Institute for Space Studies, 2880 Broadway, New York, NY 10025, USA

Alexander N. Popov Museum of Archaeology and Ethnography, Far East State University, Vladivostok 690000, Russia

Elizabeth J. Reitz Department of Anthropology, University of Georgia, Georgia, USA

James B. Richardson III Department of Anthropology, University of Pittsburgh, Pittsburgh, PA, USA

Melanie Riedinger-Whitmore Department of Environmental Science, Policy, and Geography, University of South Florida, FL, USA

Harold B. Rollins Department of Geology and Planetary Science, University of Pittsburgh, Pittsburgh, PA, USA

Michael Russo National Park Service, Southeastern Archeological Center, Tallahassee, FL 32310, USA

Daniel H. Sandweiss Department of Anthropology and Graduate School, Climate Change Institute, University of Maine, Orono, ME 04469, USA

David Sanger Department of Anthropology, University of Maine, Orono, ME 04469, USA

Calogero M. Santoro Instituto Alta Investigación, Departamento de Arqueología y Museología & Centro de Investigaciones del Hombre en el Desierto, Universidad de Tarapacá, Casilla 6-D, Arica, Chile

Kenneth E. Sassaman Department of Anthropology, University of Florida, 1112 Turlington Hall, P.O. Box 117305, Gainesville, FL 32611-7305, USA

Romuald Schild Instytut Archeologii I Ethnologii, Polskiej Akademii Nauk, Al. Solidararnosci 105, 00–140 Warszawa, Poland

James Shulmeister Department of Geological Sciences, University of Canterbury, Private Bag 4800, Christchurch, New Zealand

Vivien G. Standen Departamento de Antropología & Centro de Investigaciones del Hombre en el Desierto, Universidad de Tarapacá, Casilla 6-D, Arica, Chile

Lonnie G. Thompson Byrd Polar Research Center, Ohio State University, 1090 Carmack Road, Columbus, OH 43210, USA

Barbara Voorhies Department of Anthropology, University of California, Santa Barbara, CA 93106, USA

Fred Wendorf Department of Anthropology, Southern Methodist University, Dallas, TX 75275, USA

Cathy Whitlock Department of Earth Sciences, Montana State University, Bozeman, MT 59717, USA

Irina S. Zhushchikhovskaya Institute of History, Archaeology and Ethnography of the Peoples of the Far East, Far East Branch of the Russian Academy of Sciences, Vladivostok 690001, Russia

List of Figures

List of Tables

Preface and Acknowledgments

This volume explores climate and culture change during the middle part of the current interglacial period, the Mid-Holocene era from roughly 9000 to 5000 years ago. The original impetus for what follows was a conference held in October of 1998 at the University of Maine on "Climate and Culture at 3000 B.C." organized by Daniel H. Sandweiss and Kirk A. Maasch, and sponsored by the Foundation for Exploration and Research on Cultural Origins (FERCO). Those who spoke at the conference included Atholl Anderson, David Anderson, Daniel Belknap, Andrew Bush, Heidi Cullen, Michael Gagan, Martin Grosjean, George Jacobson, Wibjorn Karlén, Douglas Kennett, Lars Larsson, Tracey Lu, Konstantin Lutaenko, Madonna Moss, Melanie Riedinger James Richardson, Harold Rollins, David Sanger, James Shulmeister, Calogero Santoro, Lonnie Thompson, Barbara Voorhies, Harvey Weiss, Fred Wendorf, and Irina Zhushchikhovskaya. Betty Meggers was originally scheduled to speak at the conference but was unable to attend. Many subsequently contributed to this volume.

One of FERCO's founders was the anthropologist Thor Heyerdahl, perhaps best known for his research on early settlement in Oceania, recounted in books like *Kon-Tiki* and *Aku-Aku*, and whose welcoming remarks to the attendees are presented in Foreword. As Thor Heyerdahl also noted at the start of the conference, however, it was the "unselfish and profound interest of FERCO economic sponsor and co-founder Fred Olsen," a Norwegian businessman, that made FERCO possible. The editors wish to thank both of these gentlemen for inspiring, and helping finance the production of this volume.

In his later years, as the Foreword clearly demonstrates, Heyerdahl was fascinated by the processes leading to the emergence of complex societies around the world, and to the role that climate might have played in these developments. The Mid-Holocene was a time of particular interest, since it was then when the foundations of complex society or civilization were laid down in many parts of the world. Heyerdahl believed that the climate of the time, the warmest period during the Holocene, facilitated these cultural developments, and additionally offered parallels and lessons for the modern world.

Although the conference participants discussed the possibility of putting together an edited volume of papers, it was not until several years later that the idea was revisited, and the papers were finalized, most in 2006 and early 2007. This delay of a decade was fortunate, since interest in climate change has grown in recent years, and the scientific literature on the subject has multiplied dramatically. Although this volume has thus had a long gestation period, its appearance at this time is opportune, when concern with climate change is drawing even more attention than it did a decade ago.

This volume includes chapters authored by both paleoclimatologists and archaeologists, in order to confront climatic and cultural records directly. Most studies of climate and culture are carried out either by archaeologists using the paleoclimate literature or paleoclimatologists accessing the archaeological literature. Only rarely do the two sides work closely together on the final publication, and consequently misinterpretation of the other record can occur. The participants in this volume agreed that their chapters should be co-authored where possible by both kinds of scientists, to make certain that both kinds of data are properly reported and used. Most of the chapters in this volume are the result of such cooperation, often among scholars who had not previously worked together.

Many people deserve our thanks in the production of this volume. Jennifer Helé, the acquisitions editor at Academic/Elsevier, offered advice, encouragement, and suggestions throughout the preparation of the manuscript. The actual production of the manuscript was accomplished by Mrs. Linda Versteeg, Elsevier physical science books development editor, and her staff. Betsy Lightfoot, Elsevier Production Editor, Amsterdam, and Dr. M.S. Rajkumar and the production team at Macmillan India Limited, Bangalore, in particular deserve our thanks for their help with the final production of the volume. The final version of most of the artwork appearing in the volume was prepared by Kirk A. Maasch, whose skills extend well beyond those of author and editor, something the other two editors have deeply appreciated and wish to specifically acknowledge.

At each of our institutions people have provided help. At the University of Tennessee Elizabeth Martin helped with the copyediting and reference checking for each chapter. Andrew Kramer, chair of the Department of Anthropology, provided direct support as well as continual encouragement as the manuscript came together. Scott Meeks, Shane Miller, Jason O'Donoughue, and Jan Simek also provided advice and assistance in the production effort.

The editors of this volume would like to especially thank Craig Pacelli, associate editor for Education, Arts & Humanities Journals, and John A. Brown, senior sales administrator, permissions, of the Taylor & Francis Group, LLC, for permission to reprint the text of Chapter 7, "Early State Formation in Southern Mesopotamia: Sea Levels, Shorelines, and Climate Change" by Douglas J. Kennett and James P. Kennett. This chapter appeared in somewhat different form in the *Journal of Island & Coastal Archaeology* and is reproduced here thanks to the permission of the authors and the press. We also wish to thank Scott M. Fitzpatrick, the journal editor, for his help in facilitating the use of this chapter.

All three editors would like to thank their spouses for their support and patience during the long hours we spent producing this volume; even while often with them at home our minds were sometimes far in the past, in the Mid-Holocene!

Foreword

Thor Heyerdahl

Editors' note: The following piece is adapted from welcoming remarks at the FERCO Conference on Climate and Culture 3000 B.C. at the University Of Maine, 8 October 1998. The meeting led to the present volume. Thor Heyerdahl died on 18 April 2002. In fall of that year, the patrons dissolved FERCO, the organization led by Heyerdahl that sponsored the 1998 conference. However, Heyerdahl's scientific legacy continues in this volume and in the work of those who collaborated with or were inspired by him.

The choice of a topic combining culture with climate is timely indeed, in an era when the excessive strength of the El Niño Current off Peru has created severe climatic and cultural disturbances in widely scattered areas all over the world. It is a fundamental belief of the founders of the Foundation for Exploration and Research on Cultural Origins (FERCO) that climate and culture are intimately linked together, and that changes in currents and coastlines have influenced cultures and caused both involuntary and organized migrations. Man is an integral part of nature, and a reconstruction of the human past by anthropologists must be built on consultation with all the environmentally oriented sciences. As we enter a new millennium, there is an increased awareness of the need for more cross-disciplinary collaboration. We are leaving behind us a century when the enormous increase of knowledge accumulated within each discipline has forced specialization on us to the extent that the universities suffer from isolation between the different faculties.

It is our belief that one way to overcome such shortcomings may be cross-disciplinary conferences such as that now organized by FERCO with a focus on climate and culture around 3000 BC. Teamwork is bound to tighten the excessive gap between the different faculties in the future, facilitated through modern technology such as computer systems and the internet.

The need for cross-disciplinary consultation was felt early by those of us who first began to trace feasible migration routes of biological species to oceanic islands, which had originally arisen sterile from the bottom of the sea through volcanic eruption or subsequent coral growth. The combined evidence from biology and anthropology was inseparably linked together. Certain plant and animal species depended on human agency for ocean voyages, and vice-versa, so their presence on oceanic islands in pre-European times provided genetic evidence of ocean voyages in aboriginal watercraft.

Although *Homo sapiens* was studied physically together with the other animal species when the present speaker enlisted at the Faculty of Zoology at the

University of Oslo in the early 1930s, there was no other link between biologists and anthropologists. Geography happened to be added to the courses, since I was trained for subsequent work in Polynesia on animal migration to the Marquesas Islands. But only coincidence enabled me to get a full insight into contemporary knowledge and theories concerning the origins of the Polynesian people. Parallel to the University training in biology and geography, I had private access to the Kroepelien Polynesian library in Oslo at a time when Polynesian anthropology was still in the making and no more had been published by Polynesian scholars than one person could digest.

A lack of co-ordination and synthesis in Pacific migration studies was already apparent early in the twentieth century, and in my review in 1952 of more than 30 scholars who had pronounced their opinions on origins of the Polynesian people, no two had reached the same conclusion (Heyerdahl, 1952). Specialization had already forced investigators to look independently for the answer within their own field of research, ignorant of facts forthcoming within other disciplines. Such shortcomings are not permissible in our days of technological facilities, and it is our hope that organizations such as FERCO can help to bridge the gaps between the disciplines.

The peopling of the Polynesian islands took place far too late to be discussed at a conference focusing on the Mid-Holocene period. Archaeological and linguistic evidence combine to show that Polynesia remained unsettled by man until very recent times, whereas migrants from Asia reached all surrounding territories, America to the east and Melanesia to the west, probably as early as 20000 BC. A seaward migration with the Japan Current across the extreme North Pacific by-passing both Siberia and Melanesia could also have brought the Polynesian ancestry from Indonesia to Northwest America in the Mid-Holocene, and on from there to Hawaii and the rest of Polynesia still as a purely Neolithic people as late as the early part of the present millennium [second millennium AD].

Those of us who have experimented with aboriginal watercraft in the three main world oceans realize that we can only map the coastlines and the ocean bottom, for the ocean sometimes travels faster than a primitive vessel. The main currents are permanent invisible salt water rivers. We never speak of a straight voyage across the Pacific along the equatorial line, because we know that the Pacific Ocean covers half the surface of the planet, so the straight line would go through the center of the Earth. It is exactly the same distance from Southeast Asia to South America by way of the Equator as by way of the North Pole. Irrespective of the changes in coastlines and ice caps during the last 20,000 years, the mileage from Southeast Asia to America has always been the same in a curve along the Arctic coast as in the seemingly straight line in tropical latitudes.

Since human migrations and cultural changes did take place in the Mid-Holocene, any information on climatic conditions is of importance from a navigational point of view. When our experiment with the reed ship Tigris in 1977–1978 proved the feasibility of direct maritime contact between Mesopotamia, the Indus Valley, and the Red Sea, it became less probable that the almost simultaneous appearance of fully developed civilization in these three areas was purely coincidental. It has been customary to estimate the rather sudden bloom of the Sumerian, Egyptian, and Harappan civilizations to somewhere around 3100 BC and, as is well known, Sir Leonard Woolley dated a 3-meter thick layer of homogeneous mud deposit to this period which separated the foundation of the Sumerian city of Ur from earlier habitation in the same area.

Modern dating of archaeological events is approximate, and we have a tendency to dismiss written records from earlier cultures as based on mythology. Yet we forget that our modern calendar system was based on astronomical observations and mathematical calculations by people with a cultural level that impresses us. The Olmecs and Maya who helped to lay the foundation for Mesoamerican high culture practiced script centuries before many European nations. And whereas Christians, Moslems, and Buddhists began their time-reckoning with a zero year linked to the birth, enlightenment, or death of the founders of their respective faiths, the Maya began their zero year with an exact date of 4 ahan 2 cumhu which according to our calendar becomes 12 August 3113 BC.

What happened then? The Mayan astronomers had calculated one year to be 365.2329 days, which is 8.64 seconds closer to the truth than our modern calendar today. The Hindu calculated their Kali-era to begin at midnight after Ujjain, which ended on 17 February 3102 BC. That is only 11 years after the Maya began to reckon time. Another Hindu calendar system gives the beginning of the Brhaspatricaca period as 3116 BC, which is only three years before the Maya calendar.

Did the pre-European creators of these antipodal calendar systems by coincidence hit upon such remarkably similar zero years as 3102 and 3116 BC in India and 3113 BC in Mexico, respectively? Or did either flood waves or civilized human voyagers cross the ocean before the Medieval Europeans? Only by sharing knowledge and combining analytic and synthetic research on the Mid-Holocene period we may one day have the answer to this remarkable concurrence of three ancient calendar systems beginning within the same decade in the same century BC, at the end of the Mid-Holocene.

Reference

Heyerdahl, T., 1952. *American Indians in the Pacific.* George Allen & Unwin, Ltd., London.

Climate Change and Cultural Dynamics: A Global Perspective on Mid-Holocene Transitions
David G. Anderson, Kirk A. Maasch and Daniel H. Sandweiss (Editors)

Chapter 1

Climate and culture change: exploring Holocene transitions

David G. Anderson[1], Kirk A. Maasch[2], Daniel H. Sandweiss[3] and Paul A. Mayewski[2]

[1]*Department of Anthropology, University of Tennessee, TN, USA*
[2]*Department of Earth Sciences, Climate Change Institute, University of Maine, ME, USA*
[3]*Department of Anthropology, Climate Change Institute, Graduate School, University of Maine, ME, USA*

1. Introduction

Understanding climate change and its likely impacts on human culture is one of the great scientific challenges of the 21st century; responding successfully to them will be a major test for global civilization. Given current projections, the remainder of the century is likely to mark a continuing period of global climate change. The papers in this volume explore how past human cultures have responded to changes in climate and consequent changes in vegetation and precipitation patterns. Although the focus of the volume is the Mid-Holocene interval from ca. 9000 to 5000 cal yr BP, climate change and its impact on human culture throughout the Holocene is examined by most of the contributors. We believe that the research documented in this volume offers many lessons of value to scholars, politicians/planners, and the general public.

Interest in climate change has grown in recent years. The scientific literature on the subject has multiplied correspondingly, with research funding directed to the subject soaring in many countries, and major papers documenting the results of this research appearing at a rate undreamed of even a decade ago. Climate change is now explored using a wide array of data types, through multi-proxy records such as ice, lake sediment and pollen cores, tree rings, rodent and other animal nests/middens with their associated plant macrofossil and microfossil/pollen remains, paleosol and goeomorphological/geoarchaeological evidence, and archaeological and paleobiological deposits (Table 1.1). Sediment layers and tree rings, for instance, can document dramatic short-term climatic events such as floods or storms, or the impact of these events on processes such as erosion and fire frequency. Archaeological research has itself grown increasingly sophisticated and multidisciplinary, employing many of the same specializations, and a growing number of archaeologists around the world are working to unravel the relationships between climate and past human culture. Paleobiological evidence such as pollen, phytolith,

Table 1.1. A Summary of data sources used to reconstruct past climates.

Proxy data source	Some of the variables measured	Possible climatic inferences	Typical sampling interval
Ice cores	Ice chemistry, dust, $\delta^{18}O$, δD, CO_2, CH_4, tephra	Atmospheric circulation, temperature, precipitation, atmospheric composition, volcanic activity	Seasonal–annual
Tree rings	Ring width, $\delta^{18}O$, δD, $\delta^{13}C$, $\Delta^{14}C$	Temperature, precipitation (drought), solar variability	Annual
Coral, mollusks	$\delta^{18}O$, Sr/Ca, growth rate	SST, precipitation–evaporation, sea level	Monthly
Pollen	Percent, influx	Temperature, precipitation	10–100 years
Insects	Chironomid, beetle assemblages	Temperature	10–100 years
Soils and sand	Clay content, $\delta^{13}C$, dunes	Humidity, wind, CO_2	Snapshots
Closed-basin	Lake level	Precipitation–evaporation	Snapshots
Lake sediments	$\delta^{18}O$, diatoms	Temperature, salinity	10–100 years
Ice sheets	Former extent, glacial rebound	Area, thickness, bedrock depression	Snapshots
Mountain glaciers	Former extent	Snowline, air temperature	Snapshots
Marine sediments	$\delta^{18}O$, $\delta^{13}C$, foraminiferal assemblages	Global ice mass, ocean circulation, SST	100–1000 years
Raised shorelines	Elevation	Sea level, bedrock depression	Snapshots
Laminated or varved sediments	Reflectance, magnetic properties	Precipitation, wind	Annual

and fire frequency records can be critical for identifying human presence and impact on the landscape; so, too, can geoarchaeological/geophysical analyses, which can tease out anthropogenic signatures as readily as soil formation and sedimentation rates. As the papers in this volume demonstrate, the reconstruction of both past climate change and past human cultural systems is best accomplished by using data from multiple sources, or proxy records, and by specialists from different disciplines working together. Multi-proxy records reveal a more complete picture than can be obtained from individual proxy measures, just as multi-disciplinary research teams with diverse yet complementary perspectives typically attain better insights and understandings than individual scholars working alone.

As scientific knowledge of the causes of climate change has grown, so, too, has public interest in the subject. "An Inconvenient Truth," a film about global warming produced by former United States Vice President Al Gore, won an Oscar for best documentary in February 2007 (see also Gore, 2006). Earlier that same month, the United Nation's Intergovernmental Panel on Climate Change (IPCC) issued its strongest statement to date on changes that are occurring in Earth's climate (IPCC, 2007). The IPCC report noted that Earth's atmosphere now contains more carbon dioxide and methane, greenhouse gases, than at any time over the past 650,000 years, and concluded that a "warming of the climate system is unequivocal" (IPCC, 2007, p. 5). Many people alive today will likely see dramatic increases in global temperature and sea level; decreases in snow cover, sea ice, and land ice sheets and mountain glaciers; increased thawing of permafrost; more and stronger tropical storms; and changes in precipitation regimes in many parts of the planet, including probable increases in rainfall in high latitudes and decreases in lower latitudes (IPCC, 2007, pp. 8, 16). These trends are likely to continue, or perhaps accelerate, and even if greenhouse gas levels are stabilized in the decades to come, changes produced by current levels may continue for centuries.

Present-day global temperature is warmer than any time since the Medieval Warm Period, a time of slightly warmer than average Holocene temperature that occurred from ca. AD 800 to 1200 (ca. 1200–800 cal yr BP) (e.g., Broecker, 2001). Current projections for global climate around ca. AD 2100, based on a doubling of atmospheric carbon dioxide, foresee average global surface warming of between 2 and 4.5°C, with the current best estimate for an increase of about 3° (IPCC, 2007, p. 12). To determine how these changes will impact climate and biota, comparisons are sometimes made with periods in the past when planetary temperatures were higher than they are at present, such as the Eemian, or last interglacial period ca. 125,000 years ago, when polar temperatures were approximately 3–5°C warmer than at present and sea level was as much as 4–6 m above the present stand (IPCC, 2007, p. 9). While the Eemian may be an analogy for current warming, our records from that time are limited, primarily because it lies in the fairly remote past. The Medieval Warm Period is more recent, and potentially instructive, but the changes that occurred had only a few hundred years to play out, and the climate was not too different from that of the present. To understand the consequences of sustained higher than average temperatures, we believe that changes in climate and culture that occurred during the Mid-Holocene warm period (ca. 5000–9000 cal yr BP) are the best case that we can explore in detail.

2. Holocene climate change

The growth and development of modern societies occurred within the Holocene era, from ca. 11,500 cal yr BP to the present, although we are coming to realize that some of the trends toward sedentary life and agricultural intensification date well back into the last glacial era in parts of the world (e.g., Roberts, 1998; Fagan, 2004; Scarre, 2005). The period of extreme cold conditions and maximum ice sheet extent associated with the last glacial occurred about 21,000–18,000 cal yr BP. Global warming and a gradual retreat of the ice sheets began after about 15,000 cal yr BP, and proceeded with cold reversals of varying intensity until about 11,600 cal yr BP, the end of the last extended cold reversal, the Younger Dryas, which began about 12,900 cal yr BP (e.g., Bond et al., 1997, 1999; Gulliksen et al., 1998; Hughen et al., 1998, 2000; Rahmstorf, 2002). The Holocene era, the subject of this book, is assumed by scientific convention to begin at 10,000 ^{14}C yr BP, or about 11,450 cal yr BP, soon after the end of the Younger Dryas, which in climate conditions is the real boundary (Harland et al., 1989; Gibbard, 2003, p. 202).

In this volume, both radiocarbon (^{14}C yr BP) and calendar (cal yr BP) ages are employed, usually the latter, unless specific radiocarbon determinations are being reported, in which case calibrated ages are also commonly presented. Due to fluctuations in radiocarbon production and uptake, radiocarbon ages are considerably different and typically but variably younger than actual calendar ages (Fig. 1.1).

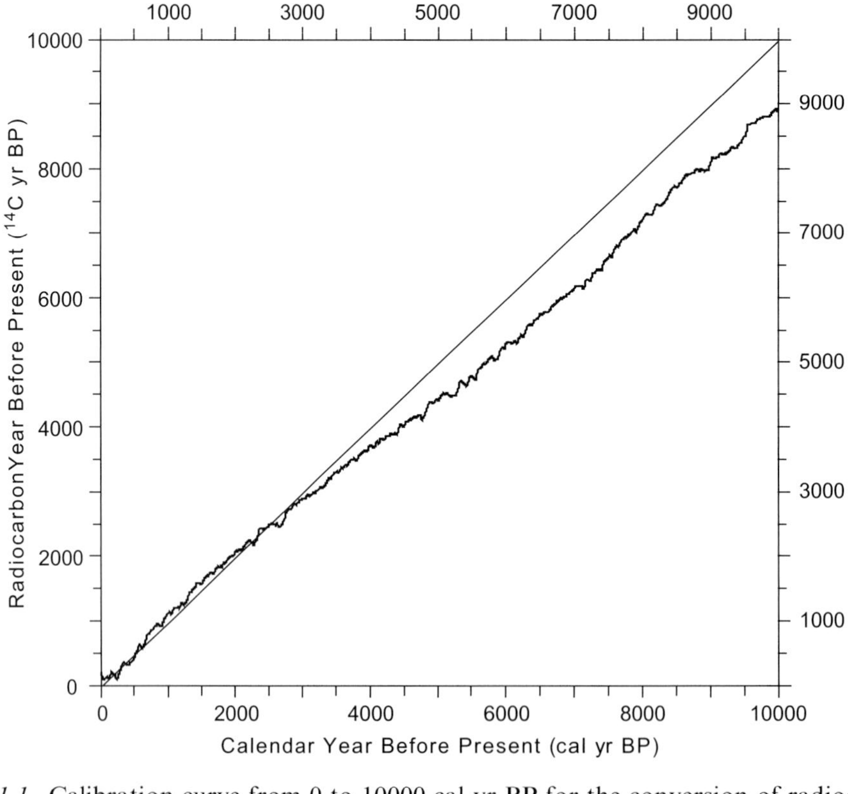

Figure 1.1. Calibration curve from 0 to 10000 cal yr BP for the conversion of radiocarbon ages to calibrated (cal) ages. The curve is based on dendrochronologically-dated tree-ring samples (after Reimer et al., 2004).

The development of the radiocarbon calibration curve has profound implications for archaeological and paleoenvironmental research (e.g., Taylor et al., 1996; Fiedel, 1999; Guilderson et al., 2005). While the offset between radiocarbon and calendar years is comparatively minor in recent millennia, it grows progressively more pronounced deeper in the past, reaching almost 1500 years at the Pleistocene/Holocene boundary 10,000 ^{14}C yr BP (11,450 cal yr BP). Calibrations have now been developed linking the radiocarbon and calendar timescales to the limits of the dating technique ca. 50,000 cal yr BP (Kitigawa and van der Plicht, 1998; Stuiver et al., 1998; Hughen et al., 2000; Reimer et al., 2004; Chiu et al., 2007). In this volume, when calendar ages based on radiocarbon dates are presented, the calibration or calibration program employed is also referenced (e.g., Reimer et al., 2004 or Stuiver et al., 1998; these are the calibrations most typically used).

Although Holocene climate is not characterized by the extreme climate fluctuations of the last glacial, it has been significantly variable. Average annual temperatures have changed by as much as a few degrees C for extended periods, sometimes with very rapid onsets and terminations, occurring on interannual to decadal scales (NRC, 2002). Holocene climate change cycles of approximately 2500 years and 1500 years are well noted in the literature (e.g., Dansgaard et al., 1971, 1993; Denton and Karlén, 1973; Pisias et al., 1973; Stuiver and Braziunas, 1989; O'Brien et al., 1995; Mayewski et al., 1997; Stager et al., 1997; Bond et al., 1997, 1999, 2001; Bianchi and McCave, 1999; Dunbar, 2000; Rahmstorf, 2002, Fleitmann et al., 2003). The pioneering work of Denton and Karlén (1973) demonstrated that globally distributed changes in glacier extent occurred throughout the Holocene about every 2500 years (Fig. 1.2). Alpine glacier extent is directly related to changes in climate, as indicated by the modern example of widespread glacier retreat coincident with climate change over the last century (e.g., IPCC, 2007). Holocene glacier advances occurred at ca. 9000–8000, 6000–5000, 4200–3800, 3500–2500, 1200–1000, and since 600 cal yr BP (Fig. 1.2) and are coincident with rapid climate changes (RCCs) observed in globally distributed proxy records of climate change (Mayewski et al., 2004). These proxy records show that Holocene climate has been dynamic at scales significant to humans and ecosystems. From the perspective of human civilization, many of these changes are fast enough (occurring over a few decades to a few hundred years) to be considered "rapid" and, as the chapters that follow demonstrate, their impacts on past societies have sometimes been quite pronounced.

The RCCs following the 9000–8000 cal yr BP event varied in geographic extent and intensity. They generally involved concomitant high-latitude cooling and low-latitude aridity, a pattern typical of long-term climate trends during the Pleistocene (e.g., Nicholson and Flohn, 1980; Maley, 1982; deMenocal et al., 2000; Gasse, 2000). The most globally extensive of the Holocene RCCs occurred about 6000–5000, 3500–2500, and after 600 cal yr BP. There were less widespread RCCs at around 4200–3800 and 1200–1000 cal yr BP. The age brackets for these RCCs were verified using the well-dated, high-resolution Greenland Ice Sheet Project Two (GISP2) chemistry series (Mayewski et al., 1997), previously correlated with the globally distributed glacier fluctuation record by O'Brien et al. (1995).

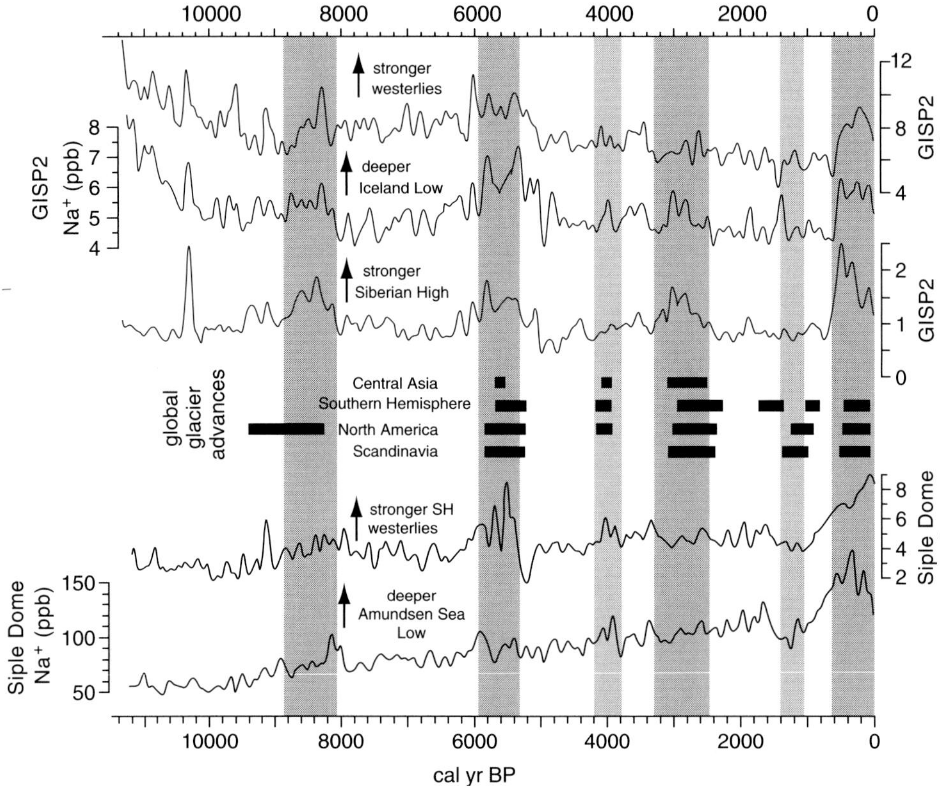

Figure 1.2. Proxy records for middle to high-latitude atmospheric circulation for the last 11,500 years obtained for the Northern Hemisphere from GISP2, Greenland, and the Southern Hemisphere from Siple Dome, Antarctica. Holocene rapid climate change (RCC) events are marked by shaded vertical bars. Changes in GISP2 Na^+ are correlated with December–January–February surface pressure over the area of the Icelandic Low such that increases (decreases) in Na^{++} coincide with decreases (increases) in pressure over this region. Increases (decreases) in GISP2 K^+ are correlated with March–April–May increases (decreases) in pressure over the region of the Siberian High (Meeker and Mayewski, 2002). Changes in GISP2 Ca^{++} are positively associated with September–October–November changes in intensity of the westerlies (Yan et al., 2006). Increases (decreases) in SD Na^+ are correlated with decreases (increases) in September–October–November surface pressure over the region of the Amundsen Sea Low. Changes in SD Ca^{++} are positively correlated with changes in the September–October–November surface mean zonal wind surrounding Antarctica, most notably the region close to 40–50°S in the Indian and Pacific Oceans. Times of distinct glacier advances are shown by horizontal black bars for Europe, North America, and the Southern Hemisphere (Denton and Karlén, 1973), and central Asia (Haug et al., 2001).

In addition to alpine glacier advances at ca. 6000–5000 and 3500–2500 cal yr BP, Northern Hemisphere RCC intervals were characterized by North Atlantic ice-rafting events (Bianchi and McCave, 1999) and strengthened westerlies over the North Atlantic and Siberia (Meeker and Mayewski, 2002; Mayewski and

Maasch, 2006). Cooling occurs over the northeast Mediterranean around 6500 and 3000 cal yr BP (Rohling et al., 2002), most likely related to winter polar air outbreaks. Westerly winds over central North America strengthen ca. 6000–5000 and 4200–3800 cal yr BP (Bradbury et al., 1993).

At lower latitudes the RCC interval ca. 6000–5000 cal yr BP marks the end of the early to Mid-Holocene humid period in tropical Africa (Gasse, 2000, 2001; deMenocal et al., 2000). Latitudinal shifts of the Atlantic Intertropical Convergence Zone expressed as changes in regional precipitation were inferred from measurements of the concentration of metals (Fe and Ti) in a marine core from the Cariaco Basin (Haug et al., 2001). A transition from wetter to drier conditions in northern South America occurred at the ca. 6000–5000 cal yr BP RCC. A proxy record for El Niño related rainstorms from a lake in Ecuador suggest that El Niño frequency increased following the ca. 6000–5000 cal yr BP RCC, and again after the RCC around 3500–2500 cal yr BP (Rodbell et al., 1999; Moy et al., 2002). This record supports western South American indicators of abrupt changes in El Niño frequency, which dominates interannual–decadal climate variability in the tropical Pacific. Paleoclimate proxy records from archaeological sites and other archives in Peru and the eastern equatorial Pacific show that El Niño activity was weak or non-existent for at least 3000 years prior to the ca. 6000–5000 cal yr BP RCC (e.g., Rollins et al., 1986; Sandweiss et al., 1996). Between ca. 5800 and 3000 cal yr BP, El Niño was present but less frequent than today. After around 3000 cal yr BP, the frequency and intensity of El Niño activity increased, becoming more similar to that of the present-day (Sandweiss et al., 2001; see Sandweiss et al., Chapter 2).

A multi-proxy climate record derived from lacustrine sediments from subtropical Chile (Jenny et al., 2002) indicates arid conditions between the ca. 9000–8000 and 6000–5000 RCCs after which time effective moisture increased progressively. Using cores from Lake Titicaca, Baker et al. (2001) have shown that maximum aridity and lowest lake level occurred between 8000 and 5500 cal yr BP. The lowest level of Lake Titicaca was reached between 6000 and 5000 cal yr BP after which lake level rose to close to its modern level.

At higher latitudes in the Southern Hemisphere, glaciers advance in the Southern Alps of New Zealand at this time (5000–6000 cal yr BP). Also, a polar ice core record from Siple Dome, Antarctica reveals that atmospheric circulation intensified at ca. 5000–6000 cal yr BP (Mayewski and Maasch, 2006; Yan et al., 2006).

3. Possible causes of Holocene climate change

Millennial scale climate variability during the Holocene may best be explained as a consequence of the dynamic balance between components of the climate system including the hydrologic cycle, heat content of the ocean, atmospheric greenhouse gas variations (including water vapor), and sea-ice extent. The forcing mechanisms likely most important in determining Holocene climate are solar variability (Denton and Karlén, 1973; O'Brien et al., 1995; Mayewski et al., 1997; Bond et al., 2001) superimposed on long-term changes in insolation, which is determined in large measure by Earth's orbital parameters. The natural feedbacks within the climate system may amplify relatively weak forcing related to fluctuations of solar output and relatively small variations in greenhouse gases (Saltzman and Moritz, 1980). The global distribution of changes in moisture balance, temperature, and atmospheric circulation for the 6000–5000 cal yr BP RCC are summarized in Figure 1.3.

Controls on Holocene climate change thus include variations in the hydrologic cycle, sea level, sea-ice extent, and forest cover. In addition, climate change can also be forced by volcanic aerosols, greenhouse gases, insolation changes, and solar variability. The hydrologic cycle that governs the latent heat distribution in the atmosphere through water vapor transport, and also the greenhouse effect, plays a major role in Holocene climate variability. This is clearly indicated by the large fluctuations in lake levels, monsoon activity, and regional precipitation patterns evident in paleoclimate records. Atmospheric methane concentrations decrease after the ca. 9000–8000 cal yr BP RCC, then steadily rise after ca. 6000–5000 cal yr BP RCC (Chappellaz et al., 1993). This, however, is likely the result rather than the cause of roughly synchronous changes in the global hydrological cycle. There are no significant systematic changes in the concentrations of volcanic aerosols (Zielinski et al., 1996; Kurbatov et al., 2006) or atmospheric carbon dioxide (Indermühle et al., 1999) over the Holocene.

Holocene climate variability, particularly during the Mid-Holocene from ca. 5000 to 9000 cal yr BP, has tended to receive relatively little research attention from integrated teams of archaeologists and climatologists. This is rapidly changing,

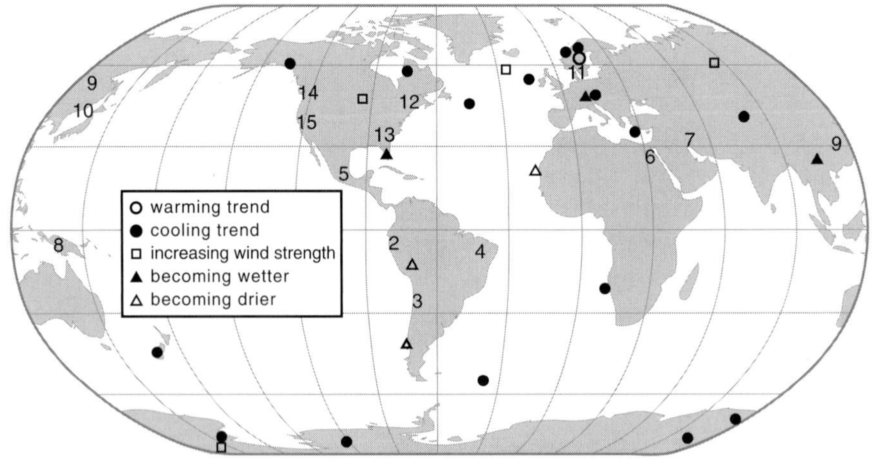

Figure 1.3. The global distribution of changes in temperature, moisture balance, and atmospheric circulation for the ca. 6000–5000 cal yr BP RCC from the Holocene proxy records summarized in Mayewski et al. (2004) are shown. The regions discussed in this book are also marked with the chapter number.

since the warmest part of the Holocene may offer a proxy of the long-term changes that are likely to occur over the next few centuries.

That the Mid-Holocene is a possible analog for future climate trends has, of course, been the subject of appreciable research, speculation, and debate for almost three decades (i.e., Budyko et al., 1978, 1987; Kellogg and Schware, 1981, Kutzbach and Guetter, 1986; Mitchell, 1990), and popular and technical articles and web sites describing global warming often note this possibility. It has long been recognized as well that the Mid-Holocene is not an exact parallel to modern circumstances, since observed Northern Hemisphere warming correlated with changing orbital parameters, specifically an increase in solar insolation, and not by an increase in greenhouse gases. Perhaps fortuitously from the perspective of resolving impacts on climate, biota, and human populations, these different potential causes have led to similar outcomes as far as projections of Northern Hemisphere and particularly high-latitude temperature and precipitation are concerned (Mitchell, 1990, pp. 1180–1183). Precipitation changes in recent years, of critical importance for sustaining agriculture and drinking water, have included "significantly increased precipitation ... in eastern parts of North and South America, northern Europe, and northern and central Asia. Drying has been observed in the Sahel, the Mediterranean, southern Africa, and parts of southern Asia" (IPCC, 2007, p. 7). Similar patterns have been noted in models of Mid-Holocene precipitation regimes (e.g., Mitchell, 1990; Ganopolski et al., 1998). The Mid-Holocene thus offers a good example of the nature and magnitude of changes in climate and biota that could occur over the long term in specific regions, and, based on what happened in the past, their possible impact on human societies in these areas.

4. Lessons from the past for the future

How does studying climate and culture change, and particularly events and processes occurring during the Mid-Holocene from ca. 9000 to 5000 years ago, help us in the modern world? In many ways, as the chapters that follow demonstrate. Lessons range from revealing the large-scale changes over time and space that may occur in variables such as vegetation cover and precipitation, and how humans responded to these changes, to developing new or improved analytical techniques and data collection strategies, as well as new ways of thinking about how we can best explore these topics. A primary lesson is that we must constantly strive to obtain the best possible temporal control of both our archaeological and paleoclimatological data, since the more precise absolute dates we have in our paleoenvironmental reconstructions and cultural sequences, the more accurately we can correlate developments over time and space, and perhaps better understand not only what was occurring, but why.

We must continually think about how to improve or develop new and innovative analytical tools to reconstruct past change. Glacial rock flour outflow sedimentation records, typically trapped in lacustrine deposits downstream, for example, are one fairly direct way to measure glacial erosion and retreat (Karlén and Larsson, Chapter 11). Plant micro and macrofossil spatial (including altitudinal) distributions are widely used to delimit the impact of climate change, or to infer that it occurred. During the Mid-Holocene, for example, the subarctic taiga/tundra boundary, or the extent of forest cover, appears to have shifted up to as much as 250 km to the north (Ganopolski, 1998, p. 1918), and pine tree remains have been shown to cluster well above their modern altitudinal range in Scandinavia (e.g., Karlén and Larsson, Chapter 11). Many of the studies herein demonstrate that changes in forest composition and distribution played a major role in Mid-Holocene culture change. A need for greater consistency and a more multivariate approach to exploring paleoclimatic/paleovegetational changes from region to region is, unfortunately, also demonstrated. In most parts of the world more pollen, fire frequency, dendroclimatology, and other proxy measures need to be obtained and examined, many of the existing records need far better temporal controls, and more of these records need to be examined collectively rather than individually. Likewise, we must be careful not to extrapolate past conditions at too great a distance from the sources of the data. For example, locally derived pollen or tree ring reconstructions of past rainfall or fire frequency may not accurately reflect regional conditions, which are moderated by ocean currents and/or other large-scale weather systems. On the other hand, ice core records from remote polar locations can be used to reliably reconstruct past changes in large-scale atmospheric circulation, and related precipitation patterns, on a continental or even hemispheric scale.

Changing climate can also lead to localized or broader scale extirpation of plant and animal species, by forcing them out of their viable ranges. As climate warms, for example, cold tolerant species may be forced to higher and higher elevations, until such refugia no longer exist; such a fate is predicted for many species in the southern Appalachians as global warming intensifies over the next few decades (e.g., Delcourt and Delcourt, 2004). Dramatic declines in eastern North American hemlock (*Tsuga canadensis*) and northern European elm (*Ulmus* spp.) that occurred toward the end of the Mid-Holocene, around 6000–5000 cal yr BP, are attributed in part to pathogen or insect infections, as well as drought, human activity, and other factors, which were in turn facilitated by warming climate (e.g., Digerfeldt, 1997; Dincauze, 2000, pp. 188–191; Bennett and Fuller, 2002; Parker et al., 2002; Foster et al., 2006). These broad changes in vegetation had major impacts on the human societies in these regions, facilitating increased hunting/gathering activities in northeastern North America and the adoption of agriculture in Scandinavia (Sanger et al., Chapter 12; Karlén and Larsson, Chapter 11). Another large-scale change in vegetation that occurred in the Mid-Holocene is the expansion of western red cedar in the Pacific northwest region of North America, a highly effective material in plank house and perhaps boat construction that is thought to have helped facilitate the emergence of complex hunting–gathering societies in this region (Moss et al., Chapter 14). Likewise, the expansion of coniferous longleaf pine in the southeastern United States, a vegetational community sustaining fewer exploitable game animals than the mixed deciduous–coniferous community that was in place previously, apparently led to a marked reduction or relocation of human populations from the Coastal Plain to the deciduous forests of the interior (Anderson et al., Chapter 13).

The Mid-Holocene thus offers several dramatic examples of how comparatively minor (i.e., no more than 2–3°C) changes in temperature can lead to the replacement of formerly dominant plant and animal species over large areas, with a concomitant impact on the human societies dependent upon them. These range changes apply to domesticates as readily as to wild species; as several authors note, small changes in temperature can markedly affect the length of the growing season for certain plant species, and hence their range of occurrence. How climate change might affect agricultural food production is a major focus of current research, since human populations worldwide are critically dependent on these resources. Small temperature changes can also affect animal populations, such as the ranges of anadromous fish and molluscan populations that have been a principal target for human populations around the world for many millennia; some species either cannot survive or actually thrive in warmer waters. As Lutaenko et al. (Chapter 10; see also Sandweiss et al., Chapter 2) demonstrate, molluscan faunal distributions, like pollen data, can help delimit the extent and impacts of warming or cooling episodes, although the authors are also careful to note that many of these species respond in an intricate way to environmental changes, so care must be taken in their interpretation.

It is also clear that the Mid-Holocene was not a period of uniform climate change, of unusual warming or drying across the world, and that it is equally dangerous to assume that climate was broadly similar even within particular regions. In the Amazon basin, the Mayan area, and the Eastern Woodlands of North America, to cite but three examples from the Americas recounted in this volume, rainfall increased in some parts of these regions and decreased in others at various times during the Mid-Holocene (Meggers, Chapter 4; Anderson et al., Chapter 13; Voorhies and Metcalfe, Chapter 5). We must recognize that there is appreciable long- and short-term climatic variability over the course of the Mid-Holocene due to as of yet incompletely understood cycles in solar output or ocean circulation.

At shorter temporal scales, increases or decreases in interannual to subdecadal climatic variability, as well as increases or decreases in seasonal (i.e., intra-annual) variability, can also have great impacts on cultural systems, perhaps as pronounced as the impacts of longer term climate trends. Thus, exploring the nature of seasonal, annual or decadal scale climatic variability, and how human cultures responded to it, is an important area for research (e.g., Kennett et al., Chapter 15). A classic example of the kind of short-term climate change that must be considered is interannual ENSO frequency, which appears to have increased after ca. 5800 cal yr BP and again after 3000 cal yr BP (Sandweiss et al., 1996, 2001, Chapter 2). Another is fire frequency, which is coming to be widely examined using tree-ring fire scar and sediment/pollen core charcoal particle records (Mohr et al., 2000; Moss et al., Chapter 14; Marlon et al., 2006).

Long-term, global scale climate change is manifest locally in different ways, and while we must not refrain from thinking in terms of global patterns or records of climate change, we also need to explore how these changes and human responses played out locally. As Lu (Chapter 9) also notes, we need to recognize and differentiate macro or global from micro or local scale climate events impacted by geography and human action.

Changes in climate and resource structure were often time transgressive, in some cases with lags between climate change and biotic response indicated; the response of vegetational communities was varied, and could be anywhere from critically dependent to minimally responsive to changes in temperature and precipitation. Local conditions and changes thus may not always proceed in lockstep or close agreement with global patterns. Vegetation changes sometimes lag behind temperature/climate changes by appreciable intervals, up to hundreds of years (Davis and Botkin, 1985); hence, pollen data may not accurately reflect when climate changed, but only responses to it.

Finally, while sudden dramatic episodes of climate change have occurred during the Holocene, such as the so-called "8200 event" (Alley et al., 1997), gradual long-term change (albeit with shorter annual to decadal fluctuations) is more typical. Culture change in many areas, at least during the Mid-Holocene, appears to have been stimulated by gradual, progressive changes in climate and biota, rather than sudden or dramatic changes at any one time.

5. Linkages between climate and culture change

Direct correlations between climate and culture change are sometimes difficult to make; assuming human societies prospered during favorable climatic periods and underwent hardship or collapse during unfavorable periods, for example, is not inevitably or invariably correct, as archaeologists, historians, and geographers have long noted (e.g., Le Roy Ladurie, 1971; DeVries, 1980; Wigley et al., 1981; Crumley, 1994; McIntosh et al., 2000; Tainter, 2000; Crumley et al., 2001; de Menocal, 2001; Redman et al., 2004; Hardesty, 2007; McGovern, 2007; Rosen, 2007). Often there were cultural responses to climate that resulted in greater organizational complexity and larger numbers of people occupying the landscape, even though environmental conditions might have been harsher than during earlier periods for certain types of subsistence activity. As discussed in this book, the Mid-Holocene record demonstrates that environmental change can trigger a range of cultural responses, from collapse, to reorganization, to expansion. In some areas, furthermore, cultural changes are observed, yet a linkage with Mid-Holocene climate is unclear, such as the emergence of agriculture in the New Guinea Highlands (Anderson et al., Chapter 8). The demonstration of spatio-temporal correlation between climate and culture change, of course, does not prove they are related. It does, however, mandate consideration of possible linkages. Among the many critical variables not in themselves directly related to climate, population size and level of socio-political complexity are particularly important. As Fagan (1999) and many others have noted, small mobile groups of foragers often have many more options for dealing successfully with climate stress on local subsistence than do large, settled populations of farmers. Below and in this volume, we tease out some of the implications of this observation in terms of subsistence stress, migration, and technological or organizational change.

We must be careful, therefore, not to assume that all or major changes in past cultural systems had a climatic trigger. Historical trends or traditions at a regional scale can significantly influence adaptations at local scales, just as individual historical events can sometimes have widespread and long lasting ramifications. To understand what is occurring in specific localities we need to recognize the cultural traditions that are in place, as well as the nature of regional political geography. A society's response may be brought about as much by its history and practices, or its location within a given region or in relation to favorable resource patches, as it is to climate change affecting temperature or vegetation. In the Mid-Holocene southeastern United States, for example, some groups occupying resource rich areas appear to have intentionally opted out of the regional trend toward increasing complexity (Anderson et al., Chapter 13). Typically these groups were located in the margins of the region, and hence were not surrounded and circumscribed spatially by other groups. In more central areas, circumscription resulted in populations quickly adapting changes their neighbors made in food production, ceremony, or warfare. Climate change might have thus forced or necessitated culture change in some areas, deliberate efforts to maintain the status quo in others, and no obvious impact in yet others. Whenever possible archaeologists should take advantage of the vast amount that has been learned through paleoenvironmental research. Unfortunately, in many parts of the world this information is underutilized by social scientists, in part because archaeologists and historians do not recognize its significance, and in part because some of them believe human agency trumps or proceeds largely unaffected by climate change (e.g., Kennett and Kennett, 2006, Chapter 7). A major lesson of this book is that climate does have a role in cultural change in most parts of the world when we look at circumstances carefully.

Direct causal links are often hard to delimit, but as the chapters that follow demonstrate, a number of major population shifts and reorganizations occurred during the Mid-Holocene that appear closely linked to concurrent changes in precipitation, sea level, growing season, or vegetation. Evidence for substantial interpersonal conflict or warfare is observed for the first time in many parts of the world during the Mid-Holocene, for instance fortification walls around settlements, burials exhibiting weapons trauma, or sites that have been razed through attack. Whether the widespread conflict is due to the climate change putting stress on people as resources declined in availability, or is a density-dependent phenomenon tied to increasing human populations, is at present unknown; both factors are assumed to have played a role in many areas (e.g., Ferguson, 1984; Haas, 1990; LeBlanc and Register, 2004; Otterbein, 2004; Gronenborn, 2005, 2007).

Both human and animal populations have tended to concentrate in well-watered areas with high exploitable biomass throughout history and prehistory. As climate and resource structure changed during the Mid-Holocene, so, too, did the size and presence of human populations in many areas; aridity commonly resulted in reduced biomass, including that of plants utilized by human populations, just as increased precipitation sometimes led to greater available biomass, and hence larger populations. As Wendorf et al. (Chapter 6) and other authors herein demonstrate, minor changes in rainfall frequency may have a much greater impact in marginal areas such as desert and grassland regions than in tropical areas/areas with much greater vegetation cover. With aridity can come heavy erosion, as vegetation cover is removed, compounding the impact on cultural systems; what rain that does occur may be more likely to run off, rather than be absorbed. In some areas such as the Atacama Desert of Peru and the Sahara of North Africa (e.g., Grosjean et al., Chapter 3; Wendorf et al., Chapter 6), exploitable subsistence resources were highly sensitive to minor changes in rainfall or temperature; when conditions changed from arid to hyperarid, human populations could no longer be maintained in some areas, resulting in depopulation or abandonment.

Human societies sometimes develop highly effective ways of dealing with climate change, especially if change leads to resource uncertainty. Technological and organizational changes are the most common strategies observed in the archaeological record from the Mid-Holocene; in the Sahara, the ability to dig deep wells, for example, allowed for use of more arid areas and for people to stay in some areas when climate became even drier (Wendorf et al., Chapter 6). Development of storage technology and organizational networks to produce and redistribute food surpluses also allowed people to buffer periods of climate-induced shortfall, at short-term scales. Long-distance trade networks appear in many areas during the Mid-Holocene, reflecting greater interaction between populations. Increased interaction between peoples in different environmental settings was likely an effective strategy to alleviate unevenly distributed subsistence stress brought on by climate change or other factors.

What kinds of sites or areas are occupied or abandoned also bears consideration. As Meggers (Chapter 4) has noted, a pronounced hiatus in rockshelter occupations occurs in the Amazonian area during the Mid-Holocene, for reasons that are not entirely clear. Locally, the period was warmer and wetter with greater biomass; use of rockshelters may have been more common when climate was colder or biomass more restricted, requiring different collection and storage strategies that may have been facilitated by the occupation of rockshelters. We need to recognize ecological and probably cultural refuges during periods of climate change/stress. Likewise, effectively delimiting anthropogenic landscape and vegetation change from climatic induced change is a continuing challenge. Changes in human settlement over time and in specific settings need to be carefully documented and tightly correlated with regional climate histories, to understand the impacts of climate change on these societies. These impacts are measured in the number of sites, the size of sites, and the density of artifacts or features such as structures or burials. We must evaluate whether hiatuses at individual sites correspond to regional changes in population numbers, or instead to population relocation, movement, or re-organization.

Another way human populations respond to climate change/stress is through migration, relocating from one area to another. The Mid-Holocene is a time of large-scale population movement, with some areas abandoned and others more densely settled. What are specific causes of migrations, and how does environmental change play a role? Unless organization and technology are capable of mitigating climate-induced reduction in subsistence resources, groups are at its mercy and must relocate from less to more favored areas, or die out. How, where, and why people moved in the past has become a major area for archaeological research (e.g., Anthony, 1990; Kelly, 2003). As Kennett et al. (Chapter 15) show, a wide range of data can be used to explore Mid-Holocene human migrations/movements, including archaeological, skeletal biological, and linguistic data. Changing biotic regimes, increases or decreases in desertification, or rising sea levels can each, in their own way, result in dramatic reductions in exploitable landmass in some areas, forcing population relocations. The locations of land masses exposed or covered by fluctuating sea level also influenced human migration patterns, by cutting off or favoring movement in certain directions. Mid-Holocene changes in winds and currents likewise impacted maritime voyaging in some areas.

Another pattern evident in many parts of the world during the Mid-Holocene, especially after ca. 6000 cal yr BP, is a dramatic increase in human use of shellfish, although it must be acknowledged up front that our knowledge of pre-Mid-Holocene use of marine resources is sparse in most coastal areas, where the ancient shorelines are submerged to varying depths. The reason we have so many surviving coastal sites from the Mid-Holocene on is tied to a global decrease in sea-level rise, as ice sheet melting slowed, shorelines reached modern levels, and estuaries formed and became stable. Voorhies and Metcalfe (Chapter 5) note, however, that sea-level stabilization since the Mid-Holocene has biased our perspective on earlier human use of coastal areas, and maritime technology in general, a point Perlman (1980) and Richardson (1981) made some decades ago. Use of coastal resources dates back almost 100,000 years in southern Africa, for example, and people reached Australia across ca. 80 km of open ocean some 45,000 years ago (Erlandson, 2002; O'Connell and Allen, 2004; Anderson et al., Chapter 8). Rising sea levels have, however, effaced much of the earlier coastal archaeological record, a problem that we will likely be facing again in the near future given global warming. Sea-level rise associated with global warming and ice sheet melting, in fact, is likely to create a vast new underwater archaeological record as well as destroy incalculable numbers of existing sites in the centuries to come, since much of the world's population, including in some of the world's largest and longest occupied cities, resides on or near the coast, and has done so for much of the Holocene. The destruction of the present and past record of human occupation and civilization along the coasts, while a major calamity for archaeology and history, is likely to be viewed as a minor concern when compared with the challenges and changes that relocating these populations will bring about.

Sea level is affected by many factors, however, so local changes may not exactly mirror global increases or decreases in ice volume. In areas of low relief, like ancient southern Mesopotamia (Kennett and Kennett, 2006, Chapter 7), comparatively minor rises in sea level resulted in major transgression as well as site burial or erosion. Arguments positing minimal human occupation in some areas may be simply an absence of evidence, because the archaeological record has been partially or totally lost to sea-level rise and associated shoreline erosion (e.g., Richardson, 1981). When examining past or possible future sea-level fluctuations, accordingly, we must recognize that this process has not been completely consistent from area to area, due to localized factors such as rebound or subsidence.

As Moss et al. (Chapter 14) note, archaeologists worldwide have argued that the rapid decrease in sea-level rise and hence coastal stabilization in the Mid-Holocene is thought to have encouraged the development and human exploitation of coastal estuarine resources. Marine resource use, and not just the development of agriculture, appears to have facilitated the development of elaborate cultures in many parts of the world during the Mid-Holocene. The Jomon culture in Japan (Lutaenko et al., Chapter 10), the Shell Mound and coastal Archaic cultures in the southeastern United States (Anderson et al., Chapter 13), and various societies along the western coast of North America (Moss et al., Chapter 14) are all examples of complex, hunting–gathering cultures that emerged at this time. Evidence for coastal resource exploitation has great antiquity, vastly predating the Mid-Holocene (Erlandson, 2002); how prevalent or effective it was during periods of comparatively greater change in sea level, as during the early part of the Holocene, is less well explored. We have to be careful not to let the expectations from such models blind us to what might actually be occurring. In both the southeastern United States and the Northwest Coast, for example, recent careful examinations have documented older and more complex sites, particularly shell middens, than we once thought existed (Anderson et al., Chapter 13; Moss et al., Chapter 14). Continuity with earlier Holocene adaptations rather than dramatic changes may be indicated, contra existing models that see an increase in the use of shellfish and anadromous fish as a response to sea-level stabilization. For instance, Terminal Pleistocene coastal groups in southern Peru targeted particular marine fish and mollusk species for intensive exploitation (Sandweiss et al., 1998), a practice that continued throughout the Holocene.

As a number of the papers in this volume also demonstrate, climate change during the Mid-Holocene helped shape the development of complex societies in several parts of the world, both through the occurrence of conditions favorable to the aggregation of larger numbers of people as well as the emergence of less favorable conditions that required new social strategies to maintain existing populations. We need to carefully consider the roles environmental variables, including changes in these variables, played in early state formation; many of the chapters herein explore these issues (e.g., Sandweiss et al., Chapter 2; Grosjean et al., Chapter 3; Voorhies and Metcalfe, Chapter 5, Wendorf et al., Chapter 6; Kennett and Kennett, 2006, Chapter 7; Lu, Chapter 9; Lutaenko et al., Chapter 10). Sea-level and climate change led to increased competition for resources, and this, coupled with "the expansion and ultimate stabilization of aquatic habitats ... favored increased population densities" (Kennett and Kennett, 2006, p. 69, Chapter 7). This process happened in several parts of the world, but primary or initial state formation occurred in only a small number of areas, including southern Mesopotamia, western South America, central China, and Mesoamerica (as opposed to in the Sea of Japan, the Pacific Northwest Coast of North America, northern Europe, the Amazon basin, or in the southeastern US), at least in part because diverse and productive suites of domestic plants and animals were also present. Sea-level stabilization in the Mid-Holocene did make the resources of coastal areas more predictable, and perhaps more bountiful, which is why complex (if not state level) societies are observed in coastal areas in many parts of the planet at this time.

Variation in rates of sea-level rise and fall can have major consequences for human societies, specifically in how such groups respond to the changes in exploitable land surface as well as in coastal/estuarine resources. The northern end of the Persian Gulf, for example, was approximately 400 km inland of its present location in the Mid-Holocene, around 6000 cal yr BP. Sites currently well removed from the ocean were quite close at the time they began their rise to prominence, such as Ubaid, Eridu, and Ur (Kennett and Kennett, 2006, pp. 74, 78, Chapter 7). Moister conditions also occurred, which likely facilitated greater crop productivity. But while the early part of the Mid-Holocene was a time of increased moisture in Mesopotamia, with higher lake levels (Kennett and Kennett, 2006, p. 76, Chapter 7), and in the Sahara (Wendorf et al., Chapter 6), just the opposite occurred in the midsouth of the United States (Anderson et al., Chapter 13). Regional trends toward greater aridity occur, but do not become pronounced in southern Mesopotamia until after ca. 6000–5500 cal yr BP, and particularly after 5000 cal yr BP, after states had emerged.

Since climate and cultural change can both occur over a wide range of timeframes, we must explore and evaluate the relationships between the two using multiscalar and multi-temporal perspectives. That is, some aspects of culture change cannot be recognized unless archaeological data are examined at a number of different spatial scales, from the site to the locality to the larger region, or over varying periods of time, from the annual or decadal to the generational and centennial scales. As Moss et al. (Chapter 14) observe, we must be careful to look to the primary data, the artifact assemblages and paleobiological remains in actual sites, rather than assume we know what is happening from later summaries or our own preconceptions of what should be present; primary reports often contain data not presented in later generalizations about these sites. Likewise, we must be ready to go back to sites and collections and re-examine them using new analytical technologies and new theoretical perspectives.

Finally, Mid-Holocene research has helped us to refine our archaeological research methods and approaches, such as procedures for site discovery, and to identify factors favoring preservation or erosion. Decreases in lake/river/spring levels during the Mid-Holocene, followed by their rise, for example, can mask archaeological sites/settlements located near their margins. Permanently submerged deposits may have better preservation than deposits characterized by fluctuating water levels. As Sanger et al. (Chapter 12) note, vegetation changes can influence erosion rates; as ground cover composition changes, so, too, will site visibility. Changes in sea level and precipitation can impact sedimentation and erosion rates, resulting in greater burial or loss of the archaeological record from certain times; Kennett and Kennett (2006, p. 81), for example, note that in Mid-Holocene Mesopotamia there may be many unrecognized Ubaid period sites "deeply buried under alluvium." We need to look in places previously ignored, and work to recover what we can from areas currently being lost or that are likely to be lost in the years to come.

6. Conclusion

The regional overviews in this volume make it abundantly clear that:

1. The Holocene was not a time of global climatic stability; change occurred at multiple spatial and temporal scales.
2. The Mid-Holocene was an era of significant social and cultural transformation in many but by no means all parts of the world.
3. Neither climatic nor cultural changes were universal or unidirectional.
4. In many parts of the globe, there were notable spatial–temporal correlations between cultural and climatic change during the Mid-Holocene. It is tempting to see correlation as causation, but that is usually an oversimplification of complex, multi-scalar, multi-modal, dynamic processes.

We have much left to learn about a critical period in human and earth history, a period that has important lessons for all of us as contemporary global change becomes our reality. It is our intent and hope that this volume will provide a global baseline for those future studies of the Mid-Holocene world.

References

Alley, R. B., P. A. Mayewski, T. Sowers, M. Stuiver, K. C. Taylor, and P. U. Clark, 1997. Holocene climatic instability: A prominent, widespread event 8200 years Ago. *Geology* 25:483–486.

Anthony, D. W., 1990. Migration in archeology: The baby and the bathwater. *American Anthropologist* 92:895–914.

Baker, P. A., G. O. Seltzer, S. C. Fritz, R. B. Dunbar, M. J. Grove, P. M. Tapia, S. L. Cross, H. D. Rowe, and J. P. Broda, 2001. The history of South American tropical precipitation for the past 25,000 years. *Science*, 291:640–643.

Bennett, K. D., and J. L. Fuller, 2002. Determining the age of the Mid-Holocene Tsuga canadensis (hemlock) decline, eastern North America. *The Holocene* 12:421–429.

Bianchi, G., and I. McCave, 1999. Holocene periodicity in North Atlantic climate and deep-ocean flow south of Iceland. *Nature* 397:515–517.

Bond, G., W. Showers, M. Cheseby, R. Lotti, P. Almasi, P. deMenocal, P. Priore, H. Cullen, I. Hajdas, and G. Bonani, 1997. A pervasive millennial-scale cycle in North Atlantic Holocene and glacial climates. *Science* 278:1257–1266.

Bond, G., W. Showers, M. Elliot, M. Evans, R. Lotti, I. Hajdas, G. Bonani, S. Johnson, 1999. The North Atlantic's 1–2 kyr climate rhythm: Relation to Heinrich events, Dansgaard/Oeschger cycle and the Little Ice Age. In *Mechanisms of Global Climate Change at Millennial Time Scales*, edited by P. Clark, R. Webb, and L. Keigwin, pp. 35–58. Geophysical Monograph Series 112. American Geophysical Union, Washington, D.C.

Bond, G., B. Kromer, J. Beer, R. Muscheler, M. Evans, W. Showers, S. Hoffmann, R. Lotti-Bond, I. Hajdas, and G. Bonani, 2001. Persistent solar influence on North Atlantic climate during the Holocene. *Science* 294:2130–2136.

Bradbury, J. P., W. Dean, and R. Y. Anderson, 1993. Holocene climatic and limnologic history of the north-central United States as recorded in the varved sediments of Elk Lake, Minnesota: A synthesis. *Geological Society of America Special Paper* 276, pp. 309–328.

Broecker, W. S., 2001. Was the medieval warm period global? *Science* 291:1497–1499.

Budyko, M. I., K. Vinnikov, O. A. Drozdov, and H. A. Efimova, 1978. Forthcoming climatic changes. *Izvestiya Rossijskoj Akademii Nauk Seriya Geografii* 6:5–20 (In Russian).

Budyko, M. I., A. B. Ronov, and A. L. Yanshin, 1987. *History of the Earth's Atmosphere.* Springer Verlag, New York.

Chappellaz, J., T. Blunier, D. Raynaud, J. Barnola, J. Schwander, and B. Stauffer, 1993. Synchronous changes in atmospheric CH4 and Greenland climate between 40 and 8 kyr b.p. *Nature* 366:443–445.

Chiu, T. C., R. G. Fairbanks, L. Cao, R. A. Mortlock, 2007. Analysis of the atmospheric ^{14}C record spanning the past 50,000 years derived from high-precision $^{230}Th/^{234}U/^{238}U$ and $^{231}Pa/^{235}U$ and ^{14}C dates on fossil corals. *Quaternary Science Reviews* 26:18–36.

Crumley, C. L., 1994. *Historical Ecology: Cultural Knowledge and Changing Landscape*. School of American Research Press, Santa Fe, NM.

Crumley, C. L., A. E. van Deventer, and J. J. Fletcher, 2001. *New Directions in Anthropology and Environment.* AltaMira Press, Walnut Creek, CA.

Dansgaard, W., S. Johnsen, H. Clausen, and C. Langway, 1971. Climatic record revealed by the Camp Century ice core. In *The Late Cenozoic Glacial Ages*, edited by K. Turekian, pp. 37–56. Yale University Press, New Haven.

Dansgaard, W., S. J. Johnsen, H. B. Clausen, D. Dahl-Jensen, N. S. Gunderstrup, C. U. Hammer, C. S. Hvidberg, J. P. Steffensen, A. E. Sveinbjornsdottir, J. Jouzel, and G. Bond, 1993. Evidence for general instability of past climate from a 250 kyr ice core, *Nature* 364:218–219.

Davis, M. B., and D. B. Botkin, 1985. Sensitivity of cool-temperature forests and their fossil pollen record to rapid temperature change. *Quaternary Research* 23:327–340.

Delcourt, P. A., and H. R. Delcourt, 2004. *Prehistoric Native Americans and Ecological Change: Human Ecosystems in Eastern North America Since the Pleistocene*. Cambridge University Press, Cambridge, UK.

deMenocal, P., J. Ortiz, T. Guilderson, J. Adkins, M. Sarnthein, L. Baker, and M. Yarusinsky, 2000. Abrupt onset and termination of the African humid period: Rapid climate responses to gradual insolation forcing. *Quaternary Science Reviews* 19:347–361.

deMenocal, P., 2001. Cultural responses to climate change during the late Holocene. *Science* 292:667–673.

Denton, G., and W. Karlén, 1973. Holocene climatic variations: Their pattern and possible cause. *Quaternary Research* 3:155–205.

DeVries, J., 1980. Measuring the impact of climate on history: The search for appropriate methodologies. *Journal of Interdisciplinary History* 10:599–630.

Digerfeldt, D., 1997. Reconstruction of Holocene lake-level changes in Lake Kalvsjön, southern Sweden, with a contribution to the local palaeohydrology at the Elm Decline. *Vegetation History and Archaeobotany* 6:9–14.

Dincauze, D. F., 2000. *Environmental Archaeology Principles and Practice*. Cambridge University Press, Cambridge, UK.

Dunbar, R. B., 2000. Climate variability during the Holocene: An update. In *The Way the Wind Blows Climate, History, and Human Action*, edited by R. J. McIntosh, J. A. Tainter, and S. K. McIntosh, pp. 45–88. Columbia University Press, New York.

Erlandson, J. M., 2002. Anatomically modern humans, maritime voyaging, and the Pleistocene colonization of the Americas. In *The First Americans: The Pleistocene Colonization of the New World*, edited by Nina G. Jablonski, Number 27, pp. 59–92. Memoir of the California Academy of Sciences, San Francisco.

Fagan, B., 1999. *Floods, Famines, and Emperors: El Niño and the Fate of Civilizations*. Basic Books, New York.

Fagan, B., 2004. *The Long Summer: How Climate Changed Civilization*. Basic Books, New York.

Ferguson, R. B., 1984. *Warfare, Culture, and Environment*. Academic Press, Orlando, FL.

Fiedel, S. J., 1999. Older than we thought: Implications of corrected dates for Paleoindians. *American Antiquity* 64:95–116.

Fleitmann, D., S. Burns, M. Mudelsee, U. Neff, J. Kramers, M. Augusto, and A. Matter, 2003. Holocene forcing of the Indian monsoon recorded in a stalagmite from southern Oman. *Science* 300:1737–1739.

Foster, D. R., W. W. Oswald, E. K. Faison, E. D. Doughty, and B. C. S. Hansen, 2006. A climatic driver for abrupt Mid-Holocene vegetation dynamics and the hemlock decline in New England. *Ecology* 87:2959–2966.

Ganopolski, A., C. Kubatzki, M. Claussen, V. Brovkin, and V. Petoukhov, 1998. The influence of vegetation–atmosphere–ocean interaction on climate during the Mid-Holocene. *Science* 280:1916–1919.

Gasse, F., 2000. Hydrological changes in the African tropics since the last glacial maximum. *Quaternary Science Reviews* 19:189–211.

Gasse, F., 2001. Hydrological changes in Africa. *Science* 292:2259–2260.

Gibbard, P. L., 2003. Definition of the middle-upper Pleistocene boundary. *Global and Planetary Change* 36:201–208.

Gore, A., 2006. *An Inconvenient Truth: The Planetary Emergency of Global Warming and What We Can Do About It*. Rodale Press, New York.

Gronenborn, D., 2005. *Klimaveranderung und Kulturwandel in neolithischen Gesellshaften Mitteleuropas, 7600–2200 cal BC*. Verlag des Romisch-Germanischen Zentralmuseums, Maines, Germany.

Gronenborn, D., 2007. Climate and socio-political crises: some cases from Neolithic central Europe. In *War and Sacrifice Studies in the Archaeology of Conflict*, edited by Tony Pollard and Iain Banks, pp. 13–32. Brill, Leiden, Netherlands.

Guilderson, T. P., P. J. Reimer, and T. A. Brown, 2005. Geoscience: The boon and bane of radiocarbon dating. *Science* 307:362–264.

Gulliksen, S., H. H. Birks, G. Possnert, and J. Mangerud, 1998. A calendar age estimate of the Younger Dryas-Holocene boundary at Kråkenes, western Norway. *The Holocene* 8:249–259.

Haas, J., 1990. *The Anthropology of War*. Cambridge University Press, Cambridge.

Hardesty, D. L., 2007. Perspectives on global change archaeology. *American Anthropologist* 109:1–7.

Harland, W. B., R. L. Armstrong, A. V. Cox, L. E. Craig, A. G. Smith, and D. G. Smith, 1989. *A Geologic Time Scale*. Cambridge University Press, Cambridge.

Haug, G., K. Hughen, D. Sigman, L. Peterson, and U. Röhl, 2001. Southward migration of the Intertropical Convergence Zone through the Holocene. *Science* 293:1304–1308.

Hughen, K. A., J. T. Overpeck, S. J. Lehman, M. Kashgarian, J. Southon, L. C. Peterson, R. Alley, and D. M. Sigman, 1998. Deglacial changes in ocean circulation from an extended radiocarbon calibration. *Nature* 391:65–68.

Hughen, K. A., J. R. Southon, S. J. Lehman, and J. T. Overpeck, 2000. Synchronous radiocarbon and climate shifts during the last deglaciation. *Science* 290:1951–1954.

Indermühle, A., T. Stocker, F. Joos, H. Fischer, H. Smith, M. Wahlen, B. Deck, D. Mastroianni, J. Tschumi, T. Blunier, R. Meyer, and B. Stauffer, B., 1999. Holocene carbon-cycle dynamics based on CO_2 trapped in ice at Taylor Dome, Antarctica. *Nature* 398:121–126.

IPCC, 2007: Summary for Policymakers. In *Clite Change 2007: The Physical Science Basis. Contribution of Working Group I to the Fourth Assessment Report of the Intergovernmental Panel on Climate Change,* edited by S. Solomon, D. Qin, M. Manning, Z. Chen, M. Marquis, K. B. Averyt, M. Tignor and H. L. Miller, pp. 1–18. Cambridge University Press, Cambridge, UK.

Jenny, B., B. L.Valero-Garcés, R. Villa-Martínez, R. Urrutia, M. Geyh, and H. Veit, 2002. Early to Mid-Holocene aridity in Central Chile and the Southern Westerlies: The Laguna Aculeo record (34°S). *Quaternary Research* 58:160–170.

Kelly, R. L., 2003. Colonization of New Land by hunter–gatherers expectations and implications based on ethnographic data. In *Colonization of Unfamiliar Landscapes: The Archaeology of Adaptation*, edited by M. Rockman and J. Steele, pp. 44–58. Routledge Taylor and Francis Group, New York.

Kellogg, J. E., and R. Schware, 1981. *Climate Change and Society: Consequences of Increasing Atmospheric Carbon Dioxide*. Westview Press, Boulder, Colorado.

Kennett, D. P., and J. P. Kennett, 2006. Early state formation in southern Mesopotamia: Sea levels, shorelines, and climate change. *Journal of Island and Coastal Archaeology* 1:67–99.

Kitigawa, H., and J. van der Plicht, 1998. Atmospheric radiocarbon calibration to 45,000 Yr B.P.: Late glacial fluctuations and cosmogenic isotope production. *Science* 279:1187–1190.

Kurbatov, A. V., G. A. Zielinski, N. W. Dunbar, P. A. Mayewski, E. A. Meyerson, S. B. Sneed, and K. C. Taylor, 2006. A 12,000 year record of explosive volcanism in the Siple Dome Ice Core, West Antarctica. *Journal of Geophysical Research* 111, D12307, doi:10.1029/2005JD006072.

Kutzbach, J. E., and P. J. Guetter, 1986. The influence of changing orbital parameters and surface boundary conditions on climate simulations for the past 18,000 years. *Journal of the Atmospheric Sciences* 43:1726–1759.

LeBlanc, S. A., and K. Register, 2004. *Constant Battles Why We Fight*. St. Martin's Press, New York.

Le Roy Ladurie, E., 1971. *Times of Feast, Times of Famine: A History of Climate Since the Year 1000*. Doubleday and Company, New York.

Maley, J., 1982. Dust, clouds, rain types, and climatic variations in tropical Africa. *Quaternary Research* 18:1–16.

Marlon, J., P. J. Bartlein, and C. Whitlock, 2006. Fire–fuel–climate linkages in the northwestern USA during the Holocene. *The Holocene* 16(8):1059–1071.

Mayewski, P. A., and K. A. Maasch, 2006. Recent warming inconsistent with natural association between temperature and atmospheric circulation over the last 2000 years. *Climate of the Past Discussions* 2:327–355.

Mayewski, P., Meeker, L., Twickler, M., Whitlow, S., Yang, Q., Lyons, W., Prentice, M., 1997. Major features and forcing of high-latitude northern hemisphere atmospheric circulation using a 110,000-year long glaciochemical series. *Journal of Geophysical Research* 102:26345–26366.

Mayewski, P. A., E. Rohling, C. Stager, K. Karlén, K. A. Maasch, L. D. Meeker, E. Meyerson, F. Gasse, S. van Kreveld, K. Holmgren, J. Lee-Thorp, G. Rosqvist, F. Rack, M. Staubwasser, R. Schneider, and S. Steig, 2004. Holocene climate variability. *Quaternary Research* 62:243–255.

McGovern, T. H., 2007. In focus: Archaeology of global change. *American Anthropologist* 109: 1–84.
McIntosh, R. J., J. A. Tainter, and S. K. McIntosh, 2000. Climate variability during the Holocene: An update. In *The Way the Wind Blows Climate, History, and Human Action*. Columbia University Press, New York.
Meeker, L., and P. A. Mayewski, 2002. A 1400 year long record of atmospheric circulation over the North Atlantic and Asia. *Holocene* 12:257–266.
Mitchell, J. F. B., 1990. Greenhouse warming: Is the Mid-Holocene a good analogue? *Journal of Climate* 3:1177–1192.
Mohr, J. A., C. Whitlock, and C. N. Skinner, 2000. Postglacial vegetation and fire history, eastern Klamath Mountains, California. *The Holocene* 10:587–601.
Moy, C. M., G. O. Seltzer, D. T. Rodbell, and D. M. Anderson, 2002. Variability of El Niño/ Southern Oscillation activity at millennial timescales during the Holocene epoch. *Nature* 420:162–165.
NRC (National Research Council Committee on Abrupt Climate Change), 2002. *Abrupt Climate Change Inevitable Surprises*. National Academy Press, Washington, DC.
Nicholson, N., and H. Flohn, 1980. African environmental and climatic changes and the general atmospheric circulation in Late Pleistocene and Holocene. *Climatic Change* 2:313–348.
O'Brien, S., P. A. Mayewski, L. Meeker, D. Meese, M. Twickler, and S. Whitlow, 1995. Complexity of Holocene climate as reconstructed from a Greenland ice core. *Science* 270:1962–1964.
O'Connell, J. F., and J. Allen, 2004. Dating the colonization of Sahul (Pleistocene Australia–New Guinea): A review of recent research. *Journal of Archaeological Science* 31:835–853.
Otterbein, K. F., 2004. *How War Began*. Texas A&M Press, College Station, TX.
Parker, A. G., D. E. Anderson, M. A. Robinson, and C. Bonsall, 2002. A review of the Mid-Holocene elm decline in the British Isles. *Progress in Physical Geography* 26:1–45.
Perlman, S. M., 1980. An optimum diet model, coastal variability, and hunter–gatherer behavior. In *Advances in Archaeological Method and Theory*, Vol. 3, edited by M. B. Schiffer, pp. 257–310, Academic Press, New York, New York.
Pisias, N., J. Dauphin, and C. Sancetta, 1973. Spectral analysis of late Pleistocene–Holocene sediments. *Quaternary Research* 3:3–9.
Rahmstorf, S., 2002. Ocean circulation and climate during the past 120,000 years. *Nature* 419:207–214.
Redman, C. L., S. R. James, P. R. Fish, and J. D. Rogers, 2004. *The Archaeology of Global Change: The Impact of Humans on the Environment*. Smithsonian Institution Press, New York.
Reimer, P. J., M. G. L. Baillie, E. Bard, A. Bayliss, J. W. Beck, P. G. Blackwell, C. E. Buck, G. S. Burr, K. B. Cutler, P. E. Damon, R. L. Edwards, R. G. Fairbanks, M. Friedrich, T. P. Guilderson, C. Herring, K. A. Hughen, B. Kromer, F. G. McCormac, S. W. Manning, C. B. Ramsey, R. W. Reimer, S. Remmele, J. R. Southon, M. Stuiver, S. Talamo, F. W. Taylor, J. van der Plicht, and C. E.Weyhenmeyer, 2004. IntCal04 terrestrial radiocarbon age calibration, 0–26 cal kyr BP. *Radiocarbon* 46(3):1029–1058.
Richardson III, J. B., 1981. Modeling the development of early complex economies on the coast of Peru: A preliminary statement. *Annals of Carnegie Museum* 50:139–150.
Roberts, N., 1998. *The Holocene: An Environmental History*. 2nd edition. Blackwell, New York.
Rodbell, D. T, G. O. Seltzer, D. M. Anderson, M. B. Abbott, D. B. Enfield, and J. H. Newman, 1999. An ~15,000-year record of El Niño-driven alluviation in southwestern Ecuador. *Science* 283:516–520.

Rollins, H. B., J. B. Richardson III, and D. H. Sandweiss, 1986. The birth of El Niño: Geoarchaeological evidence and implications. *Geoarchaeology* 1:3–15.

Rohling, E. J., P. A. Mayewski, A. Hayes, R. H. Abu-Zied, and J. S. L. Casford, 2002. Holocene atmosphere–ocean interactions: Records from Greenland and the Aegean Sea. *Climate Dynamics* 18:573–592.

Rosen A. M., 2007. *Civilizing Climate Social Responses to Climate Change in the Ancient Near East*. AltaMira Press, Plymouth, UK.

Saltzman, B., and R. Moritz, 1980. A time-dependent climatic feedback system involving sea-ice extent, ocean temperature, and CO_2. *Tellus* 32:93–118.

Sandweiss, D. H., J. B. Richardson III, E. J. Reitz, H. B. Rollins, and K. A. Maasch, 1996. Geoarchaeological evidence from Peru for a 5000 years B.P. onset of El Niño. *Science* 273:1531–1533.

Sandweiss, D. H., H. McInnis, R. L. Burger, A. Cano, B. Ojeda, R. Peredes, M. Sandweiss, and M. Glascock, 1998. Quebrada Jaguay: early maritime adaptations in South America. *Science* 281:1830–1832.

Sandweiss, D. H., K. A. Maasch, R. L. Burger, J. B. Richardson III, H. B. Rollins, and A. Clement, 2001. Variation in Holocene El Niño frequencies: Climate records and cultural consequences in ancient Peru. *Geology* 29:603–606.

Scarre, C., 2005. The human past: World prehistory and the development of human societies. Thames and Hudson, New York.

Stager, J., B. Cumming, and L. Meeker, 1997. A high-resolution 11,400-yr diatom record from Lake Victoria, East Africa. *Quaternary Research* 47:81–89.

Stuiver, M., and T. Braziunas, 1989. Atmospheric ^{14}C and century-scale solar oscillations. *Nature* 388:405–407.

Stuiver, M., P. J. Reimer, E. Bard, J. W. Beck, G. S. Burr, K. A. Hughen, B. Kromer, G. McCormac, J. van der Plicht, and M. Spurk, 1998. INTCAL98 radiocarbon age calibration, 24,000–0 cal BP. *Radiocarbon* 40(3):1041–1083.

Tainter, J. B., 2000. Global change, history, and sustainability. In *The Way the Wind Blows Climate, History, and Human Action*, edited by R. J. McIntosh, J. A. Tainter, and S. K. McIntosh, pp. 331–356. Columbia University Press, New York.

Taylor, R. E., M. Stuiver, and P. J. Reimer, 1996. Development and extension of the calibration of the radiocarbon time scale: Archaeological applications. *Quaternary Science Reviews* 15:655–668.

Wigley, T. M. L., M. J. Ingram, and G. Farmer, 1981. *Climate and History: Studies in Past Climates and their Impact on Man*. Cambridge University Press, Cambridge.

Yan, Y., P. A. Mayewski, S. Kang, and E. Meyerson, 2006. An ice core proxy for Antarctic circumpolar wind intensity. *Annals of Glaciology* 41:121–130.

Zielinski, G., P. A. Mayewski, L. Meeker, S. Whitlow, and M. Twickler, 1996. A 110,000 year record of explosive volcanism from GISP2 (Greenland) ice core. *Quaternary Research* 45:109–118.

Climate Change and Cultural Dynamics: A Global Perspective on Mid-Holocene Transitions
David G. Anderson, Kirk A. Maasch and Daniel H. Sandweiss (Editors)

Chapter 2

Mid-Holocene climate and culture change in coastal Peru

Daniel H. Sandweiss[1], Kirk A. Maasch[2], C. Fred T. Andrus[3], Elizabeth J. Reitz[4], James B. Richardson III[5], Melanie Riedinger-Whitmore[6] and Harold B. Rollins[7]

[1]*Climate Change Institute, Department of Anthropology, and Graduate School, University of Maine, Orono, ME, USA*
[2]*Climate Change Institute, Department of Earth Sciences, University of Maine, Orono, ME, USA*
[3]*Department of Geological Sciences, University of Alabama, USA*
[4]*Department of Anthropology, University of Georgia, Georgia, USA*
[5]*Department of Anthropology, University of Pittsburgh, Pennsylvania, USA*
[6]*Department of Environmental Science, Policy, and Geography, University of South Florida, Florida, USA*
[7]*Department of Geology and Planetary Science, University of Pittsburgh, Pittsburgh, PA, USA*

Abstract

In the general absence of standard, high-resolution paleoclimatic records such as lake cores or corals, archaeological remains from Mid-Holocene archaeological sites in coastal Peru provided pioneering interpretations of El Niño/Southern Oscillation (ENSO)-related paleoclimatic change in the eastern equatorial Pacific that have since been supported and amplified by multiple proxies. At the same time, archaeologists working in the region have explored the role of climatic change in cultural development, with particular attention to El Niño. In this chapter we review the history of study and the current status of Mid-Holocene climatic and cultural change along the Peruvian coast, with a focus on major transitions at ca. 5800 and 3000 cal yr BP that correlate temporally with changes in ENSO frequency.

1. Introduction

In the wake of several large-scale El Niño events over the last quarter century, archaeologists, geologists, and paleoclimatologists have shown an increasing interest in reconstructing the prehistory of this climatic anomaly. Mollusks found in archaeological sites on the north and central coasts of Peru provided the first clues that El Niño frequency had varied significantly throughout the Holocene. The totality of archaeological and paleoclimatic data available at this time support a major change in tropical Pacific climate at about 5800 cal yr BP (Rollins et al., 1986; Sandweiss et al., 1996, 2001; radiocarbon dates used in this paper were calibrated with Calib 4.3 (Stuiver et al., 1998a, b)), though it is now unclear whether El Niño was absent or just extremely rare for several millennia prior to that date. Molluscan remains from Peru also suggest that between ca. 5800 and 3000 cal yr BP, El Niño was present but less frequent than today. Modern, rapid recurrence intervals were apparently achieved only after that time (Sandweiss et al., 2001).

Here, we review available data from multiple Central Andean archives (both anthropogenic and natural) for the evolution of El Niño between ca. 9000 and 3000 cal yr BP. Turning then to the archaeological record, the onset of El Niño at 5800 cal yr BP is temporally correlated with the beginning of monumental construction on the Peruvian coast, while the apparent increase in El Niño frequency after 3000 cal yr BP is correlated with the abandonment of monumental, Initial Period temples in the same region. Is there a causal link between these processes?

2. The Peruvian archaeological record of Holocene El Niño frequency variation

In 1980, thanks to a tip from David Wilson, Sandweiss first visited the fossil beach and associated archaeological sites of the Ostra Complex (see Fig. 2.1 for the location of sites mentioned in the text). Located just north of the Santa river on the north-central Peruvian coast (9°S), these archaeological and paleontological deposits date to about 5800 to 7150 cal yr BP (Rollins et al., 1986; Perrier et al., 1994; Sandweiss et al., 1996; Andrus et al., 2003). A return visit several months later with Rollins and Richardson led to the hypothesis that the Ostra sites reflect a time when El Niño did not function as it does today (Rollins et al., 1986; Sandweiss, 1986, 1996, 2003; Sandweiss et al., 1983, 1996, 1997, 1998a, 2001).

Situated on the shores of a now-dry embayment, the principal sites of the Ostra Complex are the Ostra Base Camp (OBC), located on the southern end of the fossil bay, and the Ostra Collecting Station (OCS, Fig. 2.2), located on a rocky knoll about halfway along the shore of the fossil bay. On our first visit, we noticed that both the sites and the fossil beach contained mollusk species no longer present in the area – in fact, they are now found more than 4° of latitude to the north, near the Equator (Sandweiss et al., 1983). At Ostra, we found the same mollusks in living position in the former bay, indicating that the site's inhabitants were collecting their shellfood from the adjacent beach rather than from distant shores. Throughout this chapter, we will refer to assemblages like those from the Ostra Complex as warm-water molluscan assemblages; we will use the term "cool-water assemblages" for the Peru current-adapted species found at later sites; technically, these are "warm-tropical" and "warm-temperate", respectively. Reitz later identified similar assemblages for the fish fauna (Reitz and Sandweiss, 2001; see also Reitz et al., in press).

On the same 1980 expedition, we visited another series of sites on the shores of the Salinas de Chao, a second dry embayment 20 km further north. The earliest of these sites dated between ca. 3700 and 5350 cal yr BP and contained only cool-water mollusks characteristic of Peru and Chile today (Cárdenas, 1979, 1995; Sandweiss et al.,

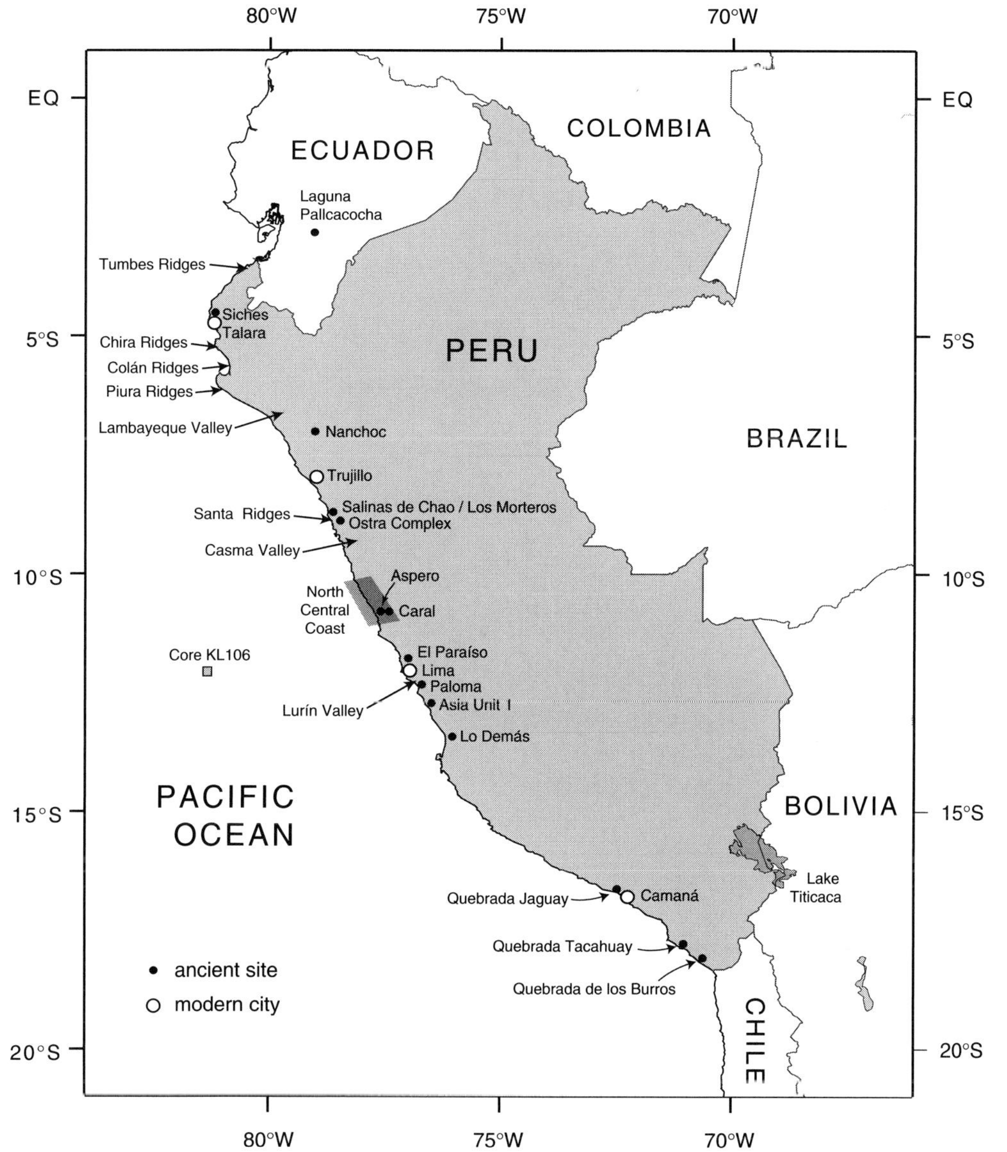

Figure 2.1. Map showing the location of sites mentioned in the text.

1983; Perrier et al., 1994; Andrus et al., 2003). Molluscan assemblages had clearly changed sometime in the centuries immediately following 5800 cal yr BP. Further north, near Talara, Peru, Richardson (1973, 1978) had observed a similar change from warm-water mangrove mollusks to cool-water species, also around 5800 cal yr BP.

We considered several hypotheses to explain these data. Was the thermally anomalous molluscan assemblage (TAMA) the result of local conditions such as solar warming of shallow embayments, or did it reflect a climatic regime different

Figure 2.2. View from the Ostra Collecting Station SSW across the fossil bay towards the Ostra Base Camp. Photo by D.H. Sandweiss.

from today? Several factors convinced us that the latter scenario was more likely correct: (1) the similar nature and timing of change at two widely separated locales in different geographic settings; (2) the restriction of warm-water molluscan assemblages to the north coast of Peru and to sites dating before 5800 cal yr BP; (3) the presence of multiple year age classes in the molluscan assemblages, both in the beach and in the sites, indicating that local conditions must have allowed sufficient exchange with the open ocean to prevent hypersalinity; and (4) the moderate diversity of the molluscan assemblages, suggesting long-term stability rather than environmental stress. We thus concluded that for some time prior to 5800 years ago, the coast of Peru north of ca. 10°S latitude was characterized by permanent warm water. From these data, we hypothesized that El Niño did not operate for some period before 5800 cal yr BP; after that time, we saw conditions as essentially the same as today (Rollins et al., 1986).

In 1990, Thomas J. DeVries and Lisa E. Wells (1990) suggested that the presence of a warm-water molluscan fauna at the Ostra sites might be due to solar warming in a completely enclosed lagoon, rather than a change in ocean circulation. The idea of anything living in a completely enclosed lagoon seemed unlikely – at this latitude and in the absence of annual rainfall, such a lagoon would rapidly go hypersaline and then dry up completely. Nevertheless, to test their idea of warm water only behind a barrier, with "normal" cold water immediately offshore of the barrier, Sandweiss returned to the Ostra sites in 1991 to recover a more extensive collection of fish as well as molluscan remains. Fish provide another source of climatic data

(Sandweiss et al., 1996; Reitz and Sandweiss, 2001; Reitz et al., in press). In 1995 and again in 2001, Richardson and Sandweiss excavated at the Siches site (4°30′S), near Talara, for the same purpose (Sandweiss et al., 1996; Sandweiss, 2003).

Sandweiss and colleagues (1996) compiled the Mid-Holocene archaeological record then available from the Peruvian coast, with particular attention to the marine fauna. These data clearly showed change at 5800 cal yr BP and north of 10°S latitude. For several millennia prior to that date, northern Peruvian sites contain predominately warm-water molluscan and (where known) fish faunas, whereas after 5800 cal yr BP for the entire Peruvian coast, and south of 10°S prior to 5800 cal yr BP, they contain predominately cool-water mollusks and fish.

Additional insight into the climatic conditions reflected by the pre-5800 cal yr BP, Mid-Holocene marine fauna in coastal sites north of 10°S came from Andrus's geochemical analyses of growth increments in fish otoliths from OBC and Siches (Andrus et al., 2002a, 2003) and in a mollusk from OBC (Andrus et al., 2005). Delta ^{18}O of the otoliths showed that in the millennium preceding 5800 cal yr BP, average sea surface temperature (SST) was about 3–4°C warmer than today, consistent with our interpretation of the marine fauna. However, the seasonal structure of SST showed a more complex picture. At Siches (4°30′S), the annual SST cycle in the Mid-Holocene paralleled that of today but was offset by 3–4°C. In contrast, at OBC winters were about as cool as today but summers were significantly warmer (Andrus et al., 2002a); the amplitude of seasonal temperature at OBC was apparently the same as the difference between normal to El Niño year SSTs today, but had an annual rather than interannual cycle. This pattern explains the difference between molluscan and fish fauna assemblages at OBC. Mollusks are sessile and therefore controlled by maximum annual temperature; OBC contained only species that can survive in warm water. Fish are mobile, so during the cool summers, cool-water fish could move north to the Ostra area while the warm-water fish would be present during the warm summers. The OBC fish fauna was dominated by warm-water species but included some cool-water fish as well (Reitz and Sandweiss, 2001).

Experiments on mollusks that survived the 1982–83 El Niño event showed that the ^{14}C content changed across the event. In growth increments deposited before and after El Niño, ^{14}C was significantly older than modern, reflecting the old, deep upwelled water of the Peru Current. During El Niño, ^{14}C gave an age close to modern, reflected the upwelling of mixed surface water resulting from the depression of the thermocline (Andrus et al., 2005). Preliminary analysis of ^{14}C in a Mid-Holocene mollusk from OBC compared to the ^{14}C age of charcoal from the same context suggests decreased upwelling compared to today, again consistent with our interpretation of Mid-Holocene climate at this locale (Andrus et al., 2002b).

In the late 1990s, further consideration of the molluscan record in Mid-Holocene Peruvian coastal sites led to additional insight (Sandweiss et al., 2001). We noticed that sites immediately post-dating the postulated onset of El Niño at 5800 cal yr BP had molluscan assemblages dominated by two species that are extremely sensitive to warm water. The large purple mussel *Choromytilus chorus* has an LT-50 (lethal temperature-50, the temperature at which 50% of the population dies in 24 h) of

28°C, based on studies in Chile (Urban, 1994). Although we do not have LT-50 data for *Mesodesma donacium*, this wedge clam was fished commercially as far north as Lima (12°S) before the 1982–83 El Niño, after which its northern limit shifted south to Lomas (15°30′S). Following the 1997–98 El Niño, the Peruvian government was forced to ban fishing of *Mesodesma* anywhere in Peru. The abundant presence of these two species in coastal sites between Lima and Trujillo (8°S) during the Late Preceramic and Initial Periods (ca. 5800–3000 cal yr BP) would not have been possible with an El Niño recurrence interval as short as it is today. The disappearance of the two mollusk species from north-central and northern Peruvian sites after 3000 cal yr BP strongly suggests an increase in El Niño frequency at that time (Sandweiss et al., 2001).

Table 2.1 summarizes the archaeological record of Terminal Pleistocene to Mid-Holocene climatic change along the Peruvian coast. The data are discussed in greater detail in Sandweiss (2003). This broad review of excavation results from multiple projects supports the outlines of Holocene change detailed from our own work (Sandweiss et al., 1996, 2001), with clearly marked transitions in the behavior of El Niño at ca. 5800 and 3000 cal yr BP.

Our most recent work on El Niño frequencies concerns the late prehispanic period. At the Inca-period fishing site Lo Demás, Sandweiss (1992) found that fish remains in the earliest deposits, ca. AD 1480–1500, were dominated by anchovies (*Engraulis ringens*), while the later deposits (ca. AD 1500–1540) contained more sardines (*Sardinops sagax*). Chavez et al. (2003) analyzed fisheries data for the Peruvian coast over the entire 20th century and compared them to Pacific climate records, finding a 50-year cycle of alternating anchovy regimes (slightly cooler average sea surface temperature (SST), less frequent El Niño) and sardine regimes (slightly warmer SST, more frequent El Niño). The faunal record from Lo Demás mirrors the regime change from anchovy to sardine, suggesting a slight change in El Niño frequency at about AD 1500 (Sandweiss et al., 2004). This transition accords with the scant Pacific Basin historical record (Quinn, 1992) and the Quelccaya ice core record (Thompson, 1992). Looking at faunal records from earlier sites, we see the potential to identify similar decadal-scale change in El Niño frequency.

3. Mid-Holocene climate from natural archives

Present-day climatic variability on interannual time scales in the tropics is dominated by El Niño/Southern Oscillation (ENSO), which involves both the atmosphere and the ocean in the tropical Pacific (e.g., Maasch, in press). Through teleconnections, extratropical climatic variability on these time scales is also impacted by ENSO.

ENSO seems to be locked in phase with the seasonal cycle, but it is not periodic (i.e., El Niño events occur at unevenly spaced intervals). The magnitude and duration of individual events vary. Over the last century (the time span of reliable instrument records) the frequency and duration of ENSO events has changed (e.g., Rajagopalan et al., 1997; Maasch, in press). Over long periods of time

Table 2.1. Climatic signals from terminal Pleistocene to Mid-Holocene archaeological sites on the Peruvian coast (after Sandweiss, 2003).

Site (S lat)	Terminal Pleistocene ~13000–11,000 cal BP	Early Holocene ~11000–9000 cal BP	Mid-Holocene ~9000–3000 cal BP		Basic references
			Mid-Holocene I ~9000–5800 cal BP	Mid-Holocene II ~5800–3000 cal BP	
Siches (4°30′S)	–	–	Warmer SSTs/seasonal precipitation/No ENSO	Cool SSTs (modern)	Sandweiss et al. (1996), Andrus et al. (2002a)
Amotape (4°40′S)	Warmer SSTs/less arid	Warmer SSTs/less arid	–	–	Richardson (1978)
Quebrada Chorrillos (6°S)	–	–	Warmer SSTs/No ENSO	–	Cárdenas et al. (1993)
Avic (6°S)	–	–	–	Cool SSTs (modern)	Cárdenas et al. (1993)
Paiján (8°30′S)	Warmer SSTs/less arid	Warmer SSTs/less arid	–	–	Chauchat et al. (1992)
Moche Valley Late Preceramic/ Initial Period sites (8°10′S)	–	–	–	Cool SSTs (modern)/low frequency ENSO	Pozorski (1979)
Salinas de Chao (8°40′S)	–	–	–	Cool SSTs (modern)/low frequency ENSO	Sandweiss et al. (1996)
Ostra (8°55′S)	–	–	Warmer SSTs/seasonal precipitation/High amplitude seasonal SST cycle/No ENSO	–	Sandweiss et al. (1996), Andrus et al. (2002a)

Continued

Table 2.1. continued

Site (S lat)	Terminal Pleistocene ~13000–11,000 cal BP	Early Holocene ~11000–9000 cal BP	Mid-Holocene ~9000–3000 cal BP		Basic references
			Mid-Holocene I ~9000–5800 cal BP	Mid-Holocene II ~5800–3000 cal BP	
Huaynuná (9°30′S)	–	–	–	Cool SSTs (modern)/low frequency ENSO	Pozorski and Pozorski (1990)
Casma Valley Late Preceramic/ Initial Period sites (9°30′S)				Cool SSTs (modern)/low frequency ENSO	Pozorski and Pozorski (1987)
Almejas (9°40′S)	–	–	Warmer SSTs/No ENSO	–	Pozorski and Pozorski (1995)
PV 35-106 (10°S)	–	–	Slightly warmer SSTs?	–	Bonavia (1996)
PV 35-6 (10°S)	–	–	–	Cool SSTs (modern)/low frequency ENSO	Bonavia et al. (1993)
Los Gavilanes (10°S)	–	–	–	Cool SSTs (modern)	Bonavia (1982)
As8 (10°45′S)	–	–	Slightly warmer SSTs?	–	Feldman (1980)
Aspero (10°45′S)	–	–	–	Cool SSTs (modern)/low frequency ENSO	Feldman (1980)

Caral (10°45′S)	–	–	–	Cool SSTs (modern)/low frequency ENSO	Shady Solís et al. (2001)
Paloma (12°30′S)	–	–	Cool SSTs (modern)		Reitz (1988)
Quebrada Jaguay (16°30′S)	Cool SSTs (modern)/ greater highland precipitation?	Cool SSTs (modern)/ greater highland precipitation?	Very arid (reduced highland precipitation?) ~8100–3500 cal BP		Sandweiss et al. (1998b)
Ring Site (17°40′S)	Cool SSTs (modern)	Cool SSTs (modern)	Cool SSTs (modern)	–	Sandweiss et al. (1989)
Quebrada Tacahuay (17°48′S)	Cool SSTs (modern)/ ENSO floods	Cool SSTs (modern)	Cool SSTs (modern)/No ENSO floods	Cool SSTs (modern)/ ENSO floods late	Keefer et al. (1998), deFrance et al. (2001)
Quebrada de los Burros (18°00′S)	–	Cool SSTs (modern)	Cool SSTs (modern)/No ENSO floods	Cool SSTs (modern)/ ENSO floods late	Fontugne et al. (1999), Lavallée et al. (1999)

(centuries-millennia), the recurrence interval and amplitude of ENSO have been even greater in magnitude than those observed in the instrument record.

Continuous natural Holocene paleoclimate archives from northern Peru, Ecuador, and the eastern tropical Pacific Ocean are difficult to find, privileging anthropogenic deposits from archaeological sites. Lakes along the desert coast are ephemeral or non-existent. Coastal waters are too cold for corals, and although sedimentation rates are high along the Peru margin, hiatuses and/or disturbance due to turbidite flows in the Holocene part of the record are common. Nevertheless, there are some high-to-medium resolution records from the region that reflect aspects of the Holocene climate of coastal Peru, as well as some low-resolution records of relevance to millennial-scale climatic variation. Although precisely dating these records is difficult, climatic change determined from them is consistent with results obtained from mollusk and fish remains from archaeological sites as described above. We discuss these regional paleoclimate records below.

3.1. Terrestrial records (low-scale resolution)

Flood deposits, soil development, lomas (xerophytic vegetation) distribution, and beach ridge morphology all offer low temporal resolution records of coastal climate in Peru. Dated flood events from the Peruvian coast reflect long-term variation in torrential rainfall events. Quebrada de los Burros (18°S) is a narrow canyon on the western slopes of the southern Peruvian Andes. Fontugne et al. (1999) identified and dated a series of debris flows in this canyon. Because it heads in the hyperarid region below the altitude of seasonal rainfall, the debris flows almost certainly represent extreme precipitation associated with El Niño events. The Quebrada de los Burros flood record has a hiatus between ca. 9600 and 3400 cal yr BP; the sedimentary record for this interval is characterized by organic layers interpreted as indicators of "a permanent water supply resulting from an increased condensation of fog at mid-altitudes" due to enhanced coastal upwelling (Fontugne et al., 1999, p. 171). The organic layers are inconsistent with El Niño activity in this region. Burros is associated with a Mid-Holocene archaeological deposit.

Half a degree north of Quebrada de los Burros, Quebrada Tacahuay is another dry canyon heading below the altitude of seasonal rainfall. There, too, a dated flood record shows a Mid-Holocene hiatus between ca. 8900 and 5700 cal yr BP (Keefer et al., 1998, 2003). This span is almost perfectly coincident with the hiatus in El Niño activity postulated from the archaeological record of coastal Peru. At Tacahuay, the flood deposits separate several episodes of human occupation from the Terminal Pleistocene to the early Mid-Holocene.

Noller's (1993) study of Quaternary soil development along the Peruvian coast shows a major disjunction at 12°S. South of that point, the absence of significant soil development and the presence of soluble minerals and salts indicate long-term hyperaridity. North of 12°S, greater soil development and the absence of significant salt accumulations document periodic rainfall events. This pattern is consistent with a period of seasonal rainfall in the Mid-Holocene; the seasonal SST structure reconstructed by Andrus et al. (2002a) for OBC would result in seasonal precipitation along the north coast during this time.

Noller's soil record is also consistent with the patterns of endemism and adaptation in lomas (fog-based) plant communities in the western foothills of the Andes, overlooking the coast. Rundel and Dillon (1998; Rundel et al., 1991) identify northern and southern Peruvian lomas-flora units with a boundary at 12°S. The southern unit, with a high degree of endemism in each lomas stand, indicates long-term hyperaridity. The northern unit shows greater similarities between now-isolated lomas stands, suggesting periods of greater moisture in the past, when the lomas were continuous. A Mid-Holocene interval of seasonal rainfall would help explain the pattern of lomas endemism.

The northern coast of Peru has five major beach ridge plains, at Santa, Piura, Colán, Chira, and Tumbes. Over the last 30 years, all but Tumbes have been studied in some detail (see Shafer et al., 2004). The Peruvian beach ridges are composed of cobbles (Santa, Colán) or sand (Piura, Chira, Tumbes) and with the exception of Colán are built by material from the four highest flow rivers of the Peruvian coast. All the ridges post-date sea level stabilization and the return of El Niño after 5800 cal yr BP.

We have hypothesized that ridges form when sediment produced by seismic activity is flushed by El Niño-caused torrential rainfall from the unvegetated desert surface of the coast and western slopes into the rivers; the increased competence of the rivers during ENSO rainfall events washes the material out to the shore, where a delta forms and then is reworked in the direction of longshore drift (north) to form the ridges (Sandweiss et al., 1983; Sandweiss, 1986). Internal ridge morphology (ibid.) and remote sensing studies of coastal change at Santa (Moseley et al., 1992) and Chira (Shafer et al., 2004) support this hypothesis.

Each ridge set contains eight or nine major ridges, though each of these is probably a composite of multiple formation events. In each case except Colán, the oldest two or three ridges are larger amplitude and better defined, while the final six ridges are lower amplitude, higher frequency, and less well defined. Based on available dates, most from the Chira ridges (Richardson, 1981, Ortlieb et al., 1993), the transition from high- to low-amplitude ridges occurs around 3000 cal yr BP, when we have identified a shift from less to more frequent El Niño events (Shafer et al., 2004). With a longer recurrence interval between rainfall episodes, there would be more time for multiple seismic events to accumulate material on the landscape, resulting in fewer but larger ridges. After 3000 cal yr BP, more frequent torrential rainfall would flush the landscape more often, leading to more but smaller ridges.

3.2. Paleolimnological data (medium-scale resolution)

Nearly continuous Holocene proxy climate records have been recovered from the Galápagos Islands and from highland lakes in Ecuador, Chile, and Bolivia, though not from the desert coast of Peru. Laguna Pallcacocha, Ecuador (2°46′S, 79°14′W and 4060 masl), analyzed by Rodbell et al. (1999) and Moy et al. (2002), contains sediments with rainstorm-related inorganic laminae. The layers containing clastic sediments that were washed into the lake during storms, measured using gray-scale and color light reflectance, match the historic record of El Niño events for the last 200 years. Using them as a proxy for El Niño, Rodbell et al. (1999) found that from Late Glacial to early Holocene (15,000 to about 7000 cal yr BP), the periodicity of elevated clastic deposition was decadal (greater than or equal to 15 years). Beginning at ~7000 cal yr BP, storm-induced clastic events came about 10–20 and 2–8.5 years apart. After ~5000 cal yr BP 2–8.5-year periodicities were most dominant, more consistent with modern El Niño frequency.

Riedinger et al. (2002) examined lithostratigraphic and mineralogic properties of sediments from hypersaline Bainbridge Crater Lake, Galápagos Islands. These laminated sediments also provide proxy evidence of past El Niño frequency and intensity. The Bainbridge record suggests that between ~7100 and 4600 cal yr BP El Niño activity was present, but infrequent. This record also indicates that the frequency and intensity of El Niño events increased at about 3100 cal yr BP.

Jenny et al. (2002) obtained a multi-proxy Holocene climate record from Laguna Aculeo in central Chile (33°50′S, 70°54′W). This record showed an arid Early to Mid-Holocene period (ca. 9500–5700 cal yr BP). After 5700 cal yr BP effective moisture increased progressively and, around 3200 cal yr BP, modern humid conditions were established. Early Holocene flood deposits, indicative of wet winters, were absent until 5700 cal yr BP. These become frequent after 3200 cal yr BP. This evidence is consistent with weak or no El Niño activity during the Early and Mid-Holocene, followed by infrequent El Niño events and then increased El Niño frequency in the late Holocene. The Laguna Aculeo chronology matches the archaeological chronology of Holocene climatic change from the Peruvian coast almost perfectly. Lake Titicaca, located between Bolivia and Peru (at about 16° to 17.50°S, 68.5° to 70°W, 3810 masl) can be used as a recorder of the precipitation over a large portion of tropical South America. Using lake cores spanning the last 25,000 years, Baker et al. (2001) have shown that maximum aridity and lowest lake level occurred in the early and middle Holocene (8000 to 5500 cal yr BP). The lowest level of Lake Titicaca was reached between 6000–5000 cal yr BP after which lake level rose to close to its modern level. During ENSO events, the Titicaca Basin tends to suffer drought; lake level rise from the mid-Holocene low stand is generally coincident with our period of infrequent ENSOs and cool coastal conditions when drought frequency may have been reduced.

3.3. Marine records (medium- to high-resolution)

Several marine sediment cores from near the coast of Peru and Chile have produced continuous climate proxy records at a resolution high enough to reconstruct Late Glacial to Holocene ENSO variability.

Rein et al. (2005) analyzed a 20,000 year-long, high-resolution marine sediment record from the El Niño region of Peru (core 106KL from 80 km off Lima/Peru; 12°03′S, 77°39.8′W, 184 m). Estimates from 106KL for past sea surface temperature, photosynthesis pigments, and a lithic proxy for El Niño flood events served as a proxy for past ENSO variability. Rein et al. (2005) concluded that an Early Holocene maximum of El Niño activity was followed by weak El Niño activity during the Mid-Holocene period (8000–5600 cal yr BP). The frequency and intensity of El Niño activity with thickest El Niño flood deposits increased after about 3000 cal yr BP. This record, too, fits our archaeologically-based reconstruction of Holocene ENSO frequency.

Lower-resolution marine records from the eastern equatorial Pacific analyzed by Loubere et al. (2003) were used to reconstruct thermocline mixed layer temperatures and nutrient contents. Stable isotopes and assemblage data from benthic and planktonic foraminifera measured in three deep-sea cores obtained off the coast of northern Peru indicate that changes in the thermocline and mixed layer consistent with increased upwelling of cooler waters began sometime after around 7000 cal yr BP.

4. Cultural records

Peruvian coastal archaeological sites contain or are associated with a variety of records pertinent to reconstructing El Niño behavior over the last 13,000 years, as reviewed above in Sections 2 and 3. These include biogeography (e.g., Reitz and Sandweiss, 2001; Reitz et al., in press; Sandweiss et al., 1996), growth increment analysis (e.g., Rollins et al., 1986, 1987), and biogeochemistry (e.g., Andrus et al., 2002a, 2002b, 2005) of mollusks and fish, differential preservation of soft organic materials, stylistic connections between distant regions sharing similar environments (Sandweiss, 1996), flood deposits (e.g., Keefer et al., 2003), and beach ridge morphology (e.g., Sandweiss, 1986; Shafer et al., 2004).

Peruvian sites also demonstrate change through time in cultural attributes that correlate temporally with the changes we have identified in El Niño frequency in the Mid-Holocene (Sandweiss et al., 2001; Sandweiss, 2003; Richardson and Sandweiss, in press). Major indicators of cultural change include settlement pattern (the distribution and function of sites across the landscape), construction style (size, form, and function of monuments as well as dwellings), subsistence practices, long-distance exchange or contact, symbolic content of artifacts and structures, and burial patterns. In this section, we focus on large-scale changes in settlement pattern, construction style, and subsistence. In terms of cultural chronology, the relevant periods are the Early Preceramic Period (ca. 13000–9000 cal yr BP), the Middle Preceramic Period (ca. 9000–5800 cal yr BP), the Late Preceramic Period (ca. 5800–4100 cal yr BP), and the Initial Period (ca. 4100–2800 cal yr BP).

Prior to 5800 cal yr BP, no large-scale monumental architecture has been identified in coastal Peru, and only a few small structures are known elsewhere in the region, as at Nanchoc on the western slopes of the northern Peruvian Andes

(Dillehay, 1992; Netherly and Dillehay, 1986). Coastal sites in the millennia preceding 5800 cal yr BP range from small fishing camps such as Early to Middle Preceramic Quebrada Jaguay (16°30′S, Sandweiss et al., 1998b) and Siches (4°30′S, e.g., Richardson, 1973, 1978; Sandweiss et al., 1996) and Middle Preceramic Ostra Base Camp (8°55′S, e.g., Sandweiss, 1996; Sandweiss et al., 1996) to villages such as Paloma (12°30′S, e.g., Benfer, 1990). Early and Middle Preceramic coastal sites had subsistence systems based on marine resources, wild plants, and occasionally early domesticated plants (Sandweiss, in press). North of 10°S, marine organisms recovered from these sites are predominately warm-water species (e.g., Sandweiss et al., 1996; Reitz and Sandweiss, 2001; Reitz et al., in press; see Section 2).

Human populations in Peru grew through time (e.g., Rick, 1987) and consequently created more and larger archaeological sites. Combined with the stabilization of sea level during the Mid-Holocene, this demographic trend resulted in an increasing number of sites preserved for analysis. In the following paragraphs, we review data for the best-known Late Preceramic and Initial Period coastal sites (see Table 1 and Moseley, 2001; Burger, 1992, *inter alia* for other sites of this time).

4.1. Late Preceramic Period

Coastal monuments first appear during the Late Preceramic Period, after the climatic transition at 5800 cal yr BP. Although Late Preceramic mounds are distributed between Lima (12°S) and the Salinas de Chao (8°40′S), it is now clear that the first florescence of monument building in coastal Peru took place on the North Central Coast (aka Norte Chico) between about 10°S and 11°S. At Aspero (10°45′S) on the shore of the Supe Valley, Feldman (1980, 1985) excavated several small, early temple mounds, but only recovered materials from the last several construction phases. These phases date to ca. 5000–4300 cal yr BP. However, Feldman also obtained one anomalously early date of ca. 5650 cal yr BP on charcoal that may have been recycled from an earlier construction phase and may therefore indicate an onset of monument building as early as that date.

Subsistence at Aspero was based on fishing, farming, and gathering. All the marine species are typical, cool-water Peru Current taxa. Among the mollusks, *Choromytilus chorus* and *Mesodesma donacium* were particularly important. The most important domesticated plants were cotton (*Gossypium barbadense*, for nets and textiles) and gourd (*Lagenaria siceraria*, for floats and containers) (Feldman, 1980), utilitarian species which Moseley (1975) calls industrial plants. Though present in Peru by the Late Preceramic Period (e.g., Perry et al., 2006), maize was not a dietary staple on the coast.

Though known for decades as Chupacigarro Grande (e.g., Kosok, 1965), the site now called Caral (10°45′S) (Fig. 2.3) was not proven to be Late Preceramic in age until recently (Shady Solís et al., 2001). A suite of radiocarbon dates, many on short-lived plants used in construction, place the site between about 4600 and 3900 cal yr BP (ibid.). Called the New World's first city, Caral is a complex settlement

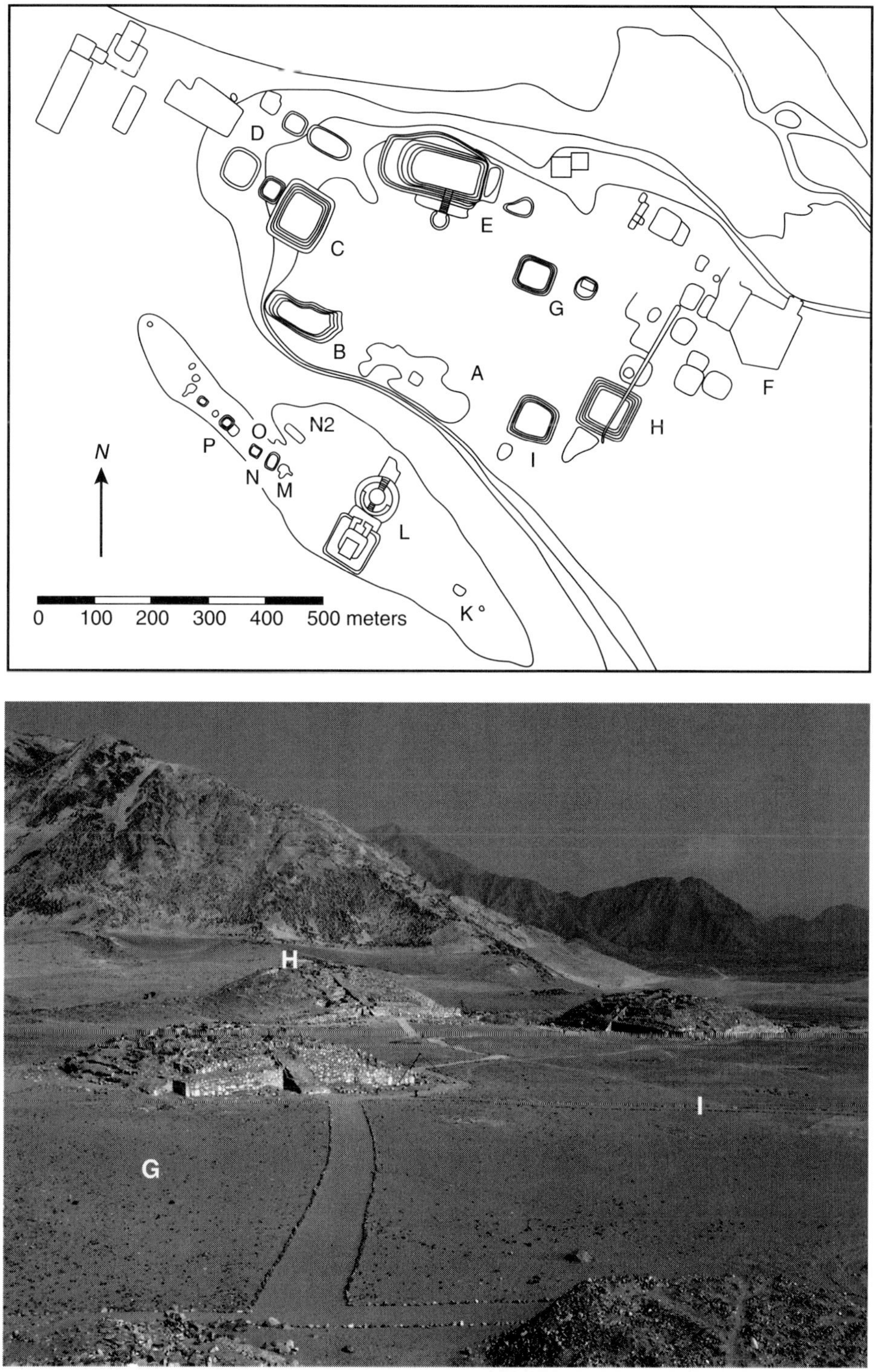

Figure 2.3. Plan of Caral (top, after Shady Solís et al., 2001) and photo of mounds G, H, and I at Caral, taken from Mound E (the Great Temple). Photo by D.H. Sandweiss.

with six large mounds and residential areas with different kinds of architecture suggesting different social classes (ibid.; Shady Solís and Leyva, 2003; Shady Solís, 2005). In contrast to Aspero, Los Morteros, and other Late Preceramic monumental sites known before 2001, Caral is located about 25 km inland, up the same valley as Aspero. Work by Shady Solís elsewhere in the Supe Valley, and more recently by Haas et al. (2004) in neighboring valleys, has uncovered more inland Late Preceramic centers with mounds.

Though subsistence data for the sites located by Haas have not been published in detail (the sites have only been tested to acquire samples for dating), Caral has been extensively excavated for over a decade, and the full panoply of remains are being analyzed by R. Shady's multidisciplinary team (e.g., Shady Solís and Leyva, 2003; Shady Solís, 2005). Despite the distance from the shore, the animal diet came almost entirely from the ocean. As elsewhere on the North Central Coast and Central Coast, *Choromytilus* and *Mesodesma* were dominant molluscan species, the most abundant fish were sardines (*Sardinops sagax sagax*) and anchoveta (*Engraulis ringens*), and the most common plants were cotton and gourd.

Los Morteros is a large mound on the fossil bay at Salinas de Chao (8°40′S). Radiocarbon dates on materials from shallow excavations date the final occupation of the structure to ca. 5500–5100 cal yr BP (Cárdenas, 1979, 1995); the structure itself is earlier, though how much earlier is unknown at this time. Molluscan remains from this site are typical cool-water Peru Current species. Los Morteros is the northernmost Late Preceramic monumental structure on the Peruvian coast.

Near Lima, El Paraíso is a large aceramic site with dates falling at the end of the Late Preceramic Period and overlapping the Initial Period (ca. 4100–3200 cal yr BP; Quilter, 1985; Quilter et al., 1991). The site covers about 58 ha and consists of six large mounds and at least five smaller structures. Though test excavations failed to find evidence for a large resident population, primary midden did provide insight into diet (Quilter et al., 1991) and climate (Sandweiss et al., 1996). Like other Late Preceramic sites, mollusks and fish provided most of the animal food, while plant food was a combination of wild and domesticated taxa. Once again, the most important crops were cotton and gourd.

Although modest-sized permanent settlements such as Asia Unit 1 (12°30′S; Engel, 1963) have been found south of Lima, El Paraíso is the southernmost Late Preceramic monumental site known to date.

Debate continues about the temporal priority of shoreline vs. inland centers in the Late Preceramic Period (Haas and Creamer, 2006; cf. Sandweiss, 2006), but the weight of evidence currently available supports a sequence beginning on the coast with fishing/farming sites, with later population growth driving expansion inland to increase production of cotton and gourds to intensify the fishing industry (Sandweiss and Rademaker, 2006; Sandweiss, in press). How complex Late Preceramic societies really were continues to be debated, but the recent work at Caral and the other North Central Coast monumental sites supports earlier arguments for social stratification, at least in the core region between about 12° and 8°S. The North Central Coast was the center of Late Preceramic development, with the greatest number, size, and complexity of monumental sites. In this pristine setting, supernatural sanctions (religion) must have played an important role in the consolidation of power in the hands of a nascent elite (Roscoe, in press).

4.2. Initial Period

During the Initial Period, the size of monumental structures increases and the geographical ranges expands south to the Lurín Valley just south of Lima (12°15′S) and north to the Lambayeque Valley (6°30′S). Like the majority of Late Preceramic monumental sites in the North Central Coast valleys, Initial Period monuments throughout the entire range tend to be located inland from the shore. Seafood is still important, but agriculture plays an increasingly significant role in subsistence (see Burger, 1992; Moseley, 2001, for a review of Initial Period coastal sites). The suite of marine species exploited during the Initial Period is substantially similar to that of the Late Preceramic Period, with *Choromytilus* and *Mesodesma* among the most important mollusks and sardines and anchoveta dominating the fish (Sandweiss et al., 2001).

Monumental construction ceased or decreased greatly in the North Central Coast valleys after the Late Preceramic Period, and the Casma Valley (9°30′S) became the focal point for Initial Period development. Among the many Casma sites of this time, Sechin Alto was the largest mound in the Americas for its epoch; like Pampa de las Llamas/Moxeke, Sechin Alto and associated sites demonstrate large-scale site planning (Pozorski and Pozorski, 1987). At Pampa de las Llamas/Moxeke, this plan extends across 2 km, uniting a temple mound (Moxeke) with a monumental storeroom (Pampa de las Llamas) along a central axis of symmetry (Pozorski and Pozorski, 1986, 1987).

A secondary focus of development occurred in the valleys around Lima (12°S), with sites such as Huaca la Florida (Patterson, 1985) and Garagay (Ravines et al., 1982) in the Rimac Valley, and a series of mound sites in the Lurín Valley (Burger, 1992). Burger's work at three of the Lurín centers, Cardal (Burger and Salazar-Burger, 1991), Mina Perdida (Burger, 1992), and Manchay Bajo (Burger, 2003), showed that these mounds were built incrementally. Burger and Salazar-Burger (1991) argue that the Lurín mounds would not have required sufficient labor and central direction to justify attributing the sites to a complex society. This view contrasts with the Pozorskis' (1986, 1987) interpretation of the Casma Initial Period sites as evidence for an early state. Given differences in the size and complexity of sites in the two valleys, social complexity may well have been unevenly distributed along the coast at this time.

Regardless of the level of social complexity in the different valleys of the Peruvian coast, people living in many of the valleys between about 6°S and 12°S built mounds during the Initial Period, continuing the tradition begun in the Late Preceramic Period. At the end of the Initial Period, the 3000-year sequence of coastal monument building came to a halt for at least several centuries at the same time that El Niño events increased in frequency (Sandweiss et al., 2001).

5. Conclusions

5.1. Summary

Drawing on the data reviewed in the preceding sections, we reconstruct the following sequence of El Niño frequency shifts and related cultural change on the Peruvian coast during the Holocene.

Before ca. 9000 cal yr BP, El Niño was present, but we do not know the frequency. People were fisher–hunter–gatherers living seasonally in small settlements such as Quebrada Jaguay.

Between ca. 9000–5800 cal yr BP, El Niño was absent or very low frequency. Fisher–hunter–gatherer lifeways continued with the addition of domesticated plants such as gourds (e.g., Erickson et al., 2005). Some settlements grew in size and may have been permanent villages such as Paloma.

Between ca. 5800–3000 cal yr BP, El Niño was present but at lower frequency than modern. Not long after the return of El Niño, people began building monumental structures on the Central and North Central Coasts. This mound-building tradition continued through this entire timespan, comprising the Late Preceramic and Initial Periods. Specific sites were built, used, and abandoned, and different valleys rose and fell in prominence, but viewed at the regional level, the tradition was unbroken.

After ca. 3000 cal yr BP, El Niño continued, but at frequencies within the modern range of variability. Shortly after this second climatic shift, the mound-building tradition stopped for hundreds of years.

5.2. Chronologies

Our chronology is built on remains found in, or in direct association with, archaeological sites. Because our ultimate goal is to help explain the cultural development of the study region, this approach gives us the most appropriate sequence. However, while the broad correlations between Mid-Holocene cultural and climatic change for the Peruvian coast are robust, we recognize that developing a detailed chronology is difficult given the multiple sources of error in age estimation from the various available archives. Most of the records we use depend on radiocarbon dating of both marine and terrestrial materials. Atmospheric ^{14}C dates must be corrected for the variable radiocarbon production rate. Further, most of the archaeological dates available for the times and places of interest are bulk dates, with the potential to be biased by old wood (e.g., Kennett et al., 2002). However, a contextual review of the available archaeological dates does not indicate a notable old wood problem for the north coast of Peru – dates tend to be in stratigraphic order and consistent across sites with similar content. For marine dates, we face the additional uncertainty of determining the appropriate reservoir correction. With variation in upwelling through the Holocene, the magnitude of the reservoir must have changed by centuries or more through time as well as through space. On-going work by Andrus and colleagues (2005) should provide a much more detailed picture of spatiotemporal change in the marine reservoir of Holocene coastal Peru.

Our chronology accords well with many natural proxy records throughout the region and the Pacific basin (see Section 3 in this chapter and other chapters in this volume). Most records indicate a period of greatly reduced interannual variability in the Pacific basin during parts of the Early and Mid-Holocene, followed by increasing interannual variability. However, not all agree with our exact timing or sequence. Laguna Pallcacocha in Ecuador (Rodbell et al., 1999; Moy et al., 2002), for instance, has an offset of approximately one to two millennia in the onset of El Niño and the later increase in ENSO frequency. At this time, we cannot say whether this discrepancy reflects a problem with chronology building or a real difference in the timing of change in the Ecuador highlands and coastal Peru. In general, global climatic change patterns have potential leads and lags from region to region; resolving those has to do with questions we are currently unable to resolve but which will be a focus of future research.

5.3. Climate and culture in Mid-Holocene Peru

Over the last 30 years, we have accumulated evidence for two major climate transitions during the Holocene on the Peruvian coast. Each of these transitions also marks a notable change in coastal societies as expressed in their settlements, subsistence, the construction (or not) of monumental architecture, and social complexity. With such temporal conjunctures, it is tempting to go from collation to correlation to causation (Sandweiss and Quilter, in press). We can easily spin a plausible story about temples to control the new climatic variability introduced with the onset of El Niño, 3000 years of success while recurrence intervals were long (50 years to a century at least), and then a crisis of faith and temple abandonment when recurrence intervals became drastically shorter (probably less than 15 years). Nevertheless, such temptation is dangerous; as Sandweiss et al. wrote in 2001

> Technology, history, cultural practices, religion, perception, and individual and group idiosyncrasies can all affect the way a society and its members respond to change. However, radical environmental change requires some response from the people who experience it.

In this chapter, we have reviewed the development of data on climatic and temporally associated cultural change during the Mid-Holocene on the Peruvian coast. We do believe that there is a relationship, though one of such complexity that it will be extraordinarily difficult to reconstruct in detail. In the final paragraphs, we point to the clearest conjunctures of climate and culture, as a guide to future research.

The most conclusive and temporally detailed evidence for a sharp climatic transition at ca. 5800 cal yr BP comes from a suite of dates on marine mollusks recovered *in situ* on paleobeaches by Perrier et al. (1994) and reproduced in Figure 2.4 (after

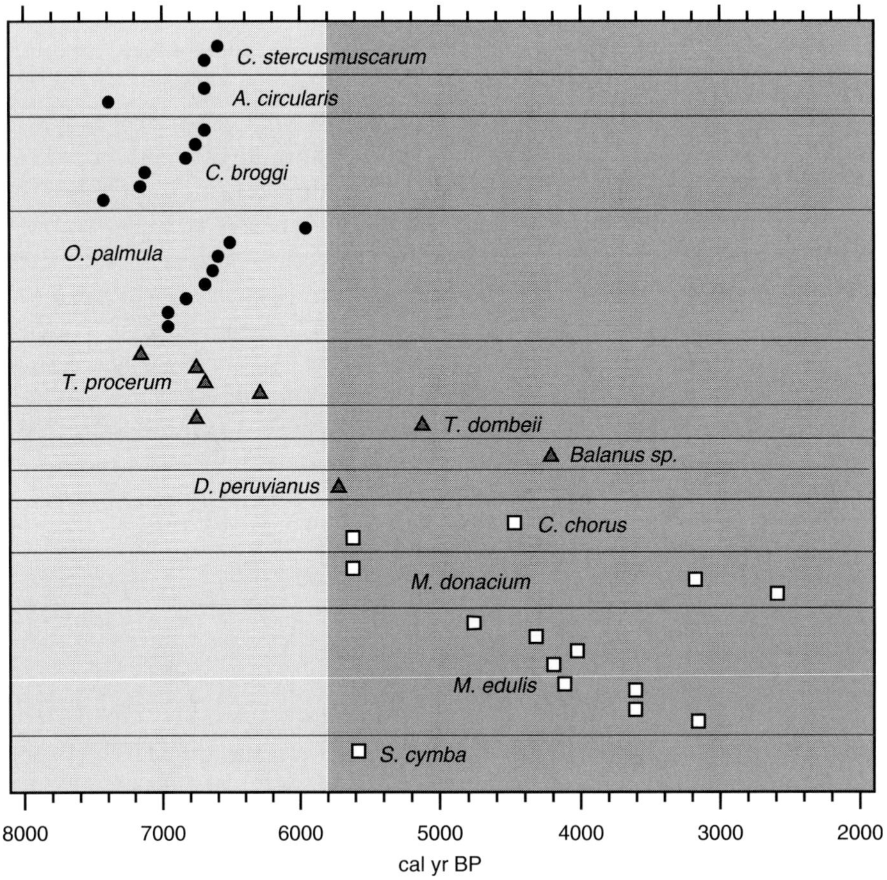

Figure 2.4. Dates on mollusks from the fossil bays at Santa and Salinas de Chao, northern Peru, arranged by water temperature requirements of the taxa. Circles are dates on species that can only tolerate warm-tropical water; triangles are dates on species that can live in both warm-tropical and warm-temperate water; and squares are species found exclusively in warm-temperate water. The difference in shading marks the postulated return of El Niño at 5800 cal yr BP. The data are from Perrier et al. (1994); the figure is after Andrus et al. (2003).

Andrus et al., 2003). All these dates should be subject to the same biases, so that even if the exact timing is offset, the direction and nature of the transition is obvious and well aligned with the less tightly constrained dates from archaeological materials (both marine and terrestrial).

The timing of change in the cultural record also fits this sequence but is less precise in chronological detail. Because early mounds tended to be built in multiple phases one on top of the other, and as yet few excavations have reached or dated the initial construction phases, is it still not possible to date the initiation of coastal temple building. The earliest dates for the use of temple mounds come from the test pits at several sites in the North Central coast valleys of Huaura, Supe, Pativilca, and Fortaleza (11°10′S to 10°40′S) (Haas et al., 2004), which lack detailed context, from the final occupation at Los Morteros (Cárdenas 1979, 1995), and from Aspero (Feldman, 1985), where the earliest date is out of context and the stratigraphically coherent suite of later dates refers to the final two construction phases (ca. 4150–5300 cal yr BP). At this time, no known dates for monumental structures fall prior to the climatic transition at 5800 cal yr BP. However, dates late in the construction sequence at Aspero and Los Morteros fall within a few centuries of 5800 cal yr BP, as do some dates from the North Central Coast sites.

The collapse of the early mound-building tradition on the Peruvian coast after about 3000 cal yr BP is apparent in the absence of radiocarbon dates on temple mounds throughout the region during a several hundred year span following approximately 3000 cal yr BP. There is one exception, Manchay Bajo in the Lurin Valley, which lasted about 100 years longer than other sites. A massive wall was built at this site, not surrounding the site as would be expected for a defensive structure, but instead protecting the monument from El Niño-related debris flows coming out of two quebradas behind the site (Burger, 2003). At Manchay Bajo, the temple leaders thus appear to have invested in El Niño mitigation strategies (Sandweiss et al., 2001; Burger, 2003). This reminds us that mound-building may be linked to climatic change, but it is ultimately the result of human decision-making embedded in historical and cultural context.

References

Andrus, C. F. T., D. E. Crowe, E. J. Reitz, and C. S. Romanek, 2002a. Otolith $\partial^{18}O$ record of Mid-Holocene sea surface temperatures in Peru. *Science* 295:1508–1511.

Andrus, C. F. T., D. E. Crowe, D. H. Sandweiss, E. J. Reitz, C. S. Romanek, and K. A. Maasch, 2003. Response to comment on "Otolith $\partial^{18}O$ record of Mid-Holocene sea surface temperatures in Peru". *Science* 209:203b.

Andrus, C. F. T., G. W. L. Hodgins, D. H. Sandweiss, and D. E. Crowe, 2002b. Molluscan radiocarbon as a proxy for El Niño-related upwelling variation in Peru. *Geological Society of America Abstracts with Program* 140:12.

Andrus, C. F. T., G. W. L. Hodgins, D. H. Sandweiss, and D. E. Crowe, 2005. Molluscan radiocarbon as a proxy for El Niño-related upwelling variations in Peru. In *Isotopic and Elemental Tracers of Cenozoic Climate Change*, edited by G. Mora and D. Surge, pp. 13–20. Geological Society of America Special Paper 395, Boulder, CO.

Baker, P. A., G. O. Seltzer, S. C. Fritz, R. B. Dunbar, M. J. Grove, P. M. Tapia, S. L. Cross, H. D. Rowe, and J. P. Broda, 2001. The history of South American tropical precipitation for the past 25,000 years. *Science* 291:640–643.

Benfer, R. A., 1990. The Preceramic period site of Paloma, Peru: bioindicators of improving adaptation to sedentism. *Latin American Antiquity* 1:284–318.

Bonavia, D., 1982. *Los Gavilanes*. COFIDE and Instituto Arqueológico Alemán, Lima.

Bonavia, D., 1996. De la caza-recolección a la agricultura: una perspectiva local. *Bulletin de l'Institut Français d'Études Andines* 25:169–186.

Bonavia, D., L. W. Johnson, E. J. Reitz, E. S. Wing, and G. H. Weir, 1993. Un sitio precerámico de Huarmey (PV35-6) antes de la introducción de maíz. *Bulletin de l'Institut Français d'Études Andines* 22:409–442.

Burger, R. L., 1992. *Chavin and the Origins of Andean Civilization*. Thames and Hudson, London and New York.

Burger, R. L., 2003. El Niño, early Peruvian civilization, and human agency, some thoughts from the Lurin Valley. In *El Niño in Perú, Biology and Culture over 10,000 years*, edited by J. Haas and M. O. Dillon, pp. 90–107. Field Museum of Natural History, Chicago, IL. Fieldiana Botany 43.

Burger, R. L., and L. Salazar-Burger, 1991. Recent investigations at the Initial Period center of Cardal, Lurin Valley. *Journal of Field Archaeology* 18:275–296.

Cárdenas M., 1979. *A Chronology of the Use of Marine Resources in Ancient Peru*. Pontificia Universidad Católica del Perú, Volkswagenwerk Stiftung, Lima.

Cárdenas M., 1995. El sitio precerámico de Los Morteros, pampa de Salinas de Chao. *Boletín de Lima* 100:45–56.

Cárdenas, M., J. Vivar, G. Olivera, and B. Huapaya, 1993. *Materiales Arqueológicos del Macizo de Illescas Sechura-Piura*. Pontificia Universidad Católica del Perú, Lima.

Chauchat, C., E. S. Wing, J. P. Lacombe, P. Y. Demars, S. Uceda, and C. Deza, 1992. *Préhistoire de la Côte Nord du Pérou: Le Paijanien de Cupisnique*. Cahiers du Quaternaire 18. CNRS-Editions, Centre Régional de Publication de Bordeaux, Bordeaux.

Chavez, F. P., J. Ryan, S. E. Lluch-Cota, and M. Ñiquen C, 2003. From anchovies to sardines and back: multidecadal change in the Pacific Ocean. *Science* 299:217–221.

deFrance, S. D., D. K. Keefer, J. B. Richardson III, and A. Umire Alvarez, 2001. Late Paleo-Indian coastal foragers: specialized extractive behavior at Quebrada Tacahuay, Peru. *Latin American Antiquity* 12:413–426.

DeVries, T. J., and L. E. Wells, 1990. Thermally-anomalous Holocene molluscan assemblages from coastal Peru: evidence for paleogeographic, not climatic change. *Palaeogeography, Palaeoclimatology, Palaeoecology* 81:11–32.

Dillehay, T. D., 1992. Widening the socio-economic foundation of Andean Civilization: prototypes of early monumental architecture. *Andean Past* 3:55–65.

Engel, F. A., 1963. *Asia Unit 1. A Preceramic settlement on the central coast of Peru*. Transactions of the American Philosophical Society 53, Part 3.

Erickson, D. L., B. D. Smith, A. C. Clark, D. H. Sandweiss, and N. Tuross, 2005. An Asian origin for a 10,000-year-old domesticated plant in the Americas. *Proceedings of the National Academy of Science USA* 102:18315–18320.

Feldman, R. A., 1980. Aspero, Peru: architecture, subsistence economy, and other artifacts of a Preceramic Maritime Chiefdom. Unpublished Ph.D. dissertation, Department of Anthropology, Harvard University, Cambridge, MA.

Feldman, R. A., 1985. Preceramic corporate architecture: evidence for development of non-egalitarian social systems in Peru. In *Early Ceremonial Architecture in the Andes*, edited by C. B. Donnan, pp. 71–92. Dumbarton Oaks, Washington, DC.

Fontugne, M., P. Usselmann, D. Lavallée, M. Julien, and C. Hatté, 1999. El Niño variability in the coastal desert of southern Peru during the Mid-Holocene. *Quaternary Research* 52:171–179.

Haas, J., and W. Creamer, 2006. Crucible of Andean civilization: the Peruvian coast from 3000 to 1800 BC. *Current Anthropology* 47:745–756, 766–775.

Haas, J., W. Creamer, and A. Ruiz, 2004. Dating the Late Archaic occupation of the Norte Chico region in Peru. *Nature* 432:1020–1023.

Jenny, B., B. L. Valero-Garcés, R. Villa-Martínez, R. Urrutia, M. Geyh, and H. Veit, 2002. Early to Mid-Holocene aridity in central Chile and the southern Westerlies: the Laguna Aculeo record (34°S). *Quaternary Research* 58:160–170.

Keefer, D. K., S. D. deFrance, M. E. Moseley, J. B. Richardson III, D. R. Satterlee, and A. Day-Lewis, 1998. Early maritime economy and El Niño events at Quebrada Tacahuay, Peru. *Science* 281:1833–1835.

Keefer, D. K., M. E. Moseley, and S. D. deFrance, 2003. A 38000-year record of floods and debris flows in the Ilo region of southern Peru and its relation to El Niño events and great earthquakes. *Palaeogeography, Palaeoclimatology, Palaeoecology* 194:41–77.

Kennett, D. J., B. L. Ingram, J. R. Southon, and K. Wise, 2002. Differences in 14C age between stratigraphically associated charcoal and marine shell from the Archaic Period site of Kilometer 4, southern Peru: old wood or old water? *Radiocarbon* 44:53–58.

Kosok, P., 1965. *Life, Land, and Water in Ancient Peru.* Long Island University Press, New York.

Lavallée, D., M. Julien, P. Béarez, P. Usselmann, M. Fontugne, and A. Bolaños, 1999. Pescadores-recolectores arcaicos del extremo sur peruano. Excavaciones en la Quebrada de los Burros (Tacna, Perú). Primeros resultados 1995–1997. *Bulletin de l'Institut Français d'Études Andines* 28:13–52.

Loubere, P., M. Richaud, Z. Liu, and F. Mekik, 2003. Oceanic conditions in the eastern equatorial Pacific during the onset of ENSO in the Holocene. *Quaternary Research* 60:142–148.

Maasch, K. A., in press. El Niño and interannual variability of climate in the Western Hemisphere. In *El Niño, Catastrophism, and Culture Change in Ancient America*, edited by D. H. Sandweiss and J. Quilter. Dumbarton Oaks, Washington, DC.

Moseley, M. E., 1975. *The Maritime Foundations of Andean Civilization.* Cummings, Menlo Park, CA.

Moseley, M. E., 2001. *The Incas and their Ancestors: The Archaeology of Peru.* 2nd edition. Thames and Hudson, London and New York.

Moseley, M. E., D. Warner, and J. B. Richardson III, 1992. Space shuttle imagery of recent catastrophic change along the arid Andean coast. In *Paleoshorelines and Prehistory: An Exploration of Method*, edited by L. L. Johnson and M. Stright, pp. 215–235. CRC Press, Boca Raton.

Moy, C. M., G. O. Seltzer, D. T. Rodbell, and D. M. Anderson, 2002. Variability of El Niño/southern oscillation activity at millennial timescales during the Holocene epoch. *Nature* 420:162–165.

Netherly, P., and T. D. Dillehay, 1986. Duality in public architecture in the upper Zaña Valley. In *Perspectives in Andean Prehistory and Protohistory*, edited by D. H. Sandweiss and P. Kvietok, pp. 85–115. Cornell University Latin American Studies Program, Ithaca.

Noller, J. S., 1993. Late Cenozoic stratigraphy and soil geomorphology of the Peruvian desert, 3 degrees-18 degrees S: a long-term record of hyperaridity and El Niño. Unpublished Ph.D. dissertation, University of Colorado, Boulder.

Ortlieb, L., M. Fournier, and J. Macharé, 1993. Beach ridge series in northern Peru: chronology, correlation, and relationship with major Late Holocene El Niño Events. *Bulletin de l'Institut Français d'Études Andines* 22:191–212.

Patterson, T. C., 1985. The Huaca La Florida, Rimac Valley, Perú. In *Early Ceremonial Architecture in the Andes*, edited by C. B. Donnan, pp. 59–70. Dumbarton Oaks, Washington, DC.

Perrier, C., C. Hillaire-Marcel, and L. Ortlieb, 1994. Paléogéographie littorale et enregistrement isotopique (13C18O) d'événements de type El Niño par les mollusques holocènes et récents du nord-ouest péruvien. *Géographie Physique et Quaternaire* 48:23–38.

Perry, L., D. H. Sandweiss, D. Piperno, K. Rademaker, M. A. Malpass, A. Umire, and P. de la Vera, 2006. Early maize agriculture and interzonal interaction in southern Peru. *Nature* 440:76–79.

Pozorski, S., and T. Pozorski, 1986. Recent excavations at Pampa del las Llamas-Moxeke, a complex Initial Period site. *Journal of Field Archaeology* 13:381–401.

Pozorski, S., and T. Pozorski, 1987. *Early Settlement and Subsistence in the Casma Valley, Peru.* University of Iowa Press, Iowa City.

Pozorski, S., and T. Pozorski, 1990. Huaynuná: A late Cotton Preceramic site on the north coast of Peru. *Journal of Field Archaeology* 17:17–26.

Pozorski, S., and T. Pozorski, 1995. Paleoenvironment at Almejas, a mid-Holocene site in the Casma Valley, Peru. *Society for American Archaeology Abstracts of the 60th Annual Meeting*, p. 154.

Pozorski S. G., 1979. Prehistoric diet and subsistence in the Moche Valley, Peru. *World Archaeology* 11:163–184.

Quilter, J., 1985. Architecture and chronology at El Paraíso, Peru. *Journal of Field Archaeology* 12:279–297.

Quilter, J., B. Ojeda, D. Pearsall, D. H. Sandweiss, J. Jones, and E. Wing, 1991. The subsistence economy of El Paraíso, Peru. *Science* 251:277–283.

Quinn, W. H., 1992. A study of southern oscillation-related climatic activity for A. D. 622–1900 incorporating Nile River flood data. In *El Niño Historical and Paleoclimatic Aspects of the Southern Oscillation*, edited by H. F. Diaz and V. Markgraf, pp. 119–149. Cambridge University Press, Cambridge, UK.

Rajagopalan, B., U. Lall, and M. A. Cane, 1997. Anomalous ENSO occurrences: an alternate view. *Journal of Climate* 10:2351–2357.

Ravines, R., H. Engelstad, V. Palomino, and D. H. Sandweiss, 1982. Materiales arqueológicos de Garagay. *Revista del Museo Nacional (Lima)* 46:135–234.

Rein, B., A. Lückge, L. Reinhardt, F. Sirocko, A. Wolf and W.-C Dullo, 2005. El Niño variability off Peru during the last 20,000 years. *Paleoceanography* 20:PA4003.

Reitz, E. J., 1988. Faunal remains from Paloma, an Archaic site in Peru. *American Anthropologist* 90:310–322.

Reitz, E. J., C. F. T. Andrus, and D. H. Sandweiss, in press. Ancient fisheries and marine ecology of coastal Peru. In *Human Impacts on Ancient Marine Ecosystems: Regional Case Studies and Global Perspectives*, edited by T. C. Rick and J. M. Erlandson. University of California Press, Berkeley.

Reitz, E. J., and D. H. Sandweiss, 2001. Environmental change at Ostra Base Camp, a Peruvian preceramic site. *Journal of Archaeological Science* 28:1085–1100.

Richardson III, J. B., 1973. The Preceramic sequence and the Pleistocene and post-Pleistocene climate of northwest Peru. In *Human Variation*, edited by D. W. Lathrap and J. Douglas, pp. 73–89. University of Illinois Press, Urbana.

Richardson III, J. B., 1978. Early man on the Peruvian north coast, early maritime exploitation and Pleistocene and Holocene environment. In *Early Man in America from a Circum-Pacific Perspective*, edited by A. L. Bryan, pp. 274–289. Occasional Papers No. 1, Department of Anthropology, University of Alberta, Edmonton, Canada.

Richardson III, J. B., 1981. Modeling the development of early complex economies on the coast of Peru: a preliminary statement. *Annals of Carnegie Museum* 50:139–150.

Richardson III, J. B., and D. H. Sandweiss, in press. Climate change, El Niño and the rise of complex society on the Peruvian coast during the middle Holocene. In *El Niño, Catastrophism, and Culture Change in Ancient America,* edited by D. H. Sandweiss and J. Quilter. Dumbarton Oaks, Washington, DC.

Rick, J. W., 1987. Dates as data: an examination of the Peruvian Preceramic radiocarbon record. *American Antiquity* 52:55–73.

Riedinger, M. A., M. Steinitz-Kannan, W. M. Last, and M. Brenner, 2002. A ~6100 14C yr record of El Niño activity from the Galápagos Islands. *Journal of Paleolimnology* 27:1–7.

Rodbell, D. T., G. O. Seltzer, D. M. Anderson, M. B. Abbott, D. B. Enfield, and J. H. Newman, 1999. An ~15,000-year record of El Niño-driven alluviation in southwestern Ecuador. *Science* 283:516–520.

Rollins, H. B., J. B. Richardson III, and D. H. Sandweiss, 1986. The birth of El Niño: geoarchaeological evidence and implications. *Geoarchaeology* 1:3–15.

Rollins, H. B., D. H. Sandweiss, U. Brand, and J. C. Rollins, 1987. Growth increment and stable isotope analysis of marine bivalves: implications for the geoarchaeological record of El Niño. *Geoarchaeology* 2:181–187.

Roscoe, P. B., in press. Catastrophe and the emergence of political complexity: a social anthropological model. In *El Niño, Catastrophism, and Culture Change in Ancient America*, edited by D. H. Sandweiss and J. Quilter. Dumbarton Oaks, Washington, DC.

Rundel, P. W., and M. O. Dillon, 1998. Ecological patterns in the Bromeliaceae of the lomas formations of coastal Chile and Peru. *Plant Systematics and Evolution* 212:261–278.

Rundel, P. W., M. O. Dillon, H. A. Mooney, S. L. Gulmon, and J. R. Ehleringer, 1991. The phytogeography and ecology of the coastal Atacama and Peruvian deserts. *Aliso* 13:1–50.

Sandweiss, D. H., 1986. The beach ridges at Santa, Peru: El Niño, uplift, and prehistory. *Geoarchaeology* 1:17–28.

Sandweiss, D. H., 1992. The archaeology of Chincha fishermen: Specialization and status in Inka Peru. *Carnegie Museum of Natural History Bulletin 29*, ix + 162 pp.

Sandweiss, D. H., 1996. Mid-Holocene cultural interaction on the north coast of Peru and Ecuador. *Latin American Antiquity* 7:41–50.

Sandweiss, D. H., 2003. Terminal Pleistocene through Mid-Holocene archaeological sites as paleoclimatic archives for the Peruvian coast. *Palaeogeography, Palaeoclimatology, Palaeoecology* 194:23–40.

Sandweiss, D. H., 2006. Comment on 'Crucible of Andean civilization: The Peruvian coast from 3000 to 1800 BC,' by J. Haas and W. Creamer. *Current Anthropology* 47:760.

Sandweiss, D. H., in press. Early fishing and inland monuments: challenging the maritime foundations of Andean civilization? In *Foundations of Andean Civilization: Papers in Honor of Michael E. Moseley*, edited by C. Marcus, C. Stanish, and R. Williams. Cotler Institute of Archaeology, Los Angeles.

Sandweiss, D. H., K. A. Maasch, D. F. Belknap, J. B. Richardson III, and H. B. Rollins, 1998a. Discussion of 'The Santa beach ridge complex,' by Lisa E. Wells. *Journal of Coastal Research* 12:1–17.

Sandweiss, D. H., K. A. Maasch, R. L. Burger, J. B. Richardson III, H. B. Rollins, and A. Clement, 2001. Variation in Holocene El Niño frequencies: climate records and cultural consequences in ancient Peru. *Geology* 29:603–606.

Sandweiss, D. H., K. A. Maasch, F. Chai, C. F. T. Andrus, and E. J. Reitz, 2004. Geoarchaeological evidence for multi-decadal natural climatic variability and ancient Peruvian fisheries. *Quaternary Research* 61:330–334.

Sandweiss, D. H., H. McInnis, R. L. Burger, A. Cano, B. Ojeda, R. Paredes, M. Sandweiss, and M. Glascock, 1998b. Quebrada Jaguay: early maritime adaptations in South America. *Science* 281:1830–1832.

Sandweiss, D. H., and J. Quilter, in press. Climate, catastrophe, and culture in the ancient Americas. In *El Niño, Catastrophism, and Culture Change in Ancient America*, edited by D. H. Sandweiss and J. Quilter. Dumbarton Oaks, Washington, DC.

Sandweiss, D. H., and K. Rademaker, 2006. First civilization in the Americas. *2007 Science Year/The World Book Annual Science Supplement* 12–25.

Sandweiss, D. H., J. B. Richardson III, E. J. Reitz, J. T. Hsu, and R. A. Feldman, 1989. Early maritime adaptations in the Andes: preliminary studies at the Ring Site, Peru. In *Ecology, Settlement, and History in the Osmore Drainage, Peru*, edited by D. S. Rice, C. Stanish, and P. R. Scarr, pp. 35–84. BAR International Series 545(i), Oxford, England.

Sandweiss, D. H., J. B. Richardson III, E. J. Reitz, H. B. Rollins, and K. A. Maasch, 1996. Geoarchaeological evidence from Peru for a 5000 years B.P. onset of El Niño. *Science* 273:1531–1533.

Sandweiss, D. H., J. B. Richardson III, E. J. Reitz, H. B. Rollins, and K. A. Maasch, 1997. Determining the beginning of El Niño: response [to comments]. *Science* 276:966–967.

Sandweiss, D. H., H. B. Rollins, and J. B. Richardson III, 1983. Landscape alteration and prehistoric human occupation on the north coast of Peru. *Annals of Carnegie Museum* 52:277–298.

Shady Solís, R., 2005. *La Civilización de Caral-Supe: 5000 Años de Identidad Cultural en el Perú*. Instituto Nacional de Cultura, Proyecto Especial Arqueológico Caral-Supe, Lima.

Shady Solís, R., J. Haas, and W. Creamer, 2001. Dating Caral, a preceramic site in the Supe Valley on the central coast of Peru. *Science* 292:723–726.

Shady Solís, R., and C. Leyva, editors, 2003. *La Ciudad Sagrada de Caral-Supe: Los Orígenes de la Civilización Andina y la Formación del Estado Pristino en el Antiguo Perú*. Instituto Nacional de Cultura, Proyecto Especial Arqueológico Caral-Supe, Lima.

Shafer Rogers, S., D. H. Sandweiss, K. A. Maasch, D. F. Belknap, and P. Agouris, 2004. Coastal change and beach ridges along the northwest coast of Peru: Image and GIS a of the Chira, Piura, and Colán beach-ridge plains. *Journal of Coastal Research* 20:1102–1125.

Stuiver, M., P. J. Reimer, E. Bard, J. W. Beck, G. S. Burr, K. A. Hughen, B. Kromer, F. G. McCormac, J. van der Plicht, and M. Spurk, 1998a. INTCAL98 radiocarbon age calibration, 24,000-0 cal BP. *Radiocarbon* 40:1041–1083.

Stuiver, M., P. J. Reimer, and T. F. Braziunas, 1998b. High-precision radiocarbon age calibration for terrestrial and marine samples. *Radiocarbon* 40:1127–1151.

Thompson, L. G., 1992. Quelccaya ice core database, 1992. *IGBP PAGES/World Data Center-A for Paleoclimatology Data Contribution Series* 92–008. NOAA/NGDC Paleoclimatology Program, Boulder, CO.

Urban, H.-J., 1994. Upper temperature tolerance of ten bivalve species off Peru and Chile related to El Niño. *Marine Ecology Progress Series* 107:139–146.

Chapter 3

Mid-Holocene climate and culture change in the South Central Andes

Martin Grosjean[1], Calogero M. Santoro[2], Lonnie G. Thompson[3], Lautaro Núñez[4] and Vivien G. Standen[5]

[1]*NCCR Climate and Institute of Geography, University of Bern, 9a Erlachstrasse, 3012 Bern, Switzerland*

[2]*Instituto Alta Investigación, Departamento de Arqueología y Museología & Centro de Investigaciones del Hombre en el Desierto, Universidad de Tarapacá, Casilla 6-D, Arica, Chile*

[3]*Byrd Polar Research Center, Ohio State University, 1090 Carmack Road, Columbus, OH, 43210, USA*

[4]*Instituto de Investigaciones Arqueológicas y Museo, Universidad Católica del Norte, San Pedro de Atacama, Chile*

[5]*Departamento de Antropología & Centro de Investigaciones del Hombre en el Desierto, Universidad de Tarapacá, Casilla 6-D, Arica, Chile*

Abstract

The South Central Andes host a wide range of different habitats from Pacific coastal areas up to extremely harsh cold and dry environments of the high mountain plateau, the altiplano *or the* puna. *Marine resources in habitats along the cold Humboldt current are abundant and very stable through time, whereas terrestrial vegetation, animal, and water resources in the habitats of the intermediate valleys, of the high valleys toward the Andes and of the high* puna *are marginal, scarce, highly variable, and hardly predictable in time. Paleoenvironmental information reveals high amplitude and rapid changes in effective moisture during the Holocene period and consequently, dramatically changing environmental conditions. Therefore, this area is suitable to study the response of hunting and gathering societies (Paleoindian and Archaic Period, between ca. 13,000 and 4500 cal yr BP; 11,000 and 4000 ^{14}C yr BP) to environmental changes, because smallest variations in the climatic conditions have large impacts on resources and the living space of humans. We analyzed environmental and paleoclimatic information from lake sediments, ice cores, pollen profiles, and geomorphic processes, and put these in relation with the cultural and geographic settlement patterns of human occupation in the different habitats in the area of southern Peru, SW Bolivia, NW Argentina, and North Chile. The time window of 5000 cal yr BP (4300 ^{14}C yr BP) considered in this context is put in perspective of the early and late Holocene in order to show a representative range of environmental and cultural changes. We found that the time broadly around 5000 cal yr BP (4300 ^{14}C yr BP) does not show significant environmental or climatic nor rapid cultural changes. The largest changes took place around 9000 cal yr BP when the humid early Holocene conditions were replaced by extremely arid but* *highly variable climatic conditions. The onset of such hostile conditions resulted in a marked decrease of human occupation, in the occupation of alternative habitats ('Ecological refuges'), in increased mobility, in a stronger orientation toward the habitats with relatively stable resources (such as the coast, the* puna seca, *and 'ecological refuges'), and in stepwise technological innovations of artifacts. In the most arid and marginal areas of the* Puna Salada *south of the Río Loa (21°S) and the adjacent valleys, the mid-Holocene aridity resulted in some sites even in a hiatus of human occupation (*'Silencio Arqueológico', *sensu Grosjean et al., 2005b). Such hostile conditions were repeatedly interrupted by sub-decadal humid spells or by short-lived extreme climatic events (floods, droughts, etc.), and lasted until ca. 3500 cal yr BP when modern conditions were established in a stepwise process. This was also the time when the* puna salada *was re-occupied at large, and irrigated agriculture emerged. Domestication of camelids in the South Central Andes (ca 5500 cal yr BP, 4800 ^{14}C yr BP) falls roughly into the time of interest around 5000 cal yr BP. Although this process is centered in the mid-Holocene harsh conditions, the climate dictate remains debatable because the onset of such harsh conditions preceded domestication by as much as 2000–3000 years.*

1. Introduction

The Atacama Desert of the South Central Andes is today an area with extremely harsh geoecological conditions and marginal resources. Thus, societies based on subsistence economies are highly susceptible to even smallest changes in the climatic and environmental settings and available resources.

As in other subtropical areas of the world, Holocene climatic changes are mainly manifested as variations in the effective moisture budget, whereas changes in temperature were relatively insignificant. This is a fundamental difference with mid- and high-latitude areas and makes the Holocene, as far as subtropical arid and semi-arid areas are concerned, one of the most interesting time windows for the study of high amplitude and abrupt climate changes.

Holocene climatic changes in the Central Andes affected primarily the water cycle (lake levels, spring flow, river discharge, groundwater tables, soil moisture, etc.) and, consequently, flora and fauna. Thus palaeo-ecological archives that record humidity, vegetation, and animal resources are the best sites to study potential impacts of climate change on early hunting–gathering subsistence societies in the Atacama Desert, which were present between ca. 13,000 and 3400 cal yr BP (11,000 and 3200 ^{14}C yr BP). Culturally, this period of time is known as the Archaic Period. It was the time when supplies of and demand for certain natural resources were in a very delicate balance, with critical implications for the demography of human societies.

It was early recognized (Le Paige, 1965; Lanning, 1967, 1973) that many archaeological sites in the Atacama Desert and elsewhere in South America are found in places with very hostile environmental conditions at present, and that paleoenvironments must have been very different from those of today. Thus, a relatively deterministic interdependence between Paleoindian/Archaic human occupation and the paleoenvironment was postulated and documented in many cases (e.g., Cardich, 1980; Massone and Hidalgo, 1981; Fernández, 1984–85; Lynch, 1990; Santoro et al., 1991; Grosjean and Núñez, 1994; Núñez and Grosjean, 1994; Núñez et al., 1994, 1996, 2001, 2002; Grosjean et al., 1997a, 2005a,b; Borrero et al., 1998; Messerli et al., 2000). In the area of the Central Andes, consensus exists that prolonged and severe droughts and arid periods had particularly strong impacts on early societies at times when buffer and storage capacities were still limited (Binford et al., 1997).

Recent advances in multidisciplinary paleoclimate research on tropical glaciers, lake sediments, geomorphologic features, paleosols, groundwater bodies, rodent middens, and pollen profiles in bogs have provided information about large scale, high amplitude, and rapid climate changes in the Central Andes during the Holocene, and have strengthened the hypothesis about the man–environment relationship. Indeed, paleoenvironments play a key-role in understanding the very complex pattern of Paleoindian and Archaic resource use in space and time, for the human occupation of different habitats from the marine coast up to the high elevation lake environments on the *altiplano* above 4500 m altitude (Grosjean et al., 2005b). The combination of archaeological and paleoenvironmental information may also shed light on the question whether climate and cultural changes were synchronous or not, whether there is a causal relationship between climate and culture, and to what extent early cultures were able to shape and manage the landscape toward more efficient resource use and for mitigating high variability or shifts in resources (Lentz, 2000). We may speculate if changes to the socio-economic and cultural patterns were adaptations to new environmental conditions and thus the result of changing environmental boundary conditions.

The aim of this chapter is to review the paleoclimate information for the mid-Holocene (between ca. 8000 and 4000 cal yr BP) in the South Central Andes, to draw a picture of the different habitats of human occupation (marine coast, valleys and *quebradas*, high elevation *puna* habitats and sites), and to compare the paleoclimate scenario with the regional archaeological information in space and time. Major research questions are:

(1) Why did people occupy or abandon specific habitats? Does a hiatus of human occupation or a change in the habitat reflect overly harsh environmental conditions whereas continuous inhabitation is indicative of stable conditions and hence resources through time?
(2) Are cultural or socio-economic changes (e.g., the beginnings of domestication or the adoption of innovative lithic industries) related to changes in the environment, or were technological and cultural changes the result of internal processes of transformation within hunting and gathering societies and aimed directly at a better management and exploitation of the environment?
(3) Was the period around 5000 cal yr BP a particularly interesting period with significant, rapid and high-amplitude climatic changes and adaptive cultural processes?

We emphasize that the unambiguous interpretation of occupations and settlement patterns is still difficult given the current state of knowledge. In some cases, our interpretations will become more complete when archaeological deposits and artifacts are better documented and dated (usually just the basal and top layers of a stratigraphic column are dated) and incomplete regional survey is clarified. In all cases, developing local and detailed environmental reconstructions, including of the geomorphological processes at every individual site, is a prerequisite to achieving a consistent and holistic view of the human–environment relationship in the past (Grosjean et al., 2005a,b).

For the purpose of this chapter, we put the mid-Holocene arid period into the perspective of the entire Holocene, starting with the swing from humid early Holocene to fully arid mid-Holocene conditions between ca. 9500 and 8500 cal yr BP, and ending with the onset of modern climatic conditions around 4000 cal yr BP. This interval brackets the time window around 5000 cal yr BP under investigation herein (see also Sandweiss et al., 1999). We delineate the research area as extending from the marine coast in the west up to the *altiplano* in the east, and from the tropical summer precipitation area in SE Bolivia and Peru at 17°S in the north to the fully arid southern margin of the *altiplano* at 28° in the south (Fig. 3.1). This area is known as the Atacama Desert, and hosts a broad range of habitats and archaeological sites with different assortments of resources such as high elevation open campsites associated with lakes on the *puna* (*dry puna* and *salt puna*), caves,

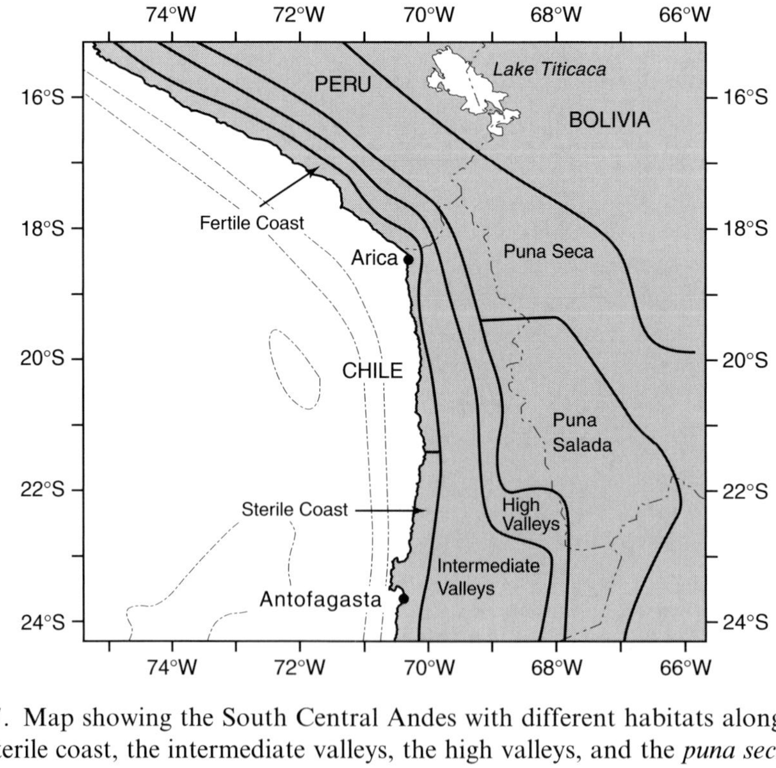

Figure 3.1. Map showing the South Central Andes with different habitats along the fertile and the sterile coast, the intermediate valleys, the high valleys, and the *puna seca* and *puna salada*.

and complex open campsites in intermediate valleys and *quebradas* (dry valleys), and densely occupied sites on the Pacific coast.

Hunters and gatherers in arid areas such as the Atacama Desert were always threatened by the extreme variability of precipitation and unpredictable periods of drought. Therefore, high mobility, and complementary and diversified use of resources in different ecological zones, was an important strategy to live in and to manage a difficult, highly variable environment. This also brought about a regional cultural development with groups that were specialized in certain habitats. However, inter-regional interaction, for instance, between the highlanders and the coastal people, were always very important. This is fundamental when patterns of concentration and dispersion of Archaic settlement in the Atacama Desert are evaluated. In this context it is important to note that a general decrease in resources during times of extreme regional aridity (such as the mid-Holocene) resulted in the formation of ecological refuges where resources were locally still available due to favorable micro-environmental conditions. This in turn led to a major concentration of animals and humans specifically around these areas despite the regional crisis and possibly also regional depopulation. The archaeological sites in these areas are thought to be the nuclei of increasing socio-economic and cultural complexity, semi-sedentarism, and the domestication of flora and fauna (Núñez, 1981; Santoro, 1989; Núñez et al., 1996; Grosjean et al., 1997a; Núñez et al., 2001).

2. The physiogeographical setting

The Andes and the high mountain plateau (*altiplano* or *puna*) form one of the most prominent mountain chains in the world. The unique physiogeographical setting with vertical gradients ranging from sea level up to peaks above 6000 m within less than 150 km horizontal distance is the result of Cenozoic tectonic uplift in the forearc region of the active tectonic convergence zone. This created a wide range of geoecological belts with extremely strong and persistent precipitation gradients between the humid windward side and the arid rain-shadow side of the N–S ranging mountain chain. The formation of the Andes led also to a broad vertical range of temperature regimes from hot climates at sea level to continuous permafrost climate above 5600 m, and to a highly variable spatial pattern of topography, slope, aspect, geological, and pedological conditions. All of these variables superposed result in a mosaic of potential habitats with characteristic local water, vegetation, and animal resources. The geoecological conditions may also involve natural hazards such as volcanism, seismic activity, tsunamis, landslides, and debris flows. Some of the variables that combined to form the living space for humans remained constant in time, some others changed very rapidly. However, it was always the humans who decided, based on their subsistence economy, technology, and ideology, whether a given living space at a specific time was regarded as favorable or hostile.

The meso- and macroscale climate of the Atacama Desert is controlled by (1) the SE Pacific Anticyclone (SPA), (2) the cold Humboldt Current, (3) the upper tropospheric Bolivian Anticyclone centered above the eastern Cordillera, and (4) the Westerly circulation belt in the mid-latitudes of Central Chile (Vuille, 1999; Garreaud et al., 2003). The SPA is a quasi-permanent dynamic high-pressure area and forms part of the southern hemisphere Hadley circulation. The all-year-round dry subsiding air masses are largely responsible for the persistent aridity in the coastal areas and the western slope of the Andes in northern Chile and Peru. The SPA also blocks the frontal systems from the zonal Westwind Drift in the mid-latitudes that bring moisture from the Pacific. Eckman upwelling of cold water in the Humboldt current off the Chilean and Peruvian coast stabilizes the SPA, and gives rise to an inversion layer at ca. 800 m altitude with the prominent coastal fog, locally known as *camanchaca*. The coastal range in northern Chile is a very effective local moisture trap for the coastal fog (Schemenauer et al., 1988) and a strong barrier against moisture transport from the Pacific into the interior of the continent. During austral summer, the *altiplano* and the western Cordillera are controlled by the 'Bolivian High' centered above the eastern Cordillera of Bolivia (Hastenrath, 1997). The 'Bolivian High' is regarded as the result of local heating of the high mountain plateau (Gutman and Schwerdtfeger, 1965; Rao and Erdogan, 1989) and latent heat release over Amazonia (Lenters and Cook, 1997). The 'Bolivian High' features strong upper tropospheric divergent flow, lively convection, easterly winds and advection of tropical Atlantic moisture from the continental lowlands east of the Andes (e.g., Hardy et al., 1998; Vuille et al., 1998). Thus the area considered here (i.e., southern Bolivia and Peru, northernmost Chile and NW Argentina) is subject to tropical summer precipitation (*Invierno boliviano*) which decreases with strong gradients from $>450\,\mathrm{mm\,yr^{-1}}$ on the Bolivian *altiplano* to $<200\,\mathrm{mm\,yr^{-1}}$ in adjacent high elevation areas to the west and south in northern Chile, and to $<20\,\mathrm{mm\,yr^{-1}}$ in areas below 2000 m elevation and the coast.

Tropical summer rainfall remains restricted to high elevations above 4000 m in the western Andes (northern Chile), while summer rainfall reaches all elevation belts in the windward very humid eastern slope of the South Central Andes (SE Bolivia and NW Argentina). The western slope of the Andes remains in the fully arid 'rain shadow' but receives some river discharge from the high Andes. A few rivers north of 22°S reach the marine coast. Frontal winter rainfall of the Westwind Drift is the common moisture source for Central Chile (*Invierno chileno*). Frontal systems further north than ca. 28°S are usually blocked by the SE Pacific Anticyclone. However, penetration of fronts is sporadically observed as far north as northernmost Chile and SE Bolivia (Vuille, 1996; Vuille and Ammann, 1997; Vuille and Baumgartner, 1998). Winter precipitation increases toward the south from ca. 100 mm in coastal areas at 27°S to >300 mm at 33°S. Thus, the Atacama Desert is currently located in the extremely dry transition zone between the tropical summer precipitation areas in the north and east (*Invierno boliviano*) and the extratropical winter precipitation areas in the south and west (*Invierno chileno*). The most arid part of this 'Arid Diagonal' crosses the Andes NW–SE at ca. 25°S (Messerli et al., 1993; Arroyo et al., 1998).

Water resources are very scarce today (Grilli, 1989). For instance, the total available water resources for the Región de Antofagasta in northern Chile (126,000 km^2) amounts to 12–18 $m^3 s^{-1}$, the larger proportion being too saline for domestic use. Also most of the endorheic lakes on the Chilean, Argentinean, and Bolivian *altiplano* are seasonally dry, very shallow, and hypersaline (Stoertz and Ericksen, 1974; Chong Diaz, 1984; Vuille and Baumgartner, 1993; Risacher et al., 2003), and the water quality in lakes, springs, groundwater, and rivers is generally affected by naturally high loads of dissolved salt, in particular arsenic. Except the two large freshwater bodies of Lake Titicaca and Lake Chungará which are located in the somewhat more humid ($P > 400\,mm\,yr^{-1}$) tropical part of the *altiplano* and have a surface or subsurface outflow, the only open water bodies with a surface of a few square kilometers are bound to active geologic fault systems with limited internal drainage (Chong Diaz, 1984; Grosjean, 1994). Small springs (discharge of some 1 $m^3 s^{-1}$) above 2500 m altitude provide water for small bogs and mires with particular ecological conditions (Ruthsatz, 1993, 1995) for animals and humans. Along the ca. 1000 km long coast of northern Chile, there are only five valleys with currently perennial or seasonal rivers connecting the *altiplano* with the Pacific. Besides these estuaries, freshwater is extremely scarce along the coast. At best, there are some small springs fed by the coastal fog, some of them being rather brackish (Núñez and Varela, 1967–68). Some of these places served also as microhabitats combined with nearby marine resources.

In contrast to the scarce terrestrial resources along the coast, the ocean offers stable and predictable resources suitable for permanent human occupation. The coast of northern Chile and southern Peru features the unique arrangement of very hostile fully arid terrestrial conditions with extremely rich marine resources of the cold Humboldt Current. High-nutrient loads of the cold water combined with high solar radiation rates sustain one of the most productive marine ecosystems and food chains in the world, and provide the base for a long tradition of marine subsistence in the coastal areas of the Atacama Desert (Llagostera, 1979, 1982; Sandweiss et al., 1996, 1998).

Terrestrial natural vegetation is an important indicator linking climatic patterns with the living space for animals and humans. Arroyo et al. (1988) show that vascular plant diversity and vegetation cover in the western Andes of northern Chile reflects well the precipitation pattern and the vegetation food resources for subsistence societies. Vegetation cover and species number are highest in the *altiplano* of northernmost Chile (18°S), decrease rapidly toward the coast (rain shadow) and toward the south (Arid Diagonal), and increase again as winter rainfall becomes stronger. In the winter rainfall areas, however, the best conditions for vegetation are found in intermediate altitudes, because low temperatures limit plant growth higher up. The occurrence of terrestrial fauna broadly follows the pattern of the vegetation. Camelids, birds, and rodents are the most important animal groups for hunting (Hesse, 1982; Santoro, 1987). Obviously, the high elevation areas and the areas with river runoff from the mountains are generally the most favorable places, whereas terrestrial resources in the low elevation areas are sparse or totally absent. However, the particular combination between the coastal fog and the moisture trapping coastal range may, in some cases, provide enough moisture to sustain surprisingly dense local vegetation (*Loma* vegetation) and the respective animals.

Natural hazards may also play a role in determining whether a given area is selected as a living space for humans. Numerous active volcanoes are found in the Western Cordillera of southern Peru, Bolivia, and northern Chile between 15°S and 27°S (Zeil, 1986). Numerous volcanic eruptions on the Atacama *altiplano* are reported for historic, Holocene, and late-glacial times (Gardeweg et al., 1984; Francis et al., 1985; Glaze et al., 1989). Some of these eruptions resulted in the collapse of large massifs, triggered immense debris flows and lahars, devastated large areas, and changed in some cases completely the hydrological drainage of a watershed. A debris avalanche, for instance, formed Lake Chungará after the late-glacial collapse of Vn. Parinacota, which dammed the earlier westward drainage (Francis and Wells, 1988). Volcanism in the Andes plays also an important role with regard to raw materials for lithic artifacts. Obsidian and basalt are usually found in the high elevation areas with Cenozoic volcanism, whereas the coastal range and the intermediate zone of the Precordillera with low-grade metamorphic rocks, Cenozoic alluvial material, and sedimentary rocks do not provide first-choice raw material for lithic artifacts. Exceptions are Devonian flint stone nodules or fine-grained sandstones.

However, earthquakes, occasional tsunamis, and volcanic eruptions are low-frequency catastrophes of rather local significance. If devastating, the impact is expected to be found in the stratigraphies of archaeological sites, which is, however, hardly observed in the Atacama (Schiappacasse and Niemeyer, 1984). Furthermore, new studies from Middle America suggest that relatively simple societies tended to recover from sudden stress of explosive volcanism more readily than complex societies (Sheets, 2001). In summary, we conclude that natural hazards and low-frequency catastrophes, although present, did not play a major role in the general regional settlement pattern over the time scales of centuries or millennia considered in this chapter.

3. Habitats for human occupation in the Atacama Desert

In order to compare the settlement patterns within and between the different sectors of the South Central Andes, we distinguish several types of habitats. Each one is characterized by a specific combination of ecological conditions (Fig. 3.1). Table 3.1 summarizes the different habitats with a qualitative index for biomass productivity and predictability (stability) of the food and water resources for human populations. We hypothesize that these two criteria were crucial when Archaic hunters and gatherers evaluated an area as a potential living space. We also expect that areas with low to medium productivity or stability were the areas which were first affected by changes in the environmental conditions, and where climatic changes had the largest impact. Thus we hypothesize that, during the mid-Holocene arid intervals, the humans not only in the *puna salada*, in the high valleys and *quebradas* but also in the intermediate

Table 3.1. Different habitats in the Atacama Desert and qualitative indices for freshwater availability, biomass production and resource stability.

Habitat	Freshwater	Biomass	Predictability
Fertile marine coast	Moderate	Marine: very high Terrestrial: very low	High
Sterile marine coast	Very low	Marine: very high Terrestrial: very low	High
Valleys, *quebradas*, oases at intermediate altitude	Low	Very low	Low
Valleys and *quebradas* towards the Andes	Moderate	Medium to high	Medium
Puna seca	Moderate	High	High
Puna salada	Moderate	Medium	Medium

areas would show the strongest impact of climate change in their cultural patterns, whereas decreasing terrestrial resources in the coastal areas and the *puna seca* did not reach the critically low levels to permanently threaten human societies.

3.1. Fertile marine coast

The coast of Peru and Chile has extremely rich and stable marine resources but terrestrial flora and fauna are very scarce. The marine food resources include a broad variety of mollusks, fish and marine mammals such as the sea lion. The marine resources are, despite variability in El Niño-Southern Oscillation (ENSO) and the ocean currents, hardly affected to the extent observed in terrestrial ecosystems, making marine resources a reliable, stable, and predictable food supply (Schiappacasse and Niemeyer, 1984; Santoro, 1987).

Only a few rivers cut through the coastal range between Majes and Pisagua (17–20°S) and convey freshwater from the mountains to the Pacific coast. Particularly favorable habitats are located around their estuaries, where the very rich and stable marine resources are complemented with fresh water, land mammals (camelids, rodents), birds, freshwater shrimp, fruits of trees (*Prosopis* sp., *Geoffrea* c.), and roots of totora (*Typha* sp.). Totora fiber was likely a very important material for construction, rope making, and cloth. The oases along such rivers reach 5–10 km inland (Fig. 3.1).

3.2. Sterile marine coast

Except the Río Loa, there is no river cutting through the coastal range between Pisagua and Chañaral (20–27°S). Thus, the Pacific coast in this area is disconnected from the high Andean freshwater resources and is fully arid. The local freshwater supply is restricted to trapped moisture from the coastal fog or small brackish groundwater wells in the interior (Núñez and Varela, 1967–68). The habitats along the sterile coast are almost exclusively based on marine resources that are as abundant and reliable as in the coastal areas further north (Fig. 3.1).

3.3. Valleys, quebradas (dry valleys), and oases at intermediate altitude

This area stretches from the coastal range to the foot zone of the high Andes and ranges between 500 and 2500 m. In the Precordillera west of the Salar de Atacama – Salar Punta Negra Graben (23–24°S), this zone reaches up to 3500 m (Fig. 3.1). The habitat is characterized by extremely arid conditions in the rain shadow of the Cordillera de los Andes. In the northern sector adjacent to the fertile coast, few oases in the interior of the transversal valleys provide living space, whereas fully arid endorheic basins and salt lakes (*Salar*) are found south of Quebrada Tiliviche (20°S, Fig. 3.1). Most of these habitats are extremely arid today. The oases are located along the few rivers from the high Andes (Arica valleys), around springs in *quebradas* (e.g., Tana and Tiliviche, Aragón, Tarapacá, and El Médano), or groundwater wells in *Salars* such as the Pampa del Tamarugal. These habitats are well-defined ecological refuges with limited resources (food and water) surrounded by extremely hostile conditions (True et al., 1970, 1971; Núñez and Zlatar, 1980; True and Gildersleeve, 1980). These habitats are fully based on scarce terrestrial resources (few camelids, rodents, birds, freshwater crustaceans, fruits of trees, *Totora* roots, etc.) and limited in their extent, which makes them highly vulnerable to climate fluctuations. Resources were very scarce and hardly predictable. However, some locations were important source areas with raw material (quartz nodules, chalcedony) for lithic artifacts.

3.4. Valleys and quebradas toward the Andes

Habitats in this area are located in higher elevation valleys and quebradas (dry valleys), between 2500 and 4000 m and connected with the high Cordillera (Fig. 3.1). In contrast to the lower elevation valleys, these sites benefit from the freshwater resources and higher precipitation rates of the high mountains. Besides many open camp sites, some of the archaeological sites are located in caves or well-protected rock shelters, on ridges, in deep valleys and near wetlands in confluence areas of rivers.

The northern part of this area is located adjacent to the more humid *Puna Seca* of southern Peru and northernmost Chile, and consists of deep valleys, steep and gentle slopes. Precipitation rates are between 200 and 300 mm yr^{-1}. A rather dense vegetation of shrubs (matorral) provides good grazing areas for camelids (guanacos and vicuñas), rodents (e.g., *Vizcacha* sp., *Ctenomis*), and taruca (*Hippocamelus* sp.), a small deer.

Geoecological conditions become harsher in valleys and quebradas toward the south, adjacent to the *Puna Salada* of northern Chile. Among the most important focal points of human occupation is a series of salars in the foot zone of the high Andes (Salar de Atacama, Salar Punta Negra) that receive fresh water from the high Cordillera. Although local precipitation is below 100 mm yr^{-1} today, many springs and groundwater wells provide favorable habitats with a rich flora and fauna. These habitats were located near the *puna* (above 4000 m) where many different geoecological zones and altitudinal belts were readily accessible and best conditions were given for a complementary use of different resources at different times of the year. A transhumant pattern of resource use (e.g., Núñez, 1981) seems obvious, and is in modern times as important as in the past. Although these habitats are favorable in many respects, the overall relative scarcity of resources (particularly in the southern part) and the relatively high variability (and low predictability) puts limits to the suitability of this area as a permanent living space for Archaic hunters and gatherers.

3.5. High puna (puna seca and puna salada)

The high elevation grasslands of the western Andes and the high mountain plateau (*altiplano*, above 4000 m) provide, as far as food and water are concerned, widespread favorable habitats for human occupation. The best places are usually found around the endorheic brackish lakes (some of them with freshwater) and salt lakes, or near the many small freshwater springs with little ponds and wetland vegetation (bogs and mires). Higher precipitation rates in the mountains ($>$200 mm yr^{-1}) provide enough moisture for disperse grass and herb vegetation (maximum cover 40–60%), and abundant animal life. In contrast to all the other habitats, the resources on the *puna* are not restricted to some favorable sites (linear or point sources), but are rather dispersed.

Following the gradients of rainfall and vegetation, the high elevation area of the South Central Andes includes the more humid *puna seca* in the north and northeast (southern Peru and Bolivia, NW Argentina and northernmost Chile), and the very arid *puna salada* in the southern part of the Atacama Desert (Arroyo et al., 1988; Santoro, 1989; Troll, 1958; Fig. 3.1). Within the *puna seca*, the areas of NW Argentina and SE Bolivia show the highest rainfall rates and the best environmental conditions regarding water and food resources. Thus we expect that these areas are the most stable human habitats (relatively speaking) with relatively low susceptibility to climatic changes, whereas the most vulnerable areas were those of the *puna salada* in the highlands of Chile south of the Río Loa.

In the *puna seca*, the list of hunted animals includes *vicuñas*, small rodents (*vizcacha, chinchilla, cholulos*), birds (ostrich, *suri*, flamingos, partridge, geese, ducks), and a wide range of plant products. The widespread grass cover (mainly *Stipa* sp. and *Festuca* sp.) provides good grazing areas for animals, and the surprisingly large wetlands (*humedales, bofedales, vegas*) are excellent habitats for camelids and birds

(Aldenderfer, 1989; Santoro, 1989). The habitats in the *puna salada* are similar to those of the *puna seca*. However, the favorable sites are more local, smaller in size, and isolated from each other.

4. Mid-Holocene aridity: Hostile conditions and scarce resources around 5000 cal yr BP

Lake sediment and ice cores, pollen profiles, plant macrofossils preserved in rodent middens, geomorphic features, and paleosol indicators provide consistent multi-proxy evidence of a dramatic decrease in average, century-to-millennial scale effective moisture during mid-Holocene times (roughly between ca. 9000 and 4000 cal yr BP). However, the issue of the mid-Holocene climate in this area has been subject to debate (Betancourt et al., 2000; Grosjean, 2001 and discussion therein; Latorre et al., 2002, 2003, 2007; Rech et al., 2002, 2003; Grosjean et al., 2003; Maldonado et al., 2005). In our view, much of this debate arose because (1) vegetation macrofossils in rodent middens record discrete and (maybe) highly variable (sub)decade-scale humid spells that are not (or poorly) recorded in lake sediments or ice cores; these in turn reflect the smoothed average mid- to long-term climate evolution (for discussion Grosjean et al., 2003) and (2) we interpret the higher groundwater tables in valleys as local features driven by geomorphic processes (Grosjean, 2001) and not as regional features driven by humid climates (Rech et al., 2002, 2003).

At multi-centennial to millennial scales, Mid-Holocene aridity was significantly greater than today, and affected the entire geo-bio-hydrosphere. Model calculations suggest that mid-Holocene annual precipitation rates in the Titicaca area were on average 18% lower than today (Talbi et al., 1999). The amplitude and rate of change at the beginning of the mid-Holocene was unique in the light of the preceding much more humid early Holocene environmental conditions.

4.1. Lake sediment and ice core records

The small endorheic lakes on the *altiplano* respond very sensitively and in a most direct way to even smallest changes in the effective moisture budget (precipitation–evaporation). Thus chemical, mineralogical, and physical properties of lake sediments and lake level changes provide information about climate change in the past. Laguna Miscanti (23°45′S, 67°45′W, 4000 m) and Laguna del Negro Francisco (27°30′S, 69°14′W, 4125 m) in northern Chile, Lake Titicaca in Bolivia, and a transect of six small Bolivian lakes between 14°S and 20°S lakes are the best studied sites regarding Holocene limnogeological changes in this part of the *altiplano*. All of the these sites show a consistent picture although the timing of the onset and the end of the mid-Holocene drought varies to some extent from site to site because of the time-space transgressive nature of climate change (Abbott et al., 1997, 2003; Grosjean et al., 2001, 2003; Tapia et al., 2003 and references therein); Laguna Miscanti is given as an example here.

Laguna Miscanti (23°44′S, 67°46′W, 4140 m) is a relatively large (15.5 km^2) 10-m deep endorheic lake with brackish (6.4–6.9 mS cm^{-1}) alkaline (pH 8.0) water. The catchment area is about 320 km^2. Limited amounts of water seep through a lava flow along the Quebrada Nacimiento Fault into Laguna Miñiques. Seismic data and information about the depositional environment, mineralogical, and chemical compositions of autigenic lake sediments provide detailed insight into the Holocene paleolake history, and the respective climatic changes (Fig. 3.2, for detailed discussion: Grosjean et al., 2001, 2003). We use the ^{14}C reservoir-corrected chronology for regional lake level changes (Geyh et al., 1999). Seismic data show four main reflectors that define three major lake sediment units (Valero-Garcés et al., 1996).

The *lowermost seismic unit* corresponds to the sediments of the late-glacial/early Holocene paleolake transgression between ca. 12,000 and 8000 ^{14}C yr BP (between 14,000 and 9000 cal yr BP). This unit consists mainly of diatomaceous mud with brackish to freshwater calcite. Gypsum concentrations are very low, and the sedimentary facies suggests pelagic conditions. These sediments correspond stratigraphically to algal bioherms and shoreline carbonates of fossil beach deposits 25 m above the current lake level. Similar paleolake features on the Bolivian *altiplano* are known as the 'Tauca' and 'Coipasa' paleolake phases (Servant and Fontes, 1978; Servant et al., 1995; Wirrmann and Mourguiart, 1995; Sylvestre et al., 1996, 1999; Bradbury et al., 2001; Placzek et al., 2006). Model calculations suggest for this time a significant increase in precipitation by a factor of 3 (annual rates of >600 mm at 23°S compared to the modern ca. 200 mm, $\Delta P = 400$ mm), a similar increase in cloudiness and reduction of evaporation rates (Grosjean, 1994). Latorre et al. (2007) found a similar factor of precipitation increase in elevations at 3000 m (from 40 to 120 mm per year). These results compare with earlier estimates for Bolivia (Hastenrath and Kutzbach, 1985; Kessler, 1991) where ΔP was estimated to ca. 200 mm yr^{-1}. Long-distance transported pollen from the east side of the Andes, the spatial pattern of the paleolakes, the gradients of equilibrium line altitudes and the geometry of glaciers in southern Bolivia and northern Chile, and the dominance of summer flowering plants in rodent middens (Markgraf, 1989, 1993; Kessler, 1991; Grosjean et al., 1995; Jenny and Kammer, 1996; Clapperton et al., 1997; Kull and Grosjean, 1998, 2000; Kull, 1999; Betancourt et al., 2000; Kull et al., 2002; Latorre et al., 2002, 2003, 2006, 2007; Maldonado et al., 2005) suggest that the increase in effective moisture was mainly due to strengthened tropical summer precipitation from the eastern side of the Andes. This in turn resulted in a strong rain shadow effect and fully arid conditions in intermediate elevations (below ca. 3000 m) on the western slope of the South Central Andes and on the Pacific coast during early Holocene times.

The *middle lacustrine seismic unit* of Laguna Miscanti (Fig. 3.2) encompasses the sediments deposited during the fully arid mid-Holocene period (between <9000 and ca. 4000 cal yr BP, Grosjean et al., 2001). The irregular and poorly stratified reflectors suggest heterogeneous deposition in a fluctuating shallow water environment. Aragonite precipitation, high gypsum contents, hardpans, and evaporite crusts suggest conditions of an ephemeral saline pan–saline lake with sub-aerial

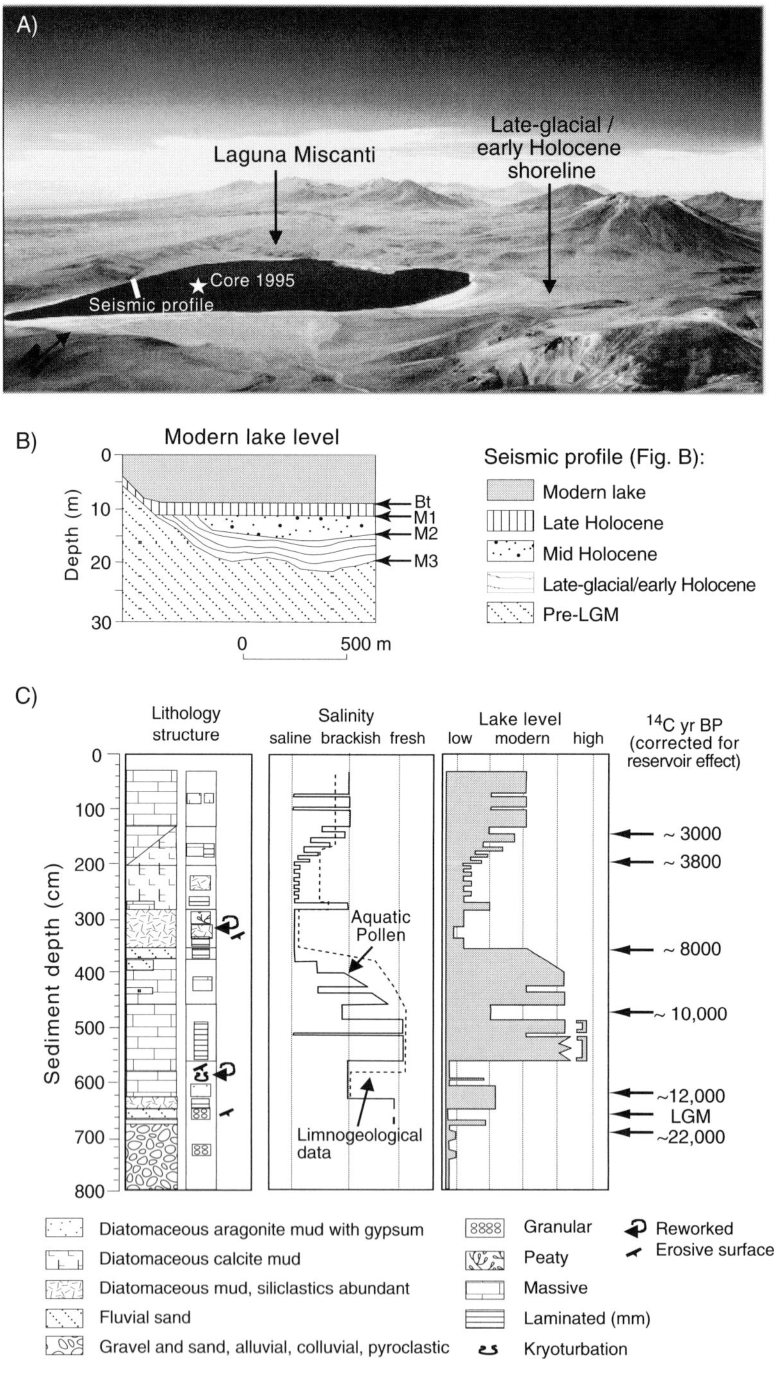

A)
Laguna Miscanti
Late-glacial / early Holocene shoreline
Core 1995
Seismic profile
B)
Modern lake level
Depth (m)
0
10
20
30
Bt
M1
M2
M3
0 500 m
Seismic profile (Fig. B):
Modern lake
Late Holocene
Mid Holocene
Late-glacial/early Holocene
Pre-LGM
C)
Lithology structure
Salinity
saline brackish fresh
Lake level
low modern high
14C yr BP (corrected for reservoir effect)
Sediment depth (cm)
0
100
200
300
400
500
600
700
800
Aquatic Pollen
Limnogeological data
~ 3000
~ 3800
~ 8000
~ 10,000
~12,000
LGM
~22,000
Diatomaceous aragonite mud with gypsum
Diatomaceous calcite mud
Diatomaceous mud, siliclastics abundant
Fluvial sand
Gravel and sand, alluvial, colluvial, pyroclastic
Granular
Peaty
Massive
Laminated (mm)
Kryoturbation
Reworked
Erosive surface

exposure of the sediments at times. The early Holocene lake sediments were exposed to erosion and truncated in the littoral part of the lake, washed into the central part of the basin or blown out. In light of the fact that levels of endorheic lakes are among the best and most direct indicators for effective moisture budgets of the past, the truncation of the sediments and the substantially lower lake level of Miscanti is one of the strongest arguments showing mid-Holocene aridity at centennial and millennial scales in this part of the Andes. However, it is important to note that the mid-Holocene sediments of Laguna Miscanti do record pronounced climate variability at the multi-decadal to decadal or shorter scale because of the inertia of the system. However, it is most likely and suggested by middens data (Latorre et al., 2003, 2006) that interannual to subdecadal variability was also high. The lake sediment data currently available are not able to provide such information. A distinct humid spell is noted around 6000–5500 cal yr BP (Grosjean et al., 2003).

Dramatic drops in lake levels are also reported for Lago Wiñaymarka, the southern sub-basin of Lake Titicaca in Bolivia. Using transfer-functions for ostracod assemblages Mourguiart and Roux (1990), Mourguiart and Carbonel (1994), and Mourguiart et al. (1998) provided quantitative evidence for extremely low lake levels (15 m lower than today) and high salinity (30 $mg\,l^{-1}$ compared to modern ca. 1 $mg\,l^{-1}$) between 9000 and 4400 cal yr BP (8100 and 3900 ^{14}C yr BP). This included also a very dry event centered around 6200–6000 cal yr BP (5300 ^{14}C yr BP) and compares favorably with earlier sedimentological evidence for low mid-Holocene lake levels in the southern basin of Titicaca (Wirrmann and De Oliveira, 1987). Pollen analysis shows that algae are almost missing in this section of the lake sediment core (Ybert, 1992). Detailed seismic profiles suggest that the level in the northern basin of Lake Titicaca dropped by as much as –85 m during this period of time (Seltzer et al., 1998; Tapia et al., 2003). Interestingly, lake sediment cores from the tropical eastern side of the Bolivian Andes (Siberia Lake 18°S, 64°45′W, 2920 m, Sifeddine et al., 1998), and the only studied lake in the southernmost *altiplano* (Laguna del Negro Francisco at 27°30′S, 69°14′W, 4125 m, Grosjean et al., 1997b), also provide convincing evidence of macro-regional mid-Holocene aridity.

The *upper lacustrine seismic unit* in Laguna Miscanti (Fig. 3.2) extends from ca. 4000 cal yr BP to the present. The sediments consist of banded to laminated diatomaceous calcitic mud rich in charophytes. Aragonite is again replaced by

---

Figure 3.2. View of Laguna Miscanti from Cerro Miñiques showing the location of the seismic profile (Fig. 3.2b) and the site of the sediment core (Fig. 3.2c). Figure 3.2b shows the schematic seismic profile with the late-glacial/early Holocene, mid-Holocene and late Holocene lake sediments (after Valero-Garcés et al., 1996). Figure 3.2c shows the sedimentology and lithology of the sediment core, the major mineralogical components (XRD), and SO_4 concentrations as an indicator of gypsum content and thus salinity. The reconstruction of the salinity and lake level changes is based on limnogeological and pollen data (Grosjean et al., 2001).

magnesian calcite, and gypsum contents are low. This suggests increasing lake levels and the formation of the modern brackish perennial 8–9 m deep lake. This is consistent with the on-lap geometry of the upper seismic sediment unit, which was deposited on top of the mid-Holocene sediments in the central part of the basin, and on top of the truncated early Holocene sediments with the erosion surface in the littoral part of the lake. The ^{14}C reservoir-corrected chronology of the lake level changes suggests that the swing from the saline mid-Holocene to the brackish late Holocene lake took place in several steps back and forth between ca. 3600 and 3000 ^{14}C yr BP (between 4000 and 3200 cal yr BP). A broadly similar timing for increasing lake levels was also found in Laguna del Negro Francisco (Grosjean et al., 1997b), in Lake Titicaca (Wirrmann and De Oliveira, 1987; Martin et al., 1993; Abbott et al., 1997; Binford et al., 1997; Mourguiart et al., 1997, 1998; Tapia et al., 2003), in the eastern Cordillera (Abbott et al., 2003) and on the eastern slope of the Bolivian Andes (lake Siberia 2900 m, Sifeddine et al., 1998) suggesting that this marked increase in lake levels and humidity was a supra-regional climate signal that started first in the northeast, extended progressively to the southwest, and terminated the mid-Holocene aridity in the South Central Andes at large.

The history of humidity changes as drawn from lake sediment records is also well reflected in the paleoclimatic archives of ice cores from tropical glaciers in the South Central Andes (e.g., Thompson et al., 1995, 1998). In particular, the ice core from Sajama (Thompson et al., 1998) shows high accumulation rates and low sulfate and chloride concentrations, which is indicative of relatively humid climatic conditions with large paleolakes and small atmospheric loads of evaporite minerals from the *altiplano* lake basins. Accumulation rates decrease and ion concentrations increase with the onset of mid-Holocene arid conditions suggesting that humidity decreased, the paleolakes disappeared, and the former paleolake basins were exposed to aeolian erosion. The mid-Holocene section of the ice core shows numerous extraordinary peaks of dust and soluble ions, again suggesting highly variable climatic conditions with extreme events at a generally very arid background climate. The time window around 5000 cal yr BP shows a significant peak in dust and nitrate. However, comparable peaks are found throughout the mid- and late Holocene period. Thus the ice core of Sajama does not seem to provide information about a significant change in climatic conditions around 5000 cal yr BP.

4.2. Vegetation records

The vegetation history as recorded in pollen profiles shows a picture of mid-Holocene aridity in the western South Central Andes and the Chilean coast between 18°S and 35°S. The database has substantially increased in recent years.

The pollen profile of Laguna Seca (18°11′S, 69°15′W, 4500 m, Fig. 3.3) in northernmost Chile shows the vegetation history of a high elevation site in the tropical summer rainfall regime. Baied (1991) described three different pollen zones. The chronology, however, is poor (two ^{14}C dates). Gramineae are dominant

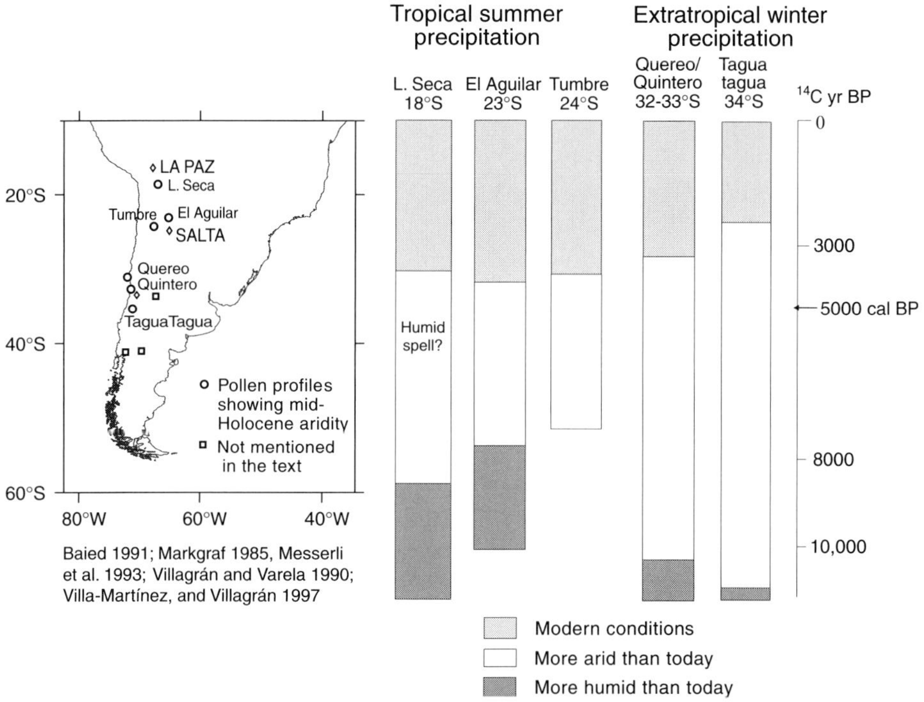

Figure 3.3. Pollen profiles in the South Central Andes and adjacent areas showing the mid-Holocene aridity. Modern vegetation patterns were established largely around 3200 cal yr BP (3000 ^{14}C yr BP). The marked shift from humid to arid conditions is observed at the end of the Pleistocene in areas with extratropical winter rainfall and around 9000 cal yr BP (8000 ^{14}C yr BP) in areas with tropical summer rainfall.

(>55%) in the late-glacial and early Holocene section (Zone 1) and long-distance transported pollen (ca. 5%) from the subtropical montane and lowland forest east of the Andes is relatively abundant. These features suggest moister conditions than today with strengthened easterly airflow and tropical summer rainfall. Long-distance pollen decreases in pollen Zone 2 (after ca. 9000 cal yr BP) suggesting increasing aridity. This trend culminated in pollen Zone 3 (after ca. 8000 cal yr BP) when the lake was replaced by peat land, and generally arid and warm mid-Holocene conditions were established. This compares with the general lake level history as described above. However, pollen from aquatic taxa and long-distance arboreal pollen from the East suggest a spell of increased moisture tentatively assigned to ca. 5800–5500 cal yr BP, which was also found in Laguna Miscanti (Grosjean et al., 2003). The first human impact on the vegetation is estimated to ca. 3500 cal yr BP (Baied, 1991), which is broadly synchronous with the rise of the lake levels.

The vegetation history of a high elevation site in NW Argentina (El Aguilar, 23°05′S, 65°45′W, 4000 m, Markgraf, 1985) suggests that relatively moist conditions lasted until ca. 8300 cal yr BP (7500 ^{14}C yr BP). Long-distance pollen from the east side of the Andes disappeared at around this time. Dry mid-Holocene conditions prevailed until ca. 4500 cal yr BP (4000 ^{14}C yr BP) when modern conditions were established.

The pollen profile at Tumbre (23°19′S, 67°47′W, 3880 m, Graf, 1992) east of the Atacama basin covers the last 8300 cal yr BP (basal date 7500±80 ^{14}C yr BP). The chronology is relatively well-constrained with seven ^{14}C dates. Graf (1992, pp. 36–106) concluded from the high percentage of Gramineae pollen (70–85%) that more humid conditions than today prevailed between 7400 and 2000 cal yr BP (between 6500 and 2000 ^{14}C yr BP). This finding is based on pollen percentages and not on pollen concentrations and disagrees with all the other available paleodata. However, based on Graf's data, we argue that the almost complete absence of wetland taxa (e.g., Cyperaceae) between 7200 and 4200 cal yr BP (6200 and ca. 3800 ^{14}C yr BP) speaks clearly for local dry mid-Holocene conditions instead, when the moisture supply for the peat bog was limited, and the wetlands were much smaller or partly absent.

Pollen in the sediments of nearby Laguna Miscanti (23°44′S, 67°46′W, 4140 m) provides a detailed high-resolution record of vegetation history covering the last 22,000 ^{14}C yr BP (Grosjean et al., 2001, analyst J. van Leeuwen). The mid-Holocene aridity is clearly found in the pollen record as aquatic freshwater taxa (*Myriophyllum* and *Ranunculus*-type) decreased gradually after ca. 9000 cal yr BP (ca. 8000 ^{14}C yr BP), and disappeared completely after ca. 8000 cal yr BP until 6900 cal yr BP, suggesting that the lake desiccated. However, Cyperaceae pollen implies patches of swamps in littoral areas and wetlands in the exposed bottom of the former lake. Subsequently, *Ruppia* returned while Cyperacea disappeared. We think that a saline shallow lake or wetlands was established (between ca. 6900 and 4000 cal yr BP) as it was also found in the limnogeological data (Fig. 3.2). The gradual initial swing to the modern aquatic vegetation is observed around or after 4000 cal yr BP.

Interestingly, the percentages of terrestrial pollen (relative abundance of the different pollen groups) in the lake sediments do not show the mid-Holocene aridity. Terrestrial pollen percentages show for some groups (e.g., Adesmia-type) the humid early and late Holocene; most other groups show no significant difference compared to the mid-Holocene. It is suggested that the overall species composition did not change much, which is consistent and expected under a persistent summer rainfall regime throughout the Holocene (e.g., Betancourt et al., 2000). However, the pollen concentration and the pollen flux rate (which we use as a proxy indicator of vegetation density and thus biomass productivity) show for most groups significantly reduced values during the mid-Holocene compared with the early and late Holocene (Grosjean et al., 2003). Pollen concentrations thus suggest overall reduced mid-Holocene vegetation cover (and biomass productivity), while the species composition remained largely the same. This would be indicative of more arid conditions relative to the present days. Rodent midden records (e.g., Betancourt et al., 2000; Latorre et al., 2002) provide vegetation information comparable to 'species compositions' and 'pollen percentages' (in the 'pollen language') but do not inform about midden production rates, rodent population density, and vegetation density (biomass productivity) as revealed by 'pollen concentration' and 'pollen flux'. It is, therefore, not surprising that the midden data show a picture comparable to that of the pollen percentages. Thus no indication for a marked dry period during the mid-Holocene is expected. But if the number of midden samples per unit time (production rate of middens) is regarded as a proxy of rodent population density, which in turn is a function of food availability, biomass productivity, and ultimately humidity (Lima et al., 2002), the midden data set (Latorre et al., 2002, 2003, 2007) shows indeed a marked mid-Holocene decline (Núñez et al., 2002).

Very interesting is the study of pollen in rodent middens along an altitudinal transect at Quabrada del Chaco (25.5°S) between 2670 and 3500 m elevation (Maldonado et al., 2005) showing that the timing of humid phases is very different in the upper elevation zone (summer precipitation regime) and the lower zone (winter precipitation regime). While the pluvial in the winter precipitation regime lasted until ca. 14,000 cal yr BP, late-glacial humid conditions related to summer precipitation prevailed exclusively in the upper elevation zone (14,000–11,000 cal yr BP). The Holocene record is very scarce at Quebrada del Chaco (Maldonado et al., 2005) suggesting overall very low primary production and, consequently, very low midden production.

Further south in the winter rainfall areas of the north-central Chilean coast 31–32°S, Cyperaceae, aquatic taxa, and arboreal pollen suggest humid conditions during late-glacial times. Aquatic taxa and arboreal pollen disappeared almost completely during the arid period between ca. 11,200–9900 and 3200 cal yr BP (10,000 and ca. 3000 ^{14}C yr BP). In some places, Cyperaceae pollen suggest that humidity returned slowly after ca. 4500 cal yr BP (4000 ^{14}C yr BP, Villagrán and Varela, 1990; Villa-Martínez and Villagrán, 1997; Maldonado and Villagrán, 2002; Maldonado and Villagran, 2006). Aquatic taxa returned by ca. 3200 cal yr BP (3000 ^{14}C yr BP), and re-colonization by forest taxa is observed after ca. 2000 cal yr BP. The period with scarce vegetation falls well into the time of coastal dune mobilization observed between 5800 and 4200 cal yr BP (5000 and 3800 ^{14}C yr BP, Villa-Martínez and Villagrán, 1997). Modern vegetation patterns were established broadly after ca. 4000 cal yr BP (around 3700 ^{14}C yr BP) at Quereo 31°55′S, Quintero 32°47′S, Quintero II 32°47′S, and Santa Julia 32°49′S (Villagrán and Varela, 1990; Villa-Martinez and Villagrán, 1997).

Particular and very powerful paleobotanic archives thus are the plant macrofossils preserved in rodent middens (Latorre et al., 2002, 2003, 2006, 2007). While the late-glacial early Holocene pluvial is consistently found in all the sites, the onset of Holocene aridity differs at the various sites, with a general trend toward progressively earlier desiccation to the south (Latorre et al., 2006). The mid-Holocene records show large variability that is not recorded in the lake sediment archives.

4.3. Geomorphic and paleosol records

Although the records of geomorphic processes and paleosol formation are often discontinuous and heterogeneous, they add important information about paleoclimatic conditions. Geomorphological records, particularly alluvial deposits, may also register short-lived climatic events of a few hours or days of duration, whereas vegetation and lake systems usually integrate seasons, years, or decades of paleoclimatic information.

Alluvial deposits in the Puripica valley of northern Chile (22°50′S, 68°04′W, 3250 m, Grosjean et al., 1997a) provide insight into mid-Holocene storm activity and climate variability. Deposits of more than 30 individual debris flows were identified between ca. 7200 and 3300 cal yr BP (between 6200 and 3100 ^{14}C yr BP). The individual deposits are interpreted as the result of low-frequency heavy storms during a hyperarid background climate with poor vegetation erosion control. The heaviest storms seem to have occurred every 500–1500 years, i.e., around 5900 cal yr BP (5080 ^{14}C yr BP), a short time before 4300–4000 cal yr BP (3790 ^{14}C yr BP), and at ca. 3500 cal yr BP (3300 ^{14}C yr BP), while moderate storms are registered every 100–200 years.

These mid-Holocene storms were of regional significance. In the Salar de Atacama, southwest of Puripica, these events are recorded in sediment profiles as fine-grained flood deposits embedded in aeolian sand. The earliest documented flood occurred at ca. 6400 cal yr BP (5605 $\pm$ 65 ^{14}C yr BP), close to the beginning of the Puripica stratigraphy (Grosjean and Núñez, 1994). Such episodic floods may also explain mid-Holocene groundwater recharge in the low-elevation areas of the Atacama Desert (Aravena, 1995), which is difficult to interpret in the light of hyperarid climate at that time. Siliciclastic inwash in the mid-Holocene sediments of Laguna Miscanti (Valero-Garcés et al., 1996) provides further evidence of regional storm activity. We also interpret the sandy matrix in the lower section of the Tumbre pollen profile (23°30′S, 3880 m, Graf, 1992) as the result of a geomorphologically unstable valley floor and alluvial activity ending around 4300 cal yr BP (3800 ^{14}C yr BP). Afterwards, peat started to dominate the Tumbre profile, suggesting that the alluvial activity decreased and stable groundwater-fed wetlands were established. The repeated short-term cycles of flooding and desiccation in the mid-Holocene sediments of Laguna del Negro Francisco (27°S) between ca. 7000 and 3900 cal yr BP (6000 and 3600 ^{14}C yr BP, Grosjean et al., 1997b) suggest a similar highly variable climate in the southwest of the *puna salada*. However, a causal link with the storms at Puripica and with the ENSO-phenomenon in general remains, in our view, inconclusive and speculative. Strong alluvial activity, scarce vegetation cover, and a hiatus in soil formation were also observed in the coastal range and the Andes of Central Chile 27–33°S between 5800 and 4000 cal yr BP (5100 and 3700 ^{14}C yr BP, Veit, 1995, 1996), and on the coast of southern Peru (Fontugne et al., 1999). Lowest lake levels in this area date to the time between 9500 and 5700 cal yr BP modern levels were reached 3200 cal yr BP (Jenny et al., 2002). Also in the *Puna Seca* of southern Peru (17°S, 3450 m), the sediment stratigraphy shows evidence of successive alluvial deposition between ca. 7800 and 6000 cal yr BP (7000 and 5200 ^{14}C yr BP, Aldenderfer, 1993).

A typical feature of mid-Holocene geomorphological activity is the infilling of steep valleys with fine-grained siliciclastic and organic (peat, diatoms) sediments. Such sediments are observed in the Salar de Atacama area (Grosjean et al., 1997a;

Rech et al., 2002, 2003) and in many other steep valleys as far north as Peru (Rech et al., 2002). Indeed, the widespread occurrence of these features suggests rather a regional climatic than a local tectonic forcing. The paleoclimatic interpretation, however, remains controversial. Whereas these deposits are mostly interpreted as a result of sediment accumulation in an overall low-energy hydrological environment with very limited surface runoff (less than today), limited river incision and thus high local aquifers but generally arid climatic conditions (humid climates would lead to surface runoff and river incision; Grosjean, 2001, P. Baker, C. Rigsby, R. Aravena, B. Warner, personal communications, 2001), and are indicative of dry climates elsewhere (e.g., in Southern Africa; I. Stengel, personal communication, 2003), Rech et al. (2002) interpret these features as a result of generally more humid climatic conditions.

In the South Central Andes, glaciers and permafrost bodies play a vital role with regard to steady freshwater supply for human consumption (e.g., Ribstein et al., 1995; Schrott, 1998). Except for the still controversial glacier advance prior to 5100 cal yr BP in Argentina 33°S (Garleff and Stingl, 1994), glaciers in Bolivia, Peru, and NW Argentina were generally at minimum extents between the late-glacial deglaciation and neoglacial advances younger than ca. 3800 cal yr BP ($<$ 3500 ^{14}C yr BP, Seltzer, 1990; Clapperton, 1993, 1994; Abbott et al., 1997; Grosjean et al., 1998).

Evidence of pronounced arid mid-Holocene conditions is also provided by terrigenous sediments in marine sediments off the Central Chilean coast 33°S (Lamy et al., 1999, Fig. 3.4). The overall low sedimentation rate throughout the early and mid-Holocene until 4160 cal yr BP suggests low river discharge rates. Clay mineralogical data show that the major river sediment sources and places of erosion were in the high Andes, whereas the coastal range remained inactive with regard to erosion. Lamy et al. (1999) also conclude for the coastal range that physical weathering was more important than chemical weathering, which is indicative of hyperarid early and mid-Holocene climatic conditions in the lower elevation areas and thus highly consistent with the findings by Maldonado et al. (2005).

Dune mobilization is a good long-term indicator for arid conditions with scarce vegetation cover. On the coast of Central Chile 32°S, dune mobilization is observed between 5800 and 4200 cal yr BP (5000 and 3800 ^{14}C yr BP, Villa-Martínez and Villagrán, 1997). Also in the tropical lowlands of the Chaco Boreal (Paraguay, Geyh et al., 1996, Fig. 3.4), TL dates of dune sands range between 7700 and 2900 TL yr BP, indicating widespread eolian processes during mid-Holocene times. The ^{14}C dates on fossil soils suggest that these dunes were stabilized after 3200 cal yr BP, right around the time when lake levels in the South Central Andes started to increase, and modern vegetation patterns were established.

Also the lack of pedogenesis on geomorphologically stable surfaces in the Salar de Atacama (e.g., at Tambillo) is a strong indicator for persistent arid conditions since ca. 9000 cal yr BP. In the more humid eastern Cordillera of the Atacama Desert 22°S (Cordillera Santa Victoria, 4500 masl), Zipprich et al. (1999) observed a mid-Holocene hiatus in soil formation (Eutric Regosol) between 9000 and 4100 cal yr BP.

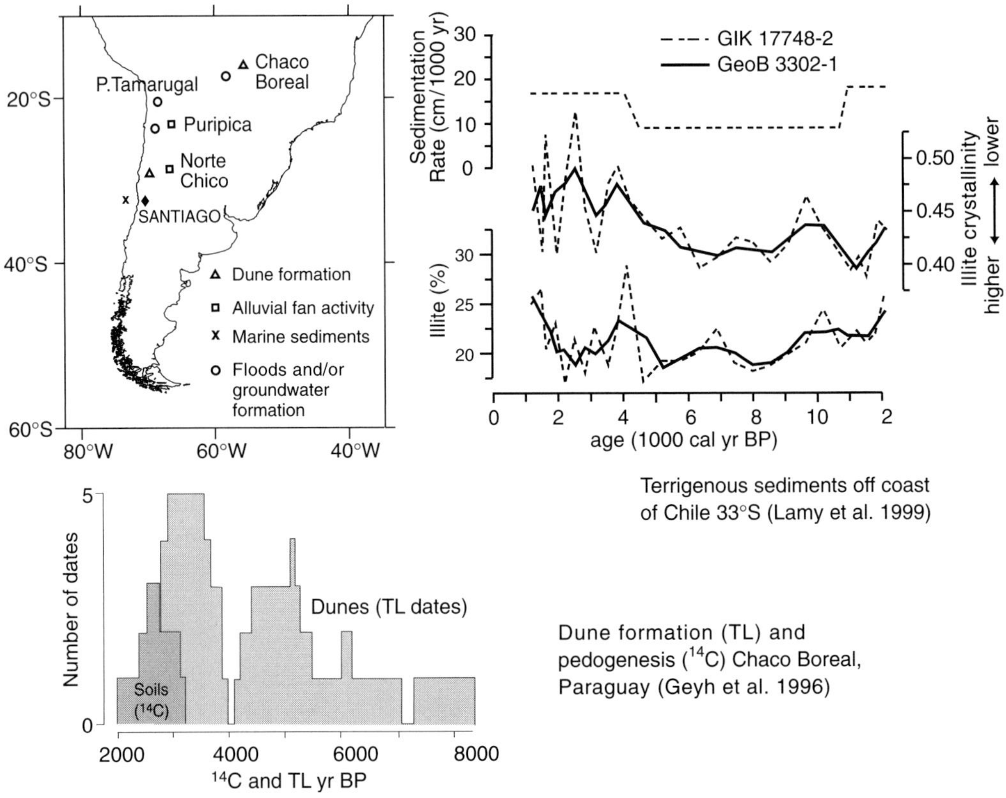

Figure 3.4. Map showing the locations with mid-Holocene geomorphological and paleosol information. In marine sediments off the Chilean coast, mid-Holocene aridity is expressed as reduced terrigenous sediment input and high Illite crystallinity. In the Chaco Boreal of Paraguay, dune mobilization is observed during the mid-Holocene until about 3200 cal yr BP (3000 ^{14}C yr BP) when the dunes were stabilized by soil formation and vegetation.

In summary, a broad range of paleolimnological, vegetation, geomorphological, and pedological information draws a consistent picture of persisting very dry climatic and harsh environmental conditions during the mid-Holocene. Such conditions are observed in the entire area of the South Central Andes at large, from the Pacific coast in the west to the high Cordillera of the Andes in the east, and from the tropical summer rainfall in the north to the extratropical winter rainfall areas in the south.

5. Archaic settlement patterns and paleoenvironmental variability

The currently available data on mid-Holocene water, floral and faunal resources show a picture of relatively very hostile environmental conditions for human societies based on pre-agricultural/pre-irrigation subsistence economies at around 5000 cal yr BP. The *puna*, a favorable living space during the early Holocene, experienced a severe persisting multi-millennia drought. The few lakes desiccated, vegetation suffered from a quantitative decrease, and glaciers reached minimum extents or disappeared. Such harsh climatic conditions resulted in a pronounced concentration of life around small very specific places, named 'ecological refuges' (Grosjean et al., 2005b), where resources were still available due to groundwater and spring discharge from regional and/or fossil Early Holocene aquifers. We hypothesize that the food and water resources were critically scarce in the *Puna* of the Atacama Desert in northern Chile (south of the Río Loa, 22°S and near the modern Arid Diagonal), whereas conditions further north in southern Bolivia, northernmost Chile, southernmost Peru, and northwestern Argentina were still rich enough to sustain a low-density, possibly highly mobile hunting and gathering population.

The fully arid conditions on the *puna salada* also affected to some extent the habitats in the intermediate zone and the longitudinal valleys because river discharge from the high Andes was reduced. Relatively stable conditions persisted in the coastal habitats where marine food resources were stable and abundant, and terrestrial food and atmospheric water supply are subordinate. Along the marine coast, the most drastic change in the habitat was most likely the rapid global rise of the sea level during the early Holocene until ca. 6000 cal yr BP when modern levels were reached (Fairbanks, 1989). A regional sea level curve is not available so far. However, the sea level rise on the order of 60–80 m during the early Holocene implies that all the archaeological sites previously located next to the beach were progressively submerged until 6000 cal yr BP and disappeared. Thus caution is needed when early and mid-Holocene coastal settlement patterns are compared.

Here, we evaluate 106 archaeological sites (with about 300 ^{14}C dates, Table 3.2; database state, 1999) in the South Central Andes in the different habitats in order to identify settlement patterns, and to relate continuous or interrupted human occupation in time to changing environments between ca. 9000 and 4500 cal yr BP (8000 and 4000 ^{14}C yr BP).

5.1. Occupation of the fertile marine coast

The 34 ^{14}C dated Archaic sites (21 in southern Peru, 13 in northern Chile) in the habitat of the fertile marine coast cover the sequence between 13,000 and 3200 cal yr BP (11,000 and 3000 ^{14}C yr BP; Figs. 3.5–3.8, Table 3.2). The initial latest late-glacial human occupation took place during a time when the humid environments in the highlands and high river discharge provided this part of the coast with fresh water. This created exceptional habitats near estuaries where complementary marine and freshwater resources were available (Núñez and Varela, 1967–68).

The first people were highly specialized on marine resources including net-fishing practice and recollection of wedge clams (*Mesodesma donacium*) as found in

Table 3.2. Uncalibrated ^{14}C dates of late Pleistocene and Holocene Archaic sites in the South Central Andes (S Peru, N Chile, and NW Argentina). Database status as of 2002.

Sites	^{14}C yr BP	Laboratory	Material	Period	Reference
A. Fertile marine coast					
Peruvian sites					
Q. Jaguay-280	11,105 ± 260	BGS-1942	Charcoal	Early Archaic	Sandweiss et al. (1998)
Q. Jaguay-280	11,088 ± 220	BGS-2024	Charcoal	Early Archaic	Sandweiss et al. (1998)
Tacaguay	10,770 ± 150	BETA-95869-C	Charcoal	Early Archaic	Keefer et al. (1998)
Q. Jaguay-280	10,770 ± 130	BGS-1702	Charcoal	Early Archaic	Sandweiss et al. (1998)
Tacaguay	10,750 ± 80	Beta-108692-A	Charcoal	Early Archaic	Keefer et al. (1998)
Q. Jaguay-280	10,725 ± 175	BGS-1937	Charcoal	Early Archaic	Sandweiss et al. (1998)
Q. Jaguay-280	10,700 ± 300	BGS-1940	Charcoal	Early Archaic	Sandweiss et al. (1998)
Q. Jaguay-280	10,600 ± 135	BGS-1939	Charcoal	Early Archaic	Sandweiss et al. (1998)
Ring site	10,575 ± 105	SI-6783	Shell	Early Archaic	Sandweiss et al. (1989)
Q. Jaguay-280	10,560 ± 125	BGS-1938	Charcoal	Early Archaic	Sandweiss et al. (1998)
Tacaguay	10,530 ± 140	Beta-108860-C	Charcoal	Early Archaic	Keefer et al. (1998)
Q. Jaguay-280	10,507 ± 125	BGS-2025	Charcoal	Early Archaic	Sandweiss et al. (1998)
Q. Jaguay-280	10,475 ± 125	BGS-1936	Charcoal	Early Archaic	Sandweiss et al. (1998)
Q. Jaguay-280	10,274 ± 125	BGS-1943	Charcoal	Early Archaic	Sandweiss et al. (1998)
Q. Jaguay-280	10,200 ± 140	NR	Charcoal	Early Archaic	Engel (1981), Sandweiss et al. (1998)
Q. Jaguay-280	10,190 ± 220	BGS-1957	Charcoal	Early Archaic	Sandweiss et al. (1998)
Q. Jaguay-280	9850 ± 170	BGS-1956	Charcoal	Early Archaic	Sandweiss et al. (1998)
Los Burros (Cañón)	9830 ± 140	10628	Charcoal	Early Archaic	Lavallée et al. (1999)
Los Burros (Test 2b)	9820 ± 80	10723	Shell	Early Archaic	Lavallée et al. (1999)
Q. Jaguay-280	9657 ± 220	BGS-2023	Charcoal	Early Archaic	Sandweiss et al. (1998)
Tacaguay	9630 ± 60	Beta-108859-A	Shell	Not cultural	Keefer et al. (1998)
Q. Jaguay-280	9597 ± 135	BGS-1960	Charcoal	Early Archaic	Sandweiss et al. (1998)
Los Burros (Cañón)	9545 ± 55	10403	Shell	Early Archaic	Lavallée et al. (1999)
Q. Jaguay-31	9393 ± 160	BGS 1966	Charcoal	Early Archaic	Sandweiss et al. (1998)
Q. Jaguay-1	9385 ± 140	BGS 1998	Charcoal	Early Archaic	Sandweiss et al. (1998)

Q. Jaguay-22	9340 ± 340	BGS 1965	Charcoal	Early Archaic	Sandweiss et al. (1998)
Q. Jaguay-1	9227 ± 110	BGS 1962	Charcoal	Early Archaic	Sandweiss et al. (1998)
Q. Jaguay-16	9200 ± 115	BGS 1967	Shell	Early Archaic	Sandweiss et al. (1998)
Q. Jaguay-280	9120 ± 300	BGS-1701	Charcoal	Early Archaic	Sandweiss et al. (1998)
Q. Jaguay-4	9115 ± 130	BGS 1993	Charcoal	Early Archaic	Sandweiss et al. (1998)
Q. Jaguay-21	9105 ± 115	BGS 1997	Charcoal	Early Archaic	Sandweiss et al. (1998)
Q. Jaguay-37	9039 ± 110	BGS 2020	Shell	Early Archaic	Sandweiss et al. (1998)
Q. Jaguay-280	9020 ± 170	BGS-1703	Charcoal	Early Archaic	Sandweiss et al. (1998)
Q. Jaguay-8	9015 ± 120	BGS 1991	Charcoal	Early Archaic	Sandweiss et al. (1998)
Q. Jaguay-1	8906 ± 115	BGS 1963	Charcoal	Early Archaic	Sandweiss et al. (1998)
Los Burros (Test 2b)	8890 ± 70	10406	Shell	Early Archaic	Lavallée et al. (1999)
Los Burros (Test 2b)	8860 ± 130	10400	Shell	Early Archaic	Lavallée et al. (1999)
Los Burros (Test 2)	8780 ± 70	10401	Shell	Early Archaic	Lavallée et al. (1999)
Q. Jaguay-20	8765 ± 180	BGS 1996	Charcoal	Early Archaic	Sandweiss et al. (1998)
P. Chira	8765 ± 160	HV 1090	Soil	Early Archaic	Ziolkowski (1993)
Q. Jaguay-43	8757 ± 110	BGS 2021	Shell	Early Archaic	Sandweiss et al. (1998)
Ring site	8755 ± 120	SI-6931	Shell	Early Archaic	Sandweiss et al. (1989)
P. Chira	8730 ± 115	BGS 1961	Shell	Early Archaic	Sandweiss et al. (1998)
Q. Jaguay-45A	8704 ± 115	BGS 2022	Shell	Early Archaic	Sandweiss et al. (1998)
Q. Jaguay-19	8615 ± 135	BGS 1995	Charcoal	Early Archaic	Sandweiss et al. (1998)
Los Burros (Cañón)	8470 ± 65	10407	Shell	Early Archaic	Lavallée et al. (1999)
Los Burros (Test 2b)	8430 ± 90	10405	Shell	Early Archaic	Lavallée et al. (1999)
Tacaguay	8430 ± 60	Beta-110330-A	Root	Not cultural	Keefer et al. (1998)
Q. Jaguay-3	8275 ± 130	BGS 1990	Charcoal	Early Archaic	Sandweiss et al. (1998)
Puyenca	8070 ± 145	Hv-1084	Charcoal	Late archaic	Ravines (1972)
Q. Jaguay-280	8053 ± 115	BGS-1944	Charcoal	Middle Archaic	Sandweiss et al. (1998)
Los Burros (Test 2b)	8040 ± 105	10634	Organic material	Early Archaic	Lavallée et al. (1999)
Kilómetro 4	8030 ± 100	Beta-77947	Charcoal	Early Archaic	Wise (1999)
Los Burros (Excavation)	8020 ± 55	10402	Shell	Early Archaic	Lavallée et al. (1999)
Tacaguay	7990 ± 30	Beta-109354-C	Charcoal	Early Archaic	Keefer et al. (1998)

Continued

Table 3.2. continued

Sites	^{14}C yr BP	Laboratory	Material	Period	Reference
Tacaguay	7920 ± 80	Beta-108861-A	Root	Not cultural	Keefer et al. (1998)
Los Burros (Excavation)	7880 ± 55	10626	Shell	Early Archaic	Lavallée et al. (1999)
Los Burros (Test 2b)	8730 ± 70	10632	Organic material	Early Archaic	Lavallée et al. (1999)
Los Burros (Profile "Capilla")	8650 ± 70	10642	Organic material	Early Archaic	Lavallée et al. (1999)
Los Burros (Test 2b)	8160 ± 70	10633	Organic material	Early Archaic	Lavallée et al. (1999)
Los Burros (Profile "Capilla")	8125 ± 30	10646	Shell	Early Archaic	Lavallée et al. (1999)
Puyenca	7855 ± 150	Hv-1086	Charcoal	Middle Archaic	Ravines (1972)
Ring site	7810 ± 105	SI-6930	Shell	Middle Archaic	Sandweiss et al. (1989)
Los Burros (Excavation)	7735 ± 40	11004	Shell	Middle Archaic	Lavallée et al. (1999)
Q. Jaguay-280	7690 ± 100	BGS-1959	Charcoal	Middle Archaic	Sandweiss et al. (1998)
Ring site	7675 ± 60	SI-4784	Shell	Middle Archaic	Sandweiss et al. (1989)
Q. Jaguay-280	7620 ± 100	BGS-1958	Charcoal	Middle Archaic	Sandweiss et al. (1998)
Q. Jaguay-17	7540 ± 110	BGS 1999	Shell	Middle Archaic	Sandweiss et al. (1998)
Q. Jaguay-280	7500 ± 130	BGS-1700	Charcoal	Middle Archaic	Sandweiss et al. (1998)
Ring Site	7415 ± 65	PITT-0142	Charcoal	Middle Archaic	Sandweiss et al. (1989)
Los Burros (Profile "Capilla")	7390 ± 50	10643	Organic material	Middle Archaic	Lavallée et al. (1999)
Los Burros (Profile "Corral")	7320 ± 80	10635	Organic material	Middle Archaic	Lavallée et al. (1999)
Q. Jaguay-5	7300 ± 105	BGS 1992	Charcoal	Middle Archaic	Sandweiss et al. (1998)
Los Burros (Excavation)	7195 ± 45	11002	Shell	Middle Archaic	Lavallée et al. (1999)
Los Burros (Profile "Capilla")	7160 ± 80	10647	Shell	Middle Archaic	Lavallée et al. (1999)
Ring site	7155 ± 180	PITT-0147	Charcoal	Middle Archaic	Sandweiss et al. (1989)

Los Burros (Profile "Capilla")	7105 ± 55	10644	Organic material	Middle Archaic	Lavallée et al. (1999)
Los Burros (Profile "Corral")	6940 ± 60	10636	Organic material	Middle Archaic	Lavallée et al. (1999)
Los Burros (Excavation)	6845 ± 30	10689	Shell	Middle Archaic	Lavallée et al. (1999)
Los Burros (Excavation)	6640 ± 50	10649	Shell	Middle Archaic	Lavallée et al. (1999)
Los Burros (Excavation)	6630 ± 70	10625/GifA 97289	Charcoal	Middle Archaic	Lavallée et al. (1999)
Los Burros (Profile "Capilla")	6595 ± 75	10645	Organic material	Middle Archaic	Lavallée et al. (1999)
Los Burros (Excavation)	6510 ± 60	10624/GifA 97288	Charcoal	Middle Archaic	Lavallée et al. (1999)
Los Burros (Excavation)	6460 ± 60	10623/GifA 97287	Charcoal	Middle Archaic	Lavallée et al. (1999)
Kilómetro 4	6220 ± 70	Beta-77951	Charcoal	Middle Archaic	Wise (1999)
Los Burros (Profile "Corral")	6180 ± 50	10637	Organic material	Middle Archaic	Lavallée et al. (1999)
Los Burros (Excavation)	6110 ± 30	10399	Shell	Middle Archaic	Lavallée et al. (1999)
Los Burros (Profile corral)	5390 ± 100	10638	Organic material	Middle Archaic	Lavallée et al. (1999)
Ring site	5060 ± 55	PITT-0144	Charcoal	Late Archaic	Sandweiss et al. (1989)
Kilómetro 4	4620 ± 90	Beta-27417	Wood-Charcoal	Late Archaic	Wise et al. (1994)
Los Burros (Profile "Corral")	4555 ± 50	10639	Organic material	Late Archaic	Lavallée et al. (1999)
Tacaguay	4550 ± 50	Beta-108536-A	Sediment	Not cultural	Keefer et al. (1998)
Carrizal	4390 ± 110	Beta-18920	Charcoal	Late Archaic	Wise (1989)
Los Burros (Profile "Corral")	4010 ± 55	10640	Organic material	Late Achaic	Lavallée et al. (1999)
Kilómetro 4	3970 ± 80	Beta-77948	Charcoal	Late Archaic	Wise (1999)
Q. Jaguay-32	3895 ± 80	BGS 1995	Shell	Late Archaic	Sandweiss et al. (1998)

Continued

Table 3.2. continued

Sites	^{14}C yr BP	Laboratory	Material	Period	Reference
Kilómetro 4	3760 ± 70	Beta-52797	Wood-Charcoal	Late Archaic	Wise et al. (1994)
Kilómetro 4	3750 ± 60	Beta-52796	Wood-Charcoal	Late Archaic	Wise et al. (1994)
Los Burros (Test 2b)	3700 ± 40	10648	Organic material	Late Archaic	Lavallée et al. (1999)
Kilómetro 4	3680 ± 70	Beta-77946	Charcoal	Late Archaic	Wise (1999)
Los Burros (Cañón)	3595 ± 90	10722	Shell	Late Archaic	Lavallée et al. (1999)
Kilómetro 4	3340 ± 70	Beta-77950	Charcoal	Late Archaic	Wise (1999)
Kilómetro 4	3240 ± 60	Beta-77943	Charcoal	Late Archaic	Wise (1999)
Los Burros (Profile "Corral")	3220 ± 50	10641	Organic material	Late Archaic	Lavallée et al. (1999)
Los Burros (Cañón)	3120 ± 80	10629	Charcoal	Late Archaic	Lavallée et al. (1999)
Los Burros (Cañón)	2825 ± 80	10631	Charcoal	Late Archaic	Lavallée et al. (1999)
Los Burros (Cañón)	2760 ± 80	10630	Charcoal	Late Archaic	Lavallée et al. (1999)
Chilean sites					
Acha-2	8970 ± 255	KE-15082	Human muscle	Early Archaic	Muñoz and Chacama (1993)
Acha-2	8900 ± 150	Teledyne SR	Charcoal	Early Archaic	Muñoz and Chacama (1993)
Acha-3	8380 ± 60	Beta-88041	Human muscle	Early Archaic	Standen and Santoro (2006)
Acha-3	8120 ± 90	Beta-40956	Human muscle	Early Archaic	Standen and Santoro (2006)
Camarones-14	7420 ± 225	I-999	Charcoal	Middle Archaic	Schiappacasse and Niemeyer (1984)
Camarones-14	7000 ± 135	I-11431	Human tissue	Middle Archaic	Schiappacasse and Niemeyer (1984)

Camarones-17	6930 ± 140	GX-15081	Wood	Middle Archaic	Muñoz et al. (1993), Aufderheide et al. (1993)
Camarones-17	6780 ± 110	GX-15080	Wood	Middle Archaic	Muñoz et al. (1993), Aufderheide et al (1993)
Camarones-14	6650 ± 155	I-9817	Charcoal	Middle Archaic	Schiappacasse and Niemeyer (1984)
Camarones-14	6615 ± 390	I-9816	Charcoal	Middle Archaic	Schiappacasse and Niemeyer (1984)
Quiani 9	6370 ± 540	GaK-8782	Charcoal	Middle Archaic	Muñoz and Chacama (1982)
Camarones Pta. Norte	6270 ± 130	Gak-7135	Charcoal	Middle Archaic	Alvarez (1980)
Camarones Pta. Norte	6240 + 160	Gak-7132	Charcoal	Middle Archaic	Alvarez (1980)
Quiani-1	6170 ± 220	I-1348	Charcoal and bone	Middle Archaic	Mostny (1964)
Quiani-9	6115 ± 280	I-11.643	NR	Middle Archaic	Muñoz and Chacama (1982)
Chinchorro-1	6070 ± 285	I-15084	Wood	Middle Archaic	Muñoz et al. (1993)
Camarones Pta. Norte	5950 + 130	Gak-7137	Charcoal	Middle Archaic	Alvarez (1980)
Camarones Pta. Norte	5880 + 160	Gak-7134	Charcoal	Late Archaic	Alvarez (1980)
Camarones Pta. Norte	5750 + 170	Gak-7133	Charcoal	Late Archaic	Alvarez (1980)
Camarones Pta. Norte	5670 + 140	Gak-7136	Charcoal	Late Archaic	Alvarez (1980)
Camarones Pta. Norte	5640 + 160	NN	Charcoal	Late Archaic	Alvarez (1980)
Camarones-Sur	5640 ± 160	GaK-8645	Charcoal	Late Archaic	Rivera (1984)
Quiani-1	5630 ± 130	I-1349	Charcoal and bone	Late Archaic	Mostny (1964)
Chinchorro-1	5560 ± 175	I-15083	Wood	Late Archaic	Muñoz et al. (1993)
Quiani-9	5250 ± 430	GaK-8781	Charcoal and bone	Late Archaic	Muñoz and Chacama (1982)

Continued

Table 3.2. continued

Sites	^{14}C yr BP	Laboratory	Material	Period	Reference
Camarones Pta. Norte	5230				Alvarez (1980)
Morro-1	5240 ± 230	GaK-9902	Wood	Late Archaic	Vera (1981)
Pisagua Viejo-4	5220 ± 245	IVIC-170	Wood	Late Archaic	Núñez (1976)
Morro-1	5160 ± 110	I-13539	Human tissue	Late Archaic	Allison et al. (1984)
Morro-1	5010 + 110	GaK-71309903	Wood	Late Archaic	Vera (1981)
Camarones-Punta	4950 ± 210	GaK-9903	Wood	Late Archaic	Alvarez (1980)
Pisagua Viejo-4	4880 ± 320	IVIC-170	Wood	Late Archaic	Núñez (1976)
Maderas Enco	4750 ± 155	GX-17464	Wood	Late Archaic	Arriaza (1995)
Camarones 8	4635 ± 90	GX-15079	Human tissue	Late Archaic	Muñoz et al. (1993), Aufderheide et al. (1993)
Morro-1	4570 ± 100	I-13542	Human tissue	Late Archaic	Allison et al. (1984)
Morro-1	4520 ± 90	Beta-40956	Wood	Late Archaic	Standen (1997)
Morro-1	4350 ± 280	I-13650	Human tissue	Late Archaic	Allison et al. (1984)
Morro-1/6	4310 ± 145	GX-	Human muscle	Late Archaic	Focacci and Chacón (1989)
Camarones 15b	4240 ± 145	GX-18256	Human tissue	Late Archaic	Muñoz et al. (1993), Rivera (1994)
Morro-1	4200 ± 100	I-13541	Human tissue	Late Archaic	Allison et al. (1984)
Morro-1	4120 ± 75	GX-17019	Human tissue	Late Archaic	Guillén (1992)
Playa Miller 8	4090 ± 105	GaK-5811	Wood	Late Archaic	Rivera (1977–78)
Morro-1	4040 ± 100	I-13543	Wood	Late Archaic	Allison et al. (1984)
Morro-1/6	4010 ± 75	GX-	Human muscle	Late Archaic	Focacci and Chacón (1989); Rivera (1994)
Camarones 15b	4010 ± 75	GX-18258	Human tissue	Late Archaic	Muñoz et al. (1993), Rivera (1994)
Lluta 13	3900 ± 100	charcoal	charcoal	Late Archaic	Santoro (1999)
Morro-1/6	3895 ± 75	GX-	Human tissue	Late Archaic	Focacci and Chacón (1989)

Morro-1/6	3880 ± 70	GX-	Human tissue	Late Archaic	Focacci and Chacón (1989)
Morro-1	3830 ± 100	I-13652	Human tissue	Late Archaic	Allison et al. (1984)
Morro-1	3790 ± 140	I-13656	Human tissue	Late Archaic	Allison et al. (1984)
Morro-1/6	3780 ± 100	I-14957	Human tissue	Late Archaic	Focacci and Chacón (1989)
Morro-1/6	3750 ± 140	GX-	Human tissue	Late Archaic	Focacci and Chacón (1989)
Morro-1	3670 ± 100	I-13651	Human tissue	Late Archaic	Allison et al. (1984)
Camarones 15d	3650 ± 200	RL-2054	Human tissue	Late Archaic	Rivera (1994)
Quiani- 7	3590 ± 100	GaK-5814	Wood	Late Archaic	Rivera (1977–78)
Morro-1/6	3560 ± 100	I- 14958	Human tissue	Late Archaic	Focacci and Chacón (1989)
Quiani- 7	3280 ± 90	I-13654		Late Archaic	Unpublished
Quiani- 7	3240 ± 90	I-13655		Late Archaic	Unpublished
Camarones-Sur	3060 ± 290	RL-2055		Late Archaic	Rivera (1994)
Camarones 15	3060 ± 100	GaK-5813	Wood	Late Archaic	Rivera et al. (1974)
B. Sterile marine coast					
Chilean sites					
La Chimba 13	9680 ± 160	P-2702	Charcoal	Early Archaic	Llagostera (1977)
La Chimba 13	9400 ± 160	P-2702	Charcoal	Early Archaic	Llagostera (1977)
La Chimba 13	9170 ± 30	TO-5631	Otoliths	Early Archaic	Costa Junqueira (2001)
Cobija-1	6030 ± 70	Beta-3933	Charcoal	Late Archaic	Bittmann (1984)
Caramucho-1	5980 ± 120	GaK-8375	Shell	Late Archaic	Sanhuesa (1980)
Cobija-13	5510 ± 50	Beta-3934	Shell	Late Archaic	Bittmann (1984)
Cobija-S1	5460 ± 140	Beta-3114	Charcoal	Late Archaic	Bittmann (1984)
Cobija-S1	5440 ± 150	Beta-3115	Charcoal	Late Archaic	Bittmann (1984)
Abtao-1	5350 ± 120	Gif-1660	NR	Late Archaic	Boisset et al. (1969)
Abtao-1	5100 ± 130	IVIC681	Shell	Late Archaic	Boisset et al. (1969)
Abtao-1	5090 ± 30	IVIC682	Shell	Late Archaic	Boisset et al. (1969)

Continued

Table 3.2. continued

Sites	^{14}C yr BP	Laboratory	Material	Period	Reference
Cobija-13	5060 ± 120	Beta-3117	Charcoal	Late Archaic	Bittmann (1984)
Abtao-2	5030 ± 70	IVIC-679	Shell	Late Archaic	Boisset et al. (1969)
Cobija-S1	4880 ± 90	Beta-3114	Charcoal	Late Archaic	Bittmann (1984)
Abtao-2	4820 ± 70	IVIC-680	Shell	Late Archaic	Boisset et al. (1969)
Abtao-1	4800 ± 70	IVIC-683	Shell	Late Archaic	Boisset et al. (1969)
Caleta Huelén-42	4780 ± 100	GaK-3546	Charcoal	Late Archaic	Núñez (1971)
Punta Guasilla-1	4730 ± 180	Beta-3121	Charcoal	Late Archaic	Bittmann (1984)
Cáñamo	3960 ± 80	GaK-102	Charcoal	Late Archaic	Núñez and Moragas (1983)
Caleta Huelén-42	3780 ± 90	GaK-3545	Wood	Late Archaic	Núñez (1971)
Abtao-1	3550 ± 100	Gif-1658	Shell	Late Archaic	Boisset et al. (1969)
Punta Guasilla-1	3490 ± 290	Beta-3112	Charcoal	Late Archaic	Bittmann (1984)
C. Valleys, *quebradas*, and oases at intermediate altitude					
Chilean sites					
Tiliviche 1(B)	9760 ± 365	SI-3116	Charcoal	Early Archaic	Núñez and Moragas (1977–78)
Aragón-1	8660 ± 230	GaK-5966	Charcoal	Early Archaic	Núñez and Zlatar (1977)
Tiliviche 1(B)	7850 ± 280	GaK-052	Vegetal fiber	Early Archaic	Núñez and Moragas (1977–78
Tiliviche 1(B)	6905 ± 65	SI-3115	Charcoal	Middle Archaic	Núñez and Moragas (1977–78)
Tarapacá 14-A	6830 ± 270	GaK-2432	Material	Middle Archaic	True et al. (1971)
Tarapacá 14-A	6430 ± 430	WSU-987	Charcoal	Middle Archaic	True et al. (1971)
Tiliviche 1(B)	6060 ± 60	SI-3114	Charcoal	Middle Archaic	Núñez and Moragas (1977–78)
Tarapacá 12	5970 ± 120	GaK-2205	Charcoal	Late Archaic	True et al. (1971)
Tarapacá 12	5250 ± 340	GaK-3895	Charcoal	Late Archaic	Tartaglia (1980)
Aragón-1	5170 ± 200	GaK-5965	Charcoal	Late Archaic	Núñez and Zlatar (1977)

Tarapacá 14-A	4780 ± 130	GaK-2529	Charcoal	Late Archaic	True et al. (1971)
Tarapacá 12	4690 ± 80	UCLA-1293	Charcoal	Late Archaic	True et al. (1971)
Tarapacá 2-A	4160 ± 80	UCLA-1834A	Wool	Late Archaic	Tartaglia (1980)
Tarapacá 12	4480 ± 170	GaK-5867	Charcoal	Late Archaic	Tartaglia (1980)
Conanoxa W(a)	4020 ± 110	IVIC-875	Charcoal	Late Archaic	Schiappacasse and Niemeyer (1969)
Conanoxa W(a)	3970 ± 120	IVIC-876	Charcoal	Late Archaic	Niemeyer and Schiappacasse (1963)
Tarapacá 18	3910 ± 170	GaK-2433	Charcoal	Late Archaic	True and Gildersleeve (1980)
Tiliviche-2	3870 ± 100	GaK-3772	Human coprolite	Late Archaic	Standen and Núñez (1984)
Conanoxa W(a)	3740 ± 130	IVIC-175	Coprolites	Late Archaic	Niemeyer and Schiappacasse (1969)
		D. Valleys and *Quebradas* towards the Andes			
Peruvian sites					
Asana	9820 ± 150	Beta 40063	NR	Early Archaic	Aldenderfer (1993)
Asana	9580 ± 130	Beta-24628	Wood	Early Archaic	Aldenderfer (1993)
Toquepala	9580 ± 160	I-1325	Charcoal	Early Archaic	Ravines (1972)
Toquepala	9490 ± 140	I-1372	Charcoal	Early Archaic	Ravines (1972)
Asana	8790 ± 170	Beta-24630	Wood	Early Archaic	Aldenderfer (1993)
Asana	8780 + 90	Beta-43920	NR	Early Archaic	Aldenderfer (1999)
Asana	8720 + 110	Beta-3303	NR	Early Archaic	Aldenderfer (1999)
Asana	8720 + 110	Beta-35599	NR	Early Archaic	Aldenderfer (1993)
Asana	8720 + 120	Beta-43922	NR	Early Archaic	Aldenderfer (1999)
Asana	8620 + 110	Beta-47057	NR	Early Archaic	Aldenderfer (1999)
Asana	8530 ± 240	Beta-18924	Charcoal	Early Archaic	Aldenderfer (1993)
Asana	8330 ± 60	Beta-43919	NR	Early Archaic	Aldenderfer (1999)
Asana	8250 ± 80	Beta-43921	NR	Early Archaic	Aldenderfer (1999)
Caru	8190 ± 130	Hv-1087	Charcoal	Early Archaic	Ravines (1967)
Asana	8080 ± 110	Beta-24627	NR	Early Archaic	Aldenderfer (1993)

Continued

Table 3.2. continued

Sites	^{14}C yr BP	Laboratory	Material	Period	Reference
Asana	8000±280	Beta-47058	NR	Early Archaic	Aldenderfer (1999)
Asana	7930±80	Beta-43923	NR	Middle Archaic	Aldenderfer (1999)
Asana	7860±110	Beta-23363	Wood	Middle Archaic	Aldenderfer (1993)
Asana	7070±110	Beta-47056	NR	Middle Archaic	Aldenderfer (1999)
Coscori	7610±130	NR	NR	Middle Archaic	Aldenderfer (1989)
Asana	7100±70	Beta 24633	NR	Middle Archaic	Aldenderfer (1993)
Asana	6850±70	Beta-25049	NR	Middle Archaic	Aldenderfer (1988 (1993
Asana	6550±110	Beta-24629	Wood	Middle Archaic	Aldenderfer (1988, 1993)
Asana	6040±90	Beta-24634	NR	Middle Archaic	Aldenderfer (1988, 1993)
Asana	5345±70	B35596/ETH6328	NR	Late Archaic	Aldenderfer (1993)
Asana	4760±90	Beta-27413	NR	Late Archaic	Aldenderfer (1993)
Asana	4640±230	Beta-27414	NR	Late Archaic	Aldenderfer (1993)
Asana	4610±60	Beta-24632	Wood	Late Archaic	Aldenderfer (1993)
Asana	4600±80	Beta-35597	NR	Late Archaic	Aldenderfer (1993)
Asana	4580±60	Beta-24631	Wood	Late Archaic	Aldenderfer (1993)
Asana	4570±60	Beta-35598	NR	Late Archaic	Aldenderfer (1993)
Asana	4330±70	Beta-43918	NR	Late Archaic	Aldenderfer (1999)
Asana	4330±130	Beta-27415	NR	Late Archaic	Aldenderfer (1993)
Asana	3640±80	Beta-23364	Charcoal	Late Archaic	Aldenderfer (1993)
Chilean sites (Arica)					
Tojotojone	9580±1950	GaK-7958	Charcoal	Early Archaic	Dauelsberg (1983)
Ticnamar	9090±75		Charcoal	Early Archaic	Rech (2001)
Patapatane	8160±160	I-12.837	Charcoal	Early Archaic	Santoro and Chacama (1984)
Toconce-Confl.	7990±125	Beta-1995	Charcoal	Middle Archaic	Aldunate et al. (1986)
Patapatane	7970±10	Beta-43019	Charcoal	Middle Archaic	Santoro et al. (2005)
Patapatane	5910±90	Beta-24634	Human bone	Middle Archaic	Standen and Santoro (1994)

Patapatane	4890 ± 130	I-12.838	Charcoal	Late Archaic	Santoro and Chacama (1984)
Guañure	4330 ± 105	I-11.873	Charcoal	Late Archaic	Santoro and Chacama (1982)
Puxuma	4240 ± 95	I-11.872	Charcoal	Late Archaic	Santoro and Chacama (1982)
Puxuma	4010 ± 100	I-11.645	Charcoal	Late Archaic	Santoro and Chacama (1982)
Quevilque	4000 ± 50	Beta-24355	Charcoal	Late Archaic	Núñez and Santoro (1988)
Piñuta	3750 ± 140	I-11.832	Charcoal	Late Archaic	Santoro and Chacama (1982)
Tojotojone	3740 ± 130	GaK-7959	Charcoal	Late Archaic	Dauelsberg (1983)
Puxuma	3510 ± 80	Beta-24357	Charcoal	Late Archaic	Santoro (1989)
Chilean sites (Puna Atacama)					
PN 99	12,251 ± 478	89-117	Obsidian Hydr.	Early Archaic	Lynch and Stevenson (1992)
PN 101	10,875 ± 450	89-121	Obsidian Hydr.	Early Archaic	Lynch and Stevenson (1992)
Tuina-1	10,830 ± 630	SI-3112	Charcoal	Early Archaic	Núñez (1983b)
San Lorenzo-1	10,400 ± 130	N-3423	Charcoal	Early Archaic	Núñez (1983b)
San Lorenzo-1	10,280 ± 120	HV-299	Charcoal	Early Archaic	Spahni (1967)
PN 71	10,154 ± 355	89–109	Obsidian Hydr.	Early Archaic	Lynch and Stevenson (1992)
Tuina-5	10,060 ± 70	Beta-107120	Charcoal	Early Archaic	Núñez et al. (2002
San Lorenzo-1	9960 ± 125	N-3423	Charcoal	Early Archaic	Núñez (1983b)
Tuina-5	9840 ± 110	Beta 107121	Charcoal	Early Archaic	Núñez et al. (2002)
PN 99	9603 ± 434	89–115	Obsidian Hydr.	Early Archaic	Lynch and Stevenson (1992)

Continued

Table 3.2. continued

Sites	^{14}C yr BP	Laboratory	Material	Period	Reference
Tambillo-2/4-a	9590 ± 110	Beta-105687	Organic sediment	No cultural	Núñez et al. (2002)
Chulqui-1	9590 ± 60	Beta-6845	Charcoal	Early Archaic	Sinclaire (1985)
PN 115a	9569 ± 445	89–113	Obsidian Hydr.	Early Archaic	Lynch and Stevenson (1992)
Tulán-68	9290 ± 100	Beta-25532	Charcoal	Early Archaic	Núñez et al. (2002)
PN 71	9127 ± 337	89–108	Obsidian Hydr.	Early Archaic	Lynch and Stevenson (1992)
Tuina-1	9080 ± 130	NR	Charcoal	Early Archaic	Lanning (1967)
Tambillo-1	9590 ± 110	Beta-105687	Charcoal	Early Archaic	Núñez et al. (2002)
Tambillo-1	8870 ± 70	Beta-63365	Charcoal	Early Archaic	Núñez (1983b)
Tambillo-1	8590 ± 130	Beta-25536	Charcoal	Early Archaic	Núñez et al. (2002)
Tulán-67	8190 ± 120	Beta-25535	Charcoal	Early Archaic	Núñez et al. (2002)
PN 115a	7961 ± 405	89–111	Obsidian Hydr.	Middle Archaic	Lynch and Stevenson (1992)
Chulqui-1	7180 ± 80	Beta-7324	Charcoal	Middle Archaic	Sinclaire (1985)
Puripica-3/P16	6460 ± 230	Beta-63366	Charcoal	Middle Archaic	Núñez et al. (2002)
PN 99	6184 ± 341	89–116	Obsidian hydr.	Middle Archaic	Lynch and Stevenson (1992)
Puripica-3/39	6150 ± 150	Beta-87200	Charcoal	Middle Archaic	Núñez et al. (2002)
Puripica-3/P13-14	6130 ± 80	Beta-63359	Charcoal	Middle Archaic	Núñez et al. (2002)
Isla Grande	6008 ± 130	NR	Charcoal	Middle Archaic	Lanning (1967)
Tulan-67	5940 ± 50	Beta-142174	Charcoal	Late Archaic	Núñez et al. (2002)
Chulqui-4	5730 ± 90	Beta-7323	Charcoal	Late Archaic	Sinclaire (1985)
Confluencia-1	5380 ± 130	NR	Charcoal	Late Archaic	Lanning (1967)
Puripica-3/P34	5130 ± 110	Beta-88951	Charcoal	Late Archaic	Núñez et al. (2002)
Calarcoco-1	5120 ±	NR	Collagen	Late Archaic	Serracino and Pereyra (1977)
Tulán-51	4990 ± 110	N-2486	Charcoal	Late Archaic	Núñez (1981)

Puripica-3/P33	4880 ± 100	Beta-45478	Charcoal	Late Archaic	Núñez et al. (2002)
Puripica-1	4815 ± 70	SI-3113	Charcoal	Late Archaic	Núñez (1980)
RanL92/Chiuchiu	4565 ± 110	I-5173	Charcoal	Late Archaic	Druss (1977)
RanL140/Chiuch.	4530 ± 110	NR	NR	Late Archaic	Druss (1977)
RanL15140/Ch.	4500 ± 116	NR	Charcoal	Late Archaic	Druss (1977)
PN 112	4387 ± 310	89–114	Obsidian hydr.	Late Archaic	Lynch and Stevenson (1992)
Kalina/Morteros-1	4370 ± 220	Beta-12977	Charcoal	Late Archaic	Aldunate et al. (1936)
Tulán-52	4340 ± 95	N-2487	Charcoal	Late Archaic	Núñez (1981)
Puripica-1	4290 ± 60	Beta-32390	Charcoal	Late Archaic	Núñez et al. (2002)
RanL92/Chiuchiu	4280 ± 170	I-7017	Charcoal	Late Archaic	Druss (1977)
Tulán-52	4270 ± 80	N-2488	Charcoal	Late Archaic	Núñez et al. (2002)
RanL133(A)/Chi.	4250 ± 105	I-5175	Charcoal	Late Archaic	Druss (1977)
Puripica-1	4160 ± 90	Beta-85226	Charcoal	Late Archaic	Núñez (1981)
Calarcoco-1	4120 ± 170	NR	Collagen	Late Archaic	Serracino (1975)
RanL4(A)/Chiu.	4115 ± 105	I-6741	Apatite	Late Archaic	Druss (1977)
RanL104(B)/Chi.	4050 ± 105	NR	NR	Late Archaic	Druss (1977)
Puripica-1	4050 ± 95	Beta-2360	Charcoal	Late Archaic	Núñez (1981)
Punta Negra-59	4040 ± 70	Beta-12908	Charcoal	Late Archaic	Lynch (1986)
Kalina/Morteros	3950 ± 50	Beta-6844	Charcoal	Late Archaic	Aldunate et al. (1936)
PN 115f	3881 ± 285	89–133	Obsidian hydr.	Late Archaic	Lynch and Stevenson (1992)
RanL118/Chiuch.	3675 ± 470	I-6742	Charcoal	Late Archaic	Druss (1977)
Tulan-67	3640 ± 120	Beta-142175	Charcoal	Late Archaic	Núñez et al. (2002)
RanL276(A)/Chi.	3625 ± 85	I-7016	Charcoal	Late Archaic	Druss (1977)
PN 36	3413 ± 111	89–132	Obsidian hydr.	Late Archaic	Lynch and Stevenson (1992)
PN 72	3257 ± 93	89–119	Obsidian hydr.	Late Archaic	Lynch and Stevenson (1992)
PN-122	3180 ± 252	89–122	Obsidian hydr.	Late Archaic	Lynch and Stevenson (1992)

Continued

Table 3.2. continued

Sites	^{14}C yr BP	Laboratory	Material	Period	Reference
PN 105	3086 ± 255	89–128	Obsidian hydr.	Late Archaic	Lynch and Stevenson (1992)
PN 36	3050 ± 142	89–131	Obsidian hydr.	Late Archaic	Lynch and Stevenson (1992)
			E. High *puna*		
Dry *puna*					
Peruvian and Chilean sites					
Hakenasa	9840 ± 40	Beta-187535	Bones	Early Archaic	LeFebvre (2004)
Las Cuevas	9540 ± 160	T-12.835	Charcoal	Early Archaic	Santoro and Chacama (1982)
Hakenasa	9520 ± 70	Beta-187534	Charcoal	Early Archaic	LeFebvre (2004)
Quebrada Blanca	9510 ± 70	Beta-139632	Charcoal	Early Archaic	Santoro and Standen (2000)
Hakenasa	9260 ± 60	Beta-187533	Charcoal	Early Archaic	LeFebvre (2004)
Hakenasa	9170 ± 70	Beta-187532	Charcoal	Early Archaic	LeFebvre (2004)
Hakenasa	8789 ± 60	Beta-187531	Charcoal	Early Archaic	LeFebvre (2004)
Hakenasa	8340 ± 300	I-13.287	Charcoal	Early Archaic	Santoro (1989)
Las Cuevas	8270 ± 250	I-13.128	Charcoal	Early Archaic	Santoro and Chacama (1984)
Quelcatani	7250 ± 170	NR	NR	Middle Archaic	Aldenderfer (1989)
Quelcatani	7100 ± 130	NR	NR	Early Archaic	Aldenderfer (1989)
Hakenasa	5140 ± 70	Beta-187530	Charcoal	Late Archaic	LeFebvre (2004)
Hakenasa	4270 ± 70	Beta-187529	Charcoal	Late Archaic	LeFebvre (2004)
Hakenasa	4380 ± 130	I-13.230	Charcoal	Late Archaic	Santoro (1989)
Hakenasa	3700 ± 60	Beta-187528	Charcoal	Late Archaic	LeFebvre (2004)
Argentine sites					
Barro Negro	12,530 ± 160	AC-735	Peat	No cultural	Fernández (1984–85)

Barro Negro	12,300 ± 170	AC-744	Peat	No cultural	Fernández (1984–85)
Barro Negro	10,740 ± 140	AC-677	Peat	No cultural	Fernández (1984–85)
Inca Cueva-4	10,620 ± 140	LP-137	Charcoal	Early Archaic	Aschero and Podestá (1986)
Leon Huasi-1	10,550 ± 300	GAK-13.402	Charcoal	Early Archaic	Fernández Distel (1980)
Cueva Yavi	10,450 ± 55	CSIC-1101	Charcoal	Early Archaic	Kulemeyer and Laguna (1996)
Huachichocana	10,200 ± 420	GAK-5847	Charcoal	Early Archaic	Fernández Distel (1986)
Barro Negro	10,200 ± 170	AC-672	Peat	No cultural	Fernández (1984–85)
Barro Negro	10,200 ± 140	AC-745	Peat	No cultural	Fernández (1984–85)
Inca Cueva 4	9900 ± 200	AC-564	Charcoal	Early Archaic	Aschero and Podestá (1986)
Cueva Yavi	9790 ± 100	CSIC-1074	Charcoal	Early Archaic	Kulemeyer and Laguna (1996)
Cueva Yavi	9760 ± 160	AC-1088	Charcoal	Early Archaic	Krapovickas (1987)
Inca Cueva-4	9650 ± 110	LP-102	Charcoal	Early Archaic	Aschero and Podestá (1986)
Huachichocana	9620 ± 130	P-2236	Charcoal	Early Archaic	Fernández Distel (1986)
Cueva Yavi	9480 ± 220	AC-1093	Charcoal	Early Archaic	Krapovickas (1987)
Quebrada Seca-3	9410 ± 120	LP-881	Charcoal	Early Archaic	Aschero (personal communication)
Quebrada Seca-3	9250 ± 100	LP-895	Charcoal	Early Archaic	Aschero (personal communication)
Inca Cueva-4	9230 ± 70	CSIC-498	Charcoal	Early Archaic	Aschero (1984)
Quebrada Seca-3	9050 ± 90	Beta-59930	Charcoal	Early Archaic	Aschero (personal communication)
Barro Negro	9200 ± 140	AC-743	Peat	No cultural	Fernández (1984–85)
Pintoscayoc	9080 ± 50	CAMS39041		Early Archaic	Hernández (2000)
Barro Negro	9050 ± 140	AC-742	Peat	No cultural	Fernández (1984–85)
Huachichocana	8930 ± 360	GAK-5847	Charcoal	Early Archaic	Fernández Distel (1986)

Continued

Table 3.2. continued

Sites	^{14}C yr BP	Laboratory	Material	Period	Reference
Quebrada Seca-3	8670 ± 350	AC-1118	Wood	Early Archaic	Aschero and Podestá (1986)
Huachichocana	8670 ± 550	P-2280	Wood	Early Archaic	Fernández Distel (1986)
Quebrada Seca-3	8660 ± 80	Beta-77747	Charcoal	Early Archaic	Aschero (personal communication)
Quebrada Seca-3	8640 ± 80	Beta-59929	Charcoal	Early Archaic	Aschero (personal communication)
Cueva Yavi	8420 ± 70	CSIC-887	Charcoal	Early Archaic	Kulemeyer and Laguna (1996)
Quebrada Seca-3	8330 ± 110	LP-267	Charcoal	Early Archaic	Aschero (personal communication)
Cueva Yavi	8320 ± 260	CSIC-908	Charcoal	Early Archaic	Kulemeyer and Laguna (1996)
Quebrada Seca-3	7760 ± 80	Beta-77746	Charcoal	Middle Archaic	Aschero (personal communication)
Cueva Salamanca-1	7410 ± 100	LP-615	Charcoal	Middle Archaic	Aschero (personal communication)
Quebrada Seca-3	7350 ± 80	Beta-59928	Charcoal	Middle Archaic	Aschero (personal communication)
Quebrada Seca-3	7220 ± 100	SMU-2364	Charcoal	Middle Archaic	Aschero (personal communication)
Quebrada Seca-3	7130 ± 110	LP-269	Charcoal	Middle Archaic	Aschero (personal communication)
Quebrada Seca-3	6160 ± 100	AC-1117	Charcoal	Middle Archaic	Aschero (personal communication)
Quebrada Seca-3	6080 ± 70	Beta-77745	Charcoal	Middle Archaic	Aschero (personal communication)
Quebrada Seca-3	5400 ± 90	LP-270	Charcoal	Late Archaic	Aschero (personal communication)

Quebrada Seca-3	5380 ± 70	Beta-59927	Charcoal	Late Archaic	Aschero (personal communication)
Inca Cueva-4	5200 ± 110	AC-1112	Charcoal	Late Archaic	Aschero and Podestá (1986)
Quebrada Seca-3	4930 ± 100	AC-1115	Charcoal	Late Archaic	Aschero and Podestá (1986)
Quebrada Seca-3	4770 ± 80	Beta-27802	Charcoal	Late Archaic	Aschero (personal communication)
Quebrada Seca-3	4510 ± 100	Beta-27801	Charcoal	Late Archaic	Aschero (personal communication)
Tomayoc	4250 ±	GIF-8710	Charcoal	Late Archaic	Lavallée et al. (1997)
Inca Cueva-7	4080 ± 80	T-1173	Charcoal	Late Archaic	Aguerre et al. (1973)
Punta de la Peña-4	4060 ± 90	Beta-77749	Charcoal	Late Archaic	Aschero (personal communication)
Peñas Chicas-1.1	3660 ± 60	LP-261	Charcoal	Late Archaic	Aschero (personal communication)
Peñas Chicas-1.1	3590 ± 55	LP-263	Charcoal	Late Archaic	Aschero (personal communication)
Tomayoc	4250 ± 50	GIF-8710	Charcoal	Late Archaic	Lavallée et al. (1997)
Tomayoc	3480 ± 40	GIF-8707	Charcoal	Late Archaic	Lavallée et al. (1997)
Tomayoc	3390 ± 50	GIF-8371	Charcoal	Late Archaic	Lavallée et al. (1997)
Tomayoc	3360 ± 50	GIF-8708	Charcoal	Late Archaic	Lavallée et al. (1997)
Tomayoc	3310 ± 40	GIF-8372	Charcoal	Late Archaic	Lavallée et al. (1997)
Tomayoc	3250 ± 60	GIF-7335	Charcoal	Late Archaic	Lavallée et al. (1997)
Salt Puna					
Chilean sites					
Aguas Calientes I	8720 ± 100	Beta-105696	Charcoal	Early Archaic	Núñez et al. (2002)
Tuyajto 1(B)	8210 ± 110	Beta-105692	Charcoal	Early Archaic	Núñez et al. (2002)
Tuyajto 1(B)	8130 ± 110	Beta-105691	Charcoal	Early Archaic	Núñez et al. (2002)

Continued

Table 3.2. continued

Sites	^{14}C yr BP	Laboratory	Material	Period	Reference
San Martín-4-a	8130 ± 50	Beta-116573	Charcoal	Early Archaic	Núñez et al. (2002)
Huasco-2	6320 ± 50	Beta-142171	Charcoal	Middle Archaic	Núñez et al. (2002)
Meniques-1	5470 ± 60	Beta-105689	Charcoal	Late Archaic	Núñez et al. (2002)
Capur-3	3390 ± 60	Beta-114536	Charcoal	Late Archaic	Núñez et al. (2002)
Capur-3	3320 ± 60	Beta-105690	Charcoal	Late Archaic	Núñez et al. (2002)
Ollague-3	3170 ± 60	Beta-114537	Charcoal	Late Archaic	Núñez et al. (2002)

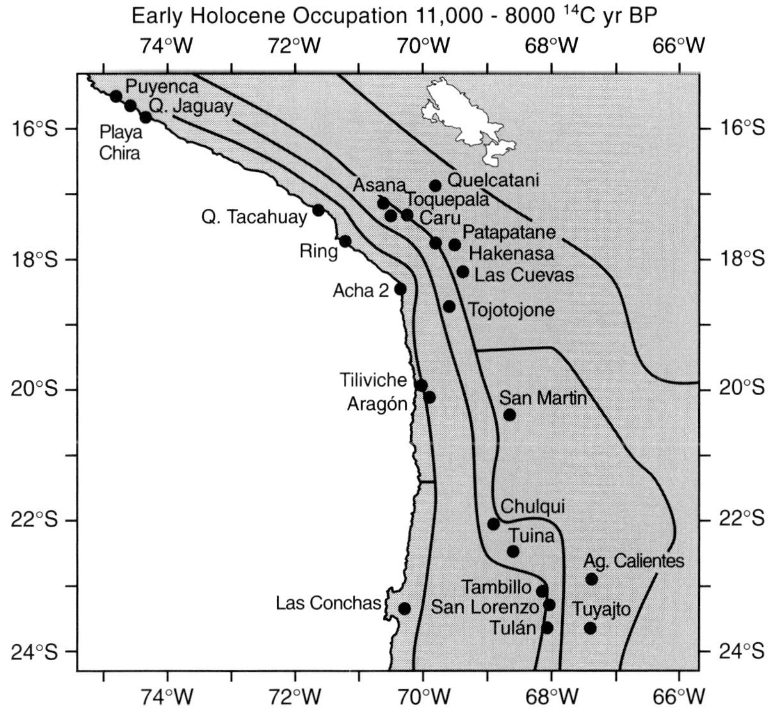

Figure 3.5. Map showing the locations of archaeological sites with early Holocene human occupation between 11,000 and 8000 ^{14}C yr BP. The black lines delineate the different habitats (Fig. 3.1).

Quebrada Jaguay (Sandweiss et al., 1998), hunting of seabirds (Quebrada Tacahuay, Keefer et al., 1998) and marine mammals. The earliest coastal sites also yielded lithic artifacts made from obsidian, which crops out 130 km inland at an altitude of 2850 m (Keefer et al., 1998; Sandweiss et al., 1998).

The Archaic settlement sequence is generally continuous. At the site-scale, however, a hiatus is observed in the south Peruvian sites between ca. 7800 and 5700 cal yr BP (7000 and 5000 ^{14}C yr BP, *Quebradas* Jaguay and Tacahuay, Ring Site, Puyenca, Figs. 3.5 and 3.8). Human occupation was restricted to ephemeral stream sites and a link with decreasing humidity in the adjacent highlands between 8000 and 3600 cal yr BP has been suggested by Sandweiss (2003). Near Arica, the sites of the Chincorro Culture (after ca. 9000 cal yr BP, Figs. 3.6 and 3.8) show considerable cultural innovations compared with the pre-Chinchorro (early) sites. Examples are the development of sophisticated procedures for human mummification, technological innovations as demonstrated in the use of harpoons, different kinds of fishhooks made of *Choromytilus* shell and bone, or the complementary use of terrestrial food resources such as camelids. This shows that major cultural changes began around 9000 cal yr BP. Also the number of coastal sites and most likely the population density increased (Fig. 3.7). However, we point to the fact that the

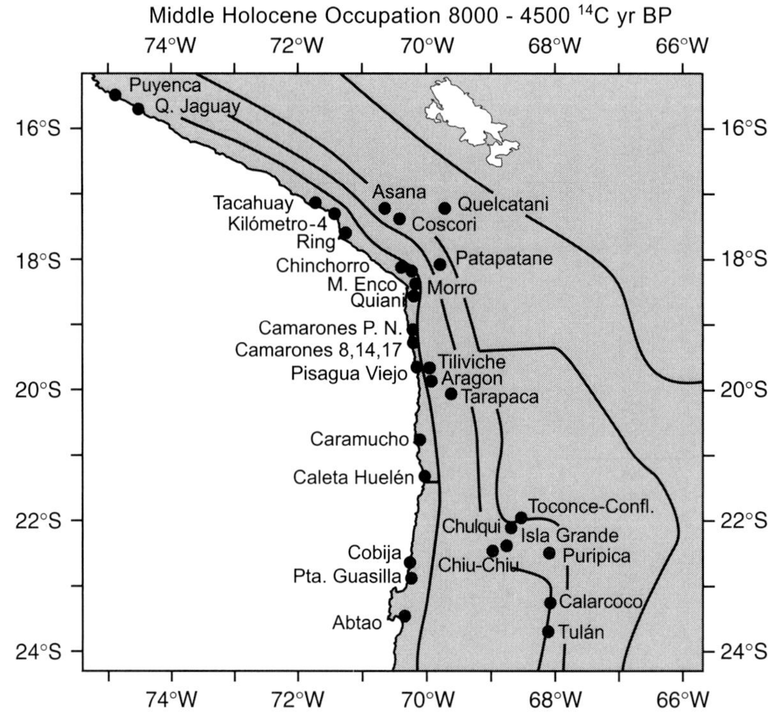

Figure 3.6. Map showing the locations of archaeological sites with mid-Holocene human occupation between 8000 and 4500 ^{14}C yr BP. The black lines delineate the different habitats (Fig. 3.1).

known early sites are all found on high terraces or in estuaries (Núñez and Moragas, 1983; Núñez and Zlatar, 1977, 1980; Múñoz et al., 1993) that were some kilometers away from the early Holocene coastline. In contrast to possibly many other unknown sites located immediately next to the previous coastline, these sites were not affected by the rising sea level prior to ca. 7000 cal yr BP. Thus, we hypothesize that the increase in the number of permanent sites after that time might also be an 'artifact' due to the stabilization of the sea level and the coast line around that time. The same observation is also made in the coastal area further to the south (sterile coast).

The coastal sites were extended open campsites with large shell middens, whereas the sites inside the *quebradas* were smaller. It is suggested that the latter ones were used sporadically, maybe as transitory logistic camps related to the rock outcrops with raw material for lithic artifacts, to collect reed fiber, or to gather terrestrial plants and hunt animals. Aragon and Tiliviche are two representatives of that type of site (Núñez and Zlatar, 1977).

The beginning of artificial mummification of the Chinchorro Culture (Guillén, 1992, 1997; Arriaza, 1994; Standen, 1997), and the diversification/intensification of resource exploitation suggests significantly increasing socio-cultural complexity on the coast

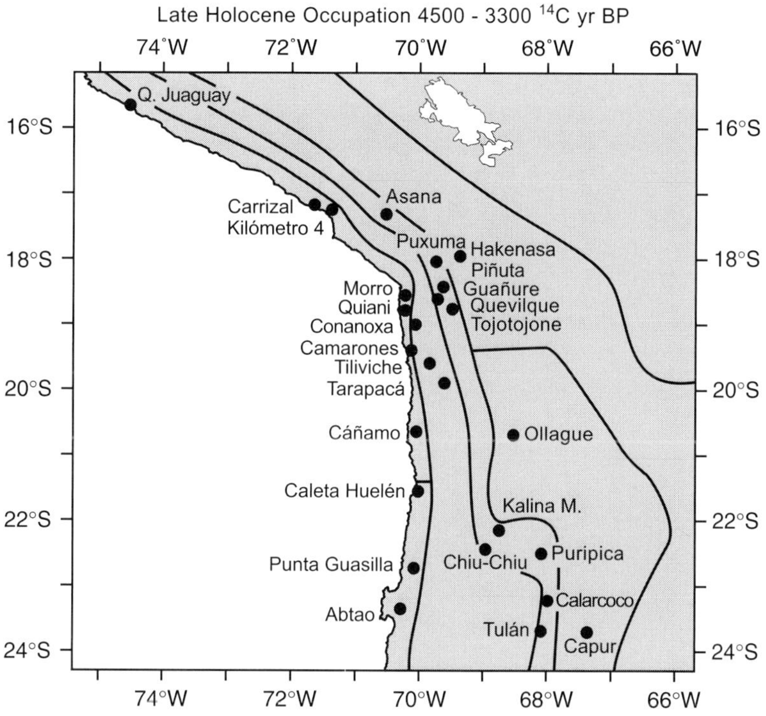

Figure 3.7. Map showing the locations of archaeological sites with late Holocene human occupation between 4500 and 3000 ^{14}C yr BP. The black lines delineate the different habitats (Fig. 3.1).

with the onset of mid-Holocene conditions. Such changes took place between ca. 9000 and 8000 cal yr BP and lasted throughout the Archaic period. There is no significant cultural or environmental change around 5000 cal yr BP (4300 ^{14}C yr BP). Interestingly, the technology and artificial mummification of the Chinchorro culture did not change much during several millennia, which suggests a rather closed, traditional, and conservative society, and is maybe even a mechanism for cohesion and sociocultural defense (Arriaza, 1995; Santoro, 1999). The Chinchorro culture disintegrated after 4000 cal yr BP when new types of burials gradually replaced mummification. Individualism (e.g., hair dress) became more important, the society more structured and new technologies with wooden tools and cotton fabric emerged. However, the marine subsistence economy and the settlement pattern on the coast remained pretty constant during the time of such transformation.

5.2. Occupation of the sterile marine coast

Las Conchas (23°33′S) is the only known and ^{14}C dated Early Holocene site along the sterile marine coast south of 22°S (Figs. 3.5 and 3.8; Table 3.2). Like the early sites further north, Las Conchas was always located several kilometers away from the

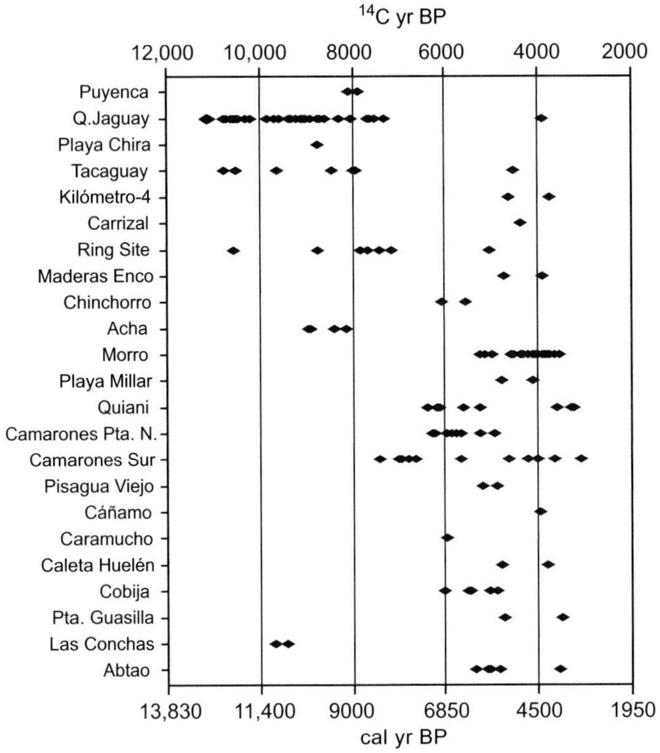

Figure 3.8. Radiocarbon chronostratigraphy of archaeological sites in the habitat of the fertile and the sterile coast. The sites are listed from north (top) to the south (bottom).

coast, and thus not affected by Holocene sea level changes. The cultural complex of this site known as the Huentelauquen Pattern (Llagostera, 1977, 1979) was based on a wide variety of marine resources (mammals, fish, and mollusks) and complementary birds and camelids. Techniques included net fishing (for Sciaenidae and Serranidae), harpoons, and collecting of mollusks (*Concholepas* and *Fissurella*). Interestingly, the variety of fish in this early site includes species of the Panamic Province that are indicative of warm water conditions along the northern Chile coast at that time (Llagostera, 1979). People seem to have lived permanently on the coast. The distinctive cultural features are geometric sandstone artifacts that are also known from areas south of Antofagasta and in Central Chile (Llagostera, 1979).

There seems to be an occupational hiatus between ca. 9000 and 6000 ^{14}C yr BP (Fig. 3.8), when a second phase of human habitation began. However, we point again to the problem of sea level rise during the early Holocene, which might have submerged many of the early coastal sites, and the fact that archaeological survey is incomplete in this area. It also remains unclear whether or not the end of Las Conchas is triggered by desiccation of the local springs at the end of the early Holocene.

Coinciding with the stabilization of sea level, the sterile coast was re-colonized by open campsites between 6700 and 3200 cal yr BP (6000 and 3000 ^{14}C yr BP;

Figs. 3.6–3.8). Although the hunting–fishing–gathering practices remained the same, the introduction of a variety of new harpoon types suggests cultural changes. Interestingly, the warm water fish species disappeared and were replaced by cold-water species (such as *Choromytilus*) by 6000 cal yr BP, whereas *Trachurus* (a warm-water fish) was absent until 4500 cal yr BP and increased afterwards (Llagostera, 1979).

Some elements of the Chinchorro culture (such as artificial mummification and technology) were introduced as far south as Antofagasta (23°S) from ca. 4500 cal yr BP (4000 ^{14}C yr BP). The presence of obsidian fragments and feathers of Andean parrots in a site at the Río Loa estuary on the one hand, and marine fish remains in sites of the middle course of the Rio Loa (Druss, 1977) and marine shells in sites of the western Andean slope (Tulán 52, Núñez and Santoro, 1988) on the other show that the cultural exchange between the coast and the high *puna* was already intensified around 5500 cal yr BP (4780 ^{14}C yr BP; Núñez et al., 1974). This falls well into the period of greatest environmental stress in the adjacent highlands. We see this as evidence of high regional mobility between the coast and the highlands and as evidence of intense exchange between peoples living in adjacent habitats largely around 5000 cal yr BP, the time window considered in the context of this chapter. However, this interpretation remains somewhat speculative because the database and the archaeological survey in this habitat are still far from complete.

5.3. Occupation of valleys, quebradas, and oases at intermediate altitude

The archaeological sites in the intermediate valleys are usually open campsites and show features of a mixed subsistence economy based on marine and terrestrial resources. The chronosequence spans between 12,000 and 4200 cal yr BP (10,000 and 3800 ^{14}C yr BP) without a significant hiatus (Fig. 3.9; Table 3.2). The early sites of Tiliviche and Aragón show strong bonds to the coast. The small size and the

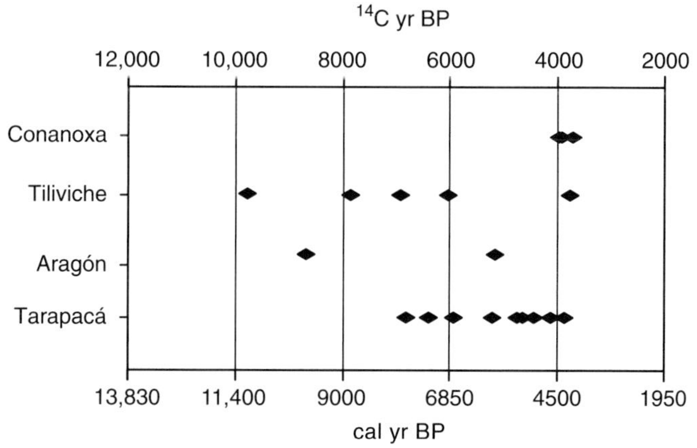

Figure 3.9. Radiocarbon chronostratigraphy of archaeological sites in the habitat of the intermediate valleys. The sites are listed from north (top) to the south (bottom).

low-density of the sites also suggest that they were rather complementary transitory camps than semi-permanent settlements observed at the coast. All of the known campsites/workshops (Figs. 3.5–3.7) are located in areas where local vegetation and water resources, reed fiber, wood, and in some cases lithic raw material were available. However, macrofossils of marine fauna and the technology of the tools suggest strong cultural bonds to the coast and the need for complementary food supplies (Niemeyer and Schiappacasse, 1963; Schiappacasse and Niemeyer, 1969; True et al., 1970, 1971; Núñez and Moragas, 1977–78; Núñez and Zlatar, 1977, 1980; Núñez et al., 1979–81, 1994; Standen and Núñez, 1984).

The (non-seasonal?) mobility pattern of the early occupants of these sites (Fig. 3.5) continues without major change until ca. 4500 cal yr BP (4000 ^{14}C yr BP), whereas major changes in the resource use are suggested. For instance, the early occupation at Aragón (pre-8600 ^{14}C yr BP, 30 km away from the coast) shows that mostly terrestrial resources were exploited (small land mammals and *Prosopis*). It is suggested that local food and river water were sufficient to sustain a (low-density?) highly mobile population, whereas the late occupation of the site after 5000 cal yr BP (4400 ^{14}C yr BP) relied strongly on marine food components. Reduced mid-Holocene river runoff from the Andes might have produced harsh conditions in these microenvironments. Other than at Aragon, the sites in Quebrada Tiliviche showed always a mixed marine-terrestrial subsistence economy throughout the early and middle Holocene (including camelids, Núñez, 1983a,b; Núñez and Moragas, 1977–78), and thus the changes were less pronounced. However, local terrestrial resources became more important after 4200 cal yr BP, (3800 ^{14}C yr BP), which coincides largely with the onset of modern more humid conditions in the high Andes, and supports the hypothesis about the importance of Andean water supply to the intermediate and coastal parts of the valleys.

Most sites at intermediate elevation are found in valleys north of the Río Loa, adjacent to the 'fertile' coast, along the waterways between the Andes and the coast. However, there are sites and lithic workshops adjacent to the 'sterile' coast ca. 40 km inland of the coast between Antofagasta and Taltal. The chronostratigraphy and paleoenvironmental context of these sites, however, is not yet known.

With regard to the mid-Holocene climate and cultural changes we emphasize the high mobility of the people (low-density population, and small transitory camps), the concentration of human activities in river oases and ecological refuges (such as Aragón, Tiliviche, and Tarapacá), and the coastal sites as a buffer zone with stable food resources and where the main camps were located. We interpret this as adaptive strategies to an environment with generally very low biomass productivity and relatively high resource variability (Table 3.1).

5.4. Occupation of valleys and quebradas toward the Andes

This habitat connects the coast with the Andes and shows also scarce precipitation, vegetation, and animal resources, but it is closer to the *puna* where the water resources are located.

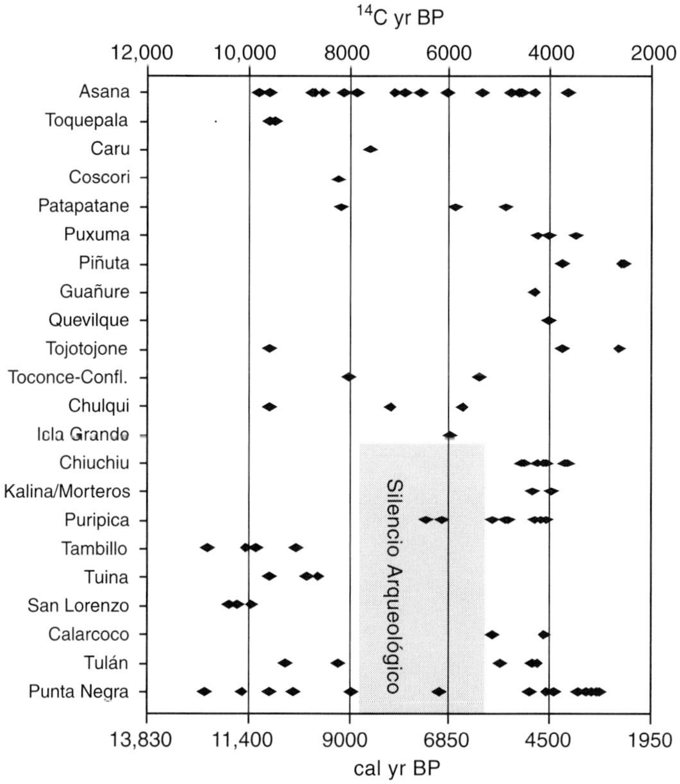

Figure 3.10. Radiocarbon chronostratigraphy of archaeological sites in the habitat of the high valleys. The sites are listed from north (top) to the south (bottom).

Asana, the most important site in southern Peru, was used continuously between 11,000 and 4000 cal yr BP (9800 and 3600 ^{14}C yr BP; Figs. 3.5 and 3.10, Table 3.2, Aldenderfer, 1993, 1999), whereas other less-well-documented sites suggest discontinuous habitation (Fig. 3.10; Table 3.2, Ravines, 1967, 1972). However, the Early Archaic occupation of Asana shows strong bonds to the lower elevation belts and the coast as suggested by the presence of lithic artifacts from outcrops at lower elevation and a settlement pattern with circular 'houses' of 2–3 m in diameter similar to what is found at coastal sites (e.g., Acha 3). At the same time, lithic materials document some links to the *puna* (Aldenderfer, 1993). Materials from the *puna* became increasingly important after ca. 10,000 cal yr BP (8800 ^{14}C yr BP) and especially between 8600 and 6900 cal yr BP (7800 and 6000 ^{14}C yr BP; Middle Archaic Period), which is thought to reflect a fundamental orientation of the mobility pattern toward the *puna*. The first architecture with circular constructions made of posts, brush-walls, and consolidated floors at wind-protected sites are dated ca. 7700 cal yr BP (6850 ^{14}C yr BP, Aldenderfer, 1988).

Aldenderfer (1993) observed a collapse in the overall number and density of artifacts (particularly the *puna* elements) particularly between 6700 and 5700 cal yr

BP (6000 and 5000 ^{14}C yr BP), which coincided with the desiccation of the local wetlands (*bofedal*) and likely with a substantial decrease in local food and water resources. Asana is thought to have been almost abandoned as it become a temporary camp site within the logistic radius of a (semi)permanent base located in the adjacent *dry puna*, possibly the Quelcatani site (Aldenderfer, 1989). However, no *puna* material has been found at Asana between 6700 and 5700 cal yr BP (6000 and 5000 ^{14}C yr BP). The major cultural change during that time is observed in the architecture, which suggests a trend toward a sporadic use of the site during short periods of time (Aldenderfer, 1993).

Intense reactivation of the site and re-establishment of strong links to the *puna* are observed around 5000 cal yr BP (4400 ^{14}C yr BP; Late Archaic). At that time, the most likely seasonal transhumant mobility pattern across various altitudinal belts and geoecological zones included pasturage of domesticated animals. Also seed processing, new oval forms, and functions of domestic architecture, ceremonial structures, and stone fences for animals suggest major cultural changes after 5000 cal yr BP (Aldenderfer, 1993).

Further to the south in northern Chile, the archaeological record of the high valleys includes six sites covering the Archaic Period between 10,900 and 2500 cal yr BP (9600 and 2400 ^{14}C yr BP; Figs. 3.5–3.7, Table 3.2). The Early Archaic sites show a highly diverse lithic industry. Faunal remains include camelids and rodents. Few marine gastropod shells (*Choromytilus*, likely of ceremonial character) suggest interaction with the coastal habitats. However, the sites are much smaller compared to Asana in Peru, and located in rock shelters and caves. Interestingly, the Middle Archaic Period starts with a 2000-year period of low human activity or even with an occupational hiatus between 9000 and 7000 cal yr BP (Fig. 3.10), which coincided with the severe regional mid-Holocene drought as recorded in nearby Laguna Seca in the *Puna* of Arica (Baied, 1991). Whereas Aldenderfer (1988) considers mainly the incomplete archaeological survey as a possible explanation, Santoro (1987, 1989) favored environmental stress, which may also have resulted in a stronger orientation toward the *puna* or the coastal sites. This might help to explain the observed increase in coastal sites, although the early Holocene sea level fluctuations remain a problem when population density on the coast is interpreted. Resolving this controversy requires a more complete archaeological survey and database.

Human occupation of the high valleys in northern Chile recovered after 6300 cal yr BP (5500 ^{14}C yr B.P; Fig. 3.10). People used a broad variety of materials and tools. Although the introduction of some tuber crops such as *ullucu* or *papalisa* (*Ullucus* sp.) and *isaño* (*Tropaelum*) is documented, the subsistence economy remained mainly based on camelids and rodents (Santoro and Núñez, 1987; Núñez and Santoro, 1988; Santoro, 1989). Other important cultural elements include rock painting. The already previously established seasonal transhumant pattern of resource use in different geoecological zones between the coast and the high *puna* remained the key-strategy for the subsistence economy.

Further to the south in the Puna de Atacama, the habitat of the high valleys is documented by 43 Archaic sites (Figs. 3.5–3.7) covering the period between 13,000 and 3200 cal yr BP (11,000 and 3000 ^{14}C yr BP; Fig. 3.10, Table 3.2). The most important archaeological areas south of the Río Loa are the Quebrada de Tulán, the Quebrada de San Lorenzo, Quebrada Puripica, and Tuina (Núñez et al., 2002). All of them connect the habitat of the *puna* (>4000 m) with the habitat in the low elevation basins of the Salar de Atacama (2500 m), the Salar Punta Negra (Lynch, 1986; Lynch and Stevenson, 1992), or with the Río Loa valley and further connections to the Pacific coast. The Early Archaic sites (Fig. 3.5) are all located in rock shelters and caves. Lithic artifacts (typical triangular points) made of exotic basalt and obsidian suggest intense transhumance with a strong orientation toward the *puna* that was readily accessible at a short distance. Many of the caves show a well-developed stratigraphy of Early Archaic archaeological deposits, and show multiple uses of these sites over a long period of time by highly mobile but likely small groups of people.

The surprisingly dense record of early sites experiences a dramatic decline with the onset of arid mid-Holocene conditions around 9000 cal yr BP (8000 ^{14}C yr BP; Fig. 3.10, Núñez et al., 2001, 2002). Due to the constant sedimentation of sterile geologic material from the ceiling, the caves of Tuina-4 and San Lorenzo are perhaps the best sites to document the mid-Holocene occupational hiatus (known as '*Silencio Arqueológico*', Núñez and Grosjean, 1994) between 9000 and <5700 cal yr BP (8000 and <5000 ^{14}C yr BP; Fig. 3.10, Table 3.2). The mid-Holocene sediments in some of the caves are totally devoid of archaeological remains, clearly separating the Early Holocene/Early Archaic archaeological strata rich in plant macrofossils, charcoal, mammal, and bird bones from the post-Archaic archaeological strata (supplement material in Núñez et al., 2002). This occupational hiatus coincides with the extremely arid mid-Holocene environmental conditions in the Salt Puna, when lake levels reached lowest stands, river discharge decreased substantially, and water resources became critically scarce. Compared with the more stable and continuously occupied high valleys further north (such as Chulqui in the Río Loa basin, Aldunate et al., 1981; Sinclaire, 1985), the valleys in the Atacama basin were always relatively poor in resources, and thus responded most sensitively to climate changes. Therefore, we think that the environmental conditions in this sector dropped below critical levels for hunting and gathering societies, as they were present during the early Holocene. This resulted in a clear hiatus in this specific area, whereas a comparable decline in water resources led to a decrease in population density, but not necessarily to a visible hiatus north of the Río Loa. This feature is typical for areas with marginal and critically scarce resources, and will repeat itself in the Salt Puna (*puna salada*), the most marginal and arid part of the *puna* (see Section 5.5.).

The mid-Holocene fully arid conditions resulted in some cases in the formation of 'ecological refuges', small atypical oases, where water was still available, resources were concentrated, and where people found the living space for discrete habitation. Such an example is documented in Quebrada Puripica (Grosjean et al., 1997a, Núñez et al., 2001). Twenty fireplaces that are physically separated by individual debris flows on the alluvial fan record in detail the stepwise cultural transformation of a hunting/gathering Early Archaic society into a very complex Late Archaic society, and thus fill the 'gap of evidence' of the regional '*Silencio Arqueológico*'. The early hunting tradition prior to 7000 cal yr BP (6200 ^{14}C yr BP) changed by 6700 cal yr BP (5900 ^{14}C yr BP) into a cultural system with large campsites, intense exploitation of wild camelids, and an innovative lithic industry with microliths and perforators, some of them made of exotic raw material. This process culminated in the Late Archaic classic site of Puripica-1 (5500 cal yr BP, 4800 ^{14}C yr BP) that showed parallel hunting and domestication of camelids, the use of local lithic materials, and the development of structured semi-sedentary settlements and naturalistic rock art (Núñez, 1981; Hesse, 1982; Grosjean and Núñez, 1994; Grosjean et al., 1997a; Núñez et al., 2001). Although human occupation at Puripica continued through the agricultural period (after ca. 3500 cal yr BP), the site lost the unique importance as an 'ecological refuge' after ca. 4200 cal yr BP (3800 ^{14}C yr BP), when modern (i.e., better) conditions were established, lake levels rose again, and widespread re-occupation of the area is observed (Tilocalar Phase beginning 3500 cal yr BP, Núñez et al., 1996).

5.5. Occupation of the high puna

The numerous sites in the habitat of the high *puna* reflect in general what has been observed in the sites of the high valleys in adjacent areas to the west. This is not surprising because the *puna* and the high valleys were always complementary ecosystems within the same economic unit. Early Archaic (<12,000 cal yr BP, 10,000 ^{14}C yr BP) human occupation in the *puna seca* is documented in three sites, Quelcatani in southern Peru, and Las Cuevas and Hakenasa in northernmost Chile (Fig. 3.11, Aldenderfer, 1989; Santoro, 1989). Interestingly, the known sites are all located in caves, which is very different from the numerous open campsites in the *puna salada* further south. The sites in the *puna seca* were repeatedly used during short intervals. As expected, there is no clear evidence of a mid-Holocene hiatus, although the density of artifacts and likely also human activity decreased significantly between ca. 9000 and 6700 cal yr BP (8000 and 6000 ^{14}C yr BP; Table 3.2). The mid-Holocene sites do not show any evidence of coastal artifacts, suggesting that the mobility pattern was restricted to the *puna* and the adjacent valleys, possibly with (or maybe due to) domesticated camelids (Núñez, 1981; Aldenderfer, 1993).

In contrast to the sites in the *puna seca* with rather continuous occupation, the sites in hydrologically sensitive areas of the *Puna Salada* of northern Chile show a distinct hiatus (*Silencio Arqueológico*) between ca. 9000 and 4500 cal yr BP (8000 and 4000 ^{14}C yr BP), which coincided with the hyperarid mid-Holocene conditions. The Early Archaic sites in the *puna salada* prior to 9000 cal yr BP (8000 ^{14}C yr BP) are small open campsites with abundant local lithic material (basalt, obsidian), typical triangular artifacts. The sites are usually strictly related to the fossil shorelines of the late-glacial/early Holocene paleolakes. The few ^{14}C dated sites (Salar Aguas Calientes I, Salar Tuyajto, Salar San Martin, Fig. 3.5; Table 3.2, Núñez

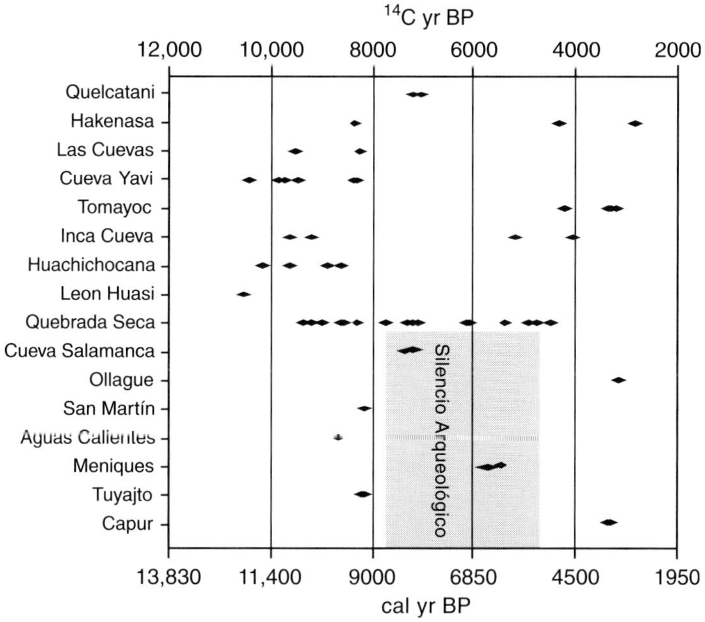

Figure 3.11. Radiocarbon chronostratigraphy of archaeological sites in the habitat of the *puna seca* and *puna salada*. The sites are listed from north (top) to the south (bottom).

et al., 2001, 2002) and the numerous sites with diagnostic triangular artifacts suggest that human occupation was widespread between 12,000 and 9000 cal yr BP (10,000 and 8000 ^{14}C yr BP), whereas there is no evidence of human occupation in the *puna salada* during the mid-Holocene between 9000 and 6200 cal yr BP (8000 and 5500 ^{14}C yr BP; Fig. 3.6). However, as a result of changing climate and geomorphological processes, alternative habitats with very favorable conditions were created in flat bottoms of desiccated lakes or in steep valleys where wetlands were formed (Grosjean et al., 2005b). In light of the well-documented paleoenvironmental scenario, we interpret this occupational re-organization (i.e. not always a hiatus) as a clear signal of extremely harsh environmental conditions. Interestingly, ^{14}C dated open camp sites document reoccupation of the lakesides at the time when regional lake levels started to increase, and modern (i.e., more humid than before) conditions were established after 4000 cal yr BP (3600 ^{14}C yr BP). The two ^{14}C dated sites showing Late Archaic reoccupation of lake shorelines are Salar Ollagüe and Salar Capur (Fig. 3.7).

Our climate–culture model for the *Puna Seca* and *Puna Salada* in Peru and Chile also applies successfully to NW Argentina (Fig. 3.11, Table 3.2). As expected, the environments in the more arid and marginal part of the NW Argentinean *Puna* were seriously affected by the mid-Holocene drought. Particularly the time between 9000 and 7000 cal yr BP (8100 and 6100 ^{14}C yr BP) was very arid (Kulemeyer et al., 1999), and resources dropped below a critical level for hunting and gathering societies. This resulted in a mid-Holocene occupational hiatus in the Archaic sites of

Inca Cueva-4, Leon Huasi, Yavi and Huachichocana, all located above 3000 m in the Argentinean *Dry Puna*. The sites of Cueva Salamanca-1 and Quebrada Seca-3 (Aschero, 1994; Rodríguez, 1999; Núñez et al., 2001) show continuous occupation.

Acknowledgments

This chapter resulted from many years of archaeological and paleoenvironmental researches in the South Central Andes supported by numerous FONDECYT (to C.S., L.N., and V.S.), Swiss NF and NCCR Climate (M.G.) and National Geographic Society projects.

References

Abbott, M. B., G. O. Seltzer, K. Kelts, and J. Southon, 1997a. Holocene paleohydrology the tropical Andes from lake records. *Quaternary Research* 47:70–80.

Abbott, M. B., B. B. Wolfe, A. P. Wolfe, G. O. Seltzer, R. Aravena, B. G. Mark, P. J. Polissar, D. T. Rodbell, H. D. Rowe, and M. Vuille, 2003. Holocene paleohydrology and glacial history of the Central Andes using multiproxy lake sediment studies. *Palaeogeography, Palaeoclimatology, Palaeoecology* 194(1–3):123–138.

Aldenderfer, M. S., 1988. Middle Archaic Period domestic architecture from southern Peru. *Science* 241:1828–1830.

Aldenderfer, M. S., 1989. Archaic Period "Complementarity" in the Osmore drainage. In *Ecology Settlement and History in the Osmore Drainage, Peru*, Part 1, edited by S. Rice, C. Stanish, and P. R. Scarr, pp. 101–128. *BAR International Series* 545(1), Oxford.

Aldenderfer, M. S., 1993. Cronología y Definición de Fases Arcaicas en Asana, Sur del Perú. *Chungara* 24/25:13–35.

Aldenderfer, M. S., 1999. *Montane Foragers: Asana and the South-Central Andean Archaic*. University of Iowa Press, Iowa.

Aldunate, C., J. Armesto, V. Castro, and C. Villagrán, 1981. Estudio Etnobotánico en una Comunidad Precordillerana: Toconce. *Boletín Museo Nacional de Historia Natural* 38.

Aldunate, C., J. Berenguer, V. Castro, L. Cornejo, J. Martínez, and C. Sinclaire, 1986. *Cronología y Asentamiento en la Región del Loa Superior*. Dirección de Investigaciones y Bibliotecas Universidad de Chile, Santiago.

Aguerre, A. M., A. Fernández Distel, and C. A. Aschero, 1973. Hallazgo de un Sitio Acerámico en la Quebrada de Inca Cueva (Provincia de Jujuy). *Relaciones Nueva Serie* 7:197–235.

Allison, M. J., G. Focacci, B. Arriaza, V. Standen, M. Rivera, and L. M. Lowenstein, 1984. Chinchorro, Momias de Preparación Complicada: Métodos de Momificación. *Chungara* 13:155–173.

Alvarez, L., 1980. Cazadores Alto-Andinos en la Costa de Arica. *Actas y Trabajos del III Congreso Peruano El Hombre y la Cultura Andina* Tomo III:1029–1031.

Aravena, R., 1995. Isotope hydrology and geochemistry of Northern Chile groundwaters. *Bulletin de l'Institut Français d'Études Andines* 24:495–503.

Arriaza, B. T., 1994. Tipología de las Momias Chinchorro y Evolución de las Prácticas de Momificación. *Chungara* 26:11–24.

Arriaza, B. T., 1995. *Beyond Death the Chinchorro Mummies of Ancient Chile*. Smithsonian Institution Press, Washington, DC.

Arroyo, M. T. K., F. A. Squeo, J. J. Armesto, and C. Villagrán, 1988. Effects of aridity on plant diversity in the northern Chilean Andes: results of a natural experiment. *Annual Missouri Botanical Garden* 75:55–78.

Arroyo, M. T. K., C. Castor, C. Marticorena, M. Muñoz, L. Cavieres, O. Matthei, F. A. Squeo, M. Grosjean, and R. Rodriguez, 1998. The flora of Llullaillaco National Park in the transitional winter–summer rainfall area of the northern Chilean Andes. *Gayana Botánica* 55/2:93–110.

Aschero, C. A., 1984. El Sitio ICC-4: Un Asentamiento Precerámico en la Quebrada de Inca Cueva (Jujuy, Argentina). *Estudios Atacameños* 7:62–72.

Aschero, C. A., 1994. Reflexiones desde el Arcaico Tardío (6000–3000 AP). *Rumitacama, Revista de Antropología* 1:13–18.

Aschero, C. A., and M. M. Podestá, 1986. El Arte Rupestre en Asentamientos Precerámicos de la Puna Argentina. *Runa* XVI:29–57.

Aufderheide, A., I. Muñoz, and B. Arriaza, 1993. Seven Chinchorro mummies and the prehistory of northern Chile. *American Journal of Physical Anthropology* 91:189–202.

Baied, C. A., 1991. Late-quaternary environments and human occupation of the South-Central Andes. Unpublished Ph.D. Dissertation, Department of Anthropology, University of Colorado, Boulder.

Betancourt, J. L., C. Latorre, J. A. Rech, J. Quade, and K. Rylander, 2000. A 22,000-year record of monsoonal precipitation from northern Chile's Atacama Desert. *Science* 289:1542–1546.

Binford, M. W., A. L. Kolata, M. Brenner, J. Janusek, M. B. Abbott, and J. H. Curtis, 1997. Climate variations and the rise and fall of an Andean civilization. *Quaternary Research* 47:235–248.

Bittmann, B., 1984. El Proyecto Cobija: Investigaciones Antropológicas en la Costa del Desierto de Atacama (Chile). *Simposio de Arqueología Atacameña, 44 Congreso Internacional de Americanistas, Manchester*. Universidad de Norte, Antofagasta.

Boisset, G., A. Llagostera, and E. Salas, 1969. Excavaciones Arqueológicas en Caleta Abtao, Antofagasta. *Actas del V Congreso Nacional de Arqueología* 1:75–112.

Borrero, L. A., M. Zárate, L. Miotti, and M. Massone, 1998. The Pleistocene-Holocene transition and human occupations in the southern cone of South America. *Quaternary International* 49/50:191–199.

Bradbury, J. P., M. Grosjean, S. Stine, and F. Sylvestre, 2001. Full- and late-glacial lake records along PEP-1 transect: their role in developing inter-hemispheric paleoclimate interactions. In *Interhemispheric Climate Linkages*, edited by V. Markgraf, pp. 265–291. Academic Press, San Diego.

Cardich, A., 1980. Origen del Hombre y la Cultura Andinos. In *Historia del Perú, Perú Antiguo, Tomo 1*, edited by J. Mejia Baca, pp. 31–156. Editorial Juan Mejia Baca, Lima.

Chong Diaz, G., 1984. Die Salare in Nordchile – Geologie, Struktur und Geochemie. *Geotektonische Forschung* 67:1–146.

Clapperton, C. M., 1993. *Quaternary Geology and Geomorphology of South America*. Elsevier, Amsterdam.

Clapperton, C. M., 1994. The quaternary glaciation of Chile: a review. *Revista Chilena de Historia Natural* 67:369–383.

Clapperton, C. M., J. D. Clayton, D. I. Benn, C. J. Marden, and J. Argollo, 1997. Late quaternary glacier advances and Palaeolake highstands in the Bolivian altiplano. *Quaternary International* 38/39:49–59.

Costa Junqueira, M. A., 2001. Modalidades de enterratorios Arcaicos en el Norte de Chile. *Chungara Revista de Antropología Chilena* 33(1):55–62.

Dauelsberg, P., 1983. Tojo-Tojone: Un Paradero de Cazadores Arcaicos. Características y Secuencias. *Chungara* 11:11–30.

Druss, M., 1977. Computer analysis of Chiu–Chiu complex settlement pattern. *El Dorado* 2:51–73.
Engel, F. A., 1981. *Prehistoric Andean Ecology, Vol. 2 of Man Settlement and Environment in the Andes*. The Deep South Humanities Press for Hunter College, New York.
Fairbanks, R. G., 1989. 17,000-year Glacio-Eustatic sea level record: influence of glacial melting rates on the younger Dryas event and deep-ocean circulation. *Nature* 342:637–642.
Fernández, J., 1984–1985. Reemplazo del Caballo Americano (Perissodactila) por Camelidos (Artiodactyla) en Estratos del Límite Pleistocénico-Holocénico de Barro Negro, Puna de Jujuy. Implicancias Paleoclimáticas Faunísticas y Arqueológicas. *Relaciones Nueva Serie* 16:137–152.
Fernández Distel, A., 1980. Los Fechados Radiocarbónicos en la Arqueología de la Provincia de Jujuy. Fechas Radiocarbónicas de la Cueva CH. III de Huachichocana, Tiuiyaco e Inca Cueva. *Argentina–Radiocarbono–en Arquelogía* 1(4/5):89–100.
Fernández Distel, A., 1986. Las Cuevas de Huachichocana, su Posición Dentro del Precerámico co n Agricultura Incipiente del Noroeste Argentino. *Beiträge zur Allgemeinen und Vergleichenden Archäologie Band* 8:353–430.
Focacci, G., and S. Chacón, 1989. Excavaciones Arqueológicas en los Faldeos del Morro de Arica Sitios Morro 1/6 y 2/2. *Chungara* 22:15–62.
Fontugne, M., P. Usselmann, D. Lavallée, M. Julien, and C. Hatté, 1999. El Niño variability in the coastal desert of Southern Peru during the mid-Holocene. *Quaternary Research* 52: 171–179.
Francis, P. W., and G. L. Wells, 1988. Landast TM observations of debris avalanche deposits in the Central Andes. *Bulletin of Volcanology* 50:258–278.
Francis, P. W., M. Gardeweg, C. F. Ramírez, and D. A. Rothery, 1985. Catastrophic debris avalanche deposit of Socompa volcano, Northern Chile. *Geology* 13:600–603.
Gardeweg, M., P. Cornejo, and J. Davidson, 1984. Geología del Volcán Llullaillaco, Altiplano de Antofagasta, Chile (Andes Centrales). *Revista de Geología de Chile* 23:21–37.
Garleff, K., and H. Stingl, 1994. Reply to Gosse & Evenson: reinterpretation of the evidence for a significant neoglacial advance in the Río Atuel valley, Mendoza province, Argentina. *Zeitschrift für Geomorphologie N. F.* 38:339–342.
Garreaud, R., M. Vuille, A. C. Clement, 2003. The climate of the altiplano: observed current conditions and mechanisms of past changes. *Palaeogeography, Palaeoclimatology, Palaeoecology* 194:5–22.
Geyh, M. A., M. Grosjean, W. Kruck, and U. Schotterer, 1996. Síncronopsis del Desarrollo Morfológico y Climático del Chaco Boreal y de Atacama en los Últimos 35.000 años AP. *Memorias del XII Congreso Geológico de Bolivia. Tarija* Tomo III:1267–1276.
Geyh, M. A., M. Grosjean, W. Kruck, and U. Schotterer, 1999. Radiocarbon reservoir effect and the timing of the late-glacial/early Holocene humid phase in the Atacama Desert, northern Chile. *Quaternary Research* 52:143–153.
Glaze, L. S., P. W. Francis, S. Self, and D. A. Rothery, 1989. The 16 September 1986 eruption of Lascar volcano, North Chile: satellite investigations. *Bulletin of Volcanology* 51:149–160.
Graf, K., 1992. Pollendiagramme aus den Anden. Eine Synthese zur Klimageschichte und Vegetationsentwicklung seit der Letzten Eiszeit. *Physische Geographie* 34:1–138.
Grilli, A., 1989. Recursos Hídricos en las Provincias de Arica y Parinacota. Direccion General de Aguas, Santiago.
Grosjean, M., 1994. Paleohydrology of Laguna Lejía (North Chilean altiplano) and climate implications for late-glacial times. *Paleogeography, Paleoclimatology, Palaeoecology* 109:89–100.
Grosjean, M., 2001. Mid-Holocene climate in the south-central Andes: humid or dry? *Science* 292:2391–2392.

Grosjean, M., M. A. Geyh, B. Messerli, and U. Schotterer, 1995. Late-glacial and early Holocene lake sediments, groundwater formation and climate in the Atacama altiplano 22–24°S. *Journal of Paleolimnology* 14:241–252.

Grosjean, M., L. Núñez, I. Cartajena, and B. Messerli, 1997a. Mid-Holocene climate and culture change in the Atacama Desert, northern Chile. *Quaternary Research* 48:239–246.

Grosjean, M., B. Valero-Garcés, B. Messerli, M. A. Geyh, B. Messerli, U. Schotterer, H. Schreier, and K. Kelts, 1997b. Mid and late Holocene limnogeology of Laguna del Negro Francisco, Northern Chile, and its paleoclimatic implications. *The Holocene* 7:151–159.

Grosjean, M., M. Geyh, B. Messerli, H. Schreier and H. Veit, 1998. A Late Holocene (<2600 BP) glacial advance in the South-Central Andes (29°S), northern Chile. *The Holocene* 8/4:473–479.

Grosjean, M., and L. Núñez, 1994. Lateglacial, early and middle Holocene environments, human occupations and resource use in the Atacama (northern Chile). *Geoarchaeology* 9:271–286.

Grosjean, M., J. N. van Leeuwen, B. Ammann, M. A. Geyh., van der W. O. Knaap, and W. Tanner, 2001. A 22,000 year sediment and pollen record from Laguna Miscanti, northern Chile, Central Andes 24°S. *Global and Planetary Change* 28:35–51.

Grosjean, M., I. Cartajena, M. A. Geyh, and L. Nunez, 2003. From proxy data to paleoclimate interpretation: the mid-Holocene paradox of the Atacama Desert, northern Chile. *Palaeogeography, Palaeoclimatology, Palaeoecology* 194(1–3):247–258.

Grosjean, M., L. A. Nuñez, and I. Cartajena, 2005a. Paleoindian occupation of the Atacama Desert, northern Chile. *Journal of Quaternary Science* 20/7–8:643–653.

Grosjean, M., L. A. Nuñez, and I. Cartajena, 2005b. Cultural response to climate change in the Atacama Desert. In *23° South: The Archaeology and Environmental History of the Southern Deserts*, edited by M. Smith and P. Hesse, pp. 156–171. National Museum of Australia.

Guillén, S. E., 1992. The Chinchorro cultura: mummies and crania in the reconstruction of preceramic coastal adapatation in the South Central Andes. Ph.D. Dissertation, Department of Anthropology, University of Michigan.

Guillén, S. E., 1997. Morro 1-5 (Arica) Momias y Sociedades Complejas del Arcaico de los Andes Centrales. *Boletín de Arqueología PUCP* 1:65–78.

Gutman, G., and W. Schwerdtfeger, 1965. The role of latent and sensible heat for the development of a high pressure system over the subtropical Andes in the summer. *Meteorologische Rundschau* 18:69–75.

Hardy, D. R., M. Vuille, C. Braun, F. Keimig, and R. S. Bradley, 1998. Annual and daily meteorological cycles at high altitude on a tropical mountain. *Bulletin of the American Meteorological Society* 79/9:1899–1913.

Hastenrath, S., 1997. Annual cycle of circulation and convective activity over the tropical Americas. *Journal of Geophysical Research - Atmosphere* 102:4267–4274.

Hastenrath, S., and J. E. Kutzbach, 1985. Late Pleistocene climate and water budget of the South American altiplano. *Quaternary Research* 24:249–256.

Hernández Llosas, M. I., 2000. Quebradas Altas de Humahuaca a través del tiempo: el caso Pintoscayoc. *Estudios Sociales del NOA* Año 4/2:167–224.

Hesse, B., 1982. Animal domestication and oscilating climates. *Journal of Ethnobiology* 2:1–15.

Jenny, B., and K. Kammer, 1996. Climate change in den Trockenen Anden: Jungquartäre Vergletscherung. *Geographica Bernensia* G46:1–80.

Jenny, B., B. Valero, R. Villa-Martinez, R. Urrutia, M. A. Geyh, and H. Veit, 2002. Early to mid-Holocene aridity in Central Chile and the southern Westerlies: the Laguna Aculeo Record (34°S). *Quaternary Research* 58:160–170.

Keefer, D. K., S. D. deFrance, M. Moseley, J. Richardson III, D. Satterlee, and A. Day-Lewis, 1998. Early maritime economy and El Niño events at Quebrada Tacahuay, Peru. *Science* 281:1833–1835.

Kessler, A., 1991. Zur Klimaentwicklung auf dem Altiplano seit dem Letzten Pluvial. *Freiburger Geographische Hefte* 32:141–148.

Kulemeyer, J. A., and L. R. Laguna, 1996. La Cueva de Yavi: Cazadores-Recolectores del Borde Oriental de la Puna de Jujuy (Argentina) entre los 12,500 y 8000 Años B.P. *Ciencia y Tecnología (Jujuy)* 1:37–46.

Kulemeyer, J. A., L. C. Lupo, J. A. Kulemeyer, and L. R. Laguna, 1999. Desarrollo Paleoecológico Durante las Ocupaciones Humanas del Precerámico del Norte de la Puna Argentina. *Bamberger Geographische Schriften* 19:233–255.

Kull, C., 1999. Modellierung Paläoklimatischer Verhältnisse, basierend auf der jungpleistozänen Vergletscherung – Ein Beispiel aus den Nordchilenischen Anden. *Zeitschrift für Gletscherkunde und Glazialgeologie* 35(1):35–64.

Kull, C., and M. Grosjean, 1998. Albedo changes, Milankovitch forcing, and late quaternary climate changes in the Central Andes. *Climate Dynamics* 14:871–881.

Kull, C., and M. Grosjean, 2000. Late Pleistocene climate conditions in the North Chilean Andes drawn from a climate-glacier model. *Journal of Glaciology* 46:622–632.

Kull, C., M. Grosjean, and H. Veit, 2002. Modeling modern and late Pleistocene glacio-climatological conditions in the North Chilean Andes (29°S–30°S). *Climate Change* 53(3):359–381.

Krapovickas, P., 1987. Nuevos Fechados Radiocarbónicos para el sector oriental de la Puna y la Quebrada Humahuaca. *Runa* XVII–XVIII:207–219.

Lamy, F., D. Hebbeln, and G. Wefer, 1999. High resolution marine record of climatic change in mid-latitude Chile during the last 28,000 years based on terrigenous sediment parameters. *Quaternary Research* 51:83–93.

Lanning, E. P., 1967. Informe Previo de las Investigaciones Realizadas por la Columbia University Field Station Durante el año 1967. *Revista de la Universidad del Norte* 2:63–68.

Lanning, E. P., 1973. Burin industries in the Pleistocene of the Andes. *Estudios Atacameños* 1:21–39.

Latorre, C., J. L. Betancourt, K. A. Rylander, and J. Quade, 2002. Vegetation invasion into absolute desert: a 45 k.y. rodent midden record from the Calama-Salar de Atacama Basins, northern Chile (lat 22–24°S). *GSA Bulletin* 114(3):349–366.

Latorre, C., J. L. Betancourt, K. A. Rylander, J. Quade, and O. Matthei, 2003. A vegetation history from the arid prepuna of northern Chile (22–23°S) over the last 13,500 years. *Palaeogeography, Palaeoclimatology, Palaeoecology* 194(1–3):223–246.

Latorre, C., J. L. Betancourt, and M. T. K. Arroyo, 2006. Late quaternary vegetation and climate history of a perennial river canyon in the Rio Salado Basin (22°S) of northern Chile. *Quaternary Research* 65:450–466.

Latorre, C., P. I. Moreno, G. Vargas, A. Maldonado, R. Villa-Martínez, J. J. Armesto, C. Villagrán, M. Pino, L. Núñez, and M. Grosjean, 2007. Quaternary environments and landscape evolution. In *Geology of Chile*, edited by W. Gibbons and T. Moreno, Chapter 8. London Geological Society Press, London.

Lavallée, D., M. Julien, C. Karlin, L. C. García, D. Pozzi-Escot, and M. Fontugne, 1997. Entre Desierto y Quebrada. Primeros Resultados de las Excavaciones Realizadas en el Abrigo de Tomayoc (Puna de Jujuy, Argentina). *Bulletin de l'Institut Français d'Études Andines* 26(2):141–176.

Lavallée, D., P. Béarez, A. Chevalier, M. Julien, P. Usselmanny M. Fontugne, 1999. Paleoambiente y ocupación prehistórica del litoral extremo-sur del Perú. Las ocupaciones del Arcaico en la Quebrada de Los Burros y alrededores (Tacna, Perú). *Boletín de Arqueología. Pontificia Universidad Católica del Perú* III:393–416.

LeFebvre, R. P., 2004. Hakenasa: the archaeology of a rock shelter in the altiplano of northern Chile. Unpublished Ph.D. Dissertation, Department of Anthropology, Rutgers, the University of New Jersey. University Microfilms International, Ann Arbor, Michigan.

Lenters, J. D., and K. H. Cook, 1997. On the origin of the Bolivian high and related circulation features of the South American climate. *Journal of the Atmospheric Sciences* 54: 656–677.

Lentz, D. L., 2000. Imperfect balance. Landscape transformations in the precolumbian Americas. Columbia University Press, New York. 547 p.

Le Paige, G., 1965. San Pedro de Atacama y su Zona. *Anales de la Universidad del Norte* 4.

Lima, M., N. C. Stenseth, F. M. Jaksic, 2002. Food web structure and climate effects on the dynamics of small mammals and owls in semi-arid Chile. *Ecology Letters* 5:273–284.

Llagostera, A., 1977. Ocupación Humana en la Costa Norte de Chile Asociada a Peces Local-Extintos y a Litos Geométricos: 9680 ± 160 A.P. *Actas VII Congreso de Arqueología de Chile* 1:93–114. Ediciones Kultrun, Santiago.

Llagostera, A., 1979. 9700 years of maritime subsistence on the Pacific: an analysis by means of bioindicators in the North of Chile. *American Antiquity* 44:309–324.

Llagostera, A., 1982. Tres Dimensiones en la Conquista Prehistórica del Mar. Un Aporte para el Estudio de las Formaciones de Pescadores de la Costa Sur Andina. *Actas del VIII Congreso de Arqueología Chilena* 217–245. Ediciones Kultrun, Santiago.

Lynch, T. F., 1986. Un Reconocimiento Arqueológico en el Salar de Punta Negra, Segunda Región. *Chungara* 16–17:75–88.

Llagostera, A., 1986. Climate change and human settlement around the late-glacial Laguna de Punta Negra northern Chile: the preliminary results. *Geoarchaeology* 1:145–161.

Llagostera, A., 1990. Quaternary climate, environment, and the human occupation of the South-Central Andes. *Geoarchaeology* 5:199–228.

Lynch, T. F., and C. M. Stevenson, 1992. Obsidian hydration dating and temperature controls in the Punta Negra region of northern Chile. *Quaternary Research* 37:117–124.

Maldonado, A., and C. Villagrán, 2002. Paleoenvironmental changes in the semiarid coast of Chile (32°S) during the last 6200 cal years inferred from a swamp-forest pollen record. *Quaternary Research* 58:130–138.

Maldonado, A., J. L. Betancourt, C. Latorre and C. Villaran, 2005. Pollen analysis from a 50,000-yr rodent midden series in the Southern Atacama Desert (25°30′S). *Journal of Quaternary Science* 20:493–507.

Maldonado, A., and C. Villagran, 2006. Variability of the northern limit of the southern west erlies over the last 9900 cal yr BP from a swamp forest pollen record along the semiarid coast of Chile (32°05′S). *Quaternary Research* 66:246–258.

Markgraf, V., 1985. Paleoenvironmental history of the last 10,000 years in northwestern Argentina. *Zentralblatt fur Geologie und Paläontologie* 11/12:1739–1749.

Markgraf, V., 1989. Palaeoclimates in Central and South America since 18,000 B.P. based on pollen and lake-level records. *Quaternary Science Reviews* 8:1–24.

Markgraf, V., 1993. Climatic history of Central and South America since 18,000 yr BP: comparison of pollen records and model simulations. In *Global Climates since the Last Glacial Maximum*, edited by H. E. Wright, pp. 357–385. University of Minnesota Press, Minneapolis.

Martin, L., M. Fournier, A. Sifeddine, B. Turcq, M. L. Absy, and J. M. Flexor, 1993. Southern oscillation signal in South American paleoclimatic data of the last 7000 years. *Quaternary Research* 39:338–346.

Massone, M., and E. Hidalgo, 1981. Arqueología de la Región Volcánica de Palli-Aike (Patagonia Meridional Chilena). *Anales Instituto de la Patagonia* 12:95–124.
Messerli, B., M. Grosjean, G. Bonani, A. Bürgi, M. A. Geyh, K. Graf, U. Schotterer, H. Schreier, and M. Vuille, 1993. Climate change and dynamics of natural resources in the atiplano of northern Chile during late glacial and Holocene times. First synthesis. *Mountain Research and Development* 13:117–127.
Messerli, B., M. Grosjean, Th. Hofer, L. Núñez, and C. Pfister, 2000. From nature-dominated to human-dominated environmental changes. *Quaternary Science Reviews* 19(1–5):459–479.
Mourguiart, P., J. Argollo, T. Corrège, L. Martin, M. E. Montenegro, A. Sifeddine, and D. Wirrmann, 1997. Changements Limnologique et Climatologiques dans le Bassin du Lac Titicaca (Bolivia), Depuis 30,000 ans. *Comptes Rendues Recherche Academie Science Paris* 325:139–146.
Mourguiart, P., and P. Carbonel, 1994. A quantitative method of paleolake-level reconstruction using ostracod assemblages: an example from the Bolivian altiplano. *Hydrobiologia* 288: 183–193.
Mourguiart, P., T. Corrège, D. Wirrmann, J. Argollo, M. E. Montenegro, M. Pourchet, and P. Carbonel, 1998. Holocene palaeohydrology of Lake Titicaca estimated from an ostracod-based transfer function. *Palaeogeography, Palaeoclimatology, Palaeoecology* 143:51–72.
Mourguiart, P., and M. Roux, 1990. Une Approche Nouvelle du Problème Posé par les Reconstructions des Paléoniveaux Lacustres: Utilisation d'une Fonction de Transfert Basée sur les Faunes D'Ostracodes. *Géodynamique* 5:151–165.
Mostny, G., 1964. Anzuelos de Concha: 6.170±220 años, *Noticiero Mensual Museo Nacional de Historia Natural* Año IX N° 98.
Muñoz, I., and J. Chacama, 1982. Investigaciones Arqueológicas en las Poblaciones Precerámicas de la Costa de Arica. *Documentos de Trabajo* 2:3–97.
Muñoz, I., and J. Chacama, 1993. Patrón de Asentamiento y Cronología de Acha-2. In *Acha-2 y los Orígenes del Poblamiento Humano en Arica*, edited by I. Muñoz, B. T. Arriaza, and A. Aufderheide, pp. 21–46. Ediciones Universidad de Tarapacá, Arica.
Muñoz, I., B. Arriaza, and A. Aufderheide, editors. 1993. *Acha-2 y los Orígenes del Poblamiento Humano en Arica.* Universidad de Tarapacá, Ediciones, Arica.
Niemeyer, H., and V. Schiappacasse, 1963. Investigaciones Arqueológicas en las Terrazas de Conanoxa, Valle de Camarones (Provincia de Tarapacá). *Revista Universitaria* 48, *Anales de la Academia Chilena de Ciencias Naturales* 26:101–166.
Núñez, L., 1971. Secuencia y Cambio en los Asentamientos Humanos de la Desembocadura del Río Loa, en el Norte de Chile. *Boletín de la Universidad de Chile* 112:3–25.
Núñez, L., 1976. Registro Regional de Fechas Radiocarbónicas del Norte de Chile. *Estudios Atacameños* 4:74–123.
Núñez, L., 1980. Cazadores Tempranos en los Andes Meridionales. Evaluación Cronológica de las Industrias Líticas del Norte de Chile. *Boletín de Antropología Americana* 2:87–120.
Núñez, L., 1981. Asentamiento de Cazadores Tardíos de la Puna de Atacama: Hacia el Sedentarismo. *Chungara* 8:137–168.
Núñez, L., 1983a. Paleoindian and Archaic cultural periods in the arid and semiarid regions of northern Chile. *Advances in World Archaeology* 2:161–222.
Núñez, L., 1983b. Paleoindio y Arcaico en Chile: Diversidad, Secuencia y Procesos. México: Ediciones Cuicuilco.
Núñez, L., and M. Grosjean, 1994. Cambios Ambientales Pleistocénico-Holocénicos: Ocupación Humana y Uso de Recursos en la Puna de Atacama (Norte de Chile). *Estudios Atacameños* 11:11–24.

Núñez, L., and C. Moragas, 1977–1978. Ocupación Arcaica Temprana en Tiliviche Norte de Chile (I Región). *Boletín Museo Arqueológico de La Serena* 16:52–76.

Núñez, L., and C. Moragas, 1983. Cerámica Temprana en Cañamo (Costa Desértica del Norte de Chile): Análisis y Evaluación Regional. *Chungara* 1:31–62.

Núñez, L., and C. Santoro, 1988. Cazadores de la Puna Seca y Salada del Area Centro-Sur Andina (Norte de Chile). *Estudios Atacameños* 9:11–60.

Núñez, L., and J. Varela, 1967–1968. Sobre los Recursos de Agua y el Poblamiento Prehispánico de la Costa Norte de Chile. *Estudios Arqueológicos* 3–4:7–42.

Nuñez, P., and V. Zlatar, 1977. Tiliviche-1 b y Aragón 1 (Estrato V): Dos Comunidades Precerámicas Coexistentes en Pampa del Tamarugal, Pisagua-Norte de Chile. *Actas y Trabajos III Congreso Peruano "El Hombre y la Cultura Andina"* Tomo 2:734–736.

Nuñez, P., and V. Zlatar, 1980. Coexistencia de Comunidades Recolectoras-Cazadoras. *Actas del V Congreso Nacional de Arqueología Argentina* Tomo 1:79–92.

Núñez, L., M. Grosjean, and I. Cartajena, 1999. Paleoenvironmental evaluation of Paleoindian-Early Archaic occupations on the Puna de Atacama (northern Chile). *Informe Final Proyecto National Geographic Society 5836-96. Universidad Católica del Norte*, Vols 1 and 2.

Núñez, L., M. Grosjean, and I. Cartajena, 2001. Human dimensions of late Pleistocene/Holocene arid events in southern South America. In *Interhemispheric Climate Linkages*, edited by V. Markgraf, Chapter 7, pp. 105–117. Academic Press, San Diego.

Núñez, L., M. Grosjean, and I. Cartajena, 2002. Human occupations and climate change in the puna de Atacama, Chile. *Science* 298:821–824.

Núñez, L., J. Varela, and R. Casamiquela, 1979–1981. Ocupación Paleoindio en Quereo (IV Región): Reconstrucción Muldisciplinaria en el Territorio Semiárido de Chile. *Boletín del Museo Arqueológico de La Serena* 17:32–67.

Núñez, L., J. Varela, R. Casamiquela, and C. Villagrán, 1994. Reconstrucción Multidisciplinaria de la Ocupación Prehistórica de Quereo, Centro de Chile. *Latin American Antiquity* 55:99–118.

Núñez, L., V. Zlatar, and P. Núñez, 1974. Caleta Huelén-42: Una Aldea Temprana en el Norte de Chile (Nota Preliminar). *Hombre y Cultura* 2:67–103.

Núñez, L., M. Grosjean, B. Messerli, and H. Schreier, 1996. Cambios Ambientales Holocénicos en la Puna de Atacama y sus Implicancias Paleoclimáticas. *Estudios Atacameños* 12:31–40.

Placzek, C., J. Quade, and P. J. Patchett, 2006. Geochronology and stratigraphy of late Pleistocene lake cycles on the southern Bolivian altiplano: implications for causes of tropical climate change. *Geological Society of America Bulletin* 118/5:515–532.

Rao, G. V., and S. Erdogan, 1989. The atmospheric heat source over the Bolivian plateau for a mean January. *Boundary-Layer Meteorology* 46:13–33.

Ravines, R., 1967. El Abrigo de Caru y sus Relaciones con Otros Sitios Tempranos del Sur del Perú. *Ñawpa Pacha* 5:39–57.

Ravines, R., 1972. Secuencia y Cambios en los Artefactos Líticos del Sur de Perú. *Revista del Museo Nacional* 38:133–184.

Rech, J., 2001. Late quaternary paleohidrology and superficial processes of the Atacama Desert, Chile: Evidence from wet land deposits and stable isotopes of soil salts. Ph.D. Dissertation, Department of Geosciences, University of Arizona.

Rech, J. A., J. Quade, and J. Betancourt, 2002. Late quaternary paleohydrology of the Central Andes (22–24°S), Chile. *Geological Society of America Bulletin* 114:334–348.

Rech, J., J. S. Pigati, J. Quade, and J. L. Betancourt, 2003. Re-evaluation of mid-Holocene deposits at Quebrada Piripica, northern Chile. *Palaeogeography, Palaeoclimatology, Palaeoecology* 194:207–222.

Ribstein, P., E. Tiriau, B. Francou, and R. Saravia, 1995. Tropical climate and glacier hydrology: a case study in Bolivia. *Journal of Hydrology* 165:221–234.
Risacher, F., H. Alonso, C. Salazar, 2003. The origin of brines and salts in Chilean Salars: a hydrochemical review. *Earth-Science Reviews* 63:249–293.
Rivera, M. A., 1977–1978. Cronología Absoluta y Periodificación en la Arqueología Chilena. *Boletín del Museo Arqueológico de La Serena* 16:13–46.
Rivera, M. A., 1984. Cuatro Fechas Radiocarbónicas para Sitios Arqueológicos del Litoral Norte de Chile. *Nuestro Norte* 2:5–11.
Rivera, M. A., 1994. Comentario sobre el Trabajo de Bernardo Arriaza: Tipología dela Momias Chinchorro y Evolución de las Prácticas de Momificación. *Chungara* 26:25–34.
Rivera, M. A., P. Soto, and D. Kushner, 1974. Aspecto sobre el Desarrollo Tecnológico en el Proceso de Agriculturización en el Norte Prehispánico Especialmente Arica (Chile). *Chungara* 3:79–107.
Rodríguez, F., 1999. Explotación de Recursos Vegetales Durante el Arcaico en la Puna Meridional Argentina. Presentación de un Caso: Quebrada Seca 3. *Actas XII Congreso Nacional de Arqueología Argentina* Tomo III:345–351.
Ruthsatz, B., 1993. Flora und ökologische Bedingungen hochandiner Moore Chiles zwischen 18°00′ (Arica) und 40°30′ (Osorno) südlicher Breite. *Phytocoenologia* 23:157–199.
Ruthsatz, B., 1995. Vegetation und Ökologie tropischer Hochgebirgsmoore in den Anden Nord-Chiles. *Phytocoenologia* 25:185–234.
Sandweiss, D. H., 2003. Terminal Pleistocene through mid-Holocene archaeological sites as paleoclimatic archives for the Peruvian Coast. *Palaeogeography, Palaeoclimatology, Palaeoecology* 194(1–3):23–40.
Sandweiss, D. H., C. A. Maasch, and D. G. Anderson, 1999. Transitions in the mid-Holocene. *Science* 283:499–500.
Sandweiss, D., J. Richardson, E. Reitz, J. Hsu, and R. Feldman, 1989. Early maritime adaptations in the Andes: preliminary studies at the ring site, Perú. In *Ecology Settlement and History in the Osmore Drainage, Peru,* Part I, edited by S. Rice, C. Stanish, and P. R. Scarr, pp. 35–84. BAR International Series 545(i), Oxford.
Sandweiss, D. H., J. B. Richardson III, E. J. Reitz, H. B. Rollins, and K. A. Maasch, 1996. Geoarchaeological evidence from Peru for a 5000 years B.P. onset of El Niño. *Science* 273:1531–1533.
Sandweiss, D. H., H. McInnis, R. L. Burger, A. Cano, B. Ojeda, R. del Carmen Paredes, M. Sandweiss, and M. D. Glascock, 1998. Quebrada jaguay: early South American maritime adaptations. *Science* 281:1830–1832.
Sanhuesa, J., 1980. Asentamiento Precerámico en la Costa Desértica de Interfluvio; Caramucho, Provincia de Iquique, I Región. Memoria para optar al Título de Arqueólogo. Universidad del Norte Antofagasta.
Santoro, C., 1987. Settlement patterns of Holocene hunting and gathering societies in the South Central Andes. Master Thesis. Department of Anthropology, Cornell University.
Santoro, C., 1989. Antiguos Cazadores de la Puna. In *Culturas de Chile desde sus Orígenes hasta los Albores de la Conquista,* edited by J. Hidalgo, H. Niemeyer, V. Schiappacasse, C. Aldunate, and I. Solimano, pp. 33–56. Andrés Bello, Santiago.
Santoro, C., 1999. MS, Cambios en los Patrones de Asentamientos del Valle de Lluta. *Informe Proyecto FONDECYT 1970597*. Manuscript in possession of the author.
Santoro, C., and J. Chacama, 1982. Secuencia Cultural de las Tierras Altas del Area Centro Sur Andina. *Chungara* 9:22–45.
Santoro, C., and J. Chacama, 1984. Secuencia Cultural de las Tierras Altas del Extremo Norte de Chile. *Estudios Atacameños* 7:85–103.

Santoro, C., and L. Núñez, 1987. Hunters of the dry Puna and the salt Puna in the northern Chile. *Andean Past* 1:57–109.
Santoro, C. M., and V. G. Standen, 2000. Ocupaciones humanas en Quebrada Blanca y cambios climáticos en el norte de Chile. Manuscript in possession of the authors.
Santoro, C., C. Baied, E. Belmonte, and E. Rosello, 1991. Evaluación de Paleomabientes Holocénicos y Adaptación de Cazadores Recolectores, Área Centro Sur Andina. *Actas del XI Congreso Nacional de Arqueología Chilena* Tomo II:25–30.
Santoro, C. M., V. G. Standen, B. T. Arriaza, and T. D. Dillehay, 2005. Andean Archaic funerary pattern, or postdepositional alteration? The Patapatane burial in the highlands of South Central Andes. *Latin American Antiquity* 16:329–346.
Schemenauer, R. S., H. Fuenzalida, and P. Cereceda, 1988. A neglected water resource: the Camanchaca of South America. *Bulletin American Meteorological Society* 69:138–147.
Schiappacasse, V., and H. Niemeyer, 1969. Comentario a Tres Fechas Radiocarbónicas de Sitios Arqueológicos de Conanoxa (Valle de Camarones, Prov. de Tarapacá). *Noticiero Mensual Museo Nacional Historia Natural* 13(151):6–7.
Schiappacasse, V., and H. Niemeyer, 1984. Descripción y Análisis Interpretativo de un Sitio Arcaico Temprano en la Quebrada de Camarones. *Publicación Ocasional* N° 41. Santiago: Museo Nacional de Historia Natural y Universidad de Tarapacá.
Schrott, L., 1998. The hydrological significance of high mountain permafrost and its relation to solar radiation. A case study in the high Andes of San Juan, Argentina. *Bamberger Geographische Schriften* 15:71–84.
Seltzer, G. O., 1990. Recent glacial history and paleoclimate of the Peruvian-Bolivian Andes. *Quaternary Science Reviews* 9:137–152.
Seltzer, G. O., P. Baker, S. Cross, R. Dunbar, and S. C. Fritz, 1998. High-resolution seismic reflection profiles from Lake Titicaca, Peru-Bolivia: evidence for Holocene aridity in the tropical Andes. *Geology* 26:167–170.
Serracino, G., 1975. Los Movimientos de los Cazadores Recolectores en la Cordillera de los Andes (entre la Latitud 21° y 26° y Longitud 67° 00′ y 70° 22’). *Estudios Atacameños* 3:17–44.
Serracino, G., and F. Pereyra, 1977. Tumbre: Sitios Estacionales en la Industria Tambilliense. *Estudios Atacameños* 5:5–17.
Servant, M., and J. C. Fontes, 1978. Les Lacs Quaternaires des Hauts Plateaux des Andes Boliviennes, Premières Interprétations Paléoclimatiques. *Cahier ORSTOM Série Géologie* X:9–23.
Servant, M., M. Fournier, J. Argollo, S. Servant-Vildary, F. Sylvestre, D. Wirrmann, and J. P. Ybert, 1995. La Dernière Transition Glaciaire/Interglaciaire des Andes Tropicales sud (Bolivie) D'Après L'étude des Variations des Niveaux Lacustres et Fluctuations Glaciaires. *Comptes Rendues Académie Science Paris* 320 IIa:729–736.
Sheets, P., 2001. The effects of explosive volcanism on simple to complex societies in Ancient Middle America. In *Interhemispheric Climate Linkages*, edited by V. Markgraf, Chapter 5, pp. 73–86. Academic Press, San Diego.
Sifeddine, A., J. Bertaux, P. Mourguiart, L. Martin, J. R. Disnar, F. Laggoun-Défarge, and J. Argollo, 1998. Etude de la Sédimentation Lacustre d'un Site de Fôret D'Altitude des Andes Centrales (Bolivie). Implications Paléoclimatiques. *Bulletin Société Géologie France* 169:395–402.
Sinclaire, A. C., 1985. Dos Fechas Radiocarbónicas del Alero Chulqui, Río Toconce: Noticia y Comentario. *Chungara* 14:71–79.
Spahni, J. C., 1967. Recherches Archéologiques à L' Embouchure du Río Loa (Côte du Pacifique - Chili). *Journal de la Société des Américanistes* 56:179–252.

Standen, V. G., 1997. Temprana Complejidad Funeraria de la Cultura Chinchorro (Norte de Chile). *Latin American Antiquity* 8:134–156.

Standen, V. G., and L. Núñez, 1984. Indicadores Antropológicos Físicos y Culturales del Cementerio Precerámico Tiliviche-2 (Norte de Chile). *Chungara* 12:135–153.

Standen, V., and C. Santoro, 1994. Patapatane-1: Temprana Evidencia Funeraria en los Andes de Arica (Norte de Chile) y sus Relaciones. *Chungara* 26:165–184.

Standen, V. G., and C. M. Santoro.,2004. Patrón funerario arcaico temprano del sitio Acha-3 y su relación con Chinchorro (costa norte de Chile). *Latin American Antiquity* 15:89–109.

Stoertz, G. E., and G. E. Ericksen, 1974. Geology of Salars in northern Chile. *US Geological Survey Professional Paper* 811:1–65.

Sylvestre, F., M. Servant, S. Servant-Vildary, C. Causse, M. Fournier, and J.-P. Ybert, 1999. Lake-level chronology on the Southern Bolivian altiplano (18°–23°S) during late-glacial time and the early Holocene. *Quaternary Research* 51:54–66.

Sylvestre, F., S. Servant-Vildary, M. Fournier, and M. Servant, 1996. Lake levels in the Southern Bolivian altiplano (19°–21°S) during the late glacial based on diatom studies. *International Journal of Salt Lake Research* 4:281–300.

Talbi, A., Coudrain-Ribstein, A., Ribstein, P. and Pouyaud, B., 1999. Computation of the rainfall on Lake Titicaca catchment during the Holocene. *Comptes Rendues Recherche Académie Science Paris* 329:197–203.

Tapia, P. M., S. C. Fritz, P. A. Baker, G. O. Seltzer, and R. B. Dunbar, 2003. A late quaternary diatom eecord of tropical climatic history from Lake Titicaca (Peru and Bolivia). *Palaeogeography, Palaeoclimatology, Palaeoecology* 194(1–3):139–164.

Tartaglia, L. J., 1980. A revised C-14 chronology for northern Chile. In *Prehistoric Trails of Atacama, Archaeology of Northern Chile, Monumenta Archaeologica* 7, edited by C. W. Meighan, and D. L. True, pp. 5–22. University of California Los Angeles, Los Angeles.

Thompson, L. G., E. Mosley-Thompson, M. E. Davis, P. N. Lin, K. A. Henderson, J. Coledai, J. F. Bolzan, and K. Liu, 1995. Late glacial stage and Holocene tropical ice core records from Huascaran, Peru. *Science* 269:46–50.

Thompson, L. G., M. E. Davis, E. Mosley-Thompson, T. Sowers, K. A. Henderson, V. S. Zagorodnov, P. N. Lin, V. N. Mikhalenko, R. K. Campen, J. F. Bolzan, J. Coledai, and B. Francou, 1998. A 25,000-year tropical climate history from Bolivian ice cores. *Science* 282:1858–1864.

Troll, C., 1958. Las Culturas Superiores Andinas y el Medio Geográfico. Translated by C. Nicholson. *Revista del Instituto de Geografía* 5:3–49.

True, D. L., L. Núñez, and P. Núñez, 1970. Archaeological investigations in northern Chile: project Tarapacá-Preceramic resources. *American Antiquity* 35:170–184.

True, D. L., L. Núñez, and P. Núñez, 1971. Tarapacá 10: a wordshop site in northern Chile. *Proceedings of the American Philosophical Society* 115:398–421.

True, D. L., and L. Gildersleeve, 1980. Archaeological investigation in northern Chile: Tarapacá 18. In *Prehistoric Trails of Atacama, Archaeology of Northern Chile, Monumenta Archaeologica* 7, edited by C. W. Meighan, and D. L. True, pp. 37–58. University of California Los Angeles, Los Angeles.

Valero-Garcés, B. L., M. Grosjean, A. Schwalb, M. Geyh, B. Messerli, and K. Kelts, 1996. Limnogeology of Laguna Miscanti: evidence for mid to late Holocene moisture changes in the Atacama altiplano (northern Chile). *Journal of Paleolimnology* 16:1–21.

Veit, H., 1995. Jungquartäre Landschafts- und Klimaentwicklung der Zentralen Andes und Ihres Westlichen Vorlandes: Kenntnisstand und Probleme. *Geomethodica* 20:163–194.

Veit, H., 1996. Southern westerlies during the Holocene deduced from geomorphological and pedological studies in the Norte Chico, northern Chile (23–27°S). *Paleogeography, Paleoclimatology, Palaeoecology* 123:107–119.
Vera, S., 1981. Momias Chinchorro de Preparación Complicada del Museo de Historia Natural de Valparaíso. *Anales del Museo de Historia Natural de Valparaíso* 14:5–18.
Villa-Martínez, R., and C. Villagrán, 1997. Historia de la Vegetación de Bosques Pantanosos de la Costa de Chile Central Durante el Holoceno Medio y Tardío. *Revista Chilena de Historia Natural* 70:391–401.
Villagrán C., and J. Varela, 1990. Palynological evidence for increased aridity on the Central Chilean Coast during the Holocene. *Quaternary Research* 34:198–207.
Vuille, M., 1996. Zur Raumzeitlichen Dynamik von Schneefall und Ausaperung im Bereich des Südlichen altiplano, Südamerika. *Geographica Bernensia* G45:1–118.
Vuille, M., 1999. Atmospheric circulation over the Bolivian altiplano during dry and wet periods and extreme phases of the southern oscillation. *International Journal of Climatology* 19:1579–1600.
Vuille, M., and C. Ammann, 1997. Regional snowfall patterns in the high, arid Andes. *Climate Change* 36:413–423.
Vuille, M, and M. F. Baumgartner, 1993. Hydrologic investigations in the North Chilean altiplano using landsat – MSS and – TM Data. *Geocarto International* 3:35–45.
Vuille, M, and M. F. Baumgartner, 1998. Monitoring the regional and temporal variability of winter snowfall in the arid Andes using NOAA/AVHRR data. *Geocarto International* 13:59–67.
Vuille, M., D. R. Hardy, C. Braun, F. Keimig, and R. S. Bradley, 1998. Atmospheric circulation anomalies associated with 1996/1997 summer precipitation events on Sajama Ice Cap, Bolivia. *Journal of Geophysical Research – Atmosphere* 103:11,191–11,204.
Wirrmann, D., and P. Mourguiart, 1995. Late quaternary spatio-temporal limnological variations in the altiplano of Bolivia and Peru. *Quaternary Research* 43:344–354.
Wirrmann, D., and L. F. De Oliveira Almeida, 1987. Low Holocene level (7700 to 3650 years ago) of Lake Titicaca (Bolivia). *Palaeogeography, Palaeoclimatology, Palaeoecology* 59: 315–323.
Wise, K., 1989. Archaic Period Research in the lower osmore region. In *Ecology, Settlement and History in the Osmore Drainage, Peru*, Part i, edited by D. S . Rice, C. Stanish, and P. R. Scarr, pp. 85–99. BAR International Series 545(i), Oxford.
Wise, K., 1999. Kilómetro 4 y la ocupación del periodo Arcaico en el área de Ilo. *Boletín de Arqueología PUCP* 3:335–363.
Wise, K., N. R. Clark, and S. R. Williams, 1994. A Late Archaic Period burial from the South-Central Andean coast. *Latin American Antiquity* 5.212–227.
Ybert, J. P., 1992. Ancient lake environments as deduced from pollen analysis. In *Lake Titicaca. A Synthesis of Limnological Knowledge*, edited by C. Dejoux, and A. Iltis, pp. 49–60. Kluwer Academic Publisher, Dordrecht.
Zeil, W., 1986. *Südamerika.* Stuttgart: Enke.
Zipprich, M., B. Reizner, W. Zech, H. Stingl, and H. Veit, 1999. Upper quaternary landscape and climate evolution in the Sierra de Santa Victoria (North-Western Argentina) deduced from geomorphologic and pedogenic evidence. *Zentralbl. für Geologie u. Paläontologie, Teil I* 7/8:997–1011.
Ziolkowski, M. S., 1993. *Andes Radiocarbon Database for Bolivia, Ecuador and Peru.* Warsaw University and Silesian Technical University, Warsaw.

Climate Change and Cultural Dynamics: A Global Perspective on Mid-Holocene Transitions
David G. Anderson, Kirk A. Maasch and Daniel H. Sandweiss (Editors)

Chapter 4

Mid-Holocene climate and cultural dynamics in Brazil and the Guianas

Betty J. Meggers
Smithsonian Institution, PO Box 37012, NMNH MRC-112, Washington, DC 20013

Abstract

It was assumed until recently that the climate and vegetation of the Neotropical lowlands were not significantly altered during the Pleistocene. This impression has been refuted by a variety of paleoclimatological evidence that indicates both Amazonia and the Brazilian coast were cooler and drier prior to ca. 7000 ^{14}C yr BP. In both regions, the forest was substantially reduced and the predominant vegetation was grass and shrubs. The similarity of the environment is reflected in the similarity of the subsistence remains and artifacts in rock shelters and open sites throughout both regions from the inception of human occupation ca. 13,000 ^{14}C yr BP until development of Holocene conditions. Simultaneously, culmination of sea level rise created new aquatic habitats along the coast and the floodplain of the Amazon with more concentrated protein resources. These ameliorations permitted larger and more sedentary communities, but inherent limitations to intensive agriculture and dependence on wild protein sources placed a ceiling on population concentration throughout the lowlands.

1. Introduction

The region considered here includes the largest extent of tropical rainforest on the planet, bounded on the south by a relatively arid coastal upland extending from tropical to temperate latitudes and on the north by lower and increasingly seasonal rainfall. At present, there is an abrupt ecological frontier between Amazonia and the Brazilian Coastal Strip, marked by the coincidence of three environmental features: (1) the boundary between the equatorial and transitional bioclimatic regions, (2) the southern margin of Amazonian rainforest vegetation, and (3) the limit of the northern physiographic zone. These differences are reflected in different types of modern land use (Fig. 4.1; Brochado et al., 1970). A similar, but less abrupt transition exists in the north, where rainforest becomes increasingly fragmented by more open vegetation culminating in the savannas of the Orinoco.

This pattern was considered to be ancient and unaffected by the climatic fluctuations that transformed temperate landscapes during the Pleistocene until the 1970s, when it was observed that the modern distributions of related species of Amazonian forest birds do not coincide with present environmental barriers to interbreeding. This suggested that the rainforest had been fragmented during the Pleistocene into "refugia" separated by more open types of vegetation that isolated ancestral populations

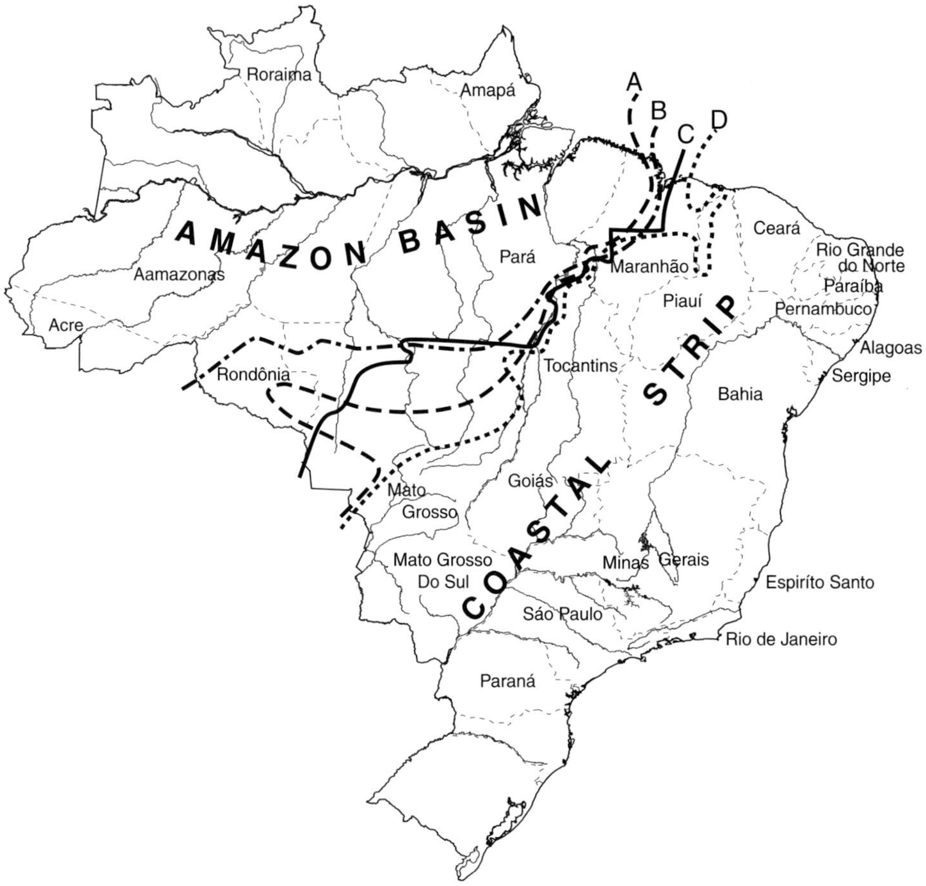

Figure 4.1. Present ecological boundary between the Coastal Strip and the Amazon Basin. A, Limit between the equatorial and transitional bioclimatic regions; B, Limit of Amazonian vegetation; C, Limit of the northern physiographic zone; D, Differences in land use (after Brochado et al., 1970, Fig. 1).

for sufficient time to permit their diversification (Haffer 1969, 1974; Prance, 1982; Hooghiemstra and van der Hammen, 1998). Although still disputed by some (e.g. Colinvaux and Oliveira, 2001; Cowling et al., 2001), the existence of fluctuations in the composition and distribution of the vegetation both in Amazonia and on the Coastal Strip during the late Pleistocene and early Holocene is increasingly supported by a variety of biogeographical, palynological, paleoecological, hydrological, geomorphological, sedimentological, and climatological data (Haffer and Prance, 2001).

Better understanding of the impact of atmospheric fluctuations such as the ENSO phenomenon and the Intertropical Convergence Zone (ITCZ), as well as identification of sea level changes and coastal dynamics, oceanic temperature variation, and marine sediment composition, increasingly supplement evidence for environmental changes based on terrestrial sources both in Amazonia and on the Coastal Strip. The geographical extent, frequency, intensity, and duration of these environmental fluctuations must be taken into consideration in reconstructing and interpreting precolumbian cultural development.

In Amazonia, recognizing significant cultural changes prior to the adoption of pottery is hampered by the combination of poor preservation of perishable remains and impermanent settlement in open locations. However, lithic camp and workshop sites and a long series of ^{14}C dates attest to the presence of humans by ca. 13,000 ^{14}C yr BP (uncal.) and indirect evidence of prehistoric population movements is provided by the disjunct distributions of languages, genetic traits, and cultural elements among surviving indigenous groups. On the Brazilian Coastal Strip, shell middens and rock shelters preserve bone and shell artifacts, burials, and subsistence remains that amplify the record left by camp sites and rock art.

In the following discussion, I will summarize the evidence for paleoclimatological fluctuations and prehistoric cultural changes since ca. 13,000 ^{14}C yr BP (uncal.) separately for the Coastal Strip and the Amazon Basin. Both the climatological and cultural data indicate that these now distinct regions experienced long- and short-term environmental fluctuations during and since the Pleistocene that favored the emergence and perpetuation of similar general cultural configurations.

2. The coastal strip

This region, known geographically as the Brazilian highlands, is dominated by the Brazilian shield. Elevation rises to between 500 and 1500 m from southern Piauí across western Bahia, Goiás, Minas Gerais, São Paulo, Paraná, and Mato Grosso do Sul, with sporadic increases to 3000 m in Minas Gerais and São Paulo. Escarpments containing rock shelters are common, especially in Goiás and Minas Gerais. As elevation increases, annual rainfall decreases from above 1550 to 400–600 mm and changes from seasonal to intermittent. Present latitudinal variations in climate create three general vegetational zones. In the north, with 5–6 month dry season, cerrado (savanna and scrub woodland) predominates. In the center, with a 2–5 month dry season, semi-deciduous forest is characteristic. In the south, with no significant dry season, *Araucaria* forest is typical (Ledru et al., 1998b). A narrow strip of rainforest extends along the Atlantic coast as far south as Espírito Santo. The only major river is the São Francisco, which flows north before turning east to empty into the Atlantic. Between Bahia and Espírito Santo, a few small rivers flow east from the coastal highland, but drainage farther south is principally toward the west into the south-flowing Paraná. A specialized lacustrine habitat that developed along the shoreline from Espírito Santo to Rio Grande do Sul as sea level rose offered unique opportunities for human exploitation.

2.1. Paleoclimatic fluctuations

Evidence comes from two principal sources: terrestrial and marine pollen cores (Fig. 4.2) and sea-level changes.

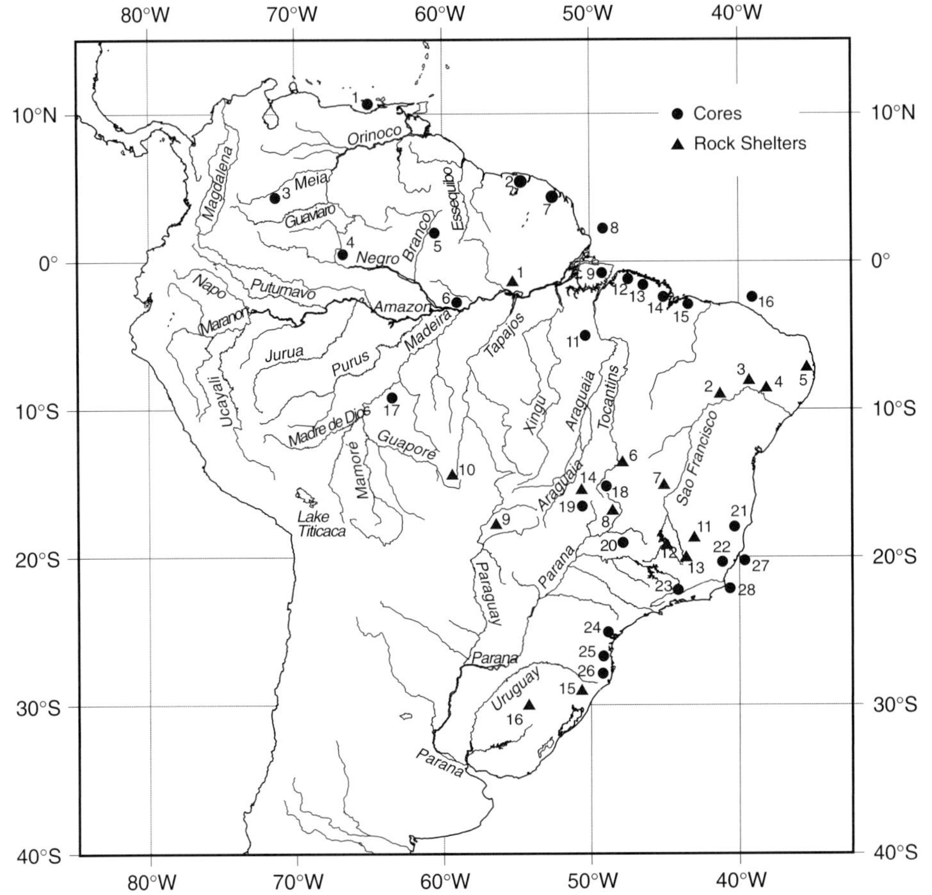

Figure 4.2. Locations of pollen cores and rock shelters mentioned in the text. Pollen cores: 1, Cariaco Basin; 2, Ogle Bridge; 3, Agua Sucia; 4, Pata; 5, Moriru; 6, Manaus; 7, Sinnamary; 8, Amazon Fan; 9, Ararí; 10, Curuá; 11, Carajás; 12, Curuça; 13, Crispim; 14, Bragança; 15, Caçó; 16, GeoB 3104-1; 17, Katira; 18, Aguas Emendadas; 19, Cromínia; 20, Salitre, Serra Negra; 21, Lago de Pires; 22, Catas Altas; 23, Itapeva; 24, Campos Gerais; 25, Poço Grande; 26, Boa Vista; 27, GeoB 3219-2; 28, GeoB 3202-1. Rock shelters: 1, Pedra Pintada; 2, Pedra Furada/Calderão do Rodriguez; 3, Sítio do Meio; 4, Gruta do Padre; 5, Pedra do Caboclo/Bom Jardim; 6, Barreiro; 7, Boquête/Boqueirão Soberbo/Varal/Pequena; 8, Gentio/Foice; 9, Santa Elina; 10, Abrigo do Sol; 11, Santana do Riacho; 12, Lagoa Santa; 13, Lapa Vermelha; 14, Serranópolis; 15, Cerrito Dalpiaz; 16, RS-TQ-58.

2.1.1. Palynological evidence

The predominance of grass and other non-arboreal taxa throughout the coast during the Late Pleistocene implies markedly drier and 5–7°C cooler conditions (Behling et al., 2001, 2002; Ledru et al., 2001; Behling, 2002b). In the south, development of modern climatic conditions with warmer temperatures and brief or no dry periods after ca. 4000 ^{14}C yr BP permitted the expansion of *Araucaria* and semi-deciduous forest (Behling, 1997; Ledru et al., 1998b). In the north, increasing seasonality, precipitation, and temperature allowed development of the heterogeneous vegetation of the cerrado (Ledru, 2002; Oliveira and Marquis, 2002). Regional diversity is implied, however, by the occurrence in Minas Gerais of landslides, peat deposits overlying erosional surfaces, and palm swamps indicative of episodes of torrential rain (Salgado-Labouriau et al., 1998), whereas drier conditions existed at Lago do Pires on the coast (Behling, 1998).

Regional variation in climate is also attested by a progressive decline in arboreal taxa between ca. 6500 and 4000 ^{14}C yr BP at Carajás. The abundance of pollen of pioneer vegetation is attributable to frequent droughts of short duration and associated fires that prevented the development of mature forest. By contrast, higher rainfall prevailed during this period in Santa Catarina (Martin et al., 1995a; Alexandre et al., 1999).

2.1.2. Sedimentological evidence

Hiatuses of 10,000 or more years in ^{14}C dates and abrupt changes in sedimentation rate and lithography in terrestrial and marine pollen cores at Salitre, Ipeva, Serra Negra, Cromínia, Aguas Emendadas, GeoB 3104-1, GeoB 3329-2, and GeoB 3202-1 also imply drought and colder temperatures during the Late Glacial Maximum (Fig. 4.3; Ledru et al., 1998a; Behling et al., 2000, 2002). Soil profiles from Paraná,

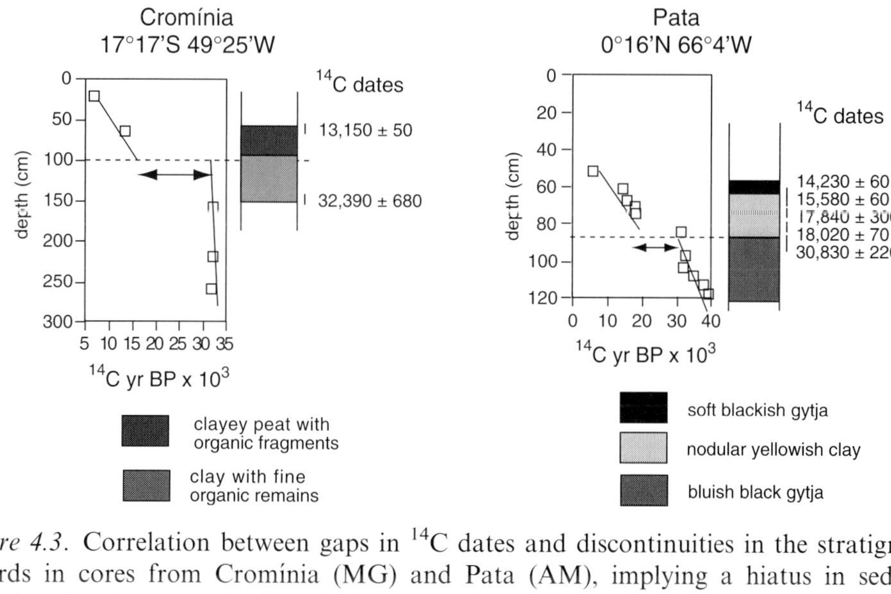

Figure 4.3. Correlation between gaps in ^{14}C dates and discontinuities in the stratigraphic records in cores from Cromínia (MG) and Pata (AM), implying a hiatus in sediment deposition lasting several millennia during the Late Glacial Maximum and accounting for the seeming continuity of forest vegetation (after Ledru et al., 1998a, Fig. 3).

São Paulo, and Minas Gerais show charcoal throughout, implying paleofires during the late Pleistocene/Holocene consistent with drier climate (Pessenda et al., 2004).

Parabolic compound dunes of eolian sand extend over an area ca. 7000 km^2 west of the middle São Francisco in Bahia, where the present climate is semi-arid and rainfall is sufficient to support caatinga vegetation. Twelve ^{14}C dates from different locations identify episodes of increased deposition ca. 4800, 3300, and 1700-850 ^{14}C yr BP (Barreto et al., 1996). This period is placed in larger temporal perspective by a pollen sequence from an adjacent peat bog, which shows a progressive decline in forest taxa from ca. 8910 to 6790 ^{14}C yr BP suggesting semi-arid conditions. Moisture increased between ca. 6230 and 4535 ^{14}C yr BP, followed by a marked decline thereafter and establishment of modern semi-arid conditions (Oliveira et al., 1999).

2.1.3. Sea level change

Systematic investigations along the Brazilian coast between Alagoas and Rio Grande do Sul have identified dead reefs, fossilized burrows of a marine arthropod, gastropod deposits, sea-urchin holes, and sandy beach deposits above present high-water level that provide the basis for detailed reconstruction of changes in sea level (Villwock et al., 1986; Suguio et al., 1988, 1991). More than 700 ^{14}C determinations permit correlating the local histories of the most recent transgression.

All of the regions sampled show a similar pattern, consisting of a rise ca. 7000 ^{14}C yr BP to ± 1 m above present mean sea level, a sudden spurt ca. 5100 ^{14}C yr BP to ± 4.8 m, followed by a rapid and then more gradual decline ca. 4100 ^{14}C yr BP to 0 or slightly below, a second rise between ca. 3800 and 3600 ^{14}C yr BP to ± 3.5 m, a brief drop, a third rise ca. 2500 ^{14}C yr BP to ± 2.5 m, followed by a slow decline to zero. Although the maximum elevation was achieved simultaneously throughout the coast, onset was delayed until ca. 6600 ^{14}C yr BP from São Paulo south. Elevation reached only ± 4 m on the coast of São Paulo and 2.5 m on the coast of Paraná (Fig. 4.4; Suguio et al., 1988 pp. 205–206; Angulo and Suguio, 1995). The substrates, locations on paleo-lagoons, and height above present sea level of selected sambaquis on the coast of São Paulo and the $\delta 13C$ values of constituent molluscs are compatible with this paleoenvironmental reconstruction (Suguio et al., 1991). Stabilization of sea level allowed the formation of lagoons and marshes with varying saline composition, with resulting diversification of flora and fauna (Lorscheitter and Dillenburg, 1998; Ybert et al., 2003).

Another perspective on shoreline fluctuation is provided by beach ridges along the central coast, which record repeated reversals in the direction of transport of sand during the past 5100 years. Observation of a reversal produced by changes in ocean-swell patterns during the 1982–1983 ENSO event suggests that "the long periods of reversed long-shore transport are associated with long periods of blocking conditions of frontal systems related to El Niño-like conditions." The much higher volumes of sand accumulated during earlier reversals imply persistence of a low Southern Oscillation index during several decades and document climatic

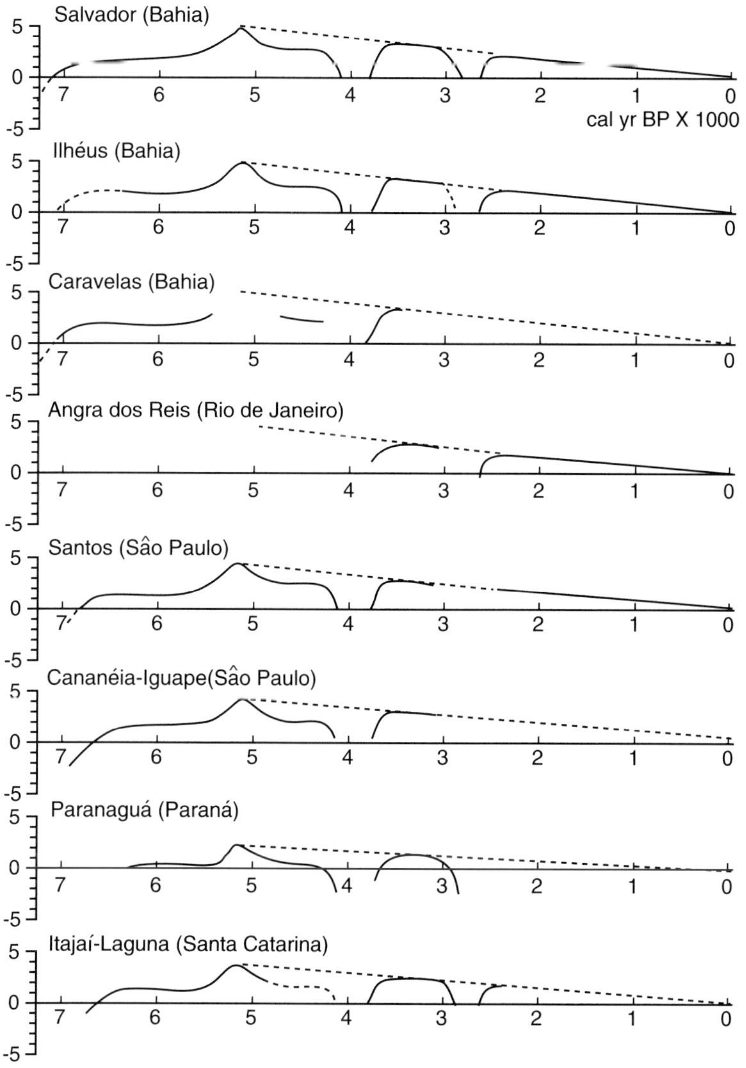

Figure 4.4. Changes in relative sea level along the Brazilian coast since ca. 7000 ^{14}C yr BP. All regions show a similar descending pattern and chronology from a maximum ca. 5100 ^{14}C yr BP. Vertical scale = meters above or below present mean sea level (after Suguio et al., 1988).

fluctuations too brief to be detectable in geological and palynological records, but sufficient to affect human populations inland as well as on the shore (McClone et al., 1992; Martin et al., 1993, p. 345, 1995a,b, 1996).

North of the mouth of the Amazon, the impact of sea level rise was minimized by the deposition of substantial amounts of Amazon sediment along the coast. About 20% of the annual discharge is carried northward along the Guianas, where it forms mud banks. Stable conditions were achieved by ca. 6000 ^{14}C yr BP and subsequent changes in shoreline fauna and flora are attributable to increased salinity rather than to sea level fluctuations (Williams, 1992; Clapperton, 1993, pp. 570–571; Nittrouer et al., 1995, p. 181; Sommerfield et al., 1995, p. 353).

2.2. *Archaeological evidence*

2.2.1. *Inland sites*

In the north, the vast majority of the known preceramic sites are rock shelters, reflecting the abundance of these features in the landscape. Preceramic open sites have been encountered where survey has been conducted in Goiás and western Bahia, but information on their extent, composition, and antiquity is slight (Schmitz et al., 1996). Rock art, executed either by engraving or painting, is also most abundant here. Regional and chronological styles have been recognized, but their social context is unknown (Prous, 1994; Aguilar, 1996; Martin, 1996; Etchevarne, 1999–2000; Martin and Asón, 2000). Open sites predominate in the south; several lithic traditions have been defined in both regions.

Rock shelters

Pedra Furada in Piauí, Santana do Riacho and Boquête in Minas Gerais, and rock shelters in the Serranópolis region of Goiás have occupations beginning ca. 11,000 ^{14}C yr BP (Fig. 4.2; Schmitz, 1987a; Schmitz et al., 1989, 2004; Prous, 1991, 1994; Barbosa, 1992; Guidon et al., 1994; Kipnis, 1998; Prous and Fogaça, 1999). Faunal remains include deer, peccary, armadillo, rodents, marsupials, and birds, as well as terrestrial and freshwater molluscs. Plant remains represent a wide variety of edible seeds, nuts, fruits, and roots. Large, elongated, unifacial blades suitable for cutting and scraping, produced by percussion, are diagnostic of the widespread Itaparica Tradition; unformalized flakes, cores, and pounders are also characteristic. About 9000 ^{14}C yr BP, bifacial tools elaborated from flakes were added, along with bone points. Where sufficient radiocarbon dates have been obtained to provide reliable evidence, they show most of the rock shelters were abandoned between ca. 7000 and 4000 ^{14}C yr BP, when resumption of warmer and wetter conditions increased the abundance and variety of terrestrial and aquatic subsistence resources (Ab'Sáber, 1980). Although some were reoccupied after ca. 2500 ^{14}C yr BP, most were used later only for burial or rock art (Fig. 4.5; Dias, 1991, pp. 69–70; Mentz Ribeiro and Ribeiro, 1999; Araujo et al., 2005).

Open sites

Rare buried sites of the Ibucuí Phase dating ca. 12,700 ^{14}C yr BP have been encountered along the left bank of the Rio Uruguai on the western boundary of Rio Grande do Sul (Miller, 1987). The following Uruguai Phase, represented at numerous locations between ca. 11,000 and 8500 ^{14}C yr BP, is characterized by

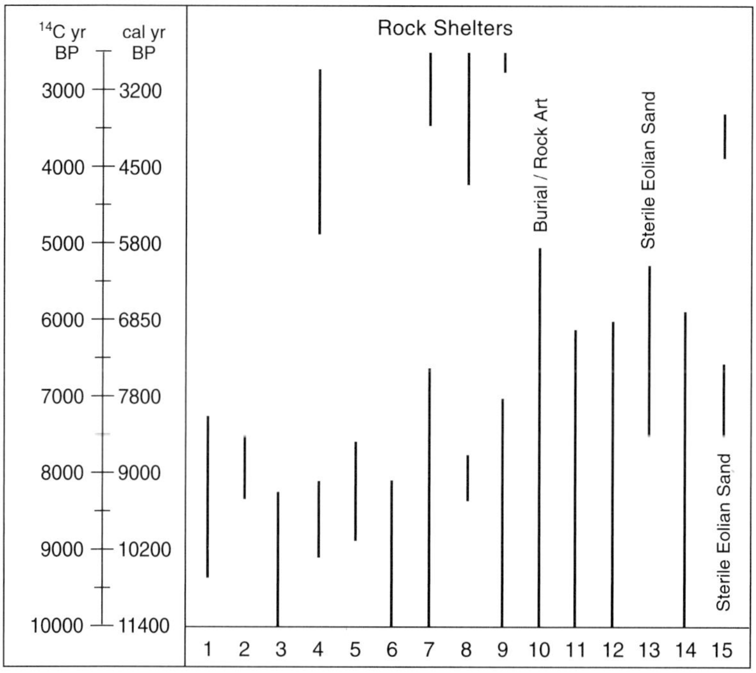

Figure 4.5. Rock shelters with sufficient ^{14}C dates to identify a hiatus between abandonment for habitation during the Pleistocene/Holocene transition and resumption of use for burial or art. 1, RS-TQ-58; 2, Lapa Pequena; 3, Lapa Varal; 4, Boqueirão Soberbo; 5, Barreiro; 6, Santana do Riacho; 7, Lapa do Gentio; 8, Lapa do Foice; 9, Boquête; 10, Lapa Vermelha; 11, Pedra Furada; 12, Calderão do Rodriguez; 13, Gruta do Padre; 14, Abrigo do Sol; 15, Pedra Pintada.

several kinds of stemmed projectile points, as well as scrapers, flakes, and expedient tools. Two contemporary traditions diverged about 7000 ^{14}C yr BP: Umbu characterized by bifacial-stemmed projectile points and Humaitá by large choppers and bifaces. Numerous habitation sites of both traditions have been recorded from Paraná to Rio Grande do Sul, those of the Humaitá Tradition associated with forested regions and those of the Umbu Tradition with more open vegetation (Schmitz, 1987b; Mentz Ribeiro, 1991; Rodríguez, 1992, Figs. 4–5).

Excavation of Cerrito Dalpiaz, a rock shelter in eastern Rio Grande do Sul occupied between ca. 5900 and 4200 ^{14}C yr BP by representatives of the Umbu Tradition, revealed a 50-cm thick stratum composed of thin lenses of ash and sand containing abundant postholes, faunal remains, artifacts, and lithic debitage. Animals of all sizes were hunted, but peccary, deer, tapir, and armadillo were most common. Fruit, snails, and molluscs were also consumed. Ovoid and pentagonal projectile points predominate initially, giving way gradually to a variety of stemmed forms with straight, convex, or concave bases. Bifacial knives, scrapers, and choppers, pitted anvil stones, small mortars with traces of red pigment, and hammer stones are also typical. Bone and horn artifacts include perforators, spatulas, pressure flakers, and fishhooks. Circular shell beads, small gastropod shells, and animal teeth were used for personal adornment (Miller, 1969).

Projectile points rarely occur in sites of the contemporary Humaitá Tradition and the diagnostic artifact is a straight or boomerang-shaped, percussion-flaked tool. Large choppers, scrapers, and bifaces predominate and are often worked only along the margins.

Although regional and chronological variations in lithics and rock art existed prior to 4000 ^{14}C yr BP, the basic settlement behavior and material culture remained generally uniform and relatively stable (Bryan and Gruhn, 1993, p. 162; also Rodríguez, 1992; Schmitz et al., 1996). The appearance of domesticated cucurbits, peanuts, beans, and cotton in several rock shelters in Minas Gerais ca. 4000 ^{14}C yr BP correlates with a shift in settlement to open locations and increased abundance of bifacial projectile points and scrapers, and polished axes (Bird et al., 1991; Dias, 1993). Several regional pottery traditions appeared between ca. 2000 and 1500 ^{14}C yr BP, associated with different types of environment and contemporary with the widespread Tupiguarani Tradition. Social and settlement behavior are similar to those in Amazonia (Brochado et al., 1970; Oliveira and Viana, 1999–2000).

2.2.2. Shoreline sites

Two kinds of shoreline sites have been distinguished: (1) shell middens, known in Brazil as sambaquis and (2) habitation sites of the Itaipu Tradition, located on coastal dunes and containing minor amounts of shell.

Sambaqui Tradition

More than 1000 shell middens have been recorded along the margins of extinct and extant shallow bays and lagoons along the coast of Pará east of the mouth of the Amazon and in scattered locations between Maranhão and Bahia, but are most abundant in the south between Espírito Santo and Rio Grande do Sul. Most attention has been focused between Rio de Janeiro and Paraná, where mangrove borders canals, lagoons, bays, and estuaries that provide a wide range of diverse habitats for aquatic flora and fauna. Tabulation of 288 ^{14}C dates from 141 southern sites by 500-year intervals shows rare occurrences prior to 5500 ^{14}C yr BP, when sea level peaked, and a decline in density after 3000 ^{14}C yr BP, when sea level stabilized leading to desiccation of lagoons and decline in subsistence resources (Ybert et al., 2003). Dimensions range from ca. 15 m long and 1 m high to 300 m long and 32 m high (Schmitz, 1998). Where conditions were most favorable, density reached 23 or more in an area of 420 km^2. In this region, isolated between the ocean and the Serra do Mar, settlement appears to have been permanent, whereas the sambaquis of Rio Grande do Sul constitute seasonal occupations by inland preceramic and ceramic groups (Schmitz, 1998; Tenório, 1998).

Some 50 species of molluscs have been identified, but four sediment-dwelling and one mangrove species comprise the vast majority (Mello, 1999). Condition varies from finely crushed to largely intact specimens. Bones of freshwater fish constitute up to 95% of the vertebrate remains in some sites; other aquatic resources include mammals (whale, dolfin), turtles, and crabs. Monkeys, armadillos, rodents, and other terrestrial fauna are abundant in some sites and absent in others. Analysis of the faunal remains at two sites along the coast of Rio de Janeiro indicated that permanent occupation could be sustained by intensive exploitation of the resources available within a radius of 2 km (Gaspar, 1995–1996; Kneip, 1998).

Cultural remains are exceeding rare in the large sambaquis on the coast of Rio Grande do Sul, the most striking objects being rare zooliths: geometric and zoomorphic ground-stone sculptures with a depression on one surface (Bryan, 1993; Gaspar, 1998; Kneip, 1998). Postholes, hearths, burials, and artifacts are often common in the smaller ones to the north. Postholes, sometimes associated with clay floors, suggest single circular or elliptical dwellings ranging from ca. 3×3 m to 14 m^2 (Gaspar, 1998, Table 2). Irregularly shaped picks, scrapers, blades, choppers, hammers, and projectile points were produced by percussion from quartz and andesite cores and flakes. Semi-polished artifacts include pestles, celts, and rubbing stones. Perforated fish vertebrae, drilled shark teeth, shell beads, and bone perforaters, spatulas, and projectile points also occur. The absence of fishhooks in sites along lagoons suggests that cast or stationary nets were used (Figuti, 1994–1995). Evidence of territoriality, contemporaneity, social inequality, and other aspects of social organization and demography remains equivocal (Blasis et al., 1998).

Whereas the southern shell middens are aceramic, several along the north coast of Brazil contain undecorated pottery. The Mina Tradition on the coast of Pará has an initial date of ca. 5000 ^{14}C yr BP, whereas the Periperi Tradition on the coast of Bahia is dated ca. 2800 ^{14}C yr BP (Calderon, 1969; Simões, 1981).

Itaipu Tradition

Populations of this tradition occupied fossil beaches or dunes accessible to forest, grassland, swamp, and lagoon habitats along the coasts of Rio de Janeiro and Espírito Santo between ca. 4500 and 1500 ^{14}C yr BP (Dias, 1992). Subsistence emphasized terrestrial resources and the faunal remains suggest the people ate anything that moved from mammals, reptiles, and birds to amphibians and insects, as well as fish and molluscs. High consumption of carbohydrates is implied by a frequency of caries exceeding that of agricultural groups and the associated unusual pattern of dental wear has been attributed to using the teeth to strip plant tissue. The high frequency of caries contrasts with their near absence among sambaqui populations (Machado, 1992). Lithic artifacts include pebble tools, grinding stones, and pitted anvil stones, as well as quartz flakes. Shell tools and ornaments are also common. Bone implements include projectile points and spatulas.

Dwellings defined by postholes vary from 3–4 m^2 to more than 30 m^2, with the larger ones typically located on the inland side of the site. The floor was hard-packed silt. Superposition of occupation levels and the abundance of burials in some sites imply their use during several millennia. An average separation of ca. 6 km between sites on the coast of Rio de Janeiro suggests they may have been central bases for macro-bands whose members dispersed periodically in family groups to exploit resources in the surrounding area and to interact with neighboring communities of the same tradition (Dias, 1992).

Excavations in the Corondó site on the coast of Rio de Janeiro produced remains of some 445 individuals. Single primary extended burials were most common, but multiple interments, secondary burials, and disarticulated bones were also encountered, the latter probably the result of disturbance by later burials. Orientation was preferably north–south. Grave goods, most often associated with adult females, included stone and shell artifacts, and bone, shell, and animal-tooth beads. Detailed analysis of the skeletal remains permitted reconstructing mortality and survivorship curves and observing a variety of fractures and degenerative conditions (Machado, 1992).

3. Amazonia

Compared to the Coastal Strip, Amazonia today is a homogeneous region topographically, climatically, and biotically. Except for protrusions of the Guayana and Brazilian shields in the northeast and southeast, elevation rarely exceeds 500 m. Rainfalls on 130 or more days per year, annual precipitation exceeds 1500 mm and reaches more than 3000 mm in the northeast, and relative humidity is normally above 80%. Daily temperature can fluctuate between 32 and 21°C, whereas annual variation averages only 3°C between the warmest and coldest month. Although a dozen major tributaries flow into the Amazon, their impact is mitigated by alternation of the influx from the northern and southern hemispheres, with the result that the normal difference between low and high water is only about 10 m. The clear and black water rivers draining the Guayana and Brazilian shields are deficient in nutrients whereas those descending from the Andes are rich in suspended sediments that are deposited on the varzea (flood plain), where they provide a diversified habitat for aquatic fauna and fertile soil for seasonal cultivation. By contrast, millennia of erosion have depleted the soil of the upland (terra firme) of soluble nutrients, creating "persisting ecological constraints on tropical agriculture" (Weischet and Caviedes, 1993).

3.1. Paleoclimatic fluctuations

Prior to the 1970s, the Amazon Basin was considered to have been unaffected by the glacial cycles that altered the flora and fauna elsewhere in the hemisphere. This assumption was called into question by Haffer's observation that the distributions of superspecies of forest-dwelling toucans do not coincide with existing disruptions of the forest, which would have prevented interbreeding (Haffer, 1969, 1974). Other biogeographers subsequently identified disjunct distributions in a variety of animals and plants that generally support the "refugia" model, whereas palynologists have produced pollen profiles that do not show discontinuities in forest vegetation during and since the Pleistocene, leading them to argue that cooler temperatures, changes in river channels, sea level rise, and other variables, rather than periodic drought, are responsible for Amazonian biodiversity (see Hoogheimstra and van der Hammen, 1998, for a review and over 500 references). New types of evidence obtained since the late 1990s and greater understanding of global climatic processes increasingly support the existence of episodes of fragmentation of the rainforest of varying durations and magnitudes during and since the Pleistocene–Holocene transition (Haffer and Prance, 2001; Rossetti et al., 2004).

3.1.1. Palynological evidence

Although pollen analysis shows continuity in rainforest vegetation at Pata in northwest Amazonia, absence of sediment deposition between ca. 31,888 and 18,000 ^{14}C yr BP and between ca. 14,000 and 5800 ^{14}C yr BP implies a decline in precipitation. A similar hiatus has been identified in cores from Carajás on the southern periphery (Fig. 4.2; Suguio et al., 1996; Ledru et al., 1998a). A pollen profile from east-central Marajó at the mouth of the Amazon, now dominated by savanna, shows repeated fluctuations between forest and savanna during the past 7000 years (Fig. 4.6; Absy, 1985) and a profile from Katira on the southwestern margin of the lowlands also shows an arid interval (Absy and Van der Hammen, 1976).

On the eastern llanos of Colombia, where the present climate favors forest, grassland predominated between ca. 9700 and 5200 ^{14}C yr BP (Behling and Hooghiemstra, 1998, p. 265). Phytoliths and charcoal from cores at a site 90 km north of Manaus suggest forest vegetation prevailed during the past 5000 years, but identify major fires between 1795 and 550 ^{14}C yr BP (Piperno and Becker, 1996). Review of 32 cores north and south of the Amazon indicates that savannas expanded during glacial periods and that the early Holocene climate was drier prior to ca. 6000–5000 ^{14}C yr BP than earlier or later (Behling and Hooghiemstra, 2001).

Another type of botanical of evidence that "the tropical forest, thought to have remained stable since the last glacial event, has in fact undergone deep modifications" is charcoal in the soil (Sanford et al., 1985; Bassini and Becker, 1990; Fearnside, 1990; Lucas et al., 1993; Meggers, 1994a; Charles-Dominique et al., 1998, p. 296; Uhl and Nepstad, 1990, p. 85). Profiles exposed along the Transamazonian Highway between Santarem and Cuiabá show sporadic pockets of charcoal dating between 6000 and 3000 ^{14}C yr BP implying, drier conditions than at present. This history of disturbance is reflected in disharmony between the present rainfall and vegetation (Soubiès, 1980). A similar inference has been drawn from the existence of enclaves of savanna in the forest of southern Venezuela, where drainage, relief, soil conditions, and climate are equivalent to those in the surrounding forest, and from the disjunct distributions of many plant species in widely separated savanna enclaves across the tropical lowlands (Eden, 1974; Harley, 1988, p. 113).

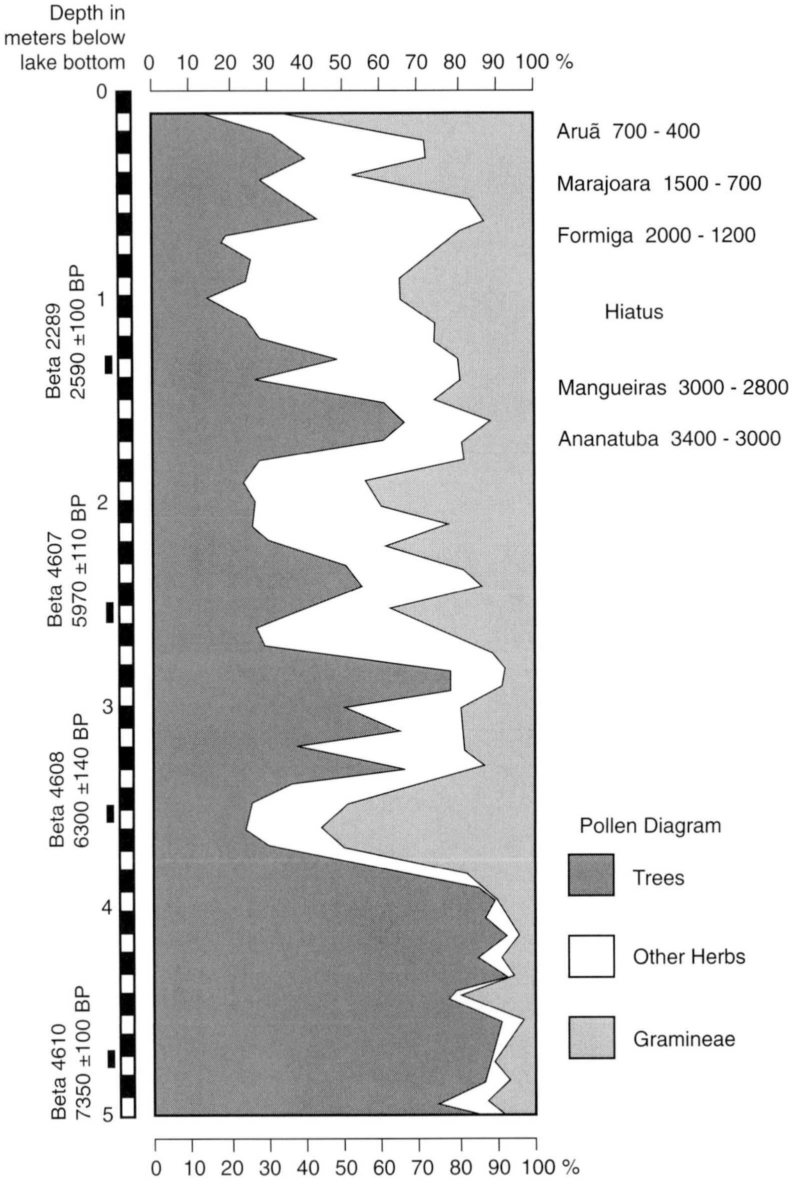

Figure 4.6. Pollen profile from Lago Ararí in east-central Marajó showing fluctuations in the extent of forest during the past 7000 years. The ^{14}C and TL dates for replacements of the archaeological phases correlate with declines in forest vegetation ca. 2800-2000, 1500, and 700 cal yr BP (after Absy, 1985, Fig. 4.9; Meggers and Danon, 1988, Fig. 2).

3.1.2. Paleoecological evidence

A detailed reconstruction of local climatic fluctuations during the past 10,000 years has been produced by a decade-long multidisciplinary investigation conducted in an undisturbed tract of rainforest 100 km inland from the coast of French Guiana, where present annual rainfall exceeds 3000 mm (Charles-Dominique et al., 1998, 2001; Ledru, 2001). The trees in primary forest and four other formations in the study area were inventoried; core samples and sediments along streams were processed for seeds, pollen, and charcoal; alluvial terraces were identified, soil profiles were analyzed for evidence of erosion, and the impact of heavy rain on the forest floor and on small agricultural clearings was observed. The geological, ecological, palynological, and climatological evidence from all sources, combined with a large set of ^{14}C dates, provide the following sequence of events:

- Major droughts between ca. 8000–7000 and 6000–4000 ^{14}C yr BP, implied by the lower diversity of primary forest species during these periods.
- More humid conditions between ca. 3000 and 2000 ^{14}C yr BP, reflected in increased taxonomic diversity.
- Drought between ca. 1800 and 1200 ^{14}C yr BP, implied by the dominance of pioneer species over large areas; an intense episode of erosion occurred ca. 1390 ^{14}C yr BP.
- More humid conditions between ca. 1200 and 900 ^{14}C yr BP, marked by expansion of the forest;
- New disturbances between 900 and 600 ^{14}C yr BP, indicated by re-emergence of pioneer assemblages.
- An episode of intense erosion ca. 530 ^{14}C yr BP.
- Consolidation of present-day vegetation ca. 300 ^{14}C yr BP.

The authors argue that "the apparitions of pioneer species during several consecutive centuries suggest that, during each of these periods, brief disturbances must have occurred every 10–30 years, impeding the establishment of mature forest species and maintaining a widespread secondary vegetation" (Charles-Dominique et al., 1998, p. 299).

3.1.3. Sedimentological evidence

A variety of geoscientific evidence supports episodes of drought and forest fragmentation during the late Pleistocene (Haffer, 1997, pp. 465–466; Haffer and Prance, 2001, pp. 582–583). Extensive deposits of eolian sand have been reported on the llanos of the Orinoco (Clapperton, 1993, pp. 199–200) and in the Negro basin (Santos et al., 1993). The Pantanal do Norte, which covers several thousand square kilometers between the Branco and Negro in north-central Amazonia, and the Parintins Formation, which extends some 400 km along the left bank of the Amazon eastward from the mouth of the Negro, have similar characteristics (Santos et al., 1993; Iriondo and Latrubesse, 1994). Four periods of eolian activity have been identified by 14 TL dates, the two most recent spanning 17,200–12,700 and 10,400–7800 BP (Filho et al., 2002). The latter interval is documented in the stratigraphy in Pedra Pintada, a rockshelter near Monte Alegre on the left bank of the lower Amazon, where the earliest human occupation ca. 10,000 ^{14}C yr BP is separated from a later occupation beginning ca. 7600 ^{14}C yr BP by an irregular deposit of wind-blown sand some 30 cm thick (Roosevelt et al., 1996). It is also represented in the earliest major period of aridity in the sequence from French Guiana and falls within the more recent hiatus in the dates from the Pata pollen core.

3.1.4. Hydrology

Eustatic changes in sea level during and since the Pleistocene have had a significant impact on the extent and composition of the Amazonian flood plain. As a consequence of the increased gradient caused by the decline of ca. 120 m in sea level during the Late Glacial Maximum, the rivers cut deeply into their beds and sediments were deposited on the Amazon fan rather than the riverbed. The conclusion that absence of an increase in grass pollen in these sediments shows "unequivocally that the Amazon lowlands were forested in glacial times as they are now" (Colinvaux, 1996, p. 389) is contradicted by hydrological research in various forested and desert landscapes in the tropics indicating that the small-scale observations of runoff relied on in climate models can be misleading. "Significantly, as the scale is increased, redistribution of overland flow becomes more dominant and takes place in the form of spatially, discontinuous surface flow ... the redistribution mechanism causes the runoff term to be virtually zero at this scale" (Bonell, 1998, p. 104). These conditions, added to the relatively small contribution (under 20%) of the eastern tributaries to the present sediment load of the Amazon (Clapperton, 1993, p. 181; Sommerfield et al., 1995, p. 353) and the massive input from increased rainfall in the eastern Andes, which produced the second and third highest water levels recorded during this century at Manaus (Molion, 1990), also discredit the reliability of pollen profiles from the Amazon fan as indicators of rainforest vegetation.

3.1.5. Specialized studies

Further support for the refugia model is provided by several kinds of specialized studies. Isotopic data suggest that reductions in atmospheric CO_2 during the Late Glacial Maximum contributed to a reduction in terrestrial biomass (Street-Perrott et al., 1997). Stable isotope ratios of soil organic carbon identify Holocene fluctuations in the forest-savanna boundary (Desjardins et al., 1996). Changes in the oxygen composition of foraminifera in Amazon fan sediments indicate a 60% reduction in Amazon flow during the Younger Dryas, implying a 40% reduction in rainfall (Maslin and Burns, 2000). In the Cariaco Basin off the east coast of Venezuela, a similar shift in foraminifera provides "a broad regional signal of water balance over tropical South America not previously identified in terrestrial records" (Peterson et al., 2000). Amazon cone isotopic stratigraphy (Showers and Beavis, 1988), sediment deposition on the Amazon continental shelf (Nittrouer et al., 1995), and carbon isotope ratios in porewaters from Amazon fan sediments (Burns, 1998) also support aridity during the Pleistocene/Holocene transition.

Fluctuations in glacial moraines and the thickness of layers in Andean ice cores provide additional evidence, since highland rainfall originates from the eastern lowlands (Heine, 2000).

Observations of the impact of recent displacements of the ITCZ and fluctuations in Sea Surface Temperature (SST) on Amazonian precipitation provide another explanation for past episodes of aridity (Peterson et al., 2000; Haug et al., 2001).

Integration of the various forms of direct and indirect evidence has led to "a majority view" that a 25–50% reduction in rainfall occurred during the Late Glacial Maximum (Heine, 2000; Thomas, 2000) and that Van der Hammen's model "appears to be more consistent with the available pollen evidence" (Haberle and Maslin, 1999, p. 36). This model is based on the current climatic and vegetational associations in northern South America, which indicate that closed forest dominates when annual rainfall exceeds 2000 mm, more deciduous forest prevails between 2000 and 1500 mm, savanna woodland or cerrado develops between 1500 and 1000 mm, and savanna/cerrado or semidesert occurs below 1000 mm. Imposing a 40% diminution on the present-day rainfall map reduces the current 1500 mm isohyet to 1000 mm, placing Carajás, Pata, and the dune fields on the upper Negro in savanna/cerrado or semidesert environments (Fig. 4.7; Van der

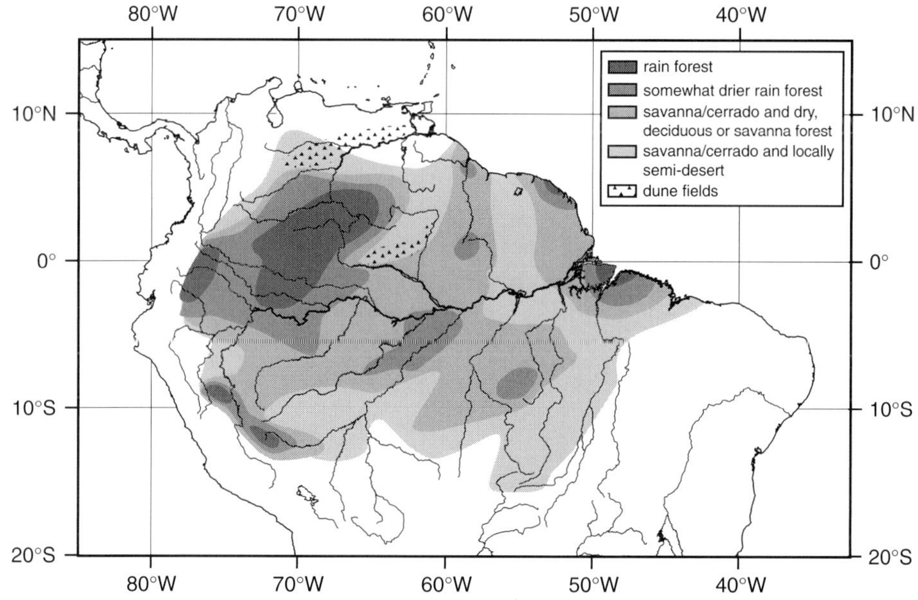

Figure 4.7. Postulated distribution of rainforest with a reduction of 40% in annual rainfall, based on the present-day isohyets and correlation of savanna and savanna woodland vegetation with rainfall below 1500 mm. A large refugium would have existed in the northwest and smaller refugia in central and eastern Amazonia (after Van der Hammen and Hooghiemstra 2000, Fig. 3).

Hammen and Hooghiemstra, 2000). Rainforest persists in a large refugium in the northwest and several smaller refugia elsewhere in the lowlands. Although, the locations of the rainfall isohyets may have been somewhat different in the past and regional variations in precipitation certainly existed, at least a 10% reduction in input from the Atlantic has been predicted by a general circulation model of atmospheric and oceanic interactions (Bush and Philander, 1998).

3.2. Holocene conditions

Geomorphological, sedimentological, and hydrological studies conducted along the Amazon above and below the Negro indicate that the varzea (floodplain) reached its present extent after stabilization of sea level and has a maximum age of 5000–6000 years (Irion et al., 1997; Behling, 2002a). The same conclusion emerged from detailed examination of the morphology of the riverbed at the mouth of the Amazon (Irion, 1984; Junk, 1984; Irion et al., 1997; Vital et al., 1998). Comparison of the configuration of the flood plain during the past two decades documents continuous cutting and depositing, shifting the locations, shapes, and magnitudes of the islands, creating and obliterating auxiliary channels, and causing minor changes in the course of the main river (Mertes et al., 1996), all of which would have affected human exploitation.

The principal source of rainfall variability during the past five millennia is the ENSO phenomenon. During the brief 1982–1983 episode, weather stations throughout Brazilian Amazonia registered precipitation 70% below normal (Nobre and Renno, 1985). During one day in February, the discharge of the Trombetas, a left-bank tributary of the lower Amazon, declined to 47 m^3 compared with a long-term average of 2100 m^3 (Molion and de Moraes, 1987). The fact that 50% or more of local rainfall originates from evapotranspiration suggests that the reduced input of moisture from the ocean would have been amplified during prolonged episodes as rainforest trees gave way to the more drought-resistant vegetation registered in pollen profiles ca. 1500, 1000, 700, and 500 ^{14}C yr BP (Absy, 1982; Salati, 1985, p. 39; Goldammer and Price, 1998, p. 278). These vegetational fluctuations are also reflected in the nitrate concentrations in late Holocene ice cores in the southern Andes (Thompson, 1995) and the fluctuating discharges in the Magdalena, Cauca, and San Jorge rivers in Colombia (Van der Hammen and Cleef, 1992).

Additional environmental uncertainty is created by the magnitude of unpredictable fluctuations in maximum and minimum water level. Records at Manaus since 1902 show the maximum to be least variable, typically ranging between 26 and 29 m above mean sea level (Fig. 4.8). Two exceptionally low crests of 25.5 and 21.5 m coincide with the 1912 and 1926 episodes of El Niño. Minimum water levels are far more erratic, ranging between 21 and 15 m, with differences of 5 m often occurring in successive years. A low of 21 m leaves the lower varzea flooded, whereas a low of 15 m drains pools and leaves fish stranded in cutoff channels. The coincidence of a low maximum and a low minimum during the 1926 episode of El Niño had a devastating impact on the aquatic fauna.

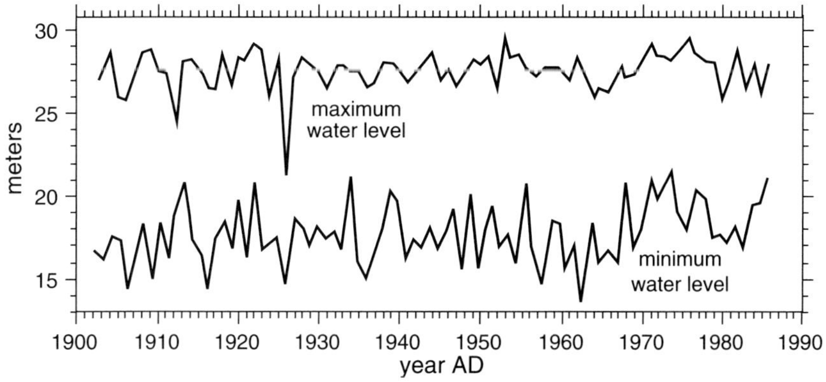

Figure 4.8. Maximum and minimum water level of the Amazon at Manaus between 1902 and 1985 (meters above mean sea level). Whereas fluctuations in maximum water level typically do not exceed 3 m, minimum water level fluctuates more than 5 m, with extremes often occurring in consecutive years. The coincidence of a low maximum and low minimum during the 1926 episode of El Niño caused high mortality of aquatic fauna (after Junk, 1989, Fig. 2).

The impact of even brief drought on the rainforest vegetation is illustrated by changes in the composition of a 50 ha plot on Barro Colorado Island in Panama as a consequence of the 1983 episode of El Niño and subsequent fluctuations in rainfall periodicity (Condit, 1998, p. 419):

> There were 37 species ... defined as moisture-demanding ... and 33 of these declined in abundance between 1982 and 1995 ... One of these moisture-specialists ... is a large, prominent canopy species that has undergone a striking crash, from 3426 stems in 1982 to 1777 stems in 1995. But it was small-stature moisture-specialists (shrubs and treelets) that were most affected: 17 of 81 declined in abundance, one went extinct ... and their total abundance fell by 35% over 13 years ... If drying trends continue, it seems likely that most of these 37 will be lost within 25 years.

The failure of many plants to flower or fruit during an abnormal weather cycle on Barro Colorado Island in 1970–1971 and the resultant famine and death among frugivores testify to the subsistence stress longer episodes would have inflicted on humans (Foster, 1982).

3.3. Archaeological evidence

The rarity of stone tools and of habitable rock shelters limits evidence of humans prior to the adoption of pottery, but camp sites and workshops encountered during environmental impact surveys document their existence throughout the lowlands by 13,000 ^{14}C yr BP at the latest.

3.3.1. Rock shelters

Although rock shelters occur in the Guayana Shield, most are too small for habitation and were used mainly for burial or temporary camps. The only known exception is Pedra Pintada near the left bank of the lower Amazon, where the earliest occupation during the 10th millennium BP is separated from two later occupations beginning ca. 7500 ^{14}C yr BP by a layer of windblown sand 20–40 cm thick containing fragments of charcoal (Roosevelt et al., 1996). The lower levels produced rare unifacial scrapers and blades, expedient tools, and abundant flakes. Subsistence remains included a variety of seeds, nuts, and palm fruits, as well as fish, mammal, reptile, and amphibian bones. Pictographs on the rear wall of the shelter have been assigned to this occupation. The similarity between this complex and the early levels at Boquête in Minas Gerais supports the existence of environmental continuity during the Pleistocene/Holocene transition (Kipnis, 1998). Two rock shelters on the southern margin of Amazonia, Abrigo do Sol and Santa Elina, attest to human presence by ca. 15,000 ^{14}C yr BP and possibly earlier (Fig. 4.2; Miller, 1987; Vilhena Vialou and Vialou, 1989; Vilhena Vialou et al., 1999).

3.3.2. Shell middens

Shell middens of the preceramic Alaka Phase appear on the northwest coast of Guyana ca. 6800 ^{14}C yr BP and were abandoned ca. 4000 ^{14}C yr BP. Area ranges from 12 × 12 to 80 × 30 m; maximum height is 1–15 m. Mammal, bird, and fish bones and crab fragments are mixed among the shells. Stone artifacts were rudimentarily shaped by percussion for use as choppers, hammerstones, picks, knives, and scrapers. Burials often occur (Evans and Meggers, 1960, pp. 38–54; Williams, 1992).

A shell midden 6 m high has been reported at Taperinha on the right bank of the Amazon below the mouth of the Tapajós. Twelve ^{14}C dates from the lower preceramic levels extend from 7090 to 5700 ^{14}C yr BP. A sterile layer separates this occupation from the upper levels, which are undated, but the presence of pottery related to the Barlovento Phase on the north coast of Colombia implies a reoccupation after ca. 3500 ^{14}C yr BP (Roosevelt et al., 1991; Meggers, 1998).

Two swampy locations on the southern margin of the rainforest were exploited for intensive shellfish gathering. Nearly 200 small middens have been documented along the Rio Paraguai and associated lakes in the pantanal of Mato Grosso do Sul. Thirteen ^{14}C dates bracket the preceramic Corumbá Phase between ca. 8300 and 2700 ^{14}C yr BP. Pottery was introduced ca. 2000 ^{14}C yr BP (Schmitz et al., 1998). Large shell middens of the Sinimbu Phase are scattered across the pantanal of the middle Guaporé. Area ranges from 25 × 20 to 210 × 130 m and height is up to 6 m. During the rainy season, they are now surrounded by water up to 1.7 m deep. Layers of crushed and calcined shell, postholes, occupation floors, and artifacts of stone, bone, and shell occur throughout, as well as lumps of clay with twig, fiber or mat impressions. Ten ^{14}C dates extend from ca. 6316 to 4300 ^{14}C yr BP, when the region was abandoned for several centuries as a consequence of drought (Miller, 1999 and n.d.).

3.3.3. Open sites

Finely chipped bifacial-stemmed projectile points resembling those from early contexts elsewhere in South America have been encountered sporadically throughout the central lowlands (Fig. 4.9; Meggers and Miller, 2003), but the only Amazonian example

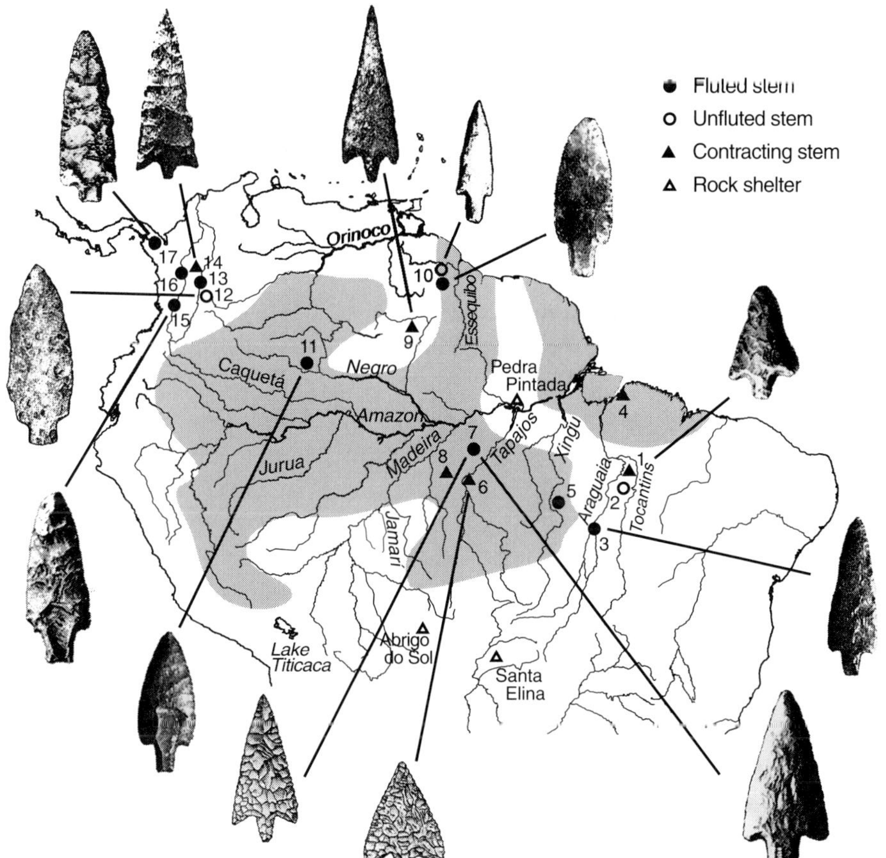

Figure 4.9. Locations of rock shelters and isolated projectile points in relation to the distribution of forest vegetation during the Late Pleistocene, assuming a 25% reduction in precipitation and a similar distribution (after Van der Hammen and Absy, 1994). 1, Itaguatins TO; 2, Darcinópolis TO; 3, Upper Araguaia TO/MT; 4, Ilha Cotijuba PA; 5, Middle Xingu PA; 6, Middle Tapajós PA; 7, Cara Preta AM; 8, Apuí AM; 9, Igarapé Murupu RR; 10, Mazaruni District, Guyana; 11, Upper Negro AM; 12, Sabana de Bogotá; 13, Middle Magdalena valley; 14, Puerto Berrio; 15, Restrepo; 16, Niquía; 17, Golfo de Urubá (after Meggers and Miller, 2003, Fig. 10.2).

from a documented context was encountered during salvage excavations along a powerline transect between the Tocantins and Araguaia, where it was associated with a variety of expedient tools and debitage. Lithics and debitage were encountered at 299 locations along this transect and 230 locations along a similar transect in northern Roraima, as well as in transects near the coast of Amapá, between the Tocantins and Tapajós, and in Rondônia between the Ji-Paraná and Rolim de Moura. Initial dates in each region range between 13,720 and 11,300 ^{14}C yr BP and many sites remained in use until 1100 ^{14}C yr BP (Meggers and Miller, 2003). A date of 14,990 ^{14}C yr BP was obtained from a site on the lower Sinnamary on the central coast of French Guiana, where rainforest would have persisted (Vacher et al., 1998).

Stratigraphic excavations in open sites along the Jamarí, a right-bank tributary of the upper Madeira in southwest Amazonia, identify three successive preceramic components: (1) the Itapipoca Phase, with dates extending from ca. 8000 to 6900 ^{14}C yr BP; (2) the Pacatuba Phase, extending from ca. 6000 to 5000 ^{14}C yr BP; and (3) the Massangana Phase, extending between ca. 4800 and 2600 ^{14}C yr BP. Artifacts of the Itapipoca Phase consist mainly of large percussion-flaked bifaces, end and side scrapers, flakes with and without retouch, and hammerstones. The succeeding Pacatuba Phase is characterized by the addition of rare small scrapers, cores, flakes, and micro-flakes, some showing micro-retouch from use (Miller et al., 1992, pp. 36–37).

In contrast to the sites of the two earlier phases, Massangana Phase occupations are associated with black soil (*terra preta*). Since black soil is characteristic of the shifting habitation sites of ceramic phases, both along the Jamarí and elsewhere in Amazonia, this situation suggests that slash-and-burn agriculture and semi-permanent settlement behavior were adopted here prior to the acquisition of pottery. This inference is supported by the addition to the lithic inventory of axes, anvil stones, small mortars and pestles, and grinding stones impregnated with hematite pigment. Cores, flakes, and micro-flakes persist from the previous phase (Miller, 1992; Miller et al., 1992, pp. 37–38). It is also compatible with genetic evidence for manioc domestication in the region (Olsen and Schaal, 2001).

Palm starch (sago) is the primary source of carbohydrate throughout southeast Asia and various kinds of evidence indicate that prehistoric Amazonian hunter-gatherers depended on the burití palm (*Mauritia flexuosa*) for starch, fruit, and grubs (Jones, 1955; Meggers, 2001b). Access to this resource increased substantially after ca. 8000 ^{14}C yr BP, when rising sea level expanded the extent of swampy habitats throughout the lowlands (Behling, 2002a). It remains a staple among several marginal groups, among them the Warao of the Orinoco delta (Heinen and Ruddle, 1994). On the western coast of Guyana, where depletion of palms and molluscs as a consequence of sea-level rise and increased aridity provoked abandonment of the shell middens for inland locations ca. 5000 ^{14}C yr BP, more intensive exploitation of terrestrial resources is reflected in the addition of axes, choppers, adzes, pitted anvil stones, and scrapers to the lithic inventory. It has been suggested that the procedures employed for processing palm starch would have preadapted the inhabitants for eliminating the toxic content of bitter manioc (Williams, 1992, pp. 238–240).

The earliest well-documented ceramic complexes are Ananatuba on Marajó, with an initial TL date of 3400 BP (Meggers and Danon, 1988), and Bacabal on the upper Guaporé, with an initial ^{14}C date ca. 3900 ^{14}C yr BP (Miller, 1999 and ms). Pottery became widespread throughout the lowlands after ca. 2000 ^{14}C yr BP, permitting construction of detailed local relative chronologies. Correlation of archaeological sequences from lowland Bolivia, the central Amazon, Marajó, northern Colombia, and Venezuela shows simultaneous discontinuities ca. 1500, 1000, 700, and 400 ^{14}C yr BP implying the dispersal and replacement of an earlier community of shifting agriculturalists by a later one as a consequence of subsistence stress inflicted by the impact of mega-Niño droughts on the biota (Fig. 4.10; Absy, 1982; Meggers, 1994a, 1996b). The extensive habitation sites along the major tributaries and sectors of the middle Amazon have been interpreted as large permanent settlements, but all those that have been investigated archaeologically are the product of multiple re-occupation by small villages during hundreds of years (Miller et al., 1992; Meggers, 2001a; Meggers and Miller, 2006).

Although it has been suggested that population pressure on the varzea was relieved by expansion up the tributaries, archaeological evidence indicates that the first rapid was a permanent ecological barrier to movement in either direction (Meggers et al., 1988). The implication that adaptations suitable for sustained exploitation of one region are not equally effective in the other is confirmed by detailed analysis of the behavior of contemporary riverine and hinterland Achuar communities in southeastern Ecuador (Descola, 1994).

3.4. Biogeographical evidence

The paucity of archaeological evidence for reconstructing indigenous movements prior to the adoption of pottery is partly compensated by the geographical distributions of languages, genetic features, and cultural traits (Meggers, 1987).

3.4.1. Linguistic distributions

The principal Amazonian language families have been assigned to two major phyla: Ge-Pano-Carib and Equatorial-Tucanoan (Greenberg, 1987). The lexicostatistical estimates for the primary separation of Ge, Pano, and Carib languages and their present isolation in non-forest habitats in eastern Brazil, eastern Peru, and the Guianas are consistent with displacement of the speakers of the proto-language from the central lowlands during consolidation of the rainforest ca. 5000 ^{14}C yr BP (Fig. 4.11; Meggers, 1994b). Similarly, the present predominance of speakers of Equatorial languages (Arawak and Tupí) in the rainforest suggests that they

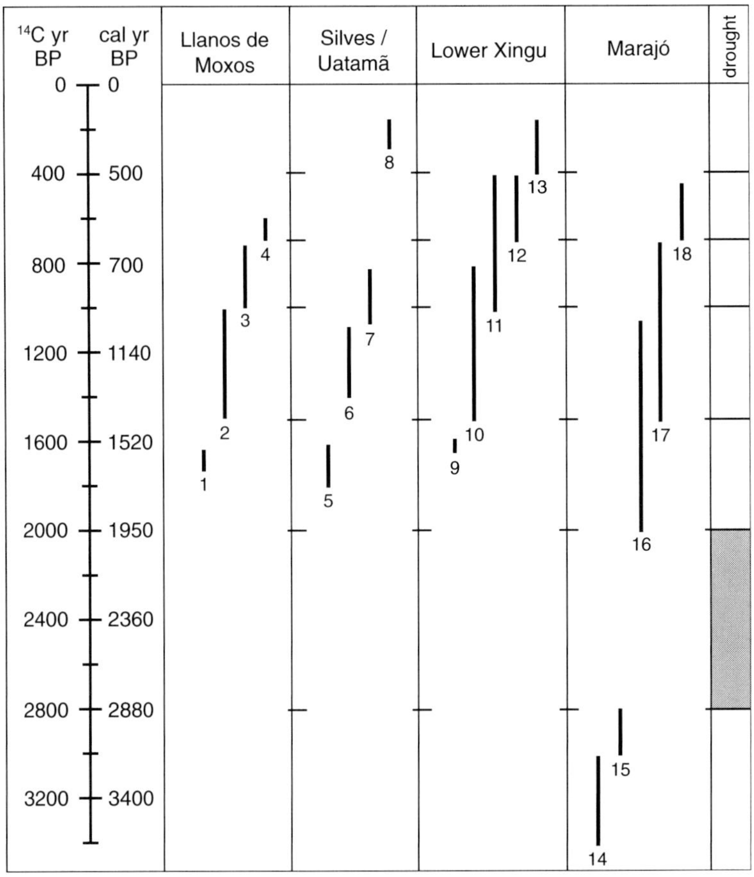

Figure 4.10. Discontinuities in well-dated ceramic sequences from lowland Bolivia (Llanos de Mojos), the central Amazon (Silves/Uatumã and Lower Xingu), and the mouth of the Amazon (Marajó), implying substitution of an earlier population by a later one ca. 1500, 1000, 700, and 400 ^{14}C yr BP. These dates coincide with severe droughts induced by mega-Niño events, which diminished the productivity of subsistence resources. 1, Casarabe Tradition; 2, Mamoré Tradition; 3, Kiusiu Tradition; 4, Ibare Tradition; 5, Polychrome Tradition, Manacapuru Subtradition; 6, Polychrome Tradition, Saracá Subtradition; 7, Incised and Punctate Tradition; 8, Uatumã Phase; 9, Macapá Phase; 10, Guará Phase; 11, Polychrome Tradition, Cacarapí Phase; 12, Pacajá Phase; 13, Curuá Phase; 14, Ananatuba Phase; 15, Mangueiras Phase; 16, Formiga Phase; 17, Polychrome Tradition, Marajoara Phase; 18, Aruã Phase (after Meggers, 1994a, Fig. 4).

expanded from a homeland in the refugium east of the Andes as the forest coalesced. The coincidence between the estimates for the subsequent diversification within Arawak and Tupí ca. 1500, 1000, and 500 ^{14}C yr BP and the mega-Niño episodes of aridity provides an explanation for the present heterogeneous linguistic, cultural element, and genetic distributions throughout Amazonia (Migliazza, 1982; Black et al., 1983).

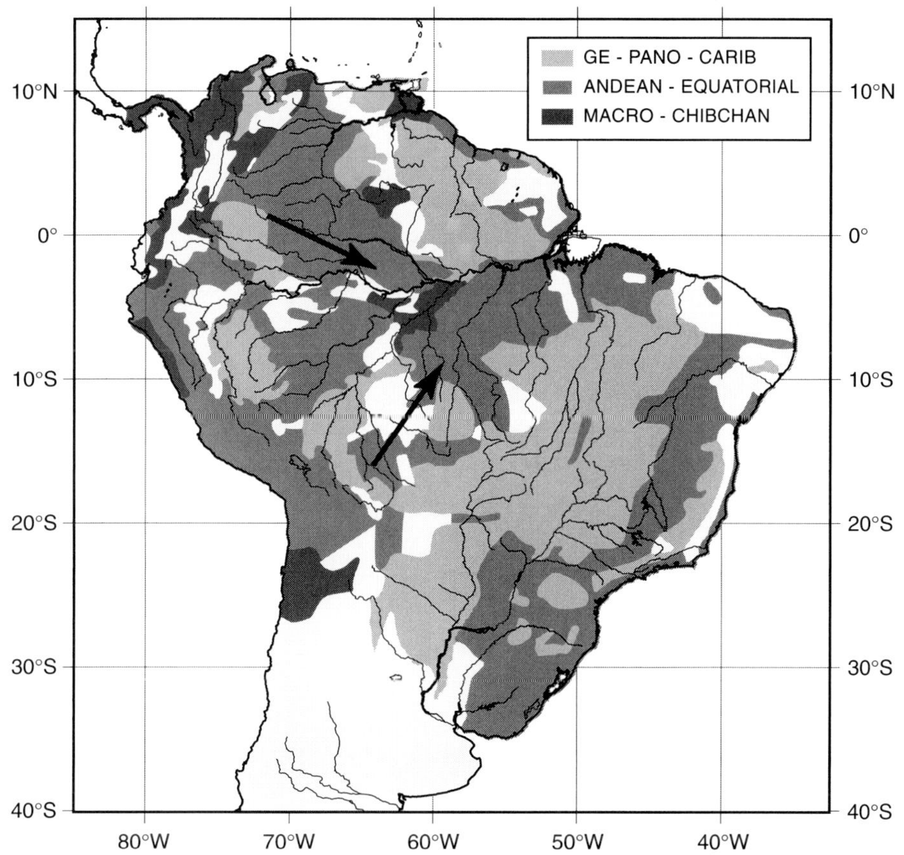

Figure 4.11. Present distributions of the principal lowland language phyla, suggesting replacement the savanna/cerrado-adapted proto-Ge-Pano-Carib speakers from the central lowlands by intrusion of forest-adapted Equatorial-Tucanoan speakers from the west. This inference is supported by the coincidence between the lexicostatistical dates for completion of primary differentiation within each phylum ca. 5000 ^{14}C yr BP and ^{14}C estimates for coalescence of the rainforest about that time, and the present-day association of Ge, Pano, and Carib speakers with open environments on the northeastern, southeastern, and western margins of the rainforest (after Meggers, 1994b, Fig. 3; Greenberg, 1987).

3.4.2. Cultural element distributions

Although the deculturation and disappearance of many indigenous Amazonian groups leaves large geographical blanks, some cultural traits free of adaptive constraints have concentric distributions compatible with the displacement of earlier by later populations in the central lowlands. Among two types of racks for roasting meat, the tripod is restricted to Amazonia whereas the tetrapod occurs in the surrounding regions, implying its greater antiquity (Fig. 4.12; Nordenskiold, 1924). Three types of finger positions for arrow releases also have circum-Amazonian

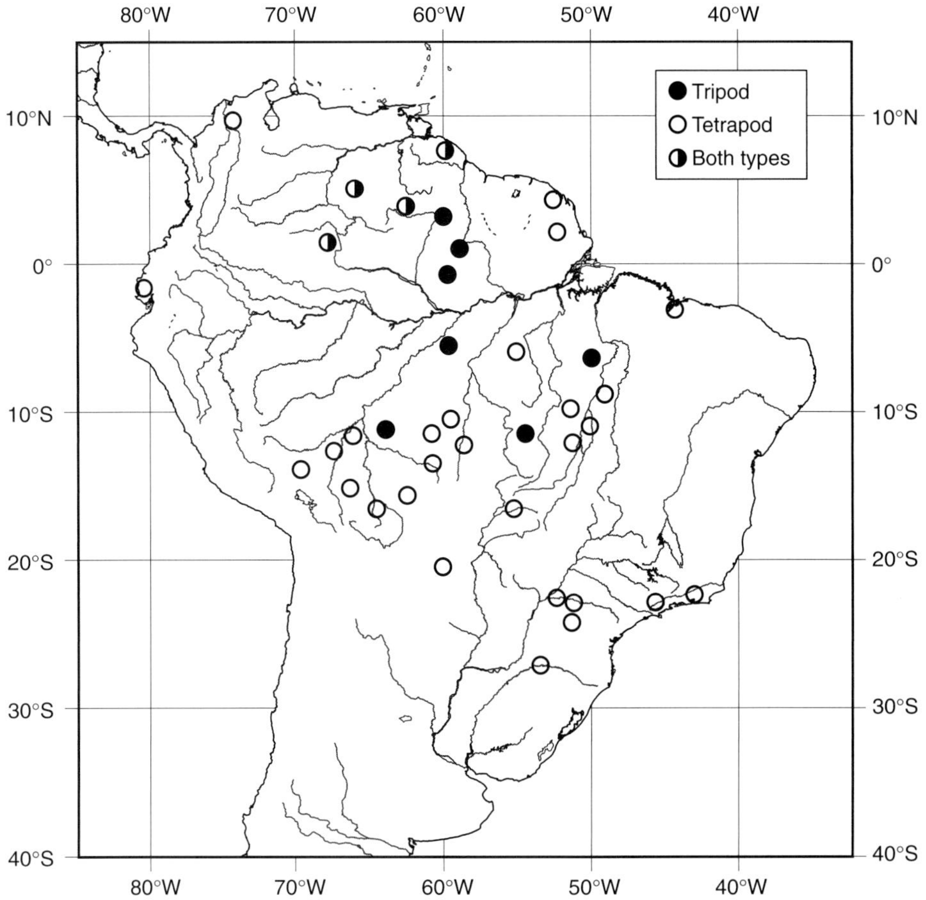

Figure 4.12. South American distributions of tripod and tetrapod racks for roasting meat. The concentric pattern is compatible with the linguistic evidence for displacement of Ge-Pano-Carib speakers by Equatorial-Tucanoan speakers in the central lowlands. Bisected circles indicate the presence of both varieties (after Nordenskiold, 1924, Map 15).

distributions, whereas the secondary position is restricted to the intervening region (Heath and Chiara, 1977, Map 5), and pole snares are more widespread than simple nooses for capturing small mammals (Ryden, 1950). Among various mythical explanations of the dark patches on the surface of the moon, that interpreting them as the face of a man is restricted to Amazonian groups (Blixen, 1992).

3.4.3. Genetic distributions

Most blood group alleles, mtDNA, and other genetic traits considered immune to natural selection have heterogeneous distributions among contemporary indigenous Amazonians. As in the case of languages, this pattern is "dramatically different from the regular clinal distributions found for North America" (O'Rourke and

Suarez, 1985, p. 24) and is attributable to the repeated fractionations and dispersals of indigenous communities triggered by mega-Niño episodes during recent millennia (Ward et al., 1975, p. 18).

An exception to this general pattern was revealed by multivariate analysis of 13 alleles among 21 indigenous populations, which produced distributions in the first three principal components suggesting dispersal from the central lowlands (Rothhammer and Silva, 1992). Examining the linguistic affiliations of the groups sampled shows 10 to be Ge-Pano-Carib speakers and 7 to be Equatorial speakers. The former are distributed peripherally to the latter with two exceptions in the eastern Guianas, which represent post-contact migrants from south of the Amazon.

3.4.4. Historical evidence

The impact on the biota and human populations in northern Amazonia of the short-term droughts associated with 20th century episodes of El Niño is proxy evidence for the situation confronted by the precolumbian inhabitants. During the 1912 event, fires burned continuously for several months on the lower Branco, the river was not navigable, and thousands of rubber gatherers are reported to have died. Between the upper Ventuari and upper Uraricuara in southern Venezuela, massive destruction of the forest by fire was followed by toppling of semi-burned vegetation. During the 1926 event, extensive fires lasting more than a month affected the entire lower Negro region, causing massive mortality among the fauna, especially large birds. The heat was sufficiently intense in some streams to kill the fish (Koch-Grünberg, 1979–1982, p. 234; Carvalho, 1952, p. 16).

The impact of the 1972–1973 drought on a Yanomami community on the Alto Siapa is illuminating:

> The Indians, accustomed to burning the dried vegetation in their fields to expand the cultivated area or remove plant residue, were careless and the flames consumed the producing plants. What happened next was inconceivable under normal conditions: the fires spread via the undergrowth and flared up on hilltops where the stony soil made vegetation sparse. Few gardens escaped the catastrophe, creating a shortage of cultivated foods. Forest resources, however, remained available. Consequently, the majority of the population abandoned their dwellings and assumed a nomadic existence, exploiting zones of forest successively. Working harder than normal, they remained constantly hungry but survived. The state of health … testified that the food shortage they had experienced was not dramatic … Nevertheless, the palmito that had replaced the plantain as their primary food was becoming increasingly scarce" (Lizot, 1974, p. 7; Meggers, 1994a, p. 332).

Recent efforts to exploit the more fertile soils of the varzea for intensive agriculture have been frustrated by the unpredictable regime of the rivers. On the Ucayali in eastern Peru, "floods annually threaten crops grown in fertile low-lying areas [and] even crops grown in all but the very highest parts of restinga are destroyed by floods once or twice every decade" (Chubnik, 1994, p. 221). The 100-year history of a varzea community on the Solimões reveals a decline in the number of settlements because of subsistence uncertainties induced by riverine fluctuations and an increase in population density on the adjacent *terra firma*, where conditions are more stable (Lima-Ayres and Alencar, 1994). The disastrous impact of the erratic flood cycle on jute production has been documented on an island near Santarem (Santos, 1982). On the middle Caquetá in eastern Colombia, where periodic losses occur as a result of untimely or excessive flooding, both indigenous groups and colonists grow commercial crops on the floodplain, but raise subsistence crops on the *terra firma* where they are less susceptible to loss (Eden, 1990, p. 124). High vulnerability to pathogens constitutes an additional hazard (Bahri et al., 1990).

4. Conclusion

The first humans arriving in lowland South America ca. 15,000 years ago would have encountered cooler and drier conditions than exist today. A 25–50% decrease in rainfall is reflected in the formation of extensive eolian sand deposits in the upper Orinoco and Negro regions, reduction of the rainforest to enclaves separated by savanna and cerrado vegetation in central Amazonia, and downcutting of the bed of the Amazon and its tributaries (Fig. 4.7). The climate on the Coastal Strip was also drier and 5–7°C cooler prior to ca. 4000 ^{14}C yr BP, reducing the availability of perennial streams and favoring development of xerophytic vegetation (Araujo et al., 2005).

The absence of an ecological barrier between Amazonia and the Coastal Strip during the Pleistocene–Holocene transition is implied by the existence of the same flora and fauna, domestic features, artifacts, and art in the early occupations at Pedra Pintada in east-central Amazonia and Boquête in Minas Gerais. Although there are local differences in lithic traditions, the characteristics of the habitation sites, subsistence remains, and artifacts throughout the lowlands prior to ca. 5000 ^{14}C yr BP imply the existence of small bands of foragers that, like surviving indigenous hunter–gatherers, moved as local resources were depleted or seasonal ones became available (Dias, 1991, 1993; Barbosa, 1992, pp. 155–159; Schmitz et al., 1996, pp. 180–184; Politis et al., 1997; Kipnis, 1998).

The achievement of Holocene climatic conditions and stabilization of sea level ca. 5000 ^{14}C yr BP had two major environmental consequences: (1) it created an ecological barrier between Amazonia and the Coastal Strip and (2) it divided each region into a narrow aquatic zone and a large hinterland. On the Coastal Strip, stabilization of sea level opened a new niche for specialized exploitation of marine resources and higher temperature and rainfall inland favored expansion of forest vegetation, the development of perennial streams, and the diversification of terrestrial and aquatic fauna. The abandonment of many rock shelters between ca. 7000 and 4000 ^{14}C yr BP reflects increased diversity and abundances of wild subsistence resources as well as the adoption of domesticated maize, cucurbits, beans, and manioc. Although communities became more sedentary and somewhat larger, a ceiling was set by dependence on wild sources of protein.

In Amazonia, consolidation of the rainforest permitted groups adapted to its exploitation to expand eastward and replace those adapted to more open environments. Simultaneously, the culmination of sea level rise allowed sediments formerly deposited on the fan to accumulate on the flood plain, creating habitats for diverse aquatic biota. Indirect evidence of agriculture is provided by the appearance in southwestern Amazonia ca. 4800 ^{14}C yr BP of habitation sites composed of black soil, diagnostic of the frequently moved settlements of contemporary shifting cultivators (Miller et al., 1992).

In contrast to other parts of the planet, Holocene climatic changes did not significantly improve the opportunities for agricultural intensification and associated cultural development. The perpetuation of the same general way of life throughout the eastern lowlands is attributable to the existence of three permanent environmental constraints: (1) edaphic and climatic impediments to agricultural intensification and to storage, (2) reliance on dispersed wild sources of protein, and (3) intermittent and unpredictable short and long-term drought. The soils of the Guayana and Brazilian Shields are among the poorest on earth as a result of millions of years of leaching and erosion. Although the flora of Amazonia is the most diverse and species-rich on the planet, it is deficient in nutrients. As a consequence, terrestrial herbivores are small, solitary, and vulnerable to over hunting (Alvard et al., 1997; Sioli, 1984; Robinson and Bennett, 2000). Any tendency to increasing population density and sedentism in both Amazonia and the Coastal Strip was truncated by the impact of periodic long-term drought on subsistence resources (Meggers, 1994a).

In Amazonia, the intensity of these constraints is reflected in the dependence of modern settlements numbering only a few hundred individuals on imported food, by the failure of well funded "development" projects to enhance sustainable agricultural productivity, and by the increasing environmental degradation that follows abandonment of indigenous practices. Many characteristics of the subsistence, settlement, and social behavior of surviving indigenous groups are intelligible as adaptations to minimize the impact of inherent ecological constraints and unpredictable subsistence stress (Meggers, 1996a, p. 192, 2001a; Meggers, 2007).

The existence of inherent limitations to sustainable intensive exploitation, population concentration, and sedentism in the Neotropical lowlands continues to be challenged by archaeologists in spite of the environmental and cultural evidence, and the failure of modern efforts at "development." They interpret prehistoric habitation sites extending a kilometer or more along the banks of the Amazon and its tributaries as confirmation of early European accounts describing large and permanent settlements organized into proto-states with powerful rulers (Erickson, 2003). They assert that "chieftaincy developed as much in the interfluvial as in the floodplain areas" and that "ancient Amerindian political and cultural life was of a level of sophistication that rivaled or even exceeded" that of contemporary Europe (Whitehead, 1990, 1994). By contrast, historians warn that "the first Europeans to set foot in Amazonia let their imaginations run away with them and claimed actually to see and hear everything they had hitherto only imagined" (Gheerbrant, 1992, p. 47; Meggers, 1993–1995). The failure of the indigenous inhabitants of

Amazonia and the Coastal Strip to develop the features we equate with "civilization" does not imply cultural stagnation. On the contrary, their success in achieving a flexible and sustainable accommodation to inherent environmental limitations and unpredictable climatic fluctuations is an accomplishment we have yet to match, much less exceed.

References

Ab'Sáber, A. N., 1980. Paleo-clima e paleo-ecologia. In *Temas de Arqueologia Brasileira I Paleo-Indio,* edited by P. I. Schmitz, A. S. Barbosa, and P. M. Ribeiro, *Anuário de Divulgação Científica* 5, 33–51. Instituto Goiano de Investigações Pré-Históricas, Goiânia.

Absy, M. L., 1982. Quaternary palynological studies in the Amazon Basin. In *Biological Diversification in the Tropics*, edited by G. T. Prance, pp. 67–73. Columbia University Press, New York.

Absy, M. L., 1985. Palynology of Amazonia: the history of the forests as revealed by the palynological record. In *Amazonia*, edited by G. T. Prance and T. E. Lovejoy, pp. 72–82. Pergamon Press, Oxford.

Absy, M. L., and T. Van der Hammen, 1976. Some palaeoecological data from Rondônia, southern part of the Amazon Basin. *Acta Amazônica* 6:293–299.

Aguilar, A., 1996. Bibliografía Sobre Registros Rupestres no Nordeste do Brasil. *CLIO, Série Arqueológica* 11:187–200.

Alexandre, A., J.-D. Meunier, A. Mariotti, and F. Soubiès, 1999. Late Holocene Phytolith and Carbon-Isotope Record from a Latosol at Salitre, South-Central Brazil. *Quaternary Research* 51:187–194.

Alvard, M. S., J. G. Robinson, K. H. Redford, and H. Kaplan, 1997. The Sustainability of Subsistence Hunting in the Neotropics. *Conservation Biology* 11:977–982.

Angulo, R. J., and K. Suguio, 1995. Re-evaluation of the Holocene sea-level maxima for the State of Paraná, Brazil. *Palaeogeography, Palaeoclimatology and Palaeoecology* 113: 385–393.

Araujo, A. G. M., W. A. Neves, K. B. Piló, and J. P. Atui, 2005. Holocene dryness and human occupation in Brazil during the "Archaic Gap." *Quaternary Research* 64, 298–307.

Bahri, S., F. Grenand, P. Grenand, J.-L. Guillaumet, and M. Lourd, 1990. La Várzea est-elle un don de l'Amazonie? Les techniques tradicionneles de mise en valeur de la várzea face à la modernité. In *Sahel, Nordeste, Amazonie*, edited by ORSTOM, pp. 105–144. UNESCO, Paris.

Barbosa, A. S., 1992. A tradição Itaparica: uma compreensão ecológica e cultural do povamento inicial do planalto central brasileiro. In *Prehistoria Sudamericana*, edited by B. J. Meggers, pp. 145–160. Taraxacum, Washington, DC.

Barreto, A. M. F., L. C. R. Pessenda, and K. Suguio, 1996. Probable drier Holocene climate evidenced by charcoal bearing middle São Francisco River Paleodunes, State of Bahia, Brazil. *Anales da Acadêmia Brasileira de Ciência* 68(Suppl.1):43–48.

Bassini, F., and P. Becker, 1990. Charcoal's occurrence in soil depends on topography in terra firme forest near Manaus, Brazil. *Biotropica* 22:420–422.

Behling, H., 1997. Late Quaternary vegetation, climate and fire history of the *Araucaria* forest and campos region from Serra Campos Gerais, Paraná State (South Brazil). *Review of Palaeobotany and Palynology* 97:109–121.

Behling, H., 1998. Late Quaternary vegetational and climatic changes in Brazil. *Review of Palaeobotany and Palynology* 99:143–156.

Behling, H., 2002a. Impact of the Holocene sea-level changes in coastal, eastern and central Amazonia. *Amazoniana* 17:41–52.

Behling, H., 2002b. South and southeast Brazilian grasslands during late Quaternary times: a synthesis. *Palaeogeography, Palaeoclimatology, Palaeoecology* 177:19–27.

Behling, H., H. W. Arz, J. Pätzold, and G. Wefer, 2000. Late Quaternary vegetational and climate dynamics in Northeastern Brazil, inferences from Marine Core GeoB 3104-1. *Quaternary Science Reviews* 19:981–994.

Behling, H., H. W. Arz, J. Pätzold, and G. Wefer, 2002. Late Quaternary vegetational and climate dynamics in Southeastern Brazil, inferences from Marine Cores GeoB 3229-2 and GeoB 3202-1. *Palaeogeography, Palaeoclimatology, Palaeoecology* 179:227–243.

Behling, H., S. G. Bauermann, and P. C. P. Neves, 2001. Holocene environmental changes in the São Francisco de Paula Region, southern Brazil. *Journal of South American Earth Sciences* 14:631–639.

Behling, H., and H. Hooghiemstra, 1998. Late Quaternary palaeoecology and palaeoclimatology from pollen records of the savannas of the Llanos Orientales de Colombia. *Palaeogeography, Palaeoclimatology, Palaeoecology* 139:251–267.

Behling, H., and H. Hooghiemstra, 2001. Neotropical savanna environments in space and time: later Quaternary interhemispheric comparisons. In *Interhemispheric Climate Linkages*, edited by V. Markgraf, pp. 307–323. Academic Press, New York.

Bird, R. McK., O. Dias, and E. T. Carvalho, 1991. Subsídios para arqueo-botânica no Brasil: o milho antigo em cavernas de Minas Gerais. *Revista de Arqueologia* 6:14–31. São Paulo.

Black, F. L., F. M. Salzano, L. L. Bermann, I. Gabbay, T. A. Weimer, and M. H. L. P. Franco, 1983. Failure of linguistic relationships to predict genetic distances between the Waiãpi and other tribes of lower Amazonia. *American Journal of Physical Anthropology* 60: 327–335.

Blasis, P. de, S. R. Fish, M. D. Gaspar, and P. R. Fish, 1998. Some references for the discussion of complexity among the Sambaqui Moundbuilders from the Southern Shores of Brazil. *Revista de Arqueología Americana* 5:75–105.

Blixen, O., 1992. Las manchas de la luna y sus explicaciones míticas en sudamérica. *Moana, Estudios de Antropología* 4(2), Montevideo.

Bonell, M., 1998. Possible impacts of climate variability and change on tropical forest hydrology. *Climatic Change* 39:215–272.

Brochado, J. P., V. Calderon, I. Chmyz, O. F. Dias, C. Evans, S. Maranca, B. J. Meggers, E. Th. Miller, N. A. Nasser, C. Perota, W. F. Piazza, J. W. Rauth, and M. F. Simões, 1970. Brazilian archaeology in 1968: an interim report on the National Program of Archaeological Research. *American Antiquity* 35. 1–23.

Bryan, A. L., 1993. The Sambaqui at Forte Marechal Luz, State of Santa Catarina, Brazil. In *Brazilian Studies*, pp. 1–114. Center for the Study of the First Americans, Corvallis.

Bryan, A. L., and R. Gruhn, 1993. Archaeological research at six cave or rockshelter sites in interior Bahia, Brazil. In *Brazilian Studies*, pp. 1–168. Center for the Study of the First Americans, Corvallis.

Burns, S. J., 1998. Carbon isotopic evidence for coupled sulphate reduction-methane oxidation in Amazon fan sediments. *Geochimica et Cosmochimica Acta* 62:797–804.

Bush, A. B. G., and S. G. H. Philander, 1998. The role of ocean-atmospheric interactions in tropical cooling during the Last Glacial Maximum. *Science* 279:1341–1344.

Calderon, V., 1969. A Fase Aratu no Recôncavo e litoral norte do Estado da Bahia. *Publicações Avulsas* 13:161–172. Museu Paraense Emílio Goeldi, Belém.

Carvalho, J. C. M., 1952. Notas de viagem ao Rio Negro. *Publicações Avulsas* 9. Museu Nacional, Rio de Janeiro.

Charles-Dominique, P., P. Blanc, D. Larpin, M.-P. Ledru, B. Riéra, C. Sarthou, M. Servant, and C. Tardy, 1998. Forest perturbations and biodiversity during the last ten thousand years in French Guiana. *Acta Oecologica* 19:295–302.

Charles-Dominique, P., P. Blanc, D. Larpin, M.-P. Ledru, B. Riéra, T. Rosique, C. Sarthou, M. Servant, and C. Tardy, 2001. Palaeoclimates and their consequences on forest composition. In *Nouragues*, edited by F. Bongers, P. Charles-Dominique, P-M. Forget, and M. Théry, pp. 79–87. Kluwer Academic Publishers, Dordrecht.

Chubnik, M., 1994. *Risky Rivers: the Economics and Politics of Floodplain Farming in Amazonia.* University of Arizona Press, Tucson.

Clapperton, C., 1993. *Quaternary Geology and Geomorphology of South America.* Elsevier, Amsterdam.

Colinvaux, P. A., 1996. Quaternary environmental history and forest diversity in the Neotropics. In *Evolution and Environment in Tropical America*, edited by J. B. C. Jackson, A. F. Budd, and A. G. Coates, pp. 359–405. University of Chicago Press, Chicago.

Colinvaux, P. A., and P. E. Oliveira, 2001. Amazon plant diversity and climate through the Cenozoic. *Palaeogeography, Palaeoclimatology, and Palaeoecology* 166:51–63.

Condit, R., 1998. Ecological implications of changes in drought patterns: shifts in forest composition in Panama. *Climatic Change* 39:413–427.

Cowling, S. A., M. A. Maslin, and M. T. Sykes, 2001. Paleovegetation simulations of lowland Amazonia and implications for Neotropical allopatry and speciation. *Quaternary Research* 55:140–149.

Descola, P., 1994. *In the Society of Nature: A Native Ecology in Amazonia.* Cambridge University Press, New York.

Desjardins, T., A. Carneiro-Filho, A. Mariotti, A. Chauvel, and C. Girardin, 1996. Changes of the forest-savanna boundary in Brazilian Amazonia during the Holocene revealed by stable isotope ratios of soil organic carbon. *Oecologia* 108:749–756.

Dias, O. F., 1991. Desenvolvimento Cultural no Horizonte 9000/4000 anos A.P. no Brasil tropical. *Revista de Arqueología Americana* 4:55–87.

Dias, O. F., 1992. A tradição Itaipu, costa central do Brasil. In *Prehistoria Sudamericana*, edited by B. J. Meggers, pp. 161–176. Taraxacum, Washington, DC.

Dias, O. F., 1993. As Origens da Horticultura no Brasil. *Revista de Arqueología Americana* 7:7–52.

Eden, M. J., 1974. Palaeoclimatic influences and the development of savanna in southern Venezuela. *Journal of Biogeography* 1:95–109.

Eden, M. J., 1990. *Ecology and Land Management in Amazonia.* Bellhaven Press, London.

Erickson, C., 2003. Historical ecology and future explorations. In *Amazonian Dark Earths*, edited by J. Lehmann, D. K. Kern, B. Glaser, and W. I. Woods, pp. 455–500. Kluwer Academic Publishers, Dordrecht.

Etchevarne, C., 1999–2000. A ocupação humana do nordeste brasileiro antes da colonização portuguesa. *Revista da Universidade de São Paulo* 44:112–141.

Evans, C., andB. J. Meggers, 1960. Archeological Investigations in British Guiana. *Bureau of American Ethnology Bulletin* 177. Smithsonian Institution, Washington, DC.

Fearnside, P. M., 1990. Fire in the tropical rain forest of the Amazon Basin. In *Fire in the Tropical Biota*, edited by J. G. Goldammer, pp. 106–116. Springer, Berlin.

Figuti, L., 1994–1995. Os sambaquis Cosipa (4200 a 1200 anos AP): estudo da subsistência dos povos pescadores coletores pré-históricos da baixada santista. *Revista de Arqueologia* 8:267–283. São Paulo.

Filho, A. C., D. Schwartz, S. H. Tatumi, and T. Rosique, 2002. Amazonian paleodunes provide evidence for drier climate phases during the Late Pleistocene-Holocene. *Quaternary Research* 58:205–209.

Foster, R. B., 1982. Famine on Barro Colorado Island. In *The Ecology of a Tropical Forest: Seasonal Rhythms and Long-term Changes*, edited by G. Leigh Jr., S. A. Rand, and D. M. Windsor, pp. 201–212. Smithsonian Institution Press, Washington, DC.

Gaspar, M. D., 1995–1996. Território de exploração e tipo de ocupação dos pescadores, coletores e caçadores que ocuparam o litoral do Estado do Rio de Janeiro. *CLIO, Série Arqueológico* 11:153–174.

Gaspar, M. D., 1998. Considerations of the *Sambaquis* of the Brazilian coast. *Antiquity* 72:592–615.

Gheerbrant, A., 1992. *The Amazon: Past, Present, and Future*. Harry N. Abrams Inc., New York.

Goldammer, J. G., and C. Price, 1998. Potential impacts of climate change on fire regimes in the tropics based on MAGICC and a GISS GCM-derived lightning mode. *Climatic Change* 39:273–296.

Greenberg, J., 1987. *Language in the Americas*. Stanford University Press, Stanford.

Guidon, N., F. Parenti, M. de F. da Luz, C. Guerin, and M. Faure, 1994. Le plus ancien peuplement de l' Amérique: le paleolithique du nordeste Brésilien. *Bulletin of Société Préhistorique Française* 91:246–250.

Haberle, S. G., and M. A. Maslin, 1999. Late Quaternary vegetation and climate in the Amazon Basin based on a 50,000 year pollen record from the Amazon fan, ODP Site 932. *Quaternary Research* 51:27–38.

Haffer, J., 1969. Speciation in Amazonian forest birds. *Science* 165:131–137.

Haffer, J., 1974. Avian speciation in tropical South America. *Publications of the Nuttall Ornithological Club* 14, Cambridge, MA.

Haffer, J., 1997. Alternative models of vertebrate speciation in Amazonia: an overview. *Biodiversity and Conservation* 6:451–476.

Haffer, J., and G. T. Prance, 2001. Climate forcing of evolution in Amazonia during the Cenozoic: on the refuge theory of biotic differentiation. *Amazoniana* 16:579–607.

Harley, R. M., 1988. Evolution and distribution of *Iriope* (Labiatae), and its relatives in Brazil. In *Proceedings of a Workshop on Neotropical Distribution Patterns*, edited by P. E. Vanzolini and W. R. Heyer, pp. 71–120. Academia Brasileira de Ciências, Rio de Janeiro.

Haug, G. H., K. A. Hughen, D. M. Sigman, L. C. Peterson, and U. Röhl, 2001. Southward migration of the intertropical convergence zone through the Holocene. *Science* 293:1304–1308.

Heath, E. G., and V. Chiara, 1977. *Brazilian Indian Archery*. The Simon Archery Foundation, Manchester Museum, Manchester.

Heine, K., 2000. Tropical South America during the Last Glacial Maximum: evidence from Glacial, Periglacial and Fluvial records. *Quaternary International* 72:7–21.

Heinen, H., and K. Ruddle, 1994. Ecology, ritual, and economic organization in the distribution of palm starch among the Warao Indians of the Orinoco Delta. *Journal of Anthropological Research* 30:116–138.

Hooghiemstra, H., and T. van der Hammen, 1998. Neogene and Quaternary development of the Neotropical rain forest: the forest refugia hypothesis, and a literature overview. *Earth Science Reviews* 44:147–183.

Irion, G., 1984. Sedimentation and sediments of Amazonian rivers and evolution of the Amazonian landscape since Pliocene times. In *The Amazon*, edited by H. Sioli, pp. 201–214. Dr. W. Junk Publishers, Dordrecht.

Irion, G., W. J. Junk, and J. A. S. N. de Mello, 1997. The large central Amazonian river floodplains near Manaus: geological, climatological, hydrological and geomorphological aspects. In *The Central Amazon Floodplain: Ecology of a Pulsing System*, edited by W. J. Junk, pp. 23–46. Springer, Berlin.

Iriondo, M., and E. M. Latrubesse, 1994. A probable scenario for a dry climate in central Amazonia during the late Quaternary. *Quaternary International* 21:121–128.

Jones, D. L., 1955. *Palms Throughout the World*. Smithsonian Institution Press, Washington, DC.

Junk, W. J., 1984. Ecology of the *várzea*, floodplain of Amazonian White-Water Rivers. In *The Amazon: Limnology and Landscape Ecology of a Mighty Tropical River and its Basin*, edited by H. Sioli, pp. 215–243. Dr. W. Junk Publishers, Dordrecht.

Kipnis, R., 1998. Early hunter-gatherers in the Americas: perspectives from central Brazil. *Antiquity* 72:581–592.

Kneip, L. M., 1998. Os pescadores, coletores e caçadores pré-históricos da área arqueológica de Saquarema, R.J. *Revista de Arqueología Americana* 15:57–73.

Koch-Grünberg, T., 1979–1982. *Del Roraima al Orinoco*. Ediciones del Banco Central de Venezuela, Caracas.

Ledru, M.-P., 2001. Late Holocene rainforest disturbance in French Guiana. *Review of Palaeobotany and Palynology* 115:161–176.

Ledru, M.-P., 2002. Late Quaternary history and evolution of the Cerrados as revealed by Palynological records. In *The Cerrados of Brazil*, edited by P. S. Oliveira and R. J. Marquis, pp. 33–50. Columbia University Press, New York.

Ledru, M.-P., J. Bertaux, A. Sifeddine, and K. Suguio, 1998a. Absence of Last Glacial Maximum records in lowland South American tropical forests. *Quaternary Research* 49:233–237.

Ledru, M.-P., M. L. Salgado-Labouriau, and M. L. Lorscheitter, 1998b. Vegetation dynamics in southern and central Brazil during the last 10,000 yr B.P. *Review of Palaeobotany and Palynology* 99:131–142.

Ledru, M.-P., R. C. Cordeiro, J. M. L. Dominguez, L. Martin, P. Mourguiart, A. Sifeddine, and B. Turcq, 2001. Late-Glacial cooling in Amazonia inferred from Pollen at Lagoa do Caçó, Northern Brazil. *Quaternary Research* 55:47–56.

Lima-Ayres, D., and E. F. Alencar, 1994. Histórico da ocupação humana e mobilidade geográfica de assentamentos na área da Estação Ecológica Mamirauá. *9° Encontro de Estudos Populacionais, Caxambu, Brasil, Memorias* 2:353–384. Associação Brasileira de Estudos Populacionais, Belo Horizonte.

Lizot, J., 1974. El río de los Periquitos: breve relato de un viaje entre los Yanomami del Alto Siapa. *Antropológica* 37: 3–23.

Lorscheitter, M. L., and S. R. Dillenburg, 1998. Holocene palaeoenvironments of the northern coastal plain of Rio Grande do Sul, Brazil, Tramandaí Lagoon Sediments. *Quaternary of South America and Antarctic Peninsula* 11:73–97.

Lucas, Y., F. Soubiès, A. Chauvel, and T. Desjardins, 1993. Estudos do solo revelam alterações climáticas da Amazônia. *Ciência Hoje* 16(93):36–39.

Machado, L. C., 1992. Biologia de grupos indígenas pré-históricos do sudeste do Brasil: as tradições Itaipu e Una. In *Prehistoria Sudamericana*, edited by B. J. Meggers, pp. 77–103. Taraxacum, Washington, DC.

Martin, G. editor, 1996. *Pré-história do Nordeste do Brasil*. Editor Universitaria da Universidade de Pernambuco, Recife.

Martin, G., and I. Asón, 2000. El horizonte "Nordeste" en el arte rupestre del Brasil. *SAGVNTVM (P.L.A.V.)* 32:67–76.

Martin, L., J. Bertaux, M.-P. Ledru, P. Mourguiart, A. Sifeddine, F. Soubiès, and B. Turcq, 1995a. Perturbaciones del régimen de las lluvias y condiciones de tipo el Niño en América del Sur tropical desde hace 7000 años. *Bulletin de l'Institut Français d'Etudes Andines* 2 4:595–605.

Martin, L., J.-M. Flexor, and K. Suguio, 1995b. Variación de la dinámica costera durante los últimos 5000 años a lo largo de la costa brasileña como indicador de condiciones de tipo El Niño. In *Climas cuaternarios en América del Sur* edited by J. Argollo and P. Mourguiart, pp. 95–109. ORSTOM, La Paz.

Martin, L., M. Fournier, P. Mourguiart, A. Sifeddine, and B. Turcq, 1993. Southern oscillation signal in South American palaeoclimatic data of the last 7000 years. *Quaternary Research* 39:338–346.
Martin, L., K. Suguio, J. M. Flexor, J. M. L. Dominguez, andA. C. S. P. Bittencourt, 1996. Quaternary sea-level history and variation in dynamics along the central Brazilian coast: consequences on coastal plain construction. *Acadêmia Brasileira de Ciências, Anais* 68(3):303–354.
Maslin, M. A., and S. J. Burns, 2000. Reconstruction of the Amazon Basin effective moisture availability over the past 14,000 years. *Science* 290:2285–2287.
McClone, M. S., A. P. Kershaw, and V. Markgraf, 1992. El Niño/Southern oscillation climate variability in Australasian and South American paleoenvironmental records. In *El Niño: Historical and Paleoclimatic Aspects of the Southern Oscillation*, edited by H. F. Diaz and V. Markgraf, pp. 435–462. Cambridge University Press, New York.
Meggers, B. J., 1987. The early history of man in Amazonia. In *Biogeography and Quaternary History in Tropical America*, edited by T. C. Whitmore and G. F. Prance, pp. 151–174. Clarendon Press, Oxford.
Meggers, B. J., 1993–1995. Amazonia on the eve of European contact: ethnohistorical, ecological, and anthropological perspectives. *Revista de Arqueología Americana* 8:91–115.
Meggers, B. J., 1994a. Archeological evidence for the impact of mega-Niño events on Amazonia during the past two millennia. *Climatic Change* 28:321–338.
Meggers, B. J., 1994b. Biogeographical approaches to reconstructing the prehistory of Amazonia. *Biogeographica* 70:97–110.
Meggers, B. J., 1996a. *Amazonia: Man and Culture in a Counterfeit Paradise.* Revised edition. Smithsonian Institution Press, Washington, DC.
Meggers, B. J., 1996b. Possible impact of mega-Niño events on precolumbian populations of the Caribbean area. In *Ponencias del Primer Seminario de Arqueología del Caribe*, edited by M. Veloz Maggiolo and A. Caba Fuentes, pp. 156–176. Altos de Chavón, República Dominicana.
Meggers, B. J., 1998. La cerámica temprana en América del Sur: invención independiente o difusión? *Revista de Arqueología Americana* 13:7–40.
Meggers, B. J., 2001a. The continuing quest for El Dorado: round two. *Latin American Antiquity* 2:304–325.
Meggers, B. J., 2001b. The mystery of the Marajoara: an ecological solution. *Amazoniana* 16:421–440.
Meggers, B. J., 2007. Sustainable intensive exploitation of Amazonia: cultural, environmental, and geopolitical perspectives. In *The World System and the Earth System: Global Socioenvironmental Change and Sustainability Since the Neolithic,* edited by A. Hornborg and C. L. Crumley, pp. 195–209. Left Coast Press, Walnut Creek, CA.
Meggers, B. J., and J. Danon, 1988. Identification and implications of a hiatus in the archeological sequence on Marajó Island, Brazil. *Journal of the Washington Academy of Sciences* 78:245–253.
Meggers, B. J., O. F. Dias, E. Th. Miller, and C. Perota, 1988. Implications of archeological distributions in Amazonia. In *Proceedings of a Workshop on Neotropical Distribution Patterns*, edited by P. E. Vanzolini and W. R. Heyer, pp. 275–294. Academia Brasileira de Ciências, Rio de Janeiro.
Meggers, B. J., and E. Th. Miller, 2003. Hunter-gatherers in Amazonia during the Pleistocene-Holocene transition. In *Under the Canopy: The Archaeology of Tropical Rain Forests*, edited by J. Mercader, pp. 291–316. Rutgers University Press, Princeton.
Meggers, B. J., and E. Th. Miller, 2006. Evidencia arqueológica para el comportamiento social y habitacional en la Amazonía prehistórica. In *Pueblos y Paisajes Antiguous de la Selva*

Amazónica, edited by G. M. Ríos, S. M. Camargo, and C. F. Calvo, pp. 325–348. Universidad Nacional de Colombia, Bogotá.

Mello, E. M. B., 1999. Fauna malacológica de sítios arqueológicos do Estado do Rio de Janeiro, Brasil: Mesogastropoda, Neogastropoda (Mollusca, Gastropoda, Prosobranchia). *Publicações Avulsas* 79:1–36. Museu Nacional, Rio de Janeiro.

Mentz Ribeiro, P. A., 1991. Os caçadores pampeanos e a arte rupestre. In *Arqueologia Pré-Histórica do Rio Grande do Sul,* edited by A. A. Kern, pp. 103–133. Porto Alegre.

Mentz Ribeiro, P. A., and C. T. Ribeiro, 1999. Escavações arqueológicas no sítio RS-TQ-58, Montenegro RS, Brasil. *Série Documento* 10. Fundação Universidade Federal do Rio Grande.

Mertes, L. A. K., T. Dunne, and L. A. Martinelli, 1996. Channel-floodplain geomorphology along the Solimões-Amazon River, Brazil. *Geological Society of America Bulletin* 108:1089–1107.

Migliazza, E. C., 1982. Linguistic prehistory and the refuge model in Amazonia. In *Biological Diversification in the Tropics,* edited by G. T. Prance, pp. 497–519. Columbia University Press, New York.

Miller, E. Th., 1969. Resultados Preliminares das Escavações no sítio pré-cerâmico RS-LN-1: Cerrito Dalpiaz (abrigo-sob-rocha). *Iheringia, Antropologia* 1:43–112.

Miller, E. Th., 1987. Pesquisas arqueológicas paleoíndigenas no Brasil Ocidental. *Estudios Atacameños* 8:37–61.

Miller, E. Th., 1992. Adaptação agrícola pré-histórica no Alto Rio Madeira. In *Prehistoria Sudamericana,* edited by B. J. Meggers, pp. 219–229. Taraxacum, Washington, DC.

Miller, E. Th., 1999. A limitação ambiental como barreira à transposição do periodo formativo no Brasil: tecnologia, produção de alimentos e formação de aldeas no sudoeste da Amazônia. In *Formativo Sudamericano,* edited by P. Ledergerber-Crespo, pp. 331–351. Abya-Yala, Quito.

Miller, E. Th., n.d. Pesquisas arqueológicas no Pantanal do Guaporé: a secuência seriada de cerâmica da Fase Bacabal. Manuscript in the possession of the author.

Miller, E. Th. et al., 1992. Arqueologia nos empreendimentos hidrelétricos da Eletronorte; resultados preliminares. Centrais Elétricas do Norte do Brasil S.A, Brasília.

Molion, L. C. B., 1990. Climate variability and its effects on Amazonian hydrology. *Interciencia* 15:367–372.

Molion, L. C. B., and J. C. de Moraes, 1987. Oscilação Sul e descarga de rios en América do Sul tropical. *Revista Brasileira de Engeneiria, Caderno de Hidrologia* 5:53–63.

Nittrouer, C. A., S. A. Kuehl, R. W. Sternberg, A. G. Figueiredo Jr., and L. E. C. Faria Jr., 1995. An introduction to the geological significance of sediment transport and accumulation on the Amazon continental shelf. *Marine Geology* 125:177–192.

Nobre, C. A., and N. O. Renno, 1985. Droughts and floods in South America due to the 1982-83 ENSO episode. *Proceedings of the 16th Conference on Hurricanes and Tropical Meteorology,,* pp. 131–133. American Meteorological Society, Houston.

Nordenskiold, E., 1924. The ethnography of South America seen from Mojos in Bolivia. *Comparative Ethnographical Studies* 3. Göteborg.

Oliveira, J.E. de, and S. A. Viana, 1999–2000. O centro-oeste antes de Cabral. *Revista USP, São Paulo* 44:142–189.

Oliveira, P. E. de, A. M. F. Barreto, and K. Suguio, 1999. Late Pleistocene/Holocene climatic and vegetational history of the Brazilian caatinga: the fossil dunes of the middle São Francisco river. *Palaeogeography, Palaeoclimatology, Palaeoecology* 152:319–337.

Oliveira, P. S., and R. J. Marquis, R.J., editors. 2002. *The Cerrados of Brazil: Ecology and Natural History of a Neotropical savanna.* Columbia University Press, New York.

Olsen, K. M., and B. A. Schaal, 2001. Microsatellite variation in Cassava (*Manihot esculenta,* Euphorbiaceae) and its wild relatives: further evidence for a southern Amazonian origin of domestication. *American Journal of Botany* 88:131–142.

O'Rourke, D. H., and B. K. Suarez, 1985. Patterns and correlates of genetic variation in South Amerindians. *Annals of Human Biology* 13:13–31.

Pessenda, L. C. R., S. E. M. Gouveia, R. Aravena, R. Boulet, and E. P. E. Valencia, 2004. Holocene fire and vegetation changes in southeastern Brazil as deduced from fossil charcoal and soil carbon isotopes. *Quaternary International* 114:35–43.

Peterson, L. C., G.H. Haug, K. A. Hughen, and U. Röhl, 2000. Rapid changes in the hydrologic cycle of the tropical atlantic during the Last Glacial Maxima. *Science* 290:1947–1951.

Piperno, D. R., and P. Becker, 1996. Vegetational history of a site in the central Amazon Basin derived from phytolith and charcoal records from natural soils. *Quaternary Research* 45:202–209.

Politis, G. G., G. A. Martínez, and J. Rodríguez, 1997. Caza, recolección y pesca como estrategia de explotación de recursos en forestas tropicales lluviosas: los Nukak de la amazonía colombiana. *Revista Española de Antropología Americana* 27:167–197.

Prance, G. T., editor, 1982. *Biological Diversification in the Tropics*. Columbia University Press, New York.

Prous, A., 1991. Fouilles de l'abri du Boquête, Minas Gerais, Brésil. *Journal de la Société des Américanistes* 77:77–109.

Prous, A., 1994. L'art rupestre du Brésil. *Bulletin de la Société Préhistorique Ariège-Pyrénees* 49:77–144.

Prous, A., and E. Fogaça, 1999. Archaeology of the Pleistocene-Holocene boundary in Brazil. *Quaternary International* 53/54:21–41.

Robinson, J. G., and E. L. Bennett, editors, 2000. *Hunting for Sustainability in Tropical Forests*. Columbia University Press, New York.

Rodríguez, J. A., 1992. Arqueología del sudeste de Sudamérica. In *Prehistoria Sudamericana*, edited by B. J. Meggers, pp. 177–209. Taraxacum, Washington, DC.

Roosevelt, A. C., M. da Costa, C. Machado, M. Michab, N. Mercier, H. Valladas, J. Feathers, W. Barnett, M. da Silva, A. Henderson, J. Silva, B. Chernoff, D. S. Reese, J. A. Holman, N. Toth, and K. Schick, 1996. Paleoindian cave dwellers in the Amazon: the peopling of the Americas. *Science* 272:373–384.

Roosevelt, A. C., R. A. Housley, M. Imazio da Silveira, S. Maranca, and R. Johnson, 1991. Eighth millennium pottery from a prehistoric shell midden in the Brazilian Amazon. *Science* 254:1621–1624.

Rossetti, D. de F., P. M. de Toledo,, H. M. Moraes-Santos, and A. E. de Santos Jr., 2004. Reconstructing habitats in central Amazonia using megafauna, sedimentology, radiocarbon, and isotope analyses. *Quaternary Research* 61:289–300.

Rothhammer, F., and C. Silva, 1992. Gene geography of South America: testing models of population displacement based on archeological evidence. *American Journal of Physical Anthropology* 89:441–446.

Ryden, S., 1950. A study of South American Indian hunting traps. *Revista do Museu Paulista* 4: 247–352.

Salati, E., 1985. The climatology and hydrology of Amazonia. In *Amazonia*, edited by G. T. Prance and T. E. Lovejoy, pp. 18–48. Pergamon Press, Oxford.

Salgado-Labouriau, M. L., M. Barberi, K. R. Ferraz-Vicentini, and M. G. Parizzi, 1998. A dry climatic event during the late Quaternary of tropical Brazil. *Review of Palaeobotany and Palynology* 99: 115–129.

Sanford, R.L., Jr., J. Saldarriaga, K. E. Clark, C. Uhl, and R. Herrera, 1985. Amazon rain-forest fires. *Science* 227:53–55.

Santos, A. M. de S., 1982. Aritapera: uma comunidade de pequenos productores na várzea amazônica (Santarem-PA). *Boletim do Museu Paraense Emílio Goeldi, Antropologia* 83. Belém.

Santos, J. O. S., B. W. Nelson, and C. A. Giovannini, 1993. Corpos de areia sob leitos abandonados de grandes rios. *Ciência Hoje* 16(93):22–25.

Schmitz, P. I., 1987a. Caçadores antigos no sudoeste de Goiás, Brasil. In *Investigaciones paleoindias al sur de la línea ecuatorial*, edited by L. Núñez and B. J. Meggers, *Estudios Atacameños* 8:16–35. San Pedro de Atacama, Universidad del Norte.

Schmitz, P. I., 1987b. Prehistoric hunter-gatherers of Brazil. *Journal of World Prehistory* 1:53–126.

Schmitz, P. I., 1998. Peopling of the seashore of southern Brazil. In *Explorations in American Archaeology*, edited by M. G. Plew, pp. 194–220. University Press of America, Lanham.

Schmitz, P. I., A. S. Barbosa, A. L. Jacobus, and M. B. Ribeiro, 1989. Arqueologia nos cerrados do Brasil central: Serranópolis I. *Pesquisas, Antropologia* 44. Instituto Anchietano de Pesquisas, São Leopoldo.

Schmitz, P. I., A. S. Barbosa, A. F. de Miranda, M. B. Ribeiro, and M. de Oliveira Barbosa, 1996. Arqueologia nos cerrados do Brasil central: Sudoeste da Bahia e leste de Goiás. *Pesquisas, Antropologia* 52. Instituto Anchietano de Pesquisas, São Leopoldo.

Schmitz, P. I., J. H. Rogge, A. O. Rosa, and M. V. Beber, 1998. Aterros indígenas no pantanal do Mato Grosso do Sul. *Pesquisas, Antropologia* 54. Instituto Anchietano de Pesquisas, São Leopoldo.

Schmitz, P. I., A. O. Rosa, and A. L. V. Bitencourt, 2004. Arqueologia nos cerrados do Brasil Central: Serranópolis III. *Pesquisas, Antropologia* 60. Instituto Anchietano de Pesquisas, São Leopoldo.

Showers, M., and W. J. Beavis, 1988. Amazon cone isotopic stratigraphy: evidence for the source of the tropical fresh-water spike. *Palaeogeography, Palaeoclimatology, Palaeoecology* 64:189–199.

Simões, M. F., 1981. Coletores-pescadores ceramistas do litoral do Salgado (Pará): nota preliminar. *Boletim do Museu Paraense Emílio Goeldi, Antropologia* 78, Museu Goeldi, Belém.

Sioli, H., 1984. Former and recent utilizations of Amazonia and their impact on the environment. In *The Amazon: Limnology and Landscape Ecology of a Mighty Tropical River and its Basin*, edited by H. Sioli, pp. 675–706. Dr. W. Junk Publishers, Dordrecht.

Sommerfeld, C. K., C. A. Nittrouer, and A. G. Figueiredo, 1995. Stratigraphic evidence of changes in Amazon shelf sedimentation during the late Holocene. *Marine Geology* 125:351–371.

Soubiés, F., 1980. Existence d'une phase sèche en Amazonie brésilienne datée par la présence de charbons dans les sols (6000-3000 ans B.P.). *Cahiers ORSTOM Série Géologie* 11:133–148.

Street-Perrott, F. A, Y. Huang, R. A. Perrott, G. Eglinton, P. Barker, L. B. Khelifa, D. D. Harkness, and D. O. Olago, 1997. Impact of lower atmosphere carbon dioxide on tropical mountain ecosystems. *Science* 278:1422–1426.

Suguio, K., L. Martin, and J.-M. Flexor, 1988. Quaternary sea levels of the Brazilian coast: recent progress. *Episodes* 11(3):203–208.

Suguio, K., L. Martin, and J.-M. Flexor, 1991. Paleoshorelines and the Sambaquis of Brazil. In *Paleoshorelines and Prehistory: An Investigation of Method*, edited by L. L. Johnson, pp. 83–99. CRC Press, Boca Raton.

Suguio, K., B. Turcq, and R. C. Cordeiro, 1996. On the paleoclimates of the last 60 ky in the eastern Amazon. *Anais da Acadêmia Brasileira de Ciência* 68:101–107.

Tenório, M. C., 1998. Abandonment of Brazilian coastal sites: why leave Eden? In *Explorations in American Archaeology*, edited by M. G. Plew, pp. 221–257. University Press of America, Lanham.

Thomas, M. F., 2000. Late Quaternary environmental changes and the Alluvial record in humid tropical environments. *Quaternary International* 72:23–36.

Thompson, L. G., 1995. Late Holocene ice core records of climate and environment from the tropical Andes. *Institut Francés d'Etudes Andines, Bulletin* 24:619–629.

Uhl, C., and D. Nepstad, 1990. Perturbaciones Naturales y Antropogénicas en la Amazonía. In *Alternativas a la deforestación*, edited by A. Anderson, pp. 45–76. Museu Goeldi and Fundación Natura, Belem.

Vacher, S., S. Jérémie, and J. Briand, editors, 1998. Amérindiens du Sinnamary (Guyane): Archéologie en fôret équatoriale. *Documents d'Archéologie Française* 70. Editions de la Maison des Sciences de l'Homme, Paris.

Van der Hammen, T., and M. L. Absy, 1994. Amazonia during the last glacial. *Palaeogeography, Palaeoclimatology, Palaeoecology* 109:247–261.

Van der Hammen, T., and A. M. Cleef, 1992. Holocene changes in rainfall in northern South America and the El Niño Phenomenon. *Erdkunde* 46:252–256.

Van der Hammen, T., and H. Hooghiemstra, 2000. Neogene and Quaternary history of vegetation, climate, and plant diversity in Amazonia. *Quaternary Science Reviews* 19:725–742.

Vilhena-Vialou, A., P. D. de Blasis, L. Figuti, P. Paillet, and D. Vialou, 1999. Art rupstre et habitats préhistoriques au Mato Grosso (Brésil). *BAR International Series* 746:9–22.

Vilhena-Vialou, A., and D. Vialou, 1989. Abrigo pré-histórico Santa Elina, Mato Grosso: habitats e arte rupestre. *Revista de Pré-História* 7:34–53. São Paulo.

Villwock, J. A., L. J. Tomazelli, E. L. Loss, E. A. Dehnhardt, N. H. Filho, F. A. Bachi, and B. A. Dehnhardt, 1986. Geology of the Rio Grande do Sul coastal province. *Quaternary of South America and Antarctic Peninsula* 4:79–97.

Vital, H., K. Stattegger, J. Posewang, and F. Theilen, 1998. Lowermost Amazon river: morphology and shallow seismic characteristics. *Marine Geology* 152:277–294.

Ward, R. H., H. Gershowitz, M. Layrisse, and J. V. Neel, 1975. The genetic structure of a tribal population, the Yanomama Indians. *American Journal of Human Genetics* 27:1–30.

Weischet, W., C. N. Caviedes, 1993. *The Persisting Ecological Constraints of Tropical Agriculture*. Wiley, New York.

Whitehead, N. L., 1990. The Mazaruni Pectoral: a golden artefact discovered in Guyana and the historical sources concerning native metallurgy in the Caribbean, Orinoco and northern Amazonia. *Archaeology and Anthropology* 7:19–38.

Whitehead, N. L., 1994. The ancient Amerindian polities of the Amazon, the Orinoco, and the Atlantic Coast: a preliminary analysis of their passage from antiquity to extinction. In *Amazonian Indians*, edited by A. C. Roosevelt, pp. 33–53. University of Arizona Press, Tucson.

Williams, D., 1992. El arcaico en el noroeste de Guyana y los comienzos de la horticultura. In *Prehistoria Sudamericana*, edited by B. J. Meggers, pp. 233–251 Taraxacum, Washington, DC.

Ybert, J.-P., W. M. Bissa, E. L. M. Catharino, and M. Kutner, 2003. Environmental and sea-level variations on the southeastern Brazilian coast during the late Holocene with comments on prehistoric human occupation. *Palaeogeography, Palaeoclimatology, Palaeoecology* 189:11–24.

Climate Change and Cultural Dynamics: A Global Perspective on Mid-Holocene Transitions
David G. Anderson, Kirk A. Maasch and Daniel H. Sandweiss (Editors)

Chapter 5

Culture and climate in Mesoamerica during the Middle Holocene

Barbara Voorhies[1] and Sarah E. Metcalfe[2]

[1]*Department of Anthropology, University of California, Santa Barbara, CA 93106, USA*

[2]*School of Geography, University of Nottingham, Nottingham, NG7 2RD, UK*

Abstract

This chapter summarizes current research on paleoclimate and prehistory in Mesoamerica during the Middle Holocene, here taken to be the period 7800–3200 cal yr BP (7000–3000 ^{14}C yr BP). Both areas of research suffer from weaknesses such as problems in chronology, precision, and uneven geographic coverage. Moreover, any attempt at synthesis is hampered by the lack of integration between paleoclimatic and prehistoric studies. Currently, the paleoclimatological data for Mesoamerica suggest that the 4600 year long time span of the Middle Holocene was relatively stable, with no major climatic fluctuations. There is, however, evidence from much of the region for a climatic oscillation (drying) around 5800 cal yr BP. In general, the Middle Holocene was warmer and wetter than the Early Holocene, and wetter and less variable than the Late Holocene. The comparable archaeological records document slow, gradual changes in subsistence, settlement and technology that have been interpreted as incremental adaptations to a slowly changing climatic/biotic/edaphic environment. The prevailing explanatory models of culture change have emphasized ecological and evolutionary processes. However, rates of change are derived from smoothed out and interpolated data and it is possible that the current view of the past is oversimplified simply because the resolution of the data is not high. The authors urge archaeologists and paleoclimatologists to work more closely together in the future so that these important issues may be properly addressed.

1. Introduction

The purpose of this chapter is to explore the paleoclimate and prehistory of Mesoamerica (i.e., the area now divided politically into Mexico and the northern Central American countries) for the Middle Holocene, between approximately 7800 and 3200 cal years ago. Our principal concern is to inquire if any major climatic fluctuations occurred at the onset and/or throughout the Middle Holocene, and, if so, to determine if prehistoric human lifeways exhibited simultaneous and possibly related changes.

In general terms the Middle Holocene (ca. 5800 cal yr BP) is seen as the time when El Niño/Southern Oscillation (ENSO) events became stronger and more frequent (Sandweiss et al., 1996, 1997; Wells and Noller, 1997; Rodbell et al., 1999; Moy et al., 2002; see also DeVries et al., 1997). Significantly, and possibly as a related phenomenon, humans developed economic systems more akin to those of today than to those of their predecessors. That is, humans in many areas transformed their ecological niches in such a way that they became even more dependent upon plants as they made the long journey from gatherers of wild foods to farmers of domesticated plants. While this is a widely acknowledged generality about the Holocene Epoch, we will inquire whether human adaptations during the Middle Holocene were in any way sharply different from those preceding that time period in Mesoamerica.

The Holocene initially was seen as a period of relative climatic stability that occurred after the dramatic changes of the Pleistocene. Early work in the tropics of the northern hemisphere focused on Africa, where many sites showed conditions significantly wetter than today during the Early and Middle Holocene. The moist conditions that spread over large areas of the Sahara (the "green Sahara") were clearly associated with a spread in Neolithic settlements (see Wendorf et al., this volume). Gasse (2000) has reviewed the range of data available for Africa and has demonstrated that this moist period can be bracketed by two dry events. The first and more pronounced occurred between 8500 and 7800 cal yr BP (the 8200 cal yr BP 'event') and corresponds to a pronounced cooling in the North Atlantic shown in the Greenland ice cores, and a second period between 4500 and 4000 cal yr BP that she identifies as the transition into the mid-late Holocene. In Africa, this later dry cool period marked the start of a long trend of increasing aridity. Some sites also show a dry episode between about 7000 and 6500 cal yr BP, which also seems to be registered in records from West Asia. An increase in precipitation associated with a strengthening of the Asian/African summer monsoon in the Early and Middle Holocene is confirmed by ice core data from the Tibetan plateau (Thompson, 2000). However, the pattern of Holocene change emerging for the Americas (Markgraf, 2001) looks rather different from the African pattern and there are clearly very pronounced differences north and south of the Equator (Markgraf et al., 2000).

2. Chronology

Paleoclimatologists and archaeologists use different temporal frameworks to organize the chronology of events in the past and here we have opted to employ both systems of classification rather than choosing one over the other. Paleoclimatologists have adopted the temporal framework of geologists who base their chronology on radiometric dating of sedimentary sequences. Radiocarbon (^{14}C) dating is the most common tool over Holocene timescales. In the literature on paleoclimate, ages may be reported in radiocarbon years, calibrated years BP, or calibrated years BC/AD. Here, we refer to calibrated years before present (cal yr BP) unless otherwise indicated. Paleoenvironmental interpretations for individual sites (based on a range of paleoecological, isotopic, and geochemical proxies) may eventually be combined to yield regional sequences. In archaeology, the basic temporal units are selected regional sequences in the prehistoric cultural record that are anchored by radiocarbon dating. The regional sequences are eventually combined to produce a general chronology for entire culture areas, such as regions or Mesoamerica as a whole. Given that the fundamental time-units are different in each of these chronological systems, it is not surprising that they divide time differently. In Figure 5.1 we show the two temporal frameworks for the period of time of interest to us here.

In the geological time scale, the upper boundary of the Pleistocene Epoch, a time when global climate was cooler and drier than today, is placed at 11,400 cal (10,000 ^{14}C) yr BP (Fig. 5.1), with the lower boundary at 2.5 million years ago. This epoch is followed by the Holocene Epoch, beginning at the Pleistocene–Holocene boundary and lasting until the present. Together these two time-units comprise the Quaternary Period.

Both epochs of the Quaternary Period are subdivided into smaller time-units, but only those of the Holocene Epoch are relevant here. There are many chronologies available for this epoch, since each is based upon interpretations of different regional geologic deposits. Here we will use the terms Early Holocene (11,400–7800 cal yr BP/10,000–7000 ^{14}C yr BP), Middle Holocene (7800–3200 cal yr BP/7000–3000 ^{14}C yr BP), and Late Holocene (3200 cal yr BP/3000 ^{14}C yr BP) until the present.

The chronology of Mesoamerican prehistory (Fig. 5.1) begins with the Paleoindian period, when the earliest people were occupying this region. The Late

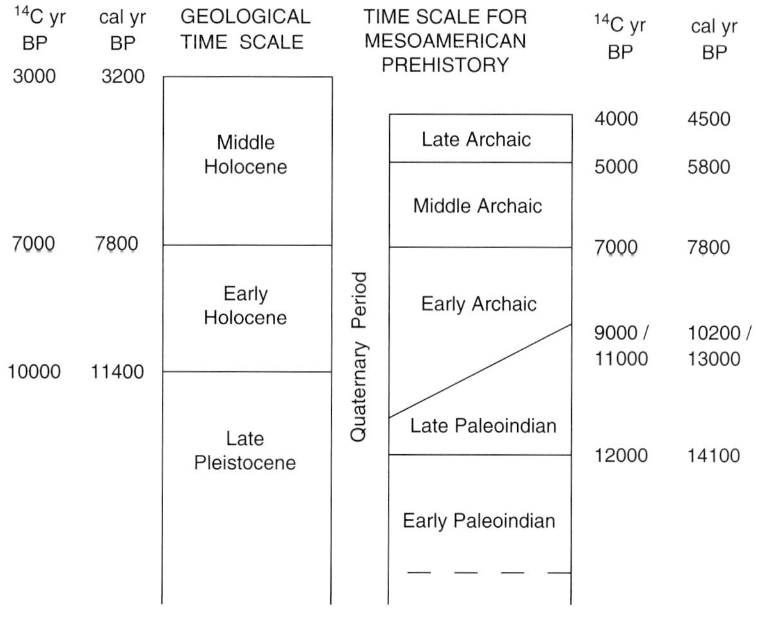

Figure 5.1. The geologic time scale compared with the time scale for Mesoamerican prehistory. Modified from Zeitlin and Zeitlin (2000, Fig. 2.1).

Paleoindian time-unit is thought to begin around 14,100 cal yr BP (12,000 ^{14}C years ago) and to extend until the Early Archaic period, which is variously placed between 13,000 and 10,200 cal years ago (11,000 and 9000 ^{14}C yr BP) (Zeitlin and Zeitlin, 2000, p. 48). The difficulty in setting this boundary stems from different views as to when a predominantly hunting way of life ended. Traditionally in North America, the Paleoindian period ends with the extinction of the horse approximately 10,200 cal BP (9000 ^{14}C yr BP) years ago (Zeitlin and Zeitlin, 2000, p. 71) but in southern Mesoamerica people were engaged in interactions with fully modern plants and animals by as early as 13,000 cal yr BP (11,000 ^{14}C yr BP). The Early Archaic ends at 7800 cal yr BP (7000 ^{14}C yr BP) and is followed by the Middle Archaic, from 7800 to 5800 cal yr BP (7000–5000 ^{14}C yr BP). Finally, the Late Archaic extends from 5800 to 4500 cal yr BP (5000–4000 ^{14}C yr BP). Accordingly, the Middle Holocene (as defined in this volume) coincides principally with the Middle and Late Archaic in the prehistoric chronology of Mesoamerica. This entire span of time, that is, earlier than the close of the Late Archaic Period, is often referred to by prehistorians as the "preceramic". This term actually connotes a developmental stage: a time prior to the use of pottery and by implication settled village life. So, the dominant lifestyle of all peoples during the Mesoamerican preceramic was that of mobile hunter-gatherers.

When radiocarbon dating was first being used by archaeologists working in Mesoamerica it was customary to report ages in their converted BC/AD forms. More recently, the practice has been for archaeologists to use calibrated dates (cal yr BP) that may often be presented as years BC/AD. For consistency and comparability with other chapters in the present work we have converted all dates into calibrated years BP, as we mentioned above. We have done this by calibrating dates that were reported as uncalibrated radiocarbon ages BP, and when warranted by restoring dates expressed as b.c./a.d. or BC/AD to their BP formats. Figure 5.1 contrasts these various dating schemes.

3. Paleoclimate

Paleoclimatic records from Mesoamerica have only recently become sufficiently numerous to attempt some sort of regional synthesis. Much of the impetus for this attempt has come through the International Geosphere Biosphere Programme (IGBP) Past Global Change Project (PAGES) and its sponsorship of a review of paleoclimatic data along a series of north–south transects called Pole-Equator-Pole (PEP). The PEP1 transect covered the Americas and focused interest on patterns of change in the Neotropics. Markgraf (2001) has summarized the work of PEP1. In the context of Mesoamerica, the chapters by Grimm et al. (2001) and Fritz et al. (2001) in the PEP1 volume provide summaries of our understanding of palynological and limnological records.

Pioneering work in Central America by Deevey (e.g., Tsukada and Deevey, 1967) and Martin (1964) revealed the challenges of attempting to reconstruct patterns of climate change in this region. In the limestone lowlands of the Yucatán peninsula (Mexico, Belize, Guatemala) late glacial and Early Holocene records show the influence of sea level change, while the problems of hardwater error in radiocarbon dating in such environments are well known (e.g., Vaughan et al., 1985). The ability to date very small organic samples using AMS has led to a considerable refinement of chronologies from these limestone areas (e.g., Islebe et al., 1996). Across the whole Mesoamerican area, long term and often intensive anthropogenic impacts also have undermined efforts to reconstruct climate (e.g., Leyden, 2002). A number of pollen records from across the region have been interpreted purely in terms of human impact (e.g., Rue 1987; Northrop and Horn, 1996; Behling, 2000). Although the magnitude of human impact was undoubtedly less for the period of interest here (7800–3200 cal years ago) than it became in the Late Holocene, it is the case that almost all records reflect some combination of climate and human disturbance. The study by Lozano-García and Vázquez-Selem (2005) from a site at 3860 m on the flanks of the volcano Iztaccíhuatl in Mexico, is an exception to this.

Perhaps the most unequivocal records of climatic change (specifically the balance between precipitation and evaporation) have come from oxygen isotope analyses of freshwater carbonates (bulk sediment or specific carbonate organisms such as gastropods). In tropical lake systems, enhanced evaporation increases the proportion of the heavier isotope ^{18}O relative to ^{16}O and this proxy can provide very clear records of periods of drought (e.g., Hodell et al., 1995; Curtis et al., 1996). Work by Rosenmeier et al. (2002), however, has shown that even this signal may be confounded by the interactions between vegetation change and catchment hydrology.

In general terms the available paleoclimatic records can be considered in three broad groups: the lowlands of the Yucatán peninsula (IIIA), the highlands of Central America (IIIB), and the highlands of the Trans-Mexican Volcanic Belt (IIIC).

3.1. Lowlands

The application of improved dating techniques and the suitability of sediments for $\delta^{18}O$ has meant that the paleoclimate records from the Yucatán lowlands are among the clearest available to us. The Late Pleistocene and the earliest Holocene were apparently cool and dry (Leyden et al., 1994). The pollen record from Lake Quexil (Guatemala) indicates that temperatures may have been 6.5–8°C colder than today at the peak of the last glaciation and that conditions were extremely arid. The general warming of the late glacial was apparently interrupted by a period of cooling (ca. 14,100–12,100 cal yr BP/12,000–10,300 ^{14}C yr BP) that may correspond with the Younger Dryas (Leyden, 1995). In spite of variations in temperature, moisture availability continued to increase and by the Early Holocene (ca. 9550 cal yr BP/ 8600 ^{14}C yr BP) the climate of lowland Guatemala may have been wetter than it is at present (Islebe et al., 1996; Wahl et al., 2006). The increasing moisture of the early Holocene was apparently interrupted by short, drier episodes (Hillesheim et al.,

2005), possibly correlated with cooling events in the North Atlantic (see above). The records from the western (Mexican) part of the Yucatán seem to indicate that the onset of moist conditions occurred somewhat later (after about 8950 cal yr BP/8000 ^{14}C yr BP). The record from Cenote San José Chulchacá in northern Yucatán indicates conditions wetter than present by about 7700 cal yr BP (Leyden et al., 1996). Generally moist conditions are recorded across the region for most of the Middle Holocene, although there are indications of some drying between 6900 and 5800 cal yr BP at San José Chulchacá and possibly in Lake Petén-Itzá, although in the latter case the change in vegetation may be due to human impact (Islebe et al., 1996). Early human impact has also been recorded for Lago Puerto Arturo, in the northern Petén, where Wahl et al. (2006) found *Zea* pollen at 4600 cal BP, although they note that there are also indicators of a drier climate for the period 4700–3400 ca. BP. After 3300 cal yr BP the climate dried, with periods of intense drought occurring in the Late Holocene (e.g., Curtis et al., 1996). The published records from Belize are sparse and of low resolution, but an unpublished study (Metcalfe et al., in preparation) shows the same pattern of Late Pleistocene–Holocene change recorded elsewhere across the Yucatán lowlands. This study also importantly illustrates how individual basins could show a very different response to the same climate forcing depending upon their hydrology. The Late Holocene drying, which is so pronounced in some records, hardly appears at all in a system that is hydrologically well buffered.

One of the major challenges in the Yucatán lowlands has been getting reliable radiocarbon dates. Vaughan et al. (1985) report that 29 radiocarbon dates of Petén lake sediments had been rejected as meaningless, with only one date on terrestrial wood from the base of a core from Lake Quexil being accepted as credible (~9450 cal yr BP/8410 $\pm$ 180 ^{14}C yr BP). Four other Holocene dates on this core were rejected. Covich and Stuiver's (1974) paper on the oxygen isotope record from Lake Chichancanab is based on four dates. Only two of these were on organic matter and the other two on marls. A correction for hardwater error (-270 years) was applied to the latter two dates based on the measured ^{14}C in the lake carbonates. This process meant that there were three dates covering the Holocene portion of the core. The study of Lake Chichancanab by Hodell et al. (1995) has a chronology based on five dates on terrestrial material and nine on the shell material of gastropods and bivalves. The comparison of aquatic and terrestrial ages indicated that modern hardwater lake error was about 1200 years. Hodell et al. (2001, 2005) report a much more tightly constrained chronology for the last 2400 years.

In some cases, it has been possible to provide independent verification of the age of the most recent sediments using ^{210}Pb dating. The core from Petén Itzá described by Islebe et al. (1996) and Curtis et al. (1998) uses ^{137}Cs, ^{210}Pb, and ^{14}C to provide a chronology for the Holocene. The authors have eight ^{14}C ages on terrestrial wood and charcoal for the period between ca. 9900 and 50 cal yr BP (8840 and 75 ^{14}C yr BP), although the youngest date was rejected when compared with the ^{210}Pb record.

This Petén Itzá core probably is the best dated Holocene sequence for the region, but as Curtis et al. (1998) discuss, the interpretation of the different environmental proxies, in terms of either climatic change or human impact, is far from straight forward. In the context of the present chapter, it is interesting to note that they identify the period 8150–5600 cal yr BP (7300–4800 ^{14}C yr BP) as the Middle Holocene, although the end date for this span is apparently pinned to one of the radiocarbon dates. The Late Holocene is marked in the Petén Itzá pollen record by a gradual increase in disturbance taxa and a decrease in lowland forest taxa (up to about 2950 cal yr BP), although other proxies show little change over this period. Around 2950 cal yr BP there seems to have been a period of major deforestation across the Petén lake district with a significant increase in erosion in these catchments.

There are now a number of sites in the Yucatán lowlands that have high-resolution proxy data, and the combination of more traditional methods (such as pollen analysis) with the results from C and O isotope studies revealing new interactions within these systems. The basic climatic pattern seems to be quite well established, with relatively little change over the period defined here as the Middle Holocene (7800–3200 cal yr BP). Major climatic variability seems to have occurred in the Late Pleistocene/Early Holocene and again in Late Holocene. The occurrence of a severe dry period spanning ca. 1700–900 cal yr BP (ca. AD 250–1050), with a major peak around AD 860, corresponding to the time of the Classic Maya collapse (AD 800–900) has attracted a lot of interest (e.g., Hodell et al., 1995; Curtis et al., 1996; Hodell et al., 2001, 2005). It is clear, however, that the apparent magnitude of this dry episode, or at least the severity of its impact, was highly variable spatially across the lowlands (Curtis et al., 1998; Metcalfe et al., in preparation).

3.2. Central American highlands

Records from the Central American highlands come mainly from the Cordillera de Talamanca in Costa Rica (e.g., Hooghiemstra et al., 1992; Horn, 1993; Islebe and Hooghiemstra, 1997). These confirm cold (possibly 8°C colder than present) and dry conditions at the last glacial maximum, with a late glacial cooling episode equivalent to the Younger Dryas. The transition into post-glacial conditions seems to have occurred relatively early here, by 12,600 cal yr BP at La Chonta (Islebe and Hooghiemstra, 1997). The early Middle Holocene was humid, although there are some signs of slightly drier conditions around 5800 cal yr BP. Although there have been suggestions that the pollen records from highland Costa Rica might indicate temperatures warmer than present in the Middle Holocene, Horn's record from the Lago de las Morrenas at 3480 masl provides no clear evidence for this, as *paramo* vegetation dominated throughout. In common with the records from the lowlands, the highland pollen records show that the Late Holocene has been characterized by drier conditions.

3.3. Highlands of the Trans-Mexican Volcanic Belt

The lake basins of the Trans-Mexican Volcanic Belt have been the focus of much research, but the climatic signals that have emerged, frankly, have been rather confusing. As the geology is predominantly volcanic, there are fewer problems with hardwater error than in the limestone region, but the effects of volcanism, tectonism, and long-term anthropogenic modifications have all left their mark. Published records covering the whole of the Holocene have come mainly from the Basin of Mexico and the states of Mexico and Michoacán to the west. The records from Central Mexico have been reviewed in some detail by Metcalfe et al. (2000) and Metcalfe (2006) and will only be summarized here. Conditions in the Basin of Mexico around the last glacial maximum seem to have been cool (or cold) and dry with layers of calcareous "caliche" being deposited in Lake Tecocomulco in the north of the basin by 18,000 cal yr BP (Caballero et al., 1999). The climate apparently became wetter in the late glacial, but much of the basin seems to have been affected by intense volcanic activity and the paleoecological records are often difficult to interpret. The diatom record from Lake Chalco in the south of the basin indicates that the Early Holocene was drier than the late glacial. The evidence for the Early Holocene is confusing with pollen data suggesting that conditions were wetter than present in the mountains (Grimm et al., 2001), while diatom and sedimentological records from the basin floor clearly indicate dry conditions (lakes Texcoco and Tecocomulco were apparently dry over much of this period). Pollen data from the Tlapacoya archaeological site in the south of the basin apparently indicated conditions both warmer and wetter than present between about 8400 and 5650 cal yr BP (González Quintero, 1986), but more recent data do not seem to support this interpretation. Lake levels apparently recovered in the south of the basin after 5800 cal yr BP and in the north after 3900 cal yr BP. Widespread human occupation seems to be recorded by pollen from about 5800 years ago and recent drainage of the basin and subsequent deflation of sediments means that much of the Late Holocene record has been lost. Lozano-García and Vázquez-Selem's (2005) high altitude pollen record indicates that cool conditions persisted until about 7200 cal BP when coniferous forest reached to, or close to, their study site. They see evidence of mid-Holocene drying, but no major change in the pollen spectra over the last 3000 years in contrast to most sites where the anthropogenic signal is strong.

The climatic history reconstructed for Lake Pátzcuaro (Michoacán) looks rather different from the Basin of Mexico (Watts and Bradbury, 1982; Bradbury, 2000; Terrett, 2000; Metcalfe et al., 2007). Here, the full glacial seems to have been cool and moist, with somewhat drier conditions in the late glacial. The Holocene has been marked by lower lake levels compared with earlier and there was some indication of slightly drier conditions around 5800 BP. The Late Holocene record shows great variability; based on the analysis of a number of short cores from across the basin, the results of O'Hara et al. (1993), Bridgwater et al. (1999), and Metcalfe et al. (2007) confirm pronounced swings between wetter and drier conditions over the last 3300 years. Lake Pátzcuaro shows clear evidence of human impact from at least 3950 years ago and there have been a number of phases of severe soil erosion over this period (O'Hara et al., 1993). The core described by Watts and Bradbury (1982) and by Bradbury (2000) only had three radiocarbon dates (bulk) within the Holocene. O'Hara's cores provided eight additional dates for the Holocene after 4000 cal yr BP; new cores have improved the chronological control for the Holocene in this basin. Results from Holocene sediments in the nearby Zacapu lake basin show fairly moist, stable conditions in the early Middle Holocene with some indications of drying around 5800 cal yr BP. Conditions then seem to have got wetter again until the Late Holocene. This site probably has the greatest number of Holocene dates of any of the studied lakes of the Trans-Mexican Volcanic Belt. There are 24 AMS dates covering the last 3800 years on cores from the modern Laguna Zacapu (Metcalfe, 1995; Leng et al., 2005) and 18 additional Holocene dates from sites on the drained floor of the main basin (Petrequin, 1994). Unfortunately, there is little overlap between the basin floor sequences and the modern lake core as the peaty sediments of the basin floor have suffered severe deflation since the area was artificially drained. Pollen (Xelhuantzi-Lopez, 1994) and diatoms (Metcalfe, 1995) have been the major sources of paleoenvironmental data for this basin. Both proxies indicate that there was only small-scale variability through the Holocene, but the record seems to have been disrupted by episodes of volcanic activity (e.g., between 7900 and 7500 cal yr BP) and possibly by tectonic uplift originating to the west of the basin. The diatom record from the lake shows substantial changes in diatom composition, but they are not easy to interpret. Only the bottom 4 m of a 14 m core lie within the Middle Holocene as defined here, with signs of somewhat shallower, warmer conditions between about 4200 and 3100 cal yr BP. The strongest climatic signal is in the Late Holocene, with a marked drying around 1050 cal yr BP.

The number of records from the Trans-Mexican Volcanic Belt has increased quite dramatically over the last decade, but a clear pattern of events has yet to emerge for either the Late Pleistocene or Holocene. There are signs that conditions across the region differed significantly in the Late Pleistocene, being quite wet in the west and dry in the east. Bradbury (1997) has suggested that the degree of penetration of mid-latitude frontal systems from the Pacific might explain this situation. Some of the records published in the 1980s from this region seemed to indicate that conditions became wetter than present in the early Middle Holocene (and probably warmer), but more recent studies have not confirmed this interpretation. Across the west central highlands there are a number of sites that show a dry episode around 5800 cal yr BP, although this event does not seem to have been as severe as those that occurred in the Late Holocene. Human impact is apparent in many records, either directly in terms of vegetation change and increases in sediment fluxes from catchments into lakes, or indirectly. Draining of a number of wetland areas (for example in the Basin of Mexico) has resulted in the loss of later Holocene sediments. There has been very limited application of stable isotope methods in the central highlands (partly because the sediments are often unsuited to conventional methods based on carbonates). We hope that the application of stable isotope analysis based on ^{18}O in diatom silica will assist in the separation of climatic from anthropogenic signals in this region (e.g., Leng et al., 2005).

3.4. Discussion

Overall, the paleoclimatic data available for the Middle Holocene (7800–3200 cal yr BP) seem to reflect warm and moist conditions across much of Mesoamerica. The early Middle Holocene seems to have been wetter than at present in the southern part of the region. It seems likely that these moist conditions were driven by changes in insolation, with a summer maximum driving enhanced summer rainfall in the Early Holocene (Leyden et al., 1996). In the Yucatán the effects of this became most evident when sea levels and regional groundwater tables had risen. There is limited evidence for this early Middle Holocene peak in available moisture in the highlands of Central Mexico.

There is some evidence for drying across Mesoamerica around 5800 cal yr BP, but aridity was much more pronounced farther north in the interior of the North American continent where the driest conditions of the Holocene were recorded between about 7900 and 5800 cal yr BP (Fritz et al., 2001). Drier and more variable climatic conditions seem to have set in across Mesoamerica from about 3300 cal yr BP, the available evidence indicates that in this tropical area the driest part of the Holocene occurred about 1000 cal yr BP. The evidence for drought is clearest for sites in the Yucatán lowlands, but there are some indications of aridity in basins of the Trans-Mexican Volcanic Belt.

Across Mesoamerica, there has been much interest in the potential interactions between climate and society. The region is prone to drought and many of the landscapes are vulnerable to disturbance, whether natural or anthropogenic. The main focus so far has been on the role of drought in bringing about the "collapse" of the Maya (e.g., Brenner et al., 2001; deMenocal, 2001). Paleoclimatic data from Lake Chichancanab (Hodell et al., 1995, 2005) and then from nearby Lake Punta Laguna (Curtis et al., 1996) were key in establishing that at least the cultural change in the Maya area occurred at the time of the most severe drought of the Holocene. Hodell et al. (2001) have looked to solar forcing to explain some of the century scale variability in Yucatán droughts.

The relationship between climatic change and settlement histories in earlier periods has not been easy to establish. In the Maya lowlands, the drying around 3300 cal yr BP may have provided conditions more favorable to sedentary populations (e.g., Rice and Rice, 1990) compared with earlier. It seems, however, that in the majority of cases there is little coherence between the available paleoclimatic data and the early archaeological sites. A study of Lago Catemaco in the Olmec region of the Sierra de los Tuxtlas in Veracruz (Byrne and Horn, 1989) was beset with dating problems. Radiocarbon dates indicating that the sediments went back more than 6900 years were rejected as being anomalously old and a tentative chronology based on tephra layers suggested that the base of the core was less than 3000 years old. As mentioned above, the records from the Basin of Mexico that should cover the main period of occupation are often severely perturbed and difficult to interpret (see Niederberger's (1987) study of the Basin of Mexico prior to the development of Teotihuacán). Coastal sites such as the Laguna de Cocos in northern Belize (Bradbury et al., 1990; Hansen, 1990) reflect complex interactions of marine, fluvial, and terrestrial systems such that direct climatic signals can be difficult to identify. More direct collaboration between the paleoclimatic and archaeological communities would be valuable in addressing the important issue of the associations between environment and society.

4. Prehistory

It is still far from clear to archaeologists when humans first arrived in the Americas, a situation that has provoked a "seemingly endless debate" on the subject (Zeitlin and Zeitlin, 2000, p. 49). In brief, there are a handful of archaeological sites throughout the Americas (see Lorenzo and Mirambell, 1999; Roosevelt, 2000; Zeitlin and Zeitlin, 2000 for recent summaries) that have been proposed to support the view that human entry into the New World occurred prior to 12,500 radiocarbon (15,100 cal) years ago. Critics of this view consider the proffered evidence as falling short of the high standards that they demand. Nevertheless, evidence appears to be accruing that humans reached the Americas some time between 35,000 ^{14}C (no calibration available) and 12,000 ^{14}C (14,100 cal yr) years ago (e.g., Zeitlin and Zeitlin, 2000, p. 62). Although the earliest proposed date for the first Americans is not universally accepted, there is a consensus among archaeologists that just prior to the end of Pleistocene, around 14,100 cal yr BP, people occupied both the American continents. We begin our overview by briefly discussing these ancient people.

The Late Pleistocene people, called Paleoindians by archaeologists, are known principally by their uniquely designed stone spear points and by the evidence that their subsistence included predation on Pleistocene megafauna. In fact, most of the known archaeological sites that are the signatures of these ancient people are places where large game animals were slaughtered and butchered. Such kill sites are especially well known from the North American plains, but one particularly important location is Santa Isabel Ixtapan in the Basin of Mexico (Fig. 5.2), where two mammoths were dispatched and butchered at the edge of the Pleistocene Lake Texcoco. Also, projectile points have been found on the surface in many places in northern Mexico (e.g., Zeitlin and Zeitlin, 2000, p. 64; Sanchez, 2001; Carpenter et al., 2005), in the Mexican central highlands and even farther south into Central America. These archaeological finds – kill sites and isolated surface discoveries of projectile points – do not provide the data needed to reconstruct ancient subsistence practices, so the role of big game in the Paleoindians' overall diet remains unknown. That is, it continues to be unclear if the Paleoindians were heavily dependent upon

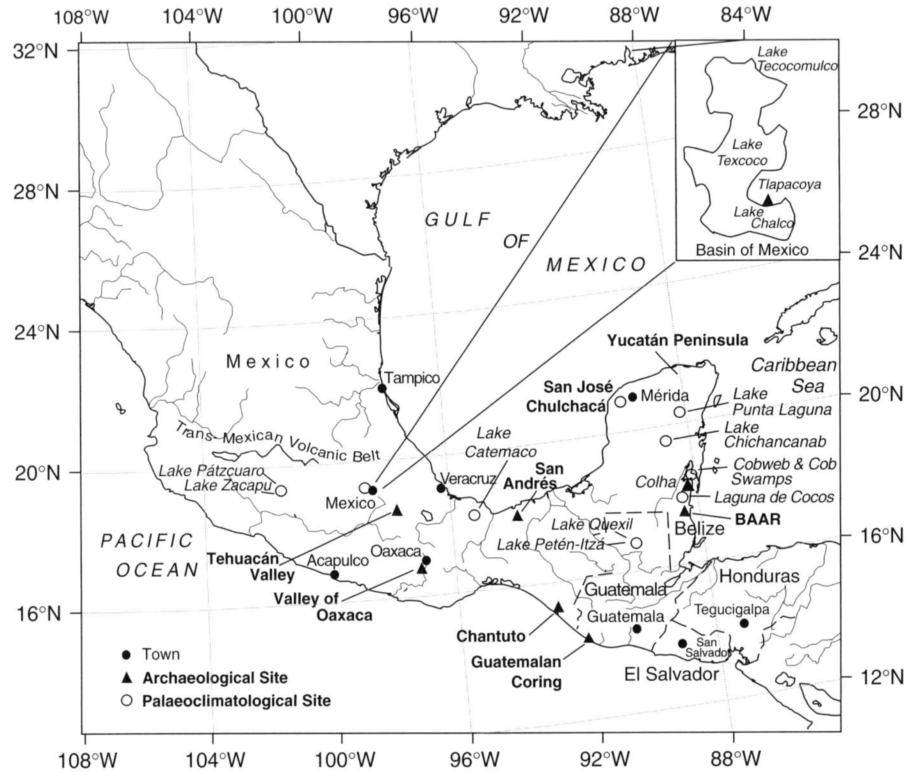

Figure 5.2. Map of Mesoamerica showing modern political boundaries, research sites, and other areas mentioned in the text.

large game, similar to people living in more recent times in high-latitude environments (e.g., the Sami, Inuit etc.), or whether these ancient people had a mixed subsistence economy of which only hunting is now archaeologically visible.

4.1. Uplands

As the Zeitlins have noted recently (Zeitlin and Zeitlin, 2000, p. 68), evidence from Mexico challenges the traditional view that Paleoindians were big-game hunters whose diet consisted principally of meat from large game animals, at least at the end of the Late Paleoindian period. This contravening evidence comes from a large, multi-disciplinary study led by Richard S. MacNeish, which focused on the prehistory of the semi-arid Tehuacán Valley in south central Mexico at the edge of the Trans-Mexican Volcanic Belt. The Tehuacán Archaeological-Botanical Project was regional in scope and spanned approximately 12,000 years (i.e., estimated as beginning at approximately 10,000 B.C.; Johnson and MacNeish, 1972, p. 18) of human occupation in the valley. It was strikingly ambitious: by the end of the project the team had investigated 456 archaeological sites (MacNeish et al., 1972, p. 344), about 20,000 non-ceramic artifacts (MacNeish et al., 1967, p. 5), 100,000 macrobotanical remains, 11,000 faunal remains, and over 100 human feces (MacNeish, 1967, p. 290).

The oldest occupation was assigned to the Ajuereado Phase, the beginning of which was not firmly dated but which ended at 9700 cal yr BP (6800 ^{14}C years BC) (Johnson and MacNeish, 1972, p. 5). In the early part of the Ajuereado Phase, at approximately 14,000 cal yr BP (10,000 ^{14}C BC), the valley was more arid and cooler than it is today, resulting in a steppe-like environment. This climatic inference, which accords with other regional paleoclimatic data, is based upon a study of ancient owl pellets. These are regurgitated cocoons of rodent hair and skeletons that are the undigested remains of owls' meals. The contents of the pellets are thought to reflect the proportions of small rodents present in an owl's diet and by extension in the surrounding environment (Flannery, 1967, p. 40ff). The small rodent population of Early Ajuereado Phase cave deposits was strikingly different from the populations of rodents dating from more recent cave deposits in the Tehuacán Valley, suggesting a very different climate at that time compared to more recently. In Flannery's (1967, p. 144) view, the Terminal Pleistocene paleoenvironment of the Tehuacán Valley resembled the very arid interior plains found today in northern Mexico and in southern Texas. These open plains, which are subject to winter frosts, are suitable habitats for the kinds of animals found in the Early Ajuereado Phase deposits, including antelope and jackrabbits, two animals with a particularly high presence in the archaeological deposits.

Most of the faunal remains recovered from Ajuereado Phase deposits that come from small- to medium-sized fauna, but fossil horse and antelope were also taken (Flannery, 1967, p. 140) by hunters using bifacially flaked projectile points of the sort believed to indicate big-game hunting (Zeitlin and Zeitlin, 2000, p. 68). Despite the technological and skeletal evidence of big-game hunting, the record is clear that smaller animals and plants greatly prevailed over the larger sized game in the overall diet. Flannery has calculated that horse and antelope constituted less than 10% of individual animals for this time period (Flannery, 1967, pp. 158, 170). Unfortunately, no one has generated estimated biomass analyses in order to compare the amount of meat represented by large vs. small animals in these deposits. Such an analysis would provide a more accurate assessment of the relative importance of big vs. small game than would simple counts of individual animals represented by each taxon. Nevertheless, the data are unambiguous that Late Pleistocene inhabitants of the Tehuacán Valley ate significant amounts of small game. Rabbits, especially jackrabbits, were particularly important, along with a wide variety of other animals that may have been trapped opportunistically. The abundance of rabbits suggests to Flannery that the late Paleoindians employed communal rabbit drives, in addition to their big game hunting techniques.

By at least 9500 cal yr BP (6500 ^{14}C yr BC), however, the fauna in the Tehuacán Valley had become completely modern (Flannery, 1967, p. 144ff) and the ecofacts recovered from archaeological sites, along with evidence from owl pellets recovered from cave deposits, suggest that the climate became more similar to that of today.

That is, the faunal evidence used to make inferences about paleoclimate indicates a fairly constant climatic regime beginning about 10,150 cal yr BP (7000 ^{14}C yr BC) and extending until the present. However, as Flannery (1967, p. 144) recognizes, faunal studies are a poorer proxy for paleoclimatic reconstructions than other standard methods such as pollen studies, and he laments that pollen was not preserved at the studied archaeological sites.

At this time in the record white-tailed deer filled the ecological niche formerly occupied by fossil horse and antelope (Flannery, 1967, p. 144). Deer are better adapted to the dense thorn forest of the valley slopes that developed in the Holocene than are horse and antelope. Also, modern forms of small animals such as fox and turtle replaced their Pleistocene counterparts. Once the modern climate regime was established in the valley, the climate appears to have remained fairly constant until the present. This inference is based, of course, upon a rather coarse-grained analysis, using the record of owl pellets from cave deposits, as well as plant and animal remains from archaeological deposits, as proxies for paleoclimate. These proxy methods are unable to detect short-term climatic variations, but they do provide compelling evidence for *overall* climatic stability during the Holocene.

Despite the evidence for a climatic change with its consequent change in plant and animal assemblages in the Tehuacán Valley at the Pleistocene–Holocene boundary, the tool assemblage of the Early Ajuereado Phase continues unchanged into the Late Ajuereado Phase. This continuity is somewhat unsettling to archaeologists because it implies that they cannot use tools to make reliable inferences about specific tool functions. Tool assemblages do change over time, of course, but this progression appears to be gradual in the Tehuacán sequence (cf. Zeitlin and Zeitlin, 2000, p. 75).

MacNeish and his collaborators also propose a gradualistic model for other aspects of the prehistory of the Tehuacán Valley. For example, based upon the analysis of food remains in many sites, the role of animals in the diet of Tehuacanos diminished steadily from 10,150 cal yr BP (7000 ^{14}C yr BC) until the arrival of the Spaniards (Fig. 5.3), whereas plants increased concomitantly (MacNeish, 1967, p. 301). Likewise, the change from wild to domesticated plants seems to be similarly gradual through the same time period, rather than exhibiting a more punctated or stepwise developmental pattern. As we discussed above, however, coarse dating resolution would preclude the detection of a stepwise trend if it existed.

Since the publication of the landmark Tehuacán Valley study, the gradualistic model of development has come to dominate archaeologists' thinking about the prehistory of Mesoamerica. For example, this model has been used to characterize early culture change in the Valley of Oaxaca, which was the focus of another ambitious archaeological project, this time directed by Kent V. Flannery. The long cultural sequence in the Valley of Oaxaca effectively begins with the Naquitz Phase (12,900 to 9550 cal yr BP/8900 to 6700 ^{14}C yr BC) (Flannery, 1986b, p. 38), coinciding with the Early Holocene, although there are hints of an earlier Late Pleistocene human presence in the valley (Marcus and Flannery, 1996, p. 45). Smith (1986, p. 265) identified over 20,000 plant remains from deposits dating to this

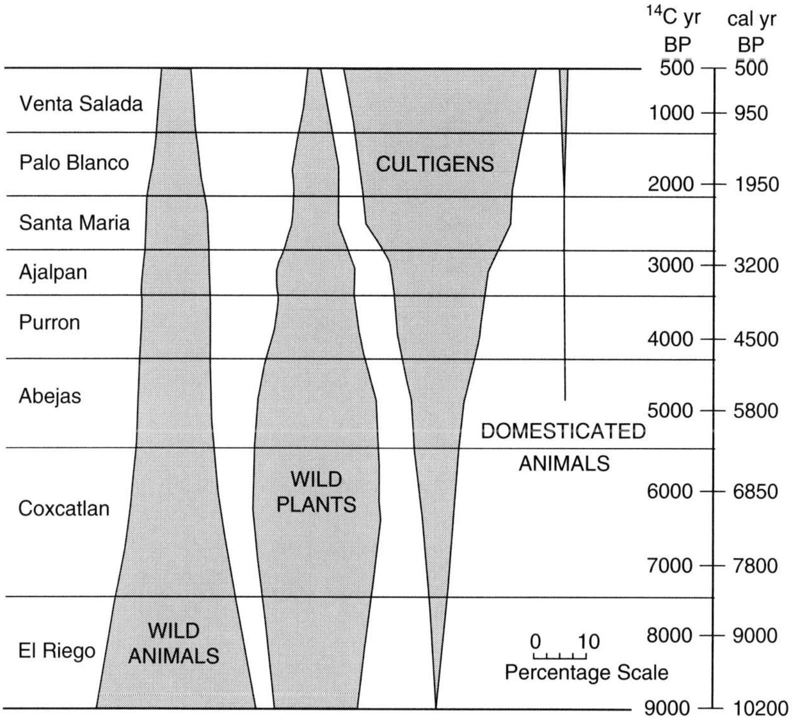

Figure 5.3. Percent of subsistence remains contributed by wild animals, wild plants, cultigens, and domesticated animals over time in Tehuacán Valley sites. Redrawn from MacNeish (1967, Fig. 186).

phase in the Guilá Naquitz cave. These remains document the use of a wide variety of wild plants, with an emphasis on acorns. Domesticated squash (*Cucurbita pepo*) was present, and pollen from domesticated corn (*Zea mays*) was identified in one sample. Beans (*Phaseolus* sp.) were well represented but still phenotypically wild (Flannery, 1986a, p. 315). These three taxa are especially important because eventually they formed the three legs of the traditional Mesoamerican diet in later, agricultural times. Nevertheless, they contributed only 4% of the plant remains by raw count in the analyzed deposits dating to the Naquitz Phase (Flannery, 1986a, p. 315). Animal remains include deer, cottontail rabbit, peccary, raccoon, turtle, and various birds, all fully modern fauna that exist in the area today.

Flannery identified changes in the frequencies of five key plants in the archaeological deposits: acorns, piñon, *susí* (a member of the spurge family), mesquite, and hackberry. The first three plants grow in the Thorn Forest on the valley slopes, whereas the last two plants listed are from the Mesquite Grassland near river drainages. Over time, mesquite remains increased greatly in frequency, whereas the other four plants declined (Flannery, 1986a, p. 316). Although it might be inferred that this cultural change was simply the result of either the spatial expansion of the Mesquite Grassland or the increasing preference for exploiting that particular biotic zone over time, Flannery believes that either view is overly simplistic because hackberry, another Mesquite Grassland plant, simultaneously decreases in importance. Flannery and his colleagues prefer the explanation that the observed changes in the use of plant taxa reflect scheduling decisions that were linked to a gradual increasing commitment to agriculture. This model considers the edaphic conditions of the Mesquite Grassland superior for agriculture compared with those of the Thorn Forest (Flannery, 1986a, p. 503). This would explain ancient Oaxacans' gradual shift away from the Thorn Forest and toward the Mesquite Grassland of the alluvial valley floor. In his model, Flannery combines the spatial component involved in reduction of search area with a selection process that narrowed the range of exploited plants (a narrowing of dietary breadth) during the transition from economic dependency upon wild to domesticated plants.

Flannery's model of how and why the ancient people of Oaxaca changed their lifestyle is grounded in ecological/evolutionary principles because his main focus, like that of MacNeish, is to investigate the origins of plant domestication and the growth of agroeconomies (i.e., farming systems) in the region. He subscribes to the process of coevolution similar to that articulated by Rindos (1984) in which changes in one population, for example, a species of plant with economic value for foraging humans, permit adaptive changes in the human populations that use them. This evolutionary approach linking biological genetic evolution to cultural processes of change has strong explanatory power and nicely fits the archaeological data.

In the final chapter of the Guilá Naquitz volume, Flannery (1986a, p. 506ff) reflects upon the different outcomes of a model that stresses human ecological adaptation, compared to those of a model that splices the archaeological record into time segments based upon changing styles of artifacts. He observes that archaeologists always use linear temporal units (e.g., phases and periods) in organizing their chronologies, as indeed was done both in the Tehuacán Valley and Oaxaca Valley studies. In these two studies, as with others on Paleoindians of the Americas, projectile point styles define regional phases. However, this practice of formulating a sequence of phases that bracket segments of time by its very nature tends to emphasize discontinuities rather than continuities in prehistory.

The Oaxacan preceramic sequence does not end with the Early Archaic but continues on through the Middle and Late Archaic periods. Unfortunately, however, the archaeological data from these later times are less robust, and therefore not equivalent to those from the Early Archaic. The Middle Archaic Jícaras Phase is known only from one open-air campsite (Gheo-Shih) on the valley floor where neither animal bones nor macrobotanical remains were preserved. These preservational problems preclude any reliable reconstruction of the subsistence base at the campsite. The site also lacked charcoal for radiocarbon dating. Cross-dating artifacts with those from sites with known ages provided the approximate age estimate of 7800–4500 cal yr BP (5000–4000 ^{14}C yr BC; Flannery, 1986b, p. 38). *Zea* pollen was present, which suggested to the excavators that maize horticulture might have been carried out during the wet season. The site is inferred to have been a seasonal macroband encampment.

Finally, Late Archaic Period occupations in Oaxaca are known from several sites: a winter deer-hunting camp, a probable maguey roasting site, and a rock shelter that served as a short-term plant processing station. The functional inferences about these sites are drawn primarily from the nature of the respective tool assemblages, which are not yet published in detail.

Lately, investigators have been increasingly turning to analyses of sediment cores to address questions about early human–plant interactions, especially the onset of plant cultivation and dependency on agroeconomies. One on-going study of particular interest in the Mexican uplands is in the Central Balsas drainage, a region now thought to be where maize was domesticated and perhaps early agriculture took root. This region, in the western state of Guerrero, is currently being studied by Piperno and her colleagues. They are finding that intensive forest clearing has considerable antiquity and that by four thousand years ago the landscape had been modified substantially by anthropogenic fires used to clear land for cultivation (Piperno, 2006). The evidence for forest clearing is accompanied by the presence of maize pollen.

4.2. Lowlands

Archaeologists have also adopted a gradualistic model as the dominant paradigm to explain changes in the archaeological record of the Mesoamerican lowlands. However, the quality and quantity of archaeological data from the lowlands still lags behind that of the highlands. Archaeologists confront substantial difficulties in studying preceramic occupants of the lowlands because of low site visibility and poor organic preservation at open-air sites, which until recently have been the most frequently investigated site type. Lately, however, we have made considerable strides in our understanding of the preceramic on both the Atlantic and Pacific coasts.

On the Atlantic side, the most intensive efforts to study the preceramic have been carried out in northern Belize. For example, investigations of the lithic industry at the Classic Maya site of Colha revealed preceramic deposits predating the earliest Maya presence at approximately 2850 cal yr BP (900 cal yr BC) (Andrews and Hammond, 1990, p. 580). Soil humates date the earliest of two preceramic deposits to 5350–4850 cal yr BP (3400–2900 cal yr BC) (Iceland, 1997, p. 11), that is, during the Late Archaic Period. This deposit was interpreted as a quarry-production locale (Iceland, 1997, p. 94) for a chert core-blade industry. Projectile points, considered to be possibly Lowe style, were produced at this location (Iceland, 1997, p. 134). Unfortunately, the lithics in this early deposit had no associated features or organic remains (Iceland, 1997, p. 11). A younger occupation, also designated as "preceramic", follows the Late Archaic Period at Colha (Iceland, 1997, p. 11). This later "preceramic" occupation is too recent to concern us here (3450–2850 cal yr BP/1500–900 cal yr BC).

MacNeish and colleagues mounted an important regional project (called BAAR) in present day Belize where Paleoindian and Archaic projectile points had previously been collected from the surface. The research team found at least nine preceramic sites in northern Belize and excavated six of them. They used stylistic variations of stone tools to define three preceramic phases, but the lack of datable materials prevented the establishment of an absolute chronology (Lohse et al., 2006; Zeitlin and Zeitlin, 2000, p. 86). In MacNeish's interpretation, the three-phase sequence spanned the entire Early to Late Archaic Period. Other investigators (e.g., Kelly, 1993; Lohse et al., 2006), however, think that the archaeological assemblages are much more recent than the ages ascribed to them by MacNeish. Kelly (1993, p. 215), for example, argues that the specific projectile point type diagnostic of MacNeish's earliest phase (Lowe point) now has associated radiocarbon dates that suggest this spear point was in use between 2500 and 1900 years B.C. (ca. 4450–3850 cal yr BP). This inference is bolstered by Pohl and her colleagues who dated a Lowe projectile point at Pulltrouser Swamp to 4160 cal yr BP (reported as 2210 cal yr BC) (Pohl et al., 1996, p. 363). This places the points and their associated material within the Late Archaic (for an additional discussion see Zeitlin and Zeitlin, 2000, p. 87), rather than the Early Archaic as MacNeish inferred. A recent reevaluation by Lohse and colleagues (2006) place the entire known Archaic record of Belize between 3400 and 900 B.C.

Both the Colha and BAAR projects produced data that indicate people were present in northern Belize during the Late Archaic Period. Recent work at wetland sites employing standard archaeological excavation techniques and the extraction of sediment cores has confirmed and amplified this result. Cultigens, including maize and manioc, have been found in pollen assemblages from Cobweb swamp (Jones, 1991, 1994) and at Cob swamp (Pohl et al., 1996, p. 363) where they occur prior to 4950 cal yr BP (cited as 3000 cal yr BC) and perhaps as early as 5350 cal yr BP (cited as 3400 cal yr BC). Evidence for massive forest clearing in the vicinity of Cob swamp appears around 4450 cal yr BP (cited as 2500 cal yr BC), suggesting the presence of full-blown agroeconomies. Although these Belizean data could be interpreted as either the result of an endogenous development or the incursion of agricultural people into the Belizean lowlands in Late Archaic times, MacNeish subscribed to a gradualistic view: "The implication drawn by MacNeish is that Early Archaic foraging groups in Belize shifted toward a more extensive use of plant foods, ultimately becoming dependent on agricultural produce" (Zeitlin and Zeitlin, 2000, p. 88). Recent work by Pohl and Jones and their collaborators supports this view. Kelly (1993, p. 225), in contrast, sees a sharp discontinuity between the preceramic projectile points and those of the later Maya, starting perhaps around 1200 B.C. These contrasting interpretations nicely illustrate Flannery's observation, mentioned above, that stylistic studies of artifacts emphasize discontinuities, whereas ecological studies emphasize continuities of developmental change.

At the Early to Middle Formative archaeological site of San Andrés, Tabasco, Mexico, Pope et al. (2001) used traditional archaeological excavation methods combined with sediment cores to reconstruct the site's paleoenvironmental and paleoecological history. They found that pollen from plants of the genus *Zea* occurs as early as 7050 cal yr BP (5100 cal yr BC). The morphological characteristics suggest that the pollen is from teosinte, a wild form of *Zea*. Because of its clear association with evidence of forest clearing the investigators think that the pollen came from wild plants under cultivation. Pollen that is typical of domesticated maize appears in the record only a mere hundred years later. Both these dates are well within the Middle Archaic Period, when the site area was situated within a paleoestuary. Recently, other investigators (Sluyter and Dominguez, 2006), who analyzed a similar pollen sequence from the Veracruz coastal plain, have raised questions about accepting such an early date for maize in the Gulf Coast lowlands. Their concern is based on methodological considerations connected with the possibility of bioturbation in the sampled sediments from San Andrés.

Returning to the record from San Andrés, a pollen grain dated to 6550 cal yr BP (cited as 4600 cal yr BC) may be from domesticated manioc (*Manihot*). By this time, the evidence suggests that extensive land clearing and maize cultivation were under way. Evidence of domesticated sunflower (*Helianthus annuus*) and cotton (*Gossypium* sp.) appear about 4450 cal yr BP (cited as ca. 2500 cal yr BC), when the area had become a paleolagoon at the end of the Late Archaic Period. The microbotanical evidence for the progression of wild to domesticated *Zea* appears to record the gradual domestication process in this coastal wetland setting. Since the small site of San Andrés is located only 5 km from the major Middle Formative Olmec site of La Venta, these data show that by the Late Archaic/Early Formative boundary several cultigens (maize, manioc, sunflower, and cotton) were being farmed in the Olmec heartland of Tabasco. Bones and shells recovered from the site from about 5350 cal yr BP (cited as 3400 cal yr BC) also indicate human presence, but so far no discrete archaeological features have been identified.

Voorhies and her colleagues have also adopted a gradualistic development model for Archaic Period archaeological data from the Chantuto region in the south Pacific coast of Mexico. Six shellmounds and one open-air site on the inner coastal plain provide a regional chronology anchored by multiple radiocarbon dates. The earliest subphase, Chantuto A, dates between 7500 and 6000/5500 cal yr BP (Voorhies et al., 2002), within the Middle Archaic, whereas the Chantuto B subphase falls within the Late Archaic, between 6000/5550 and 3500 cal years ago. These bracketing dates for the two subphases are different from those previously published (Blake et al., 1995) because Voorhies obtained more radiocarbon dates since that article was written (see Voorhies et al., 2002, 2004).

Only one of the investigated archaeological sites on the south Pacific coast of Mexico, a shellmound, dates to the Middle Archaic Period. In terms of site structure and contents it strongly resembles the nearby later shellmounds dating to the Late Archaic (cf. Voorhies, 1996, 2004; Voorhies et al., 1991), and like them Voorhies interprets it as a resource processing camp where animals from the estuarine-lagoon habitat were cooked and probably sun dried prior to transport inland. The aceramic deposits at these sites consist of couplets of burned and unburned shell layers that are interpreted as the archaeological signatures of clambakes. Voorhies suspects that in addition to clam meat, fish and shrimp were also processed in large quantities at all the shellmound sites, but the evidence for this activity is indirect and not conclusive.

The structure (Michaels and Voorhies, 1999; Voorhies, 2004) and faunal contents (Wake et al., 2004) of all six shellmounds are so similar from bottom to top, and to each other, that it is tempting to see them as portraying a stable, unchanging human adaptation to an estuarine-lagoon system throughout the Middle and Late Archaic Periods. However, seasonality studies of clam harvesting, the artifact record, and the microbotanical record all suggest to the investigators that a model of gradual change is more appropriate. Kennett used oxygen isotopic analysis of clam shells (*Polymesoda radiata*) to determine whether clams were harvested during the wet or dry season (Kennett and Voorhies, 1996; Voorhies et al., 2002) throughout the Middle to Late Archaic Period. During the Middle Archaic and early Late Archaic clams were harvested year-round, with a slight preference for the dry season. However, toward the end of the Late Archaic Period, a gradual yet striking change in seasonality occurred, with a conspicuous shift toward clam harvesting only during the wet season. This narrowing of the window of time spent at the shellmound sites led to the cessation of clam shell build-up some time around 3500 cal yr BP.

The artifacts at the shellmounds also reflect change, but the tempo of the change is impossible to detect because the overall frequency of artifacts is extremely low at these sites. The only artifacts in the lower Chantuto A deposits are large ark shells (a type of bivalve, *Anadara grandis*) and cooking stones, but a unique stratum immediately overlying the Chantuto A strata contained turtle shell fishhooks and other possible components of fishing tackle. Both artifact assemblages from the Middle Archaic shellmound contrast with those in the shellmounds dating to the Late Archaic Period. The investigators found crude obsidian flakes (Clark, 1989), milling stones, and several other stone tool types in the upper levels of the Archaic deposits in all the investigated Late Archaic shellmounds. Since the archaeologists excavated a greater volume of these upper Chantuto B deposits compared with the lower ones, it is difficult to determine precisely where and when the first examples of a particular tool type appear in the archaeological record. Clearly, however, the diversity of tools increases from early to late in the shellmound deposits dating to the Chantuto B subphase. The greatest diversity of surviving tools, however, was found at the inland open-air site, which is not only functionally different from the shellmound sites but is possibly late in the Late Archaic Period, based upon a single radiocarbon date. The precise dating of this site is problematic, however.

Finally, phytoliths from both Middle and Late Archaic shellmound contexts, as well as at the Late Archaic inland site, show a progressive decline of forest taxa and their replacement by disturbance vegetation (Jones and Voorhies, 2004). Along with this trend, *Zea* phytoliths have been recovered from the upper Late Archaic Period deposits at one shellmound and from the coeval inland site. These data show that domesticated maize was definitely present in the diet of the Chantuto people during the last stages of the Late Archaic Period, at the same time that visits to the coastal wetlands were becoming less frequent. To Jones, the observed changes in the phytolith records appear to be overwhelmingly anthropogenic, rather than climatically induced, but he had no means to segregate the possible effects of these two agents of change. Recent cores from several locations away from archaeological sites on the Chiapas coast are providing new insights about its paleoenvironment and the earliest appearance of cultigens in the region (Kennett et al., 2007).

Another source of information about the paleoenvironment of the shellmound builders comes from the oxygen isotope studies of clam shells mentioned above. Kennett and Voorhies (1995) found that the range of $\delta^{18}O$ values in the archaeological shells is similar to that of today's lagoonal waters and modern clam shells, suggesting that ancient lagoons of the south Pacific coast of Mexico were similar to those of today. This result implies that rainfall regimes during the Middle and Late Archaic Periods were not radically different from those of today.

Farther south along the Pacific coast, Hector Neff and his research associates (Neff et al., 2001) have recently initiated a research project to investigate the early prehistory and paleoenvironment along the Guatemalan coastal margin. They took sediment cores from several different locations near early archaeological sites, which in this region date to approximately 4000 cal years ago. The investigators are examining these cores to determine the geomorphological processes of coastal formation and to search for evidence of human occupation that predates the archaeologically known sites. Radiocarbon dates tie the inferred processes to an absolute chronology.

In most places investigated by Neff and his team, wave action associated with rising sea level caused erosion to dominate the coastal processes until marine transgression ceased about 5000–4000 cal years ago. Afterward, mangrove-lined estuaries expanded as coastal erosion gave way to shoreline progradation. In one protected location, however (Manchón core MAN015), mangrove peats began developing shortly after 7000 cal yr BP (Neff et al., 2006a). Decisive evidence of early human impacts comes from the Sipacate zone (cores SIP001 and SIP014), where mangroves became established shortly after 6000 cal years ago. Here, a dramatic change in vegetation occurred around 5500 cal years ago, as documented in the record of fossil pollen obtained from the sediment core (Jones et al., 2001; Neff et al., 2002, 2006b) that shows mangrove and swamp forest pollen disappearing in favor of pollen from freshwater plants. This shift from brackish to freshwater may have resulted from the outlet of a paleolagoon being closed off from the sea. Whatever the cause, this change took place at the same time when *Zea* pollen (Jones et al., 2001) first occurred in the record and when a bloom in charcoal suggests forest clearance. Even more intriguing is the presence, throughout the bottom section of the core, of phytoliths similar to those observed in maize (Collins et al., 2003). However, they may be from domesticated maize or from a wild grass. Positively identified maize phytoliths occur at about the same level as the *Zea* pollen noted above.

Neff's research team found microbotanical evidence of human presence that significantly predates the known archaeological sites along the Pacific Guatemalan coast (Neff et al., 2006b). Apparently, coeval archaeological sites have low visibility and are yet to be identified. This research complements that of Voorhies and her group by providing the first solid data about coastal geomorphological history in the region of the south Pacific coast of Mesoamerica. Voorhies' archaeological data are unequivocal in indicating that while Archaic people sojourned in the coastal wetlands they focused entirely on estuarine-lagoonal resources. Neff's data indicate that these coastal features formed as early as 7000 cal years ago but did not become prevalent until shoreline progradation replaced marine transgression between 5000 and 4000 cal years ago.

4.3. Discussion

This brief overview of the prehistory of Mesoamerica with emphasis on the Middle Holocene/Middle and Late Archaic is not intended to be comprehensive, but only to point to several important studies that illustrate the types of data that are currently available. For a truly comprehensive overview of Late Pleistocene to Holocene human lifeways in Mesoamerica, see the article by Zeitlin and Zeitlin (2000) and earlier reviews such as those of MacNeish (1986), MacNeish and Nelken-Turner (1983), and Stark (1981). Our primary objective is to evaluate the current status of archaeological research for this time span in Mesoamerican prehistory.

First, we note that there are relatively few archaeological studies of Mesoamerican prehistory during the Middle Holocene, and even maps that identify sites from that time span (e.g., Zeitlin and Zeitlin, 2000, Map 2.3) are somewhat misleading because of the great unevenness of the retrieved data. However, interest in the early prehistory of Mesoamerica seems to be swelling, which encourages us to expect breakthroughs in the foreseeable future.

Second, archaeological sites in several lowland regions (e.g., in Tabasco and in the Petén and Pacific coast of Guatemala) have eluded detection by archaeologists so paleobotanical remains provide the only archaeological data for reconstructing human presence during the Middle Holocene. We believe that this lack of evidence is due to several factors, including low initial archaeological site visibility coupled with high rates of sedimentation and erosion due to geologic processes. Their exceptionally high archaeological visibility makes shellmound sites the exception, but in the lowlands Cerro de las Conchas is the only intensively investigated Middle Holocene site.

Third, it is worth emphasizing that most Mesoamerican research projects formulated expressly to investigate Archaic lifeways have focused principally upon documenting changes in subsistence patterns. We note that the emphasis on subsistence change in these studies is perfectly reasonable because this was the most significant developmental process that was underway in the region at that time. However, in some studies it has not yet been possible to investigate other aspects of human life.

Fourth, even in the most thorough and detailed studies the subsistence data are not so finely resolved that we can say with confidence that the rate of change was slow and gradual, but merely that it appears to us that way. Archaeological data in general are notorious for having low resolution and usually are better for portraying sweeping trends in prehistory rather than accurate portraits of short-term events. Take for example, an idea expressed by Flannery (1986a, p. 504) about the possible impetus in the prehistoric past that might have encouraged ancient people to either experiment or not experiment with their farming activities. Flannery and his colleague Robert Reynolds (1986) note that among modern Oaxacan farmers, farming behavior is linked to assessments about whether a particular growing season is likely to be a wet one or a dry one. If they anticipate dry conditions, farmers behave conservatively, whereas their behavior is more experimental and innovative when they expect especially rainy conditions. Accordingly, the behavior of Oaxacan farmers in their subsistence activities is very much conditioned by annual variations in the climate. However, these annual climatic variations are not really predictable, despite the fact that farmers do their best to anticipate them. Our point here is that with the currently available data, researchers can rarely detect such small-scale (annual) changes either in the archaeological or the climatological records. So, we think it is fair to say that even if the subsistence changes made by Mesoamericans during preceramic times actually occurred in small steps, it is unlikely that we would be able to detect this pattern with our current analytical tools.

An important aspect of this problem derives from dating methods. Radiocarbon dates are sometimes unavailable (as in the BAAR study), but more often are too few in number to permit precision and accuracy in interpretations of rate of change. The adoption of a two-sigma range on calibrated radiocarbon dates also widens the range of estimates on BC/AD dates. Essentially, archaeologists fit the data points to a continuous curve that portrays developmental change as gradual, whether or not it really was. Currently, we simply lack the fine temporal resolution necessary to determine whether subsistence change was really gradual or punctated.

However, the theoretical models currently embraced by researchers give weight to the gradualistic view. Co-evolutionary models that emphasize the symbiosis between certain human and plant populations generally reinforce the idea of a slow and gradual pace of change. So, too, do models that emphasize the manipulation of plant populations by nonagricultural peoples in order to enhance certain characteristics of economic plants while they are still genetically wild. That is, current models emphasize continuities rather than discontinuities.

5. Conclusions

It has only recently become possible to attempt a regional synthesis of Middle Holocene paleoclimate and prehistory in Mesoamerica. Although relevant studies are not yet abundant, an increasing number of investigations involve more sites dispersed geographically across the region. In addition, there are more varied approaches to the reconstruction of the past, particularly in the reconstruction of past climates, which may be accomplished using a variety of different proxy data.

Clearly, the more independent lines of evidence that are brought to bear on the subject, the more accurate will be the final interpretation of the past. However, the basis of any synthesis is good data, and we think that the relevant data suffer from several weaknesses. These include problems in chronology, precision, and integration between paleoclimatic and prehistoric studies.

Sound temporal frameworks are the basis for comparison and correlation of paleoclimate and prehistory. However, our studies are bedeviled by problems in dating that include the hard water effects in limestone areas on the accuracy of radiometric dates, the uncertainties inherent in radiocarbon dating and calibration (Pilcher, 1993), archaeological sites dated only by relative dating methods, and too few dates to allow fine temporal resolution of local paleoclimate and archaeological sequences. Many of the paleoclimatological studies, even those reliant on small sampling intervals, have insufficient radiocarbon dates to permit dating small-scale climatic events evident in proxy records. We desperately need high precision radiocarbon dating in all local sequences. Without enough precise temporal data, we cannot detect rates of change, or correlations, much less relatively short-term events. For example, ENSO-related events are short-term, as they occur in the global climate system on time scales of months to several years. However, few of the data sets currently available are precise enough to allow the detection of such events during the Middle Holocene in Mesoamerica. For this record we must depend upon annually laminated proxy data such as speleothems, high-resolution lake and ocean sediments, and coral sections (Baumgartner et al., 1989). Such high-resolution data are only beginning to accumulate for the area and time under present consideration (see also Poore et al., 2003).

Closely tied to problems of dating are those of sampling precision. In paleoclimatological studies, close-interval sampling is needed to detect small-scale changes in the proxies being used for inferences about paleoclimate. It is even more imperative that archaeological sampling techniques become more precise, to reveal the record in greater detail. In addition to dating problems, the archaeological data suffer from use of large sampling intervals that necessarily tend to gloss over small-scale perturbations in the data. High-resolution prehistory is very much needed.

Finally, we are struck by the lack of integration between studies of Mesoamerican paleoclimate and prehistory for the Middle Holocene. In regions where the paleoclimatic data are especially good there are no associated archaeological records. This, for example, is the current situation in the central Yucatán Peninsula. The opposite situation also obtains. For example, in the Tehuacán Valley the archaeological record is good but the paleoclimatic data are based on low-resolution proxies. We currently lack truly integrated studies based upon direct collaboration between paleoclimatologists and archaeologists. We hope to see such studies formulated in the near future. Acknowledging these problems, we offer the following summary.

The paleoclimatological data for Mesoamerica currently suggest that the four thousand year long time span of the Middle Holocene was a relatively stable period without major climatic fluctuations. In general, it was warmer and wetter than the Early Holocene, and wetter and less variable than the Late Holocene. Slightly drier conditions may have occurred in some areas around 5800 cal yr BP, the time when sea level transgression ceased and shoreline progradation began. This change may have resulted from the decline of direct insolation forcing in the northern hemisphere (decreasing seasonality), with the ITCZ retreating from its very northerly early Holocene position (Tedesco and Thunell, 2003). The comparable archaeological records document slow, gradual changes in subsistence, settlement, and technology that have been interpreted as incremental adaptations to a slowly changing climatic/biotic/ edaphic environment. Prevailing explanatory models of cultural change have emphasized ecological and evolutionary processes. However, rates of change are derived from smoothed and interpolated data, and our current view of the past may be oversimplified simply because the resolution of our data is not high.

Acknowledgments

Metcalfe wishes to thank Sarah Davies and Ann Breen for their contributions in Mexico and Belize, respectively, and Vera Markgraf for her support through PEP1. Voorhies thanks Linda A. Brown and Michelle Woodward for field assistance at Cerro de las Conchas, where the fieldwork was supported by the H. John Heinz III Charitable Fund Grant Program for Latin American Archaeology. Subsequent laboratory analyses of plant and animal remains from that site were supported by the Foundation for Exploration and Research on Cultural Origins (FERCO). Additionally, we are grateful to several colleagues for their help in the preparation of this chapter. David Hodell and Mark Brenner kindly shared with us two very useful papers that were unpublished at the time we began this endeavor. Hector Neff, Mary Pohl, and Kevin O. Pope read the passages that pertained to their own research and offered suggestions that we gratefully acknowledge. Other readers who offered constructive comments on earlier drafts include Alan P. Covich, George H. Michaels, Signe Nygaard, and Michael R. Talbot. We greatly appreciate their expert advice and encouragement.

References

Andrews, E. W., and N. Hammond, 1990. Redefinition of the Swasey phase at Cuello, Belize. *American Antiquity* 55:570–584.

Baumgartner, T. R., J. Michaelsen, L. G. Thompson, G. T. Shen, A. Soutar, and R. E. Casey, 1989. The recording of interannual climatic change by high-resolution natural systems: tree-rings, coral bands, glacial ice layers and marine varves. In *Aspects of Climate Variability in the Pacific and Western Americas*, edited by D. W. Peterson, pp. 1–14. *Geophysical Monograph*, 55. American Geophysical Union, Washington, DC.

Behling, H., 2000. A 2860-year high-resolution pollen and charcoal record from the Cordillera de Talamanca in Panama: a history of human and volcanic forest disturbance. *The Holocene* 10:387–393.

Blake, M., J. E. Clark, B. Voorhies, G. Michaels, M. W. Love, M. E. Pye, A. A. Demarest, and B. Arroyo, 1995. Radiocarbon chronology for the Late Archaic and Formative periods on the Pacific Coast of Southeastern Mesoamerica. *Ancient Mesoamerica* 6:161–183.

Bradbury, J. P., 1997. Sources of glacial moisture in Mesoamerica. *Quaternary International* 43/44:97–110.

Bradbury, J. P., 2000. Limnologic history of Lago de Pátzcuaro, Michoacán, Mexico for the past 48,000 years: impacts of climate and man. *Palaeogeography, Palaeoclimatology, Palaeoecology* 163:69–95.

Bradbury, J. P., R. M. Forester, W. A. Bryant, and A. P. Covich, 1990. Paleolimnology of Laguna de Cocos, Albion Island, Rio Hondo, Belize. In *Ancient Maya Wetland Agriculture. Excavations on Albion Island, Northern Belize*, edited by M. D. Pohl, pp. 119–154. Westview Press, Boulder.

Brenner, M., D. A. Hodell, J. Curtis, M. Rosenmeier, M. W. Binford, and M. Abbott, 2001. Abrupt climate change and pre-Columbian cultural collapse. In *Interhemispheric Climate Linkages*, edited by V. Markgraf, pp. 87–103. Academic Press, San Diego.

Bridgwater, N., T. H. Heaton, and S. L. O'Hara, 1999. A Late Holocene palaeolimnological record from Central Mexico, based on faunal and stable isotope analysis of ostracod shells. *Journal of Paleolimnology* 22:383–397.

Byrne, R., and S. P. Horn, 1989. Prehistoric agricultural forest clearance in the Sierra de los Tuxtlas, Veracruz, Mexico. *Palynology* 13:181–193.

Caballero, M., S. Lozano, B. Ortega, J. Urrutia, and J. L. Macias, 1999. Environmental characteristics of Lake Tecocomulco, Northern Basin of Mexico, for the last 50,000 years. *Journal of Paleolimnology* 22:399–411.

Carpenter, J. P., G. Sanchez, and C. M. E. Villapando, 2005. The Late Archaic/early agricultural period in Sonora Mexico. In *Late Archaic Across the Borderlands*, edited by B. J. Vierra, pp.13–40. University of Texas Press, Austin.

Clark, J. E., 1989. Obsidian tool manufacture. In *Ancient Trade and Tribute: Economies of the Soconusco Region of Mesoamerica*, edited by B. Voorhies, pp. 215–228. University of Utah Press, Salt Lake City.

Collins, S., D. M. Pearsall, and N. Duncan, 2003. Prehistoric agriculture on the Pacific Coast of Guatemala: insights from phytolith analysis. Paper Presented at the 68th Annual Meeting of the Society for American Archaeology, Milwaukee, Wisconsin.

Covich, A., and M. Stuiver, 1974. Changes in oxygen 18 as a measure of long-term fluctuations in tropical lake levels and Molluscan populations. *Limnology and Oceanography* 19:682–691.

Curtis, J. H., D. A. Hodell, and M. Brenner, 1996. Climate variability on the Yucatán Peninsula (Mexico) during the last 3500 years, and implications for Maya cultural evolution. *Quaternary Research* 46:37–47.

Curtis, J. H., M. Brenner, D. A. Hodell, R. A. Balser, G. A. Islebe, and H. Hooghiemstra, 1998. A multi-proxy study of Holocene environmental change in the Maya lowlands of Petén, Guatemala. *Journal of Paleolimnology* 19:139–159.

deMenocal, P. B. 2001. Cultural responses to climate change during the Late Holocene. *Science* 292:667–673.

DeVries, T. J., L. Ortlieb, and A. Díaz, 1997. Determining the early history of El Niño. *Science* 276:965–966.

Flannery, K. V., 1967. The vertebrate fauna and hunting patterns. In *The Prehistory of the Tehuacan Valley Environment and Subsistence*, edited by D. S. Byers, Vol. 1, pp. 132–177, University of Texas Press, Austin.

Flannery, K. V., 1986a. Food procurement area and Preceramic diet at Guilá Naquitz. In *Guilá Naquitz: Archaic Foraging and Early Agriculture in Oaxaca, Mexico*, edited by K. V. Flannery, pp. 303–317. Academic Press, Orlando.

Flannery, K. V., 1986b. Guilá Naquitz in spatial, temporal and cultural context. In *Guilá Naquitz: Archaic Foraging and Early Agriculture in Oaxaca, Mexico*, edited by K. V. Flannery, pp. 31–42. Academic Press, Orlando.

Fritz, S., S. E. Metcalfe, and W. Dean, 2001. Holocene climate patterns in the Americas inferred from paleolimnological records. In *Interhemispheric Climate Linkages*, edited by V. Markgraf, pp. 241–263. Academic Press, San Diego.

Gasse, F. 2000. Hydrological changes in the African tropics since the Last Glacial Maximum. *Quaternary Science Reviews* 19:189–21.

González-Quintero, L., 1986. Análisis polínicos de los sedimentos. In *Tlapacoya: 35,000 años de historia del Lago de Chalco Colección Científica, Serie Prehistoria*, pp. 157–166. Instituto Nacional de Antropología e Historia, México.

Grimm, E., S. Lozano-García, H. Behling, and V. Markgraf, 2001. Holocene vegetation and climate variability in the Americas. In *Interhemispheric Climate Linkages*, edited by V. Markgraf, pp. 325–370. Academic Press, San Diego.

Hansen, B., 1990. Pollen stratigraphy of Laguna de Cocos. In *Ancient Maya Wetland Agriculture. Excavations on Albion Island, Northern Belize*, edited by M. D. Pohl, pp. 155–186. Westview Press, Boulder.

Hillesheim, M.B., D. A. Hodell, B. W. Leyden, M. Brenner, J. H. Curtis, F. S. Anselmetti, D. Ariztegui, D. G. Back, T. P. Guilderson, M. F. Rosenmeier and D. W. Schnurrenberger, 2005. Climate change in lowland Central America during the Late Deglacial and Early Holocene. *Journal of Quaternary Science* 20: 363–376.

Hodell, D. A., J. H. Curtis, and M. Brenner, 1995. Possible role of climate in the collapse of classic Maya civilization. *Nature* 375:391–394.

Hodell, D. A., M. Brenner, J. Curtis, and T. Guilderson, 2001. Solar forcing of drought frequency in the Maya lowlands. *Science* 292:1367–1370.

Hodell, D. A., M. Brenner, and J. H. Curtis, 2005. Terminal Classic Drought in the Northern Maya lowlands Inferred from Multiple Sediment Cores in Lake Chichancanab (Mexico). *Quaternary Science Reviews* 24:1413–1427.

Hooghiemstra, H., A. M. Cleef, G. Noldus, and M. Kappelle, 1992. Upper Quaternary vegetation dynamics and palaeoclimatology of the La Chonta Bog area (Cordillera de Talamanca, Costa Rica). *Journal of Quaternary Science* 7:205–225.

Horn, S. P., 1993. Postglacial vegetation and fire history in the Chirripó Paramo of Costa Rica. *Quaternary Research* 40:107–116.

Iceland, H. B., 1997. The preceramic origins of the Maya: the results of the Colha preceramic project in Northern Belize. Unpublished Ph.D. dissertation, Department of Anthropology, The University of Texas at Austin.

Islebe, G. A, and H. Hooghiemstra, 1997. Vegetation and climate history of Montane Costa Rica since the Last Glacial Maximum. *Quaternary Science Reviews* 16: 589–604.

Islebe, G. A., H. Hooghiemstra, M. Brenner, D. A. Hodell, and J. Curtis, 1996. A Holocene vegetation history from Lowland Guatemala. *The Holocene* 6:265–271.

Johnson, F. and R. S. MacNeish, 1972. Chronometric dating. In *The Prehistory of the Tehuacan Valley, Volume Four: Chronology and Irrigation*, edited by F. Johnson, pp. 3–55. University of Texas Press, Austin.

Jones, J. G., 1991. Pollen evidence for early settlement and agriculture in Northern Belize. Unpublished Ph.D. dissertation, Department of Anthropology. Texas A&M University, College Station.

Jones, J. G., 1994. Pollen evidence for early settlement and agriculture in Northern Belize. *Palynology* 18:205–211.

Jones, J. G., D. Pearsall, and C. Vientamilla, 2001. Pollen and phytolith evidence for settlement and subsistence in Pacific Coastal Guatemala. Paper Presented at the 66th Annual Meeting of the Society for American Archaeology, New Orleans, Louisiana.

Jones, J. G., and B. Voorhies, 2004. Human and Plant Interactions. In *Coastal Collectors in the Holocene: The Chantuto People of Southwest Mexico*, edited by B. Voorhies senior author, pp. 300–343. University Press of Florida, Gainesville.

Kelly, T. C., 1993. Preceramic projectile-point typology in Belize. *Ancient Mesoamerica* 4:205–227.

Kennett, D. J., and B. Voorhies, 1995. Middle Holocene periodicities in rainfall inferred from oxygen and carbon isotopic fluctuations in prehistoric tropical estuarine Mollusc shells. *Archaeometry* 37:157–170.

Kennett, D. J., and B. Voorhies, 1996. Oxygen isotopic analysis of archaeological shells to detect seasonal use of wetlands on the Southern Pacific Coast of Mexico. *Journal of Archaeological Research* 23:689–704.

Kennett, D. J., B. Voorhies, J. G. Jones, H. Neff, D. R. Piperno, T. A. Wake , K. L. Blake, B. Culleton, J. A. Gómez-García, and N. Martínez-Taqueña, 2007. Informe Técnico Final del Proyecto Arcaico-Formativo:Costa De Chiapas (Temporada 2005). Submitted to the Instituto Nacional de Antropología e Historia, Mexico.

Leng, M.J., S. E. Metcalfe, and S. J. Davies, 2005. Investigating Late Holocene climatic variability in Central Mexico using carbon isotope ratios in organic materials and oxygen isotope ratios from diatom silica within lacustrine sediments. *Journal of Paleolimnology* 34:413–431.

Leyden, B. W., 1995. Evidence of the Younger Dryas in Central America. *Quaternary Science Reviews* 14:833–839.

Leyden, B. W., 2002. Pollen evidence for climatic variability and cultural disturbance in the Maya lowlands. *Ancient Mesoamerica* 13:85–101.

Leyden, B. W., M. Brenner, D. A. Hodell, and J. Curtis, 1994. Orbital and internal forcing of climate on the Yucatán Peninsula for the past ca. 36,000 ka. *Palaeogeography, Palaeoclimatology, Palaeoecology* 109:193–210.

Leyden, B. W., M. Brenner, T. Whitmore, J. H. Curtis, D. Piperno, and B. H. Dahlin, 1996. A record of long and short-term climatic variation from Northwest Yucatán: Cenote San José Chulchaca. In *The Managed Mosaic: Ancient Maya Agriculture and Resource Use*, edited by S. L. Fedick, pp. 30–50. The University of Utah Press, Salt Lake City.

Lohse, J.C., J. Awe, C. Griffith, R. M. Rosenswig, and F. Valdez Jr., 2006. Preceramic occupations in Belize: updating the Paleoindian and Archaic record. *Latin American Antiquity* 17:209–226.

Lorenzo, J. L., and L. Mirambell, 1999. The inhabitants of Mexico during the upper Pleistocene. In *Ice Age People of North America: Environments, Origins and Adaptations*, edited by R. Bonnichsen and K. L. Turnmire, pp. 482–496. Oregon State University, Corvallis.

Lozano-García, S. and L. Vázquez-Selem, 2005. A high elevation Holocene pollen record from Iztaccíhuatl Volcano, Central Mexico. *The Holocene* 15:329–338.

MacNeish, R. S., 1967. A summary of the subsistence. In *The Prehistory of the Tehuacan Valley, Volume 1: Environment and Subsistence*, edited by D. S. Byers, Vol. 1, pp. 290–309. University of Texas Press, Austin.

MacNeish, R. S., 1986. The preceramic of Middle America. *Advances in New World Archaeology* 5:93–129.

MacNeish, R. S., and A. Nelken-Turner, 1983. The preceramic of Mesoamerica. *Journal of Field Archaeology* 10:71–84.

MacNeish, R. S., A. Nelken-Turner, and I. W. Johnson, 1967. *The Prehistory of the Tehuacan Valley: Nonceramic Artifacts*. Vol. 2. University of Texas Press, Austin.

MacNeish, R. S., F. A. Peterson, and J. A. Neely, 1972. The archaeological reconnaissance. In *The Prehistory of the Tehuacan Valley, Volume 5: Excavations and Reconnaissance*, edited by R. S. MacNeish, M. L. Fowler, A. García Cook, F. A. Peterson, A. Nelken-Turner, and J. A. Neely, Vol. 5, pp. 341–495. University of Texas Press, Austin.

Marcus, J., and K. V. Flannery, 1996. *Zapotec Civilization: How Urban Society Evolved in Mexico's Oaxaca Valley*. Thames & Hudson, London.

Markgraf, V., 2001. *Interhemispheric Climate Linkages*. Academic Press, San Diego.

Markgraf, V., T. R. Baumgartner, J. P. Bradbury, H. F. Díaz, R. Dunbar, B. Luckman, G. O. Seltzer, T. W. Swetnam, and R. Villalba, 2000. Paleoclimate reconstruction along the pole-equator-pole transect of the Americas (PEP 1). *Quaternary Science Reviews* 19:125–140.

Martin, P. S., 1964. Paleoclimatology and a tropical pollen profile. *Proceedings of the 6th Congress of the International Quaternary Association (INQUA), Warsaw,* 1961:319–323.

Metcalfe, S. E., 1995. Holocene environmental change in the Zacapu Basin, Mexico: a diatom-based record. *The Holocene* 5:196–208.

Metcalfe, S. E., 2006. Late Quaternary environments of the northern deserts and central trans-volcanic belt of Mexico. *Annals of the Missouri Botanical Garden* 93:258–273.

Metcalfe, S., A. Breen, M. Murray, P. Furley, A. Fallick, and A. McKenzie, in preparation, Holocene environmental change in Northern Belize: records from the New River Lagoon.

Metcalfe, S. E., S. J. Davies, J. Braisby, M. J. Leng, A. J. Newton, N. L. Terrett, and S. L. O'Hara, 2007. Long and short term change in the Pátzcuaro Basin, Central Mexico. *Palaeogeography, Palaeoclimatology, Palaeoecology* 247:272–295.

Metcalfe, S. E., S. L. O'Hara, M. Caballero, and S. S. Davies, 2000. Records of late Pleistocene-Holocene climatic change in Mexico – a review. *Quaternary Science Reviews* 19:699–721.

Michaels, G. H., and B. Voorhies, 1999. Late Archaic period coastal collectors in Southern Mesoamerica: the Chantuto people revisited. In *Pacific Latin America in Prehistory: The Evolution of Archaic and Formative Cultures*, edited by M. Blake, pp. 39–54. Washington State University Press, Pullman.

Moy, C. M., G. O. Seltzer, D. T. Rodbell, and D. M. Anderson, 2002. Variability of El Niño/Southern oscillation activity at millennial timescales during the Holocene Epoch. *Nature* 420:162–165.

Neff, H., D. E. Friedel, and J. G. Jones, 2001. Humans on the changing Holocene Landscape of the Guatemalan Pacific Coast. Paper Presented at the 66th Annual Meeting of the Society for American Archaeology, New Orleans, Louisiana.

Neff, H., B. Arroyo, J. G. Jones, D. M. Pearsall, and D. E. Freidel, 2002. Nueva evidencia pertinente a la ocupación temprana del sur de Mesoamérica. Paper presented at the XII Encuentro Internacional: Los Investigadores de la Cultura Maya, Campeche, November 10–14, México.

Neff, H., D. M. Pearsall, J. G. Jones, B. Arroyo, and D. E. Freidel, 2006a. Climate change and population history in the Pacific lowlands of Southern Mesoamerica. *Quaternary Research* 65:390–400.

Neff, H., D. M. Pearsall, J. G. Jones, B. Arroyo, S. Collins, and D. E. Freidel, 2006b. Early Maya adaptive patterns: Mid – Late Holocene paleoenvironmental evidence from Pacific Guatemala. *Latin American Antiquity*. 17:287–315.

Niederberger Betton, C., 1987. Paleopaysages et archeologie pre-urbaine du Bassin de Mexico (Mexique). Collection Études Mésoaméricaines Vol XI, 1 & 2, Centro de Estudios Mexicanos y Centro-Americanos, México D.F.

Northrop, L. A., and S. P. Horn, 1996. Precolumbian agriculture and forest disturbance in Costa Rica: Palaeoecological evidence from two lowland rainforest lakes. *The Holocene* 6:289–299.
O'Hara, S. L., F. A. Street-Perrott, and T. P. Burt, 1993. Accelerated soil erosion around a Mexican highland lake caused by prehispanic agriculture. *Nature* 362:48–50.
Petrequin, P., editor. 1994. 8000 años de la Cuenca de Zacapu. Evolución de los paisajes y primeros desmontes. *Cuadernos de Estudios Michoacanos* 6. CEMCA, México D.F.
Pilcher, J. R., 1993. Radiocarbon dating and the palynologist: a realistic approach to precision and accuracy. In *Climate Change and Human Impact on the Landscape*, edited by F. M. Chambers, pp. 23–32. Chapman & Hall, London.
Piperno, D. R., 2006. Quaternary environmental history and agricultural impact on vegetation in Central America. *Annals of the Missouri Botanical Garden* 93:274–296.
Pohl, M. D., K. O. Pope, J. G. Jones, J. S. Jacob, D. R. Piperno, S. D. deFrance, D. L. Lentz, J. A. Gifford, M. E. Danforth, and J. K. Josserand, 1996. Early agriculture in the Maya lowlands. *Latin American Antiquity* 7:355–372.
Poore, R. Z., H. J. Dowsett, S. Verardo, and T. M. Quinn, 2003. Millennial- to century-scale variability in Gulf of Mexico Holocene climate records. *Paleoceanography* 18(2):1048.
Pope, K. O., M. D. Pohl, J. G. Jones, D. L. Lentz, C. von Nagy, F. J. Vega, and I. R. Quitmyer, 2001. Origin and environmental setting of ancient agriculture in the lowlands of Mesoamerica. *Science* 292:1370–1373.
Reynolds, R. G., 1986. An adaptive computer model for the evolution of plant collecting and early agriculture in the eastern valley of Oaxaca. In *Guilá Naquitz: Archaic Foraging and Early Agriculture in Oaxaca, Mexico*, edited by K. V. Flannery, pp. 439–500. Academic Press, Orlando.
Rice, D. S., and P. M. Rice, 1990. Population size and population change in the central Petén Lake region, Guatemala. In *Pre-Columbian Population History in the Maya Lowlands*, edited by T. P. Culbert and D. S. Rice, pp. 123–148. University of New Mexico Press, Albuquerque.
Rindos, D. A., 1984. *The Origins of Agriculture: An Evolutionary Perspective.* Academic Press, New York.
Rodbell, D. T., G. O. Seltzer, D. M. Anderson, M. B. Abbott, D. B. Enfield, and J. H. Newman, 1999. An ~15,000-year record of El Niño-driven alluviation in Southwestern Ecuador. *Science* 283:516–521.
Roosevelt, A. C., 2000. Who's on first? *Natural History* 109:76–79.
Rosenmeier, M. F., D. M. Hodell, M. Brenner, J. H. Curtis, J. B. Martin, F. S. Anselmetti, D. Ariztegui, and T. P. Guilderson, 2002. Influence of vegetation change on watershed hydrology: implications for paleoclimatic interpretation of lacustrine $\delta^{18}O$ records. *Journal of Paleolimnology* 27, 117–131.
Rue, D. J., 1987. Early agriculture and early postclassic Maya occupation in western Honduras. *Nature* 326:285–286.
Sanchez, M. G., 2001. A synopsis of Paleoindian Archaeology in Mexico. Manuscript prepared for the 3rd Annual Julian D. Hayden Student Paper Competition. University of Arizona, Tucson.
Sandweiss, D. H., J. B. Richardson III, E. J. Reitz, H. B. Rollins, K. A. Maasch, 1996. Geoarchaeological evidence from Peru for a 5000 years B.P. Onset of El Niño. *Science* 273:1531–1533.
Sandweiss, D. H., J. B. Richardson III, E. J. Reitz, H. B. Rollins, K. A. Maasch, 1997. Determining the early history of El Niño. *Science* 276:965–967.
Sluyter, A. and G. Dominguez, 2006. Early maize (*Zea mays* L.) cultivation in Mexico: dating sedimentary pollen records and its implications. *Proceedings of the National Academy of Sciences* 103:1147–1151.

Smith, C. E., Jr., 1986. Preceramic plant remains from Guilá Naquitz. In *Guilá Nauitz: Archaic Foraging and Early Agriculture in Oaxaca, Mexico*, edited by K. V. Flannery, pp. 265–274. Academic Press, Orlando.

Stark, B., 1981. The rise of sedentary life. In *Supplement to the Handbook of Middle American Indians*, edited by J. Sabloff, Vol. 1, pp. 245–272. University of Texas Press, Austin.

Tedesco, K., and R. Thunell, 2003. High resolution tropical climate record for the last 6,000 years. *Geophysical Research Letters* 30:1–4.

Thompson, L. G., 2000. Ice core evidence for climate change in the tropics: implications for our future. *Quaternary Science Reviews* 19:19–35.

Tsukada, M., and E. S. Deevey, 1967. Pollen analysis from four lakes in the Southern Maya area of Guatemala and El Salvador. In *Quaternary Palaeoecology*, edited by E. J. Cushing and H. E. Wright, pp. 303–331. Yale University Press, New Haven.

Vaughan, H. H., E. S. Deevey, and S. E. Garrett-Jones, 1985. Pollen stratigraphy of two cores from the Petén lake district. In *Prehistoric Lowland Maya Environment and Subsistence Economy*, edited by M. Pohl, Vol. 77, pp. 73–89. Papers of the Peabody Museum, Harvard University Press, Cambridge.

Voorhies, B., 1996. The transformation from foraging to farming in the lowlands of Mesoamerica. In *The Managed Mosaic: Ancient Maya Agriculture and Resource Use*, edited by S. L. Fedick, pp. 17–29. University of Utah Press, Salt Lake City.

Voorhies, B., 2004. *Coastal Collectors in the Holocene: The Chantuto People of Southwest Mexico*. The University Press of Florida, Gainesville.

Voorhies, B., G. H. Michaels, and G. M. Riser, 1991. Ancient shrimp fishery. *National Geographic Research and Exploration* 7:20–35.

Voorhies, B., D. J. Kennett, J. G. Jones, and T. A. Wake, 2002. A Middle Archaic archaeological site on the West Coast of Mexico. *Latin American Antiquity* 13:179–200.

Wahl, D., R. Byrne, T. Schreiner, and R. Hansen, 2006. Holocene vegetation change in the Northern Peten and its implications for Maya prehistory. *Quaternary Research* 65:380–389.

Wake, T. A., N. Anikouchine, and B. Voorhies, 2004. Food procurement and processing: fish and game remains at the shellmound sites. In *Coastal Collectors in the Holocene: The Chantuto People of Southwest Mexico*, edited by B. Voorhies, Senior Author, pp. 158–206. The University Press of Florida, Gainesville.

Watts, W. A., and J. P. Bradbury, 1982. Paleoecological studies at Lake Patzcuaro on the West-Central Mexican plateau and at Chalco in the basin of Mexico. *Quaternary Research* 17:56–70.

Wells, L. E, and J. S. Noller, 1997. Determining the early history of El Niño. *Science* 276:966.

Wendorf et al., this volume. Middle Holocene Environment of North and East Africa, with Special Emphasis on the African Sahara.

Xelhuantzi-Lopez, M. S. 1994. Estudio palinológico de cuatro sitios ubicados en la Cuenca de Zacapu: fondo de cienega, contacto Lomas-Cienega, pantano interno y Loma Alta. In *8000 años de la Cuenca de Zacapu. Evolución de los paisajes y primeros desmontes*, edited by P. Petrequin. *Cuadernos de Estudios Michoacanos* 6, 81–93. CEMCA, México D.F.

Zeitlin, R. N., and J. F. Zeitlin, 2000. The Paleoindian and Archaic cultures of Mesoamerica. In *The Cambridge History of the Native Peoples of the Americas, Vol. II: Mesoamerica*, Part 1, edited by R. E. W. Adams and M. J. MacLeod, pp. 45–121. Cambridge University Press, Cambridge.

Chapter 6

Middle Holocene environments of north and east Africa, with special emphasis on the African Sahara

Fred Wendorf[1], Wibjörn Karlén[2] and Romuald Schild[3]
[1]*Department of Anthropology, Southern Methodist University, Dallas, TX 75275, USA*
[2]*Geografiska Annaler Ser. A, Institutionen fur Kulturgeografi, Box 513, 751 20 Uppsala, Sweden*
[3]*Instytut Archeologii I Ethnologii, Polskiej Akademii Nauk, Al. Solidararnosci 105, 00–140 Warszawa, Poland*

Abstract

In this chapter, we discuss climatic fluctuations in northern and eastern Africa during the Holocene and in particular the middle Holocene. The major emphasis is on the Sahara and the mountains of eastern Africa, because the climatic changes in these areas were very dramatic, and the evidence for those changes is the most visible. In addition, for the Sahara there are numerous radiocarbon age determinations tied to the climatic events in that area. Within the Sahara, we treat the eastern part in the greatest detail, primarily because extensive detailed work has been done there and the chronology of the entire sequence is controlled by a large series of radiocarbon dates from charcoal. Discussions of the cultural phenomena that accompanied the middle Holocene climatic fluctuations focuses on the Eastern Sahara and the Nile.

1. Introduction

Most of Africa is within 30° of the equator, and temperatures are generally above the global mean. With high evaporation throughout most of the continent, precipitation is the most important climatic element. In the Holocene there were significant changes in rainfall and this review focuses on this aspect of Africa's past climates. The greatest rainfall occurs in an east–west band extending from 10° to 15° on either side of the equator (Fig. 6.1). Most of these rains are monsoonal, and come from moist air masses moving from the south Atlantic into the lowland areas between May and September. Other rains come in the mountainous areas of eastern Africa between October and April, mostly from the Indian Ocean (Grove, 1993, p. 32). Poleward from the tropics the climate during most of the year is dominated by descending air and is arid, except for the east side of southern Africa. Modern records of rainfall indicate that fluctuations seem to cluster over two time scales, one over a period of a decade or two, and other departures from the mean last about a century. The cause or causes of these variations in rainfall is unknown (Grove, 1993).

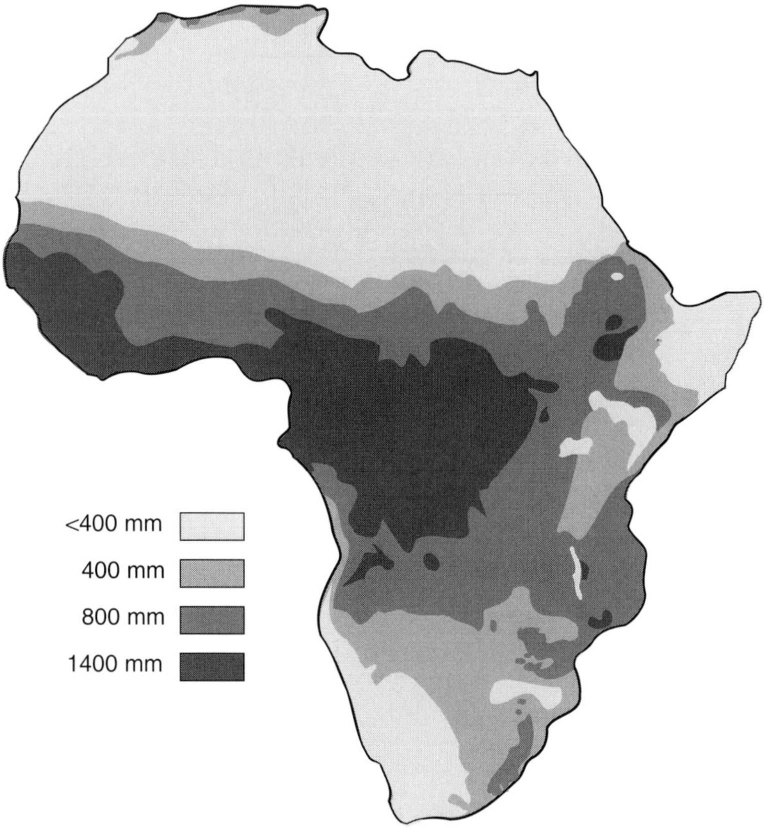

Figure 6.1. Modern annual distribution of precipitation in Africa (in mm) (redrawn from Grove, 1993, p. 40).

The initial attempt to correlate changes in climate in Africa with climate changes in higher latitudes was made by Nilsson (1940), who was testing a hypothesis by Wayland (1934) that periods of higher rainfall, or pluvials, were coincident in Africa with periods of glacial advance in Europe. Working with lakes near the still glaciated areas in the highlands of East Africa and Ethiopia, Nilsson associated higher lake levels with moraines of more extensive and older glaciers, and correlated these with glacial advances in Europe. This pluvial theory was rapidly adopted by most students of past climates in Africa and it became the unifying hypothesis for correlations between areas (see Cole, 1954, pp. 34–63; Clark, 1950). Although questions were raised about using climate for stratigraphic correlations (Cooke, 1957; Flint, 1959), it was not until radiocarbon dates became widely available that the pluvial hypothesis was finally abandoned.

The model currently dominating paleoclimatic studies in Africa and elsewhere is the astronomical theory proposed by Croll (1864) and elaborated by Milankovitch (1941). It is now generally understood that the alternating periods of aridity and increased moisture in tropical Africa and the Sahara are closely related to past variations in the Intertropical Convergence (ITC, or sometimes referred to as the Intertropical Convergence Zone or ITCZ), one of the most important influences on the climate of Africa. Today, (and also throughout the Late Pleistocene and Holocene, and probably for some time before then) the ITC controls both the expansion and northward movement of the summer monsoon rains and their retreat at the end of the summer. The most important link in this chain of evidence was the observation that low lake levels in tropical Africa were contemporary with glacial conditions in higher latitudes (not interglacials, as previously thought), that postglacial warming coincided with high lake levels, and that these have a nonlinear relationship with increases in the maximum solar radiation in the northern hemisphere (Kutzbach, 1981; Gillespie et al., 1983; Street-Perrott and Roberts, 1983; Kutzbach and Street-Perrott, 1985). Also, there is evidence for several abrupt periods of aridity in the records of East African lakes. It is interesting to note that one of these occurs shortly after an initial rise of lake levels at around 12,500 ^{14}C yr BP (14,650 cal yr BP) and seems to coincide with the Younger Dryas dating between 10,800 (12,730 cal yr BP) and 9700 ^{14}C yr BP (10,950 cal yr BP) (Street-Perrott and Perrott, 1990; Lamb et al., 1995; Grove, 1993; Hassan, 1997; Haynes, 1997).

Other important evidence on climatic change in tropical eastern and northern Africa has been obtained from analyses of cores taken from the Persian Gulf south of the Arabian peninsula. These indicate that the major transition between glacial and Holocene conditions occurred around 13,060 ^{14}C yr BP (15,530 cal yr BP) (Sirocko et al., 1993). A series of rapid and abrupt transitions that strengthened the monsoons were noted prior to and at the onset of the Holocene. According to the evidence from these cores, the transitions occurred at 14,300 ^{14}C yr BP (17,140 cal yr BP), 13,500 ^{14}C yr BP (16,160 cal yr BP), 13,060 ^{14}C yr BP (15,530 cal yr BP) and 9900 ^{14}C yr BP (11,000 cal yr BP). Contemporary with an interval when the continental heat flow was strongest, between 8850 and 7850 ^{14}C yr BP (9880 and 8560 cal yr BP), there was also a long wet period when the Arabian peninsula was vegetated sufficiently to reduce markedly the amount of wind-blown sand.

2. Tropical Africa

Convincing evidence for climatic change in tropical Africa during the final Pleistocene and Holocene has been obtained from numerous pollen and sediment studies on cores taken at crater lakes and other closed basins from Ethiopia to West Africa.

2.1. East Africa

Important data on rainfall in tropical Africa come from an exceptionally well-dated sequence of changes in lake level in the Lake Ziway-Shala Basin in central Ethiopia.

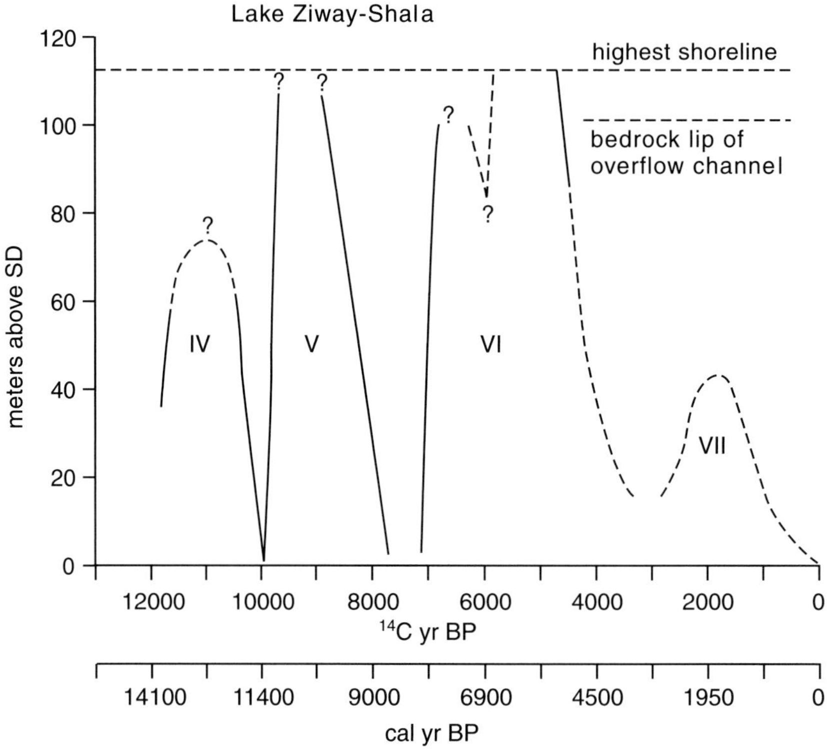

Figure 6.2. Holocene lake levels in the Ziway-Shala basin, Ethiopia (redrawn from Gillespie et al., 1983, p. 682).

As seen in Fig. 6.2, this lake was very low in the Final Pleistocene until around 12,000 ^{14}C yr BP (13,990 cal yr BP) when it expanded and then fell again between 10,400 (12,300 cal yr BP) and 9950 ^{14}C yr BP (11,060 cal yr BP). This was immediately followed by a rapid rise in the level of the lake, followed by a drying episode that began around 8500 ^{14}C yr BP (9470 cal yr BP) during which the lake reached its lowest level between 7800 and 7200 ^{14}C yr BP (8520 and 7950 cal yr BP). The lake then rose abruptly and reached its highest level between 7000 and 5200 ^{14}C yr BP (7790 and 5940 cal yr BP), broken by a brief, abrupt arid phase around 5900 ^{14}C yr BP (6730 cal yr BP) (Gillespie et al., 1983). Around 5000 ^{14}C yr BP (5730 cal yr BP) a major recession in the level of the lake occurred, and dry conditions have prevailed since then.

Pollen studies at two equatorial forest crater lakes are also very interesting. They are: Lake Bosumtwi, a meteorite impact crater in the forest zone of southern Ghana (6° 30′N), and Lake Barombi Mbo, a volcanic crater in west Cameroon (4° 40′N). The cores from both of these Ghana and Cameroon lakes go back to about 25,000 years ago (Maley, 1997).

2.2. Lake Bosumtwi

The vegetation in the area around Bosumtwi crater is today a closed tropical forest. In the final Pleistocene the level of the lake in the crater was much lower than today; however, it was much higher than now between 13,000 and 4000 ^{14}C yr BP (15,440 and 4440 cal yr BP), with three sub-peaks: between (1) 12,500 and 10,500 ^{14}C yr BP (13,810 and 12,420 cal yr BP); (2) 10,000 and 8200 ^{14}C yr BP (11,270 and 9124 cal yr BP); and (3) 7500 and 4000 ^{14}C yr BP (8260 and 4420 cal yr BP). The lake during the last peak was over 100 m higher than today and was probably overflowing (Fig. 6.3). Two brief arid intervals with lower lake levels separate these sub-peaks. There was a major abrupt episode of aridity shortly after 4000 ^{14}C yr BP and another about 500 years ago. The lake was much lower than today during both of these last arid periods. Studies of fossil pollen from the core show that the maximum forest extension began here around 9500 (10,510 cal yr BP) years ago, and that the first retreat of the northern edge of the forest may have occurred when the lake level fell between 4000 and 3000 ^{14}C yr BP (4440 and 3190 cal yr BP) (Talbot et al., 1984; Maley, 1991).

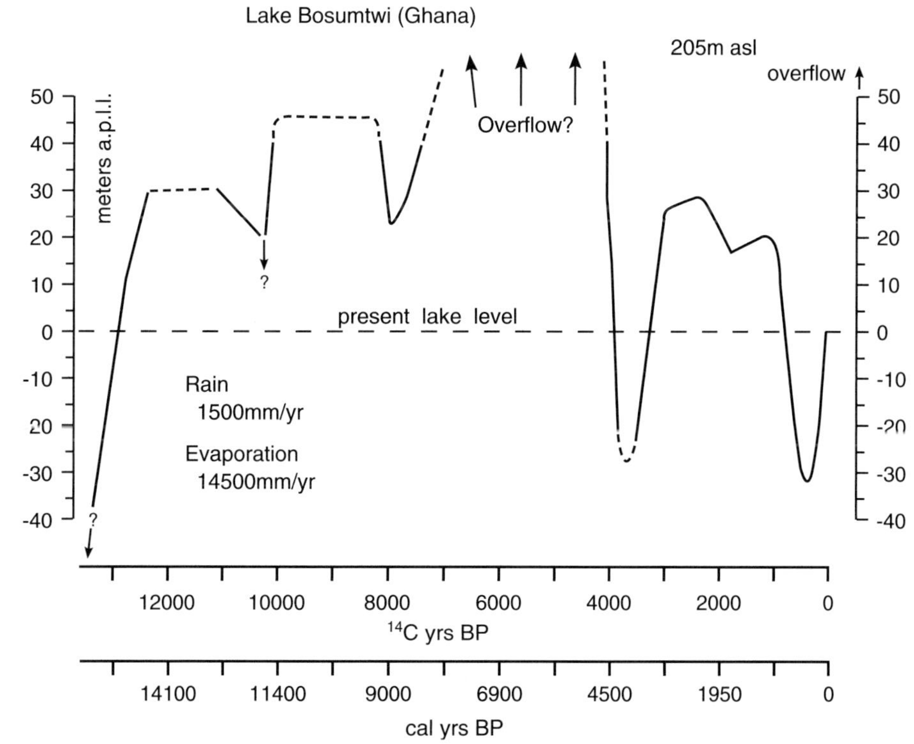

Figure 6.3. Holocene lake levels in the Bosumtwi Crater, Ghana (redrawn from Talbot et al., 1984, p. 188).

2.3. Lake Barombi Mbo

In western Cameroon in the area around Barombi Mbo the modern vegetation is composed of two main elements, an evergreen rainforest and a semi-deciduous rain forest, with a small area of relic savanna. Pollen frequencies come from a core taken in the center of Lake Barombi Mbo, and dated by 15 radiocarbon determinations. Taxa of arboreal pollen typical of evergreen and wetter forests reached their maximum values between ca. 9500 and 3000 ^{14}C yr BP (10,510 and 3190 cal yr BP), and characteristic elements of montane cloud forests show a strong maximum between 3800 and 3400 ^{14}C yr BP (4150 and 1680 cal yr BP). The area of the forest changed abruptly around 2800 ^{14}C yr BP (2870 cal yr BP), with the maximum retreat between 2500 and 2000 ^{14}C yr BP (2610 and 1940 cal yr BP) (Giresse et al., 1994). This phenomenon of rapid forest retreat at this time has also been seen in several other lakes in central Atlantic Africa and in the southwest Congo (Giresse et al., 1994; Maley, 1997, pp. 613–614).

The degree of open vegetation in the area around Barombi Mbo is indicated by relatively high values for grass pollen, which in the horizons dated to before 20,000 ^{14}C yr BP are around 10–15%, and increase to 20–30% until 12,500 years ago (14,650 cal yr BP), with a peak at 53% at 15,100 ^{14}C yr BP (18,020 cal yr BP). After 12,500 ^{14}C yr BP grass pollen frequencies erratically decrease until 9500 ^{14}C yr BP (10,510 cal yr BP), and then remain very low until 3000 ^{14}C yr BP, from none to 2%, followed by an increase to between 20 and 40% and between 3000 and 2000 ^{14}C yr BP (3160 and 1940 cal yr BP), at which point grass pollen declines to the modern level of 7%. The inferred changes in the landscape are as follows (Giresse et al., 1994):

> It is interesting to note that neither of these lakes yields any evidence of change in climate that might be identified with the middle Holocene. There is no change in vegetation or lake level that might distinguish the middle from the early Holocene. The dry phase seen at Lake Bosumtwi just prior to 3000 ^{14}C yr BP (3160 cal yr BP), might be a local marker for the end of the middle Holocene, but it is not evident at Barombi Mbo. On the other hand, the marked dry phase dated between 2500 and 2000 ^{14}C yr BP (2610 and 1940 cal yr BP) at Barombi Mbo occurs in cores from several lakes in the Congo and East Africa. In so far as we know, no explanation has been offered as to the cause of these regional differences.

3. The Sahara

To the north of the belt of intense tropical rains on either side of the equator, mean values of rainfall abruptly decline. The tropical forest is replaced by woodlands and savanna, and by the true desert of the Sahara farther north. In these areas precipitation is often less than 20 mm per year, and in a few places, like the Eastern Sahara of Egypt and Sudan, and the Tanezrouft in southernmost Algeria, precipitation is less than 5 mm per year (Fig. 6.4). In the climatic model outlined above, periods of higher temperatures should mean greater penetration of the ITC into the

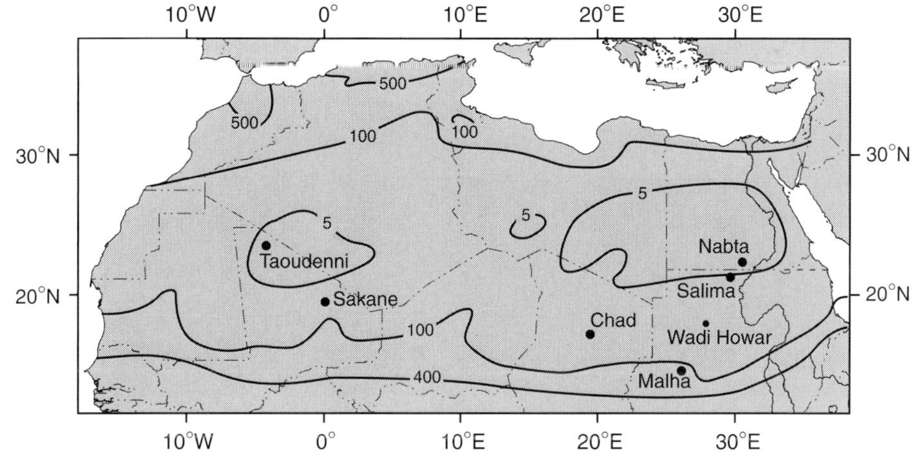

Figure 6.4. Map of North Africa showing locations of areas discussed in text and distribution of annual rainfall (in mm) (redrawn from Haynes and Mead, 1987, p. 88, with additional data from Petit-Maire et al., 1990, p. 336).

Sahara with maximum rain for some distance equatorward of the ITC, with precipitation diminishing between there and the stationary band of sea breeze related to coastal rainfall (Ilesanmi, 1971). Thus rainfall is variable behind the ITC, depending on relative location and longitude (Bryson, 1992; Bryson and Bryson, 1996). The correlation is not precise, because as expected, the climatic changes differ from area to area in intensity and duration. In many respects the Sahara seems to have a very complex climatic record, sometimes more sensitive than the lakes in tropical Africa. This may be because minor declines in monsoon intensity that are invisible in the tropical lake sediments have a greater impact in the marginal areas. In the Saharan seasonal lakes (or playas), abrupt declines in rainfall are recorded either by erosional events, or by reduced ground cover and an increase in wind-blown sand. It is only in the last 15 or 20 years, however, that enough detailed work had been done in a variety of areas within the Sahara, with accompanying suites of radiocarbon dates, to permit correlation between wet and dry episodes here with broader climatic phenomena of regional or even global significance.

The almost lifeless arid and hyper-arid areas of the Sahara have not always been as dry as they are today. The evidence for more moist periods is widespread throughout North Africa, and consists of old and now dry and sand-filled stream channels, numerous internallydrained basins with remnants of deposits testifying to the former presence of permanent or seasonal lakes, abundant fossil faunal remains of aquatic and terrestrial species that today live only in better-watered areas farther south, and carbonized plant remains of types which today occur only in wetter areas several hundred km to the south. Some of this evidence of a more humid past refers to wet intervals during the Last Interglacial or before, but the more obvious and better-preserved evidence is of Holocene age. This evidence also records an unstable climate, with wet periods frequently interrupted by episodes of aridity that profoundly impacted those human groups living in the Sahara.

3.1. Western Sahara

3.1.1. Erg Ine Sakane and vicinity, Mali

The dune and basin area in northeastern Mali and adjacent Algeria has yielded several well-dated Holocene paleoclimatic sequences. One of the most interesting of these is the one from the Ine Sakane sand sea in northeastern Mali (Petit-Maire and Riser, 1981, 1983; Hillaire-Marcel et al., 1983) where five climato-stratigraphic episodes have been defined. The sequence is dated by 30 radiocarbon age determinations, all on fossil mollusk shells. These dates probably should be viewed with caution, because of the tendency of mollusk shell to absorb both older and more recent carbon. The area lies between 20° and 21°N latitude, and today receives precipitation of around 50–60 mm per year. The first phase in the paleoclimatic sequence is a period of hyper-aridity that coincides with the final Pleistocene and the early part of the Holocene. This is followed by two lacustrine episodes separated by a semi-arid interval.

During the first lacustrine episode the lakes were extensive and deep, with well-established shorelines 13 m above the modern floor of the depression. One of the lakes covered an area of around 300 km^2. There are numerous archeological sites associated with the second lake, but surprisingly, they seem to be much less frequent with the earlier lake. The fauna are not identified as to which lacustrine episode they are associated with, but include numerous large animals typical of the African savanna, such as elephant, rhino, hippo, wild buffalo, Barbary sheep, Dorcas gazelle, and zebra. Both young and adult elephants and Barbary sheep are present, suggesting the presence of herds rather than isolated animals. In addition, there are Nile perch, which require deep fresh water, crocodiles, and several species of mollusks, both aquatic and terrestrial. A well-watered savanna landscape is indicated for most of the Holocene in this area.

3.1.2. Taoudenni

Slightly north of Ine Sakane, between 22° and 23°N, in northernmost Mali, an area that now receives less than 5 mm of precipitation per year, is the hydrologically isolated depression of Taoudenni. A chain of lakes existed in this basin during the Holocene. Extensive deposits of salt are an integral part of the lake sediments. Two lacustrine phases are recorded, the first beginning around 9000 ^{14}C yr BP (9980 cal yr BP), with the greatest moisture between 8300 and 6700 ^{14}C yr BP (8740 and 7510 cal yr BP); and the second between 6700 and ca. 4000 ^{14}C yr BP (7510 and 4460 cal yr BP), with effective aridity present by 4500 ^{14}C yr BP (5200 cal yr BP) (Fabre and Petit-Maire, 1988). The chronology of the sequence is based on 22 radiocarbon dates, many of them on wood or charcoal.

One of the most interesting features of the Taoudenni sequence is the evidence that climatic instability prevailed in this area throughout the Holocene. There were four regressions in lake level and a minimum of two lake expansions between 8800 and 8300 ^{14}C yr BP (9780 and 9310 cal yr BP), or one cycle about every 250 years. There were four more cycles between 8300 and 6700 ^{14}C yr BP (9310 and 7510 cal yr BP), and either eight or nine cycles between 6760 and 3840 ^{14}C yr BP (7550 and 4230 cal yr BP). On an even smaller scale, microcycles of alternating clay and halite lamina are also evident. These are estimated to represent events of about 50 years duration, or possibly short events with a 50-year return interval.

These two major wet intervals also seem to be represented farther north, in the northwestern Sahara, where two intervals of humid conditions are separated by periods of aridity in the Mellala Sebkha and inter-dunal ponds in the Grand Erg Occidental (Fontes and Gasse, 1989). There the early humid period is dated between 9300 and 7200 ^{14}C yr BP (10,280 and 7960 cal yr BP), and the second between 6200 and 4500 ^{14}C yr BP (7090 and 5150 cal yr BP).

3.2. Central Sahara

3.2.1. Chad basin

The central southern Sahara has one of the most important and complete records of the Holocene environment available for Africa. The Chad basin can be fed by runoff from the central Saharan highlands, by local rainfall, and by the influx of water from rivers flowing from the south. Paleoenvironmental research on this basin has been concerned with the complex interaction of these three sources of water, each of which carries a distinctive suite of pollen. The current rainfall in the Chad basin is 350 mm per year, and it has been known since the early 1900s that Lake Chad has experienced major oscillations in the past. At one time it was as large as the Caspian Sea. Extensive geological and palynological investigations in the Chad basin have permitted detailed environmental reconstructions for this area during most of the Holocene (Maley, 1977a,b; Rognon and Williams, 1977).

The geological, palynological, and diatom studies resulted in the subdivision of the early Holocene into several episodes (Fig. 6.5). The detailed chronology proposed by Maley is based in large part on the projection of radiocarbon dates from many localities onto two main pollen diagrams, and may, therefore, be subject to some error. There may also be other problems with the chronology because some of the dates are from carbonates. After the late Pleistocene period of hyper-aridity, there were three weak, but distinct expansions of the lake, each separated by arid episodes, sometimes accompanied by eolian action. There was an abrupt regression of the lake accompanied by dune migration and erosion for which there are two dates of 10,900–10,300 ^{14}Cyr BP. This arid interval may correlate with the Younger Dryas.

Following these early events is a complex period of multiple transgressions extending from about 9200 ^{14}C yr BP to the present. There are five transgressions during this period, not all of which were accompanied by local rainfall.

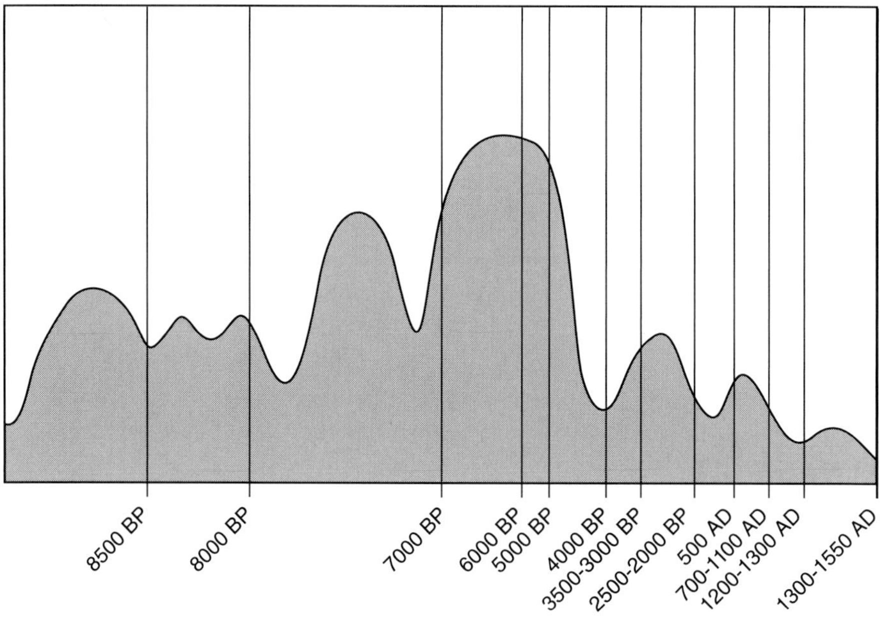

Figure 6.5. Holocene lake levels in Chad (redrawn from Maley, 1977a, p. 574).

During Stage II_1 there was local rainfall as well as influxes of water from the central Saharan highlands and from the south, followed by an abrupt interval of local aridity (8500–8400 ^{14}C yr BP; 9460–9410 cal yr BP). Calculations of combined hydrological and energy balances of the paleolake yielded an estimate of at least 650 mm per year of precipitation during this period (Kutzbach, 1980). Another interruption occurs between 8200 and 8100 ^{14}C yr BP (9120 and 8990 cal yr BP). The Stage II_2 transgression was caused by water coming from the south, contemporaneous with local arid conditions. A major period of aridity occurred between Stages II_3 and II_4 (7300–7100 ^{14}C yr BP; 8060–7870 cal yr BP) when Lake Chad almost went dry and there was no influx of water from either the north or south. Local monsoon rains and water from both the highlands and the south occurred with the Stage II_4 transgression, with the maximum highland input between 7200 and 6600 ^{14}C yr BP (7960 and 7450 cal yr BP). This seems to represent a middle Holocene interval of increased rainfall in the central Sahara. A long regression in the lake occurred after Stage II_4 between 5000 and 3900 ^{14}C yr BP (5730 and 4350 cal yr BP), and several minor fluctuations characterize Stage II_5.

Recent work in the Nigerian Chad basin by Breunig and Neumann (2002) note that the earliest evidence of human presence in the basin during the Holocene is a dugout canoe that is dated 8100 ^{14}C yr BP. Somewhat later, two sites with Saharan-type pottery, and located on the Bama Ridge (the southwestern shoreline of the early and mid-Holocene "Mega Chad") are dated on charcoal between 6600 and 6000 ^{14}C kyr BP. The Mega Chad apparently disappeared during a period of aridity between 6000 and 3600 ^{14}C yr BP. The exposed lake floor was colonized around 3500 ^{14}C yr BP by cattle pastoralists, who may have come into the area from the north, where pastoralists with similar pottery are known. Two phases are represented, and intensive plant collecting occurred in both phases. The plant remains, known only from plant impressions in pottery, consist of several varieties of wild grasses, rice and Paniceae in Phase 1 and early Phase 2. Grains of domestic *pennisetum* appear around 3000 ^{14}C yr BP, but represent only a minor component of the grass identified, and only become dominant near the end of Phase 2.

3.2.2. Southwestern Fezzan, Libya

An important body of new evidence for a complex sequence of abrupt Holocene climatic changes in the central Sahara comes from extensive geoarcheological studies of the southwestern Fezzan in Libya (Cremaschi, 1998, 2002; Cremaschi and di Lernia, 1998; Garcea, 2001; di Lernia, 2002). Wetter phases are indicated by a variety of proxy data, including alluvial deposits, rock shelter and cave sediments, travertines, paleosols, and lacustrine sediments. The oldest stratigraphic unit found in shelters and caves is reddish Aeolian sand that is correlated with fossil dunes observed outside the caves and thought to indicate a period or periods of aridity. This arid period must have been broken by brief intervals of moisture, because middle Paleolithic Aterian occupations occur in the red sand in two of the caves. Several thermoluminiscence (TL) and Optically Stimulated Luminescence (OSL) dates place the red sand and the Aterian occupations between 61,000 and 90,000 cal yr BP.

Terminal Pleistocene and early Holocene wet conditions are indicated by travertines deposited along bedding planes of the sandstone bedrock (Cremaschi, 2002). These travertines record a high precipitation rate that recharged the hydrographic net inside the mountain. Several U/Th dates indicate that they were deposited from 15,600 to 9700 cal yr BP. These dates indicate that the final Pleistocene in the Fezzan was much wetter than any interval in the Holocene. Also, the absence of travertine as young as 9000 cal BP is regarded as evidence that the onset of aridity was underway by that date. It should be noted, however, that lacustrine deposits that could correlate with this period of maximum wetness are not recorded in the surrounding and immediately adjacent lowland and dune areas, as would be expected if the final Pleistocene was even as wet as the early Holocene. The suggestion that the onset of aridity began around 9000 cal yr BP also conflicts with the evidence for increased rainfall and large lakes in the lowland and dune areas in the early Holocene, and the presence of numerous Epipaleolithic sites (elsewhere dated between 9800 and 8500 ^{14}C yr BP) along the shores of these fossil lakes. It seems to us that there may be some problem with these U/Th dates. They may be too recent. Despite the impressive series of U/Th dates, some consideration should be given to the possibility that the deposition of these travertines may have occurred much earlier, possibly during one or more of the wet phases of the Last Interglacial.

Before 9000 ^{14}C yr BP, southwestern Libya was occupied by pottery-using hunter-gatherers. The early Holocene human settlements in the caves and shelters record two cycles of occupation, each of which begins in an interval of relative wetness and gradually becomes more arid. The first cycle begins around 9800 ^{14}C yr BP with the deposition of the initial organic rich anthropogenic deposits. The latest occupation in this cycle is dated 8000 ^{14}C yr BP. The second cycle begins around 7200 ^{14}C yr BP, or perhaps a little later. Well defined geological proxy in the caves as well as in the lakes in the dune area, indicate a major period of aridity during the period between 8000 and 7500 ^{14}C yr BP. There are very few settlements known that date during this period. The second cycle begins about 7200 ^{14}C yr BP and ends around 3700 ^{14}C yr BP. Shortly after the onset of the second interval of increased wetness cattle and sheep/goat pastoralists appear in the area and occupy some of the caves and lake shores (di Lernia, 2002). Fossil pollen and a break in the radiocarbon sequence indicate another abrupt dry interval that coincided with an abandonment of the area, between 6400 and 6100 ^{14}C yr BP. Severe dry conditions seem to have began around 5000 ^{14}C yr BP, as indicated by the collapse of shelter roofs and the end of desert varnish deposition.

Other climatic and cultural data for the southwestern Fezzan come from the extensive lakes that covered a third of the whole area. These were formed mostly inside the dune areas and are related to a rise of the water table. Numerous Epipaleolithic and pastoral sites occur along the shores of these lakes. As was noted in the caves, two cycles are represented; the first is dated between 8400 and 7400 ^{14}C yr BP (the excavations in the lake sites have been limited, and it is likely that some of the early Epipaleolithic sites date earlier, perhaps as early as 9000 or 9800 ^{14}C yr BP). A brief period of aridity follows the first lake cycle, with the second cycle beginning around 7300 ^{14}C yr BP. The first pastoral sites (with cattle and ovicaprines) are apparently associated with this second lacustrine cycle. Those sites with sheep/goat are unlikely to be any older than 7000 ^{14}C yr BP, because that is about the oldest date for sheep/goat in Egypt, the most likely source for both the cattle and the sheep/goat. The end of this second lake event is not well dated, but it most likely ended after 5200 ^{14}C yr BP, the most recent date for a pastoral site in the lake area.

3.3. East Africa

As mentioned above, precipitation is the climatic factor of importance for changes in vegetation boundaries in most of Africa, and most studies therefore focused on changes in precipitation. However, there are some studies that have also yielded information about temperature.

A few of these are from Mount Kenya in East Africa. Coetzee (1967) studied the pollen assemblage in a lake near the present tree limit (2400 m a.s.l.) and found evidence of an amelioration after 14,050 + 360 ^{14}C yr BP (16,850 cal yr BP). The conditions remained the same up to about 12,600 ^{14}C yr BP (14,800 cal yr BP). A distinct minimum occurred around 11,000 ^{14}C yr BP (12,920 cal yr BP) but the climate had returned to previous conditions by 10,560 ^{14}C yr BP (11,950 cal yr BP).

A distinct warm and wet period is dated to an interpolated age of 6400 ^{14}C yr BP (7300 cal yr BP). The midpoint of an approximately 1000-year long cold event is dated to 5860 ^{14}C yr BP (6700 cal yr BP). This was followed by the warmest period indicated in the entire pollen diagram at 4000 ^{14}C yr BP (4400 cal yr BP), a period that was also wet. A cooler and dryer period appears to have begun around 2400 ^{14}C yr BP (2400 cal yr BP). Coetzee presents additional information retrieved from Cherangi Hills, where relatively cool and dry periods are dated to between 6330 and 4820 ^{14}C yr BP (7200–5550 cal yr BP). A cold period at Lake Rutundi is dated to 2090$\pm$215 ^{14}C yr BP (2000 cal yr BP). A pollen study from Rusaka, Burundi indicates cool periods around 7400 ^{14}C yr BP (8150 cal yr BP), and after a warm period around 6800 ^{14}C yr BP (7600 cal yr BP) another cooling at about 5100 ^{14}C yr BP (5900 cal yr BP). Around 4500 ^{14}C yr BP (5150 cal yr BP) the climate was again relatively warm. A trend toward cooler climate began 3900 ^{14}C yr BP (4350 cal yr BP) and the lowest point was reached around 2000 ^{14}C yr BP (1950 cal yr BP). After a warm period 1900–1100 (1850–1000 cal yr BP) the climate became cooler and remained cool up to recent time (Bonnefille et al., 1995). A short cool event around 1200 ^{14}C yr BP (1150 cal yr BP) was superimposed on this general trend.

A 3.3 m sediment core from a pro-glacial lake on Mount Kenya yields continuous information about rock-flour influx during the last 6000 years (Karlén et al., 1999). This rock-flour is mainly coming from the erosion of two glaciers, which have retreated rapidly during the last 100 years. Several short events were superimposed on the long-term trends of the mid-Holocene record (Fig. 6.6). The glaciers were large prior to 5000 ^{14}C yr BP (5700 cal yr BP). After a major recession the glacier again expanded after about 4000 ^{14}C yr BP (4500 cal yr BP) and remained active until about 2700 ^{14}C yr BP (2800 cal yr BP). Minor fluctuations occurred during this period but a major recession took place first between 2700 ^{14}C yr BP and 2200 ^{14}C yr BP (2800 and 2200 cal yr BP). After a 500 years period of active glacier erosion centered around 2100 ^{14}C yr BP (2000 cal yr BP), the size of the glacier appears to have diminished. However, short expansions occurred around 1400 (1300 cal yr BP), 600 (600 cal yr BP), and 350 ^{14}C yr BP (400 cal yr BP). The sediments between 2700 and 2200 ^{14}C yr BP (2800–2300 cal yr BP) are interpreted as indications of cold-based glacier conditions. The climate is likely to have been relatively cold during periods of advancing glaciers and particularly cold during the period of cold-based glaciers.

Oxygen isotopes from a lake on the eastern side of Mount Kenya, Small Hall Tarn (Barker et al., 2001; Gasse, 2002), like the data from the Hausberg Tarn located on the western side of the mountain, indicate a major change in climate around 5850 ^{14}C yr BP (6650 cal yr BP). At least partly this change in ^{18}O composition is likely to have been caused by increased precipitation. However, a lowering of the amount of arboreal pollen in a bog in Burundi (Rusaka, Bonnefille et al., 1995) and several lake levels indicate a slightly dryer climate around this date. The Small Hall Tarn record indicates heavy precipitation between 2900 ^{14}C yr BP (3000 cal yr BP) and 2050 ^{14}C yr BP (2000 cal yr BP), a period for which the Hausberg Tarn data indicate

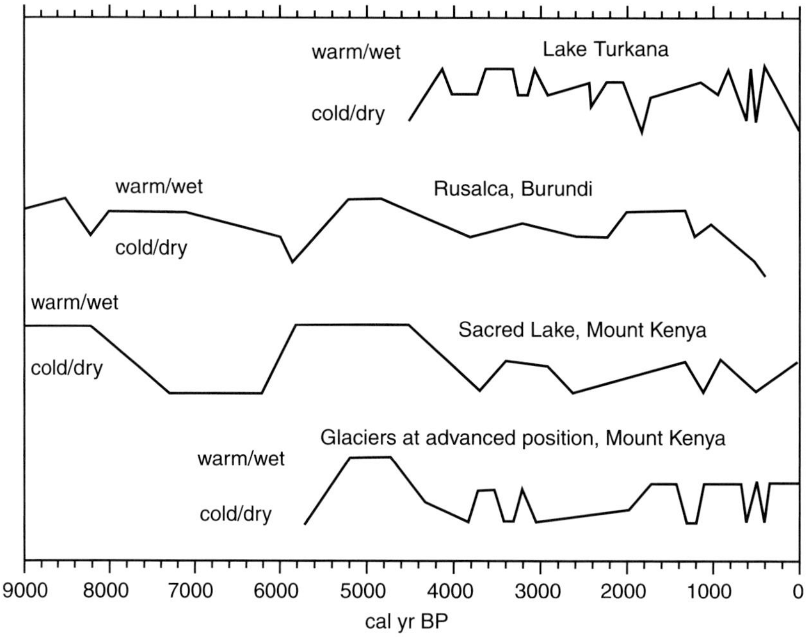

Figure 6.6. Generalized description of changes in the mid-Holocene climate. Several records indicate cold/dry conditions around 5250 ^{14}C yr BP (6000 cal yr BP). Warm/wet conditions dominated the climate around 4450 ^{14}C yr BP (5000 cal yr BP). A long dry phase between 4150 and 2050 ^{14}C yr BP (4500 and 2000 cal yr BP) was briefly interrupted around 3450 and 3150 ^{14}C yr BP (3700 and 3200 cal yr BP) and possibly also around 2500 ^{14}C yr BP (2500 cal yr BP) (Bonnefille et al., 1995; Karlén et al., 1999; Gasse, F., 2000).

both episodes of wet and dry climate. The water level of studied lakes in East Africa is likewise not entirely consistent with the results from Hausberg Tarn. Possibly, shifts in the monsoon circulation can have a distinct effect on the local climate.

Most climatic records are based on the interpretation of proxy data, which only yields indirect information about climate such as vegetation, lake level, or glacial response to a change in several parameters of the climate. Frequently the age models are based on only a few dates and because rate of sedimentation varies, presented dates may be off by several hundred years. Considering the often relatively low resolution, it is impossible to claim simultaneous changes in climate. However, large-scale variations in climate seem similar for several records from East Africa. Climate was cool and dry around 5250 ^{14}C yr BP (6000 cal yr BP), and it became substantially warmer and wetter around 4450 ^{14}C yr BP (5000 cal yr BP), but after this date there was a trend toward cooler and drier conditions. Except for two short reversals around 3450 and 3050 ^{14}C yr BP (3700 and 3200 cal yr BP), the trend continued up to about 2750 ^{14}C yr BP (2800 cal yr BP). Around this date the climate improved and was relatively warm during the following 600 years. A large expansion is dated to around 2100 ^{14}C yr BP (2000 cal yr BP) and minor advances are dated to 1400 ^{14}C yr BP (1300 cal yr BP), and around 600 and 350 ^{14}C yr BP (600 and 400 cal yr BP).

Records based on variations in the size of glaciers on Mount Kenya indicate several cool periods during the last 6000 years. Glaciers respond sensitively to changes in temperature and solid precipitation (Paterson, 1994). Paleorecords of glacier variations therefore yield information on changes in climate. While dates on the morphology of fronting glaciers reveal discontinuous information about major events, pro-glacial lacustrine sediments can provide continuous records on changes in glacial erosion. Because the rock-flour production for wet-based glaciers is proportional to the size of the glacier and the velocity with which it moves over the substrate, the content of rock-flour in relation to the organic carbon in pro-glacial sediments reveals variations in the activity of wet-based glaciers (Karlén, 1976, 1981; Leonard, 1986a,b; Gilbert and Desloges, 1992; Matthews and Karlén, 1992; Leemann and Niessen, 1994). The size that an advancing glacier will attain depends on its size at the onset of the climatic event, as well as on the severity and the duration of the climatic event. For an advance lasting long enough to permit a glacier to approach its steady state (in which there is a balance between mass balance and glacier size), the climate and the morphology in the area of the glacier snout will determine the length of the advance. Since even relatively large glaciers have a response time of less than 100 years (Nye, 1960; Paterson, 1994), the small glaciers on Mount Kenya are likely to approach a steady state situation even when responding to short-term climate changes. In contrast to wet-based glaciers, a cold-based glacier is frozen to the substrate and the ice does not erode the substrate. A cold-based glacier therefore does not release any rock flour; it is also relatively unresponsive to climatic change.

Several ^{14}C dates on sediments from a small pond dammed between a bedrock ridge and a distinct end moraine, formed by the expansion and confluence of Lewis, Darwin, and Tyndall Glaciers, indicate a major glacier advance shortly before 4950 ^{14}C yr BP (5700 cal yr BP). Even dates from a nearby lake indicate a glacier retreat from an advanced position before 4750 ± 90 ^{14}C yr BP (5710 cal yr BP). Modeling of glacier response on changes in climate shows that an advance of the discussed size could be caused by a lowering of the temperature by 0.5°C during a few hundred years. An advance could be a result of increased precipitation, but this is not likely for this particular event because several records from the surrounding area indicate that the climate during this period was relatively dry.

3.4. Eastern Sahara

Today the Eastern Sahara is one of the driest places on earth. It is the driest part of the Sahara, with annual precipitation of less than 1 mm per year. During the early and middle Holocene, however, there were intervals of significant increase in rainfall, at times possibly as much as 200 mm per year in southern Egypt west of the Nile, much of which came as summer monsoons. Even during the wet periods, however, it remained an unstable environment with frequent droughts. Except for these abrupt droughts, of which five, possibly six are recorded (Schild and Wendorf,

2002), the area was relatively moist between 10,000 and 4500 ^{14}C yr BP (11,270 and 5150 cal yr BP), and again briefly around 3800 ^{14}C yr BP (4150 cal yr BP).

In the atmospheric circulation model discussed above, rainfall during periods of monsoon expansion is weaker at higher latitudes than in the tropics. This monsoon gradient hypothesis can be evaluated by looking at the data from a series of localities arranged in a rough south–north line from northern Sudan to southern Egypt. The southernmost locality is Malha Crater, which today is near the southern edge of the desert area of northern Sudan. The most northerly localities are Nabta Playa and Bir Kiseiba both located in southern Egypt, near the northern limit of the monsoon rains during most of the Holocene. The intervening localities are the Wadi Howar, which is about 300 km north of Malha Crater, and the Oyo and Selima basins, both situated between the Wadi Howar and Nabta Playa.

3.4.1. Malha Crater

In northern Darfur, a paleolimnological study of a crater lake has yielded a rainfall record for that area which has interesting similarities and differences as compared to locations farther north at Nabta and Bir Kiseiba. Malha is located near 15°N latitude, some 7°S of Nabta and Bir Kiseiba, in an area that today has a mean precipitation of slightly less than 115 mm per year. A 9.25 m core was taken from the lake and analyzed, and a radiocarbon date of 8290 ^{14}C yr BP (9310 cal yr BP) was obtained from the base of the core (Mees et al., 1991; Dumont and El Moghraby, 1993).

Primarily on the basis of changes in salinity, six periods were defined, each separated from the others by intervals much drier than today, and sometimes the lake was completely dry (Fig. 6.7). The ages of these wet and dry intervals have been estimated, but not directly dated. The overall trend during the last 8300 (9310 cal yr BP) years has been "progressive aridification, punctuated by brief, intensely dry intervals" (Dumont and El Moghraby, 1993, p. 385). At present Malha lake is permanent and saline, but the lake was fresh and deep during Stages I and III. It was brackish during Stages II and IV. Compared to previous stages, it has been generally drier since ca. 3000 ^{14}C yr BP (3160 cal yr BP) (Stages V and VI), and much drier than today around 2000 ^{14}C yr BP (1940 cal yr BP), followed by an increase in precipitation that has continued until today. After ca. 7200 ^{14}C yr BP (7950 cal yr BP) the indicated rainfall fluctuates between more arid than today (punctuated by five episodes of hyper aridity) and intervals of slightly more moisture than at present. The period between ca. 7000 and 3000 ^{14}C yr BP (7770 and 3160 cal yr BP), the middle Holocene, seems to be set apart by abrupt periods of hyper-aridity at ca. 7000 and 3000 ^{14}C yr BP, of which, the ca. 7000 ^{14}C yr BP arid episode might be contemporary with a similar event in the Egyptian Sahara that is dated between 7300 and 7100 ^{14}C yr BP (8060 and 7870 cal yr BP).

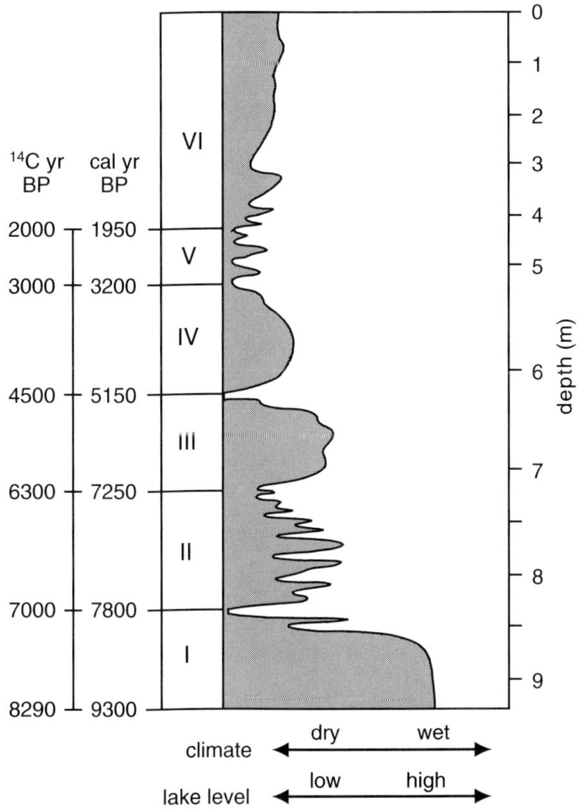

Figure 6.7. Reconstructed lake levels in Malha Crater, northern Sudan (redrawn from Dumont and El Moghraby, 1993, p. 384).

3.4.2. Wadi Howar

In northern Sudan, between 17° and 18°N, is the now dry Wadi Howar, a long west–east drainage net that at one time flowed into the Nile near the Dongola Bend. Although now a hyper-arid region, at times during the Holocene there were numerous marshes and large, deep (more than 10 m) freshwater lakes that formed behind dunes on the floor of the wadi. Radiocarbon dates indicate moist conditions prevailed in this area from before 8585 ^{14}Cyr BP (9500 cal yr BP) until after 3825 ^{14}C yr BP (4170 cal yr BP) (Kropelin, 1993). Numerous remains of large savanna animals, including elephant, rhino, and hippo, have been found associated with the lacustrine sediments, and suggest that the wadi was surrounded by a highly developed savanna ecosystem. Although detailed reports on the ceramics from sites along the margins of Wadi Howar have been published (Keding, 1997; Jesse, 2003), only preliminary reports of the paleoenvironmental research are available (Kuper, 1986, 1989).

3.4.3. Selima Oasis and Oyo

These are two basins located in the hyper-arid region of northern Sudan, an area that today probably receives between less than 1 and 5 mm of precipitation per year, usually during the summer. Today the only water available is in shallow wells. At Selima the water level is less than 1 m below the surface in the lowest part of the depression, however, a marsh was seen there in 1821 (Haynes et al., 1989). Both basins contained large, deep lakes in the Holocene, and the chronologies of both sequences are controlled by large suites of radiocarbon dates.

Pollen and sedimentary analyses of the lake deposits at Selima indicate that in the Early Holocene a permanent lake was present from around 9700 to after 9000 ^{14}C yr BP (10,930 to after 9980 cal yr BP), followed by a period of aridity and deflation. The lake returned around 8400 ^{14}C yr BP (9410 cal yr BP) and continued until ca. 7000 ^{14}C yr BP (7790 cal yr BP), with maximum wet conditions around 8000 ^{14}C yr BP (8740 cal yr BP), at which time the lake was about 20 m deep. Pollen recovered from the lake sediments indicates the vegetation at that time was an open scrub savanna, consisting of sparse woodlands and scrub mixed with grasslands on pediment slopes and sand sheets (Haynes et al., 1989 p. 132). During this period Selima was a savanna, but it was not far from the savanna-desert transition area. The vegetation changed around 7000 ^{14}C yr BP (7790 cal yr BP) to a semi-desert steppe until ca. 6000 ^{14}C yr BP (6830 cal yr BP), when increasing aridity is evident. The water table continued to drop and all surface water apparently was gone by 4000 ^{14}C yr BP (4460 cal yr BP). A similar sequence occurs at Oyo where between 8000 (8840 cal yr BP) and 6100 ^{14}C yr BP (6950 cal yr BP) there was a deep lake, and the pollen indicates an annual monsoon rainfall of around 400 mm per year (Ritchie et al., 1989).

3.4.4. Nabta Playa and Bir Kiseiba

The Southwestern Desert of Egypt is one of the driest places on earth. Today it receives less than 1 mm of rain per year, and it is seemingly devoid of life. It has not always been this dry, but the last time it was really wet with permanent lakes and flowing streams was in the final phase of the Last Interglacial, about 70,000 years ago, when there was about 500 mm of rain per year. From then until the Final Pleistocene, around 12,000 to 13,000 ^{14}C yr BP (14,000– 15,200 cal yr BP) the Southwestern Desert was hyper-arid (Wendorf et al., 1984, 1993). At that time the desert began to receive some rainfall, but not very much, probably around 100 mm per year, and mostly during the summer months, as the tropical monsoons of central Africa expanded and began to move northward into what is now southernmost Egypt. These rains were sufficient only to form seasonal lakes, or playas that left fluvial sediments on the floors of the deflational basins that had formed during preceding intervals of aridity.

The evidence that the monsoon rains began in this area before the beginning of the Holocene and during the Final Pleistocene comes from two basins, one known as El Adam Playa, near Bir Kiseiba, and the other, El Gebal El Baid Playa located about 50 km north of Nabta Playa. In both playas there are sites of the earliest known Holocene archeological entity in the area, and with several dates on charcoal between 9800 and 8870 ^{14}C yr BP (11,200 and 9700 cal yr BP), and below these sites in both basins there are from 5 to 6 m of sandy playa sediments with a band of sandy silts that suggest an even wetter episode. The lithostratigraphy of the playa deposits is known only from boreholes; therefore, no clear unconformities have been observed. A break, however, perhaps recording a period of aridity, has been postulated above the silty sand lens. This inferred break is correlated with the dry phase of the Younger Dryas. If this can someday be confirmed, the onset of monsoon rains in this area may have begun before 13,000 ^{14}C yr BP (15,100 cal yr BP), and might be correlated with the Allerod Oscillation. There is, however, no reliable basis for estimating how rapidly the early playa sediments seen in the boreholes accumulated.

The lithostratigraphic evidence for Holocene and Final Pleistocene climatic events in the Nabta-Kiseiba area is based on signals that are rather monotonous and limited to the accumulation of suites of clastic sediments in several enclosed, internally drained basins by aeolian, lacustrine, and alluvial phenomena, interrupted by major unconformities caused by aeolian erosion. The periods of accumulation are generally coeval with wet interphases and local rainfall, while the intensive aeolian erosion usually marks hyper-arid phases. Numerous radiocarbon measurements (170) as well as the association of the deposits with well-dated prehistoric entities support a relatively precise chronological placement of the deposits (Fig. 6.8).

A different archeological entity, with different lithic typology and sometimes technology, characterizes each Holocene age wet interphase. These entities disappear from the desert with the onset of the following hyper-arid phase. These abrupt dry periods must have been very severe, because each one not only saw the entire area abandoned, but also the basins were scoured out by wind deflation, dunes were deposited, and the water table was lowered to levels close to those of today. Then, with the onset of the following wet interphase the desert was reoccupied. Sometimes the new groups share many features with the preceding entity, however, more often they are very different, and may indicate that a new population entered the area. Based on extensive new work at Nabta and the surrounding basins, the following revised sequence of arid phases and humid interphases has been defined for the Nabta-Kiseiba area (Schild and Wendorf, 2001a,b; Wendorf and Schild, 2001; Fig. 6.9):

Pre-El Adam Humid Interphase (ca. 13,000–11,000 ^{14}C yr BP). No known occupation. Allerod Oscillation?
Inferred Pre-El Adam Arid Phase (11,000–10,000 ^{14}C yr BP). No known occupation. Younger Dryas?
El Adam Humid Interphase (9800–8700 ^{14}C yr BP). El Adam Early Neolithic.
Post-El Adam Arid Phase (8700–8600 ^{14}C yr BP). No occupation.
El Ghorab Humid Interphase (8550–8200 ^{14}C yr BP). El Ghorab Early Neolithic.
Post-El Ghorab Arid Phase (8200–8100 ^{14}C yr BP). No occupation.

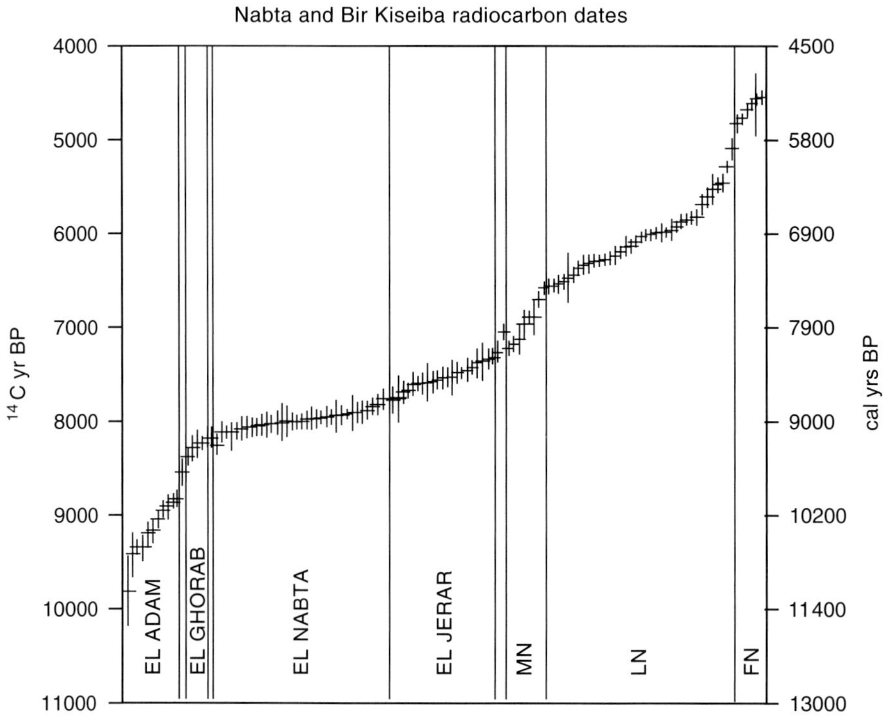

Figure 6.8. Plot of uncalibrated radiocarbon age determinations from Nabta and Bir Kiseiba that are directly tied to the stratigraphic sequences of those areas (data from Schild and Wendorf, 2001b).

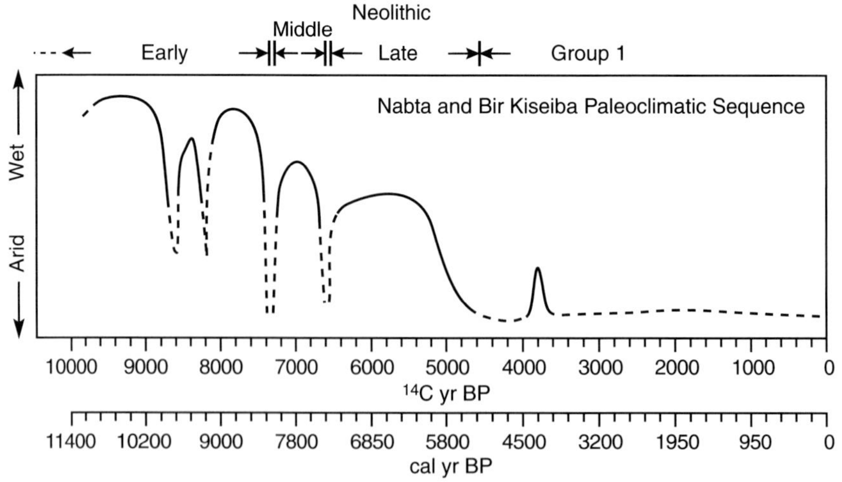

Figure 6.9. Indicated changes in precipitation at Nabta Playa during the Holocene.

El Nabta/Al Jerar Maximum Humid Interphase (8050–7300 ^{14}C yr BP). El Nabta/ Al Jerar Early Neolithic.
Post-Al Jerar Arid Phase (7250–7100 ^{14}C yr BP). No occupation.
Ruat El Ghanam Humid Interphase (7100–6600 ^{14}C yr BP). El Ghanam Middle Neolithic.
Post-Ru'at El Ghanam Arid Phase (6600–6500 ^{14}C yr BP). No occupation.
Ru'at El Baqar Humid Interephase (6500–5800 ^{14}C yr BP). El Baqar Late Neolithic.
Post-Ru'at El Baqar Arid Phase (5800–5500 ^{14}C yr BP). No occupation.
Bunat El Asnam Humid Interphase (5500–4500 ^{14}C yr BP). El Asnam Final Neolithic
Modern interval of aridity (after 4500 ^{14}C yr BP, broken by short moist phases around 3800–3600 ^{14}C yr BP, C-Group settlement).

There is almost no direct evidence of the amount of rainfall during any of these Holocene wet episodes, except that the rains were sometimes heavy, as indicated by the extensive beach and shore features that in some wet interphases relate to large, deep (5 + m) seasonal lakes. The major source of paleoenvironmental data for this interval comes from proxy data recovered from the numerous archeological sites that have been excavated and studied in detail (Hester and Hobler, 1969; Schild and Wendorf, 1977, 1981, 2001a,b,c; Gautier, 1980, 1984, 2001; Wendorf and Schild, 1980, 2001; Wendorf et al., 1984; Kuper, 1989; Neumann, 1989, 1993; McDonald, 1991, 1993, 1998; Wendorf and Close, 1992; Wasylikowa et al., 1993, 1995, 1996, 2001; Barakat, 1995). Additional valuable proxy data have come from several nonarcheological stratigraphic, paleontological, and palynological studies by earth scientists (Haynes et al., 1989; Pachur and Kropelin, 1989; Pachur and Hoelzmann, 1991).

El Adam Early Neolithic

The first known settlements associated with playa deposits of Final Pleistocene–early Holocene age have 13 associated radiocarbon dates on charcoal between 9800 and 8750 ^{14}C yr BP (11,200 and 9700 cal yr BP; Schild and Wendorf, 2001b,c). These earliest known sites are very small, probably occupied by a few individuals for only short periods of time. Some of these localities show evidence of multiple reuse, but even these are relatively small, with thin cultural debris confined to areas rarely exceeding 100 m^2, again suggesting brief seasonal occupations (Connor, 1984; Wendorf et al., 1984).

These earliest known sites are located in the lower parts of deflational basins that were seasonally flooded during and immediately after the summer rains. Because none of these early sites show traces of water wells, it is presumed that they were occupied only during the early part of the dry season, when water was still available in the deepest part of the basins. Without water wells, it is unlikely that any of the known early sites could have been occupied during the driest time of the year, winter and spring, when there was no surface water.

In the area between Nabta Playa and Bir Kiseiba, from ca. 100 to 250 km west of the Nile, and from the Eocene plateau on the north, at about 23°N latitude, south to the Sudan border, and probably beyond, these earliest small settlements are believed to have been occupied by cattle herders, some of whom may have been special task groups (Gautier, 1980, 1984, 1987, 2001; Wendorf and Schild, 1994, 2001). Almost all of these early camps have yielded rare bones of domestic cattle. They are among the oldest domestic cattle known, and they indicate a possible independent center for cattle domestication in northeast Africa. This interpretation is also supported by mtDNA analyses of modern cattle (Bradley et al., 1996). The fact that cattle remains are rare in these sites has led to the suggestion that the modern African pattern of cattle use, where cattle are exploited mostly for milk and blood, and are rarely killed for meat, evolved in this or a nearby desert area. The cattle by-products may have been an important source of food to these earliest groups using this area, and was a kind of "walking larder" (Close, 1990).

Other fauna found in these sites are all small, desert-adapted species that do not require surface water, but obtain moisture from vegetation and dew. They are mostly small gazelle, hare, and a few small carnivores. The character and species paucity of the fauna clearly indicates an environment with very low carrying capacity. Although preservation in these open sites is poor, a few carbonized plant remains, mostly grass seeds, have been recovered from these earliest sites. The presence of grinding stones in almost all of these localities indicates that plant foods may have been a relatively consistent food resource. Characterized as "hunter-gatherer-cattle herders," it is obvious that the known sites represent but one segment of their seasonal round. The presence of cattle, which need to drink at least every other day, suggests that there must be other sites of this entity in an adjacent area where water was available in the dry season from dug wells, or where there was permanent water on the surface. In either instance, there also had to be adequate dry-season forage for the cattle.

The archeology from these earliest hunter-gatherer-cattle herder sites includes a characteristic lithic tool assemblage and rare, well-made, pottery, made with apparently local crushed granite, and with distinctive rocker stamped designs (Nelson, 2001). These sites are grouped into a taxonomic entity known as El Adam Early Neolithic, and the interval of relatively moist conditions from around 10,000 to 8750 ^{14}C yr BP (11,500 to 9700 cal yr BP) is termed the Adam Humid Interphase (Schild and Wendorf, 2001a,b; Wendorf and Schild, 2001).

El Ghorab Early Neolithic

There was a brief, sharp arid episode after 8750 and before 8550 ^{14}C yr BP (9700 and 9500 cal yr BP). It is identified as the Post-El Adam Arid Phase, and it coincides with the disappearance of the Adam entity from the Southwestern Desert of Egypt. In the subsequent moist interval, the Ghorab Humid Interphase, the area was reoccupied by small groups who also apparently lacked water wells and used the area only seasonally, much as did El Adam groups. A few El Ghorab sites have yielded traces of small, shallow basin-shaped hut floors. They also were hunter-gatherer pastoralists, and share many elements with the earlier El Adam. The Ghorab phase lasted until around 8250 ^{14}C yr BP (9200 cal yr BP) when there was another abrupt, intense arid interval, the Post-El Ghorab Arid Phase, when the area was again abandoned.

El Nabta/Al Jerar Early Neolithic

The beginning of El Nabta/Al Jerar Humid Interphase, dated between 8050 and 7300 ^{14}C yr BP (9500 to 8000 cal yr BP; Schild and Wendorf, 2001c), marks the interval of the highest rainfall in the Southwestern Desert since the Last Interglacial. This was the Holocene Maximum in this part of Egypt, and the period when there are more and larger sites and presumably the largest population at any time during the entire Holocene. Annual rainfall may have been around 200 mm, or slightly more. The increased rainfall supported a wide variety of grasses, herbs, legumes, tubers, shrubs, and trees. Over 20,000 charred plant remains belonging to 130 taxa, including 10 varieties of trees and shrubs, were recovered from one El Nabta site, E-75-6. The most frequent were edible plants, including sorghum and two varieties of millets, all of which were wild, although there are suggestions that the sorghum may have been cultivated (Barakat, 2001; Krolik and Schild, 2001; Wasylikowa et al., 2001). All but a few of these identified plants grow today in southwestern Egypt, but only as scattered small patches and never in dense stands. To find dense stands one would have to look farther south in northern Sudan, in the northern section of the Sahelian belt.

It is also interesting that during this more than 800 year Humid Interphase at Nabta Playa, rates of sedimentation were sharply reduced, with less than 2 m of total playa accumulation, in spite of the considerable influx of water that flooded the lower part of the basin. This strongly indicates a relatively dense grass cover over the entire, huge catchment area. In contrast, over 3 m of playa was rapidly deposited in the same area during a brief (ca. 100 years), relatively dry phase that immediately followed, when there were still some rains, but the plant cover over the uplands was apparently significantly reduced (Schild and Wendorf, 2001a).

The groups who reoccupied the Nabta-Kiseiba area around 8050 ^{14}C yr BP (9500 cal yr BP) were economically still hunter-gatherer-cattle pastoralists, but they came with, or rapidly developed, a different economy, a new settlement system, and several new technologies, all of which functioned as an integrated system. These new technologies made it possible to live in the desert throughout the year, including the driest period, and to exploit the available resources more efficiently and intensively. Among the new technologies was the ability to dig large, deep wells to provide water for themselves and for their cattle. There was also a change in the way plant foods were used, from immediate consumption to intensive harvesting and storage in pits for use when other food resources were not available. There were a few other minor, but interesting changes as well. Although a few sherds of well-made, rocker stamped pottery occurred in the earliest El Adam settlements, and in all subsequent Early Neolithic variants, and the temper in these sherds suggests that it was made locally, or in another area with similar rocks. In Al Jerar sites, however, the pottery becomes abundant, and firing failures and unfinished vessels indicate that it was undoubtedly locally made.

With water wells they could stay in the area from late fall until the onset of the summer rains. The edible plants, mostly grass seeds, tubers and fruits, also matured during the early part of this period, in the fall and early winter. Storage facilities justified an *intensification* of effort on the abundant plants, and provided the resources that enabled them to stay in the area during the late winter and spring when other food was not available. But why did they make this change in their food economy? An analysis of modern hunter-gatherers living in similar marginal environments suggests that the use of intensification and storage occurs only when the group is *circumscribed*, that is, when they are restricted to a limited area (Binford, 1990, 2001, p. 238f). This suggests that it is highly likely the Nabta/Jerar economic system developed in response to restriction on group mobility, and not because there was more rainfall and more abundant plants. They did not intensify their collection of plants and store them because they were there, but because they had to, if they were to survive in this area during the dry season (Wendorf and Schild, 2002a). For some reason, they could not move to another area where food and water was available in the winter and spring. It is unlikely that these new groups came from the Nile Valley or from farther north, or west; there are no known settlements in those areas with similar features. The most likely hypothesis is that they came from the south, in what is now northern Sudan, and if so they may already have been collecting and storing plant foods and digging water wells before they moved. They, or other similar groups, also may have been the "home bases" for the people in the El Adam and El Ghorab herding camps. Unfortunately, our knowledge of the archeology in this area is extremely limited.

Post-Al Jerar Arid Phase

The El Nabta/Al Jerar Humid Interphase was followed by a sharp, brief period of hyper-aridity between 7200 and 7100 ^{14}C yr BP (7950 and 7850 cal yr BP) when the previously deposited playa deposits were extensively eroded, much of the modern topography of the basins was formed, and the desert again was abandoned. Moisture returned to the area around 7100 ^{14}C yr BP (7850 cal yr BP), but the rainfall was never again as much as during the previous Humid Interphases. From this point on the deposition of playa-like deposits was limited to the mouths of some tributary drainages, and no playa sediments accumulated in the center of the basin.

Ru'at El Ghanam Middle Neolithic Humid Interphase

Another increase in rainfall occurred shortly after 7100 ^{14}C yr BP (7850 cal yr BP), and the area was once again reoccupied, this time by groups identified as Ru'at

El Ghanam Middle Neolithic (Fig. 6.9). Rainfall was much reduced during this Humid Interphase and settlements were restricted to a few localities with perched water tables or other unusual settings where water could be obtained in the dry season by digging wells (Wendorf and Schild, 2001). One of the most important of these localities was Site E-75-8 where, during the preceding period of hyper-aridity, wind deflation had hollowed out a depression through the earlier playa silts down into the bedrock. With the return of moisture, vegetation began to grow in the bottom of this hollow and a phytogenic dune began to accumulate here, with some interfingering of reworked silts (Schild and Wendorf, 2001b).

Middle Neolithic settlements are very rare in the Nabta/Kiseiba area. The only large site of this period is at E-75-8. Here they dug numerous deep walk-in wells to get water from the perched water at the south end of the site, and upslope in the accumulating dune, they dug numerous large and deep storage pits, and presumably had a food economy closely similar to that of the preceding Nabta/Al Jerar Early Neolithic groups. It seems likely that the Ghanam Middle Neolithic were also constricted in their range and were forced to store food in the fall and early winter for consumption during the period of scarcity, in the late winter and spring.

Perhaps the most important new feature in the Ru'at El Ghanam Middle Neolithic was the introduction of domestic sheep/goat, almost certainly coming from the Red Sea Hills or the Nile Valley, and ultimately from Southwest Asia. The integration of sheep/goat gave them a mixed herding economy that may have impacted how they handled their cattle herds.

Other interesting changes were in the pottery. The ceramics in the earliest of these Middle Neolithic sites is similar to that in the preceding Al Jerar sites, that is, it is decorated with impressed, rocker stamped designs, but they used combs with larger teeth and the designs were not as carefully executed as it had been previously. Slightly later in this interval they began to partially obliterate the impressed designs while the clay was still soft, then, still later, the exterior was simply roughened (Nelson, 2001).

The similarity of this pottery to earlier ceramics in this area does not necessarily indicate that the Ghanam Middle Neolithic people were the same as those who were present during the preceding Al Jerar occupation. Closely similar rocker stamped designs occur widely across the Sahara at about this time, and a very different population may be represented. Indeed, this is suggested by the dramatically different lithic technology, as evidenced in tool style, and raw material preference. Flakes rather than blades were the dominant tool blanks, and there was a shift from flint as the preferred raw material for retouched tools to even more use of local chert, and other metamorphosed rock. There was an accompanying increase in the frequency of expedient tools with little or no retouch, and the most frequent raw material, quartz, was rarely retouched. Among the retouched tools there were also important changes, including the appearance of wide, thick "segmental pieces," and flakes with retouched bases that were naturally pointed or partially backed to a point at the tip.

Post-Ru'at El Ghanam Arid Phase

The weak humid interval of the Middle Neolithic was terminated by another sharp episode of hyper-aridity between 6700 and 6600 ^{14}C yr BP (7600 and 7450 cal yr BP). Significant remodeling of the landscape also occurred during this dry phase, and the desert was almost certainly abandoned. When the rains returned around 6550 ^{14}C yr BP (7380 cal yr BP), fine-grained silts and sands were deposited in a few sub-basins, but most of the sediments of this interval consist of finely laminated beds of small gravels, sands, and silts that accumulated as fans in the mouths of the major wadis entering the basin from the surrounding uplands. It is also here on these fans that most of the Late and Final Neolithic sites are located.

Ru'at El baqar Late Neolithic Humid Interphase

This humid interphase is even drier than the previous interphase associated with the Ghanam Middle Neolithic, but clearly a better adjustment to this aridity was made, because El Baqar settlements are numerous and widely dispersed over the area. Even so, the settlements are thin and usually poor in artifacts. Radiocarbon dates place this interphase from around 6.55 to about 5800 ^{14}C yr BP (7400 to 6700 cal BP). Another short arid phase separates the Baqar Late Neolithic Humid Interphase from the following humid interval associated with the Asnam Final Neolithic that begins around 5700^{14}C yr BP (6500 cal yr BP) and lasts until about 4500 ^{14}C yr BP (5200 cal yr BP). After 4500 ^{14}C yr BP there seems to have been a major desertification of the area that was only interrupted by a weak humid spell. It was during this last, weak humid period that C-Group settlers penetrated the desert at Dungul, and built several villages that are dated at 3600^{14}C yr BP (3900 cal yr BP; Hester and Hobler, 1969), and as far west as Nabta where a three-room house has two dates of 3800 and 3100 ^{14}C yr BP (4200 and 3300 cal yr BP; Applegate and Zedeno, 2001).

Site E-75-8, with its long sequence of rich settlements, of which some may have been fairly large, seems to be unique for this period in the Nabta-Kiseiba area. On the other side of the basin, and elsewhere in the Nabta-Kiseiba area, there are only short-lived camps with only a few artifacts. Site E-75-8 is distinguished also by the presence of simple houses or shelters, one of which was excavated, a round, shallow basin hut with a central dish-shaped hearth. Numerous wells were dug during this period at Site E-75-8, as were large, bell-shaped storage pits, suggesting that they, like the preceding Nabta/Al Jerar Early Neolithic and the Ghanam Middle Neolithic, were restricted in the area they were able to exploit, but by storing food they could live in the desert during the dry season, and perhaps year-round. Despite considerable searching for carbonized plant foods, however, the only evidence of plants recovered were from the fiber tempered pottery and consisted of a few seed casts of wild millet and *Setaria*, both commonly used for food today by groups living in the Sahelian zone. Nevertheless, the numerous grinding stones in all of these sites indicate that plant foods were an important component of the Late Neolithic diet.

Faunal samples from Baqar Late Neolithic sites are dominated by sheep/goat and dorcas gazelle, followed by hare and cattle. Dog also occurs for the first time.

The importance of sheep/goat and cattle clearly indicates that there was a mixed herding economy based on both cattle and sheep.

While there is a resemblance in the pottery throughout the Early and Middle Neolithic, the Baqar Late Neolithic pottery is technically radically different; neither stamping nor rough textured surface treatment occurs. The Baqar vessels are usually very thin, and some are burnished on both the exterior and interior surfaces. Almost none are decorated. Many of them have "black tops" and smudged interiors, and some have burnished red slips. Vessel forms are restricted to small simple bowls and tall, slender jars with conical bases. These vessels are made of alluvial clay that appears to have been refined. The temper is also different, it is sand rather than crushed granite, and some of the pottery is fiber tempered. Surface treatments of the Baqar Late Neolithic pottery, particularly the "black top" and red slipped burnished wares, are closely similar to those found on some Baderian pottery from the Nile Valley. They are, however, about 500–700 years older in the desert. While these new ceramic technologies may have originated in the desert, it seems more likely that they came from still earlier, as yet unknown Pre-Baderian Neolithic settlements embedded in the floodplain silts of the Nile.

The most dramatic change evident in the Baqar Late Neolithic is in the area of ritual and cosmology (Wendorf and Schild, 2002b). These new "ritual" features appear to represent the roots of some of the most important Ancient Egyptian beliefs, rituals and religion. It has been suggested that Nabta Playa was a regional ceremonial center that fostered integration among the various related, but usually widely dispersed pastoralist groups who would periodically gather at Nabta for ceremonial, social and political purposes. The Nabta locality may have functioned much like the regional ceremonial centers that are still found today among many Sub-Saharan cattle pastoralists. Among the ritual features at Nabta Playa is a "calendar circle" that marked the position of the rising sun at the summer solstice around 6000 years ago (Malville et al., 1998), and a group of stone covered tumuli located along the northwestern edge of a major wadi where it enters the Nabta Basin, and containing offerings of partially disarticulated remains of cattle and sheep/goat, usually with parts or all of several different animals in the same tumulus. The largest of these tumuli also covered a chamber containing a fully articulated young adult cow. A stick from the roof covering the chamber gave a radiocarbon date of 6470 ^{14}C yr BP (7340 cal yr BP). Yet another stone covered tumuli covered an articulated human skeleton without the skull (Applegate et al., 2001).

Perhaps the most intriguing ritual features at Nabta Playa are the four groups of complex structures or shrines that were built on flattish playa remnants near the center and western edge of the basin. These are more than 50 of these complex structures known, two of which were completely excavated, a third was tested, and two others were drilled to bedrock. All of them share the same basic features. The surface component of these complex structures consist of an oval or rectangle, from 6 to 8 m long and from 5 to 6 m wide, and made with large, roughly shaped blocks of quartzitic sandstone, set upright. In each instance, before the surface component was built, the builders first found, presumably by probing or digging, a table rock or mushroom rock buried below 4–5 m of playa deposits. They then dug a pit from 5 to 6 m in diameter and from 4 to 5 m deep to expose the table rock and the bedrock area around it. In one instance they carefully shaped the top and sides of the table rock, leaving a projection at the north end, while in another they merely removed three flakes from a northern projection of the table rock.

The pit of the complex structure with the carefully shaped table rock was then partially refilled and a large shaped stone "sculpture," 1.7 m long, 1.25 m high, and 0.5 m thick was carefully placed over the center of the table rock and blocked into position with small slabs, with the "head" oriented slightly west of north. The pit was then refilled to the surface and the oval of rough upright stones put into place, and two large stelae were placed close together standing upright near the southern edge oval, facing north. In most of the other complex structures the surface architecture included only one large stelae, but all of them were otherwise similar. Earlier reports on the complex structures (Wendorf and Schild, 1998) stated that these large central stelae had been placed horizontally, but further study indicates that they had been placed upright in shallow pits. The amount of labor involved in excavating the pits to expose the buried table rocks, and in quarrying and shaping the large surrounding blocks and the stelae, clearly indicate the presence of an authority that had control over manpower and resources for significant periods.

Other important ritual features at Nabta Playa include three groups of megalithic alignments that radiate out like spokes on a wheel from the largest complex structure described above (the one with the "shaped sculpture"). The alignments consist of 8–12 large to medium sized, carefully shaped stelae, some with anthromorphic outlines, while others were shaped into simple rectangles. All are made of quartzitic sandstone blocks. The largest of these stelae are estimated to weigh from 8 to 10 tons, the smallest, around 100 kilos. These alignments of stelae have astronomical implications; they mark the positions of rising stars of importance in Ancient Egyptian cosmology. One set of these alignments were toward *Dubhe*, the brightest star in the Big Dipper, another aligns with the two brightest stars in the *Orion* belt, *Alnilam* and *Alnilak*, and the third with *Sirius* (Malville et al., 1998; Malville in Wendorf and Malville, 2001). The calculated dates for these alignments range from 4775 to 3800 cal yr BC. The stelae in the alignments were carefully shaped, and had been originally placed standing upright in shallow pits. A few are still in place, but most have fallen over and are broken, but with the aid of a computer, several of the broken stelae have been virtually reassembled (Schild and Wendorf, 2004).

3.5. Nile Valley

The major source of water in the Nile is in the uplands of eastern Africa. Thus, the flow of the Nile varied with the strength of the monsoons farther south. Unlike the Eastern Sahara, however, the Nile did not go dry, and human habitation was always possible even during the driest part of the Late Pleistocene. The problem is that we know very little about the archeology in the early and middle Holocene in the Valley, and what little is known comes from Catfish Cave in Egyptian Nubia, dated at 7060 ^{14}C yr BP (7860 cal yr BP) (Wendt, 1966), the Qarunian site of E29G1 in the Fayum dated to 7140 ^{14}C yr BP (7920 cal yr BP) (Wendorf and Schild, 1976, p. 312), and the so-called Khartoum Variant localities near Wadi Halfa, dated at 6540 ^{14}C yr BP $\pm$ 110 years (7390 cal yr BP) (Shiner, 1968; Close, 1980). The date for this last complex is almost surely too recent, because the ceramics in the Khartoum Variant sites are stylistically closely similar to the pottery in the Al Jerar settlements in the Nabta/Kiseiba area. The Jerar ceramics are about 700 years older than the Khartoum Variant sites near Wadi Halfa. There are, however, thick wide crescents like those in the Middle Neolithic at Nabta in the Khartoum Variant sites.

The lithic industry in the Fayum site of E29G1 is Terminal Paleolithic in character and includes numerous bladelets produced from single platform cores. The most frequent tools are arch-backed and straight-backed bladelets, followed by notches, denticulates, and a few geometrics, mostly triangles and trapezes. Bone points made from catfish pectoral spines also occur. Grinding stones are present, but they are not numerous. Neither Catfish Cave nor the Qarunian sites contain pottery, domestic animals, or domestic plants. Both Catfish Cave and the Qarunian sites were primarily fishing localities and fish remains dominate the associated faunal assemblages in both sites (Wendorf and Schild, 1976). These sites probably represent a part of the economy that developed when the Nile began to down cut through the Late Pleistocene valley fill after 12,000 ^{14}C yr BP (13,880 cal yr BP) and terminated the complex Final Pleistocene plant gathering economy (Wendorf et al., 1989; Wendorf and Schild, 1989).

The first Neolithic settlements known in the Nile Valley are dated to around 6400 ^{14}C yr BP (7270 cal yr BP), and are all in the Fayum. It is highly likely, however, that there was a Neolithic presence somewhere in the Valley before 7100 ^{14}C yr BP (7870 cal yr BP); how else could the sheep, a Southwest Asian domesticate, have reached the Egyptian Sahara by that date? The Fayum settlements are very large communities with burnished and smudged ceramics, a complex lithic technology, and a mixed economy of wheat and barley cultivation, fishing, and sheep and cattle husbandry that undoubtedly owes a great deal to stimulus or perhaps migration from Southwest Asia. Shortly thereafter, there are Neolithic settlements all along the Nile from the Delta to Nubia, all sharing many similar elements. It is these groups that are the most likely source of the ceramics and, perhaps other features, seen in the Late Neolithic of the Sahara. Somewhat later, between 5500 and 5000 ^{14}C yr BP (6290 and 5730 cal yr BP) the Neolithic had become sufficiently complex in the Nile Valley that the onset of the Egyptian State is clearly anticipated. These later Neolithic groups are identified as Predynastic.

4. Discussion

As evident in Fig. 6.10, the early Holocene in most areas of Africa north of the equator was characterized by high permanent lake levels, or in drier areas such as

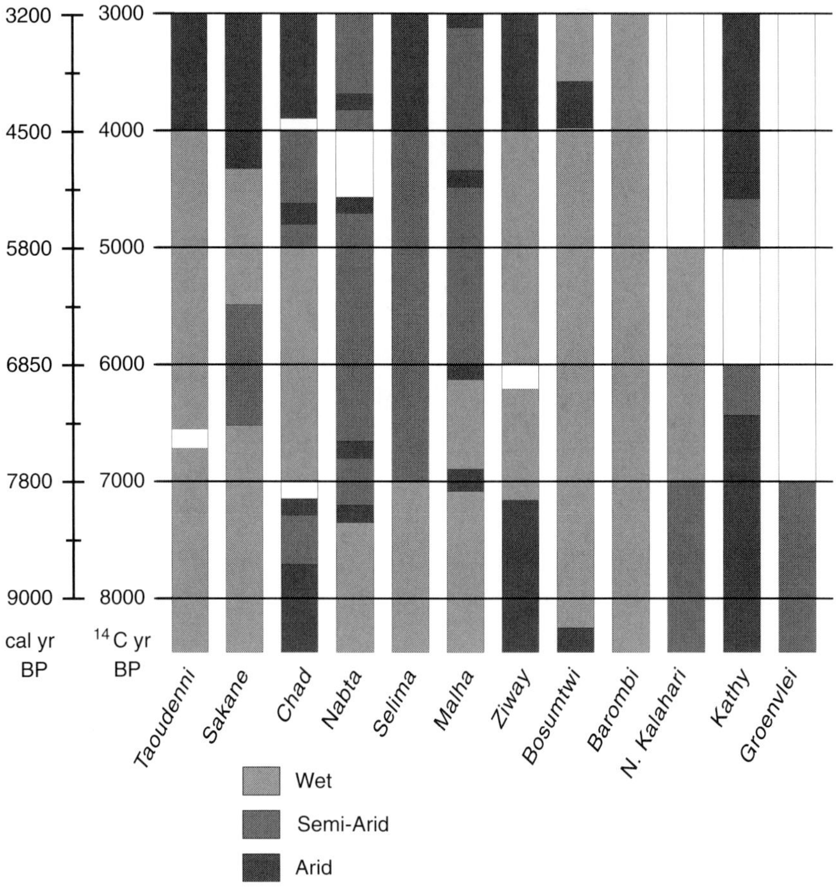

Figure 6.10. Summary of climatic proxy data discussed in text. All calibrated radiocarbon dates are the midpoint of the one-sigma range rounded to the nearest 10 years. They are taken from University of Washington, Quaternary Isotope Laboratory Radiocarbon Calibration Program, Revision 3.0.3c (Stuiver et al., 1993).

Nabta, by seasonal lakes or playas. Increased rainfall is recorded for the period from before 8000 ^{14}C yr BP (8740 cal yr BP) for the basins in the tropics (Ziway-Shala, Bosumtwi, and Barombi Mbo), as well as for the Sahelian and hyper-arid zones (Malha, Selima, Nabta, Sakane, and Taoudenni). The exception is Chad, which stands out as an aberration. The problem may be with some of the radiocarbon dates, many of which are on carbonates.

In most of these sequences, the end of the early Holocene seems to coincide with an abrupt reduction of rainfall around 7000 ^{14}C yr BP (7790 cal yr BP). The broader climatic significance of this abrupt arid interval is not well understood, but an earlier and similar episode seems to coincide with the worldwide drop in temperatures known as the Younger Dryas. The subsequent abrupt declines in rainfall in the Sahara also may be associated with declines in temperature. We should note, however, that the timing of this arid phase is not consistent across the entire Sahara.

In the western Sahara, if the radiocarbon dates are correct, the break between two lacustrine events occurred later, between 6500 and 5500 ^{14}C yr BP (7380 and 6290 cal yr BP), while at Bosumtwi in Ghana, the decline is earlier, and dated to about 8000 ^{14}C yr BP (8740 cal yr BP). Obviously, the relationship between temperature and precipitation in the Sahara is complex, and varies across the continent.

The middle Holocene is generally regarded as a period of higher temperatures, particularly summer temperatures in the Northern Hemisphere that were at a maximum (winter temperatures continued to rise until recently). But was the middle Holocene really a period of warmer temperatures in Africa? Probably. The model outlined in the first part of this chapter assumes a nonlinear relationship between mean temperatures and variations in the monsoons. In its simplest form, colder temperatures are thought to be related to a reduction in the monsoons, and conversely, higher temperatures result in greater rainfall in the tropics and a northward expansion of the rain belt. Thus, if it was warmer in the middle Holocene, as is widely believed, then there should be more rain in both the tropics and in the Sahara. As may be seen in Figures 6.2, 6.3, and 6.6, in the tropics the highest lake levels, even higher than those in the early Holocene, occur in the middle Holocene levels at Ziway-Shala, Bosumtwi, and Chad, and the maximum forest development at Barombi Mbo continues through this interval.

A mid-Holocene warm period is also known from pollen studies on Mount Kenya by Coetzee (1967), who found that the forest reached its Holocene maximum extent between about 4850 and 4150 ^{14}C yr BP (5500 and 4500 cal yr BP) (Fig. 6.9). These dates coincide with dates indicating high organic content in a pro-glacier lake at Mount Kenya. This is interpreted as an indication of relatively warm climate and only small glaciers.

These long-range climatic records are based on the interpretation of proxy data, which only yields indirect information about climate such as vegetation, lake level, or glacial response to changes in several parameters of the climate. Frequently, the models are based on only a few dates, and because rate of sedimentation varies, the presented dates may be off by several hundred years. Considering the often relatively low resolution, it is impossible to claim simultaneous changes in climate over all of North Africa. However, large-scale variations in climate seem similar for several records from East Africa. Climate was cool and dry around 5350 ^{14}C yr BP (6000 cal yr BP), and it became substantially warmer and wetter around 4450 ^{14}C yr BP (5000 cal yr BP). Shortly after 5350 ^{14}C yr BP (5000 cal yr BP) a trend toward cooler and drier began. Except for two short reversals around 3450 and 3150 ^{14}C yr BP (3700 and 3200 cal yr BP), the trend continued up to about 2150 ^{14}C yr BP (2000 cal yr BP). Around this date the climate improved and a temporary warm and wet event occurred around 1600 ^{14}C yr BP (1500 cal yr BP). Distinct but short cooler events occurred around 1300 and 400 ^{14}C yr BP (1200 and 500 cal yr BP).

In the Sahara, the proxy data and the temperature-rainfall model are in general agreement. Modern climatic data show that the position of the ITC is not a straight line across Africa, but varies with longitude (Bryson, 1992, p. 253). We would expect the same in the past, and this is precisely what we see in the proxy data from the Eastern and Western Sahara for the middle Holocene. The proxy data from Malha (Fig. 6.6), Selima (Fig. 6.10), and Nabta (Fig. 6.7) are consistent, and all indicate much less rainfall in this area during the 7000–3000 ^{14}C yr BP interval than during the early Holocene. In contrast, in the Western Sahara, at Sakane, it was drier between 6500 and 5500 ^{14}C yr BP, but the second lacustrine event at Taoudenni, which existed during part of the middle Holocene, besides being out of phase, is weaker than the earlier lake event. The modern period of hyper-aridity in the Sahara also appears to have a chronological slope that is both north to south and east to west between 5000 and 3800 ^{14}C yr BP (5730 and 4150 cal yr BP). In the east, hyper-aridity began at Nabta around 5000 ^{14}C yr BP, while at Selima, just 200 km farther south, after a long interval of declining rainfall, the onset of hyper-aridity is dated around 4000 ^{14}C yr BP (4420 cal yr BP) when the last surface water disappeared. Very dry conditions also were present in the Western Sahara at Sakane and Taoudenni at about the same time, at 4500 and 4000 ^{14}C yr BP (5200 and 4420 cal yr BP). Some of these differences between the two areas are probably related to variations in the position of the coastal rainfall belt in far western Africa. This period of dry climate in the Western Sahara coincides with the period of clearly warm and wet conditions on Mount Kenya.

In so far as the Eastern Sahara is concerned, the data might be interpreted as indicating that throughout this period there were brief northward pulsations and retreats resulting in highly unstable rainfall frequencies. These northward pulsations became weaker through time, and at Nabta and Kiseiba ceased altogether around 4500 ^{14}C yr BP (5200 cal yr BP), except for a brief moist interval around 3800 ^{14}C yr BP (4150 cal yr BP).

Acknowledgement

We wish to express our thanks to Dr. Reid A. Bryson for his very helpful comments on an earlier version of this chapter. We greatly benefited from his suggestions, but any errors or omissions are entirely our own responsibility.

References

Applegate, A., A. Gautier, and S. Duncan, 2001. The north Tumuli of the Nabta late Neolithic ceremonial complex. In *Holocene Settlement of the Egyptian Sahara, Volume 1, The Archaeology of Nabta Playa*, edited by F. Wendorf, R. Schild, and Associates, pp. 468–488. Kluwer Academic/Plenum Publishers, New York.

Applegate, A., and N. Zedeno, 2001. Site E-92-8: A late Prehistoric C-group component at Nabta Playa. In *Holocene Settlement of the Egyptian Sahara, Volume 1, The Archaeology of Nabta Playa*, edited by F. Wendorf, R. Schild, and Associates, pp. 529–533. Kluwer Academic/Plenum Publishers, New York.

Barakat, H., 1995. Contribution Archeobotanique a l'Histoire de la Vegetation dans le Sahara Oriental et Soudan Central. Unpublished Ph.D. Dissertation, University de Droit, d'Economie et des Sciences d'Aix-Marseille.

Barakat, H., 2001. Anthracological studies in the Neolithic sites of Nabta Playa, south Egypt. In *Holocene Settlement of the Egyptian Sahara, Volume 1, The Archaeology of Nabta Playa*, edited by F. Wendorf, R. Schild, and Associates, pp. 592–600. Kluwer Academic/Plenum Publishers, New York.

Barker, P. A., F. A. Street-Perrott, M. J. Leng, P. B. Greenwood, D. L. Swain, R. A. Perrott, R. J. Telford, and K. J. Ficken, 2001. A 14,000-year oxygen isotope record from diatom Silica in two Alpine Lakes on Mt. Kenya. *Science* 292:2307–2310.

Beaumont, P. B., E. M. Van Zinderen Bakker, and J. C. Vogel, 1984. Environmental Changes Since 32,000 BP at Kathu Pan, Northern Cape. In *Late Cenozoic Palaeoclimates of the Southern Hemisphere*, edited by J. C. Vogel, pp. 329–338. Balkema, Rotterdam.

Binford, L. R., 1990. Mobility, housing, and environment: a comparative study. *Journal of Anthropological Research* 46:119–152.

Binford, L. R., 2001. *Constructing Frames of Reference: An Analytical Method for Archaeological Theory Building Using Hunter-Gatherer and Environmental Data Sets*. University of California Press, Berkeley.

Bonnefille, R., G. Riollet, G. Buchet, M. Icole, R. Lafont, M. Arnold, and D. Jolly, 1995. Glacial/interglacial records from intertropical Africa, high resolution pollen and carbon data at Rusaka, Burundi. *Quaternary Science Review* 14:917–936.

Bradley, D. G., D. E. Machugh, P. Cunningham, and R. T. Loftus, 1996. Mitochondrial diversity and the origin of African and European cattle. *Proceedings of the National Academy of Sciences USA* 93:5131–5135.

Breunig, P., and K. Neumann, 2002. From hunters and gatherers to food producers: new archaeological and archaeobotanical evidence from the West African Sahel. In *Droughts, Food and Culture. Ecological Change and Food Security in Africa's Later Prehistory*, edited by F. A. Hassan, pp. 123–155. Kluwer Academic/Plenum Publishers, New York.

Bryson, R. A., 1992. A macrophysical model of the Holocene intertropical convergence and jetstream positions and rainfall for the Saharan region. *Meteorology and Atmospheric Physics* 47:247–258.

Bryson, R. A., and R. U. Bryson, 1996. Macrophysical climatic modeling of Africa's late Quaternary climate: site-specific, high-resolution applications for archaeology. *African Archaeological Review* 14(3):143–160.

Clark, J. D., 1950. A review of prehistoric research in northern Rhodesia and Nyasaland. In *Proceedings of the Third Pan-African Congress on Prehistory, Livingstone 1955*, edited by J. D. Clark, pp. 412–432. Chatto & Windus, London.

Close, A. E., 1980. Recent radiocarbon dates from northern Africa. *Journal of African History*, 21:145–167.

Close, A. E., 1990. Living on the edge: Neolithic herders in the eastern Sahara. *Antiquity*, 64:79–96.

Coetzee, J. A., 1967. Pollen analytical studies in east and southern Africa. *Paleoecology Africa and the Surrounding Islands and Antarctica* 3:1–140.

Cole, S., 1954. *The Prehistory of East Africa*. Pelican, Penguin Books, Harmondsworth, England.

Connor, D. R., 1984. Report on Site E-79-8. In *Cattle-Keepers of the Eastern Sahara: The Neolithic of Bir Kiseiba*, edited by F. Wendorf and R. Schild, Assem; A. E. Close, pp. 217–250. Department of Anthropology, Institute for the Study of Earth and Man, Southern Methodist University, Dallas.

Cooke, H. B. S., 1957. The problem of quaternary Glacio-Pluvial correlation in east and southern Africa. In *Proceedings of the Third Pan-African Congress on Prehistory, Livingston 1955*, edited by J. D. Clark, pp. 51–55. Chatto & Windus, London.

Cremaschi, M., 1998. Late Quaternary geological evidence for environmental changes in south-western Fezzan (Libyan Sahara). In *Wadi Teshunat. Paleoenvironment and Prehistory in South-western Fezzan (Libyan Sahara)*, edited by M. Cremaschi and S. di Lernia, pp. 13–47. Quaderni di Geodinamica Alpina e Quaternaria 7, C.N.R., Milan.

Cremaschi, M., 2002. Late Pleistocene and Holocene climatic changes in the central Sahara. The case study of the southwestern Fezzan, Libya. In *Droughts, Food and Culture. Ecological Change and Food Security in Africa's Later Prehistory*, edited by F. A. Hassan, pp. 65–82. Kluwer Academic/Plenum Publishers, New York.

Cremaschi, M., and S. di Lernia, editors, 1998. *Wadi Teshunat. Paleoenvironment and Prehistory in South-western Fezzan (Libyan Sahara)*. Quaderni di Geodinamica Alpina e Quaternaria 7, C.N.R., Milan.

Croll, J., 1864. On the physical cause of the change of climate during geological epochs. *Philosophical Magazine* 28:121–137.

di Lernia, S., 2002. Dry climatic events and cultural trajectories: adjusting middle Holocene pastoral economy of the Libyan Sahara. In *Droughts, Food and Culture. Ecological Change and Food Security in Africa's Later Prehistory*, edited by F. A. Hassan, pp. 225–250. Kluwer Academic/Plenum Publishers, New York.

Dumont, H. J., and A. I. El Moghraby, 1993. Holocene evolution of climate and environment and stone "City" iuins in northern Darfur, Sudan: is there a relationship? In *Environmental Change and Human Culture in the Nile Basin and Northern Africa Until the Second Millennium B.C.*, edited by L. Krzyzaniak, M. Kobusiewicz, and J. Alexander, pp. 381–397. Poznan Archaeological Museum, Poznan.

Fabre, J., and N. Petit-Maire, N., 1988. Holocene climatic evolution at 22–23°N from two palaeolakes in the Taoudenni area (Northern Mali). *Palaeogeography, Palaeoclimatology and Palaeoecology* 65:133–148.

Flint, R. F., 1959. On the basis of Pleistocene correlation in East Africa. *Geological Magazine* 96:265–284.

Fontes, J., and F. Gasse, 1989. On the ages of humid Holocene and late Pleistocene phases in North Africa – remarks on "Late Quaternary Climatic Reconstruction for the Maghreb (North Africa)" by P. Rognon. *Palaeogeography, Palaeoclimatology and Palaeoecology* 70:393–398.

Gasse, F., 2000. Hydrological changes in the African tropics since the Last Glacial Maximum. *Quaternary Science Reviews* 19:189–211.

Gasse, F., 2001. Hydrological changes in Africa. *Science* 292:2259–2260.

Garcea, E. A. A., editor, 2001. Uan Tabu in the settlement history of the Libyan Sahara. In *Arid Zone Archaeology, Monographs* 2. Edizioni All'insegna del Giglio, Gruppo Editoriale Internazionale, Rome.

Gautier, A., 1980. Appendix 4. Contributions to the archaeozoology of Egypt. In *Prehistory of the Eastern Sahara*, edited by F. Wendorf and R. Schild, pp. 317–344. Academic Press, New York.

Gautier, A., 1984. Archaeozoology of the Bir Kiseiba region. Eastern Sahara. In *Cattle Keepers of the Eastern Sahara: The Neolithic of Bir Kiseiba*, edited by F. Wendorf and R. Schild, Assem; A. E. Close, pp. 49–72. The Department of Anthropology, Southern Methodist University, Dallas.

Gautier, A., 1987. Prehistoric man and cattle in North Africa: a dearth of data and a surfeit of models. In *Prehistory of Arid North Africa. Essays in Honor of Fred Wendorf*, edited by A. E. Close, pp. 163–188. Southern Methodist University Press. Dallas.

Gautier, A., 2001. The early to late Neolithic archeofaunas from Nabta and Bir Kiseiba. In *Holocene Settlement of the Egyptian Sahara, Volume 1, The Archaeology of Nabta Playa,* edited by F. Wendorf, R. Schild, and Associates, pp. 609–635. Kluwer Academic/Plenum Publishers, New York.

Gilbert, R., and J. R. Desloges, 1992. The late Quaternary sedimentary record of Stave Lake, southwestern British Columbia. *Canadian Journal of Earth Sciences* 29:1997–2006.

Gillespie, R., F. A. Street-Perrott, and R. Switsur, 1983. Post-Glacial Arid Episodes in Ethiopia have Implications for Climate Prediction. *Nature* 306:680–683.

Giresse, P., J. Maley, and P. Brenac, 1994. Late Quaternary palaeoenviornments in the Lake Barombi Mbo (West Cameroon) deduced from pollen and carbon isotopes of organic matter. *Palaeogeography, Palaeoclimatology and Palaeoecology* 107:65–78.

Grove, A. T., 1993. Africa's climate in the Holocene. In *The Archaeology of Africa, Food, Metals and Towns*, edited by T. Shaw, P. Sinclair, B. Andah, and A. Okpoko, pp. 32–42. Routledge, London.

Hassan, F. A., 1997. Holocene paleoclimates of Africa. *African Archaeological Review* 14(4): 213–230.

Haynes, C. V., Jr., 1997. Geoarchaeological Manifestations of Younger Dryas and Later Climate Change in the Darb el Arba'in Desert, Eastern Sahara. Oral presentation at The Paleo-monsoon Workshop, Siwa Oases, Egypt, 11–22 January sponsored by the International Geosphere-Biosphere Program (IGBP) of INQUA.

Haynes, C. V., Jr., and A. R. Mead, 1987. Radiocarbon dating and paleoclimatic significance of subfossil limiclaria in northwestern Sudan. *Quaternary Research* 28:86–99.

Haynes, C. V., Jr., C. H. Eyles, L. A. Pavlish, J. C. Ritchie, and M. Rybak, 1989. Holocene palaeoecology of the eastern Sahara: Selima Oasis. *Quaternary Science Reviews* 8:109–136.

Hester, J. J., and P. M. Hobler, 1969. Prehistoric settlement patterns in the Libyan Desert. University of Utah Anthropological Papers No. 92, Nubia Series 4, University of Utah Press, Salt Lake City.

Hillaire-Marcel, C., J. Riser, P. Rognon, N. Petit-Maire, J. C. Rosso, and I. Soulie-Marche, 1983. Radiocarbon chronology of Holocene hydrologic changes in northeastern Mali. *Quaternary Research* 20:145–164.

Ilesanmi, O. O., 1971. An empirical formulation of an ITD rainfall model for the tropics. A case study of Nigeria. *Journal of Applied Meteorology* 10:882–891.

Jesse, F., 2003. Rahib 80/87. *African Praehistorica* 16. Heinrich Barth Institut, Koln.

Karlén, W., 1976. Lacustrine sediments and the limit variations as indicators of Holocene climatic fluctuations in Lappland: Northern Sweden. *Geografiska Annaler* 58A(12): 1–34.

Karlén, W., 1981. Lacustrine sediment studies. *Geografiska Annaler* 63A(3–4).273–281.

Karlén, W., J. L. Fastook, K. Holmgren, M. J. Malmstrom, J. A. Matthews, A. E. Odada, J. Risberg, G. Rosqvist, P. Sandgren, A. Shemesh, and L. O. Westerberg, 1999. Holocene glacier fluctuations on Mount Kenya, East Africa, between 6000 cal. years BP and the present. *Ambio* 28:409–418.

Keding, B., 1997. Djabarona 84/13. *African Praehistorica* 9. Heinrich Barth Institut, Koln.

Krolik, H., and R. Schild, 2001. Site E-75-6: an El Nabta and Al Jerar village. In *Holocene Settlement of the Egyptian Sahara, Volume 1, The Archaeology of Nabta Playa*, edited by F. Wendorf, R. Schild, and Associates, pp. 111–146. Kluwer Academic/Plenum Publishers, New York.

Kropelin, S., 1993. The Gilf Kebir and lower Wadi Howar: contrasting early and mid-Holocene environments in the eastern Sahara. In *Environmental Change and Human Culture in the Nile Basin and Northern Africa Until the Second Millennium B.C.*, edited by L. Krzyzaniak, M. Kobusiewicz, and J. Alexander, pp. 249–258. Poznan Archaeological Museum, Poznan.

Kuper, R., 1986. Wadi Howar and Laquiya. Recent field studies into the early settlement of northern Sudan. In *Nubische Studien. Proceedings of the Symposium of the International Society of Nubian Studies, Heidelberg 1982*, edited by M. Krause, pp. 129–136. Mainz.
Kuper, R., editor, 1989. Forschungen zur Umweltgeschichte der Ostsahara. *Africa Praehistorica* 2. Köln.
Kutzbach, J. E., 1980. Estimates of past climate at Paleolake Chad, North Africa, based on a hydrological and energy-balance model. *Quaternary Research* 14:210–223.
Kutzbach, J. E., 1981. Monsoon climate of the early Holocene: climate experiment with the Earth's orbital parameters for 9000 years ago. *Science* 214:59–61.
Kutzbach, J. E., and F. A. Street-Perrott, 1985. Milankovitch forcing of fluctuations in the level of tropical lakes from 18 to 0 kyr B.P. *Nature* 317:130–134.
Lamb, H. F., F. Gasse, A. Benkaddour, N. El Hamouti, S. van der Kaars, W. T. Perkins, N. J. Pearce, and C. N. Roberts, 1995. Relations between century-scale Holocene arid intervals in tropical and temperate zones. *Nature* 373:134–137.
Leemann, A., and F. Niessen, 1994. Varve formation and the climatic record in an Alpine proglacial lake: calibrating annually laminated sediments against hydrological and meteorological data. *The Holocene* 4:1–8.
Leonard, E. M., 1986a. Varve studies at Hector Lake, Alberta, Canada, and the relationship between glacial activity and sedimentation. *Quaternary Research* 25:199–214.
Leonard, E. M., 1986b. Use of lacustrine sedimentary sequences as indicators of Holocene glacial history, Banff National Park, Alberta, Canada. *Quaternary Research* 26:218–231.
Lorius, C., L. Merlivat, J. Jouzel, and M. Pourchet, 1979. A 30,000-yr isotope climate record from Antarctic Ice. *Nature* 280:644–648.
Maley, J, 1977a. Palaeoclimates of central Sahara during the early Holocene. *Nature* 269: 573–577.
Maley, J, 1977b. Analyses Polliniques et Paléoclimatologie des Douze Derniers Millenaires du Bassin du Tchad (Afrique Centrale) Recherches Françaises sur le Quaternaire (INQUA 1977), pp. 187–197. *Supplement au Bulletin AFEQ*, No. 50, Paris.
Maley, J, 1991. The African rain forest vegetation and palaeoenvironments during late Quaternary. *Climatic Change* 19:79–98.
Maley, J, 1997. Middle to late Holocene changes in tropical Africa and other continents: paleomonsoon and sea surface temperature variations. In *Third Millennium BC Climate Change and Old World Collapse*, edited by H. N. Dalfes, G. Kukla, and H. Weiss, NATO ASI Series, Vol. I 49, pp. 611–640. Springer, Berlin Heidelberg.
Malville, J. McK., F. Wendorf, A. Mazhar, and R. Schild, 1998. Megaliths and Neolithic astronomy in southern Egypt. *Nature* 392:488–492.
Matthews, J. A., and W. Karlén, 1992. Asynchronous neoglaciation and Holocene climatic change reconstructed from Norwegian glaciolacustrine sedimentary sequences. *Geology* 20:991–994.
McDonald, M., 1991. Technological organization and sedentism in the epipalaeolithic of Dakhleh Oasis, Egypt. *African Archaeological Review* 9:81–109.
McDonald, M., 1993. Cultural adaptations in Dakhleh Oasis, Egypt, in the early to mid-Holocene. In *Environmental Change and Human Culture in the Nile Basin and Northern Africa Until the Second Millennium B.C.*, edited by L. Krzyzaniak, M. Kousiewicz, and J. Alexander, pp. 199–209. Poznan Archaeological Museum, Poznan.
McDonald, M., 1998. Early African Pastoralism: view from Dakhleh Oasis (South Central Egypt). *Journal of Anthropological Archaeology* 17:124–142.
Mees, F., D. Verschuren, R. Nijs, and H. J. Dumont, 1991. Holocene evolution of the Crater Lake at the Malha, northwest Sudan. *Journal of Palaeoclimatology* 5:227–253.

Milankovitch, M., 1941. *Kanon der erdbestrahlung und seine. Anwendung auf das Eiszeiten Problem*. Serbian Academy. Special Publication. Belgrade. (English translation published by Israel Program for Scientific Translations, Jerusalem, 1969. Available from U.S. Department of Commerce, Washington, DC).

Nelson, K., 2001. The pottery of Nabta Playa: a summary. In *Holocene Settlement of the Egyptian Sahara, Volume 1, The Archaeology of Nabta Playa*, edited by F. Wendorf, R. Schild, and Associates, pp. 534–543. Kluwer Academic/Plenum Publishers, New York.

Neumann, K., 1989. Zur Vegetationsgeschichte der Ostsahara im Holozän. Holzkohlen aus Prähistorischen Fundstellen. Mit Einem Exhurs über die Holzhohlen von Fachi-Dogonbolo Nigher. In *Forschungen zur Umwelgeschichte des Ostsahara*, edited by R. Kuper, pp. 13–181. African Praehistorica 2. Heinrich Barth Institut, Koln.

Neumann, K., 1993. Holocene vegetation of the eastern Sahara: charcoal from prehistoric sites. In *Environmental Change and Human Culture in the Nile Basin and Northern Africa Until the Second Millennium B.C.*, edited by L. Krzyzaniak, M. Kobusiewicz, and J. Alexander, pp. 153–169. Poznan Archaeological Museum, Poznan.

Nilsson, E., 1940. Ancient changes of climate in British East Africa and Abyssinia. *Geografriska Annaler* (Stockholm) 22:1–79.

Nye, J., 1960. The response of glaciers and ice sheets to seasonal and climatic changes. *Proceedings of the Royal Society* 256A(1287):559–584.

Pachur, H.-J., and P. Hoelzmann, 1991. Paleoclimatic implications of late Quaternary lacustrine sediments in western Nubia, Sudan. *Quaternary Research* 36:257–276.

Pachur, H.-J., and S. Kröpelin, 1989. L'Aridification du Sahara Oriental à l'Holocène Moyen et Supéreur. *Bulletin de la Société Géologique de France* Serie 8, 5(1):99–107.

Paterson, W. S. B., 1994. *The Physics of Glaciers*. Pergamon, London.

Petit-Maire, N., and J. Riser, J., 1981. Holocene Lake deposits and palaeoenvironments in central Sahara, northeastern Mali. *Palaeogeography, Palaeoclimatology, Palaeoecology* 35: 45–61.

Petit-Maire, N., and J. Riser, editors, 1983. *Le Sahara malien à l'Holocène*. C.N.R.S., Paris.

Petit-Maire, N., D. Commelin, J. Fabre, and M. Fontugne, 1990. First evidence for Holocene rainfall in the Tanezrouft hyperdesert and its margins. *Palaeogeography, Palaeoclimatology, Palaeoecology* 79:333–338.

Ritchie, J. C., C. H. Eyles, and C. V. Haynes, 1989. Sediment and pollen evidence for an early to mid-Holocene humid period in the eastern Sahara. *Nature* 314:352–355.

Rognon, P., and M. A. J. Williams, 1977. Late Quaternary climatic changes in Australia and North Africa: a preliminary interpretation. *Palaeogeography, Palaeoclimatology, Palaeoecology* 21:285 327.

Schild, R., and F. Wendorf, 1977. *The Prehistory of Dakhla Oasis and Adjacent Desert*. Polish Academy of Sciences, Ossolineum, Warsaw.

Schild, R., and F. Wendorf, 1981. *The Prehistory of an Egyptian Oasis*. Polish Academy of Sciences, Ossolineum, Warsaw.

Schild, R., and F. Wendorf, 2001a. Geoarchaeology of the Holocene climatic optimum at Nabta Playa, Southwestern Desert, Egypt. *Geoarchaeology* 16:7–28.

Schild, R., and F. Wendorf, 2001b. Geomorphology, lithostratigraphy, geochronology and taphonomy of sites. In *Holocene Settlement of the Egyptian Sahara, Volume 1, The Archaeology of Nabta Playa*, edited by F. Wendorf, R. Schild, and Associates, pp. 111–146. Kluwer Academic/Plenum Publishers, New York.

Schild, R., and F. Wendorf, 2001c. Combined prehistoric expedition's radiocarbon dates associated with Neolithic occupations in the Southern Western Desert of Egypt. In *Holocene Settlement of the Egyptian Sahara, Volume 1, The Archaeology of Nabta Playa*, edited by

F. Wendorf, R. Schild, and Associates, pp. 111–146. Kluwer Academic/Plenum Publishers, New York.

Schild, R., and F. Wendorf, 2002. Paleo-ecologic and paleo-climatic background to socio-economic changes in the South Western Desert of Egypt. In *Egypt and Nubia: Gifts of the Desert*, edited by R. Friedman, pp. 21–27. British Museum Press, London.

Schild, R., and F. Wendorf, 2004. The megaliths of Nabta Playa. *Academia* 1(1):10–15.

Shiner, J. L., 1968. The Khartoum Variant industries. In *The Prehistory of Nubia*, edited by F. Wendorf, 2 vols. and Atlas, pp. 768–790. Fort Burgwin Research Center and Southern Methodist University Press, Dallas.

Sirocko, F., M. Sarnthein, H. Erleneeuser, Hl. Lange, M. Arnold, and J. C. Duplessy, 1993. Century-scale events in monsoonal climate over the past 24,000 years. *Nature* 364:322–324.

Street-Perrott, F. A., and R. A. Perrott, 1990. Abrupt climate fluctuations in the tropics: the influence of Atlantic Ocean circulation. *Nature* 343:607–612.

Street-Perrott, F. A., and R. A. Roberts, 1983. Fluctuations in closed basin lakes as an indicator of past atmospheric circulation patterns. In *Variations in the Global Water Budget*, edited by F. A. Street-Perrott, M. Beran, and R. D. Ratfcliffe, pp. 331–345. Reidel, Dodrcht.

Stuiver, M. and P. J. Reimer, 1993. Extended 14C data base and revised calib 3.0 14C age. *Radiocarbon* 35: 215–230.

Talbot, M. R., D. A. Livingstone, P. G. Palmer, J. Maley, J. M. Melack, G. Delibrias, and S. Gulliksen, 1984. Preliminary results from sediment cores from Lake Bosumtiwi, Ghana. *Palaeoecology of Africa and the Surrounding Islands* 16:173–192.

Wasylikowa, K., J. R. Harlan, J. Evans, F. Wendorf, R. Schild, A. E. Close, and H. Królik, 1993. Examination of botanical remains from early Neolithic houses at Nabta Playa, Western Desert, Egypt, with special reference to sorghum grains. In *The Archaeology of Africa: Food, Metal and Urbanisation in Africa*, edited by T. Shaw, P. Sinclair, B. Andah, and A. Okpoko, pp. 154–164. Routledge, London.

Wasylikowa, K., J. Mitka, A. Walanus, F. Wendorf, and R. Schild, 1996. Distribution of plant macrofossils within a settlement: a case study on plant exploitation manners in a hunter-gatherer site at Nabta Playa, south Egypt. In *Bulletin of the XIII Congress of the International Union of Prehistoric and Protohistoric Sciences*, pp. 47–56. Forli, Italy.

Wasylikowa, K., R. Schild, F. Wendorf, H. Krolik, L. Kubiak-Martens, and J. R. Harlan, 1995. Archaeobotany of the early Neolithic Site E-75-6 at Nabta Playa, Western Desert, south Egypt (preliminary results). *Acta Palaeobotanica* 35:133–155.

Wasylikowak, W., H. Barakat, L. Boulos, A. Butler, J. Dahlberg, J. Hather, and J. Mitka, 2001. Site E-75-6: vegetation and subsistence of the early Neolithic at Nabta Playa, Egypt, reconstructed from charred plant remains. In *Holocene Settlement of the Egyptian Sahara, Volume 1, The Archaeology of Nabta Playa*, edited by F. Wendorf, R. Schild, and Associates, pp. 544–591. Kluwer Academic/Plenum Publishers, New York.

Wayland, E. J., 1934. Outline of prehistory and Stone Age climatology in the Bechuanaland protectorate. *Memoirs d'Académie Royale des Sciences Coloniales* 25:1–42.

Wendorf, F., and A. E. Close, 1992. Early Neolithic food-economies in the eastern Sahara. In *The Followers of Horus: Studies in Memory of Michael Allen Hoffman*, edited by R. Friedman, and B. Adams, pp. 155–162. Oxbow Monographs 20. Oxford.

Wendorf, F., and J. Malville, 2001. The megalithic alignments. In *Holocene Settlement of the Egyptian Sahara, Volume 1. The Archaeology of Nabta Playa*, edited by Fred Wendorf and Romuald Schild and Associates. Kluwer Academic/Plenum Publishers. New York.

Wendorf, F., and R. Schild, 1976. *Prehistory of the Nile Valley*. Academic Press, New York.

Wendorf, F., and R. Schild, 1980. *Prehistory of the Eastern Sahara*. Academic Press, New York.

Wendorf, F., and R. Schild, 1989. Conclusions. In *The Prehistory of Wadi Kubbaniya, Volume 3. Late Paleolithic Archaeology*, edited by F. Wendorf, and R. Schild, Assem.; A. Close, pp. 768–824. Southern Methodist University Press, Dallas.
Wendorf, F., and R. Schild, 1994. Are the early Holocene cattle in the eastern Sahara domestic or wild. *Evolutionary Anthropology* 3(4):118–128.
Wendorf, F., and R. Schild, 1998. Nabta Playa and its role in northeastern African prehistory. *The Journal of Anthropological Archaeology* 17:97–123.
Wendorf, F., and R. Schild, 2001. Conclusions. In *Holocene Settlement of the Egyptian Sahara, Volume 1, The Archaeology of Nabta Playa*, edited by F. Wendorf, R. Schild, and Associates, pp. 648–675. Kluwer Academic/Plenum Publishers, New York.
Wendorf, F., and R. Schild, 2002a. The role of storage in the Neolithic of the Egyptian Sahara. In *Tides of the Desert – Gezeiten der Wuste, Contributions to the Archaeology and Environmental History of Africa in Honour of Rudolph Kuper*, edited by Jennerstrasse 8. Heinrich-Barth-Institut. Cologne, Germany.
Wendorf, F., and R. Schild, R., 2002b. Implications of incipient social complexity in the late Neolithic in the Egyptian Sahara, In *Egypt and Nubia: Gifts of the Desert*, edited by R. Friedman, pp. 13–20. British Museum Press, London.
Wendorf, F., and R. Schild (Assemblers), and A. E. Close, editor, 1984. *Cattle Keepers of the Eastern Sahara: The Neolithic of Bir Kiseiba*. The Department of Anthropology, Institute for the Study of Earth and Man, Southern Methodist University, Dallas.
Wendorf, F., and R. Schild (Assemblers), and A. E. Close, editor, 1989. *The Prehistory of Wadi Kubbaniya, Volumes 2 and 3*. Southern Methodist University Press, Dallas.
Wendorf, F., and R. Schild (Assemblers), and A. E. Close, editor, 1993. *Egypt During the Last Interglacial. The Middle Paleolithic of Bir Tarfawi and Bir Sahara East*. Plenum Press, New York.
Wendt, E., 1966. Two prehistoric archaeological sites in Egyptian Nubia. *Postilla* 102:1–46.

Chapter 7

Influence of Holocene marine transgression and climate change on cultural evolution in southern Mesopotamia☆

Douglas J. Kennett[1,*] and James P. Kennett[2]

[1]*Department of Anthropology, University of Oregon, Eugene, OR 97403-1218, USA*
[2]*Department of Geological Sciences and Marine Science Institute, University of California, Santa Barbara, CA 93106, USA*

Abstract

The evolution of the earliest complex state-level societies and cities, from small sedentary communities, took place in southern Mesopotamia between 8000 and 5000 cal yr BP during the 'Ubaid and Uruk Periods. Attempts to explain this transition often discount the role of environmental change and evaluate available archaeological evidence for urban-based state development either within a static environmental context or under conditions similar to those of the present. This is no longer tenable given newly available paleoenvironmental records for the region. Postglacial sea-level rise resulted in the inundation and creation of the Arabo-Persian Gulf and rich coastal and aquatic habitats formed in southern Mesopotamia as the marine transgression slowed in the Middle Holocene. These habitats favored the establishment and growth of 'Ubaid Period communities and the efficient transport of goods, ideas, and people throughout the region. High water tables also promoted early experimentation with irrigation agriculture and the expansion of these systems as populations grew and the humid conditions of the Early Holocene gave way to increasing aridity. We argue that the critical confluence of eustatic and climatic changes unique to this circumscribed region favored the emergence of highly centralized, urban-based states.

1. Introduction

Southern Mesopotamia was the site of the earliest large, highly integrated political systems marked by administrative hierarchies and rulers with significant power and authority – so called state-level societies (Rothman, 2001, 2004). As states developed in this region, people became differentiated socially, more specialized economically, and highly integrated and centralized politically (Flannery, 1972). This process resulted in greater interdependence and cooperation between members of society, but it

☆ This chapter appeared in somewhat different form in the *Journal of Island & Coastal Archaeology* 1:67–99, and is used with permission of the Taylor & Francis Group, LLC.
*Corresponding author. *E-mail address:* dkennett@uoregon.edu (D.J. Kennett).

also required a majority of people to relinquish autonomy and allow others to have greater economic, social, and political benefits. How and why this occurred in this region first, and later in several other locations around the world during the Middle and Late Holocene, remains a central question in anthropological archaeology (e.g., Postgate, 1992; Blanton et al., 1993; Flannery, 1998; Marcus and Feinman, 1998; Stein, 1998, 2001; Pollock, 1999; Adams, 2000a; Feinman, 2000; Feinman and Manzanilla, 2000; Algaze, 2001a; Rothman, 2004; Yoffee, 2005).

Although there is some evidence for occupation earlier (see below), the first traces of permanent human settlement in southern Mesopotamia occurred in the beginning of the ‘Ubaid Period at ~8000 cal yr BP. Between 8000 and 5500 cal yr BP human populations increased and aggregated into small towns and nucleated villages. These communities ultimately provided the foundation for integrated state-level societies and urban centers between 5500 and 5000 cal yr BP, with an associated complex of technological innovations including large-scale irrigation agriculture and writing. A number of theories have been proposed for the earliest known state-level societies including: (1) technological and agricultural innovation (Adams, 1981, 2000b); (2) bureaucratic development necessary to build, maintain, and manage large-scale irrigation necessary for agriculture (Wittfogel, 1957, 1981); (3) information processing and the development of centralization and political hierarchies (Wright and Johnson, 1975, 1985; Wright, 1977, 1994, 1998, 2001); (4) increasing warfare in an environmentally and socially circumscribed area (Carneiro, 1988); (5) increasing intra- and interregional exchange, colonialism, and cross-cultural contact (Wright and Johnson, 1975; Algaze, 1993, 2001b); and (6) religious ideology and the control or mobilization of labor (Hole, 1983). Proponents of multicausal models suggest that state development resulted from several interrelated factors, including characteristics of the regional environment (Flannery, 1972; Redman, 1978; Adams, 1981; Crawford, 1991; Hole, 1994; Stein, 1994, 2001; Rothman, 1994, 2004; Kouchoukos, 1999; Algaze, 2001a; Pournelle, 2003; Wilkinson, 2003). Others have suggested that state formation emerged as a result of intrinsic human social interrelations independent of environmental factors (Pollock, 1992).

The processes leading to the emergence of state-level societies in southern Mesopotamia were multivariate, but we argue that these developments should be considered within the context of environmental change (Fairbanks, 1989; Sanlaville, 1989; Eisenhauer et al., 1992; Petit-Maire, 1992; Sirocko et al., 1993; Teller et al., 2000; Aqrawi, 2001). Changes in environmental conditions are often thought to have played a major role in cultural demise (Weiss et al., 1993; Hodell et al., 1995; Issar, 1995; Weiss, 1997). In contrast, the potential importance of climate and related environmental change is less accepted as a critical variable in the development of cultural complexity, but has recently received increased attention (Hole, 1994; Spier, 1996; Sandweiss et al., 1999; Kennett and Kennett, 2000). Except for Hole (1994; also see Nützel, 2004), however, there has been limited consideration of the potential role of environmental change in the evolution of the state in southern Mesopotamia, where the environment has been incorrectly characterized by some archaeologists as stable during the Holocene (Pollock, 1992). Evaluation of the available paleoenvironmental data demonstrates distinct correlations between the timing of Holocene environmental and cultural changes. These correlations, by themselves, do not prove causality but must be considered when evaluating the timing and nature of emergent cultural complexity in this region.

We propose that this development was stimulated, in part, by increased competition for resources caused by successive changes in sea level, shorelines, and climate specific to this region. In particular, the expansion, and ultimate stabilization of aquatic habitats associated with the marine transgression and more productive compared with today, favored increased population densities and early group formation, community stability, enhanced maritime trade, and the emergence of social hierarchies. The natural diversity of resources in coastal/aquatic habitats, in combination with newly domesticated plants and animals, provided the economic foundation for these developing communities, as they did elsewhere during the Early and Middle Holocene (Binford, 1968; Moseley, 1975; Clark and Blake, 1994). In this contribution we describe these changes within the context of cultural development and discuss the possible implications of these interrelationships.

2. Postglacial environmental change in southern Mesopotamia

Southern Mesopotamia lies in present-day southern Iraq at the head of the Arabo-Persian Gulf. Modern climatic conditions are arid to semiarid with a mean annual rainfall of 139 mm (ranging from 72 to 316 mm; Adams, 1965). Severe dust storms occur during the summer months due to semipermanent low-pressure zones over the Gulf that draw hot, dry winds across the alluvial plain. Because of extreme aridity, agriculture is largely limited to the floodplains of the Tigris–Euphrates–Karun Rivers that converge in an extensive wetland region associated with the El Schatt Delta.

The Arabo-Persian Gulf is roughly 1000 km long and ranges from 350 km to as little as 5 km wide at the Straits of Hormoz, where it joins the Gulf of Oman in the northern Indian Ocean. This is the shallowest inland sea of significant area in the world, the bathymetry reflecting a gently inclined basin with a mean depth of only 40 m and almost nowhere exceeding 100 m except near the Straits of Hormoz (Fig. 7.1c) (Seibold and Vollbrecht, 1969; Sarnthein, 1971; Purser and Seibold, 1973). Late Quaternary changes in sea level played a major role in shaping the environment of this region (Gunatilaka, 1986; Cooke, 1987; Sanlaville, 1989; Teller et al., 2000). Although southern Mesopotamia is located on a subsiding sedimentary basin (Lees and Falcon, 1952), tectonic influences on eustacy, including subsidence, are considered to have been relatively minor compared with glacioeustatic effects during the last 15,000 yr (Macfadyen and Vita-Finzi, 1978; Gunatilaka, 1986; Cooke, 1987; Sanlaville, 1989; Lambeck, 1996; Aqrawi, 2001). This differs from an earlier, widely accepted view that the balance between

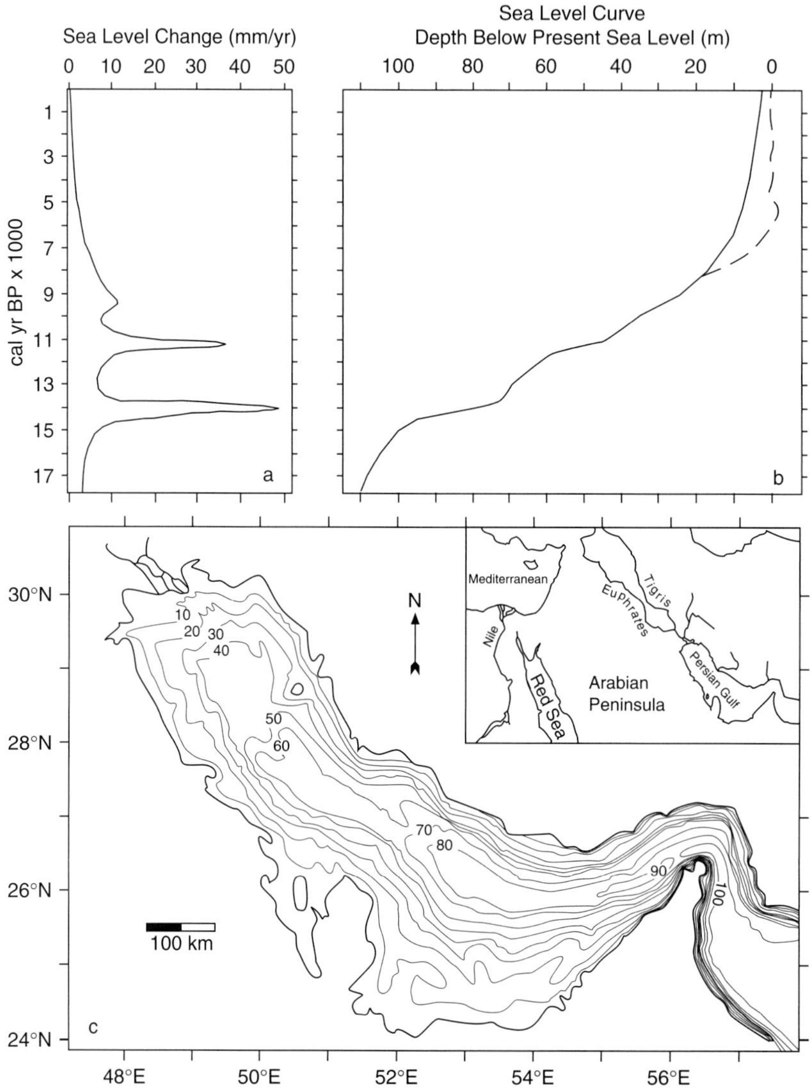

Figure 7.1. (a and b) Sea-level change during the last 18,000 yr (Lighty et al., 1982; Fairbanks, 1989; Sanlaville, 1989; Yafeng et al., 1993). Solid line in (b) shows well-established sea-level curve based on Caribbean corals (Lighty et al., 1982; Fairbanks, 1989); dashed line is local sea-level curve estimated for northern Arabo-Persian Gulf (Sanlaville, 1989; Yafeng et al., 1993). All radiocarbon dates have been calibrated to calendar years (cal yr BP). (c) Modern bathymetry of the Arabo-Persian Gulf (adapted from Sarnthein, 1972).

sedimentation and subsidence rates in southern Mesopotamia maintained the Gulf shoreline and delta close to their present-day positions throughout the Holocene and that inland incursions of the ocean resulted from subsidence (Lees and Falcon, 1952).

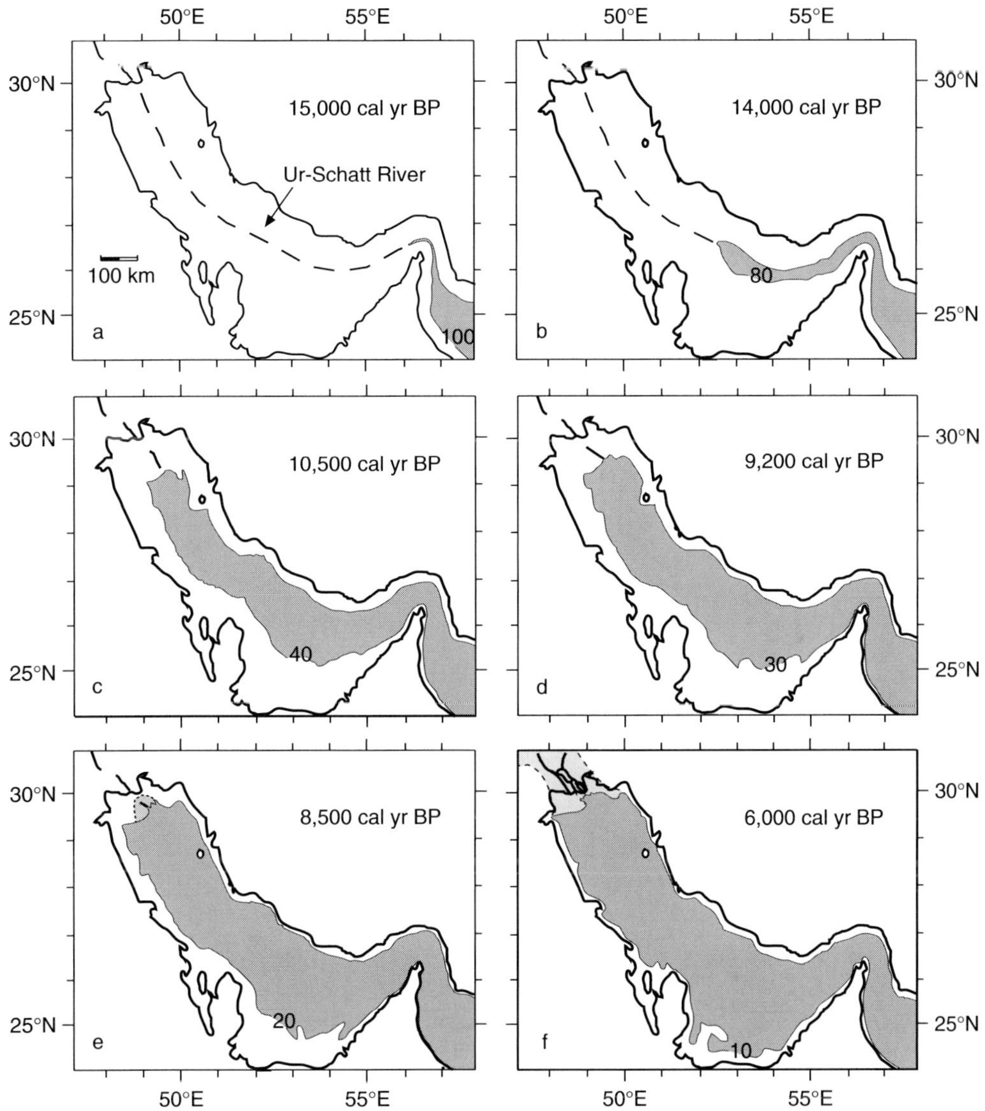

Figure 7.2. Maps of successive time intervals showing the marine transgression into the Arabo-Persian Gulf during Late Pleistocene to Early/Middle Holocene. Sea-level estimates from (Fairbanks, 1989) and based on modern bathymetry (Sarnthein, 1972). Position of the Ur–Schatt River (ancient Schatt River) estimated from bathymetry (Sarnthein, 1972). Marine transgression in southern Mesopotamia at 6000 cal yr BP (F) adapted from (Sanlaville, 1989) and shown in detail in Fig. 7.4.

Environmental conditions during the latest Pleistocene through Middle Holocene were different from those of today in southern Mesopotamia and the Gulf region. About 15,000 cal yr BP, global sea level was still 100 m below present (Fig. 7.1a, b; Fairbanks, 1989). Owing to the shallowness of the Arabo-Persian Gulf,

Late Quaternary marine transgression associated with deglaciation was only beginning to enter this dry, subaerial, basin through the Straits of Hormoz (Fig. 7.2a; Vita-Finzi, 1978; Lambeck, 1996). Calcareous detritus in pericoastal dunes in the United Arab Emirates was wind-transported from the exposed Gulf floor during the Late Pleistocene (100,000 to 12,000 cal yr BP; Teller et al., 2000). At this time the Tigris–Euphrates–Karun River system flowed into the Gulf of Oman as the Ur–Schatt (ancient Schatt) River (Seibold and Vollbrecht, 1969; Gunatilaka, 1986) that traversed the full length of the Gulf in its deepest present-day sector. The Ur–Schatt River flowed in an incised canyon, now completely submerged, but still evident in the present-day bathymetry of the middle to lower Gulf (Seibold and Vollbrecht, 1969; Sarnthein, 1971). This canyon was formed by

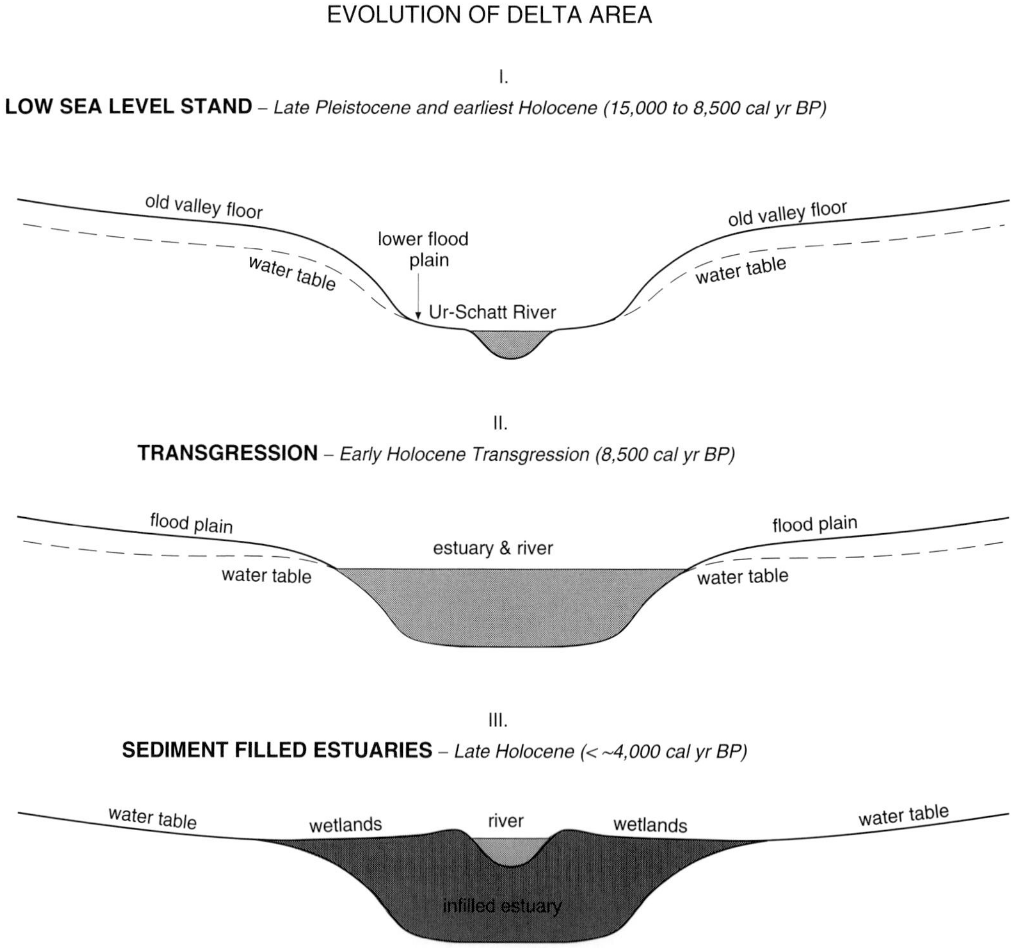

Figure 7.3. Schematic cross sections showing inferred three stages in evolution of the delta region of southern Mesopotamia during the latest Quaternary.

downcutting during low sea levels of the last glaciation. A deep sea canyon extending southwards from the head of the Gulf of Oman (Seibold and Ulrich, 1970) was almost certainly cut by turbidity currents carrying sediments southwards from the head of the Gulf of Oman. Large volumes of sediments appear to have been transported to the Gulf of Oman by the Ur–Schatt River during Quaternary low sea-level stands, implying substantial river flow. At this time, the modern delta did not exist in southern Mesopotamia (Fig. 7.3) and narrow floodplains were restricted to the incised river canyons. Severe aridity at this time is indicated by the presence of drowned ridge and trough features in the northern Arabo-Persian Gulf, interpreted as fossil sand-dune fields (Sarnthein, 1971), in combination with sedimentological (Sarnthein, 1972; Diester-Haass, 1973) and oxygen isotopic data (Sirocko et al., 1993).

Following 15,000 cal yr BP, marine transgression formed the Arabo-Persian Gulf (Fig. 7.2b–f). Sea-level rise during this interval was highly variable, but averaged ~1 cm/yr until ~9000 cal yr BP (Fig. 7.1a), after which the rate of rise slowed (Lighty et al., 1982; Fairbanks, 1989; Warne and Stanley, 1993). The pattern of global sea-level change after 9000 cal yr BP has yet to be firmly established. Sea-level curves from the western Atlantic (Lighty et al., 1982) and southeastern

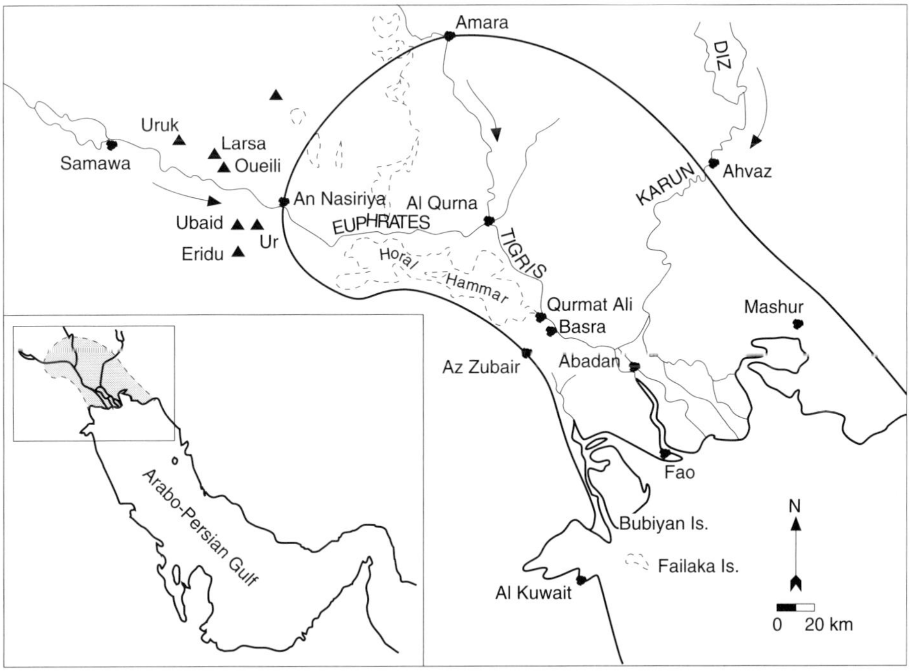

Figure 7.4. Estimated shoreline at 6000 cal yr BP in southern Mesopotamia superimposed on present-day geography. Triangles show locations of early settlements and dots indicate modern cities (adapted from Sanlaville, 1989).

Mediterranean (Warne and Stanley, 1993) show a slow rise with a steadily decreasing rate (average ~0.3 cm/yr) from 7000 cal yr BP to the present (Fig. 7.1b). In these curves, rates of rise are especially slow after 5500 cal yr BP. This pattern compares with sea-level rise estimates based on western Australian evidence of ~0.7 cm/yr from 9800 cal yr BP to a maximum high stand at 6300 cal yr BP, followed by a slight decrease in sea level inferred to be associated with a cessation of polar ice sheet melting (Eisenhauer et al., 1992). In spite of these differences, it is clear that the rate of global sea-level rise decreased significantly between 6300 and 5500 cal yr BP. Rapid rise in sea level during the Early Holocene (Siddall et al., 2003) of ~1 cm/yr created a lateral marine transgression in the Arabo-Persian Gulf of ~110 m/yr, one of the highest rates known for any region. The early stages of this transgression (~15,000 and 11,000 cal yr BP) mainly filled the deeply incised canyon of the Ur–Schatt River in its lower to middle reaches. The transgression later inundated the broader, shallower Gulf region (Fig. 7.4). Most notable rapid rises in sea level in the Arabo-Persian Gulf occurred between 12,000 and 11,500 cal yr BP and again from 9500 to 8500 cal yr BP and during these periods the lateral transgression probably exceeded 1 km/yr (Teller et al., 2000, p. 306).[1]

Inundation of the Arabo-Persian Gulf coincided with an interval of increased seasonal rainfall across the Arabian Peninsula and southern Mesopotamia between ~10,000 and 6,000 cal yr BP. This interpretation is based on a variety of indicators including sedimentological evidence for increased river runoff into the Arabo-Persian Gulf (Diester-Haass, 1973), speleothem records from Israel and Oman (Bar-Matthews et al., 1997; Burns et al., 1998, 2001), pollen evidence for more widespread, less arid vegetation, and the presence of interdune lakes and on the Arabian Peninsula

[1] This manuscript was initially prepared in response to a talk given by Walter Pitman linking biblical flood mythology to the rapid inundation of the Black Sea during the Holocene, an idea he later published in a book entitled "Noah's Flood: The New Scientific Discoveries About the Event That Changed History" (Ryan and Pitman, 1998). Although it is inherently difficult to link past cultural and environmental developments with mythology, we instead argue based on the environmental and cultural history of southern Mesopotamia that the Sumerian flood myth (as recorded in the Gilgamesh Epic) and the succeeding biblical flood narrative, most likely have their origins in the glacioeustatic latest Quaternary transgression in the Arabo-Persian Gulf; the largest, shallowest inland sea contiguous with the ocean. Flood mythology among maritime societies is virtually universal (http://www.talkorigins.org/faqs/flood-myths.html) and most likely linked to eustatic rises in sea level during the Late Pleistocene and Early Holocene. As Teller et al. (2000) point out, a rapid marine transgression occurred in the Arabo-Persian Gulf during this time and likely displaced people living along this waterway. Such momentous events were surely passed down orally for generations. What makes southern Mesopotamia unique from other parts of the world is that writing developed early, associated with the formation of state-level societies, and this story was ultimately recorded as in the well-known Epic of Gilgamesh. Given Occam's razor, we hypothesize, as do Teller et al. (2000; also see Potts, 1996), that the origins of Biblical flood mythology is most likely in southern Mesopotamia rather than in the vicinity of the Black Sea.

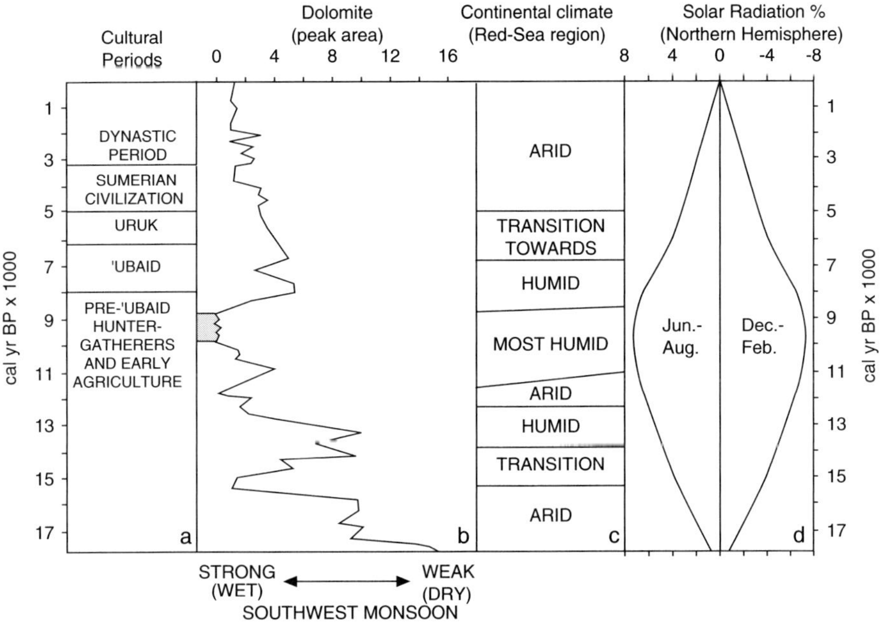

Figure 7.5. Correlations between Late Quaternary climatic changes (1000 cal yr BP = ka; Almogi-Labin et al., 1991; Sirocko et al., 1993) and major cultural periods (a) in southern Mesopotamia (Adams and Nissen, 1972; Wright, H. T. and Rupley, 2001; Rothman, 2004; also see Endnote 2). Fluctuations in dolomite abundance (peak area) values during Late Quaternary in a sediment core from Arabian Gulf (b) reflect changes in aridity in Arabia–southern Mesopotamia region (Sirocko et al., 1993). The gray band on the dolomite curve represents interval of highest humidity. Shown at right (d) is the percent change in solar radiation in the Northern Hemisphere summer (June–August) and winter (December–February) during the last 18,000 yr, resulting from the earth's orbital perturbations (Kutzbach and Gallimore, 1988; Kutzbach and Guetter, 1986). This caused summers to be warmer and winters colder in Arabia and southern Mesopotamia, with seasonal differences and inferred strength of summer monsoons peaking at ~9–10,000 cal yr BP. Note relations with humid–arid cycles in the Red Sea region (c) representing a synthesis of changes in lake levels, vegetation history, and sediment data (Almogi-Labin et al., 1991).

(McClure, 1976; Street-Perrott and Roberts, 1983; Rossignol-Strick, 1987; Roberts and Wright, 1993; Lézine et al., 1998) and southern Mesopotamia (Wright, 1993; Yan and Petit-Maire, 1994) between 9000 and 6000 cal yr BP. Moister conditions are also inferred from an extensive network of ephemeral channels (wadis) over the Arabian Peninsula that run into the Arabo-Persian Gulf and Arabian Sea (Höltz et al., 1984; Dabbagh et al., 1998) and appear to have been more active during the Late Pleistocene and Early Holocene (Wilkinson, 2003). Increased rainfall in this region between ~10,000 and 8,600 cal yr BP is also inferred from an absence of dolomite at this time in a sediment core from the Gulf of Oman (Sirocko et al., 1993) (Fig. 7.5b). Under arid conditions dolomite is formed in coastal, supra-tidal evaporitic environments (sabkas) in the Gulf region and wind-transported to the Gulf of Oman.

Overall, Early Holocene climatic conditions in Arabia were semiarid (~250–300 mm annual rainfall) compared with the aridity of today (50–100 mm) (Whitney et al., 1983).

Paleoceanographic and terrestrial climatic records in the Red Sea region indicate relatively humid conditions between ~12,000 and 6,000 cal yr BP (Almogi-Labin et al., 1991), with the wettest interval occurring between ~10,000 and 7,800 cal yr BP (Haynes et al., 1989; Street-Perrott and Perrott, 1990; Fig. 7.5c). Significant regional increase in rainfall between 11,700 and 5,400 cal yr BP has also been inferred from a decrease in oxygen isotopic values in planktonic foraminifera and pteropods in a Red Sea core (21°N), interpreted to reflect decreased surface water salinities (Rossignol-Strick, 1987). Continental freshwater runoff at this time was sufficiently large to decrease surface water salinities in the Red Sea relative to present-day values. Inferred low salinities peaked between ~8500 and 6700 cal yr BP, then increased to 5400 cal yr BP when average postglacial values were reached (Rossignol-Strick, 1987). Wet (humid) conditions between 9200 and 7250 cal yr BP in this region are confirmed based on increases in terrigenous sediment input and decreased surface water salinities as reflected by oxygen isotopes in foraminifera species in sediment cores from the northernmost Red Sea (Arz et al., 2003; see Fig. 7.6). These data are consistent with pollen records and evidence for high lake levels indicating Early Holocene (10,000–5,500 cal yr BP) increases in precipitation and a pluvial maximum (~10,000–7,000 cal yr BP) in the Levant (Issar, 2003) and throughout sub-Saharan Africa (Kutzbach and Street-Perrot, 1985; Ritchie et al., 1985; Street-Perrott et al., 1985; Petit-Maire, 1986, 1990, 1992; Haynes and Mead, 1987; Pachur and Kröpelin, 1987, 1989; COHMAP, 1988; Haynes et al., 1989; Gasse et al., 1990, 1991; Street-Perrott and Perrott, 1990; Ambrose and Sikes, 1991), coinciding with the so-called hypsithermal climatic interval (Lamb, 1977; Kutzbach and Street-Perrot, 1985; Petit-Maire, 1986; COHMAP, 1988; Gasse et al., 1991). Pollen (Rossignol-Strick, 1987; Roberts and Wright, 1993;Lézine et al., 2002), lake level (McClure, 1976), and marine sediment (Diester-Haass, 1973; Sirocko et al., 1993) data indicate that humid conditions persisted until ~6000 cal yr BP, although a gradual decrease in humidity had begun after ~8000 cal yr BP (Vita-Finzi, 1978, Ritchie et al., 1985). Analysis of paleoclimatic data suggests that Southwest Indian monsoon strength for the broader Asia–East African region was greatest between 11,000 and 5,000 cal yr BP (Overpeck et al., 1996), but in the Middle East, maximum activity seems to have occurred between 9000 and 7000 cal yr BP (Yan and Petit-Maire, 1994; Bar-Matthews et al., 1997; Lézine et al., 1998).

The marine transgression reached the present-day northern Gulf area between 9000 and 8000 cal yr BP (Gunatilaka, 1986; Lambeck, 1996; Aqrawi, 2001), inundating the entrenched Ur–Schatt River valley and forming an extensive marine estuary in the location of the present delta area. The modern delta has since filled the estuary (Cooke, 1987; Aqrawi, 2001). Movement of the coastline associated with the marine transgression was so rapid that sedimentation would have been minimal in the newly developing, open estuary and no delta would have formed

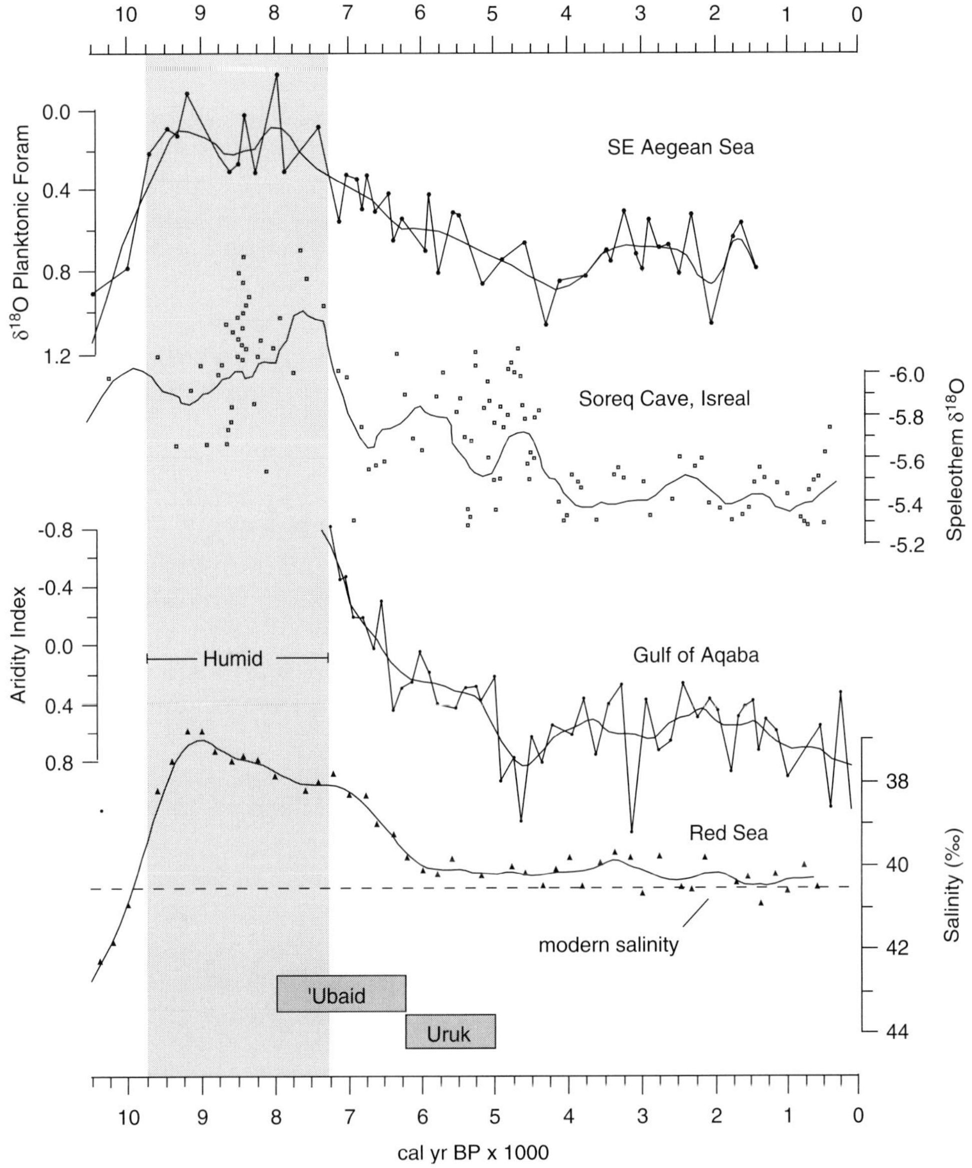

Figure 7.6. Comparison of four climate records from the eastern Mediterranean/Red Sea regions showing Early Holocene humidity and the onset of regional drying after ~7000 cal yr BP (adapted from Arz et al., 2003). Major cultural periods shown at the bottom of the figure.

(Cooke, 1987). From ~9000 cal yr BP onwards, and particularly after ~6000–5500 cal yr BP, sea-level rise slowed globally (Fig. 7.1a,b; Lighty et al., 1982; Fairbanks, 1989) or possibly reached a Holocene maximum at ~6000 cal yr BP (Eisenhauer et al., 1992; Yafeng et al., 1993). In the northern Arabo-Persian Gulf, sea level reached its current elevation at 6000 cal yr BP, was possibly 2.5 m higher between 6000 and 5000 cal yr BP or longer (Lambeck, 1996), and thereafter was relatively stable close to present-day levels (Al-Asfour, 1978; Sanlaville, 1989). Sediments from boreholes in the Tigris–Euphrates Delta indicate marine-brackish conditions in the vicinity of Fao at ~9000 cal yr BP, near Basra at 8000 cal yr BP, and as far north as Nisiriyya by 6000 cal yr BP (Aqrawi, 2001, p. 275). By ~6000 cal yr BP estuaries had expanded to their northernmost limits. Marine deposits of 7000 cal yr BP age have been reported inland (80–100 km) at Lake Hammar and estuarine deposits of this age are also known in cores 40 km farther to the northwest (Hudson et al., 1957; Sanlaville, 1989). The northernmost known extent of estuarine sediments in boreholes of Holocene age is 400 km inland from the present head of the Gulf (Cooke, 1987). Estuarine sediments from bore holes containing foraminifera, marine mollusks, and other marine fossils, reflect brackish to marine conditions during the Middle Holocene near the ancient city of Ur on the Euphrates (Sanlaville, 1989; Aqrawi, 2001) and at Amarah on the Tigris (Macfadyen and Vita-Finzi, 1978; Aqrawi, 2001). Estuaries probably extended even farther northwards following the Tigris–Euphrates–Karun River canyons and formed a variety of productive wetland habitats throughout this region (Pournelle, 2003). Extensive floodplains and associated high water tables would have formed during the Middle Holocene with deceleration of sea-level rise as it began to approach present levels, with the consequent development of a complex mosaic of aquatic habitats (estuaries, rivers, wetlands, and marshes).[2]

By the Middle Holocene, regional climatic conditions had become more arid. This change is indicated by a diversity of evidence including an abrupt increase in aeolian sediment and dune formation on the southern periphery of the Arabo-Persian Gulf at 6000 BP (United Arab Emirates; Bray and Stokes, 2004). Evidence from the Red Sea region indicates a marked trend towards increasing aridity between 7000 and 5000 cal yr BP (Almogi-Labin et al., 1991; Arz et al., 2003). Pronounced regional desiccation by 5000 cal yr BP dried up lakes in the region encompassing Arabia, the Red Sea, and sub-Saharan Africa (McClure, 1976; Ritchie et al., 1985; Pachur and Kröpelin, 1989; Almogi-Labin et al., 1991; Roberts and Wright, 1993; Sukumar et al., 1993). Developing aridity also led to increased dust transport evident in the Arabian Sea after 5300 cal yr BP (Sirocko et al., 1993). In sub-Saharan Africa, lake level and pollen records also indicate that regional desiccation became widespread between 6000 and 5000 cal yr BP (Ritchie et al., 1985; Roberts, 1989; deMenocal et al., 2000).

Major deceleration of sea-level rise after the Middle Holocene led to sediment infilling of the estuary, accelerated by erosion (Melguen, 1973; Cooke, 1987).

[2] These habitats were highly productive and not comparable to the infilled wetlands of lower productivity inhabited historically by "Marsh Arabs" (Ochsenschlager, 2004). We argue that this lifeway has been used incorrectly as ethnographically analogous to early human populations at the head of the Gulf.

Southeastward progradation of the Mesopotamian Delta commenced, eventually extending ~200 km to the southeast. This led to the development of extensive wetland areas (Larsen and Evans, 1978; Sanlaville, 1989). The trend towards increasing aridity in the broad region continued through the Late Holocene. Aeolian deposits within deltaic sediments were most pronounced in southern Mesopotamia between 5000 and 4000 cal yr BP and point to severe aridity during this interval (Aqrawi, 1993, 1995, 2001). Archaeological evidence for the abandonment of settlements also suggests that southern Mesopotamia became extremely arid during the Late Holocene (Wright, 1981; Nissen, 1988; Weiss et al., 1993). Flood levels in the Nile also declined between 5000 and 4000 cal yr BP (Roberts, 1989) and central African rift lakes desiccated completely between 3400 and 3000 cal yr BP (Ambrose and Sikes, 1991).

The changes in Holocene precipitation and associated shifts of the Sudanian–Sahelian vegetation belt over north Africa (Roberts, 1989), the Arabian Peninsula, and southern Mesopotamia have been linked to changing intensification of summer monsoons related to northward shift of the Intertropical Convergence Zone and the influence of a more moist westerly airstream affecting the Mediterranean region (Kutzbach, 1983; Street-Perrott and Roberts, 1983; Kutzbach and Guetter, 1986; COHMAP, 1988; Roberts and Wright, 1993; Clemens et al., 1996). Holocene fluctuations in the strength of the South Asian monsoon in the Middle East region resulted from differential thermal response of land and ocean surfaces due to orbitally caused (Milankovitch) changes in the strength of the seasonal cycle and solar radiation in the Northern Hemisphere (Kutzbach and Guetter, 1986; Kutzbach and Gallimore, 1988; Clemens et al., 1996). At 9000 cal yr BP, orbital perturbations of the earth were such that the perihelion occurred in July rather than January, as it is today, and the axial tilt of the earth was greater than it is now. As a result, July (summer) average solar radiation in the Northern Hemisphere was ~7% higher than that of today (Kutzbach and Guetter, 1986; Kutzbach and Gallimore, 1988). This caused summers to be warmer and winters colder in Arabia and southern Mesopotamia, with seasonal differences peaking at ~9,000–10,000 cal yr BP. Resulting increase in airflow over Arabia from the Indian Ocean, associated with southwest summer monsoon, led to higher seasonal precipitation (Kutzbach and Guetter, 1986) and upwelling near south Arabia that peaked about 9000 cal yr BP (Prell, 1984). Northern summer monsoons were stronger between 12,000 and 6,000 cal yr BP in conjunction with increased summer radiation (COHMAP, 1988). Associated increased summer monsoon precipitation over tropical lands were ~10–20% (Kutzbach and Guetter, 1986). The broad arid–humid–arid cycle of the Middle East region during the last 18,000 yr, including southern Mesopotamian (Roberts and Wright, 1993), was largely controlled by this change in the seasonal cycle. Increased monsoonal strength during the Early Holocene led to the higher annual precipitation as far north as southern Mesopotamia (Whitney et al., 1983; Kutzbach and Guetter, 1986; Kutzbach and Gallimore, 1988; Roberts and Wright, 1993). Conditions further north remained relatively dry (Kutzbach and Guetter, 1986; Roberts and Wright, 1993). After

~5500 cal yr BP, weakening of the monsoons led to increasing aridity over the Arabian Peninsula (Sirocko et al., 1993).

3. State development in southern Mesopotamia

States with well-developed urban centers and administrative hierarchies first appeared in southern Mesopotamia about 5000 cal yr BP (Late Uruk or LC5; Johnson, 1973; Wright and Johnson, 1975; Adams, 1981; Nissen, 1988, 2001; Postgate, 1992; Pollock, 1999; Rothman, 2004; Yoffee, 2005; Fig. 7.5a). The people of southern Mesopotamia were on the leading edge of what Childe (1950) referred to as the second great revolution – the integration of large numbers of people into one social, economic, and political system ruled by an elite class with the help of an elaborate administrative hierarchy. Archaic states spread quickly in the Near East after this time and developed independently in other parts of the world, including Mesoamerica, South America, and China (Feinman and Marcus, 1998). The development of social differentiation, economic specialization, and ultimately political centralization culminating in the emergence of state-level societies is a central archaeological research question addressed from a variety of perspectives (Flannery, 1972; Blanton et al., 1993; Feinman and Marcus, 1998; Pollock, 1999). Southern Mesopotamia has played an important role in modeling state origins and emergence of centralized administrative hierarchies because of the long tradition of research in the region (Rothman, 2004; Yoffee, 2005).

Origins of the first city dwellers in southern Mesopotamia have long been debated and are crucial for understanding the social, economic, and political processes culminating in the state (see Meissner, 1920; Speiser, 1930; Frankfort, 1956; Kramer, 1963; Oates, 1991; Potts, 1997; Bottéro, 2001). Similarities in architecture and ceramic styles in southern Mesopotamia from the Early 'Ubaid through Late Uruk Periods (~8000 to 5000 cal yr BP) suggest a certain degree of demographic and cultural continuity, rather than the intrusion of outside peoples, prior to the emergence of the state (Oates, 1960). Population continuity is also indicated by a series of superimposed temples at the site of Eridu through this interval (Oates, 1960; Potts, 1997, p. 47). This was the period when some small villages grew in size relative to neighboring settlements and ultimately became the first state centers in which social and political hierarchies developed.

The settlement history of people in southern Mesopotamia prior to 8000 cal yr BP is far from clear. A transition from mobile hunting and gathering to sedentary agriculture is evident in west Asia between 12,000 and 8,000 cal yr BP (Flannery, 1969; Bar-Yosef and Belfer-Cohen, 1989; Henry, 1989; McCorriston and Hole, 1991; Bar-Yosef and Meadow, 1995; Harris, 1996, 1998; Meadow, 1996; Garrard, 1999; Zeder and Hesse, 2000). This was a time of early development of agriculture and animal domestication (Smith, 1998), emergence of small villages, some of which were fortified, and pottery manufacturing in various locations (Redman, 1978). Early agricultural villages in northern Mesopotamia (e.g., Tell es-Sawwan) were generally limited in size to a few hundred people engaged in simple agriculture and animal domestication (Moore, 1985). We suspect that at this time people were living along the Ur–Schatt and associated floodplains, a natural corridor connecting northern Mesopotamia with the interior shallow basin known today as the Arabo-Persian Gulf. This basin was then undergoing rapid marine transgression and stories resulting from this inundation may be the source of biblical flood mythology (Potts, 1996; Teller et al., 2000). Evidence for any settlements along the Ur–Schatt River has since been obscured by flooding and/or covered by sediments.

Some archaeological evidence exists for a pre-'Ubaid occupation along the northeast coast of the Arabo-Persian Gulf. Furthermore, Neolithic stone tool assemblages (Arabian bifacial tradition; Potts, 1997, p. 52) dating to between 9600 and 5500 cal yr BP are also common near now desiccated inland lakes across the Arabian Peninsula (Zarins et al., 1981; Edens, 1982; Uerpmann, 1992; Potts, 1993, 1997; Edens and Wilkinson, 1998) and at several locations along the western edge of the Arabo-Persian Gulf by at least 7300 cal yr BP (Glover, 1998; Beech et al., 2005; Connan et al., 2005). Visible architecture is rarely encountered at Arabian Neolithic sites, with the exception of small stone structures reported at several coastal locations (e.g., Kuwait, Connan et al., 2005; Qatar, Inizan, 1988, Marawah Island, United Arab Emirates; Beech et al., 2005). Artifact and faunal/floral assemblages from these sites suggest that people combined sheep and goat herding with small game hunting, the collection of local grasses, and even periodic cereal crop farming (Potts, 1993, 1997). This was combined with fishing, shellfishing and other marine resources (e.g., sea turtles; Beach et al., 2005) at several sites along the Arabo-Persian Gulf (Glover, 1998; Connan et al., 2005). These data indicate a mixed foraging and food producing strategy comparable to what Smith (2001) has described as low-level food production. Stone tool assemblages dating to ~8000 cal yr BP are also known from wadi systems in western Iraq that flow into the Euphrates River (Zarins, 1990). Similar stone tool assemblages, although poorly described, are documented in southern Mesopotamia (Potts, 1997, p. 53; Zarins, 1992) where they were found at Ur (Woolley, 1955), near Eridu (Potts, 1997), and at Tell Oueli and Tello (Cauvin, 1979; Inizan and Tixier, 1983). Preceramic sites on the eastern fringe of southern Mesopotamia (Tell Rihan III in the Hamrin and Choga Banut in Kuzistan) dating to ~8500 cal yr BP also suggest linkages between people living in the Zagros and the southern alluvium (Aurenche, 1987). Overall, the archaeological record for this period suggests that significant populations lived across the desert regions of Arabia and Mesopotamia, practicing a diverse range of subsistence strategies tied to a variety of aquatic habitats (e.g., lakes, springs, creeks, rivers) resulting from the northward shift of the south Asian monsoon belt during the Early to Middle Holocene.

The presence of stone tools comparable to the southern Arabian biface tradition is consistent with the idea of a pre-Ubaid occupation earlier than 8000 cal yr BP in southern Mesopotamia (see Oates, 1960, 2004). Exploration for pre-Ubaid sites is impeded by thick alluvial sediments deposited since the near stabilization of sea level at ~6200 cal yr BP and a water table that is higher than when sites of this age were occupied. Even early 'Ubaid sites in the region (e.g., Hajji Muhammad) are deeply buried under alluvium (Huot, 1989), so it is likely that sites of this age are more numerous. Several sites are now known to have slightly earlier ceramics comparable in form to Samarran and Choga Mami assemblages from central and northern Mesopotamia (WS 298, Adams and Nissen, 1972; Tell Oueli, Calvet, 1989; Huot, 1989; see Potts, 1997). Tell Oueli is the most impressive of these sites with early levels, now designated as Ubaid 0 (~8000 cal yr BP; Oates, 2004) that are 5 m thick in places and contain ceramics and cigar-shaped building bricks reminiscent of cultural traditions in northern and central Mesopotamia (Huot, 1989). This is consistent with the idea of a small, relatively sedentary population living around wetlands at the head of the Arabo-Persian Gulf as sea-level rise slowed. Whether they moved into this area as sea level was stabilizing or were pushed from the south with rapid sea-level rise remains unresolved, but if biblical flood mythology originates in southern Mesopotamia it is likely that at least a portion of the resident population on the floodplains of the Ur–Schatt River were displaced to the north because of this marine transgression (Potts, 1996; Teller et al., 2000).

By the beginning of the 'Ubaid Period (~8000 cal yr BP), small villages and towns were more common across greater Mesopotamia. Much of this region was linked through social networks, and similarities in artifacts indicate widespread exchange of goods and knowledge. The archaeological record suggests distinctive demographic trends in southern Mesopotamia and the Susiana Plain during this time. Larger numbers of known settlements dating to the Early 'Ubaid Period result from a combination of greater archaeological visibility (larger sites visible above alluvium) and increases in regional population. At this time the communities of Eridu, 'Usaila, Ur, and Tell al- 'Ubaid were small, averaging about 1 ha in size with estimated populations seldom exceeding 1000 people. These small communities were widely dispersed and lacked the linear distribution typical of settlements dependent on irrigation canals (Adams, 1981, p. 59), although irrigation agriculture was practiced elsewhere in Mesopotamia and was employed in some parts of the southern alluvial plain (Oates and Oates, 1976; Wilkinson, 2003). The carbonized remains of *Triticum monococcum* (Einkorn) and *Hordeum vogare* (Barley) in 'Ubaid 0 (~8000–7500 BP) levels at Tell Oueli also suggest that some form of irrigation was in use (Huot, 1989, 1996), or that the higher humidity evident in Early Holocene climate records was sufficient to sustain rain-fed agriculture, perhaps in combination with opportunistic use of seasonally receding flood zones or the higher water tables close to the head of the Gulf (Kouchoukos, 1999). Early settlements were located on slight rises (turtle backs) within aquatic habitats resulting from seasonal monsoonal rains or in the wetlands at the head of the Arabo-Persian Gulf as sea-level rise slowed (Pournelle, 2003). Such locations were at the interface between fresh and salt water and were optimal for fresh water accessibility, hunting/fishing, transportation, and irrigation agriculture (Oates, 1960). Within this aquatic context, a broad-spectrum economy developed during the ‘Ubaid Period, based upon small-scale agriculture with an emphasis on salt-tolerant crops, animal domestication (e.g., oxen), hunting, fishing, and trade (Woolley, 1929; Huot, 1989; Sanlaville, 1989, p. 14).

By Middle ‘Ubaid times (‘Ubaid 2–3) some communities in southern Mesopotamia grew larger than their neighbors, a two-tiered settlement system that often marks the emergence of hierarchically organized (nonstate) societies (Wright, 1981; Stein, 1994). Important centers included Eridu, Ur, and Uquir (Adams, 1981; Wright, 1981; Wright and Pollock, 1986). Eridu and Ur were particularly large by this time, both having grown to between 9 and 10 ha in size with estimated populations of 2000–3000 people (Adams, 1981). A similar pattern occurred on the adjacent Susiana Plain (southwestern Iran) where Choga Mish expanded rapidly to 11 ha, dwarfing other agricultural communities in the region (Wright and Johnson, 1985). As some communities expanded in size within the Mesopotamian heartland, ‘Ubaid Period ceramics appear in Neolithic settlements and shell middens along the east coast of the Arabian Peninsula (Masry, 1974; Oates et al., 1977; Zarins et al., 1981; Potts, 1993; Uerpmann and Uerpmann, 1996; Glover, 1998). The appearance of ‘Ubaid Period ceramics is associated with the first evidence for a maritime trade network suggested by the remains of a barnacle-covered reed and bituminous boat from coastal Kuwait (Site H3, As-Sabiyah; ‘Ubaid 3, 7300–6900 cal yr BP; Carter, 2002, 2003; Connan et al., 2005). Population expansion continued across the southern Mesopotamian alluvium in Late ‘Ubaid times (‘Ubaid 3–4) and a network of large and small settlements developed with economies based on irrigation agriculture (Wright, 1981; Wright and Pollock, 1986). With this expansion, elements of material culture associated with the ‘Ubaid Period (e.g., uniform pottery style, Berman, 1994; other clay objects-cone-head figurines, sickles, and “nails,” Wright and Pollock, 1986; architecture, Roaf, 1984; Huot, 1989) first appear in northern Mesopotamia (Tobler, 1950; Stein, 1994; Rothman, 2002), a pattern interpreted as political integration centered on the southern alluvium (Algaze, 1993) or cultural replication due to contact with people to the south (Stein, 1994). Interestingly, there is little evidence for warfare in southern Mesopotamia until the end of the ‘Ubaid Period. Settlements were not fortified and ‘Ubaid seals do not show war-related depictions. In contrast, warfare in northern Mesopotamia during this period is suggested by the presence of fortified settlements and interpreted as evidence for intrusive ‘Ubaid expansion from the south (Stein, 1994).

The distribution of wealth items within ‘Ubaid Period sites is suggestive of differential access to economic benefits (Stein, 1994). Economic and political differentiation is also indicated by the hierarchical distribution of settlements, a pattern first visible during the Middle ‘Ubaid Period (Wright, 1981). Unlike hierarchical societies that developed in other parts of the world (Flannery, 1968; Earle, 1987; Clark and Blake, 1994; Clark and Pye, 2000), little evidence exists for elite control of long-distance exchange systems and centralized control of high status craft production (Stein, 1994). Mortuary studies also provide little evidence for social ranking and depictions of rulers are rare (Stein, 1994; Wright and

Pollock, 1986). Instead, 'Ubaid Period society was centered on the temple complex and ideology appears to have played an important organizing role in these communities (Hole, 1994, p. 139; Stein, 1994). These temples occur at focal settlements throughout the region, a pattern that remained remarkably stable for 1500 yr (Stein, 1994). Temples were rectangular in form, oriented to the cardinal directions, and contained altars and offering tables. A series of superimposed temples excavated at the important center of Eridu suggests continuity in settlement and social organization at this location throughout the 'Ubaid Period (Oates, 1960). The first temple in this sequence was constructed during the 'Ubaid 1 phase and subsequent temples were larger and more elaborate. Offerings associated with these temples indicate that aquatic resources (e.g., fish) played a central role in this society (Bottéro, 2001). The economic importance of estuarine and riverine fish is also indicated by faunal remains in 'Ubaid 4 levels at Tell Oueli (Huot, 1989). It is possible that the ideological system represented by the 'Ubaid temple complex was used to legitimize differential access to key elements of the farming system (e.g., water, land, and labor; Stein, 1994) as people began to intensify agricultural production to sustain larger populations in the region. Based on the large size of territories during the Late 'Ubaid Period and some evidence for centralized storage facilities at Tell Oueli (Huot, 1989), Stein (1994) has argued that ideological manipulation was used to mobilize food surpluses to storage facilities at focal communities. Regardless, by the end of the 'Ubaid Period (~6300 cal yr BP)[3] it is clear that some communities were substantially larger than their neighbors, were ruled by hereditary leaders, and were administered by institutionalized administrative organizations.

The economic, social, and political complexity evident at the end of the 'Ubaid Period culminated during the Uruk Period (6300–5000 cal yr BP) with the development of the first urban-based states at ~5000 cal yr BP (Late Uruk or LC5; Johnson, 1973; Wright and Johnson, 1975; Adams, 1981; Rothman, 2004). A significant population expansion occurred in southern Mesopotamia during the 'Ubaid to Uruk Period transition (~6300 cal yr BP), but some areas saw population decline (northern alluvium) as settlements became more concentrated in the southern alluvium (Adams, 1981, pp. 60–61). The city of Uruk-Warka, with deposits extending back into the 'Ubaid Period, was the largest and certainly the most prominent on the southern alluvium through the Uruk Period (Nissen, 2001). This community grew to 250 ha by the end of the Uruk Period and the urban core expanded to 100 ha in size (Finkbeiner, 1991; Nissen, 2001), with an estimated 10,000 inhabitants (Redman, 1978). Most of the elaborate public buildings at Uruk-Warka date to the Late Uruk Period (3–5) and reflect a development of complex

[3] The boundary between the 'Ubaid and Uruk Periods is provisionally placed at ~6300 cal yr BP based on the recalibration of radiocarbon dates for 'Uruk-related assemblages in greater Mesopotamia (Wright and Rupley, 2001) and personal communication with Joan Oates who has suggested that the boundary date between 'Ubaid and Uruk is sometime before ~6200 cal yr BP. However, the boundary could be as early as 6700 cal yr BP in some locations (Hole, 1994).

administrative systems over the course of about 600 yr (Nissen, 2001). The increased size of Uruk-Warka resulted from indigenous population growth (Johnson, 1988–89) or migration of people from adjacent areas. This occurred as communities were abandoned further to the north in southern Mesopotamia and the adjacent Susiana Plain (Johnson, 1973; Adams, 1981), perhaps fostered by increasing aridity evident throughout the region or due to deltaic progradation.

Uruk-Warka was the largest urban center in southern Mesopotamia by the Late Uruk Period (Nissen, 2001), dwarfing even the large settlement of Susa on the adjacent Susiana Plain (~18 ha; Hole, 1987). The urban core of this city was surrounded by a defensive wall and divided into two discrete areas, each containing large free-standing buildings visible from a considerable distance (Heinrich, 1982; Nissen, 2001). Some of these large structures are interpreted as public buildings representing temples, "cult houses," or assembly halls (Schmid, 1980; Heinrich, 1982; Nissen, 2001). Cylinder seals and clay tablets found in dumps behind administrative buildings represent the first writing systems and appear to have been used primarily for information storage and accounting purposes (Nissen et al., 1993). A vigorous economy is suggested by the remains of workshops and kilns in the city center, and substantial evidence exists for craft specialization, with major advances in metallurgy and pottery manufacture visible in the record (e.g., fast wheel; Adams and Nissen, 1972). Artistic achievement also flourished. Clear social and political hierarchies are indicated by artistic representations and by a clay tablet containing a "standard professions list," a categorization of professions from rulers to laborers (Nissen et al., 1993; Nissen, 2000). This list indicates that a strong division of labor was well established and that the hierarchical elements of society likely had roots earlier in the 'Ubaid and Uruk Periods. A strong political authority is also suggested by hints of forced labor and the control of an extensive agricultural irrigation system. This is also indicated by rank-size differences between communities and evidence from cylinder seals showing strong economic, political, and social integration between cities and smaller communities in the region (Wright and Johnson, 1975). Other large cities in the region dating to this time include Kish, Nippur, Gersu, and Ur (Nissen, 2001).

The southern cities of Mesopotamia were linked to other communities in greater Mesopotamia via exchange networks. These networks were crucial for state development because people living on the southern alluvium were able to acquire goods unavailable locally (e.g., high grade wood and metal; Algaze, 1993). By the Late Uruk Period, the movement of goods to southern cities was facilitated by outposts strategically located in areas containing valuable resources or close to natural trade routes affording control over the distribution of these materials (e.g., Godin Tepe, Habub Kabira, Hacinebi Tepe, Jabal Aruda; Algaze, 2001b; Rothman, 2001). The records at these sites suggest that colonists/merchants from southern Mesopotamia were able to gain access to critical materials and wealth objects necessary to support the emerging social hierarchy in the heartland. This occurred during the Late Uruk Period between about 5500 and 5000 cal yr BP and is known as the Uruk expansion (Algaze, 1993, 2001b; Wright and Rupley, 2001). Algaze (2001b) argued convincingly that the process of state development was linked to broader economic interactions that partially enabled leaders to sustain the development of economic, social, and political hierarchies in southern Mesopotamia.

4. Discussion

In southern Mesopotamia the cultural changes leading to integrated state level societies occurred during a 3000 yr period between the beginning of the 'Ubaid Period at ~8000 cal yr BP and the end of the Uruk Period at about 5000 cal yr BP. Although states flourished after 5000 cal yr BP, during the Sumerian Period, the foundations were built during the 'Ubaid (8000–6300 cal yr BP) and Uruk (6300–5000 cal yr BP) Periods. This was a dynamic interval of human cultural evolution, a punctuated series of events, in a long (~80,000 yr; Klein, 1999) history of human hunting and gathering and early agricultural economies (Early Holocene; Moore, 1985). The first urban-based states appeared in southern Mesopotamia and the Susiana Plain, although early sedentary agricultural communities and hunter-gatherers were then well distributed throughout much of the region. We suggest that state development in southern Mesopotamia resulted from human responses stimulated, in part, by a particular succession and confluence of environmental changes unique to this region during the Early and Middle Holocene. Rapid and extensive marine transgression, coupled with higher rainfall, was followed by stabilization of sea level and increasing aridity. Interrelated human responses resulted in intense and increasing competition for favorable resources that were becoming increasingly circumscribed due to aridity and population expansion (Flannery, 1972; Service, 1978; Carneiro, 1988). These changing conditions favored population aggregation, intensified agricultural production, economic specialization, and the formation of social and political hierarchies founded on and reinforced by new or existing ideological systems. In this context, the first urban centers emerged and controlled adjacent communities with elaborate and centralized administrative hierarchies led by a small elite class.

Our hypothesis includes elements of previous models that incorporated several interrelated factors: population increase, environmental and social circumscription, increased warfare, development of extensive irrigation agriculture, and favorable conditions for trade (Haas, 1982; Rothman, 2004). Nevertheless, we argue that any explanation is incomplete if it considers these factors, alone or in combination, in the absence of environmental change. Population increase was a necessary component for the development of state-level societies, but by itself was inadequate (Wright and Johnson, 1975; Adams, 1981). Likewise, state development was possibly the cause rather than the result of large-scale irrigation agriculture and increasing interregional trade (Adams, 1974; Wright and Johnson, 1975). However, large-scale irrigation could not have occurred in this region without sufficiently high, near-stable sea levels necessary for the development of extensive floodplains at river level and a high water table.

Relatively few researchers have stressed climate change as an important factor in contributing to the development of city-based states in southern Mesopotamia. Exceptions are Hole (1994), Nissen (1988), and Sirocko et al. (1993), who suggested that state development was tied to increased regional aridity during the Middle Holocene. Hole (1994) stressed the importance of climatic instability as a major trigger in cultural development, suggesting that short-term environmental shocks encouraged collective action to mitigate them. Sea-level change has also been considered to have caused human migrations and strongly influenced the development of agrarian economies (Van Andel, 1989; Stanley and Warne, 1993; Ryan and Pitman, 1998). Recent work in Egypt has linked deceleration of global sea-level rise between 8500 and 7500 cal yr BP with the formation of the Nile Delta and the initiation of farming settlements (Stanley and Warne, 1993, 1997; Stanley and Chen, 1996). Hole (1994) considered sea-level rise to be insignificant for cultural development in southern Mesopotamia, however, arguing that sea-level rise was too slow for river aggradation to keep pace, but stability of the land surface fostered the extensive use of canal systems and aggregated settlement.

Our view is that climatically induced environmental change is one of several variables leading to the emergence of centralized states in this region. Given the coincidence of major climatic shifts and the emergence of social hierarchies and centralized states, these physical factors should weigh more heavily than other variables that are more difficult to document quantitatively. We argue that certain behaviors were favored (probabilistically, not deterministically) within this dynamically changing environmental and social context that had major evolutionary implications for the formation of social hierarchies and ultimately the institutionalized administrative hierarchies characteristic of urban-based states (see Winterhalder and Goland [1997] for a similar argument with respect to agricultural origins). Therefore, we emphasize decision making or human responses to a changing set of environmental and social circumstances in southern Mesopotamia in three distinct, but continuous stages.

4.1. Stage I: 9000–8000 cal yr BP

Little is known about cultural development during the pre-'Ubaid Period of southern Mesopotamia because populations were likely small and dispersed and the archaeological record is either covered by water and/or buried by alluvium deposited during floodplain aggradation. We suggest that Early Holocene climatic conditions between 9000 and 8000 cal yr BP were an important catalyst for later cultural developments. Climatically this period was the most propitious for people living in this region because of higher humidity and more reliable summer monsoon rainfall (Arz et al., 2003; Petit-Maire et al., 1997). Widely distributed lakes in northern Arabia and southern Mesopotamia, created by seasonal monsoonal conditions, would have favored dispersed settlement and seasonal mobility. The wadi systems of northern Iraq and the floodplains along the Ur–Schatt River provided localized aquatic resources attractive to early populations.

Sparsely distributed stone tools characteristic of the Arabian bifacial tradition provide tantalizing evidence for pre-'Ubaid occupation in the southern Mesopotamia Delta region (Woolley, 1955; Cauvin, 1979; Inizan and Tixier, 1983; Zarins, 1992). These data are consistent with other evidence from along the northeast coast of the Arabo-Persian Gulf and more broadly across the Arabian Peninsula for small preceramic settlements near now-dry lakebeds dating to between ~9600 and 7000 cal yr BP (Potts, 1993). Evidence suggests that people practiced a range of mixed foraging-farming strategies combining the herding of sheep and goats with hunting for wild game and the collection of local grasses and various aquatic resources (Potts, 1997; Beech et al., 2005). We suspect that small groups of semi-nomadic people, practicing similar subsistence strategies, also lived along the Ur–Schatt River during this interval. As the rate of global sea-level rise decreased during the Early Holocene, the extent of transgression increased in the Arabo-Persian Gulf because of the low topographic gradients. Rapid transgression of the shoreline (~110 m per annum) caused continuous displacement of peoples and competition for optimal locations on the shifting boundary between fresh and salt water systems. It also compressed the number of people living along this watercourse into a smaller area, may have stimulated group formation or movement of people into adjacent, uninhabited environments of equal or greater economic potential.

4.2. Stage II: 8000–6300 cal yr BP – 'Ubaid Period

We suggest that the appearance of relatively large communities in southern Mesopotamia at Eridu, Uruk, and other locations reflects an aggregation of sedentary populations adjacent to the newly formed estuaries and associated wetlands (Aqrawi, 2001; Sanlaville, 1989). As Oates (1960, 2004) noted, these habitats would have been ideal locations for access to fresh water, hunting/fishing of terrestrial and aquatic animals, and transportation (also see Pournelle, 2003). The potential for irrigation agriculture also improved with increasing stabilization of sea level by the end of this period. In addition, it is possible that the Arabo-Persian Gulf was less saline and more productive compared to later in time, an idea based on data from the Red Sea suggesting less saline conditions related to the displacement of the monsoonal belt at this time (Arz et al., 2003). Increases in the size and number of settlements in southern Mesopotamia during the Early 'Ubaid Period (Phases 0–1) suggest that regional populations were expanding, a product of indigenous population growth or immigration from elsewhere in greater Mesopotamia. Several focal communities started to emerge at this time (e.g., Eridu, Tell Oueli), positioned on elevated landforms at the head of the Arabo-Persian Gulf. Settlement at several of these communities was remarkably stable through the 'Ubaid Period, a point that is best illustrated by a series of superimposed temples at Eridu (Oates, 1960).

The quantity of fish bone found as offerings in these temples highlights the importance of aquatic habitats at least at this location (Bottéro, 2001).

The nonlinear distribution of these settlements suggests that irrigation agriculture (Adams, 1981) was not an essential or dominant element in the subsistence economy, although it was used elsewhere in greater Mesopotamia (Oates and Oates, 1976; Wilkinson, 2003). Given the humid conditions at the beginning of the 'Ubaid Period and the initial formation of productive wetlands, it seems reasonable to assume that these communities were founded upon more diverse mixed economies that included hunting, fishing, herding, and farming (dry and irrigated fields). Given low regional populations and favorable environmental conditions, we suggest that irrigation agriculture was not essential for the formation and stability of these early 'Ubaid settlements. The more dispersed nature of settlement in the earliest 'Ubaid suggests that populations were less restricted or circumscribed than later in time and that a range of habitats could sustain populations using a mixture of subsistence practices. This economic base would have been facilitated, in part, by varied habitats associated with the newly created wetland areas at the head of the Gulf and lacustrine habitats scattered across the region related to the humid conditions of this period. In other words, as groups increased in size, impacting local habitats and competing for localized resources, there were other economically viable options available. This type of environmental context would have favored splintering of communities rather than integration into larger groups.

During the Middle 'Ubaid Period ('Ubaid 2–3) certain communities grew larger in size relative to surrounding communities and some evidence exists for differential access to resources and control of food stores by elite group members (Stein, 1994). The aggregation of people at focal communities located near optimal wetland locations occurred during a period of rapid regional drying, related to the southern retreat of the South Asian monsoonal belt (Arz et al., 2003), that would have reduced the extent of lacustrine habitats in southern Mesopotamia. Sea level also continued to rise during this interval (7 m or more, Fairbanks, 1989; Eisenhauer et al., 1992) and the expansion of maritime trade along the western margin of the Arabo-Persian Gulf occurred within this environmental context (Carter, 2002, 2003; Connan et al., 2005). At the head of the Gulf, this rise caused a marine inland transgression of at least 200 km, or ~100 m/yr. In this case, low-lying areas were inundated to create a continually changing mosaic of aquatic habitats (Pournelle, 2003), and all but the most stable landforms (those occupied by communities like Eridu or Ur) were flooded. Optimal freshwater and estuarine environments continued to shift inland displacing human populations. This dynamic mosaic would have stimulated increased competition for localized and circumscribed resources and the need to constantly redefine territorial boundaries and village locations as rapidly as within a single generation.

Larger groups formed at optimal locations, where environmental conditions created opportunities for ambitious individuals to exploit competitive advantages. These advantages ultimately formed the basis for the social and political hierarchies that emerged by the end of the 'Ubaid Period (~6300 cal yr BP). It appears that people compared the costs and benefits of joining a larger group and that the benefits (or perceived benefits) were greater than the available alternatives, even at the cost of economic and social subjugation. Under these conditions, the use of ideology, centered on the temple, became increasingly important for legitimizing the emerging economic/social inequities and the status of hereditary leaders. The expansion of Late 'Ubaid settlements down the western margin of the Arabo-Persian Gulf could be related to environmental, demographic, or social pressures in southern Mesopotamia that stimulated migration. It could also be related to new economic opportunities created by major improvements in irrigation agriculture technology or the persistence of seasonal monsoonal rains farther to the south as conditions continued to dry in southern Mesopotamia.

4.3. Stage III: 6300–5000 cal yr BP – Uruk Period

The Uruk Period represents the culmination of earlier developments leading to the first fully urban state-level society by ~5000 BP (Late Uruk or LC5; Wright and Johnson, 1975; Nissen, 1988, 2000, 2001; Crawford, 1991; Rothman, 2004). Populations increased in southern Mesopotamia during the 'Ubaid to Uruk Period transition (~6300 cal yr BP). This demographic trend tracks the expansion of irrigation agriculture as indicated by the greater occurrence of canals dating to this time and of linear settlement patterns suggesting that communities were becoming more reliant on these important systems (Adams, 1981). Increased food surpluses provided a firmer basis for expanding populations and a growing, socially stratified society that included administrators, craftspeople, and other specialists. We also suggest that population expansion during the Uruk Period was no coincidence: demographic growth was favored by improving irrigation technology coupled with high water tables and the expanding floodplains linked to the deceleration of sea-level rise.

The near-stabilization of sea level also favored further enlargement of towns optimally located on the margins of the expanding wetlands at the head of the Arabo-Persian Gulf. Textual evidence indicates that Eridu and Ur were located on the coast about 5000 cal yr BP (Larsen and Evans, 1978; Sanlaville, 1989; Lambeck, 1996). Archaeological evidence suggests that during the 'Ubaid Period communities in the Eridu and Tell Oueli regions became increasingly maritime (Huot, 1989). Even the later Sumerians (~4500 cal yr BP) are considered to have been a maritime culture (Falkenstein, 1951; Jacobsen, 1960). Boat transportation significantly increases the efficiency with which people can deliver resources (agricultural or otherwise) to central places (Ames, 2002). Maritime transport would therefore have helped mobilize food surpluses to focal communities such as Eridu or Ur. The combination of fishing and farming also provides a powerful economic engine for population expansion, while maritime voyages to distant locations facilitated the exchange of ideas and provided exotic goods consumed by emerging elites. Exotic materials served as important status markers and reinforced the existing social and political hierarchies.

In this environmental, demographic, and social context, the settlement of Uruk-Warka emerged as a dominant regional center. First established in the 'Ubaid Period, this community grew in size during the Uruk Period with the most pronounced growth between ~5500 and 5000 cal yr BP (Nissen, 2001). Public architecture and the first writing systems, used primarily for information storage and accounting purposes, appear at this time. These traits point to the existence of an elaborate administrative hierarchy governed by a supreme religious elite that controlled the economic, social, and political affairs of neighboring communities – the first urban-based state in the world. The relatively rapid growth at Uruk-Warka over 500–600 yr parallels the partial abandonment of communities in northern portions of the alluvium and on the adjacent Susiana Plain. Movement of people from hinterland communities suggests that the economic and social opportunity at cities like Uruk-Warka was attractive relative to that available elsewhere. We argue that the foundation for these opportunities was provided, in part, by the expansion of irrigation systems afforded by the near-stabilization of sea level coupled with the infrastructural improvements developed and maintained by the new administrative hierarchy. Decreased agricultural production (e.g., dry farming) on the margins of the southern alluvium, due to increasingly dry conditions throughout Mesopotamia (Arz et al., 2003), would have attracted people to cities then supported by well-developed irrigation systems. Aridification throughout greater Mesopotamia also led to continued environmental and social circumscription in these southern cities, which heightened competition for increasingly localized resources associated with aquatic habitats. The defensive wall around the city of Uruk-Warka is suggestive of social and political instabilities that would have further favored the formation of larger groups for defensive purposes. Social instabilities and the threat of war likely contributed to the decision of many to move from hinterland to urban settings – the best available alternative even with the severe economic and social inequities inherent in the new social order. Under these conditions people were attracted to these new urban centers and accepted the ideological system established by the ruling elite to legitimize their elevated positions. Population expansion and the replication of the Uruk social system fostered its expansion into northern Mesopotamia where Uruk outposts were established to gain additional resources and where similarly organized state-level societies emerged.

5. Conclusions

We argue that urbanism and cultural complexity in southern Mesopotamia resulted from a series of decisions by many people over several millennia under continuously changing environmental, demographic, economic, social, and political conditions. The aggregate effect of these decisions culminated in the first state-level societies in the region. Global eustatic and climate changes influenced dynamic environmental conditions in southern Mesopotamia, along with interrelated changes in human demography, economy, and sociopolitical organization.

Coastal and aquatic habitats played a critical role in shaping the cultural evolutionary history of the region. The earliest stable settlements were established adjacent to newly formed, productive estuaries and associated wetlands as the marine transgression slowed between 8000 and 6300 cal yr BP – the 'Ubaid Period. These locations were optimally located near fresh water and diverse aquatic habitats, and were ideally positioned for efficient transport of goods and people using watercraft. Small-scale irrigation agriculture could also have been employed along wetland margins where water tables were higher. People living in more marginal parts of southern Mesopotamia moved into larger communities because this action either maintained or improved their economic well-being, even in the face of subjugation by emerging elites.

Continued population growth occurred in southern Mesopotamia during the 'Ubaid to Uruk transition (~6300 cal yr BP) as sea-level rise decreased significantly between 6300 and 5500 cal yr BP. A deceleration in marine transgression stimulated the expansion of floodplains in southern Mesopotamia and the high water table necessary for large-scale, floodplain irrigation agriculture. Increased food surpluses resulting from irrigation agriculture provided the basis for expanding populations and the developing sociopolitical structure. Increasing aridity in southern Mesopotamia between 6200 and 5000 cal yr BP contributed to further circumscription of populations, increased competition for favorable land and water resources, aggregation of populations in centers close to the rivers and estuaries, greater reliance on irrigation agriculture, and the need for the development of larger population centers for defensive purposes. It is within this context that the important city of Uruk-Warka emerged at the center of the first urban-based state between 5500 and 5000 cal yr BP.

Acknowledgments

We appreciate the critical support of the National Science Foundation (Archaeology, and Marine Geology and Geophysics). We thank Dale Krause for encouragement and valuable suggestions; Walter Pitman for initial stimulation; David Hodell for suggestions regarding climate/orbital relations; and Frank Sirocko for providing sediment data. We also thank Robert McC. Adams, Mark Aldenderfer, Geoff Bailey, Jon Erlandson, Brian Fagan, Scott Fitzpatrick, Michael Glassow, Michael Jochim, Sarah McClure, Joan Oates, John Orbell, Dan Potts, Katharina Schreiber, Bruce Winterhalder, and Herbert Wright for useful information and/or suggestions that helped improve this paper. Several anonymous reviewers also provided valuable constructive criticism.

References

Adams, R. McC., 1965. *Land Behind Baghdad*. University of Chicago Press, Chicago.

Adams, R. McC., 1974. Anthropological perspectives on ancient trade. *Current Anthropology* 15:239–258.

Adams, R. McC., 1981. *Heartland of Cities: Surveys of Ancient Settlement and Land Use on the Central Floodplain of the Euphrates.* University of Chicago Press, Chicago.
Adams, R. McC., 2000a. Scale and complexity in archaic states. *Latin American Antiquity* 11:187–193.
Adams, R. McC., 2000b. Accelerated technological change in archaeology and ancient history. In *Cultural Evolution: Contemporary Viewpoints*, edited by G. Feinman and L. Manzanilla, pp. 95–117. Kluwer Academic, New York.
Adams, R. McC., and H. J. Nissen, 1972. *The Uruk Countryside: The Natural Setting of Urban Societies.* University of Chicago Press, Chicago.
Al-Asfour, T., 1978. The marine terraces of the Bay of Kuwait. In *The Environmental History of the Near and Middle East Since the Last Ice Age*, edited by W. C. Brice, pp. 245–254. Academic Press, New York.
Algaze, G., 1993. *The Uruk World System: Dynamics of Expansion of Early Mesopotamian Civilization.* University of Chicago Press, Chicago.
Algaze, G., 2001a. Initial social complexity in Southwestern Asia: the Mesopotamian advantage. *Current Anthropology* 42:199–233.
Algaze, G., 2001b. The prehistory of imperialism: The case of Uruk Period Mesopotamia. In *Uruk Mesopotamia and Its Neighbors: Cross-Cultural Interactions in the Era of State Formation*, edited by M. S. Rothman, pp. 27–83. School of American Research Press, Santa Fe, NM.
Almogi-Labin, A., C. Hemleben, D. Meischner, and H. Erlenkeuser, 1991. Paleoenvironmental events during the last 13,000 years in the Central Red Sea as recorded by Pteropoda. *Paleoceanography* 6:83–98.
Ambrose, S. H., and N. E. Sikes, 1991. Soil carbon isotope evidence for Holocene habitat change in the Kenya Rift Valley. *Science* 253:1402–1404.
Ames, K. M., 2002. Going by boat: the forager-collector continuum at sea. In *Beyond Foraging and Collecting: Evolutionary Change in Hunter-gatherer Settlement Systems*, edited by B. Fitzhugh and J. Habu, pp. 19–52. Kluwer Academic/Plenum Publishers, New York.
Aqrawi, A. A. M., 1993. Palygorskite in the recent fluvio-lacustrine and deltaic sediments of southern Mesopotamia. *Clay Minerals* 28:153–158.
Aqrawi, A. A. M., 1995. Brackish-water and evaporitic Ca-Mg carbonates in the Holocene lacustrine/deltaic deposits of southern Mesopotamia. *Journal of the Geological Society of London* 152:259–268.
Aqrawi, A. A. M., 2001. Stratigraphic signatures of climatic change during the Holocene evolution of the Tigris-Euphrates delta, lower Mesopotamia. *Global and Planetary Change* 28:267–283.
Arz, H. W., F. Lamy, J. Pätzold, P. J. Müller, and M. Prins, 2003. Mediterranean moisture source for an Early-Holocene humid period in the Northern Red Sea. *Science* 300: 118–121.
Aurenche, O., 1987. Remarques sur le peuplemente de la Mésopotamia. In *La Préhistoire de la Mésopotamia*, edited by J.-L. Huot, pp. 85–89. Editions du CNRS, Paris.
Bar-Matthews, M., A. Ayalon, and A. Kaufman, 1997. Late Quaternary paleoclimate in the Eastern Mediterranean Region from stable isotope analysis of speleothems at Soreq Cave, Israel. *Quaternary Research* 47:155–168.
Bar-Yosef, O., and A. Belfer-Cohen, 1989. The origins of sedentism and farming communities in the Levant. *Journal of World Prehistory* 3:447–491.
Bar-Yosef, O., and R. H. Meadow, 1995. The origins of agriculture in the Near East. In *Last Last Hunters-First Farmers: New Perspectives on the Prehistoric Transition to Agriculture*, edited by T. D. Price and A. B.. Gebauer, pp. 39–94. School of American Research Press, Santa Fe, NM.

Beech, M., R. Cuttler, D. Moscrop, H. Kallweit, and J. Martin, 2005. New evidence for the Neolithic settlement of Marawah Island, Abu Dhabi, United Arab Emirates. *Proceedings of the Seminar for Arabian Studies* 35:37–56.

Berman, J., 1994. The ceramic evidence for sociopolitical organization in 'Ubaid southwestern Iran. In *Chiefdoms and Early States in the Near East: The Organizational Dynamics of Complexity*, edited by G. Stein and M. S. Rothman, pp. 23–33. Prehistory Press, Madison, WI.

Binford, L. R., 1968. Post-Pleistocene Adaptations. In *New Perspectives in Archaeology*, edited by S. R. Binford and L. R. Binford, pp. 313–341. Aldine, Chicago.

Blanton, R., S. Kowalewski, G. Feinman, and L. Finsten, 1993. *Ancient Mesoamerica: A Comparison of Change in Three Regions.* (second edition). Cambridge University Press, Cambridge.

Bottéro, J., 2001. *Everyday Life in Ancient Mesopotamia.* The John Hopkins University Press, Baltimore, Maryland.

Bray, H. E., and S. Stokes, 2004. Temporal patterns of arid-humid transitions in the south-eastern Arabian Peninsula based on optical dating. *Geomorphology* 59:271–280.

Burns, S. J., D. Fleitmann, A. Matter, U. Neff, and A. Mangini, 2001. Speleothem evidence from Oman for continental pluvial events during interglacial periods. *Geology* 29:623–626.

Burns, S. J., A. Matter, N. Frank, and A. Mangini, 1998. Speleothem-based paleoclimatic record from northern Oman. *Geology* 26:499–502.

Calvet, Y., 1989. La phase 'Oueili de l'époque d'Obeid. In *Larsa, travaux de 1985*, edited by J.-L. Huot, pp. 129–151. Editions Recherche sur les Civilisations, Paris.

Carneiro, R. L., 1988. The circumscription theory. *American Behavioral Scientist* 31:497–511.

Carter, R., 2002. Ubaid-period boat remains from As-Sabiyah: excavations by the British Archaeological Expedition to Kuwait. *Proceedings of the Seminar for Arabian Studies* 32:13–30.

Carter, R., 2003. The Neolithic origins of seafaring in the Arabian Gulf. *Archaeology International* 2002/2003:44–47.

Cauvin, M.C., 1979. Tello et l'origine de la houe au Proche Orient. *Paléorient* 5:193–206.

Childe, V. G., 1950. The urban revolution. *Town Planning Review* 21:3–17.

Clark, J., and Blake, M., 1994. The power of prestige: competitive generosity and the emergence of rank societies in lowland Mesoamerica. In *Factional Competition and Political Development in the New World*, edited by J. W. Fox and E. M. Brumfiel, pp. 17–30. Cambridge University Press, Cambridge.

Clark, J., and M. Pye, editors. 2000. *Olmec Art and Archaeology in Mesoamerica.* Yale University Press, New Haven.

Clemens, S. C., D. W. Murray, and W. L. Prell, 1996. Nonstationary phase of the Plio-Pleistocene Asian monsoon. *Science* 274:943–948.

COHMAP, 1988. Climatic changes of the last 18,000 years: observations and model simulations. *Science* 241:1043–1052.

Connan, J., R. Carter, H. Crawford, M. Tobey, A. Charrié-Duhaut, D. Jarvie, P. Albrecht, and K. Norman, 2005. A comparative geochemical study of bituminous boat remains from H3, As-Sabiyah (Kuwait), and RJ-2, Ra's al-Jinz (Oman). *Arabian Archaeology and Epigraphy* 16:21–66.

Cooke, G. A., 1987. Reconstruction of the Holocene coastline of Mesopotamia. *Geoarchaeology* 2:15–28.

Crawford, H., 1991. *Sumer and the Sumerians.* Cambridge University Press, New York.

Dabbagh, A. E., K. G. al-Hinai, and M. A. Khan, 1998. Evaluation of the Shuttle Imagining Radar (SIR-C/X-SAR) data for mapping paleo-drainage systems in the Kingdom of Saudi-Arabia. In *Quaternary Deserts and Climatic Change*, edited by A. S. Alsharhan, K. W. Glennie, G. L. Whittle, and C. G. St. C. Kendell, pp. 483–493. Balkema/Brookfield, Rotterdam.

deMenocal, P., J. Ortiz, T. Guilderson, J. Adkins, M. Sarnthein, L. Baker, and M. Yarusinsky, 2000. Abrupt onset and termination of the African Humid Period: Rapid climate responses to gradual insolation forcing. *Quaternary Science Review* 19:347–361.

Diester-Haass, L., 1973. Holocene climate in the Persian Gulf as deduced from grain-size and Pteropod distribution. *Marine Geology* 14:207–223.

Earle, T., 1987. Chiefdoms in archaeological and ethnohistorical context. *Annual Review of Anthropology* 16:279–308.

Edens, C., 1982. Towards a definition of the Rubc al Khali "neolithic." *Atlal* 6:109–124.

Edens, C., and T. J. Wilkinson, 1998. Southwest Arabia during the Holocene: recent archaeological developments. *Journal of World Prehistory* 12:55–119.

Eisenhauer, A., J. Chen, and G. J. Wasserburg, 1992. Holocene sea level determination relative to the Australian continent. *Abstracts from American Geophysical Union Spring Meeting, Montreal*, p. 172.

Fairbanks, R. G., 1989. A 17,000-year glacio-eustatic sea level record: influence of glacial melting rates on the Younger Dryas event and deep-ocean circulation. *Nature* 342:637–642.

Falkenstein, A., 1951. Die Eridu-Hymme. *Sumer* 7:119–125.

Feinman, G. M., 2000. Cultural evolutionary approaches and archaeology: past, present, and future. In *Cultural Evolution: Contemporary Viewpoints*, edited by G. M. Feinman and L. Manzanilla, pp. 3–11. Kluwer Academic/Plenum Publishers, New York.

Feinman, G. M., and J. Marcus, editors. 1998. *Archaic States*. School of American Research, Santa Fe, NM.

Feinman, G. M., and L. Manzanilla, editors. 2000. *Cultural Evolution: Contemporary Viewpoints*. Kluwer Academic/Plenum Publishers, New York.

Finkbeiner, U., 1991. Uruk Kampagne 35–37, 1982–84: Die archäologische Oberflächenuntersuchung. *Ausgrabungen in Uruk-Warka Endberichte* 4. von Zabern, Mainz.

Flannery, K. V., 1968. The Olmec and the valley of Oaxaca: a model for interregional interaction in Formative times. In *Dumbarton Oaks Conference on the Olmec*, edited by E. Benson, pp. 79–110. Dumbarton Oaks, Washington, D.C.

Flannery, K. V., 1969. Origins and ecological effects of early domestication. In *Iran and the Near East in the Domestication and Exploitation of Plants and Animals*, edited by P. J. Ucko and G. W. Dimbley, pp. 73–100. Aldine-Atherton, Chicago.

Flannery, K. V., 1972. The cultural evolution of civilizations. *Annual Review of Ecology and Systematics* 3:399–426.

Flannery, K. V., 1998. The ground plans of Archaic States. In *Archaic States*, edited by G. M. Feinman and J. Marcus, pp. 15–58. School of American Research, Santa Fe, NM.

Frankfort, H., 1956. *The Birth of Civilization in the Near East*. Double Day, Garden City.

Garrard, A., 1999. Charting the emergence of cereal and pulse domestication in south-west Asia. *Environmental Archaeology* 4:67–86.

Gasse, F., M. Arnold, J. C. Fontes, M. Fort, E. Gibert, A. Huc, L. Bingyan, L. Yuanfang, L. Qing, F. Melieres, E. V. Campo, W. Fubao, and Z. Qingsong, 1991. A 13,000-year climate record from Western Tibet. *Nature* 353:742–745.

Gasse, F., R. Téhet, A. Durand, E. Gibert, and J.-H. Fontes, 1990. The arid-humid transition in the sahara and the sahel during the last deglaciation. *Nature* 346:141–146.

Glover, E., 1998. Mangroves, molluscs and man: Archaeological evidence for biogeographical changes in mangrove around the Arabian Peninsula. In *Arabia and its Neighbours. Essays on Prehistorical and Historical Developments Presented in Honour of B. de Cardi*, edited by C. S. Phillips, D. T. Potts, and S. Searight, pp. 221–232. Brepols, Leuven.

Gunatilaka, A., 1986. Kuwait and the Northern Arabian Gulf: a study in Quaternary sedimentation. *Episodes* 9:223–231.

Haas, J., 1982. *The Evolution of the Prehistoric State*. Columbia University Press, New York.
Harris, D.R., 1996. The origins and spread of agriculture and pastoralism in Eurasia: an overview. In: *The Origins and Spread of Agriculture and Pastoralism in Eurasia*, edited by D. R. Harris, pp. 552–573. Smithsonian Institution Press, Washington, D.C.
Harris, D.R., 1998. The origins of agriculture in southwest Asia. *The Review of Archaeology* 19:5–11.
Haynes, C. V., C. H. Eyles, L. A. Pavlish, J. C. Ritchie, and M. Rybak., 1989. Holocene palaeoecology of the Eastern Sahara; Selima Oasis. *Quaternary Science Reviews* 8:109–136.
Haynes, C. V. J., and Mead, A. R., 1987. Radiocarbon dating and paleoclimatic significance of subfossil limicolaria in Northwestern Sudan. *Quaternary Research* 28:86–99.
Henrich, E., 1982. Die Tempel und Heiligtümer im Alten Mesopotamien. Walter de Gruyter, Berlin.
Henry, D., 1989. *From Foraging to Agriculture. The Levant at the End of the Ice Age.* University of Pennsylvania Press, Philadelphia.
Hodell, D. A., J. H. Curtis, and M. Brenner, 1995. Possible role of climate in the collapse of Classic Maya civilization. *Nature* 375:391–394.
Hole, F., 1983. Symbols of religion and social organization at Susa. In *Beyond the Hilly Flanks*, edited by T. C. Young, P. Smith, and P. Mortensen, pp. 315–344. Oriental Institute, Chicago.
Hole, F., editor. 1987. *The Archaeology of Western Iran: Settlement and Society from Prehistory to Islamic Conquest.* Smithsonian Institution Press, Washington, D.C.
Hole, F., 1994. Environmental instabilities and urban origins. In *Chiefdoms and Early States in the Near East: The Organizational Dynamics of Complexity*, edited by G. Stein and M. S. Rothman, pp. 121–143. Prehistory Press, Madison, WI.
Hötzl, H., A. R. Jado, H. Moser, W. Rauert, and J. G. Zötl, 1984. Climatic fluctuations in the Holocene. In *The Quaternary Period in Saudi Arabia*, edited by A. R. Jado and J. G. Zötl, Vol. 2, pp. 301–314. Springer-Verlag, New York.
Hudson, R. G. S., F. E. Eames, and G. L. Wilkins, 1957. The fauna of some recent marine deposits near Basrah, Iraq. *Geological Magazine* 94:393–401.
Huot, J.-L., 1989. 'Ubaidian Villages of Lower Mesopotamia. In *Upon these Foundations: The 'Ubaid Reconsidered, Proceedings from the 'Ubaid Symposium, Elisnore*, edited by E. F. Henrickson and I. Thuesen, pp. 19–42. Museum Tusculanum Press, University of Copenhagen, Institute of Ancient Near Eastern Studies, Copenhagen.
Huot, J.-L., 1996. *Oueli: Travaux de 1987 et 1989*. Éditions Recherche sur les Civilisations, Paris.
Inizan, M.-L., editor. 1988. *Préhistoire à Qatar. Mission archéologique francaise. ii.* Éditions Recherche sur les Civilisations, Paris.
Inizan, M.-L. and J. Tixier, 1983. Tell el 'Oueli: le material lithique. In *Larsa et Oueili, travaux de 1978–1981*, edited by J.-L. Huot, pp. 164–175. Éditions Recherche sur les Civilisations, Paris.
Issar, A., 1995. Climatic change and the history of the Middle East. *American Scientist* 83:350–355.
Issar, A., 2003. *Climate Changes During the Holocene and Their Impact on Hydrological Systems.* Cambridge University Press, Cambridge.
Jacobsen, T., 1960. The waters of Ur. *Iraq* 22:174–185.
Johnson, G. A., 1973. *Local Exchange and Early State Development in Southwestern Iran.* Anthropological Papers 51, Museum of Anthropology, University of Michigan, Ann Arbor.
Johnson, G. A., 1988–89. Late Uruk in greater Mesopotamia: expansion or collapse. *Origini* 14:595–611.
Kennett, D. J., and J. P. Kennett, 2000. Competitive and cooperative responses to climatic instability in Southern California. *American Antiquity* 65:379–395.
Klein, R. G., 1999. *The Human Career: Human Biological and Cultural Origins* (second edition). The University of Chicago Press, Chicago.
Kouchoukos, N., 1999. *Landscape and Social Change in Late Prehistoric Mesopotamia.* Unpublished Ph.D. dissertation, Yale University.

Kramer, S. N., 1963. *The Sumerians.* University of Chicago, Chicago.

Kutzbach, J. E., 1983. Monsoon rains of the Late Pleistocene and Early Holocene: patterns, intensity and possible causes of changes. In, *Variations in the Global Water Budget*, edited by A. Street-Perrott, M. Bern, and R. Ratcliffe, pp. 371–389. Reidel, Dordrecht.

Kutzbach, J. E., and R. G. Gallimore, 1988. Sensitivity of a coupled atmosphere/mixed layer ocean model to changes in orbital forcing at 9000 years B.P. *Journal of Geophysical Research* 93(D1):803–821.

Kutzbach, J. E., and P. J. Guetter, 1986. The influence of changing orbital parameters and surface boundary conditions on climate simulations for the past 18,000 years. *Journal of Atmospheric Science* 43:1726–1759.

Kutzbach, J., and F. A. Street-Perrott, 1985. Milankovitch forcing of fluctuations in the level of tropical lakes from 18 to 0 kyr BP. *Nature* 317:130–134.

Lamb, H. H., 1977. *Climatic History and the Future.* Methuen and Co. Ltd., London.

Lambeck, K., 1996. Shoreline reconstructions for the Persian Gulf since the Last Glacial Maximum. *Earth and Planetary Science Letters* 142:43–57.

Larsen, C. E., and G. Evans, 1978. The Holocene geological history of the Tigris-Euphrates-Karun delta. In *The Environmental History of the Near and Middle East Since the Last Ice Age*, edited by W. C. Brice, pp. 227–244. Academic Press, New York.

Lees, G. M., and N. L. Falcon, 1952. The geographical history of the Mesopotamian plains. *The Geographical Journal* 118:24–39.

Lézine, A., J. Saliège, R. Mathieu, T.-L. Mery, S. Tagliatela, V. Charpentier, and S. Cleuziou, 2002. Mangroves of Oman during the late Holocene: climatic implications and impact on human settlements. *Vegetation History and Archaeobotany* 11:221–232.

Lézine, A., J. Saliège, C. Robert, F. Wertz, and M. Inizan, 1998. Holocene lakes from Ramlat as-Sab'atayn (Yemen) illustrate the impact of monsoon activity in Southern Arabia. *Quaternary Research* 50:290–299.

Lighty, R. G., I. G. Macintyre, and R. Stuckenrath, 1982. Acropora palmata reef framework: a reliable indicator of sea level in the Western Atlantic for the past 10,000 years. *Coral Reefs* 1:5–130.

Macfadyen, W. A., and C. Vita-Finzi, 1978. Mesopotamia: the Tigris-Euphrates delta and its Holocene Hammar fauna. *Geology Magazine* 115:287–300.

Marcus, J., and G. M. Feinman, 1998. Introduction. In *Archaic States*, edited by G. M. Feinman and J. Marcus, pp. 3–13. School of American Research Press, Santa Fe, NM.

Masry, A., 1974. *Prehistory in Northeastern Arabia: The Problem of Interregional Interaction.* Field Research Projects, Coconut Grove, FL.

McClure, H. A., 1976. Radiocarbon chronology of late Quaternary lakes in the Arabian Desert. *Nature* 263:755–756.

McCorriston, J., and F. Hole, 1991. The ecology of seasonal stress and the origins of agriculture in the Near East. *American Anthropologist* 93:46–62.

Meadow, R.H., 1996. The origins and spread of agriculture and pastoralism in northwestern South Asia. In *The Origins and Spread of Agriculture and Pastoralism in Eurasia*, edited by D. R. Harris, pp. 390–412. Smithsonian Institution Press, Washington, D.C.

Meissner, B., 1920. *Babylonien und Assyrien I.* Winters, Heidelberg.

Melguen, M., 1973. Correspondence analysis for recognition of facies in homogeneous sediments off an Iranian river mouth. In *The Persian Gulf*, edited by B. H. Purser, pp. 99–113. Springer-Verlag, Berlin.

Moore, A. M. T., 1985. The development of Neolithic societies in the Near East. In *Advances in World Archaeology*, edited by F. Wendorf and A. Close, Vol. 4, pp. 1–69. Academic Press, New York.

Moseley, M., 1975. *The Maritime Foundations of Andean Civilization.* Cummings, Menlo Park, CA.

Nissen, H.-J., 1988. *The Early History of the Ancient Near East, 9000–2000 BC.* University of Chicago Press, Chicago.

Nissen, H-J., 2000. A Mesopotamian hierarchy in action in ancient Uruk. In *Hierarchies in Action: Cui Bono?*, edited by M. W. Diehl, pp. 210–217. Occasional Paper 27, Center for Archaeological Investigations, Southern Illinois University, Carbondale.

Nissen, H.-J., 2001. Cultural and political networks in the ancient Near East during the fourth and third millennia BC. In *Uruk Mesopotamia and Its Neighbors: Cross-Cultural Interactions in the Era of State Formation*, edited by M. S. Rothman, pp. 149–180. School of American Research, Santa Fe, NM.

Nissen, H.-J., P. Damerow, and R. K. Englund, 1993. *Archaic Bookkeeping.* University of Chicago Press, Chicago.

Nützel, W., 2004. *Einführun de Geo-Archäologie des Vorderen Orients.* Reichert Verlag, Wiesbaden.

Oates, J., 1960. Ur and Eridu, the prehistory. *Iraq* 22:32–50.

Oates, J., 1991. Babylonia and Elam in prehistory. In *Mésopotamie et Elam*, L. de Meyer and H. Gasche, pp. 23–26. Mesopotamian History and Environment Occasional Paper 1, Ghent.

Oates, J., 2004. Ubaid Mesopotamia revisited. In *From Handaxe to Khan: Essays Presented to Peder Mortensen on the Occasion of his 70th Birthday*, edited by K. von Folsach, H. Thrane, and I. Thuesen, pp. 87–104. Aarhus University Press, Aarhus.

Oates, D., and J. Oates, 1976. Early irrigation agriculture in Mesopotamia. In *Problems in Economic and Social Archaeology*, edited by G. Sieveking, I. H. Longworth, and K. E. Wilson, pp. 109–135. Duckworth, London.

Oates, J., T. E. Davidson, D. Kamilli, and H. McKerrell, 1977. Seafaring Merchants of Ur? *Antiquity* 51:221–234.

Ochsenschlager, E. L., 2004. *Iraq's Marsh Arabs in the Garden of Eden.* University of Pennsylvania Museum of Archaeology and Anthropology, Philadelphia.

Overpeck, J., D. Anderson, S. Trumbore, and W. Prell, 1996. The Southwest Indian monsoon over the Last 18,000 years. *Climate Dynamics* 12:213–225.

Pachur, H. J., and S. Kröpelin, 1987. Wadi Howar: paleoclimatic evidence from an extinct river system in the Southeastern Sahara. *Science* 237:298–300.

Pachur, H. J., and S. Kröpelin, 1989. L'aridification du Sahara Oriental á l'Holocéne moyen et supérieur. *Bulletin of the Society for Géology, France* 237:99–107.

Petit-Maire, N., 1986. Paleoclimates in the Sahara of Mali: a multidisciplinary study. *Episodes* 9:7–16.

Petit-Maire, N., 1990. Will greenhouse green the Sahara? *Episodes* 13:103–107.

Petit-Maire, N., 1992. Lire l'avenir dans les archives géologiques. *La Recherche* 243:566–570.

Petit-Maire, N., L. Beufort, and N. Page, 1997. Holocene climate change and man in the present day Sahara. In *Third Millenium BC Climate Change and Old World Collapse*, edited by H. Nuzhet Dalfes, G. Kukla, and H. Weiss, pp. 289–308. NATO ASI Series I: Global Environmental Change 49. Springer-Verlag, Berlin.

Pollock, S., 1992. Bureaucrats and managers, peasants and pastoralists, imperialists and traders: Research on the Uruk and Jemdet Nasr Periods in Mesopotamia. *Journal of World Prehistory* 6:297–336.

Pollock, S., 1999. *Ancient Mesopotamia: The Eden that Never Was.* Cambridge University Press, Cambridge.

Postgate, J. N., 1992. *Early Mesopotamia: Society and Economy at the Dawn of History.* Routledge, New York.

Potts, D. T., 1993. The late prehistoric, protohistoric, and early historic periods in Eastern Arabia ca. 5000–1200 B.C. *Journal of World Prehistory* 7:163–212.

Potts, D. T., 1996. Dilmun, Le paradis et la transgression Flandrienne. *Collected Papers of the Society for Near Eastern Studies* 5:21–27.

Potts, D. T., 1997. *Mesopotamian Civilization: The Material Foundations*. Cornell University Press, Ithaca, New York.

Pournelle, J. R., 2003. *Marshland of Cities: Deltaic Landscapes and the Evolution of Early Mesopotamian Civilization*. Unpublished Ph.D. dissertation, University of California, San Diego.

Prell, W. L., 1984. Monsoonal climate of the Arabian Sea during the late Quaternary; a response to changing solar radiation. In *Milankovitch and Climate*, edited by A. Berger, J. Imbrie, J. Hays, G. Kukla, and B. Saltzman, pp. 349–366. NATO ASI Series C: Mathematical and Physical Sciences 126. Riedel Publishing Co., Dordrecht.

Purser, B. H., and E. Seibold, 1973. The principal environmental factors influencing Holocene sedimentation and diagenesis in the Persian Gulf. In *The Persian Gulf: Holocene Carbonate Sedimentation and Diagenesis in a Shallow Epicontinental Sea*, edited by B. H. Purser et al., pp. 1–9. Springer-Verlag, New York.

Redman, C. L., 1978. *The Rise of Civilization: From Early Urban Society in the Ancient Near East*. W. H. Freeman and Company, San Francisco.

Ritchie, J. C., C. H. Eyles, and C. V. Haynes, 1985. Sediment and pollen evidence for an Early to Mid Holocene humid period in the Eastern Sahara. *Nature* 314:352–355.

Roaf, M., 1984. Ubaid houses and temples. *Sumer* 43:80–90.

Roberts, N., 1989. *The Holocene: An Environmental History*. Basil Blackwell, New York.

Roberts, N., and H. E. Wright Jr., 1993. Vegetational, lake-level, and climatic history of the Near East and Southwest Asia. In *Global Climates Since the Last Glacial Maximum*, edited by H. E. Wright, J. E. Kutzbach, T. Webb III, W. F. Ruddiman, F. A. Street-Perrott, and P. J. Bartlein, pp. 194–220. University of Minnesota Press, Minneapolis, MN.

Rossignol-Strick, M., 1987. Rainy periods and bottom water stagnation initiating brine accumulation and metal concentrations. *Paleoceanography* 2:333–360.

Rothman, M. S., 1994. Introduction Part I: evolutionary typologies and cultural complexity. In: *Chiefdoms and Early States in the Near East: The Organizational Dynamics of Complexity*, edited by G. Stein and M. S. Rothman, pp. 1–22. Prehistory Press, Madison, WI.

Rothman, M. S., 2001. The local and the regional: introduction. In *Uruk Mesopotamia and Its Neighbors: Cross-Cultural Interactions in the Era of State Formation*, edited by M. S. Rothman, pp. 3–26. School of American Research, Santa Fe, NM.

Rothman, M. S., 2002. *Tepe Gawra: The Evolution of a Small Prehistoric Center in Northern Iraq*. University of Pennsylvania, Museum of Archaeology and Anthropology, Philadelphia.

Rothman, M. S., 2004. Studying the development of complex society: Mesopotamia in the Late Fifth and Fourth Millennia BC. *Journal of Archaeological Research* 12:75–119.

Ryan, W., and W. Pitman, 1998. *Noah's Flood: The New Scientific Discoveries about the Event that Changed History*. Simon and Schuster, New York.

Sandweiss, D. H., K. A. Maasch, and D. G. Anderson, 1999. Transitions in the Mid-Holocene. *Science* 283:499–500.

Sanlaville, P., 1989. Considérations sur l'evolution de la Basse Mésopotamie au cours des derniers millénaires. *Paléorient* 15:5–27.

Sarnthein, M., 1971. Oberflächensedimente im Persischen Golf und Golf von Oman II. Quantitative Komponentenanalyse der Grobfraktion. In *Meteor Forschungsergebnisse, Herausgegeben von der Deutschen Forschungsgesellschaft*, edited by E. Seibold and H. Closs, pp. 1–82. Gebrüder Borntraeger, Berlin.

Sarnthein, M., 1972. Sediments and history of the postglacial transgression in the Persian Gulf and Northwest Gulf of Oman. *Marine Geology* 12:245–266.

Schmid, J. G., 1980. *Die Madrasa des Kalifen al-Mustansir in Baghdad.* Bagh. Forschungen 4. von Zabren, Mainz.

Seibold, E., and J. Ulrich, 1970. Zur Bodengestalt des nordwestlichen Golfs von Oman. In *Meteor Forschungsergebnisse: Herausgegeben von der Deutschen Forschungsgesellschaft*, edited by E. Seibold and H. Closs, pp. 1–14. Gebrüder Borntraeger, Berlin.

Seibold, E., and K. Vollbrecht, 1969. Die Bodengestalt des Persischen Golfs. In *Meteor Forschungsergebnisse: Herausgegeben von der Deutschen Forschungsgesellschaft*, edited by E. Seibold and H. Closs, pp. 31–56. Gebrüder Borntraeger, Berlin.

Service, E. R., 1978. Classical and modern theories of the origins of government. In *Origins of the State*, edited by R. Cohen and E. R. Service, pp. 21–34. Institute for the Study of Human Issues, Philadelphia.

Siddall, M., E. J. Rohling, A. Almogi-Labin, C. Hemleben, D. Meischner, I. Schmelzer, and D. A. Smeed, 2003. Sea-level fluctuations during the last glacial cycle. *Nature* 423:853–858.

Sirocko, F., M. Sarnthein, H. Erlenkeuser, H. Lange, M. Arnold, and J. C. Duplessy, 1993. Century-scale events in monsoonal climate over the past 24,000 years. *Nature* 364:322–324.

Smith, B. D., 1998. *The Emergence of Agriculture.* Scientific American Library, New York.

Smith, B. D., 2001. Low-Level food production. *Journal of Archaeological Research* 9:1–43.

Speiser, E. A., 1930. *Mesopotamian Origins.* University of Pennsylvania, Philadelphia.

Spier, F., 1996. *The Structure of Big History.* Amsterdam University Press, Amsterdam.

Stanley, D. J., and Z. Chen, 1996. Neolithic settlement distributions as a function of sea level-controlled topography in the Yangtze delta, China. *Nature* 24:1083–1086.

Stanley D. J., and A. G. Warne, 1993. Sea level and initiation of Predynastic culture in the Nile Delta. *Nature* 363:435–438.

Stanley D. J., and A. G. Warne, 1997. Holocene sea-level change and early human utilization of deltas. *GSA Today: A Publication of the Geological Society of America* 7(12):1–7.

Stein, G., 1994. Economy, ritual, and power in 'Ubaid Mesopotamia. In *Chiefdoms and Early States in the Near East: The Organizational Dynamics of Complexity*, edited by G. Stein and M. S. Rothman, pp. 35–46. Prehistory Press, Madison, WI.

Stein, G., 1998. Heterogeneity, power, and political economy: some current research issues in the archaeology of Old World Complex Societies. *Journal of Archaeological Research* 6:1–44.

Stein, G., 2001. Understanding ancient state societies in the Old World. In *Archaeology at the Millennium: A Sourcebook*, edited by G. Feinman and D. Price, pp. 353–379. Kluwer Academic, New York.

Street-Perrott, F. A., and R. A. Perrott, 1990. Abrupt climate fluctuations in the tropics: the influence of Atlantic Ocean Circulation. *Nature* 343:607–612.

Street-Perrott, F. A., and N. Roberts, 1983. Fluctuations in closed basin lakes as an indicator of past atmospheric circulation patterns. In *Variations in the Global Water Budget*, edited by A. Street-Perrott, M. Beran, and R. Ratcliffe, pp. 331–345. Reidel, Dordrecht.

Street-Perrott, F. A., N. Roberts, and S. Metcalfe, 1985. Geomorphic implications of late Quaternary hydrological and climatic changes in the Northern Hemisphere tropics. In *Environmental Change and Tropical Geomorphology*, edited by I. Douglas and T. Spencer, pp. 165–183. George Allen and Unwin, London.

Sukumar, R., R. Ramesh, R. K. Pant, and G. Rajagopalan, 1993. A δ^{13} C record of late Quaternary climate change from tropical peats in Southern India. *Nature* 364:703–706.

Teller, J., K. W. Glennie, N. Lancaster, and A. K. Singhvi, 2000. Calcareous dunes of the United Arab Emirates and Noah's Flood: The postglacial reflooding of the Persian (Arabian) Gulf. *Quaternary International* 68–71:297–308.

Tobler, A., 1950. *Excavations at Tepe Gawra*, Vol. 2. University Museum Monograph, University of Pennsylvania, Philadelphia.

Uerpmann, M., 1992. Structuring the Late Stone Age of southeastern Arabia. *Arabian Archaeology and Epigraphy* 3:65–109.

Uerpmann, M., and H.-P. Uerpmann, 1996. Ubaid pottery in the Eastern Gulf–new evidence from Umm al-Qaiwan (U.A.E.). *Arabian Archaeology and Epigraphy* 7:125–139.

Van Andel, T. H., 1989. Late Quaternary sea-level changes and archaeology. *Antiquity* 63:733–745.

Vita-Finzi, C., 1978. Recent alluvial history in the catchment of the Arabo-Persian Gulf. In *The Environmental History of the Near and Middle East Since the Last Ice Age*, edited by W. C. Brice, pp. 255–261. Academic Press, New York.

Warne, A. G., and D. J. Stanley, 1993. Archaeology to refine Holocene subsidence rates along the Nile delta margin, Egypt. *Geology* 21:715–718.

Weiss, H., 1997. Late Third Millennium abrupt climate change and social collapse in West Asia and Egypt. In *Third Millennium BC Climate Change and Old World Collapse*, edited by H. Nuzhet Dalfes, G. Kukla, and H. Weiss, NATO ASI Series Vol. I, 49, pp. 711–723. Springer-Verlag, Berlin.

Weiss, H., M.-A. Courty, W. Wetterstrom, F. Guichard, L. Senior, R. Meadow, and A. Curnow, 1993. The genesis and collapse of Third Millennium North Mesopotamian civilization. *Science* 261:995–1004.

Whitney, J. W., D. J. Faulkender, and M. Rubin, 1983. The environmental history and present condition of Saudi Arabia's northern sand seas. *Open File Report, U.S. Geological Survey*. U.S. Geological Surveys, Washington, D.C, pp. 83–749.

Wilkinson, T. J., 2003. *Archaeological Landscapes of the Near East*. The University of Arizona Press, Tucson.

Winterhalder, B., and C. Goland, 1997. An evolutionary ecology perspective on diet choice, risk, and plant domestication. In *People, Plants, and Landscapes: Studies in Paleoethnobotany*, edited by K. J. Gremillion, pp. 123–160. University of Alabama Press, Tuscaloosa, AL.

Wittfogel, K. A., 1957. *Oriental Despotism: A Comparative Study of Total Power*. Yale University Press, New Haven, CT.

Wittfogel, K. A., 1981. *Oriental Despotism*. Vintage, New York.

Woolley, C. L., 1929. *The Sumerians*. Clarendon Press, Oxford.

Woolley, C. L., 1955. *Ur Excavations IV. The Early Periods*. The British Museum and the University Museum, London and Philadelphia.

Wright, H. T., 1993. Environmental determinism in Near Eastern prehistory. *Current Anthropology* 34:458–469.

Wright, H. T., 1977. Recent research on the origin of the state. *Annual Review of Anthropology* 6:379–397.

Wright, H. T., 1981. The southern margins of Sumer: archaeological survey of the area of Eridu and Ur. In *Heartland of Cities*, edited by R. McC. Adams, pp. 295–324. University of Chicago Press, Chicago.

Wright, H. T., 1994. Prestate political formations. In *Chiefdoms and Early States in the Near East*, edited by G. Stein and M. S. Rothman, pp. 67–84. Prehistory Press, Madison, WI.

Wright, H. T., 1998. Uruk states in southwestern Iran. In *Archaic States*, edited by G. M. Feinman and J. Marcus, pp. 173–197. School of American Research Press, Santa Fe, NM.

Wright, H. T., 2001. Cultural action in the Uruk World. In *Uruk Mesopotamia and Its Neighbors: Cross-Cultural Interactions in the Era of State Formation*, edited by M. S. Rothman, pp. 123–148. School of American Research Press, Santa Fe, NM.

Wright, H. T., and G. A. Johnson, 1975. Population, exchange, and early state formation in Southwestern Iran. *American Anthropologist* 77:267–289.

Wright, H. T., and G. A. Johnson, 1985. Regional perspectives on southwest Iranian state development. *Paléorient* 11/2:25–30.

Wright, H. T., and S. Pollock, 1986. Colloques internationaux CNRS, Préhistoire de la Mésopotamie, 17–19 December 1984. Editions du CNRS, Paris, pp. 317–329.

Wright, H. T., and E. S. A. Rupley, 2001. Calibrated radiocarbon age determinations of Uruk-related assemblages. In *Uruk Mesopotamia and Its Neighbors: Cross-Cultural Interactions in the Era of State Formation*, edited by M. S. Rothman, pp. 85–122. School of American Research Press, Santa Fe, NM.

Yafeng, S., K. Zhaozheng, W. Sumin, T. Lingyu, W. Fubao, Y. Tandong, Z. Xitao, Z. Peiyuan, and S. Shaohua, 1993. Mid-Holocene climates and environments in China. *Global & Planetary Change* 7:219–233.

Yan, Z., and N. Petite-Maire, 1994. The last 140 ka in the Afro-Asian arid/semi-arid transitional zone. *Palaeogeography, Palaeoclimatology, Palaeoecology* 110:217–233.

Yoffee, N., 2005. *Myths of the Archaic State: Evolution of the Earliest Cities, States, and Civilizations.* Cambridge University Press, Cambridge.

Zarins, J., 1990. Early pastoral nomadism and the settlement of lower Mesopotamia. *Bulletin of the American Schools of Oriental Research* 280:31–65.

Zarins, J., 1992. The early settlement of southern Mesopotamia: a review of the recent historical, geological, and archaeological research. *Journal of the American Oriental Society* 112:55–77.

Zarins, J., A. Murad, and K. al-Yish, 1981. Comprehensive Archaeological Survey Program – the second preliminary report on the Southwestern Province. *Atlal* 5:9–42.

Zeder, M. A., and B. Hesse, 2000. The initial domestication of goats (Capra hircus) in the Zagros Mountains 10,000 years ago. *Science* 287:2254–2257.

Climate Change and Cultural Dynamics: A Global Perspective on Mid-Holocene Transitions
David G. Anderson, Kirk A. Maasch and Daniel H. Sandweiss (Editors)

Chapter 8

Mid-Holocene cultural dynamics and climatic change in the Western Pacific

Atholl Anderson[1,*], Michael Gagan[2] and James Shulmeister[3]

[1]*Research School of Pacific and Asian Studies, Australian National University, ACT 0200, Australia*

[2]*Research School of Earth Sciences, Australian National University, ACT 0200, Australia*

[3]*Department of Geological Sciences, University of Canterbury, Private Bag 4800, Christchurch, New Zealand*

Abstract

Human settlement in the Southwest Pacific began 40,000 or more years ago, but until about 3000 cal yr BP it was restricted to the region of Near Oceania that consists primarily of Australia and New Guinea. In Australia, there was a set of cultural changes during the mid- to late Holocene, which included the introduction of the dingo, the florescence of an assemblage of composite tools, movement of residential settlement into new environments, and a significant and sustained increase in the occupancy of archeological sites. These changes have been attributed most often to demographic and social factors, but some of them might be responses to relative resource scarcity induced by mid-Holocene climatic changes. Evidence from across the Western Pacific region indicates that significant climatic changes occurred during the mid-Holocene. These changes were neither all uniform nor simple but they appear to reflect generally strengthened circulation systems, including the onset of modern ENSO periodicities about 500 yr BP. In New Guinea, the Holocene development of agriculture in the Highlands is generally attributed to endogenous processes, but climatic change, particularly in relation to ENSO conditions, might have been influential. A relationship between climatic and cultural change is therefore plausible in several cases, but further investigation of these propositions will need much better comparative data on demographic trends that, it can be assumed, were also significant in mid-Holocene cultural dynamics.

1. Introduction

The Western Pacific is a region of dramatic contrasts in human history. On the ancient and geologically stable continent of Australia there has been habitation for

*Corresponding author. *E-mail address:* atholl.anderson@anu.edu.au (Atholl Anderson).

more than 40,000 years (Allen and O'Connell, 2003) and archeological remains of similar antiquity occur in New Guinea and its nearest offshore islands – New Britain and New Ireland. On outlying islands, such as Manus, settlement could have only occurred through open sea crossings of greater than 150 km, accomplished more than 20,000 years ago. Yet that early promise in maritime capability was not widely fulfilled until very much later (Anderson, 2000, 2003). There were probably people in the Southeast Solomons by the late Pleistocene but there is no evidence of human settlement any further to the north, east, or Southeast in Oceania until the late Holocene, about 3000 cal yr BP. The Pacific divides, therefore, into two broad regions: "Near Oceania" and "Remote Oceania", with the dividing line at the eastern-end of the larger Solomon islands (Fig. 8.1). This paper focuses upon Near Oceania where cultural sequences span the mid-Holocene, but before discussing all those, it is worth outlining the spread of Neolithic cultures across both regions during the late Holocene because that, too, might have been initiated in some way by mid-Holocene climatic changes.

Older, aceramic cultures of coastal and insular Near Oceania were largely replaced about 3200–3300 cal yr BP by a material culture that included red-slipped, dentate-stamped pottery known as Lapita ware (Specht and Gosden, 1997). Lapita expanded rapidly eastward, reaching New Caledonia by about 3000 cal yr BP, and Fiji, Tonga, and Samoa by 2800–2900 cal yr BP (Anderson, 2001a). Other

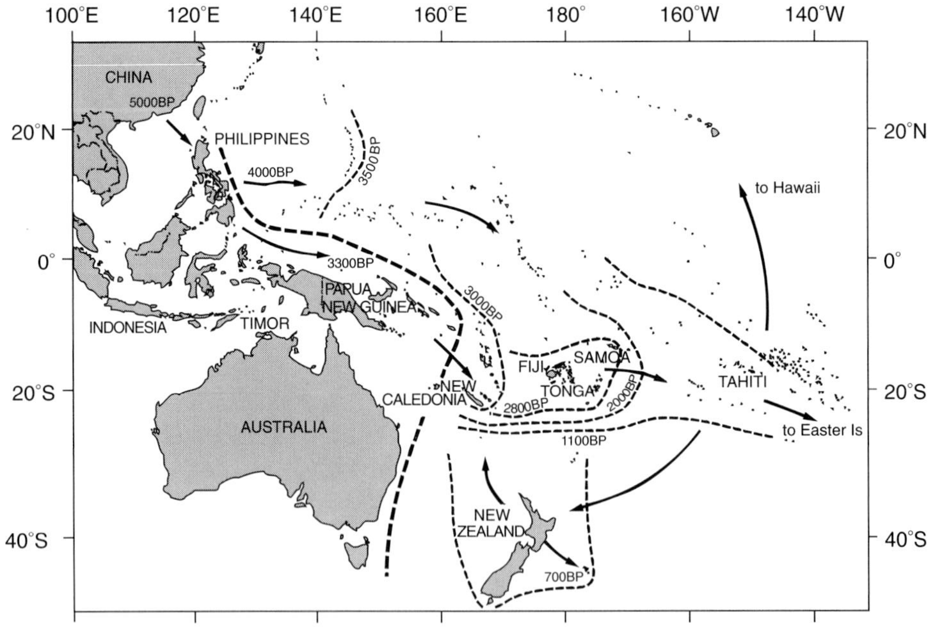

Figure 8.1. The Southwest Pacific region showing the main landmasses, the boundary (heavy dashed lines) between Near Oceania to the west and Remote Oceania to the east, with approximate isochrons (narrow dashed lines) of initial island colonization.

red-slipped ceramics, polished stone adzes and chisels, slate and shell tools, and ornaments and fish hooks, sometimes associated with remains of pigs, dogs, and chickens, are found in mid- to late-Holocene sites throughout island Southeast Asia. They probably reflect various mainland Southeast Asian sources but one route of dispersal was almost certainly through Taiwan. Corded-ware ceramics arrived there about 6000 cal yr BP from coastal south Chinese traditions dated to 7000–5000 cal yr BP, and the development of those in Yuanshan and other ceramic traditions after 4500 cal yr BP marks the point of rapid expansion toward Oceania (Spriggs, 1999). Neolithic culture reached the Philippines before 4000 cal yr BP, the Marianas by 3500 cal yr BP, and Western Melanesia, in the form of Lapita ware, by 3300 cal yr BP. There were some important changes along the way, notably the loss of rice agriculture that reached the Philippines but failed to penetrate any further to the southeast where taro, yams, and tree crops prevailed.

The spread of Neolithic culture into island Southeast Asia and Oceania during the period 5000–3000 cal yr BP raises the question of a climatic impetus, strengthening El Niño-Southern Oscillation (ENSO) conditions and some changes in monsoon circulation at about 5000 cal yr BP. One suggestion arises from Liu (2000), who attributes major population shifts and socio-political realignment concurrent with the collapse of Lungshan culture before 4000 cal yr BP to lower temperatures and increased flooding in Northeast China. Wu and Liu (2004) present a similar and more wide-ranging hypothesis which suggests that significantly wetter and cooler conditions in South China, beginning about 4500 cal yr BP led, as did drought elsewhere in China, to the widespread collapse of Neolithic cultures, 4200–4000 cal yr BP. One potential adjustment may have been increasing migration, and as the impact of ENSO increased during the late Holocene, then the strengthening frequency of offshore winds could have produced a significant increase in sea travel toward island Southeast Asia, perhaps accelerated in the 4th millennium BP by the addition of sails to watercraft (Anderson, 2000, 2004; McGrail, 2001, p. 356; Anderson et al., 2005).

Turning to Near Oceania, here including Australia, the mid-Holocene appears as a significant hinge in late Quaternary environmental and cultural history. Rising post-glacial sea levels severed the Sahul landmass into its three main constituents of New Guinea, Australia, and Tasmania, in the period 11,000–8000 cal yr BP, and by 5000 cal yr BP the loss of more than two million km^2 of late Pleistocene landmass had reduced the land area in the Southwest Pacific to its smallest extent within the last 100,000 years. Land areas and coastal complexity increased somewhat in the late Holocene as a result of a slight retreat from a post-glacial high sea-level, in some areas, after 4000 cal yr BP and the catch-up of coral reef growth as sea-levels stabilized (Dickinson, 2001). The mid-Holocene, therefore, was probably the least suitable period for island or coastal settlement in the Western Pacific since the last glacial era (Enright and Gosden, 1992). From an archeological point of view, sea level changes had an equally dramatic effect because they wiped out or concealed most of the potential evidence of coastal and insular settlement prior to the mid-Holocene. To consider the potential relationship of cultural dynamics to climatic change at that time, it is desirable therefore to review archeological data that are not wholly bound to coastal occupation.

We will look at two cases, pitched at different geographical scales and representative of the fundamental cultural juxtaposition in the region: (1) the suite of cultural changes which occurred in hunter–gatherer societies during the mid- to late Holocene in Australia, and (2) those changes which appear to document the early history of agriculture in Highland New Guinea. Before examining these case studies, we briefly review the modern climate and inferred mid-Holocene climate changes.

2. Regional climatology

Three major zones of atmospheric circulation are represented in the Southwest Pacific region: the Monsoonal Belt (5–15°S), the Sub-tropical Anticyclonic Belt (20–35°S) with the southeast trade-winds on its northern flank, and the Westerly Belt (40–60°S). Because the Southern Hemisphere is dominated by oceans, zonal atmospheric circulation is less modified than in the Northern Hemisphere. As warm air rises at the equator, it generates very humid climates, especially where the uplift is accentuated by tall mountain ranges, as in New Guinea. Areas that receive this moisture for part of the year fall within the Monsoonal Belt. The air that rises over the equator travels to higher latitudes at high elevations and, in the Southern Hemisphere, it cools and sinks at about 30°S. Thus, in the Sub-tropical Anticyclonic Belt, where the air sinks, there is little surface air movement and little precipitation. The air that sinks at 30°S either flows southwards and is deflected by the spin of the earth into the westerly wind belt that dominates the mid-latitudes, or it turns northwards to be deflected into the southeast trade-wind belt. All of these zones are displaced southward in the Austral summer and move equatorward during the winter (Figs. 8.2 and 8.3).

In addition to the basic zonal pattern of atmospheric circulation, regional and local factors play a major role in controlling the distribution of rainfall. The Walker Circulation, a Pacific-wide zonal circulation in tropical latitudes, is thought to result primarily from strong atmospheric convection over the Western Pacific Warm Pool which drives air eastward at high altitude to sink in the East Pacific Dry Zone. The seasonal transit of the Intertropical Convergence Zone southwards from the Asian continent in the Austral spring and summer brings monsoonal rainfall to the tropical regions of the Southwest Pacific. Another feature of regional significance in terms of rainfall is the South Pacific Convergence Zone that connects the convective region over the Warm Pool with the higher latitudes of the southwestern Pacific (Allan et al., 1997). At higher latitudes, onshore winds crossing warm water on the eastern side of temperate Australia make that area much moister than equivalent latitudes in Western Australia. Mountain ranges, especially in New Guinea and New Zealand, also dramatically modify the local climates by blocking prevailing winds and concentrating rainfall on windward slopes.

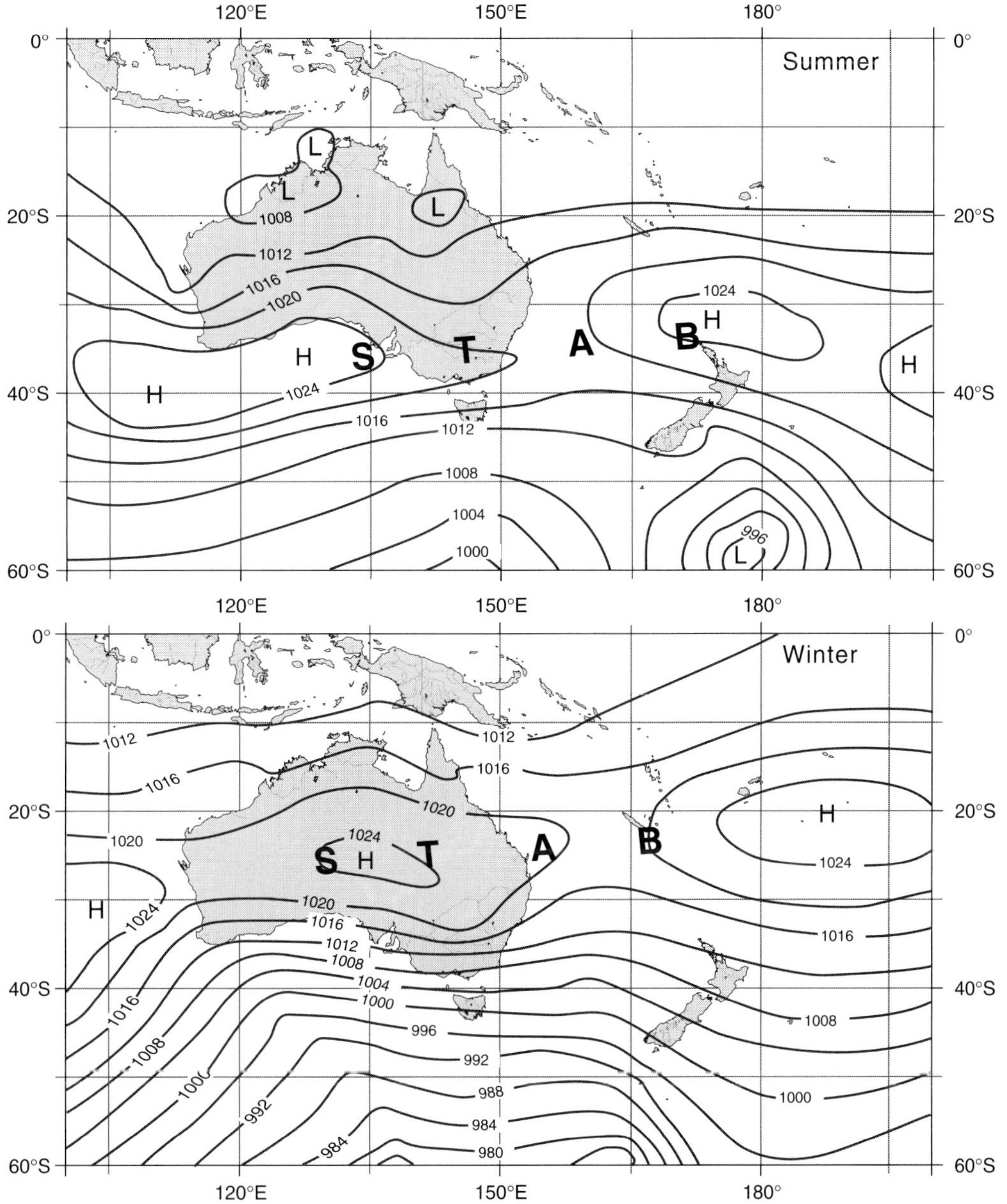

Figure 8.2. Position of the sub-tropical anticyclonic belt (STAB) over the Southwest Pacific region in summer and winter. This belt separates the mid-latitude westerlies to the south from the intertropical convergence zone (ITCZ) to the north (modified from Sturman and Tapper, 1996).

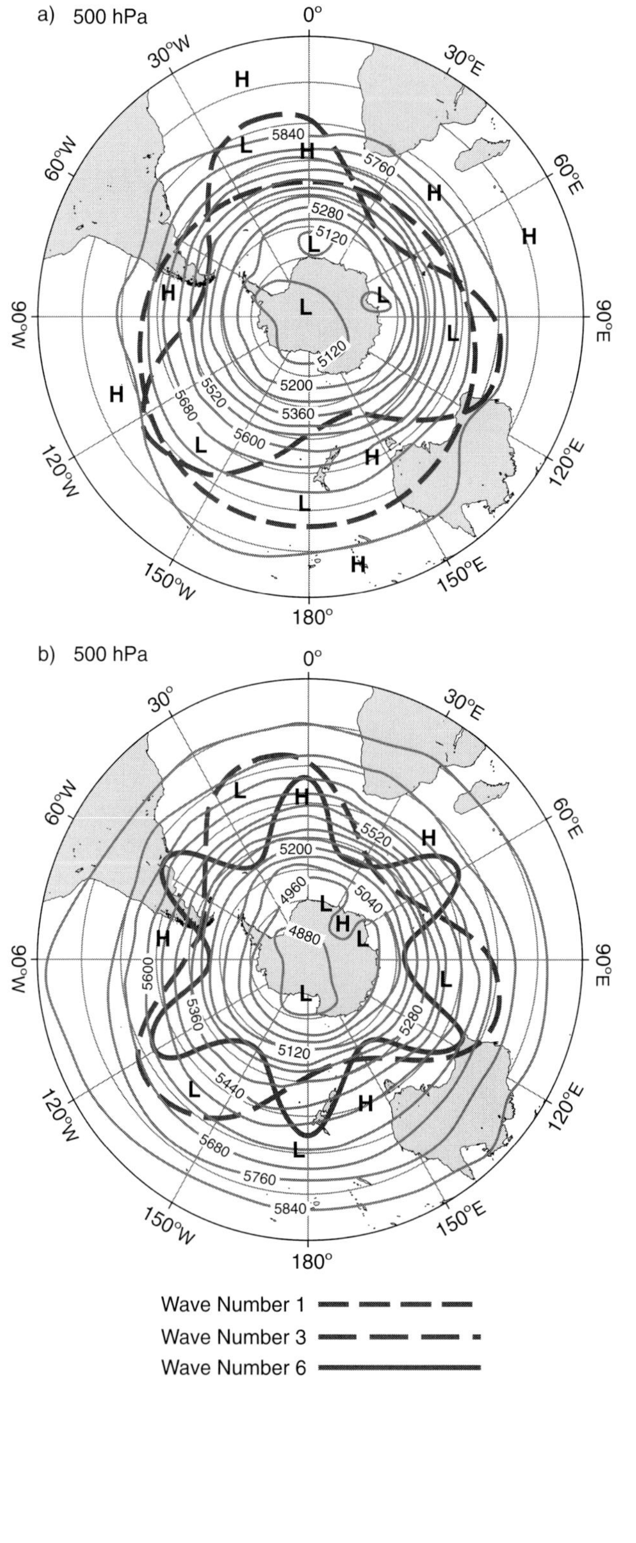
a) 500 hPa
0°
30°W
30°E
60°W
60°E
90°W
90°E
120°W
120°E
150°W
150°E
180°
5840
5760
5280
5120
5200
5360
5520
5680
5600
b) 500 hPa
0°
30°
30°E
60°W
60°E
90°W
90°E
120°W
120°E
150°W
150°E
180°
5520
5200
4960
5040
4880
5600
5360
5280
5120
5440
5680
5760
5840
Wave Number 1
Wave Number 3
Wave Number 6

Precipitation patterns in New Guinea are particularly complex due to the orographic effects of an east–west blocking range (the highest point, Mount Djaja, is 5029 m high) and the interactions of two major moisture sources. The most important of these moisture sources are monsoonal rains that occur during the austral summer (January–April). These are sourced from the northwest and while they provide significant rainfall to most areas, a rain shadow affects some lowland southeastern areas, such as the Fly River delta. Southeasterly trade winds, which flow during May to October, are modified as they pass over the Coral Sea, southeast of New Guinea, and provide substantial rainfall to parts of the southeast, notably the Fly River delta, and to the mountains, while a rain shadow occurs on the northwestern side of the island and in the southernmost lowlands. Mountain areas also generate much convective rainfall activity throughout the year. The Highlands of New Guinea consequently have high rainfalls with a fairly even distribution of rain throughout the year, but both sources of rainfall are highly vulnerable to circulation changes associated with an ENSO.

ENSO is embedded within the Walker Circulation, which is the most significant modifier of inter-annual climates in the southwestern Pacific. On a 2–7 year basis, the Walker Circulation weakens and the convective center over Indonesia moves eastward where it generates anomalous rainfall from the central equatorial Pacific to the Peruvian coast. This is an El Niño. The effects of El Niños are well known in the Southwest Pacific where droughts occur in Southeast and northern Australia, Papua New Guinea, and the far western portion of Oceania. An excellent review of the oceanic, atmospheric, and hydrologic responses to ENSO in the southwestern Pacific region can be found in Allan et al. (1997). By contrast, it is less well known that the impacts of ENSO are equally severe in New Zealand. In particular, El Niño years are associated with anomalously strong southwesterly flow over New Zealand (Gordon, 1985) and this pattern is linked to changes in the Southern Hemisphere polar-jet stream, which controls the angle of approach of Southern Ocean cyclonic systems on to New Zealand (Fig. 8.3).

A detailed analysis of historical climate records spanning the last century has produced a new view of the spatial patterns of ENSO variability through time (Allan et al., 1997). This analysis suggests that rainfall, sea-surface temperatures (SSTs), and wind-field patterns associated with ENSO differ strongly among events, and the centers of action also shift. The cause of these shifts is not clear; leading candidates include a change in the basic state of the tropical climate system (Federov and Philander, 2000) or a modulation of ENSO by decadal-scale

---

Figure 8.3. Idealized Rossby waves nos. 1 and 3 (Fig. 8.3a) and nos. 3 and 6 (Figure 8.3b) in the Southern Hemisphere, at 500 hPa elevation. Low-Rossby numbers are associated with zonal flow and high numbers with meridional flows. Unlike the Northern Hemisphere, where the polar-jet is anchored to the Rocky Mountains, the Southern Hemisphere polar-jet is less restrained and modification of the Rossby waves affects the track of cyclonic systems approaching New Zealand. The dominance of southwesterly flows over New Zealand during El Niño years may be related to shifts in position or wave number of the Rossby wave.

variability. There is increasing recognition of decadal oscillation signals in the Southwest Pacific (e.g., Salinger et al., 2001) which compound and/or confound ENSO signals. These signals do not, however, offer a complete explanation of variation.

3. Climatic and environmental change in the mid-Holocene

3.1. Terrestrial reconstructions

Before considering the archeological data for putative interactions between climate and culture, it is necessary to determine which climate changes occurred during the mid-Holocene in the Western Pacific. As might be expected in such a large area, these changes are not consistent across the region. In addition, one of the primary problems of examining pre-instrumental climate changes in areas of long-term human habitation is that many of the indicators that are used to identify climate change are also susceptible to anthropogenic modification. A classic example, and one that is noted again in the context of New Guinea, is fire. Increased burning, as evidenced by charcoal in soils, lake beds, and other sediments, may be the result of increasing aridity due to climate change, or of changes in human land use, either in type or intensity. To determine whether there were any climatic changes of consequence, it is necessary to consider a situation where anthropogenic landscape modification post-dates the mid-Holocene. In this context, the best paleoenvironmental records regionally come from New Zealand.

In New Zealand, the early part of the Holocene was somewhat warmer than the present day, with a climatic optimum at ca. 8000 cal yr BP (e.g., McGlone et al., 1993). After this time a gradual increase in frosts and droughts occurred. Although a number of vegetation changes occurred during the Holocene throughout New Zealand, McGlone (1988) and McGlone et al. (1993) have not ascribed these to any distinct changes in climate, except at about 3000 cal yr BP when southern New Zealand became wetter and cooler in response to enhanced westerly circulation. Shulmeister (1999) noted, however, that there is strong evidence from the resurgence of glacial activity in the Southern Alps (Gellatly et al., 1988) and from the reduction or elimination of frost intolerant taxa from sites in central New Zealand (McGlone and Moar, 1977) for a significant climatic change at ca. 5000 cal yr BP. Glacial advances at that time have been attributed also to strengthening westerly or southwesterly flows (Fitzharris et al., 1992; Shulmeister et al., 2004).

From these observations, we can deduce that the dominant climate signal in New Zealand is the gradual strengthening of westerly zonal circulation through the Holocene (Shulmeister et al., 2004). All the zonal circulations are linked and, in general, as one circulation strengthens or weakens, the others will respond in a similar manner. Thus, strengthening the westerlies in New Zealand should also mean increasing the southeast trades and the monsoon winds in the sub-tropics and tropics as the Holocene progressed. Shulmeister (1999) provides evidence that

ENSO is linked to this circulation change and that it was either absent or substantially reduced prior to about 5000 cal yr BP.

In northern Australia, analysis of fossil pollen assemblages in sediment cores indicates that after a gradual increase in moisture through the early Holocene, there appears to be an effective precipitation (EP) maximum between 5000 and about 3500 cal yr BP (e.g., Kershaw, 1983; Shulmeister and Lees, 1995). This period was up to 50% wetter than modern (Kershaw and Nix, 1989), whereas the EP maximum in Southeast Australia is inferred to be only 5–10% above modern values (Dodson, 1998) and occurred between about 7000 and 5000 cal yr BP. In Southeast Australia, an arid phase began after 5000 cal yr BP.

Holocene environmental change has been investigated also in Papua New Guinea. Analysis of pollen assemblages in sediment cores (below) suggests that rapid post-glacial warming had crossed into a regime slightly warmer and wetter than today by 9000 cal yr BP, and there is evidence that the treeline climbed from 2700 to 4000 m. In the period 8500–5000 cal yr BP, alpine ice fields disappeared from Mt. Wilhelm and the treeline was 100–200 m higher than at present. At the end of this "hypsithermal interval", the treeline retreated to its current position, temperatures cooled slightly, and from 3500 cal yr BP to the present there were at least four small glacial advances (Haberle, 1994, p. 178, 1996).

3.2. Ocean–atmosphere reconstructions

Documenting past changes in the Western Pacific Warm Pool (mean SST $>28^\circ$C) is particularly important for understanding Holocene climate change. The relatively warm, fresh, "skin" of buoyant Warm Pool water occupies only 0.05% of the total ocean-water mass, yet it plays a leading role in driving atmospheric circulation (Webster, 1994). Even small changes in tropical SST can lead to a marked increase in surface–ocean evaporation and water vapor in the atmosphere (Flohn et al., 1990). Thus, any change in the temperature, size, or positioning of the Warm Pool in the past would have had a profound effect on global climate (Cane and Clement, 1999).

Recent research has revealed new insights into the temperature, size, and variability of the Warm Pool since the Last Glacial Maximum (LGM). Studies of foraminiferal Mg/Ca and alkenones from deep-sea sediment cores provide long, continuous histories of changes in mean SST (Pelejero et al., 1999; Lea et al., 2000; Kienast et al., 2001; Koutavas et al., 2002; Stott et al., 2002; Rosenthal et al., 2003; Visser et al., 2003). These reconstructions generally agree that SSTs in the Warm Pool region during the LGM were $\sim$2–4°C cooler than at present. In contrast, the tropical East Pacific exhibits a much smaller cooling of $\sim$1°C during the LGM (Koutavas et al., 2002). Interestingly, records near the equator or in the Southern Hemisphere tropics show a rapid rise to SSTs 0–1°C higher than modern values during the early-middle Holocene ($\sim$10,000–4000 cal yr BP).

New coral Sr/Ca paleothermometry records from northern Australia and Indonesia are in good agreement with the foraminiferal Mg/Ca estimates of early-middle Holocene warming of tropical Western Pacific SSTs (Gagan et al., 2004). New coral records show that SSTs reached modern values by ~8500 cal yr BP. Mid-Holocene SSTs in southern Indonesia fall within 0.5°C of modern values, whereas corals from the inshore Great Barrier Reef, Australia, indicate SSTs ~1°C warmer than the present (Gagan et al., 1998). Taken together, the foraminiferal Mg/Ca and coral Sr/Ca records indicate a general cooling of SSTs since the early Holocene that has not reversed until the 20th century.

To improve our understanding of tropical ocean–atmosphere interactions in the past, the Mg/Ca and Sr/Ca paleothermometers have been used to remove the temperature component of the oxygen isotope signal in biogenic carbonates, and thereby reveal changes in seawater ^{18}O concentrations as a proxy for surface–ocean salinity (e.g., Gagan et al., 1998, 2000; Lea et al., 2000; Hendy et al., 2002; Stott et al., 2002; Rosenthal et al., 2003). In the tropics, changes in surface–ocean salinity primarily reflect changes in the balance between evaporation and precipitation at the ocean surface. An important new finding is that the foraminiferal and coral records both show significant freshening of the ocean surface in the tropical Western Pacific region since ~8000–6000 cal yr BP (Gagan et al., 1998; Stott et al., 2002).

In summary, the oceanic climate reconstructions indicate a semi-permanent La Niña-like state in tropical Pacific SSTs during the early-middle Holocene (Koutavas et al., 2002; Stott et al., 2002). A prolonged, westward-concentrated La Niña during the early-middle Holocene agrees with terrestrial paleoclimate records indicating warmer and wetter conditions in the tropical Western Pacific region, as discussed earlier. Moreover, the Holocene cooling to a more El Niño-like state in Pacific SSTs explains why Near Oceania became progressively drier in the late Holocene.

The climate-change scenario described above provides interesting possibilities for enhanced eastward-directed sea travel to Remote Oceania during the late Holocene. Geoarcheological evidence of the late Holocene emergence of habitable atolls, as the result of a 1–2 m drawdown in sea level, has been linked by Dickinson (2001) and Kerr (2003) to the pattern of human colonization. The impact of emerging atolls on migration patterns, however, was probably rather slight because most Oceanic atolls lie in Micronesia and the Tuamotus, well to the north of the main routes known, from archeological evidence, to have been used between archipelagos in Melanesia and Polynesia. Very few atolls emerged on those routes. There were none in the critical passage between the main Solomon Islands and the Reef Islands – the boundary between Near and Remote Oceania, very few between West and East Polynesia, and none between East and South Polynesia.

More persuasive is the potential impact of climatic change on wind systems that were critical for voyaging at a time when sailing technology probably did not permit passages against the wind (Anderson, 2000, 2001b). The transition to a more El Niño-like state in Pacific SSTs during the late Holocene would have reduced the average strength of the Pacific tradewinds, as indicated by climate models (Clement et al., 2000; Liu et al., 2000). Indeed, the frequency of "windows of opportunity" to travel east on frequent westerly wind reversals, against the prevailing southeasterly trades, would have increased as El Niño events became more common in the late Holocene. Anderson (2004) and Anderson et al. (2005) suggest that quasi-cyclic enhancement of westerlies at a millennial scale may have provided climatic forcing of eastward migration in Remote Oceania, especially around 3000 and 1000 cal yr BP.

The opposing changes of the surface–ocean and terrestrial freshwater balances during the Holocene are also intriguing. During the La Niña-like state of the early-middle Holocene, strong tradewinds and westward transport of moisture would have increased rainfall on land in Near Oceania, as indicated by the terrestrial paleoclimate records. In contrast, the freshening of the surface–ocean (and drying of Near Oceania) toward the late Holocene indicates that the focus for rainfall shifted eastward, toward the open ocean, as occurs during modern El Niño events.

Taken together, the paleoclimate evidence suggests that relatively weak and variable Pacific tradewinds, interspersed with frequent westerlies, could have promoted eastward sea travel to Remote Oceania during the late Holocene. It is possible that rainfall shifted toward Remote Oceania during the late Holocene, thereby increasing the survival rate of early seafarers, if not also providing an incentive for them to disperse from drought-prone western islands (Anderson, 2004). Such a possibility leads us to examine what is known about changes in the frequency of El Niño events during the Holocene, and any bearing that might have on human settlement patterns in the tropical Western Pacific.

3.3. Holocene evolution of ENSO

The most continuous, high-resolution record of ENSO for the early Holocene comes from laminated-clastic deposits in a high-altitude lake, Laguna Pallcacocha, in Ecuador (Rodbell et al., 1999; Moy et al., 2002). Today, these clastic laminae record anomalously high rainfall during El Niño events. Interestingly, the sedimentary record shows a clear suppression of ENSO variability, with periodicities of ~15 years, from 12,000 cal yr BP (the beginning of the record) to 7000 cal yr BP (Fig. 8.4). The result is in good agreement with pollen records from South America, New Zealand, and Australia, which indicate that early Holocene vegetation did not include types adapted to the periodic droughts associated with ENSO (McGlone et al., 1992; Markgraf and Diaz, 2000). By comparing early Holocene vegetation, lake level, and fire history records from these regions, McGlone et al. (1992) concluded that the circum-Pacific precipitation patterns, reduced environmental variability, and absence of fire all suggest the suppression of ENSO between 10,000 and 8000 cal yr BP. A coral record from Papua New Guinea also shows a suppression of ENSO variability at 6500 cal yr BP (Tudhope et al., 2001).

Several lines of paleoclimate evidence suggest that the onset of modern ENSO variability occurred between 7000 and 4000 cal yr BP. Spectral analysis of the

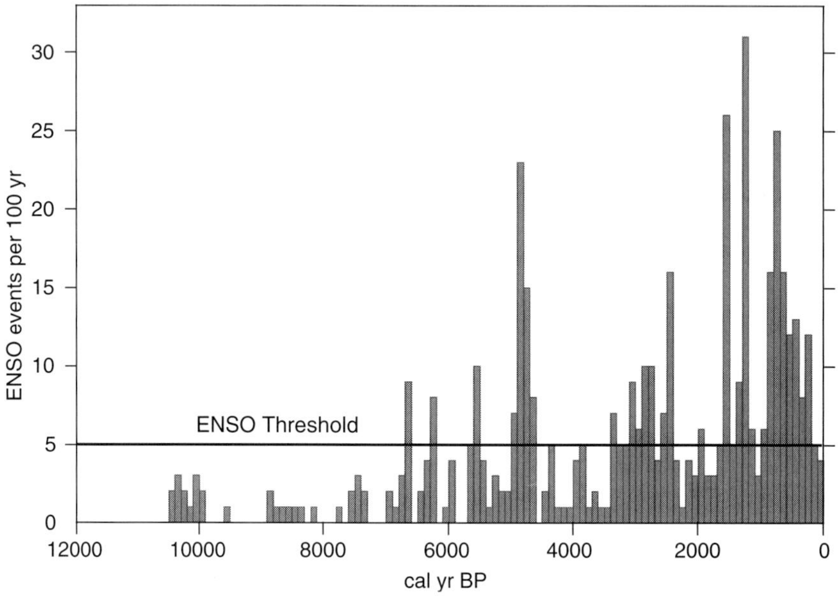

Figure 8.4. Summary of the number of moderate–strong El Niño events in 100-year windows since 12,000 cal yr BP, based on the analysis of clastic laminae in lake Laguna Pallcacocha, southern Ecuador (after Moy et al., 2002). The solid line indicates minimum number of events (~5) required to produce ENSO-band variance. Data available from the NOAA/NGDC Paleoclimatology Program website: http://www.ngdc.noaa.gov/paleo/pubs/moy2002.

15,000-year high-resolution record of storm-derived clastic sedimentation in Laguna Pallcacocha, Ecuador (Rodbell et al., 1999; Moy et al., 2002) shows that the transition to modern ENSO periodicities (2–8 yr) began ~7000–5000 cal yr BP (Fig. 8.4). Sandweiss et al. (1996) and Rollins et al. (1986) reached a similar conclusion based on their analysis of fossil mollusk and fish assemblages and other geoarcheological evidence from coastal Peru. On the western side of the Pacific basin, the first occurrence of drought-adapted pollen taxa in lake sediment cores from tropical northern Australia indicates ENSO onset at ~4000 cal yr BP (Shulmeister and Lees, 1995). A composite charcoal abundance record derived from 10 lake and wetland records from eastern Indonesia and Papua New Guinea reflects changes in the pattern of regional burning from the LGM to the present (Haberle et al., 2001). Higher charcoal concentrations from the mid- to late Holocene (5000 cal yr BP to the present) are interpreted to reflect higher precipitation variability associated with the onset of modern ENSO variability.

The picture emerging for the most recent 5000 cal yr of ENSO history indicates that it began to operate as it does now, but with variability on millennial timescales, and a peak in ENSO frequency and magnitude at ~1800–1200 cal yr BP (Fig. 8.4). The record of clastic sedimentation from Laguna Pallcacocha, Ecuador, shows that El Niño events became more frequent over the Holocene until about 1200 BP, and then declined toward the present (Moy et al., 2002). Superimposed on this long-term trend are periods of relatively high and low ENSO activity, alternating at a time scale of about 2000 years. Geoarcheological evidence from the coast of Peru indicates an increase in ENSO event frequency after 3200–2800 cal yr BP (Sandweiss et al., 2001). Isotope records from massive fossil coral microatolls at Christmas Island in the central equatorial Pacific also indicate that ENSO variability reached, or exceeded, modern values from 2500 to 1700 cal yr BP (Woodroffe and Gagan, 2000; Woodroffe et al., 2003).

In summary, the most continuous, high-resolution record of ENSO (Moy et al., 2002) indicates a significant suppression of El Niño during the early Holocene, with less than five moderate–strong events/century recorded (Fig. 8.4). In contrast, several different paleo-ENSO records indicate a maximum in ENSO magnitude and frequency during the late Holocene. The Moy et al. (2002) reconstruction indicates up to 30 moderate–strong events/century for several periods during the late Holocene, which is significantly more than were observed during the 20th century. Cultural developments during the mid- to late Holocene in Near Oceania can now be considered in relation to the long-term history of climatic and environmental change.

4. Cultural change in mid- to late Holocene Australia

Australia is so large and environmentally variable that it cannot be expected that a single set of cultural changes would occur in the mid- to late Holocene. Yet, leaving aside some regional contradictions and differences, a group of features with at least very broad distribution have been proposed (Williams, 1987; Bird and Frankel, 1991; Lourandos and Ross, 1994; Lourandos, 1997). These are as follows:

(1) The addition to long-established stone tool types, and their partial replacement by, a new assemblage known as the "small tool tradition" (mainland, but not Tasmania), referring essentially to small stone elements of various kinds that had been parts of composite or hafted tools. The most important of the newly-abundant tool types were backed artifacts, bifacial points, and tula adzes.
(2) Evidence of increasing site numbers and increasing intensity of site occupation, with inferred settlement patterns that include greater mobility in many areas, but also signs of logistically-organized habitation in large base camps, the increasing use of defined cemeteries, notably in the lower Murray valley, and an efflorescence in painted rock art which is also differentiated into regional styles.
(3) Expansion of settlement into new habitats including the sandridge deserts, subalpine areas, tropical rainforest, and offshore islands with concomitant changes in resource utilization including greater use of marine resources, the processing of toxic plants for food, and the specialized grinding of grass-seeds.
(4) The arrival of a domestic animal, the dog or dingo (excluding Tasmania), which drove its indigenous competitors on the mainland, *Thylacinus* sp. (Marsupial Tiger) and *Sarcophilis* sp. (Tasmanian Devil), into extinction.

Of the new tool types, the backed artifacts and bifacial points were already in existence but scarce by the early Holocene (Hiscock and Attenbrow, 1998) and there is evidence of occasional cycad processing and grass-seed grinding which dates to the late Pleistocene (Lourandos, 1997; Gorecki et al., 1997). But whether quantitatively or qualitatively, most of the inferred changes are dated to later than 4000 cal yr BP. There is no archeological evidence of the dingo earlier than about 3500 cal yr BP (Lourandos, 1997), systematic processing of toxic cycads may go back to about 4000 cal yr BP (Beaton, 1982), although the data are not as clear as might be wished (Hiscock, personal communications), and specialized grass-seed grinding to 3500 cal yr BP in the semi-arid zone. Data on site usage patterns suggest that the point of inflection toward more intensive occupation was generally at 2500–3500 cal yr BP.

The general similarity of these inferred cultural changes to events occurring during the early Holocene in many other parts of the world, where they have been characterized as the development of complex foraging, has not escaped notice, but whether, in Australia, they fit together as pieces of a single jigsaw puzzle, and to what they might be attributed, has been argued for more than 20 years. The null hypothesis, still not clearly rejected in most regions, is that with the exception of some newly-introduced elements, notably the dingo, the increasing diversity and frequency of mid- to late Holocene archeological remains presents an impression of change which is conveyed largely by the relative survivability of a greater range and quantity of material toward the present.

Many other explanations have invoked an hypothesis of land-use and resource procurement intensification, the general form of which is that, by the mid- to late Holocene, demographic imperatives, especially population growth, had begun to convert some mobile socio-economic strategies into more complex and sedentary systems. Changes in technology and social organization including more extensive exchange and kinship networks and more elaborate ceremonial behavior were associated with increased occupation of hitherto marginal environments, a broadening of the resource base, and sharpened differentiation of territoriality and associated behaviors such as the production of rock art. It is worth looking at this proposition and some alternatives in a little more detail by reference to one of the regional cases, Southeast Australia.

4.1. Southeast Australia

Modern vegetation patterns were re-established in Southeast Australia by the early Holocene, following late Pleistocene aridity. A maximum warming and moisture regime, with rainfall 5–10% higher and temperatures 1–2°C higher than today, was then attained in the period 8000–6000 cal yr BP (Dodson, 1998). Since 5000 cal yr BP there has been a small decrease in temperature and precipitation and some expansion of alpine grasslands from about 4000 cal yr BP. There was a period of drier conditions during which peat formation succeeded shallow lake environments in some catchments and in both inland and coastal lakes water levels declined substantially until about 2000 cal yr BP (Dodson et al., 1992).

Throughout Southeast Australia there is a slight increase in the rate of archeological site occupation during the early and mid Holocene with the frequency increasing dramatically after 4000 cal yr BP. While factors, such as the loss of mid-Holocene coastal sites by erosion, 6000–4000 cal yr BP, and the possibility that at least some of the variation reflects population re-distribution in the landscape, might have contributed to this pattern (Head, 1986); there seems little doubt that significant overall population growth had also occurred. The upward trend encompassed an expansion of settlement into environments not previously occupied to any extent. For example, from about 2500 cal yr BP, large earth mounds which served as foundations for settlements appear in the swamplands of southwestern Victoria (Williams, 1987), along with extensive drainage and fish trap systems. Habitation also expanded in the semi-arid mallee scrublands of northwest Victoria from about 2500 cal yr BP (Ross et al., 1992) and in the Southern Highlands. In Tasmania, isolated from the mainland for thousands of years, settlement expanded into the western forests from about 4000 cal yr BP, which might reflect the increasing efficiency of forest firing, and site occupancy generally and the use of coastal resources also increased noticeably.

The takeoff in archeological site occupancy at 4000 cal yr BP in Southeast Australia is not contemporary with the initiation of mid-Holocene climatic change, nor with the early stages of its impact on lake levels, especially in the coastal areas where most sites are located (Fig. 8.5). Proponents of intensification models argue, therefore, that this process had little, if anything, to do with climatic change; that not only is the chronology mismatched between climatic and cultural change, but also that if climate was an influential variable then there ought to have been significant increases in site occupation during the period of higher humidity and greater terrestrial bio-production in the early Holocene, not during the more stressful late Holocene (Lourandos, 1997).

In regard to the first argument, there is no reason to expect that cultural dynamics would necessarily track changes in climate. Lowered precipitation and temperatures could have had varied and unexpected impacts on sedimentary and biological systems, while to the inherent complexity of those reactions must be added the resilience and diversity of cultural behavior. In addition, the relationship of cultural to natural systems is mediated by other processes that, at some scales of analysis, may be regarded as independent of change in either, notably the timing and strength of population increase or of immigration. These various considerations may be regarded as "buffering" the relationship between climatic and cultural change and therefore as capable of delaying the manifestation of a cultural response in the archeological record.

The second issue of why the strong increase in site occupancy, and presumably in population density, occurred during a period when there might have been greater resource stress than earlier is more difficult to answer but several potential explanations have been canvassed. As indicated above, one is simply that the initial

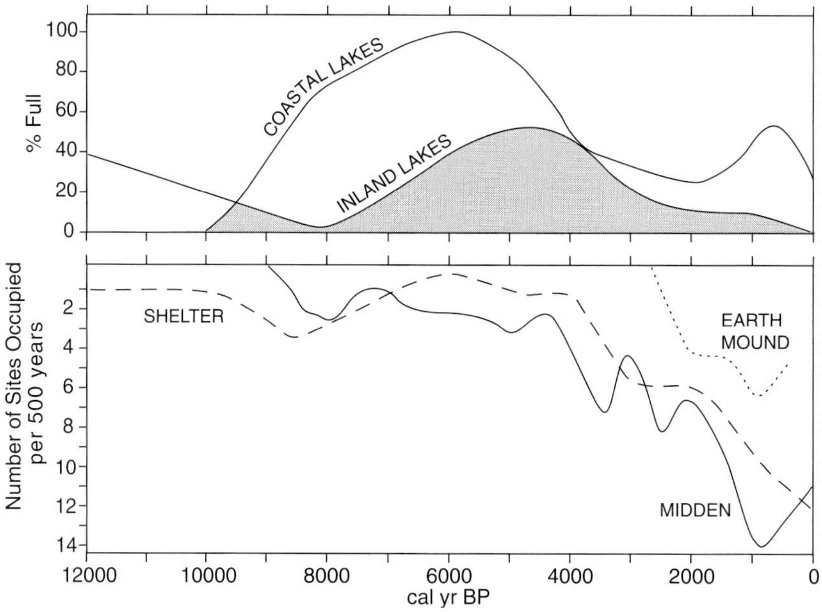

Figure 8.5. Variation in the level of coastal and inland lakes in Southeast Australia (above the time line) compared to the trends in occupancy of different sites in the same region (below the time line). Information from Dodson et al. (1992) and Lourandos and Ross (1994).

assumption may be invalid. Slightly cooler and drier conditions, and possibly greater climatic instability, might actually have produced greater rather than less access to economic resources. An opening up of the humid forest could have facilitated the use of fire to create richer, habitat mosaics, as seems evident from late-Holocene palynological profiles, while slightly lower late-Holocene sea levels seem, in fact, to have increased the extent of productive coastal freshwater wetlands in Southeast Australia (Head, 1986).

4.1.1. Climatically-induced stress

Other potential explanations have assumed the validity of increased stress on resources. Haberle and David (2004) argue that increased post-glacial population density by 6000 cal yr BP in northern Queensland, together with climatic change, led to a significant decline in bio-production which, in turn, resulted in a broadening of the diet spectrum and increasing territoriality and regionalization of material culture after 3700 cal yr BP.

Hiscock (1994) has developed an hypothesis which proposes that the mid-Holocene proliferation of small tools in mainland Australia generally, represents a solution to the economic risk of scheduling uncertainties in changing environments. Each of the main types was easily transportable. In addition, the bifaces combined the advantages of cores producing a consistent flake form, of being easily re-sharpened and of combining useful edges and points; tula adzes were also capable of multiple uses and extended use life, while backed artifacts appear to have been hafted as multiple points in spears and knives. This was a uniform, easily portable, reliable, and functionally flexible toolkit, which required relatively small quantities of raw material. It was especially suited to high residential mobility in environments where resource distributions were relatively unpredictable and foraging expeditions thus required reliable multi-purpose technologies to maximize extraction opportunities as these were encountered. Similar arguments regarding the use of multiple elements to increase redundancy and thereby lower the risk of implement failure, and other propositions about the relationship of composite tools to environmental uncertainty, have been raised in European, African, and American prehistory and continue under development in Australia (Rowland, 1999; Hiscock, 2002).

Given that the mid-Holocene was a period of climatic change which increased the relative uncertainty of many resource environments, then the emergence of the composite or small tool tradition, as well as the refinement of other technologies such as toxic nut and grass seed processing, made it possible to advance residential settlement into some of the least predictable environments in Australia, notably the sand-ridge deserts. By the late Holocene, when reduced mobility and increased population density is apparent in many coastal districts, the backed artifacts and points become less abundant and undifferentiated flake tools more prominent.

A similar kind of argument has been advanced to account for the noticeable increase in the use of marine resources, not only in the southern mainland of Australia, but also in Tasmania, after about 4000 cal yr BP. This is manifested both in the use of offshore islands, mainly from about 3000 cal yr BP (O'Connor, 1992), with implied developments in watercraft, and in the frequency of coastal middens. On the mainland there is evidence of extensive fishing and shellfishing. In Tasmania, fishing ceased, but it was replaced by intensive exploitation of other sub-tidal resources, notably abalone and crayfish, as well as culling of seal and muttonbird colonies. Since the same general trend appears on both sides of Bass Strait, it is difficult to explain it by reference to a coincidence in demographic or societal trends, unless population growth, for instance, was following the same post-glacial trajectory in each case, which is possible. The environmental hypothesis proposes that decreased precipitation and increased seasonality with frequent summer droughts and forest fires associated with the onset of intensive ENSO conditions led to increased stress on terrestrial resources and a re-orientation of the resource base toward a greater emphasis upon the large, reliable, year-round protein sources of the coast (Sim, 1998). That the climatic switch might have been quite severe locally is suggested by the extinction on Flinders Island, some time after about 4500 cal yr BP, of a human population that had survived there since 8000 cal yr BP. Sim (1998) argues that an already stressed population could not tolerate the increased incidence of drought and forest fires in the late Holocene and, having no means of reaching the Tasmanian mainland, must simply have perished, a fate probably coincident with the similar extinction of a long-established population on Kangaroo Island in South Australia. It should be acknowledged, however, that patterns of offshore island habitation in Holocene Australia are complex and enigmatic and no single explanation of them is apparent, particularly none of biogeographic origin (Bowdler, 1995).

The various explanations of mid- to late-Holocene cultural dynamics that have been advanced in Australian archeology find similar expression in the rather different circumstances of New Guinea, notably in the archeology of the Highlands.

5. Highland New Guinea agriculture

As part of the Pleistocene landmass known as Sahul, New Guinea has probably been occupied for as long as Australia. There are archeological remains, notably waisted blades recovered from the Huon peninsula, which probably date to about 40,000 cal yr BP. Holocene research on mainland New Guinea has concentrated on the great Highlands valleys in the central cordillera. These are located between 1300 and 2500 m above sea level and historically have been home to dense populations (up to 300 people per km^2) whose subsistence revolved around intensive gardening, notably of sweet potato (*Ipomoea batatas*, of American origin and introduced by the late 16th century AD) and pig husbandry (Golson and Gardner, 1990). The antiquity of Highlands habitation, and of agriculture there, has been the subject of considerable investigation.

There is little direct evidence of the early arrival of people in the New Guinea pollen records. An increasing frequency of the edible nut genus, *Pandanus*, in late Pleistocene to early Holocene spectra, and a second rise in its frequency from the mid-Holocene might reflect its intentional management at Highlands sites. More important as a potential cultural indicator is the frequency of sustained forest firing. Some swamp cores show that there was frequent or continuous forest firing during the late Pleistocene from about 30,000 cal yr BP and that this declined dramatically or ceased about 8000 cal yr BP (as in cores from Kosipe, Haeapugua, and Telefomin). The Holocene records are variable. At the lowland Hordorli site a continuous fire record exists from about 12,000 cal yr BP. In the Highlands at Kelela forest firing begins at about 7000 cal yr BP, at Telefomin it resumes at 5000 cal yr BP, at Noreikora it begins about 4500 cal yr BP, and at Sirunki at 4300 cal yr BP, but there is then reforestation and continuous clearance from 2000 cal yr BP (Haberle, 1993, 1994, 1995, 2000). Variable as these records are, and indicative of quite localized differences in cultural manipulation of forests across altitudes ranging from 500 to 3500 m, they disclose nonetheless evidence of a substantial increase in burning after 6000 cal yr BP and a tendency for peak Holocene charcoal counts to be attained within the period 4500–1000 cal yr BP (Haberle et al., 2001). It is reasonable to assume that this reflects both a greater incidence of firing by growing Highlands populations and a greater vulnerability to fire of late Holocene forests as a result, primarily, of ENSO-influenced climatic conditions. It is uncertain to what extent ENSO conditions may have distorted the fire signal, producing records that appear to be of cultural frequency and intensity, but the reduction of Highlands valleys to grassland in the mid-Holocene was probably mainly anthropogenic. Increased firing after 5000 cal yr BP can be inferred from an increase in grassland pollen and in accumulation of organic sediments in many Highlands valleys beginning at about 5000–4000 cal yr BP (Haberle, 1994, p. 191).

In order to discuss these changes in relation to cultural dynamics, we need to examine the sedimentary and structural sequence at the Kuk site, situated at 1580 m altitude in the Wahgi Valley (Fig. 8.6). This shows that at about 10,000 cal yr BP, a channel and associated depressions and holes (Phase 1) were formed on the surface of a dark gray, slightly organic clay. Whether some of these can be distinguished convincingly from natural features, especially of fluvial origin, is uncertain and it is important to resist the temptation of a retrospective credibility being afforded by later and indisputable agricultural features having been constructed at higher levels in the same place. Denham et al. (2003) note that starch grains and raphides of *Colocasia*, cf. *esculenta* (taro) occur on the worked edge of a stone tool from Phase 1 and that there are also microfossils of *Musa* spp. (banana) in early Holocene contexts. It is not clear that either of these plants were domesticated, indeed *Eumusa* (wild banana) seeds in Phase 1 suggest it was not, and the situation is uncertain in regard to taro (Haberle personal communications). At the same time, however, the abundance of *Pandanus* (edible nut) and Musaceae which developed in the local environment during the early to mid-Holocene might suggest that some encouragement of preferred plants was by then underway.

An inorganic brown-gray clay was deposited upon the Phase 1 surface, and this clay continued to accumulate through the period 9000–6000 cal yr BP. Phase 2 begins in the period 7000–6400 cal yr BP (whether ditches were constructed immediately on the Phase 2 surface is not clear, according to Denham, personal communications) and Phase 3 about 4300–3800 cal yr BP, both within a partly organic dark brown to black clay (Denham et al., 2003). At about 3000 cal yr BP, another drainage system was established (Phase 4), and there were also later phases of structural construction and refurbishment.

The early drainage features in general have been interpreted as representing wetland garden systems of various kinds, often associated with dryland forest-fallow gardening, but some might be natural waterways, especially in Phase 1. Formal ditch construction is not evident until about the fifth millennium BP. The main crops may have been bananas (Musaceae), sugar cane (*Saccharum officinarum*), yams (*Dioscorea alata*), and especially taro (*Colocasia esculenta*), varieties of which, amongst other crops (e.g., Pacific almond, *Canarium* sp., betel nut, *Areca catechu*, and sago palm, *Metroxylon* spp.), might have been domesticated within New Guinea (Golson, 1991a, 1991b; Yen, 1991; Haberle, 1995; Hope and Golson, 1995). Bananas are represented by phytoliths throughout the archeological record at Kuk and, by some stage, at least some taxa were probably domesticates. Taro

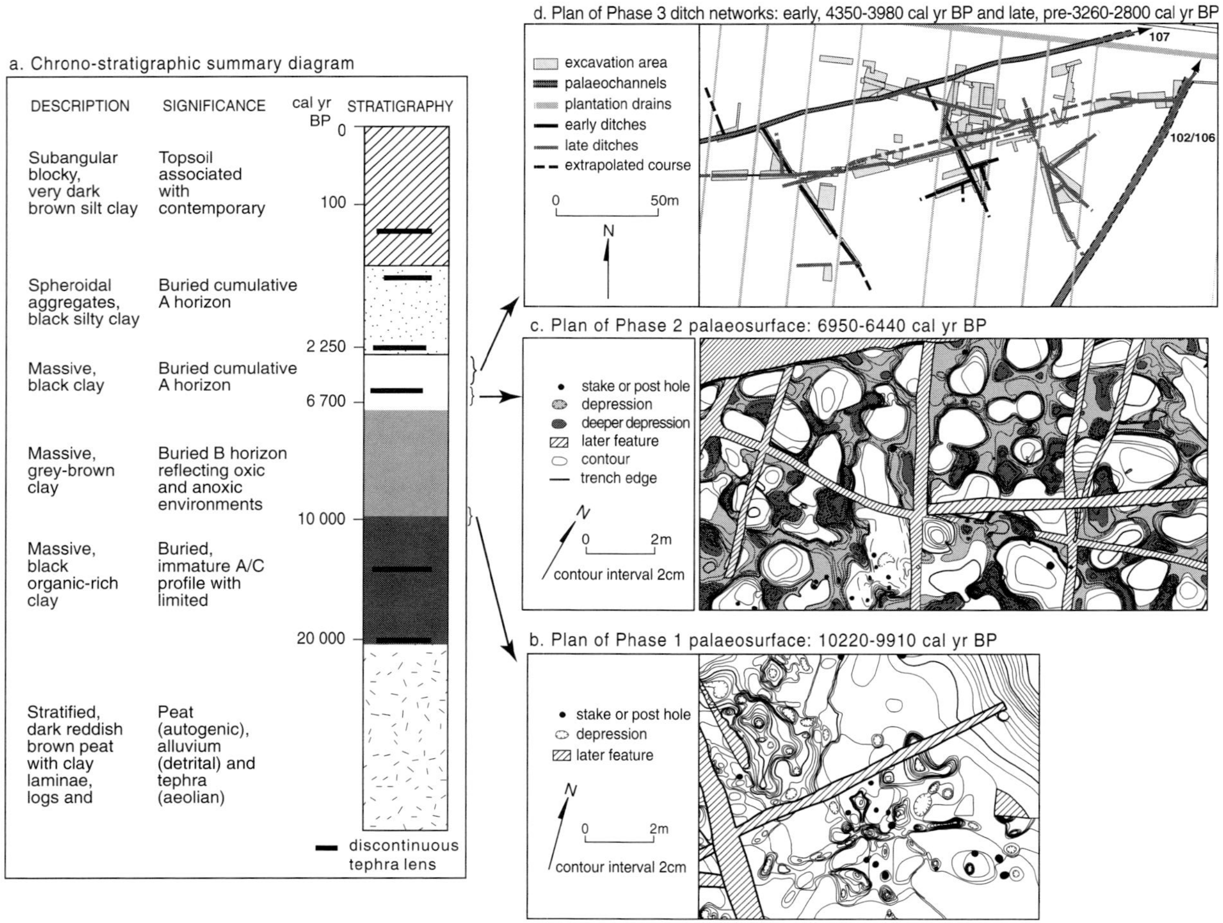

Figure 8.6. Archeo-stratigraphic representation of Phases 1, 2, 3 at Kuk (Reprinted with permission from Denham et al., 2003; Copyright 2003 AAAS).

and yams are also native to New Guinea. Yen (1995) suggested that their local domestication was no more strongly supported than is the introduction of cultivars from Asia during the Holocene, but Lebot (1999) has reviewed biomolecular data indicating that cultivars of taro, yam, sugarcane, breadfruit, plantain, and banana may have local origins in New Guinea. When and where they became domesticated, however, remains uncertain.

The sequence of subsistence development which has been constructed for the Wahgi Valley, and more generally for Highland New Guinea, begins with encouragement of useful forest taxa, notably *Pandanus*, as early as 30,000 cal yr BP (evidenced at Kosipe, but not yet in the Wahgi valley). It is argued that by the early Holocene there was construction of informal drainage features, possibly associated with incipient development of Highlands agriculture. The data, though, might be interpreted equally as re-orientation of non-agricultural settlement patterns and harvesting activities associated with environmental changes during the Holocene transition. Haberle and David (2004) suggest that taro, sugarcane, and gourd had dispersed naturally from lowlands to Highland valleys by the early Holocene and that there was no significant forest clearance until after about 8000 cal yr BP. From a wider review of palaeoecological sites from Highland valleys, Haberle (2003) concluded that either the Phase 1 agriculture argued to exist at Kuk was highly localized, or the antiquity of early agriculture in New Guinea needs re-assessment. Denham (2003), in fact, has rejected earlier claims that the Kuk wetlands were modified for agricultural purposes in Phase 1, although he accepts that the archeological evidence supports plant exploitation practices at that time. As Ballard (2003) emphasizes, however, there is no ultimate truth to be sought here, rather more, and more illuminating, archeological narratives. For example, the almost exclusive emphasis on plants in the discussion of Highlands subsistence development during the Holocene ignores the likelihood that swamplands in forested country, being relatively open and rich in herbs and invertebrates, were also highly attractive to game and their hunters; human activities on swamp surfaces might just as often have resulted from those practices, now no longer represented by bone which would not survive long in acidic soils. Indeed, the modified Phase 1 surface, with pits, post-holes, and a scatter of stone artifacts, some evidently used to cut or scrape plants, is reminiscent of inland hunter-gatherer camps in southern New Zealand, which lay at all times beyond the range of agriculture (Anderson, 1989).

More convincing evidence of agricultural practice occurs in Phase 2 at Kuk and also at Kana in the Highlands (Muke and Mandui, 2003) where it is argued that some features are contemporary with Kuk Phase 2. With extensive forest clearance from 7000 cal yr BP, possibly reflected in an increasing prominence of ground-stone axe-adzes (Denham, 2003), phases of wetland agriculture became more frequent. It is not yet clear that the number of Highlands archeological sites had increased significantly by the mid-Holocene, and therefore whether there occurred a similar vulnerability, through implied population density, to the impacts of climatic change that have been identified in Australian cases (above).

Drainage ditches and intervening gardening structures became progressively patterned and formal from Phase 2 onward and by about 3000 cal yr BP (also in the Baliem valley), forest-fallow gardening had largely collapsed. The valley bottoms were mostly in grassland, and by 2500 cal yr BP, agriculture was dominated by soil tillage with intensive, rotational gardening in wetter areas (involving systematic fallowing to clear out taro beetle (*Papuana* spp.) infestations), established by Phase 4 (Golson, 1990, 1991a, 1991b, 1997; Golson and Gardner, 1990; Bayliss-Smith and Golson, 1992; Bayliss-Smith, 1996).

The gathering diversity and complexity of Highlands agriculture from about the mid-Holocene is probably marked in other ways. There are the stone axe-adzes, and after 2500 cal yr BP, regional exchange networks emerged involving specialized production of high-quality stone implements and other products. There are examples of specialized agricultural implements from late-Holocene swamps, including a ditching spade dated 4564–4130 cal yr BP (ANU-2282: Golson, 1996). There is evidence of significantly declining diversity in wild fauna from about 5000 cal yr BP, and the pig (*Sus scrofa*) may have been husbanded in the Highlands by the mid-Holocene. That matter, however, is very uncertain and vigorously debated. Neither dog nor chicken remains are found in Highlands sites until after about 2000 cal yr BP, although the former, together with pig bone, is argued to occur in coastal middens from 6000 cal yr BP (Bulmer, 1998; Spriggs, 1999).

5.1. Relevance of climatic change

To what extent is climatic change implicated in this process of adopting agriculture, if at all? Golson and his colleagues (Golson, 1977; Golson and Gardner, 1990; Bayliss-Smith and Golson, 1992) thought that the interaction of cultural dynamics and environmental change in Highland New Guinea's prehistory reflected alternate innovations in dryland and wetland agriculture, each of which had environmental consequences. Thus, the 9000–6000 cal yr BP gray-clay unit at Kuk, which accumulated at a rate indicative of increasing erosion of open ground under agricultural conditions, was suggested as representing a dryland forest environment in which soil structures and fertility declined under repeated firing and forest-fallow cultivation. In turn, this land-use pattern produced costs of agricultural efficiency that eventually exceeded those of the labor-intensive construction and maintenance of large-scale wetland systems. Preference for the latter then increased as grassland came to dominate the landscape of Highland valleys and wetland agriculture intensified into a soil tillage and pig husbandry complex. There are alternative explanations of the prehistoric sequence of Highlands agriculture (e.g., by Kelly (1988) who emphasizes the increasing needs of pig rearing and by Gorecki (1986) who suggests mobile use and re-use of swamps), with population growth, as well as increasing pig production, becoming major considerations by the late Holocene. However, the common basis of all of these explanations is that none see mid-Holocene change in climatic variables as essential to explanation.

Nevertheless, Golson and Gardner (1990) conceded the potential significance of Brookfield's (1989) observation that the gray-clay unit corresponded fairly precisely with the hypsithermal interval and that sedimentary and cultural changes following this period might have reflected, to some extent, mid-Holocene climatic changes, notably an increasing impact under ENSO conditions of drought and frost on swamp management. Haberle and Chepstow-Lusty (2000) have elaborated on this suggestion. They point out that the upper margins of the probable natural ranges of the main cultigens extended into the Highlands and that initial domestication may have been a response to periodic environmental stress on these plants. Similarly, when ENSO conditions set in by the mid-Holocene, those may have stimulated further human action. One response might have been to move from drier ground on to swamp surfaces where a water supply was more assured. Another may have been closer control of water management systems. The earliest plausible evidence of agriculture, represented by ditch construction in Phase 2 at Kuk, dates to after about 7000 cal yr BP. How long after is unknown, except that it must have been before about 4300 cal yr BP when Phase 3 constructions began. That interval encompasses the period when El Niño activity first became elevated during the Holocene.

Golson (1990, p. 145) had noted of the Phase 1–3 systems that "their structural features are not linear and uniform but consist in part of small basins and interconnecting runnels which can admit and circulate water, as well as dispose of its excess." He thought that this system indicated broadly-based mixed gardening involving irrigation as well as drainage which, in Phases 4–6, was progressively replaced with staple cropping, initially of taro, in carefully-gridded ditch and raised-bed systems. This process culminated in the adoption of sweet potato cultivation, that cultigen becoming available and having superior qualities of production at the time of the Little Ice Age, when severe El Niño events may have occurred frequently (Haberle, 2000).

Adaptive water management is, at least, a plausible argument in the light of the known impact of ENSO climatic variability, and especially of El Niño conditions, on modern, indigenous agriculture in the New Guinea Highlands. Bourke's (2000) analysis of the 1997 El Niño event shows that in Papua New Guinea generally, 1.2 million people, about 40% of the total rural population, were suffering a severe, and in places life-threatening, food shortage, and that 400,000 had a grossly inadequate water supply by December 1997. The impact was greatest on small islands, where drought devastated the food crops and so weakened the populations that the effort of open sea fishing became too great. In the Highlands valleys, frosts were experienced down to 1450 m in altitude (in the 1972 El Niño they were so severe that forest and grassland as well as crops were damaged, Waddell, 1989) and there was prolonged drought often associated with bushfires (see also Johns, 1989). In addition to crop failure, the dry conditions affected hunting success for wild game. People turned to traditional famine foods and there was substantial migration out of the Highlands valleys to lower altitudes and to urban areas. Health declined significantly with increasing incidence of diarrhea, dysentery, malaria, pneumonia, typhoid, and skin diseases, and death rates by burning or starvation increased significantly, at least 500 attributed to the drought. In Irian Jaya (West Papua), too, about 700 deaths were attributed to the 1997 drought, many of them in the Baliem valley. It is not difficult to imagine that if El Niño conditions had begun to occur after 5000 cal yr BP with anything like the frequency of the recent past: 1956, 1964–1965, 1972, 1982, 1987, and 1997, or with the greater frequency now documented (Fig. 8.4), then there might have been a powerful incentive toward measures of risk aversion. However, it should also be noted that significant famines and episodes of crop failure have occurred in years without El Niño drought, in fact at times of excessive rainfall.

Perhaps the essential point here is that it is not so much the one case or the other as the unpredictable instability of ENSO conditions as a whole which may have constituted an imperative for ameliorating cultural measures. Chappell (2001) makes a more general case for linking the ultimate development of agriculture in New Guinea to declining frequency and magnitude of environmental disturbance at the Holocene transition, and the general significance of stability to agricultural development is a point well taken. An increasing emphasis on wetland gardening in systems which attempted to stabilize the management of agricultural water through periods of climatic instability would have been one appropriate response. Dewar's (2003) recent hypothesis about the significance of rainfall variability is pertinent here; his emphasis was geographical but it might just as easily have referred to the age-frequency distribution of ENSO effects. Whether the cultural changes in prehistory were actually impelled in this implied way remains, however, beyond our current ability to determine.

6. Some conclusions

For both Australia and New Guinea, it is possible to construct plausible propositions that link mid- to late-Holocene climatic changes to cultural dynamics, largely through the mechanisms of technological and subsistence pattern responses to increased uncertainty about access to or control of critical resources. There are several alternatives, however, that remain potentially significant. One is the possibility that cultural change in the mid- to late Holocene was generated significantly by new technologies and associated items arriving from Southeast or southern Asia. This looks especially likely in the case of Australia. Introduction of the dog and of new plant technologies might reflect the wider Neolithic dispersal in the region after 4000 cal yr BP, and some backed artifacts and microliths are similar to contemporary Indonesian industries such as the Toalian (Bellwood, 1997, p. 191). Agriculture in Highland New Guinea is older, dating fairly clearly to Phase 2 at Kuk, and as such it precedes the Southeast Asian agricultural dispersal, although it might have been influenced after 4000–3000 cal yr BP by contact with agricultural systems of Southeast Asian origin.

In referring regional culture change to the significant impact of exogenous agencies, the climatic and immigration hypotheses discussed here are opposed by those that, in varying degrees, substitute social and demographic imperatives. Intensification hypotheses, including in Highland New Guinea, play down the potential influence of contact with other societies and explicitly reject a causative role for climatic change in particular, although they incorporate long-term ecological processes. Thus, Golson (1997, p. 40) states that "though climate was by no means stable through the Holocene...the fluctuations are not likely to have affected the processes under review here in significant ways." Such propositions of endogenous process have to recognize at some level, however, that population growth – and its consequences for density distribution, territoriality, and so on – escape at a regional and larger scale from the intervention of cultural direction. Exogenous models are obliged similarly to concede that there are cultural mechanisms which mediate the impact of environmental change. Yet, still there remains a philosophical difference between perceptions of culture change processes as determined mainly by intrinsic or extrinsic factors.

In this matter, we are mindful of Norman Yoffee's (1985, p. 47) comment on Holocene cultural dynamics in Australia that "...one cannot simply derive economic and social institutions from environmental factors, no matter how simple the society." We would argue that the same is true for demographic factors. Now that the effort invested in recent palaeoenvironmental research has given us a much better grasp of how climate might be related to mid-Holocene cultural dynamics, it is becoming critical for the relative testing of those hypotheses, as for its own sake, to invest a similar effort in the elucidation of palaeodemographic processes and trends.

Reconstructing palaeodemography may be much more difficult, but confidence in the influence of climatic variability as one of the potential prime movers of cultural change during the mid-Holocene depends very substantially upon our ability to deal with the other, palaeodemography. Our guess is that with better data we would see continuous population growth from the late Pleistocene onward, and perhaps that it was exponential in character with rapid growth occurring from about the mid-Holocene. It is implicit in this conjecture, of course, that the long-term population curve would have tracked the broad pattern of late Quaternary climatic change and to that extent a significant mid-Holocene population growth inflection would be a predictable result of earlier climatic amelioration. If so, the mid- to late-Holocene cultural changes would reflect only the longer term climatic pattern delivered through demographic shifts. Consequently, whether the cultural dynamics of that period reflect the contemporaneous smaller-scale climatic movements, including the quasi-cyclical strengthening of ENSO conditions, remains a difficult question and probably one in which the answers are commensurately regional or local in scale. Nevertheless, the mid-Holocene changes in Australian material culture and socio-settlement patterns and the development of water management and agriculture in the New Guinea Highlands appeal as promising cases in the Western Pacific for continuing analysis.

Acknowledgments

We thank ANU colleagues, Jim Allen, Tim Denham, Jack Golson, Simon Haberle, Peter Hiscock, and Geoffrey Hope, for providing unpublished information and comments on earlier drafts of this paper. Tim Denham and the American Association for the Advancement of Science kindly permitted publication of Figure 8.6. We also thank the organizers of the FERCO conference for their support and assistance, especially Kirk Maasch and Dan Sandweiss, and other participants in the symposium, for their comments and suggestions. We are indebted particularly to David Anderson for bringing our paper to publication.

References

Allan, R., J. Lindesay, and D. Parker, 1997. *El Niño Southern Oscillation and climatic variability*. CSIRO, Australia.

Allen, J., and J. F. O'Connell, 2003. The long and short of it: Archaeological approaches to determining when humans first colonised Australia and New Guinea. *Australian Archaeology* 57:5–19.

Anderson, A. J., 1989. *Prodigious birds: Moas and Moa-hunting in Prehistoric New Zealand*. Cambridge University Press, Cambridge.

Anderson, A. J., 2000. Slow boats from China: Issues in the prehistory of Indo-Pacific Seafaring. In *East of Wallace's Line: Studies of Past and Present Maritime Cultures in the Indo-Pacific Region*, edited by P. Veth, and S. O'Connor, pp. 13–50. Modern Quaternary Research in Southeast Asia 16. Balkema, Rotterdam.

Anderson, A. J., 2001a. Mobility models of lapita migration. In *The Archaeology of Lapita Dispersal in Oceania*, edited by G. Clark, A.J. Anderson, and T. Vunidilo. Pandanus Press, Canberra.

Anderson, A. J., 2001b. Towards the sharp end: The type and performance of prehistoric polynesian voyaging canoes. In *Pacific 2000: Proceedings of the Fifth International Congress on Easter Island and the Pacific*, edited by C. M. Stevenson, G. Lee, and F. J. Morin, pp. 29–37. The Easter Island Foundation, Los Osos.

Anderson, A. J., 2003. Initial human dispersal in remote oceania: Pattern and explanation. In *Pacific Archaeology: Assessment and Prospects*, edited by C. Sand, pp. 71–84. Museé Territorial de Nouvelle Caledonié, Noumea, New Caledonia.

Anderson, A. J., 2004. Taking to the boats: The Prehistory of Indo-Pacific colonization. National Institute for Asia and the Pacific Lectures Series, NIAP, Canberra.

Anderson, A. J., J. Chappell, M. Gagan, and R. Grove, 2005. Prehistoric maritime migration in the Pacific Islands: An hypothesis of ENSO forcing. *The Holocene* 16:1–6.

Ballard, C., 2003. Writing (Pre)History: Narrative and archaeological explanation in the New Guinea Highlands. *Archaeology in Oceania* 38:135–148.

Bayliss-Smith, T., 1996. People-plant interactions in the New Guinea highlands: Agricultural heartland or horticultural backwater? In *The Origins and Spread of Agriculture and Pastoralism*, edited by D. R. Harris, pp. 499–552. UCL Press, London.

Bayliss-Smith, T., and J. Golson, 1992. A Colocasian revolution in the New Guinea highlands? Insights from phase 4 at Kuk. *Archaeology in Oceania* 27:1–21.

Beaton, J. M., 1982. Fire and water: Aspects of Australian aboriginal management of cycads. *Archaeology in Oceania* 17:59–67.

Bellwood, P., 1997. *Prehistory of the Indo-Malaysian Archipelago*. University of Hawaii Press, Honolulu.
Bird, C. F. M., and D. Frankel, 1991. Problems in constructing a prehistoric regional sequence: Holocene South-East Australia. *World Archaeology* 23:179–192.
Bowdler, S., 1995. Offshore Islands and maritime explorations in Australian prehistory. *Antiquity* 69:945–958.
Bourke, R. M., 2000. Impact of the 1997 drought and frosts in Papua New Guinea. In *El Niño – History and Crisis*, edited by R. H. Grove and J. Chappell, pp. 149–170. White Horse Press, Cambridge.
Brookfield, H., 1989. Frost and drought through space and time, Part III: What were conditions like when the high valleys were first settled? *Mountain Research and Development* 9: 306–321.
Bulmer, S., 1998. At "The Edge of the Range": Re-Thinking the Origins of Agriculture in the New Guinea high valleys. Unpublished paper, 16th Congress of the Indo-Pacific Prehistory Association, Melaka, Malaysia.
Cane, M. A., and A. C. Clement, 1999. A role for the tropical Pacific coupled ocean-atmosphere system on Milankovitch and millennial timescales: Part II: Global impacts. mechanisms of global climate change at millennial time scales. *Geophysical Monograph* 112:373–383.
Chappell, J., 2001. Climate before agriculture. In *Histories of Old Ages: Essays in Honour of Rhys Jones*, edited by A. J. Anderson, I. Lilley, and S. O'Connor, pp. 171–183. Pandanus Press, Canberra.
Clement, A. C., R. Seager, and M. A. Cane, 2000. Suppression of El Niño during the mid-holocene by changes in the earth's orbit. *Paleoceanography* 15:731–737.
Denham, T., 2003. Archaeological evidence for mid-holocene agriculture in the interior of Papua New Guinea: A critical review. *Archaeology in Oceania* 38:159–176.
Denham, T. P., S. G. Haberle, C. Lentfer, R. Fullagar, J. Field, M. Therin, N. Porch, and B. Winsborough, 2003. Origins of agriculture at Kuk Swamp in the highlands of New Guinea. *Science* 301:189–193.
Dewar, R. E., 2003. Rainfall variability and subsistence systems in Southeast Asia and the Western Pacific. *Current Anthropology* 44:369–388.
Dickinson, W. R., 2001. Paleoshoreline record of relative holocene sea levels on Pacific Islands. *Earth-Science Reviews* 55:191–234.
Dodson, J. R., 1998. Timing and response of vegetation change to milankovitch forcing in temperate Australia and New Zealand. *Global and Planetary Change* 18:161–174.
Dodson, J., R. Fullagar, and L. Head, 1992. Dynamics of environment and people in the forested crescents of temperate Australia. In *The Naive Lands; Prehistory and Environmental Change in Australia and the Southwest Pacific*, edited by J. Dodson, pp. 115–159. Longman Cheshire, Melbourne.
Enright, N. J., and C. Gosden, 1992. Unstable Archipelagos-South-West Pacific environment and prehistory since 30,000 BP. In *The Naive Lands; Prehistory and Environmental Change in Australia and the Southwest Pacific*, edited by J. Dodson, pp. 160–198. Longman Cheshire, Melbourne.
Federov, A. V., and S. G. Philander, 2000. Is El Niño Changing? *Science* 288:1997–2002.
Fitzharris, B. B., J. E. Hay, and P. D. Jones, 1992. Behaviour of New Zealand glaciers and atmospheric circulation changes over the past 130 years. *The Holocene* 2:97–106.
Flohn, H., A. Kapala, H.-R. Knoche, and H. Machel, 1990. Recent changes of the tropical water and energy budget and of Midlatitude circulations. *Climate Dynamics* 4:237–252.

Gagan, M. K., L. K. Ayliffe, J. W. Beck, J. E. Cole, E. R. M. Druffel, R. B. Dunbar, and D. P. Schrag, 2000. New views of tropical paleoclimates from corals. *Quaternary Science Reviews* 19:45–64.

Gagan, M. K., L. K. Ayliffe, D. Hopley, J. A. Cali, G. E. Mortimer, J. Chappell, M. T. McCulloch, and M. J. Head, 1998. Temperature and surface-ocean water balance of the mid-holocene tropical Western Pacific. *Science* 279:1014–1018.

Gagan, M. K., E. J. Hendy, S. G. Haberle, and W. S. Hantoro, 2004. Post-Glacial evolution of the Indo-Pacific Warm Pool and El Niño-Southern Oscillation. *Quaternary International* 118/119:127–143.

Gellatly, A. F., T. J. H. Chinn, and F. Rothlisberger, 1988. Holocene Glacier cariations in New Zealand: A review. *Quaternary Science Reviews* 7:227–242.

Golson, J., 1977. No Room at the top: Agricultural intensification in the New Guinea highlands. In *Sunda and Sahul: Prehistoric Studies in Southeast Asia, Melanesia and Australia*, edited by J. Allen, J. Golson, and R. Jones, pp. 601–638. Academic Press, London.

Golson, J., 1990. Kuk and the development of agriculture in New Guinea: retrospection and introspection. In *Pacific Production Systems: Approaches to Economic Prehistory*, edited by D. E. Yen and J. M. J. Mummery, pp. 139–147. Occasional Papers in Prehistory (ANU) 18, Canberra.

Golson, J., 1991a. Bulmer Phase II: Early Agriculture in the New Guinea Highlands. In *Man and a Half: Essays in Pacific Anthropology and Ethnobiology in Honour of Ralph Bulmer*, edited by A. Pawley, pp. 484–491. The Polynesian Society, Auckland.

Golson, J., 1991b. The New Guinea highlands on the eve of agriculture. *Bulletin of the Indo-Pacific Prehistory Association* 11:82–91.

Golson, J., 1996. The Tambul Spade. In *Work in Progress: Essays in New Guinea Highlands Ethnography in Honour of Paula Brown Glick*, edited by H. Levine and A. Ploeg, pp. 152–171. Peter Lang, Frankfurt Am Main.

Golson, J., 1997. From Horticulture to Agriculture in the New Guinea Highlands: A Case Study of People and their Environments. In *Historical Ecology in the Pacific Islands: Prehistoric Environmental and Landscape Change*, edited by P. V. Kirch and T. L. Hunt, pp. 39–50. Yale University Press, New Haven.

Golson, J., and D. S. Gardner, 1990. Agriculture and sociopolitical organisation in new guinea highlands prehistory. *Annual Review of Anthropology* 19:395–417.

Gordon, N. D., 1985. The Southern oscillation: A New Zealand perspective. *Journal of the Royal Society of New Zealand* 15:137–155.

Gorecki, P., 1986. Human occupation and agricultural development in the Papua New Guinea Highlands. *Mountain Research and Development* 6:159–166.

Gorecki, P., M. Grant, S. O'Connor, and P. Veth, 1997. The morphology, function and antiquity of australian grinding implements. *Archaeology in Oceania* 32:141–150.

Haberle, S. G., 1993. Pleistocene Vegetation Change and Early Human Occupation of a Tropical Mountainous Environment. In *Sahul in Review: Pleistocene Archaeology in Australia, New Guinea and Island Melanesia*, edited by M. A. Smith, M. Spriggs, and B. Fankhauser, pp. 109–122. Occasional Papers in Prehistory (ANU) 24, Canberra.

Haberle, S. G., 1994. Anthropogenic Indicators in Pollen Diagrams: Problems and Prospects for late Quaternary Palynology in New Guinea. In *Tropical Archaeobotany: Applications and New Developments*, edited by J. G. Hather, pp. 172–201. Routledge, London.

Haberle, S. G., 1995. Identification of cultivated Pandanus and Colocasia in pollen records and implications for the study of early agriculture in New Guinea. *Vegetation History and Archaeobotany* 4:195–210.

Haberle, S. G., 1996. Palaeoenvironmental changes in the eastern highlands of Papua New Guinea. *Archaeology in Oceania* 31:1–11.

Haberle, S. G., 2000. Vegetation response to climatic variability. A palaeoecological perspective on the ENSOpPhenomenon. In *El Niño: History and Crisis*, edited by R. Grove and J. Chappell, pp. 66–78. White Horse Press, Cambridge.

Haberle, S. G., 2003. The emergence of an agricultural landscape in the highlands of New Guinea. *Archaeology in Oceania* 38:149–158.

Haberle, S. G., and A. Chepstow-Lusty, 2000. Can climate shape cultural development? A view through time. *Environment and History* 6:349–369.

Haberle, S. G., and B. David, 2004. Climates of change: Human dimensions of holocene environmental change in low latitudes of the PEPII transect. *Quaternary International* 118/119:165–179.

Haberle, S. G., G. S. Hope, and S. van der Kaars, 2001. Biomass burning in Indonesia and Papua New Guinea: Natural and human induced fire events in the fossil record. *Palaeogeography, Palaeoclimatology, Palaeoecology* 171:259–268.

Head, L., 1986. Palaeoecological contributions to Australian prehistory. *Archaeology in Oceania* 21:121–129.

Hendy, E. J., M. K. Gagan, C. Alibert, M. T. McCulloch, J. M. Lough, P. J. Isdale, 2002. Abrupt shift in tropical sea surface salinity at end of little ice age. *Science* 295: 1511–1514.

Hiscock, P., 1994. Technological responses to risk in holocene australia. *Journal of World Prehistory* 8:267–292.

Hiscock, P., 2002. Pattern and Context in the Holocene Proliferation of Backed Artefacts in Australia. In *Thinking Small: Global Perspectives on Microlithization*, edited by S. Kuhn and R. Elston, pp. 163–177. Archaeological Papers of the American Anthropological Association (AP3A), number 12.

Hiscock, P., and V. Attenbrow, 1998. Early Holocene backed artefacts from Australia. *Archaeology in Oceania* 33:49–62.

Hope, G., and J. Golson, 1995. late Quaternary change in the mountains of New Guinea. *Antiquity* 69:818–830.

Johns, R. J., 1989. The influence of drought on tropical rainforest vegetation in Papua New Guinea. *Mountain Research and Development* 9:248–251.

Kelly, R. C., 1988. Etoro Suidology: A Reassessment of the pig's role in the prehistory and comparative ethnology of New Guinea. In *Mountain Papuans: Historical and Comparative Perspectives from New Guinea Fringe Highlands Societies*, edited by J. F. Weiner, pp. 116–186. Michigan University Press, Ann Arbor.

Kerr, R. A., 2003. Geological society of america meeting: Pacific migration arrested by meltdown's high waters. *Science* 302:1888–1889.

Kershaw, A. P., 1983. The vegetation record from Northeastern Australia 7 ± 2 ka. In *CLIMANZ 1*, edited by J. Chappell and A. Grindrod, pp. 100–101. Australian National University, Canberra.

Kershaw, A. P., and Nix, H. A., 1989. The use of bioclimatic envelopes for estimation of quantitative paleoclimatic values. In: *CLIMANZ 3*, edited by T. H. Donnelly and R. J. Wasson, pp. 78–85. Division of Water Resources, CSIRO, Canberra.

Kienast, M., S. Steinke, K. Stattegger, and S. E. Calvert, 2001. Synchronous tropical South China Sea SST change and greenland warming during deglaciation. *Science* 291: 2132–2134.

Koutavas, A., J. Lynch-Stieglitz, T. M. Marchitto Jr., and J. P. Sachs, 2002. El Niño-like pattern in Ice Age tropical Pacific Sea surface temperature. *Science* 297:226–230.

Lea, D. W., D. K. Pak, and H. J. Spero, 2000. Climate impact of late quaternary equatorial pacific sea surface temperature variations. *Science* 289:1719–1724.
Lebot, V., 1999. Biomolecular evidence for plant domestication in sahul. *Genetic Resources and Crop Evolution* 46:619–628.
Liu, L., 2000. The Development and decline of social complexity in North China: Some environmental and social factors. *Bulletin of the Indo-Pacific Prehistory Association* 20: 14–34.
Liu, Z., Kutzbach, J., Wu, L., 2000. Modeling climate shift of El Niño variability in the holocene. *Geophysical Research Letters* 27:2265–2268.
Lourandos, H., 1997. Continent of hunter-gatherers: New perspectives in Australian prehistory. Cambridge University Press, Cambridge.
Lourandos, H., and Ross, A., 1994. The great "Intensification" debate: Its history and place in Australian archaeology. *Australian Archaeology* 39:54–62.
Markgraf, V., and H. F. Diaz, 2000. The past ENSO record: A synthesis. In *El Niño and the Southern Oscillation: Multiscale Variability and Global and Regional Impacts*, edited by H. F. Diaz and V. Markgraf, pp. 465–488. Cambridge University Press, Cambridge.
McGlone, M. S., 1988. New Zealand. In *Vegetation History*, edited by B. Huntley and T. Webb III, pp. 557–599. Kluwer Academic Publishers, Rotterdam.
McGlone, M. S., A. P. Kershaw, and V. Markgraf, 1992. El Niño/Southern Oscillation Climatic Variability in Australasian and South American Paleoenvironmental Records. In: *El Niño. Historical and Paleoclimatic Aspects of the Southern Oscillation*, edited by H. F. Diaz and V. Markgraf, pp. 435–462. Cambridge University Press, Cambridge.
McGlone, M. S., and N. T. Moar, 1977. The *Ascarina* decline and post-glacial climatic change in New Zealand. *New Zealand Journal of Botany* 15:485–489.
McGlone, M. S., M. J. Salinger, and N. T. Moar, 1993. Paleovegetation Studies of New Zealand's Climate Since the Last Glacial Maximum. In *Global Climates since the Last Glacial Maximum*, edited by H. E. Wright Jr., J. E. Kutzbach, T. Webb III, W. F. Ruddiman, F. A. Street-Perrott, P. J. Bartlein, pp. 294–317. University of Minnesota Press, Minneapolis.
McGrail, S., 2001. *Boats of the World: From the Stone Age to Medieval Times*. Oxford University Press, Oxford.
Moy, C. M., Seltzer, G. O., Rodbell, D. T., and Anderson, D. M., 2002. Variability of El Niño/ Southern Oscillation Activity at Millennial Timescales During the Holocene Epoch. *Nature* 420:162–165.
Muke, J., and H. Mandui, 2003. In the shadows of Kuk: Evidence for prehistoric agriculture at Kana, Wahgi Valley, Papua New Guinea. *Archaeology in Oceania* 38:177–185.
O'Connor, S., 1992. The timing and nature of prehistoric Island use in Northern Australia. *Archaeology in Oceania* 27:49–60.
Pelejero, C., J. O. Grimalt, S. Heilig, M. Kienast, and L. Wang, 1999. High-resolution $U^{k}{}_{37}$ temperature reconstructions in the South China sea over the past 220 kyr. *Paleoceanography* 14:24–231.
Rodbell, D.T., G. O. Seltzer, D. M. Anderson, M. B. Abbott, D. B. Enfield, and J. H. Newman, 1999. An ~15,000 year record of El Niño-Driven Alluviation in Southwestern Ecuador. *Science* 283:516–519.
Rollins, H. B., J. B. Richardson III, and D. H. Sandweiss, 1986. The birth of El Niño: Geo-archaeological evidence and implications. *Geoarchaeology* 1:3–15.
Rosenthal, Y., D. W. Oppo, and B. K. Linsley, 2003. The Amplitude and Phasing of Climate Change During the Last Deglaciation in the Sulu Sea, Western Equatorial Pacific. Geophysical Research Letters 30, 1428, doi:10.1029/2002GL016612.

Ross, A., T. Donnelly, and R. Wasson, 1992. The Peopling of the Arid Zone: Human-Environment Interactions. In *The Naive Lands; Prehistory and Environmental Change in Australia and the Southwest Pacific*, edited by J. Dodson, pp. 76–114. Longman Cheshire, Melbourne.

Rowland, M., 1999. Holocene environmental variability: Have its impacts been underestimated in Australian prehistory. *The Artefact* 22:11–48.

Salinger, M. J., J. A. Renwick, and A. B. Mullan, 2001. Interdecadal Pacific oscillation and South Pacific climate. *International Journal of Climatology* 21:1705–1721.

Sandweiss, D. H., K. A. Maasch, R. L. Burger, J. B. Richardson III, H. B. Rollins, and A. C. Clement, 2001. Variation in Holocene El Niño frequencies: Climate records and cultural consequences in ancient Peru. *Geology* 29:603–606.

Sandweiss, D. H., J. B. Richardson III, E. J. Reitz, H. B. Rollins, and K. A. Maasch, 1996. Geoarchaeological evidence from Peru for a 5000 years B.P. Onset of El Niño. *Science* 273:1531–1533.

Sim, R., 1998. The Archaeology of Isolation? Prehistoric Occupation of the Furneaux Group of Islands, Bass Strait, Tasmania. Unpublished PhD dissertation, Australian National University.

Shulmeister, J., 1999. Australasian evidence for Mid-Holocene climate change implies precessional control of walker circulation in the pacific. *Quaternary International* 57/58:81–91.

Shulmeister, J., and B. G. Lees, 1995. Pollen evidence from tropical Australia for the onset of an ENSO-dominated climate at c. 4000 BP. *The Holocene* 5:10–18.

Shulmeister, J., I. Goodwin, J. Renwick, K. Harle, L. Armand, M. S. McGlone, E. Cook, J. Dodson, P. P. Hesse, P. Mayewski, and M. Curran, 2004. The Southern Hemisphere Westerlies in the Australasian Sector During the Last Glaciation Cycle: A Synthesis. *Quaternary International* 118/119:23–53.

Specht, J., and C. Gosden, 1997. Dating Lapita Pottery in the Bismarck Archipelago, Papua New Guinea. *Asian Perspectives* 36:175–199.

Spriggs, M. T. J., 1999. Archaeological dates and linguistic sub-groups in the settlement of the Island Southeast Asian-Pacific region. *Bulletin of the Indo-Pacific Prehistory Association* 18:17–24.

Stott, L., C. Poulsen, S. Lund, and R. Thunell, 2002. Super ENSO and global climate oscillations at millennial time scales. *Science* 297:222–226.

Tudhope, A. W., C. P. Chilcott, M. T. McCulloch, E. Cook, J. Chappell, R. M. Ellam, D. W. Lea, J. M. Lough, and G. B. Shimmield, 2001. Variability in the El Niño Southern oscillation through a glacial-interglacial cycle. *Science* 291:1511–1517.

Visser, K., R. Thunell, and L. Stott, 2003. Magnitude and timing of temperature change in the Indo-Pacific warm pool during deglaciation. *Nature* 421:152–155.

Waddell, E., 1989. Observations on the 1972 Frosts and subsequent relief programme among the Enga of the Western highlands. *Mountain Research and Development* 9: 210–223.

Webster, P. J., 1994. The role of hydrological processes in ocean-atmosphere interactions. *Reviews of Geophysics* 32:427–476.

Williams, E. 1987. Complex hunter-gatherers: A view from Australia. *Antiquity* 61:310–321.

Woodroffe, C. D., M. Beech, and M. K. Gagan, 2003. Mid-Late Holocene El Niño Variability in the Equatorial Pacific from Coral Microatolls. *Geophysical Research Letters* 30:1358, doi:10.1029/2002GL015868.

Woodroffe, C. D., and M. K. Gagan, 2000. Coral microatolls from the central pacific record late Holocene El Niño. *Geophysical Research Letters* 27:1511–1514.

Wu, W., and T. Liu, 2004. Possible role of the "Holocene event 3" on the Collapse of Neolithic cultures around the central plain of China. *Quaternary International* 117:153–166.

Yen, D. E., 1991. Domestication: The Lessons from New Guinea. In *Man and a Half: Essays in Pacific Anthropology and Ethnobiology in Honour of Ralph Bulmer*, edited by A. Pawley, pp. 558–569. The Polynesian Society, Auckland.

Yen, D. E., 1995. The Development of Sahul Agriculture with Australia as Bystander. *Antiquity* 69:831–847.

Yoffee, N., 1985. Perspectives on trends towards social complexity in prehistoric Australia and Papua New Guinea. *Archaeology in Oceania* 20:41–48.

OTHER SELECTED REFERENCES

Modern climates

Sturman, A., and N. Tapper, 1996. The Weather and Climate of Australia and New Zealand. Oxford University Press, Oxford.

Climate Change and Cultural Dynamics: A Global Perspective on Mid-Holocene Transitions
David G. Anderson, Kirk A. Maasch and Daniel H. Sandweiss (Editors)

Chapter 9

Mid-Holocene climate and cultural dynamics in eastern Central China

Tracey L.-D. Lu

Department of Anthropology, The Chinese University of Hong Kong, Shatin, N. T. Hong Kong

Abstract

The Holocene Optimum (~7000–4500 cal yr BP) witnessed the prosperity of many prehistoric cultures in eastern Central China, the majority of them farming societies. The discovery of hundreds of archaeological sites dated to this period demonstrates increased populations and social complexity. Climatic fluctuations and consequent changes of natural resources apparently produced significant impacts on the development of the Neolithic cultures in the middle and lower Yellow and Yangzi River Valleys. However, environmental factors should not be considered as the only causal factors of cultural changes, and we must use caution in interpreting scientific and archaeological data to reconstruct past climates and cultural developments.

1. Introduction

The term 'eastern Central China' here refers to the middle and the lower Yellow and Yangzi Valleys in China. This vast landmass encompasses the area approximately between latitudes 25° and 40° and longitudes 103° and 123°E (Fig. 9.1). The north border of this area lies along the Yan Mountains, the south border is the Five Mountain Range. The western part of this area reaches the Wei River Valley, and the eastern boundary is the sea (Fig. 9.1).

This landmass is the core area of Chinese civilization and one of the evolutionary centers of human beings and cultures in the world. Since the 1950s, rich archaeological data have been found in this area and studied by different scientific approaches. The outcomes of these studies facilitate a tentative reconstruction of the paleoclimates and cultural dynamics in eastern Central China during the middle Holocene.

In this paper, 'mid-Holocene' refers to the period from approximately 7000 to 4500 cal yr BP, when significant natural and cultural changes occurred. The climate during this period was very warm with high precipitation, and sea level rose to its Holocene maximum of about 2–4 m higher than present at around 5000 cal yr BP (Shi, 1993; Wang et al., 1995; Yi et al., 2003). The climate after approximately 4500 cal yr BP became colder but with higher precipitation (Liu, 2000; Wu and Liu, 2004).

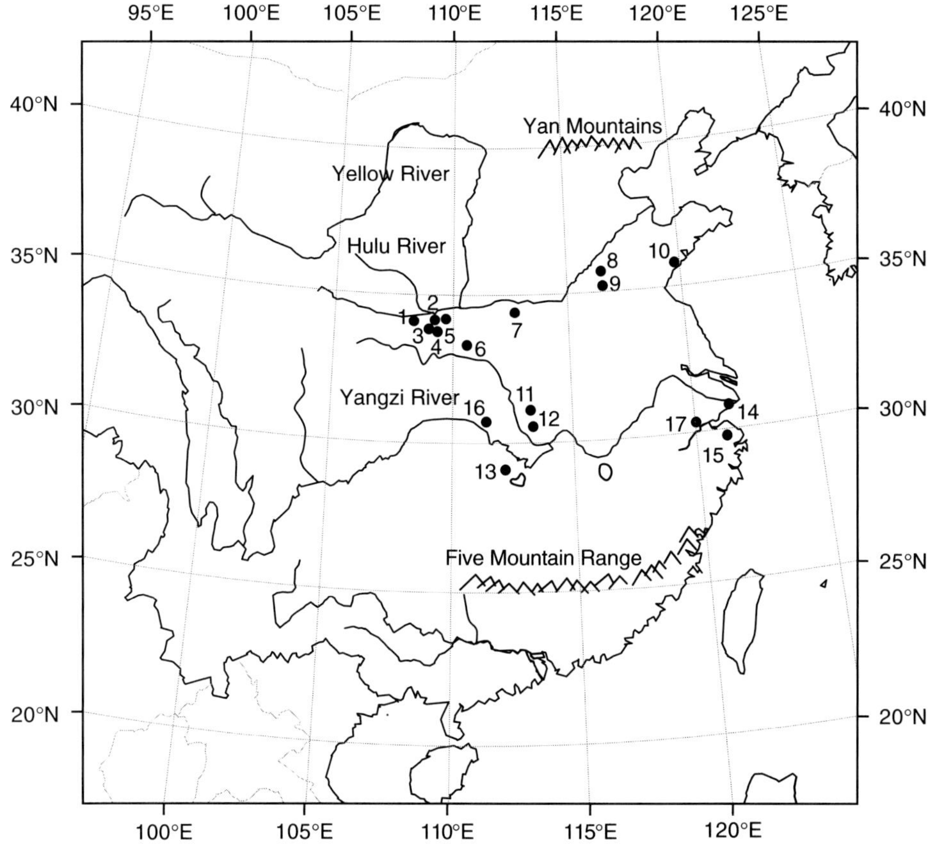

Figure 9.1. Sites mentioned in the text. 1. Anban 2. Jiangzhai 3. Banpo 4. Xiehu 5. Huaxian 6. Xiawanggang 7. Dahecun/Xishan 8. Dawenkou 9. Xikangliu 10. Sanlihe 11. Diaolongbei 12. Qujialing 13. Chengtoushan 14. Songze 15. Hemudu 16. Daxi 17. Liangzhu.

Archaeologically, the period from 7000 to 4500 cal yr BP was transitional in terms of cultural development in both the Yellow and the Yangzi Valleys. The mid-Holocene witnessed the emergence of new, more complex, and less egalitarian cultures in both river valleys. This paper synthesizes the paleoclimates, paleoenvironments, and Neolithic cultures in eastern Central China during the period from approximately 7000 to 4500 cal yr BP.

2. Paleoclimate of the mid-Holocene in eastern Central China

Today this area is a temperate to sub-tropical zone, with steppe and deciduous forest in the Yellow Valley, and broad-leafed trees in the Yangzi Valley. The average yearly temperature ranges from 6 to 12°C in the Yellow Valley, and from 14 to 16°C in the Yangzi Valley (The Editing Committee of the Physiography of

China, 1984). The precipitation in the Yellow Valley averages between 500 and 900 mm; in the Yangzi Valley it ranges from 1000 to 1600 mm (The Editing Committee of the Physiography of China, 1984).

Current fauna in the Yellow Valley is adapted to a mild climate, and in the Yangzi Valley, to a warm and moist environment. Generally speaking, the East Asian monsoon has a primary influence on the current climate in this area (Zhou et al., 1996; Wu and Liu, 2004). During winter, the East Asian continent is cooler than the Pacific Ocean. There is a high surface pressure over the continent and a low surface pressure over the ocean in the winter; as a result, the winter monsoon blows from the continent to the ocean, bringing cold and dry weather as well as dusts (An et al., 1993). In summer, the continent is warmer and the ocean cooler; thus the pressure gradient is reversed, and there is a steady blow from the ocean to the continent, bringing warm and wet weather to the land (An et al., 1993, p. 45). However, the climates of different areas are also influenced by local factors such as latitude and elevation (He et al., 2004).

Since the 1960s there has been much effort to reconstruct quaternary paleoclimates in China. It is now proposed that paleoclimate in China was controlled by the movements of the East Asian paleomonsoon (e.g., An et al., 1993; Zhou et al., 1996; Wu and Liu, 2004), causing variations and changes of temperature, precipitation, sediment accumulation, and consequently of flora and fauna.

Based upon the geological data gathered by An and his colleagues from the Baxie Basin located in the Loess Plateau of the middle Yellow Valley (Fig. 9.1) (An et al., 1993), the summer monsoon seems to have been quite strong in the early Holocene, but weakened after 6000 cal yr BP. The stratigraphic profile from Baxie shows that paleosol accumulated during the period from approximately 10,600 to 6200 cal yr BP, but loess accumulated again after that time (An et al., 1993). This would indicate a cooler and drier paleoclimate after 6200 cal yr BP. Analysis of the MS (magnetic susceptibility) signal further supports this climatic pattern (An et al., 1993).

Similar climatic changes may also have occurred in the nearby Hulu River (Fig. 9.1). In this region, sedimentary and pollen analyses suggest that the Holocene Climate Optimum occurred between 7800 and 5500 cal yr BP, but that the climate became cooler and drier after 5500 cal yr BP (Mo et al., 1996).

The pollen profile from the Banpo Neolithic site in the middle Yellow Valley (Fig. 9.1) indicates that forest-steppe dominated the vegetation at approximately 6000–5600 cal yr BP, but shows a change from forest-steppe to mainly grassland with spare trees after 5600 cal yr BP (Ke and Sun, 1990). Drought-resistant grasses such as Artemisia, Chenopodiaceae, and Compositae dominated the latter period (Ke and Sun, 1990). This pollen profile suggests that paleoclimate after 5600 cal yr BP was drier and cooler, but human activities might have also played a role in creating this signal.

In the middle Yangzi Valley, analysis of the geological stratum of local Neolithic cultures and their location above the current water level of the Yangzi River suggests that there were several flooding phases between 8000 and 5500 cal yr BP and between 4700 and 3500 cal yr BP (Zhu et al., 1997). Theoretically, the flooding would have occurred during periods of higher precipitation, while the time of less frequent flooding would represent a period of lower precipitation. Thus, Zhu and his colleagues argue that paleoclimate in the middle Yangzi Valley between 8000 and 5500 cal yr BP was very moist, but that between 5500 and 4700 cal yr BP it was relatively dry (Zhu et al., 1997). Similar climatic patterns in the lower Yangzi Valley are associated with sea level changes (Wu, 1983).

Geological and paleontological data, as well as information collected from lake and oceanic cores, imply that the paleoclimate in China was warm and moister before approximately 5000 cal yr BP, but much colder after 4500 years ago (Wu and Liu, 2004). Generally speaking, pollen profiles found in eastern Central China indicate a significant increase of broad-leafed trees and thermophilic and hygrophilous plants such as ferns, Typha, and Polypodiaceae, between approximately 7000 and 5000 cal yr BP, especially in the Yangzi Valley (Yi et al., 2003). It has been estimated that the annual mean temperature between 7000 and 6000 years ago in Eastern China was roughly 1–3°C higher than the present (Shi et al., 1993). After approximately 5000 cal yr BP, conifers increased and evergreens decreased (Yi et al., 2003), suggesting climatic cooling.

Faunal data also supports the claim of a climatic optimum by 5000 cal yr BP. Substantial quantities of remains of bamboo rat (*Rhizomys sinensis*) and Chinese water deer (*Hydropotes inermis*) were found in the Banpo assemblage in the Yellow Valley dated to between 7000 and 6100 years ago (Li and Han, 1963) (Fig. 9.1; Table 9.1). Giant panda (*Ailuropoda melanoleuca*), Sumatran rhinoceros (*Rhinoceros sumatrensis*), muntjac (*Muntiacus* sp.), and sika deer (*Cervus nippon*) were recovered from the early stratum of the Xiawanggang assemblage, Henan province, dated between 6888 and 4500 cal yr BP (Fig. 9.1; Tables 9.1 and 9.2). Large quantities of Chinese water deer and sika deer occurred in the first phase of the Jiangzhai assemblage, dated to approximately 6900 to 6400 cal yr years ago (Fig. 9.1; Table 9.1) (Banpo Museum et al., 1988). Remains of David's deer (*Elaphurus davidianus*) and Chinese alligator (*Alligator sinensis*) appear frequently in the Dawenkou assemblages of the lower Yellow Valley, dated between 6700 and 5770 years ago (Fig. 9.1; Table 9.1) (The Institute of Archaeology CASS, 1984). These animals are adapted to a warm climate with abundant water and rich grassland. They are now extinct in the Yellow River Valley and are only found in the Yangzi Valley and South China. The presence of these animals portrays a favorable environment with abundant water and vegetation in the Yellow Valley by 5000 cal yr BP. The presence of bamboo rat and giant panda suggests that there must have been dense stands of bamboo in these areas. As bamboo is a sub-tropical to tropical plant and grows only in part of the Yangzi Valley and South China today, the northward expansion of bamboo in the mid-Holocene indicates higher temperature and precipitation. In the Yangzi Valley, more animal species of sub-tropical to tropical zones have been discovered. For example, Hemudu has more than 40 species, including Asian elephant, Sumatran rhinoceros (*R. sumatrensis*), Chinese alligator, and several species of deer (Fig. 9.1) (Zhejiang Provincial Institute of Antiquity and Archaeology, 2003).

Table 9.1. Radiocarbon dates of some Neolithic cultures in eastern Central China.

Site	Location	Sample no.	Stratum	^{14}C date (half-life 5730)	^{14}C date[a] calibrated (cal yr BP)	References
Banpo	Middle Yellow Valley	ZK38	Phase I	6065 ± 110	7143–6780	Xia (1977)
		ZK121		5905 ± 105	6859–6636	
		ZK122		5840 ± 105	6780–6497	
		ZK127		5585 ± 105	6470–6289	
		ZK148		5490 ± 140	6412–6115	
Jiangzhai	Middle Yellow Valley	BK77041	Phase I	5970 ± 110	6901–6676	Banpo museum et al. (1988)
		ZK265		5835 ± 70	6857–6419	
		ZK264		5745 ± 140	6727–6406	
		ZK454	Phase II	5030 ± 85	5902–5657	
		ZK157-0	Phase III	4890 ± 150	5848–5468	
Xiawanggang	Middle Yellow Valley	GY ?		5875 ± 175	6888–6479	Tong (1998)
		?		4270 ± 170	5035–4564	
Xiehu	Middle Yellow Valley	ZK916-1		4995 ± 100	5894–5616	
		ZK-2176		4335 ± 185	5285–4648	IA, CASS (1989)
		ZK-2209		4970 ± 90	5885–5603	
Dahecun	Middle Yellow Valley	ZK520		5740 ± 130	6721–6401	IA, CASS (1992)
		BK76003		4800 ± 90	5613–5335	
		ZK185		5025 ± 100	5906–5650	
		BK76001		4550 ± 100	5441–4997	
		BK76004		4500 ± 140	5320–4871	
Dawenkou	Lower Yellow Valley	BK79016	Layer 6	5810 ± 90	6734–6490	Tong (1998)
		BK79012		5710 ± 130	6670–6319	
		BK79014	Layer 5	5450 ± 100	6311–6115	
		BK79019		5480 ± 100	6401–6187	
		ZK469	Layer 4A	5505 ± 105	6408–6198	
		BK79071		5590 ± 90	6458–6295	
		BK79015		5390 ± 80	6287–6040	

Continued

Table 9.1. continued

Site	Location	Sample no.	Stratum	^{14}C date (half-life 5730)	^{14}C date[a] calibrated (cal yr BP)	References
		BK70918	Layer 4B	5520±100	6410–6208	
		BK79010		5410±90	6296–6055	
		BK79017		5350±90	6278–5990	
		BK79013		5180±90	6024–5773	
Daxi	Middle Yangzi Valley	ZK892		5940±260	7158–6448	IA, CASS (1984)
		ZK832		5330±145	6287–5928	
Diaolongbei	Middle Yangzi Valley	ZK-2577		5190±100	6167–5775	IA, CASS (1991)
		ZK2578		5135±95	5988–5749	
		ZK2579		5265±110	6195–5915	
		ZK2580		5280±90	6181–5929	
		ZK2581		4880±85	5661–5492	
		ZK2582		5020±95	5907–5650	
Qujialing	Middle Yangzi Valley	ZK2397		4975±140	6216–5378	
		ZK0124		4145±100	4918–4466	
		ZK0125		4195±160	5018–4490	
Hemudu	Lower Yangzi Valley	Zk0263		6060±120	6859–6559	IA, CASS (1991)
		Zk0263(2)		6085±100	6859–6646	
		Zk0590		6200±85	6986–6753	
		BK78105		5560±80	6580–6569	
		BK78110		5310±90	6331–5971	
Liangzhu	Lower Yangzi Valley	ZK0049		4700±100	5696–5101	IA, CASS (1991)
		ZK0097		4695±90	5658–5104	
		ZK0044		4335±85	5338–4697	
		ZK0047		4245±85	5085–4577	

[a] Data calibrated by Stuiver et al. (1998).

Table 9.2. The Xiawanggang fauna in the middle Yellow Valley.

Species	Quantity of bones from each phase					M.N.I.					Habitat
	I	II	III	IV	V	I	II	III	IV	V	
Reptilia (reptile)											
Amyda sp.	+					1					Fresh water
Testudinidae	+	+				1	1				Terrestrial, humid area
Pisces (fishes)											
Cyprinus sp.			+					1			Fresh water
Aves (birds)											
Pavo sp.	+					1					Subtropical area
Primate											
Macaca mulatta	+					1					(Sub)tropical hills
Rodentia											
Hystrix sp.		+						1			Various: forest, steppe
Carnivora											
Canis familiaris	+	+		+	+	1	1		1	1	
Nyctereutes procyonoides			+					1			Forest, dense bush, close to water
Canidae int.		1	1								
Ursus thibetanus	+					1					Moist deciduous forest
Ailuropoda melanoleuca		1					1				Mild climate; bamboo
Meles meles		+					1				Forest, dense vege
Actonyx collaris		+		+	+		1		1	1	Forest
Panthera tigris	+					1					Various
Felis bengalensis		+					1				Forest
Lutra sp.		+					1				Close to fresh water

Continued

Table 9.2. continued

Species	Quantity of bones from each phase					M.N.I.					Habitat
	I	II	III	IV	V	I	II	III	IV	V	
Proboscidea											
Elaphas maximus	1					1					(Sub)tropical ecozone
Perissodactyla											
Rhinoceros sumatrensis	6	1				1	1				(Sub)tropical area
Artiodactyla											
Sus scrofa domesticus	+	+	+	+	+	1	1	1	1	1	
Sus scrofa	+	+				1	1				Forest and open land
Moschus moschiferus	+					1					
Muntiacus sp.	2	1				1	1				
Cervus nippon	+	+	+	+	+	1	1	1	1	1	Forest, grassland
Capreolus capreolus				+	+				1	1	Forest and grassland in cold area
Bubalus sp.		+					1				(Sub)tropical areas

Chronology: phase I is dated to around 6800 cal yr BP; phase II around 6500 cal yr BP; phase III approximately 5500–5000 cal yr BP; phase IV about 5000–4500 cal yr BP.
Source: IA, Henan Province and the Yangzi Archaeology Team (1989).

The paleoclimate became cold and dry after 4500 cal yr BP in eastern Central China, causing local floral and faunal changes, although these changes varied from region to region (He et al., 2004). Generally speaking, most of the animal species adapted to sub-tropical or tropical environments seem to have withdrawn from the Yellow Valley after 5000 cal yr BP. In Jiangzhai (Fig. 9.1), the quantities of Chinese water deer and sika deer reduced significantly after 5000 cal yr BP (Banpo Museum et al., 1988). It is not yet clear, however, whether this reduction was caused by the climatic change, or by over-hunting of human beings, or by both processes.

During the Climatic Optimum by 5000 cal yr BP, peat and mud were widespread throughout eastern Central China (Gao and Li, 1986). Sea level also rose. It has been inferred that the maximum Holocene transgression occurred in Eastern China at approximately 6000 years ago (Uehara et al., 2002). Naturally deposited marine shells dated between approximately 6000 and 5000 cal yr BP have been widely found in the lower Yellow and the lower Yangzi Valley, some of the shells located in areas now more than 30 km away from the coast (Yan and Huang, 1987), indicating marine transgression during this period. It is argued that sea level in Eastern China was about 1–3 m higher than the present from 6000 to 5000 cal yr BP, but retreated to close to its current position after 5000 cal yr BP (Shi et al. 1993).

Between 7000 and 5000 cal yr BP, the mean yearly temperature was about 2–4°C higher than present, and the precipitation was about 200–400 mm higher than present (An et al., 1990). This paleoclimatic change seems consistent with the global pattern. The termination of the Holocene Climatic Optimum is considered to be after 5000 cal yr BP, followed by a progressive cooling period (Roberts, 1998, p. 162). Such a global climatic pattern would have caused the weakening of summer monsoon and the strengthening of the winter monsoon on the East Asian continent, thus a cool and dry paleoclimate (Wu and Liu, 2004).

3. Cultural dynamics in eastern Central China

According to archaeological data, the period from 7000 to 5000 cal yr BP witnessed the prosperity of many Neolithic cultures in China. Numerous prehistoric groups were scattered over eastern Central China during this time, most of them are agricultural societies developing from their respective predecessors in their own regions.

Archaeological and archaeobotanic data indicate that agricultural societies were established in both the Yellow and the Yangzi Valleys by no later than 8400 cal yr BP (Lu, 1999), and by 5000 cal yr BP the Neolithic assemblages in eastern Central China numbered in the thousands (The Institute of Archaeology CASS, 1984). The period between 5500 and 4700 cal yr BP saw the diminishing of some archaeological cultures and the emergence of others. In the middle Yellow Valley, the Yangshao culture reached its final phase at around 5000 cal yr BP and was succeeded by the Longshan culture. In the lower Yellow Valley, the Longshan culture replaced the Dawenkou culture at approximately 4600 cal yr BP. In the middle Yangzi Valley, the Daxi culture was succeeded by the Qujialing culture at approximately 5200 cal yr BP, and the Hemudu and Majiabang culture in the lower Yangzi Valley gave way to the Liangzhu culture at more or less the same time (The Institute of Archaeology CASS, 1984; Cultural Relics Publishing House, 1999). What were the natural and cultural dynamics in eastern Central China during this period of cultural transition? It is impossible to discuss this issue comprehensively in a paper of this length, so I will briefly investigate it from selected aspects in terms of distribution and continuity changes of human settlements, subsistence strategies, burials, the occurrence of walled towns and prestige goods, and cultural interaction and human migration.

3.1. Human settlements in eastern Central China

3.1.1. Distribution of human settlements

Generally speaking, prehistoric settlements dated between 7000 and 4500 cal yr BP in the Yellow Valley are located on either river/stream terraces or open plains. Shi (1993) has argued that the Neolithic cultures dated to the period of 5500–5000 cal yr BP in the middle Yellow Valley were usually located closer to the river than their predecessors. He further suggested that this change of location might have been caused by the sudden climatic changes from warm and wet to cold and dry, the latter forcing people to move closer to water resource. Though theoretically correct and possible, this hypothesis is not always consistent with archaeological data. Surveys conducted in the Wei River and in Henan Province do not illustrate a significant change of locations of archaeological assemblages in the middle Yellow Valley during the mid-Holocene (IA, Henan Province, 1991).

On the other hand, the lower Yellow River Valley was a land of swamps, lakes, mud, delta deposits, and lagoons during the mid-Holocene (Wang et al., 1995). Marine transgression occurred in this area and the eastern edge of the modern Yellow River delta was inundated at approximately 6000–5400 cal yr years ago (Wang et al., 1995). To date, hundreds of local human settlements dated to the mid-Holocene have been located on inland hilly flanks, small plains, or river terraces (Zhang, 2001). Many of these human settlements are very large, measuring up to hundreds of thousands of square meters (Zhang, 2001), indicating dense populations successfully adapted to a warm and wet climate.

In the Yangzi Valley, prehistoric settlements have also been found on river terraces, open plains, and hilly flanks in the mid-Holocene (Zhu et al., 1997). The middle and lower reaches of the Yangzi Valley today are plains with lakes and many streams, as well as hills and eroded plateaus (Wang et al., 1995). Neolithic settlements dated to as early as 8500 cal yr BP have been found on plains or river terraces in the middle Yangzi Valley. Archaeological sites dated to the mid-Holocene clustered along the tributaries of the Yangzi River (Jin et al., 1997), the majority located on river terraces or small hills with streams nearby (Tan, 1991; Wang, 2002). Wang (2002) has argued that there were floods during the period from

5800 to 5500 cal yr BP and from 5000 to 4800 cal yr BP in the middle Yangzi Valley, and that some early Neolithic settlements were completely destroyed by flood.

The climatic changes of the mid-Holocene and related marine transgression also significantly affected the lower Yangzi Valley. Based on sediment, pollen, and fauna analyses, it has been argued that there were frequent marine transgressions and regressions between 7000 and 6500 years ago; archaeological sites dated to this period are located on hills and plateaus in areas at least some 30 km from the present coast (Wu, 2000). However, as the sea level gradually approached its current position after 6000 years ago (Hori et al., 2001), human beings also moved towards the coast and populations seem to have been very dense during the period from 5300 to 4000 years ago in the Yangzi River delta (Wu, 2000).

3.1.2. Continuity of occupation

The continuity of human occupation varies in different areas. Published data indicate different occupation sequences within the mid-Holocene Neolithic cultures in eastern Central China.

Continuous occupation

Continuous occupation means that local human occupation seems to have been stable in the mid-Holocene. This type of occupation sequence can be further divided into four sub-categories:

1. *Stable cultural development*, documented by similar sizes of residential areas and similar richness of archaeological remains through different temporal phases. The Xiehu assemblage found in Shaanxi province (The 6th Shaanxi Archaeological Team, 1989), and the Dahecun assemblage found in Henan province (The Institute of Archaeology CASS, 1984) (Fig. 9.1) illustrate such constant continuity.
2. *Unstable cultural development*, characterized by a reduced residential area and a quantitative decline of cultural remains at around 6000–5500 cal yr BP, and an increase in both measures after 5000 cal yr BP. In the Jiangzhai assemblage of the Wei River Valley, the early Neolithic village dated to between 6700 and 6000 cal yr BP measured over 20,000 m^2 (Table 9.1) (Banpo Museum et al., 1988) and the archaeological remains were very abundant. Yet after 6000 cal yr BP, both the size of occupation and the richness of archaeological remains were significantly reduced, even after making allowance for the effect of post-depositional destruction. After 5000 cal yr BP, the site was still occupied for another several hundred years and the quantity of cultural remains increased (Banpo Museum et al., 1988). However, it is not clear whether the apparent cultural decline after 6000 cal yr BP was caused by climatic changes or human activities, or both.
3. *Cultural collapse* in some regions. A comprehensive study on the paleogeography and prehistoric culture in the Hulu River, middle Yellow Valley, reveals that the local Neolithic culture dated to 5000–4800 cal yr BP shrank, as measured by smaller size of occupations and very rare occurrence of painted pottery (Mo et al., 1996).

4. *Progressive cultural development*, manifested by enlarged residential areas and richer cultural remains after 5300 cal yr BP and exemplified by the Diaolongbei assemblage found in the middle Yangzi Valley (Fig. 9.1) (Diaolongbei Archaeology Team, 1991). This sequence suggests an ongoing and successful cultural development process.

Discontinuous occupations

Discontinuous occupations have been found in some regions. An archaeological survey conducted in the Wei River basin (about 22,033 km^2) identified a total of 121 Neolithic sites; among them, only 5 sites, or less than 6% of the total, show cultural continuity from the late Yangshao to the early Longshan period from approximately 5500 to 4700 cal yr BP onwards (Table 9.3; Fig. 9.2) (Wei River Archaeology Team, 1991, 1992).

The Wei River Valley belongs to the current administrative region of Shaanxi province, where other provincial archaeological surveys also revealed interesting data. More than 4000 Neolithic sites have been found in Shaanxi. Remains of the Yangshao culture, a farming culture dated from approximately 7000 to 5000 cal yr BP, have been identified in approximately 2040 sites; remains of the Longshan culture, which is also a farming culture dated from approximately 5000 to 4000 cal yr BP, have been found in over 2200 sites (IA, Shaanxi Province, 1997). During the Yangshao cultural period of 7000 to 5000 years ago, 60% of the archaeological assemblages were found in the middle Shaanxi province, 34% in the north and 6% in the south. But the succeeding Longshan cultural period between approximately 5000 and 4000 years ago had only 34% of sites in the middle region, 64% in the north, and only 2% in the southern area of the Shaanxi province. The surveyors also stated that only 22% of the Longshan assemblages have a continuous occupation from the Yangshao to the Longshan phases in Shaanxi province (IA, Shaanxi Province, 1997).

These data raise two points. First, the focus of human occupation seems to have shifted from the middle to the north in Shaanxi after approximately 5000 years ago. Archaeological data suggest that human occupations in the middle Shaanxi region from 7000 to 5000 cal yr BP were much denser than those after 5000 cal yr BP. It seems that after 5000 cal yr BP there was a sudden decline of occupations in the middle area, and a significant increase of human occupations in the northern area. Is this phenomenon an indicator of prehistoric human migration? If so, what were the causes of such migration? If not, how can this archaeological data be interpreted? Answering these questions will require further work.

Second, a substantial proportion of the human occupations in Shaanxi seems discontinuous. The above survey data suggest that 78% of the Yangshao occupations in Shaanxi province were not continuously occupied into the Longshan cultural period. Why did this occur? What caused these prehistoric people to abandon their homes? Was it climatic changes? Was it over-exploitation of natural resources and land depletion, or other reasons? Some have argued that the cold spell after 4500 cal yr BP in China caused the collapse of many Neolithic cultures (Wu and

Table 9.3. Cultural sequences and size of mid-Holocene archaeological sites in the Wei River Valley.

Types of cultural remains found in each site	Size (m^2) of sites	Thickness of deposit (m)	Quantity of sites of this culture	Percentage of total sites of all periods in this valley	Note
Early Yangshao	Most between 20 and 30 km^2; maximum 300 km^2	1–3	10	10.75	
Middle Yangshao	Most between 30 and 50 km^2; maximum 1200 km^2	Most 1–2; maximum 5	31	33.33	
Late Yangshao	Most between 20 and 100 km^2; maximum 150 km^2	Most 1–2; maximum 4	15	16.13	
Middle Yangshao and Longshan culture	Range from 75 to 600 km^2	1–4	12	12.90	Two sites also contain early Zhou remains
Late Yangshao and Longshan	Range from 14.4 to 200 km^2	1–2.5	5	5.38	Three sites also contain early Zhou remains
Early Yangshao and Bronze Age	Range from 4.8 to 61.6 km^2	1–2.5	7	7.53	
Middle Yangshao and Bronze Age	Range from 12 to 40 km^2	1–3	5	5.38	
Late Yangshao and Bronze Age	Range from 8 to 600 km^2	Most 1–2; maximum 6	8	8.60	
Total			93	100.00	

Sources: The 6th Shaanxi Archaeological Team (1989) and Wei River Archaeology Team (1991, 1992).

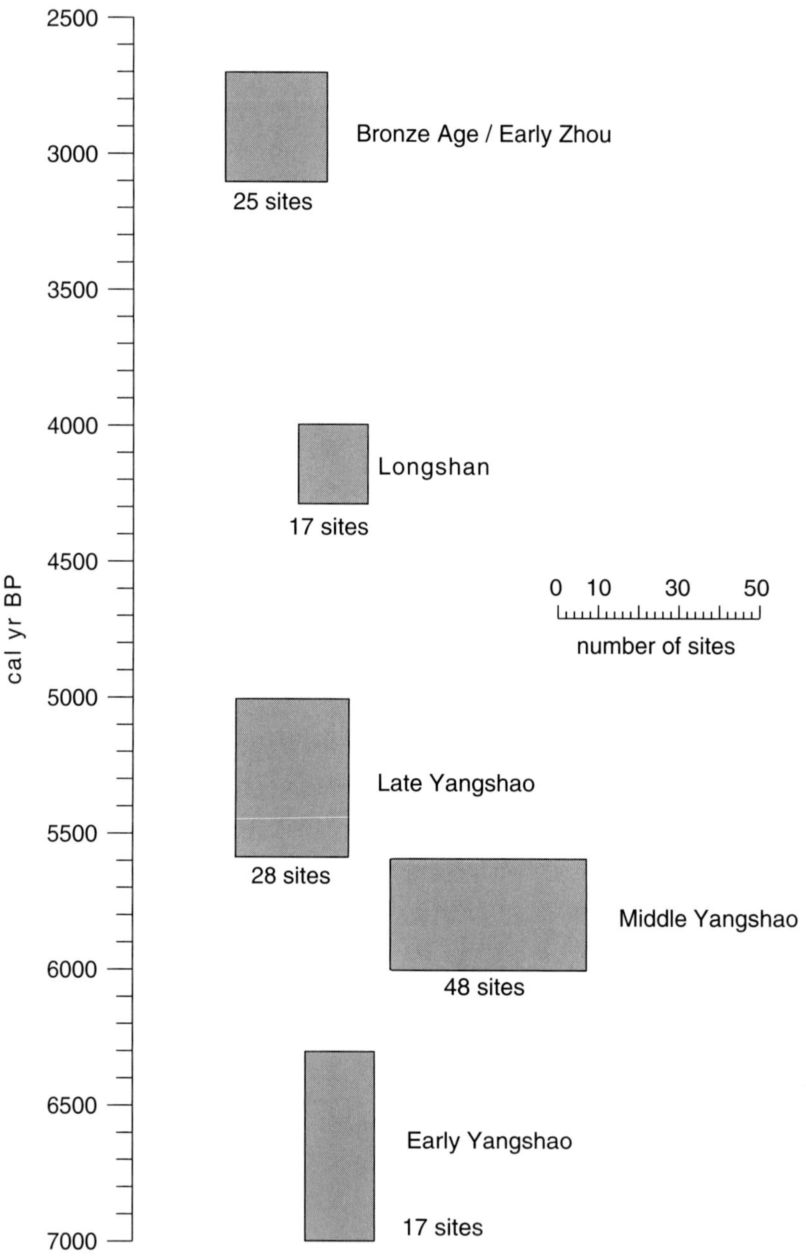

Figure 9.2. Density and length of human occupations of different periods found in the Wei River Valley.

Liu, 2004), while others believe that the climatic changes might have triggered human competition and the development of more complex societies (Liu, 2000). However, no conclusion should be reached before carefully investigating the natural and cultural elements of these archaeological assemblages.

Recently, He et al. (2004) have argued that there were asynchronous variations of climatic changes in China, and that the Holocene Optimum commenced earlier and lasted longer in the higher latitudes than in the lower latitudes. If the current archaeological data reflect the prehistoric human movements in the present Shaanxi province, which is in higher latitudes, and if the local climate was still mild after 4500 cal yr BP according to He and his colleagues, then the movements of the farming societies from the central and south to the north area might have resulted from natural resource depletion due to continuous farming activities, and the need to search for new land and resources, other than from dramatic climatic changes.

Animal remains found in Jiangzhai in Shaanxi may provide some clues. Very rich remains of wild animals occur in the first cultural phase at Jiangzhai (Table 9.4). This phase was also the period when the Jiangzhai society reached its peak, demonstrated by expanded village size and rich archaeological remains. However, animal remains diminish significantly after this phase and are almost non-existent in phase III (Table 9.4) (Banpo Museum et al., 1988). Meanwhile, the size of the village and the richness of material culture during phases II and III also decreased significantly (Banpo Museum et al., 1988).

As the bamboo rat (*Rhizomys sinensis*) and several species of deer of the Jiangzhai fauna were adapted to a warm and moist environment (Table 9.4), their presence in phases I and II and absence in phase III may suggest a warm and moist climate from approximately 6900 to 5600 cal yr BP, and a sudden climatic change after 5600 cal yr BP (phase III) (Table 9.1). On the other hand, judging from the animal remains found in phase I, hunting must have been a very frequent activity then. Did the frequent hunting over-exploit faunal resources and consequently reduce one of the major components of the Jiangzhai diet, thus forcing a part of this society to move to other areas in search of new natural resources? This is also possible.

Another important factor for human movement is the depletion of land. Chen (1991) argues that slash and burn was the major cultivation technique used by the Neolithic farmers in the Yellow Valley. Ethnological data from southwest China indicate that this technique causes rapid deforestation and land depletion in about 10 years, so that farmers have to move to other areas as the land around their settlements loses fertility (Chen, 1991). Might a similar situation have occurred in prehistoric Jiangzhai and other Neolithic sites in the Yellow River Valley, and was it the major force for migration? This hypothesis needs further study.

Human occupation also seems discontinuous in other areas. An archaeological survey in Henan Province in the middle Yellow Valley indicates a similar phenomenon. For example, out of 50 Neolithic sites found around the present Zhenzhou City and Xinyang County, only 18 sites, or about 27.7%, were occupied continuously from the Yangshao to the Longshan phases (IA, Henan Province, 1991).

River floods and marine transgression might be responsible for a similar discontinuity reported in the middle and lower Yangzi Valley (Wu, 1983; Yu et al., 1999). Details of the majority of Neolithic sites in the Yangzi Valley are not available, but Pei (2000) reports that some Neolithic sites dated to approximately 7000 years ago in the middle Yangzi Valley are found beneath mud deposits or in modern lakes,

Table 9.4. The Jiangzhai fauna in the middle Yellow Valley.

Species	Quantity of bones from each phase					M.N.I.				Habitat
	I	II	IV	V	Total	I	II	IV	V	
Pisces(fishes)										
Cyprinus sp.	3				3	2				Fresh water
Ctenopharyngodon sp.	4				4	2				
Pisces indet.	62				62	n.a.				
Aves(birds)										
Pelecanus sp.	1				1	1				Close to fresh water
Aquila sp.	1				1	1				
Grus sp.	1				1	1				Close to fresh water
Gallus sp.	3				3	1				
Incectivora										
Erinaceus europaeus	1				1	1				Forest, grassland etc.
Scaptochirus inoschatus	7				7	1				Dry steppe
Primate										
Macaca mulatta[a]	2				2	1				(Sub)tropical hills
Rodentia										
Myospalax fontanieri	14				14	4				Steppe, hills etc.
Rhizomys sinenesis[a]	5	5	4		14	2	2	2		Bamboo
Lepus sp.	1	1	5		7	1	1	2		Various
Carnivora										
Canis familiaris	7		5	19	31	2		2	1	Various
Cuon alpihus[a]	1				1	1				Forest, dense shrubs
Nyctereutes procyonoides	13	6	10		29	5	1	4		Forest, dense vege. close to water

Ursus thibetanus	3				3	2				Moist deciduous forest.
Meles meles	12		1	1	14	4		1	1	Forest, dense vege
Actonyx collaris	2		1		3	2		1		Warm and moist areas
Panthera tigris	1				1	1				
Felis sp.	1		1		2	1		1		
Artiodactyla										
Sus domesticus	521	45	88	20	674	85	8	12	4	
Moschus moschiferus	6				6	3				Hilly areas
Hydropotes inermis[a]	168	19	78	8	273	21	4	16	1	Tall reeds/rush along rivers
Cervus nippon	651	26	311	101	1089	48	7	19	11	Forest and grassland
Cervus sp.	311	78	58	40	487	19	7	5	6	
Produrcas gutturosa	7		4	4	15	2		1	1	
Bos sp.	72	15		2	86	3	2		1	
Artiodactyla indet.	168	21	20	50	258	n.a.	n.a.	n.a.	n.a.	
Mammalia indet.	190	129	2	80	407	n.a.	n.a.	n.a.	n.a.	
Vertebrate indet.	39				39	n.a.				

Chronology: phase I dated to around 6000 cal yr BP; phase II at 5800 cal yr BP; phase III estimated to be 5500 cal yr BP; phase IV estimated to be 5000–4800 cal yr BP; phase V to be 4300–4000 cal yr BP.

Source: Banpo Museum et al. 1989. (According to the excavation report, there were discontinuity between phases II and III, III and IV, and IV and V.)

[a] Species extinct in the Yellow Valley today.

indicating that floods or rising water levels might have destroyed these human settlements. In the lower Yangzi Valley, several flooding periods, dated to approximately 6500, 6000, 5500, 5000, and 4000 cal yr years ago, might have destroyed many local Neolithic settlements (Yu et al., 2000).

3.2. Subsistence strategies

The discovery of stone and organic tools and the remains of foxtail and broomcorn millets, rice and other cultivars shows that agriculture was the major subsistence strategy in eastern Central China during the mid-Holocene. Domesticated foxtail and broomcorn millets have been widely found in mid-Holocene Neolithic assemblages in the Yellow Valley, and rice has been found in many archaeological sites in the Yangzi Valley (Lu, 2004), showing that agriculture remained a major activity from approximately 7000 to 4500 years ago in these areas. Rice remains have also been found in Anban, Huaxian and other archaeological sites in the Yellow Valley during the mid-Holocene (Fig. 9.3), indicating the northward expansion of rice farming from the Yangzi Valley. Remains of domesticated pig, dog, and chicken have also been widely found. In the lower Yellow Valley the Dawenkou culture (Fig. 9.1) often used pig mandibles as grave goods (The Institute of Archaeology CASS, 1984), probably as an indicator of individual wealth.

Nevertheless, hunting, gathering, and fishing remained important subsistence strategies. Many mid-Holocene archaeological sites in eastern Central China contain arrowheads and harpoons, as well as remains of nuts, seeds of wild grasses and fruits, terrestrial animals, birds, fish, and fresh water and marine shells. In the lower Yellow Valley, alligator bones often occurred in archaeological deposits before 5000 cal yr BP (Gao and Hu, 1991). Yuan and An (1998) have argued that the mid-Holocene archaeological assemblages have a higher proportion of domesticated animals within total animal remains than do the early Holocene assemblages. While this is the case in many Neolithic assemblages found in eastern Central China, the faunal data from the Jiangzhai and Xiawanggang assemblages in the Yellow Valley show the contrary (Fig. 9.1; Tables 9.2 and 9.4). The Hemudu assemblage of the lower Yangzi Valley had over 40 species of wild animals, associated only with domesticated dog and pig (The Institute of Archaeology CASS, 1984). Faunal data found in a succeeding Songze site of the lower Yangzi Valley dated to 6000–5000 cal yr years ago illustrate that deer still dominated the animal remains (Shanghai Management Committee of Antiquity, 1987), indicating the importance of hunting; but the quantity and diversity of other wild animals reduced significantly compared to that found in Hemudu.

Undoubtedly the ratio of wild to domesticated animals is an indicator of the subsistence strategies of different groups in different areas. However, the subsistence strategy of a particular group would have depended on the natural (faunal and floral) resources available in a given area. An abundance of deer and other wild animals in the Jiangzhai vicinity by 5500 cal yr BP would have encouraged hunting, and resulting in the greater presence of bones of wild animals found in the archaeological deposits.

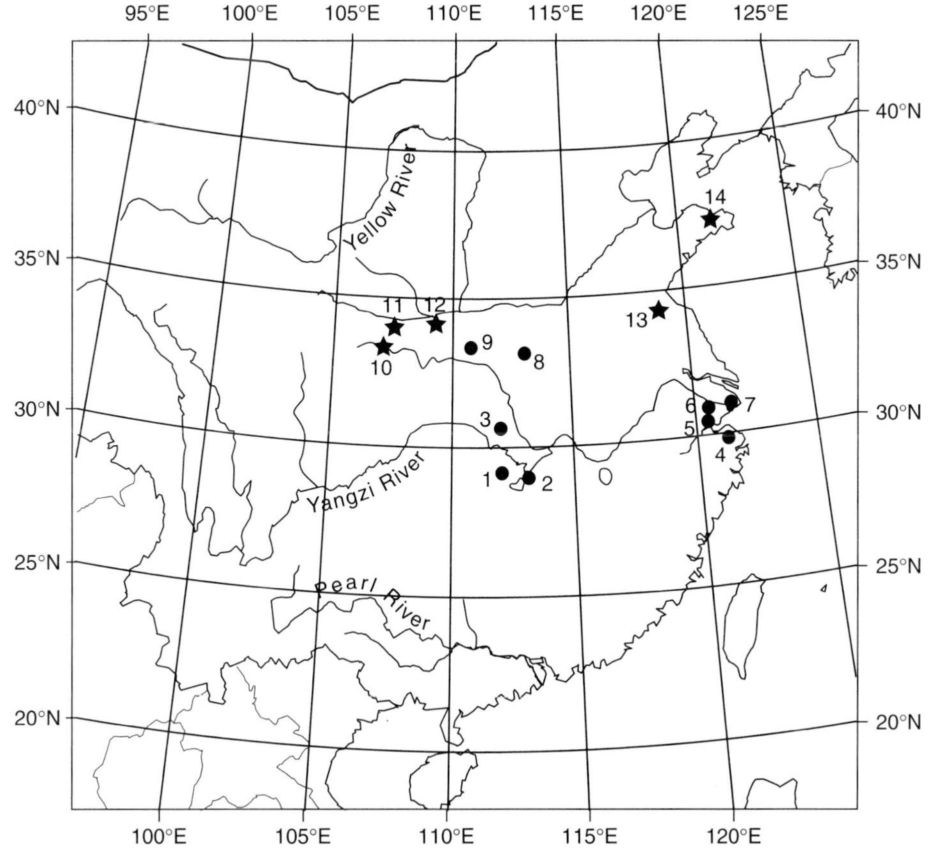

Figure 9.3. The Expansion of rice farming in the mid-Holocene in eastern Central China.
• Loci of cultivated rice
★ Loci of cultivated rice and foxtail millet
Archaeological sites with rice and/or millet remains: (1) Pengtoushan (9000–8000 cal yr BP), (2) Fenshanbao (approximately 7500 cal yr BP), (3) Daxi Culture (more than one loci) (6400–5300 cal yr BP), (4) Hemudu (7000–6000 cal yr BP), (5) Luojiajiao (7100–6899 cal yr BP), (6) Caoxieshan (6400–6115 cal yr BP), (7) Songze (5850–5250 cal yr BP), (8) Jiahu (8400–7600 cal yr BP), (9) Xiawanggang (7210–6729 cal yr BP), (10) Lijiacun/Hejiawan (7179–6796 cal yr BP), (11) Anban (4500–4000 cal yr BP), (12) Huaxian (5000–4000 cal yr BP), (13) Lianyungang (approximately 7000 cal yr BP), and (14) Qixia (4873–4502 cal yr BP).
Sources: Institute of Archaeology, Chinese Academy of Social Science (1991) and Chen (1994).

In the Jiangzhai assemblage, Qi (1988) reports that the average age of death for deer was very low. Qi examined 30 individuals of sika deer and discovered that 40% of these individuals were killed between 2.5 and 3 years old, 20% were killed between 2 and 2.5 years old, and 17% were killed before reaching 2 years old. As the average life expectancy of sika deer is between 15 to 20 years, Qi (1988) conjectured that the lower age of death reflects a practice of deer herding. However, it is also possible that young deer were more vulnerable to human hunting than mature individuals. Meanwhile, while the high frequency of sika deer and other deer found in Jiangzhai suggests the substantiality and importance of these animal resources, the reduced quantity of animal remains after the early period may indicate over-hunting. Faunal details for many other archaeological sites are not yet available, so it is not clear whether a similar decline of faunal resources also occurred in other areas.

3.3. *Burials*

Archaeological data to date suggest that single and primary burials were the most common during the mid-Holocene in the Yellow River Valley, although multiple and secondary burials also occurred. Generally speaking, during the period from approximately 7000 to 5000 cal yr BP, dead adults in the Yellow Valley were buried in earth pits, while children were often placed in large urns (The Institute of Archaeology CASS, 1984). Grave goods were commonly found, the majority being pottery and stone tools. However, craft and prestige goods such as jade ornaments have been reported from the Dawenkou culture in the lower Yellow Valley dated to 6600–5770 cal yr BP (Table 9.1), clearly indicating the occurrence of social stratification (The Institute of Archaeology CASS, 1984; Cultural Relics Publishing House, 1999).

After 4800 cal yr BP, burials of the Longshan culture in both the middle and the lower Yellow River Valley have greater differences in terms of grave structure and the quantity and quality of grave goods, demonstrating increased social complexity. While many burials dated to this period contained only a few or even no grave goods, a smaller proportion of burials were richly furnished, with wood coffins and prestige goods, including the delicately-made chalky white ceramic and egg-shell cups (see the following section) (Cultural Relics Publishing House, 1999).

Similar phenomena have also been discovered in the Yangzi Valley. Burials of the Daxi culture dated between 7000 and 6000 years ago in the middle Yangzi Valley (Table 9.1) already demonstrate quantitative and qualitative differences in terms of grave goods, suggesting decreased social egalitarianism (Cultural Relics Publishing House, 1999). Jade and other prestige goods were found in burials dated from 6000 to 4500 cal yr BP in both the middle and the lower Yangzi Valley (Cultural Relics Publishing House, 1999). Some of the nearly one hundred Liangzhu burials found in the lower Yangzi Valley were richly furnished. Wang (1998) reports that 70 jade items were found in a Liangzhu tomb in the lower Yangzi Valley, indicating that the occupant was a special individual. Several human-built mounds found recently in the lower Yangzi River Valley have burials containing numerous jade items and associated with remains of large houses and ritual activities (Cultural Relics Publishing House, 1999).

3.4. *The occurrence of prestige goods and walled towns*

Prestige goods are usually valuable objects symbolizing the social status of their owners. When deciding what was valuable in prehistoric contexts, Renfrew and Bahn (2000) suggested that valuables were probably objects which take a long time to make or which use exotic or less accessible materials. In China's context, the mid-Holocene witnessed the occurrence of many valuable objects, including the egg-shell pottery produced by the Longshan people in the lower Yellow Valley. The walls of these vessels are as thin as 0.5 to 1 mm, hence the nickname of "egg-shell pottery" (Feng et al., 1987). Undoubtedly, such delicate items could only be made by specialized craftsmen. Similar egg-shell pottery with painting has also been found in the Qujialing culture in the middle Yangzi Valley dated to about 4550–4195 cal yr BP (Feng et al., 1987).

Jade objects comprise another important category of valuable items. As mentioned above, many jade items have been found in the lower Yellow and the Yangzi Valley from approximately 6500 cal yr years onward. While jade items found in the middle to later phases of the Dawenkou culture dated prior to 5500 cal yr BP in the lower Yellow Valley were mainly body ornaments, many of the jade items found in the Liangzhu culture after 5000 cal yr BP in the Yangzi Valley are probably ritual implements, as they are heavy, large, and have mysterious, incised animal motifs (The Institute of Archaeology CASS, 1984).

Walled towns also occurred from 6000 to 4500 cal yr BP in eastern Central China. Towns with protective ditches, walls, and gates have been discovered both in the Yellow and the Yangzi Valleys. The earliest town in the Yellow Valley is the Xishan town of approximately 30,000 m^2 in Zhenzhou, Henan province, dated to between 5300 and 4800 cal yr BP (Fig. 9.1) (Ren, 1998). Remains of another town called Xikangliu of 35,000 m^2, dated to around 5000 cal yr BP, have been found in Shandong province, the lower Yellow Valley (Fig. 9.1) (Ren, 1998). In the middle Yangzi Valley, the Chengtoushan town dated to approximately 6000 cal yr BP has been found in Hunan province (Fig. 9.1) (Ren, 1998), and five other towns dated between 5000 and 4600 cal yr BP have been discovered in the present Hunan and Hubei provinces (Ren, 1998). The emergence of towns and the surrounding archaeological remains suggest increased group conflicts, social complexity, and hierarchy, paving the way for the establishment of Chinese civilization (Liu, 2000).

3.5. Cultural interaction and human movements

Cultural interactions and human migrations in prehistory have long been documented in China. As mentioned above, archaeological survey conducted in the Wei River Basin of the Yellow Valley suggests possible human movements in the mid-Holocene. Around 5000 cal yr BP, a Qujialing culture from the Yangzi Valley occupied the Xiawanggang site in the middle Yellow Valley (Fig. 9.1), which had been previously occupied by the Yangshao group of the Yellow Valley. This phenomenon might have been the result of human migration, or the outcome of a tribal conflict or/and conquest. On the other hand, it is also evident that the Neolithic culture in the Yangzi Valley received cultural influences from the Yangshao culture of the Yellow Valley, indicated by the decorations and colors of pottery found in the Diaolongbei assemblage (Fig. 9.1) (Hubei Archaeology Team CASS, 1992).

4. Discussion: paleoclimates, paleoenvironments and cultural dynamics in the mid-Holocene in eastern Central China

4.1. Issues in the reconstruction of the past climates and environments

The study of paleoclimatic changes and human adaptation (or failure of such adaptation) and cultural developments in mainland China often consists of two steps. The first step is to reconstruct the paleoclimates and paleoenvironments by various scientific approaches; the second step is to analyze and identify (or attempt to identify) human reactions to climatic changes as derived from archaeological data. However, there are several questions on this practice.

Regarding the reconstruction of paleoclimates, first, the issue of macro- versus microclimates should be noticed. The macroclimate here is defined as a global climatic pattern, which has/had a broad impact, such as the Last Glacial Maximum. The microclimate is defined as regional or local climatic cycles not only controlled by the global pattern, but also affected by local geographic settings (e.g., high or low latitudes, coast or mountains, loess or desert), and even human activities. In other words, there might have been different microclimates in different areas even within one macroclimate cycle. For example, the climatic changes in areas at lower latitudes might not have been as significant as those at higher latitudes (He et al., 2004). Furthermore, the geographic location of each small area and the vegetation of that area would also have affected the climate. In China, temperature curves of the last 10,000 years have been established based upon pollen profiles and other methods in different areas, but these local curves are not always consistent with the global climatic cycle (Shi et al., 1993). Apart from analytic errors, this discrepancy may reflect different microclimates in different areas. A recent study of paleoclimates in China has demonstrated this regional variation (He et al., 2004).

The second issue on paleoclimatic reconstruction is the accuracy of data. Many climatic reconstructions have been based on pollen profiles. However, like all research approaches, pollen analysis has its limitations. First, some plant species produce little or no catchable pollen. Those are the "low-presenting" or "non-presenting" species in pollen profiles, and very often they are species of sub-tropical to tropical ecozones (Li, 1998). Therefore, the retrieved pollen profiles are bound to be incomplete, even biased. A study based on pollen from surface soil in the middle Yellow Valley reveals that the pollen assemblage inaccurately reflects the local vegetation (Yao, 1989). Our recent comparative study of pollen from the surface soil and living vegetation in South China shows a similar discrepancy between pollen profile and plants (Lu, 2003). This limitation of pollen analysis should be taken into account when we reconstruct past climates. To overcome the problem of data inaccuracy, the reconstruction of past climates should be based on an integrated multi-disciplinary approach.

Another important, probably more crucial issue is the impact of human beings on past vegetation and fauna. For example, as mentioned above, the pollen profile of the Banpo assemblage indicates that there was an increase of grasses and a decrease of trees in the late phase of the Banpo culture. *Prima facie*, this can be interpreted as evidence of a cooling period. But it could also have been the result of fire continuously set by human beings (i.e., slash and burn). Substantial quantities of charcoal have been reported from the Banpo assemblage (Ke and Sun, 1990). According to ethnological data from Australia, fire can reduce trees and stimulate the growth of grasses (Nicholson, 1981). Therefore, human activities could have caused the decline of trees and the increase of grasses at Banpo during the mid-Holocene. As most people in mid-Holocene eastern Central China were farmers, their farming activities would have significantly accelerated deforestation, in turn causing erosion of the land and flooding, reduction of trees and increase of grasses, and higher levels of evaporation. Consequently, these processes would have changed the environment and even affected the microclimate (e.g., higher evaporation) in that area. When we seek to reconstruct the past climatic changes, we must also examine human impacts on climatic changes.

The fourth issue on reconstruction of paleoclimate is the accuracy of absolute dating, which involves two problems. First, local vegetation compositions may not reflect the climatic changes until about 100–200 years later (Davis and Botkin, 1985). Therefore, pollen data were likely lagging behind the actual paleoclimatic deteriorations. Second, some pollen profiles in China are not dated, which means the present cycle of climatic changes is neither completed nor accurate enough. Consequently, it can be difficult to correlate climatic and cultural changes.

In summary, there are many issues on the reconstruction of past climates and environments in China. If archaeologists want to integrate the past climatic and environmental changes with cultural changes, and if natural scientists want to reconstruct past natural changes with cultural evolution accurately, they must work together to carry out collaborative projects to tackle the aforementioned problems.

4.2. The interpretation of cultural changes

Since the 1980s, some natural scientists have been studying the past human cultural changes in China based on climatic and other natural changes, such as sea level changes and flooding (e.g., Wu, 1983; Zhu et al., 1997; Lu and Wu, 1999; Yu et al., 1999; Wu and Liu, 2004). While their enthusiasm should be much appreciated, many of these discussions tend to overemphasize the impact of the natural changes upon human cultures, and to ignore human dynamics in cultural changes and developments. Some described the cultural changes that occurred between 5000 and 4000 cal yr BP as a period of "collapse of Neolithic Culture" (Wu and Liu, 2004) and totally ignored the fact that the foundation of Chinese civilization was established exactly in this period (Liu, 2000). These discussions tend to draw a direct and straightforward correlation between climatic and cultural changes, and to reach the conclusion that human cultures prospered when there were favorable climates, and collapsed when there were bad climates (Lu and Wu, 1999, p. 70). Such a discussion of environmental determinism is not only biased and misleading, but also intellectually unsound and oversimplified.

As discussed above, there are still many issues that affect the accuracy of reconstructing past climate and environments. Therefore, first, we must be cautious when trying to rebuild the paleoclimatic framework of any given area. Second, human beings are not passive adapters to their environments. Different human cultures have different reactions towards environmental changes, and these differences reflect their own technologies, skills, subsistence strategies, as well as various cultural heritage assets each group inherited from its ancestors. In addition, the cultural dynamics between and within each human group should not be ignored. Liu (2000) has argued that differences in intra- and intergroup dynamics, in social, political and economic systems, and in human actions all played their roles in cultural changes.

We should also consider whether human reaction to climatic changes can be identified from archaeological remains, and if it can, how much information can be extracted, and how accurate is the information? Climatic changes include annual cycles, and many basic attributes such as temperature and precipitation, etc. can differ appreciably from one year to another. How do we accurately identify such fine scale climatic changes in the past? Even the high-resolution approaches in use today cannot fully solve this problem. On the other hand, cultural changes represented in archaeological remains are the results of continuous activities of human beings. These cultural changes are often measured by tens or even hundreds of years in prehistory. In addition, human reactions/adaptations towards natural changes are often belated. Furthermore, such reaction/adaptations are often affected and decided by other cultural and social elements of different groups, not just climatic changes. All these considerations make the attempt to correlate natural and cultural changes more complicated.

In summary, there can be data bias and inaccuracy in paleoclimatic reconstructions, and the analysis of cultural dynamics in relation to environmental change is not a straightforward matter. On the other hand, there were indeed climatic changes during the mid-Holocene in eastern Central China. What would have been the impacts of these changes upon human beings? How did the prehistoric groups adapt to these climatic changes? It is possible to investigate the climatic and cultural dynamics from a broader prospective.

4.3. Climatic and cultural dynamics

Theoretically, the warm and moist paleoclimate present by 5500 cal yr BP would have been favorable to the agricultural societies in many ways. First, a warmer and moister climate would have produced a higher effective temperature (Kelly, 1995), which would have shortened the growing cycles of cultivated cereals. Second, a warm and moist climate would have increased water levels in rivers and small lakes, and would have made the task of watering crops (if not yet irrigation) easier. Third, higher temperatures and precipitation levels would have enabled the cultivation of more cultivars, some of which could not have been cultivated in certain areas before due to insufficient water and lower temperatures. As mentioned above, it was during this period that rice was cultivated in the Yellow Valley (Fig. 9.3). Finally, higher temperature and precipitation levels would have also produced richer and more diversified wild vegetation, and consequently, richer animal resources. Thus on the one hand, the output of agriculture would have increased. On the other hand, more nuts, fruits, and animal resources would have become available.

However, a warmer paleoclimate with higher precipitation might have had some disadvantages. For example, the water level of rivers could have increased substantially, causing floods from time to time, which might have posed severe problems for prehistoric farmers. The warm climate also seems to have caused marine transgression along the eastern coast of China (Shi and Wang, 1982; Yu et al., 1999), inundating human habitats. This would have forced human beings to retreat from some areas.

When the climate changed from warm and moist to cool and dry, this shift might have caused problems, particularly in inland and/or continental areas, such as the loess plateau in the middle Yellow Valley, where the summer monsoon does not easily penetrate, but the spell of the winter monsoon is often strong. Once the climate became dry and cool, the loess began to accumulate again, precipitation levels declined, temperatures lowered, and consequently, fewer trees grew and grasses began to predominate. Animals adapted to rich water areas would have emigrated. Thus the availability of floral and faunal resources would have decreased, crop watering would have been more difficult due to the lower water levels of the rivers and lakes, and settlement sites would have moved to shorten the distance between water resources and residence. In addition, the lower effective temperature might have lengthened the growing season for cultivated plants and generally reduced the productivity of agriculture as well. In some areas the reduced quantities of water and other natural resources might have forced human beings to leave their original homes in search of more hospitable lands.

On the other hand, the cool and dry climate might have been good for those who lived in the Yangzi Valley, particularly those along the Yangzi River. According to Zhu and his colleagues, this climatic change might have reduced flooding, and provided a more favorable environment for the prehistoric occupants (Zhu et al., 1997).

How did human beings in prehistoric eastern Central China adapt to these climate changes? Several attributes provide some clues.

4.3.1. Location of settlements

As discussed above, when the climate was warm and wet, prehistoric settlements should have been located at places of higher elevation. On the other hand, during the cool and dry climate period with reduced precipitation, it is expected that human beings would have moved their settlements to areas of lower elevation to facilitate access to the lowered water levels.

Archaeologically, there are some evidence of such migration along river terraces in both the Yellow and the Yangzi Valleys (e.g., Zhu et al., 1997; Zhang, 2001), but these movements are not visible in all areas. Survey in Henan province indicates that most of the Neolithic archaeological assemblages found in plain areas were located near modern villages, many of which are close to small rivers or streams. This seems to suggest that the geographic locations of local human settlements remained little changed for the last few thousand years, although the occupation might not have been continuous.

How do we interpret the relatively stable settlements in plain areas in the mid-Holocene in eastern Central China? One possibility is that the climatic changes during the mid-Holocene were not severe enough to force humans to move their settlements, particularly those well-established agricultural societies. Another explanation is that there might have been other methods used by humans to adapt to lowered water levels, for example, digging wells. Remains of a well have been found in the lower Yangzi Valley dated to around 5000 cal yr BP (Shi, 1993). In the Yellow Valley, wells have been found during the Longshan period from 4800 to 3900 cal yr BP (The Institute of Archaeology CASS, 1984).

4.3.2. Continuity or discontinuity of occupation

This is another possible indicator of environmental and cultural changes. While continuity of occupation might suggest a successful human adaptation and/or a favorable environment, discontinuity of human occupation may have been caused by many factors, from sudden climatic deterioration to over-exploitation of local natural resources, or even social conflicts (Liu, 2000). Different cultural elements should be taken into account when we study discontinuities in human occupation; discontinuous occupation in an area does not necessarily indicate cultural collapse. In the Wei River Valley, for example, although occupation seems to have been discontinuous, the quantity of archaeological remains and sizes of Neolithic sites dated between 5500 and 4500 cal yr BP in the whole area remained similar, when significant climatic changes seem to have occurred (Table 9.3).

4.3.3. Cultivation activities

The occurrence of different cultivars in different areas over different periods may indicate human adaptations to fluctuating climatic conditions. From approximately 8000 to 7000 years ago the major cultivars in the Yellow Valley were foxtail and broomcorn millets. The discovery of cultivated rice in the Yellow Valley in the mid-Holocene indicates that prehistoric farmers were exploiting the warm and moist climate to increase the diversity of their cultivars (Fig. 9.3). As the output of rice farming should have been higher than millet farming,[1] prehistoric farmers during the mid-Holocene should have been able to obtain more food.

Other cultivars have also been discovered. Chinese cabbage (*Brassica pekinensis*) seeds have been found in Banpo, in the middle Yellow River Valley, dated to around 6800 cal yr years ago (Fig. 9.1) (Chen, 1994). Remains of gourd have also been found in Hemudu and Luojiajiao in the lower Yangzi Valley, both dated to about 7000 cal yr BP (Fig. 9.3) (Chen, 1994). These data suggest that horticulture was practiced during the mid-Holocene in eastern Central China. However, much work is required to investigate the occurrence of horticulture in China and its relation to climatic changes.

4.3.4. House structures

The majority of house remains dated from 7000 to 5500 cal yr BP in the Yellow Valley are semi-subterranean, which is suitable for people living in a cold and dry ecozone, as this structure will keep the house warm in winter and cool in summer. However, above-ground houses occurred after 6400 cal yr BP (Gong and Wang, 1991). On the other hand, the early houses found in the Hemudu assemblage in the lower Yangzi Valley (Fig. 9.3) were pile-dwelling houses (The Institute of Archaeology CASS, 1984). During the mid-Holocenc, both pile-dwelling and above-ground house structures have been found in the Yangzi Valley (The Institute of Archaeology CASS, 1984).

These different types of house structures may represent human adaptations to different climates and environments. Generally speaking the climate in the Yellow Valley was, and has been, much drier and cooler than that in the Yangzi Valley even during the period of warm and moist climate. Thus the semi-subterranean and subterranean houses in the Yellow Valley would be a better solution in a cold and windy climate. On the other hand, precipitation levels were much higher in the Yangzi Valley; and flooding would have been a constant threat. In such an environment, the pile-dwelling structure would be a better solution.

5. Conclusions

In summary, current data suggest that the warm and moist paleoclimate between approximately 7000 and 5500 cal yr BP provided favorable environments and

[1] Based on agronomic data, in the 1950s, the output of paddy rice is about 1890 $kg\,ha^{-1}$, broomcorn millet about 615 $kg\,ha^{-1}$, and foxtail millet about 700 $kg\,ha^{-1}$ (Cai, 1999; Chinese Academy of Agronomy, 1986; Shanxi Academy of Agronomy, 1987). With less-developed techniques and fertilizer, it can be inferred that the output of rice and millets in Neolithic period must have been lower, but the output of rice should still be higher than millet.

abundant natural resources for human beings in eastern Central China, although it might have caused problems in some aspects. Different Neolithic cultures in this landmass adapted to the climatic changes by different methods.

The mid-Holocene saw the prosperity and termination of some Neolithic cultures such as the Yangshao and the Dawenkou cultures in the Yellow Valley, and the Songze and Daxi cultures in the Yangzi Valley (Fig. 9.1). It was also during the mid-Holocene that the successors of these cultures emerged with advanced technologies of agriculture, house construction, pottery, and prestige goods. The emergence of walled towns indicates an increase in social complexity. Following this transitional period came another important prehistoric epoch in China, the latter dated between 4500 and 4000 cal yr BP. This period was associated with large tombs and precious grave goods, suggesting that civilization, or state level societies, were beginning to emerge in the Yellow and the Yangzi Valleys.

Regarding the dynamics between nature and human cultures, while climatic changes can produce significant impacts on human beings by altering the

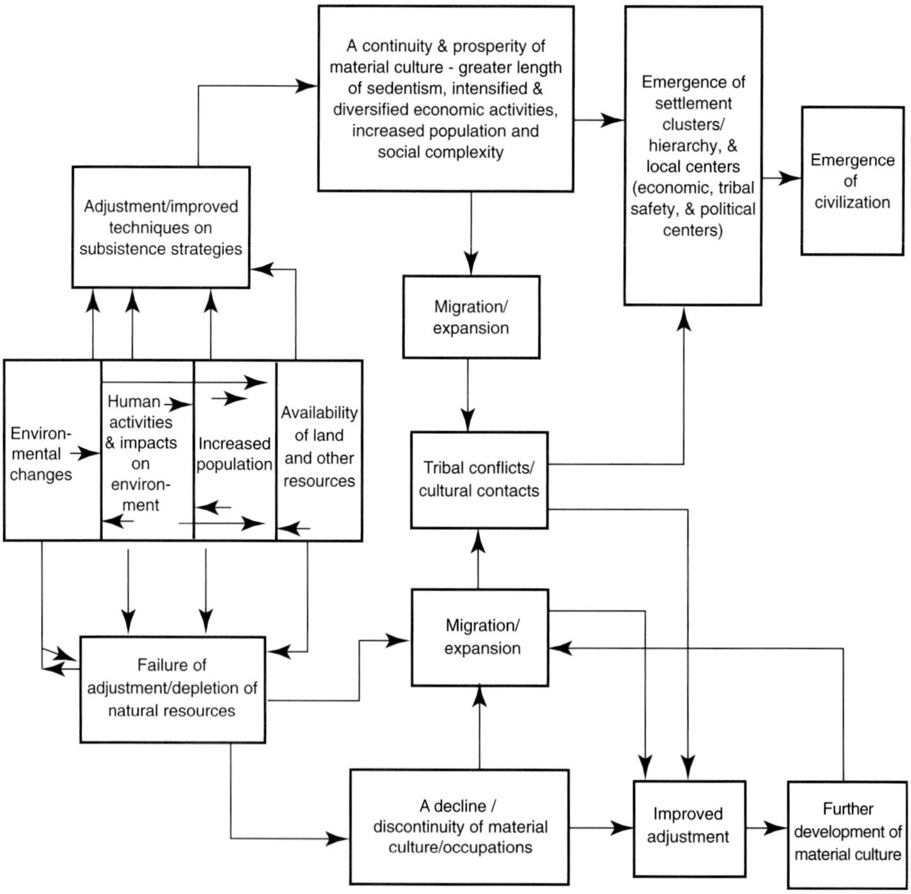

Figure 9.4. Dynamics between nature, culture and human beings.

availability of water, land, floral and faunal resources, human beings are also active agents, and can make adjustments. Successful adjustments facilitate continuous cultural development and prosperity, often represented by continuous or even extended occupations, rich artifact assemblages, probably increased social complexity, and other measures. Less successful adjustments may result in the decline of local cultures, reflected in a decrease of the quantity and quality of artifacts and shrinkage or abandonment of settlements. Such difficulties may have even impacted the viability of entire cultures. Archaeological data in eastern Central China today seem to suggest the co-occurrence and co-existence of both successful and less successful groups in the mid-Holocene.

It is hypothesized that some of these unsuccessful groups might have had to move to other areas to search for new resources. The emigrating groups might have come in contact with other groups, including those that had been able to adapt more successfully to their changed environments. These cultural contacts would have been channels for less unsuccessful groups to learn adjustments and techniques and to develop more sustainable subsistence strategies (Fig. 9.4). Though some of these meetings might have involved conflict, cultural interaction could have occurred as well. However, for groups living in more circumscribed social and environmental circumstances, their situation might have been more difficult.

The study of cultural and natural dynamics in prehistoric China is far from completed. As discussed above, many questions remain to be answered, both on the reconstruction of paleoclimates and paleoenvironments, and on the examination of past cultural developments. Natural scientists and archaeologists need to work together to tackle these questions.

References

An, Z. S., X. H. Wu, Y. T. Lu, D. E. Zhang, X. J. Sun, and G. R. Dong, 1990. Zujin Er'wannian Zhongguo Guchuanjing Bianqian de Chubu Yanjiu (A Study on paleoenvironmental changes in China since of the last 20,000 years). In *Huangtu, Disiji Dizhi-Quanqiu Bianhua* (*Loess, Quaternary and Global Changes*) edited by D. S. Liu, Vol. 2, pp. 1–29. Science Press, Beijing.

An, Z. S., S. C. Porter, W. J. Zhou, Y. C. Lu, D. J. Donahue, M. J. Head, X. H. Wu, J. Ren, and Z. Hongbo, 1993. Episode of strengthened summer monsoon climate of Younger Dryas age on the loess plateau of Central China. *Quaternary Research* 39:45–54.

Banpo Museum, Shaanxi Institute of Archaeology, and Lintong County Museum, 1988. *Jiangzhai-Xinshiqi Shidai Yizhi Fajue Baogao* (*The Excavation Report of the Jiangzhai Neolithic Site*). Cultural Relics Publishing House, Beijing.

Cai, Y., 1999. *Mizi* (*Broomcorn Millet*). Chinese Agronomy Press, Beijing.

Chen, W. H., 1991. *Zhongguo Nongye Kejishi Tuji* (*Atlas of the History of Agricultural Technology in China*). Agriculture Press, Beijing.

Chen, W. H., 1994. *Zhongguo Nongye Kaogu Tulu* (*Atlas of the Agricultural History of China*). Jiangxi Science and Technology Press, Nanchang.

Chinese Academy of Agronomy, 1986. *Zhongguo Daozuoxue* (*Rice Cultivation in China*). Agriculture Press, Beijing.

Cultural Relics Publishing House, Ed., 1999. *Xinzhongguo Kaogu Wushinian* (*Fifty Years of Chinese Archaeology*). Cultural Relic Publishing House, Beijing.

Davis, M. B., and D. B. Botkin, 1985. Sensitivity of cool-temperature forests and their fossil pollen record to rapid temperature change. *Quaternary Research* 23:327–340.

Diaolongbei Archaeology Team, 1991. Diaolongbei Yizhi Fajue Huo Zhongda Shouhuo (Significant discoveries from the Diaolongbei assemblage). *Zhongguo Wenwu Bao* (*Chinese Antiquity News*), February 10, front page.

Feng, X. M., Z. M. An, J. H. An, B. Q. Zhu, and Q. Z. Wang, 1987. *Zhongguo Taocishi* (*The History of Chinese Ceramics*). Cultural Relic Publishing House, Beijing.

Gao, G. R., and B. H. Hu, 1991. A preliminary study on ecological environment in Neolithic Shandong. *Huanjing Kaogu* (*Research on Environmental Archaeology*) 1:140–142.

Gao, F. Q., and H. S. Li, 1986. Disiji Nitan Chengtanqi de Huafen yu Qihou Bodong (Periodicity of Quaternary peat formation and climatic fluctuation). *Hiyang Dizhi yu Disiji Dizhi* (*Marine Geology and Quaternary Geology*) 6(2):95–103.

Gong, Q. M., and S. J. Wang, 1991. Jiangzhai Zaoqi Shengtai Huanjing de Yanjiu (A study on ecological environment of the early period of the Jiangzhai assemblage). *Huanjing Kaogu* (*Environmental Archaeology*) 1:78–84.

He, Y., W. H. Theakstone, Z. L. Zhang, D. Zhang, T. D. Yao, T. Chen, Y. P. Shen, and H. X. Pang, 2004. Asynchronous Holocene climatic changes across China. *Quaternary Research* 61:52–63.

Hori, K., Y. Saito, Q. H. Zhao, X. R. Cheng, P. X. Wang, Y. Sato, and C. X. Li, 2001. Sedimentary facies and Holocene progradation rates of the Changjiang (Yangtze) delta, China. *Geomorphology* 41:233–248.

Hubei Archaeology Team CASS, 1992. Hubei Zaoyangshi Diaolongbei Xinshiqi Shidai Yizhi Shijue Jianbao (The preliminary report of the trial excavation at Diaolongbei, Hubei province). *Kaogu* (*Archaeology*) 7:589–606.

IA, Henan Province, 1991. *Henansheng Wenwu Dituji* (*Atlas of the Antiquities in Henan Province*). The Publishing House of Chinese Map, Beijing.

IA, Henan Province and the Yangzi Archaeology Team, 1989. *Xichuan Xiwanggang* (*The Xiawanggang Assemblage of Xichuan County*). Cultural Relics Publishing House, Beijing.

IA, Shaanxi Province, 1997. Shaanxi Wenwudian Fenbu (The distribution of antiquities in Shaanxi province). *Wen Bo* (*Relics and Museums*) 3:10–24.

Jin, Z. G., C. X. Xiong, and P. Y. Xie, 1997. *Hunan Wenwu Dituce* (*Atlas of Antiquities of the Hunan Province*). Hunan People's Press, Changsha.

Ke, M. H., and J. H. Sun, 1990. Xi'an Banpo de Guzhibei yu Guqihou (The paleovegetation and paleoclimate of Banpo, Xi'an). *Kaogu* (*Archaeology*) 1:87–93.

Kelly, R. L., 1995. *The Foraging Spectrum: Diversity in Hunter-Gatherer Lifeways*. Smithsonian Institution Press, Washington, DC.

Li, W. Q., 1998. *Zhongguo Disiji Zhibei yu Huanjin* (*Quaternary Vegetation and Environment in China*). Science Press, Beijing.

Li, Y. H., and D. F. Han, 1963. Banpo Xinshiqi Shidai Yizhi zhi Shoulei Guluo (Animal remains found in the Banpo assemblage). In *Xi'an Banpo*, edited by The Institute of Archaeology CASS and Banpo Museum, pp. 255–269. Cultural Relics Publishing House, Beijing.

Liu, L., 2000. The development and decline of social complexity in north China: some environmental and social factors. *Indo-Pacific Prehistory Association Bulletin* 20:14–34.

Lu, T. L.-D., 1999. The Transition from Foraging to Farming and the Origin of Agriculture in China. BAR International Series No. 774, Cambridge.

Lu, T. L.-D., 2003. Palaeoenvironment and resources at Zengpiyan. In *Zengpiyan-A Prehistoric Cave in South China*, edited by X.G. Fu, pp. 251–269. Cultural Relics Publishing House, Beijing.

Lu, T. L.-D., 2004. The Origin and dispersal of agriculture and human diaspora in East Asia. In *The peopling of East Asia: Putting together Archaeology, Linguistics and Genetics*, edited by L. Sagart, pp. 51–62. Routledge Curzon, London.

Lu, W., and B. L. Wu, 1999. Zhongyuan Xingshiqi Wenhua yu Guqihou de Guanxi (Relation between Neolithic cultures and paleoclimates in the Central Plain). *Scientia Geographica Sinica* 1(1):69–72.

Mo, D. W., F. Li, S. C. Li, and Z. C. Kong, 1996. Gansu Huluhe Liuyu Zhong Quanxinshi Huanjing Yanhua Jiqi dui Renlei Huodong de Yingxiang (The Paleoenvironment of the middle Holocene in the Hulu River area in Gansu Province and its effects on human activities). *ACTA Geographica Sinica* 51(1), 59–69.

Nicholson, P. H., 1981. Fire and the Australian Aborigines – an enigma. In *Fire and the Australian Biota*, edited by A. M. Gill, R. H. Groves, and I. R. Noble, pp. 55–77. Australian Academy of Science, Canberra.

Qi, G. Q., 1988. Jiangzhai Xinshiqi Shidai Yizhi Dongwuqun de Fenxi (Faunal analysis of the Neolithic Jiangzhai assemblage. In *Jiangzhai-Xinshiqi Shidai Yizhi Fajue Baogao* (*The Excavation Report of the Jiangzhai Neolithic Site*), edited by the Banpo Museum, pp. 504–538. Cultural Relics Publishing House, Beijing.

Pei, A. P., 2000. Lun Zaoshi Xiaceng Wenhua (Archaeological assemblage found in the lower layer of Zaoshi). In *Subingqi yu Xiandai Zhongguo Kaoguxue* (*Su Bingqi and Archaeology in China*), edited by B. Su, pp. 414–431. Cultural Relics Publishing House, Beijing.

Ren, S. N., 1998. Zhongguo Shiqian Chengzhi Kaocha (A study on prehistoric cities in China). *Kaogu* (*Archaeology*) 1:1–16.

Renfrew, C., and P. Bahn, 2000. *Archaeology: Theories, Methods and Practice*. 3rd edition. Thames and Hudson, London.

Roberts, N., 1998. *The Holocene – An Environmental History*. Blackwell Publishers Ltd., Oxford.

Shanghai Management Committee of Antiquity, 1987. *Songze* (*Songze – A Neolithic Site*). Cultural Relics Publishing House, Beijing.

Shanxi Academy of Agronomy, 1987. *Zhongguo Guzi Zaipeixue* (*Foxtail Millet Cultivation in China*). Agronomy Press, Beijing.

Shi, S. H., 1993. Zhongguo Quanxinshi Gaowenqizhong de Qihou Tubian Shijian Jiqi dui Renlei de Yingxiang (Climatic changes and their impacts on human beings during the Holocene Megathermal in China). *Haiyang Dizhi yu Disiji Dizhi* (*Marine Geology and Quaternary Geology*) 13(4):65–73.

Shi, Y. F., and J. T. Wang, 1982. Zhongguo Wan Disiji de Qihou, Bingchuan he Haipingmian Bianhua (The changes of palaeoclimate, glacier and sea levels during the upper Quaternary in China). In *Disanjie Qaunguo Disiji Xueshui Huiyi Luwenji* (*Treaties of the 3rd National Conference on Quaternary Research*), edited by The Research Committee of the Quaternary Research in China, pp. 111–121. Science Press, Beijing.

Shi, Y. F., Z. Z. Kong, S. M. Wang, L.Y. Tang, F. B. Wang, T. D. Yao, X. T. Zhao, P. Y. Zhang, and S. H. Shi, 1993. Mid-Holocene climates and environments in China. *Global and Planetary Changes* 7:219–233.

Stuiver, M., P. J. Reimer, E. Bard, J. W. Beck, G. S. Burr, K. A., Hughen, B. Kromer, F. G. McCormac, J. v. d. Plicht, and M. Spurk, 1998. INTCAL98 Radiocarbon Age Calibration, 24,000–0 cal BP. *Radiocarbon* 40:1041–1083.

Tan, S. P., 1991. Ningzhen Diqu Taixing Yizhi de Juluo Fenbu yu Gudili Huanjing (The location of prehistoric settlements and the paleo-geographic environments in Ningzhen areas). *Huanjing Kaogu* (*Environmental Archaeology*) 1:143–147.

The Editing Committee of the Physiography of China, 1984. *Zhongguo Ziran Dili – Qihou* (*Physiography of China – The Climate*). Science Press, Beijing.

The 6th Shaanxi Archaeological Team, 1989. Shaanxi Lantian Xiehu Xinshiqi Shidai Yizhi Fajue Jianbao (A preliminary report on the excavation of Xiehu assemblage in Lantian county, Shaanxi province). *Kaogu* (*Archaeology*) 6:497–504.

The Institute of Archaeology, CASS (IA, CASS), 1984. *Xingzhongguo de Kaogu Faixian he Yanjiu* (*Archaeological Discoveries and Research in New China*). The Cultural Relics Publishing House, Beijing.

The Institute of Archaeology, CASS (IA, CASS), 1989. Fangshexing Tanshisi Niandai Ceding Baogao (C14 dating report). *Kaogu* (*Archaeology*) 7:657–661.

The Institute of Archaeology, CASS (IA, CASS), 1991. *Zhongguo Kaoguxue zhong Tanshisi Niandai Shujuji* (*C14 dates of Chinese Archaeology*). Cultural Relics Publishing House, Beijing.

The Institute of Archaeology, CASS (IA, CASS), 1992. Fangshexing Tanshisi Niandai Ceding Baogao (C14 dating report). *Kaogu* (*Archaeology*) 7:655–662.

Tong, Z. C., 1998. *Zhongguo Xinshiqi Yanjiu* (*Chinese Neolithic Stone Artifact Research*). Bashu Publishing, Chengdou.

Uehara, K., Saito, Y., and Hori, K., 2002. Paleotidal regime in the Changjiang (Yangtze) Estuary, the East China Sea, and the Yellow Sea at 6 ka and 10 ka estimated from a numerical model. *Marine Geology* 183:179–192.

Wang, J. F., Buhesiao'er, K. Y. Yang, Y. F. Ma, X. Z. Wang, S. L. Zhang, J. Y. Liu, T. S. Li, F. Zhu, G. N. Ju, W. Lu, R. J. Wang, Q. C. Ma, Q. Wan, X. N. Zheng, J. Y. Liu, C. Y. Wang, S. Li, C. A. Shi, and B. Liu, 1995. *Rendi Guanxi Yanjin jiqi Tiaokong* (*The Evolution and Adjustment of the Relationship between Human and Environment*). Science Press, Beijing.

Wang, H. X., 2002. Changjiang Zhongyou Chenghao Juluo de Yuanqi, Gongneng jiqi dui Wenming Jincheng de Zuoyong (The origin and development of settlements with ditches in the middle Yangzi Valley and their impacts on the origin of civilization). *Zhongguo Wenwu Bao* (*Chinese Antiquity News*), October 11, p. 7.

Wang, Z. G., 1998. Liangzhu Wenhua Yuqi Zonglun (A discussion on the jade items of the Liangzhu culture). In *East Asian Jade*, edited by T. Chung, pp. 251–264. The Chinese University of Hong Kong Press, Hong Kong.

Wei River Archaeology Team, 1991. Weishui Liuyu Yangshao Wenhua Yizhi Diaocha (The survey of the Yangshao sites in the Wei River Valley). *Kaogu* (*Archaeology*) 11:961–982.

Wei River Archaeology Team, 1992. Weishui Liuyu Longshan Wenhua Yizhi Diaocha (The survey of the Yangshao sites in the Wei River Valley). *Kaogu* (*Archaeology*) 12: 1057–1067.

Wu, J. M., 2000. Jiangsu Xingshiqi Shidai Yizhi Fenbu yu Huanjing Yanbian (The distribution of Neolithic sites and environmental changes in Jiangsu Province). *Huanjing Kaogu Yanjiu* (*Environmental Archaeology*) 2:104–110.

Wu, W. T., 1983. Cong Xinshiqi Shidai Wenhua Yizhi kan Huangzhouwan Liangan de Quanxinshi Gudili (Study on the Holocene geography in the Huangzhou Bay based upon the distribution of Neolithic sites). *Dili Xuebao* (*ACTA Geographica Sinica*) 38(2):113–127.

Wu, W. X., and T. S. Liu, 2004. Possible role of the "Holocene Event 3" on the collapse of Neolithic Cultures around the Central Plain of China. *Quaternary International* 117:153–166.

Xia, N, 1977. Tanshisi Ceding Niandai he Zhongguo Shiqian Kaoguxue (C14 dating and the prehistoric archaeology in China). *Kaogu* (*Archaeology*) 4:217–232.

Yan, Q. S., and S. Huang, 1987. Hangjiahu Pingyuan Quanxinshi Chenji Huanjing Yanbian (Evolution of Holocene sedimentary environment in the Hangjiahu Plain). *Dili Xuebao* (*ACTA Geographica Sinica*) 42:1–15.

Yao, Z. J., 1989. Shanxi Zhongtiaoshan Diqu Biaotu Huafen Fenxi (Surface pollen analysis in Zhongtiao Mountain, Shanxi Province). *Dili Xuebao* (*ACTA Geographica Sinica*) 44(4):469–478.

Yi, S. H., Y. Saito, Q. Zhao, and P. Wang, 2003. Vegetation and climate changes in the Changjiang (Yangtze River) Delta, China during the last 13,000 years inferred from pollen records. *Quaternary Science Reviews* 22:1501–1519.

Yu, S. Y., C. Zhu, and W. Z. Qu, 1999. Taihu Dongan Pingyuan Zhongquanxinshi Qihou Zhuanxin Shijian yu Xinshiqi Wenhua Zhongduan (Mid-Holocene climatic transition and rapid interruption in Neolithic cultures in the eastern plain of the Taihu Lake). *Dili Kexue* (*Scientia Geographica Sinica*) 19(6):549–554.

Yu, S. Y., C. Zhu, F. B. Wang, and H. Li, 2000. Taihu Liuyu Quanxinshi Qihou-haipinmian Duanqi Zhendang Shijian Jiqi dui Xinshiqi Wenhua de Yingxiang (The Neolithic climate in the Tai Lake area, sea level fluctuations and their impacts on Neolithic cultures). *Dili Kexue* (*Scientia Geographica Sinica*) 20(4):331–336.

Yuan, J., and J.Y. An, 1998. On two topics in the zoo-archaeological research of China. *EAANnouncements* 25:8–10.

Zhang, X. H., 2001. Shandong Shiqian Juluo Shikong Guanxi Hongguan Yanjiu (Macro-study on the time-space relation of the prehistoric settlements in Shandong). In *Su Bingqi and Modern Chinese Archaeology*, edited by B. Su, pp. 227–245. Science Press, Beijing.

Zhejiang Provincial Institute of Antiquity and Archaeology, 2003. *Hemudu – Excavation Report of a Neolithic Site*. Cultural Relics Publishing House, Beijing.

Zhou, W. J., D. J. Donahue, S. C. Porter, T. A. Jull, X. Li, M. Stuiver, Z. An, E. Matsumoto, and G. Dong, 1996. Variability of monsoon climate in East Asia at the end of the Last Glaciation. *Quaternary Research* 46:219–229.

Zhu, C., S. Y. Yu, and C. C. Lu, 1997. Changjiang Sanxia ji Jianghan Pinyuan Diqu Quanxinshi Huanjing Kaogu yu Yichang Honglao Zaihai Yanjiu (The study of Holocene environmental srchaeology and extreme flood disaster in the Middle Yangzi Valley). *Dili Xuebao* (*ACTA Geographica Sinica*) 52(3):8–277.

Climate Change and Cultural Dynamics: A Global Perspective on Mid-Holocene Transitions
David G. Anderson, Kirk A. Maasch and Daniel H. Sandweiss (Editors)

Chapter 10

Mid-Holocene climatic changes and cultural dynamics in the basin of the Sea of Japan and adjacent areas

Konstantin A. Lutaenko[1], Irina S. Zhushchikhovskaya[2], Yuri A. Mikishin[3] and Alexander N. Popov[4]

[1]*Institute of Marine Biology, Far East Branch of the Russian Academy of Sciences, Vladivostok 690041, Russia*

[2]*Institute of History, Archaeology and Ethnography of the Peoples of the Far East, Far East Branch of the Russian Academy of Sciences, Vladivostok 690001, Russia*

[3]*Coastal Research Center, Far East State University, Vladivostok 690000, Russia*

[4]*Museum of Archaeology and Ethnography, Far East State University, Vladivostok 690000, Russia*

Abstract

Mid-Holocene climatic/paleogeographical changes, cultural dynamics, and their relations in the basin of the Sea of Japan are reviewed. The sequence and chronology of warmings/coolings are reconstructed as follows: the first warming of the mid-Holocene took place between 7200 and 6200 ^{14}C yr BP in Sakhalin, although the maximum warming was between 5000 and 6000 ^{14}C yr BP; in Japan, the peak of warmth might be traced at the level of 6500–5500 or 6500–5000 ^{14}C yr BP; a significant cooling in Japan is detected between 4000 and 4500 ^{14}C yr BP, which corresponds to the Early Subboreal cooling in Sakhalin Island about 4400 ^{14}C yr BP, to that in Primorye about 4200–4700 ^{14}C yr BP, and in the Kurile Islands approximately 4500–4800 ^{14}C yr BP. The Tsushima Current's inflow in the Sea of Japan started after 9000–9500 ^{14}C yr BP, but the influx of the current on a full scale might occur approximately 8000 ^{14}C yr BP. It is suggested that a period of high storm frequency in Korea and Primorye, corresponding to maximum warming in Japan and Sakhalin, took place 5300–6000 ^{14}C yr BP, Climatic estimates show that the "climatic optimum" (time span between 5000 and 6000 ^{14}C yr BP) is characterized in the Russian Far East by the annual average temperature excess of 2–6°C and more humid conditions. Annual average sea surface temperatures (SSTs) in the Sea of Japan were 1–2°C higher compared to the present. Archaeological records from Japan, Korea, and the southern Russian Far East (i.e., evidence for material and spiritual culture, and the subsistence patterns of the prehistoric populations) indicate important changes were occurring around 5000 ^{14}C yr BP. In general, the main cultural changes in the Japan Sea basin correspond to the period after *the climatic optimum of the Holocene. This conclusion reinforces the need for further investigation of the dynamics of climate and culture in the mid-Holocene.*

1. Introduction

The basin of the Sea of Japan is the most studied area in the northwestern Pacific from the viewpoint of both Holocene climate and archaeological evidence. For that reason we selected this area for analytical review of mid-Holocene changes in climate and cultural dynamics. The study area includes the Russian Far East, Japan, and Korea (Fig. 10.1). The southern part of the Russian Far East consists of the territories of Sakhalin Island and two areas – Primorsky Krai (or Primorye) and Khabarovsky Krai ("krai" is an administrative unit of "provincial" level in the Russian Federation) located along the coasts of the Japan and Okhotsk Seas. The southern part of the Khabarovsky Krai is also called Priamurye (the area lying in the basin of Amur River, the largest river in the Russian Far East). The southern part of the Russian Far East covers an area between ~55°N and 42°N. Precise paleoclimatic data for the northern Okhotsk Sea, Kurile Islands, Kamchatka, and Chukotka Peninsulas are still scarce.

Some geographical features of the research area are of great importance for the understanding of past cultural–historical processes. The territories of the research

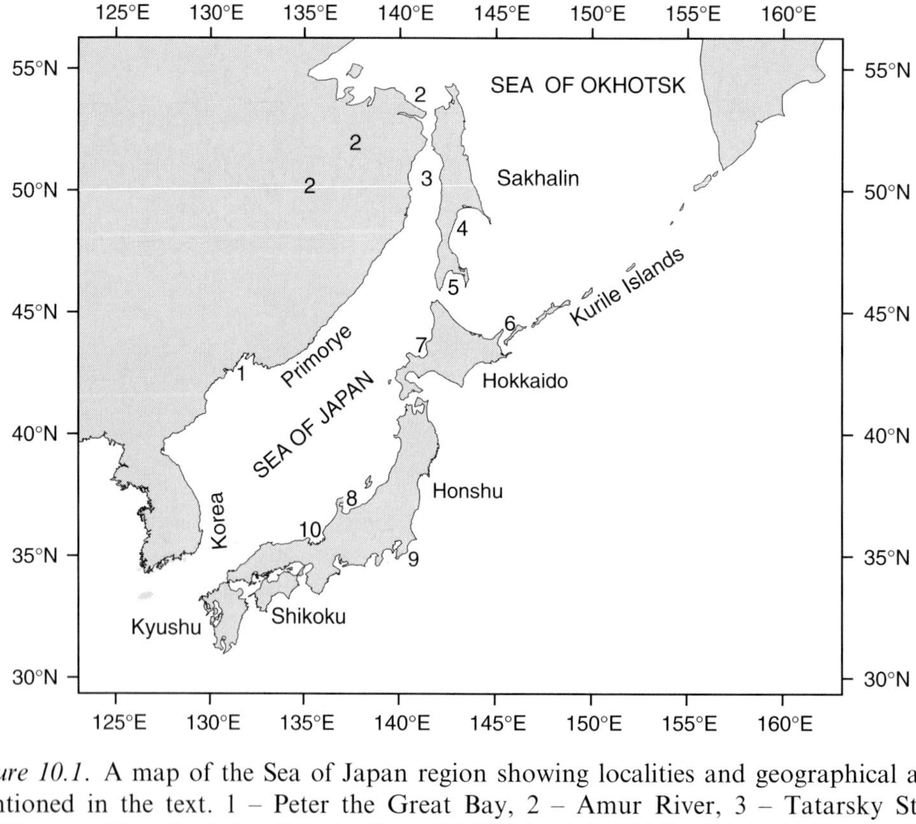

Figure 10.1. A map of the Sea of Japan region showing localities and geographical areas mentioned in the text. 1 – Peter the Great Bay, 2 – Amur River, 3 – Tatarsky Strait, 4 – Terpenya Bay, 5 – Aniva Bay, 6 – Kunashir(i) Island, 7 – Ishikari Bay, 8 – Toyama Bay, 9 – Tokyo Bay, and 10 – Wakasa Bay.

area form a circular pattern surrounding the Sea of Japan. Some territories are mainland, such as the Korean Peninsula, Primorye, and Lower Amur River regions, while the others are insular, such as the Japanese Archipelago and Sakhalin. Neighboring territories separated by the sea are located at short distances one from another (Fig. 10.1). Mainland regions of the research area representing the eastern outskirts of the Eurasian continent are linked with more distant western territories by two natural "roads" – the Amur River streaming in a west-to-east direction, and a "steppe corridor" passing through southern Siberia, Mongolia, and Northeastern China. These geographical peculiarities influenced to a certain extent the past cultural history of the research area, especially processes of cultural contacts, migrations and diffusions, and the continuity and discontinuity of prehistoric population development.

The chronological framework of our investigation span the interval from 8000 to 3000 yr BP; all dates used in the text are non-calibrated radiocarbon ages (hereafter "^{14}C yr BP"). According to the classic Blytt–Sernander scheme, the mid-Holocene includes the Atlantic and the Subboreal (8000–2200 ^{14}C yr BP). The most important event of the mid-Holocene is believed to be the post-glacial climatic optimum and high stand of the sea level, whose traces are recognized in many areas of the World Ocean, and by biotic events such as the northward shift of the boundary of broad-leaved forests, the northward migrations of warm-water faunal elements (i.e., as documented by the appearance of so-called thermally anomalous faunas in coastal areas), an increase in the bioproductivity of some marine waters, and so on. It should be noted that the definition of the "climatic optimum" is not generally accepted. Optimum in itself is a combination of most favorable conditions, and this definition is frequently intuitively used in paleoecology. But what is optimum for thermophilous organisms is not so for organisms living in cold or even temperate environments. Hereafter, when we use the term "climatic optimum", we merely mean the maximum warm climatic conditions of the Middle Holocene.

The main purpose of this paper is to delimit climatic patterns and analyze cultural dynamics during the mid-Holocene in the basin of the Sea of Japan, to correlate climatic, environmental, and cultural changes. Explanatory models are proposed to show probable relationships between culture and climate in the past.

Detailed climatic and environmental change research has been underway in Japan and the southern Russian Far East since the 1960s, resulting in a significant understanding of the region's Holocene paleogeography. The approaches of the Russian and Japanese authors are sometimes sufficiently different that it is difficult to directly correlate the paleoclimatic data. For instance, Russian authors widely accept and make use of the Blytt–Sernander climatostratigraphic scheme, whereas Japanese paleogeographers mainly use cultural periodization. The details of Holocene vegetation changes are the main "fine" indicator of climatic changes. Radiocarbon determinations are much more common and better known in Japan compared to Korea and the Russian Far East. The Russian literature dealing with Holocene paleoenvironmental changes is difficult for many scholars to access given the language barrier, and is rarely cited or critically analyzed by Japanese authors.

At the same time, the extensive Japanese literature is itself difficult to review by non-Japanese-speaking authors.

The primary evidence for past cultural dynamics is obtained from the archaeological record. In all parts of the research area, the mid-Holocene chronological limits correspond to what is known broadly as the Neolithic period. The amount of archaeological research in these different regions encompassing this period is uneven.

In Japan, large-scale systematic archaeological investigation has occurred on many sites of this period. Dozens of sites have been completely excavated, dated, and published in detail. The archaeological record includes sites of the famous Neolithic Jomon culture, some stages of which correlate with the mid-Holocene temporal limits of 8000–3000 ^{14}C yr BP. Jomon culture has been examined in detail for about a century, but it is only after World War II, with the advent of radiocarbon dating, that the "golden age" of Jomon studies started. At present many Jomon sites are being found and excavated in all part of the Japanese Archipelago, from Hokkaido in the north to Okinawa in the south. The cultural complexes of the Neolithic have been most completely worked out in Japan, where there are far more radiocarbon dates than in other parts of the research area.

The Korean Peninsula has witnessed intensive archaeological investigations since the 1950s and 1960s, and many Neolithic sites are known that date from 8000 to 3000 ^{14}C yr BP. The basic cultural sequence of the Korean Neolithic is now known, although a current problem is to attain a more detailed systematization of the archaeological contexts. Radiocarbon dating is increasingly commonplace in Korean archaeology, and large numbers of absolute dates exist for the Neolithic, more than for most other cultural periods.

Periodic archaeological investigations in the Russian Far East began in the 1950s. At present, the general time-space systematics of cultural remains from the Paleolithic to the Early States period are known and outlined. Increasing amounts of information are being recovered, and recent discoveries are correcting and adding to our conceptions of the past.

The Neolithic of the Russian Far East closely correlates with the mid-Holocene period. All recognized Neolithic cultures fall within or partially overlap with the chronological limits under consideration herein, from ca. 8000 to 3000 cal yr BP. The Russian Far East has fewer completely excavated and securely dated Neolithic sites than Korea and especially Japan. The lack radiocarbon dates is most pronounced for the Lower Amur region than for other parts of the Russian Far East. The Neolithic cultural sequence of the Lower Amur region is based, in large part, on relative dating.

To reconstruct past cultural dynamics, we must distinguish the changing components of culture and their archaeological indicators. The former include information about subsistence, demography, social structure, basic cultural context, material culture, and spiritual culture. The latter are the remains and evidences of settlements and dwellings patterns, subsistence patterns, burial and ritual objects patterns, and tools and ceramic patterns. Obviously, the availability of certain indicators of cultural dynamics depends primarily on the degree of preservation, and the volume and completeness of the archaeological record. The sets of archaeological indicators of past cultural processes in the various regions of the research area are different. They are most variable and representative for Japan and Korea and more restricted for the Russian Far East.

2. Climatic and environmental changes

Climatically, the structure of the mid-Holocene was heterogenous. Although it is generally believed that a warming occurred in the mid-Holocene, there were several warm epochs alternating with cold ones. This has been shown most clearly for the Russian territory in its European and Siberian parts, and only recently for Sakhalin Island.

Khotinsky (1982) recognized in the "Middle Holocene" three thermal maxima: the Boreal maximum (8300–8900 ^{14}C yr BP), the Late Atlantic maximum (4700–6000 ^{14}C yr BP), and the Middle Subboreal maximum (3200–4200 ^{14}C yr BP). The Boreal thermal maximum was most pronounced in Siberia and the Far East; the Middle Subboreal in the north of Russian Plain, and the Late Atlantic in the greater part of the forest zone of Northern Eurasia. The Late Atlantic maximum most closely corresponds to the global climatic optimum of the Holocene. The warming in the Boreal was no less significant, when compared to that during the Atlantic, but the Boreal warming was not expressed widely and was not lengthy (Khotinsky, 1982, p. 145). These conclusions are based on generalized schemes of paleoclimates deduced from pollen analysis.

There are several other reconstructions by Russian authors providing maps of paleotemperatures and precipitation for the entire territory of the former Soviet Union, including areas adjacent to the Sea of Japan basin. According to a quantitative reconstruction of Burashnikova et al. (1982), during the period from 5000 to 6000 ^{14}C yr BP a gradual increase in temperatures compared to present-day occurred north of ca. 48°N, from 1–2°C in the south to 3–4°C in the north, northeast, and in Kamchatka. At a latitude of 40–50°N, there was a "neutral belt" where July temperatures at 5000–6000 ^{14}C yr BP were not significantly different from recent ones. Thus, the changes in temperatures during the climatic optimum were geographically heterogenous – the greatest increases in temperature took place in the northern and northeastern parts of the research region in the former USSR, where temperature differences reached 4°C above present-day temperatures (Burashnikova et al., 1982). In a north–south direction, there was also a gradual lowering of precipitation, and in the Russian Far East the climate was more dry than today. This indicates metachronous climatic variations during the climatic optimum in different geographical zones and regions.

Borzenkova (1990), analyzing the paleotemperature curves for northwestern Europe and western Siberia, suggested about 30 global warmings and coolings took place during the Holocene. He emphasized the importance of distinguishing two warmings, the Early Boreal (9200–8900 ^{14}C yr BP) and the Late Atlantic. The first warming was evidenced by the changes of vegetation in the continental (non-glacial) regions of the Northern and Southern Hemispheres, and the second one by maximum temperatures (i.e., a warming up) of the upper layer of the ocean. The Early Boreal warming was thought to be characterized by the summer temperatures 5–6°C above present-day ones in the northeast of the former USSR (i.e., north-western Pacific areas), in Alaska, and in temperate latitudes of the Southern Hemisphere, and by the maximum northward shifting of forests during the Holocene (Borzenkova, 1990).

Climatic reconstructions for the entire territory of the former USSR and Northern Hemisphere were undertaken, respectively, by Khotinsky and Savina (1985), Klimanov (1989a), and Velichko and Klimanov (1990), inclusive of areas adjacent to the coast of the North Pacific Ocean. According to Velichko and Klimanov (1990), the warmest period of the Holocene history occurred about 5000–6000 ^{14}C yr BP, and it corresponds to Atlantic-3, or climatic optimum. These authors presented maps of the distribution of annual average temperatures, average January and July temperatures, total amounts of annual precipitation and deviations of these parameters from recent ones. The authors noted that their schemes agreed quite well with reconstructions produced by Khotinsky and Savina (1985). When determining the quantitative characteristics of climate, Velichko and Klimanov (1990) used the information-statistical method, which is based on the statistical relationship between the composition of Recent spore-pollen spectra and present-day climatic conditions. According to their reconstructions, average July temperatures were higher by 2°C or more compared to recent average July temperatures in the Russian Far East territory, eastern China, and Japan. This is in agreement with the data of Burashnikova et al. (1982) for Sakhalin and Kamchatka, and with those of Klimanov (1989b) for Primorye where average July temperatures were approximately 2°C higher than at present. Klimanov (1989b) estimated that average January temperatures during the climatic optimum were higher than present ones throughout the territory of the former Soviet Union. In the Far East, wet conditions dominated, and precipitation increased by 25–50 mm (Klimanov, 1989b); this viewpoint, however, is in conflict with the reconstruction of Burashnikova et al. (1982).

Using the zonal formational method for the climatic reconstructions at about 5000 ^{14}C yr BP (i.e., the establishment of relations between Recent vegetation zones, groups of vegetational formations, and present-day climatic characteristics), Khotinsky and Savina (1985) argued for an increase in annual amounts of precipitation in the northern, southern, and partly eastern regions of the former USSR. The authors also reconstructed January and July temperatures in the form of maps.

All of the above-mentioned reconstructions were based on either statistical methods or other methods of analysis of pollen data. Punning and Raukas (1985), in their extensive review of paleogeography of Northern Europe, correctly noted that the higher the latitude, the less palynological information there is available; thus, the reliability of reconstructions based on this kind of data is decreased in these areas. This is, probably, a major type of error associated with paleoclimatic reconstructions based on this method – an inadequate number of paleoclimatic sites are averaged to indicate climate at a large-scale (Liao et al., 1994).

2.1. Sakhalin and Kurile Islands

Three monographs published during the last 20 years by different groups of paleogeographers summarize much of the available evidence on climatic and vegetational changes in the Holocene in this area (Svitoch et al., 1988; Mikishin and Gvozdeva, 1996; Korotky et al., 1997), studies augmented by recent work by Y. Igarashi and collaborators (Igarashi, 1997; Igarashi and Igarashi, 1998; Igarashi et al., 2000) (Tables 10.1–10.3).

According to Svitoch et al. (1988), warming of climate in the Boreal–Atlantic caused a wide distribution of forest landscapes, which were dominated by birch (*Betula*)–broad-leaved vegetation with considerable admixture of oak (*Quercus*) in southern Sakhalin and mainly dark-coniferous (*Picea* and *Abies*) forests in the middle part of the island (western coast). Khotinsky (1977) suggested that the most favorable conditions for the flourishing of dark coniferous and broad-leaved forests occurred in Sakhalin Island between 8900 and 8300 ^{14}C yr BP. In this case, climatic optimum shifts from the Atlantic to the Boreal. However, in a later paper Khotinsky (1982, p. 147) spoke about this suggestion with some uncertainity: "only the Late Atlantic phase... can be considered as the climatic optimum of the Holocene of whole Eurasia. In Siberia, the Boreal thermal maximum can be regarded as the climatic optimum." The assumption about shifting of the optimum to the Boreal was criticized by Korotky et al. (1988), Mikishin (1996), Mikishin and Gvozdeva (1996), and Mikishin et al. (1997). Korotky et al. (1988) noted that postglacial transgression, in this case, reached its maximum (high stand of sea level) about 8500–9000 ^{14}C yr BP, although the coincidence of peaks of both warmth and sea level rise was established for this region of the North Pacific. Mikishin et al. (1997) re-studied several Holocene sections with peat layers where thermophilous pollen assemblages had been detected (in the area near Uandi Cape) and regarded them as a proof of Boreal thermal maximum. These authors found that the

Table 10.1. The structure of the mid-Holocene climatic changes in southwestern Sakhalin Island (based on Mikishin and Gvozdeva, 1996).

The Holocene stages	Chronological levels (^{14}C yr BP)	Climatic episodes
Subboreal	About 4400	Cooling
Atlantic	After 4900	Warming
	5100–4800/4900	Cooling
	5800/5900–5100	Maximum warming
	6200/6300–5800/5900	Cooling
	7200–6200/6300	Warming

Table 10.2. Climatic parameters of the Atlantic in Sakhalin Island.

Area	Average annual temperature (°C)	July/August temperature (°C)	January temperature (°C)	Precipitation (mm/year)	Source
Sakhalin Island	0	15 (July)	−24	700 or 500	Korotky et al. (1996a, 1997)
Southeastern Sakhalin Island	5–7,5	18–21 (August)	−5–7	Up to 1000–1200 or more than 800	Mikishin and Gvozdeva (1996)
Western Sakhalin Island (coastal area of Tatarsky Strait)	–	17 (July)	–	500	Burashnikova et al. (1982)

Table 10.3. Development of vegetation and climatic changes in southern Sakhalin during the mid-Holocene.

Age (^{14}C yr BP)	Blytt–Sernander scheme	Vegetation	Climate
3000–2300	Subboreal	Predominance of dark-coniferous and small-decidious forests with admixture of various broad-leaved species	Similar to the present, but a little warmer
3000		Motley grass meadows with predominance of Umbelliferae; heather vegetation	Colder and significantly drier than the present
4100–3000(?)		Elm–oak, partly small-decidious and dark-coniferous forests	Much warmer and drier than the present
4400–4100		Maximum distribution of dark-coniferous taiga	Similar to the present in its temperature parameters, but more wet
4800–4400 (?)		Predominance of small-decidious forests and distribution of frigid shrubs	Colder and, possibly, drier than the present
4800–4900	Atlantic	Elm–oak, partly small-decidious forests. Maximum distribution of oak	Much warmer than the present
5100		Significant distribution of dark-coniferous; in a lesser degree-small-decidious forests	Colder and wetter than the preceeding stage; close to the present
5800–5100		Elm–oak and oak–hazel forests. Maximum distribution of hazel (*Juglans*)	Much warmer (maximum of warmth), but wetter than the present
6300–5800		Predominance of small-decidious and dark-coniferous forests, decreasing broad-leaved forests	Relatively cold, but warmer than the present
7200		Elm–oak; in a lesser degree, dark-coniferous forests	Warm and wet, much warmer than the present

Continued

Table 10.3. continued

Age (^{14}C yr BP)	Blytt–Sernander scheme	Vegetation	Climate
7800	Boreal–Atlantic	Distribution of Sphagnum swamps, sedge meadows, frigid shrub with predominance of *Alnaster*; in a lesser degree, birch forests	Cold and wet
9300–8100		Decreasing birch forests, distribution of dark-coniferous and broad-leaved species. Appearance of *Aralia, Fraxinus*, and fern *Osmunda*	Close to the present, but wetter and slightly colder
		Predominance of birch forests, decreasing frigid shrub. Appearance of oak, disappearance of fern; distribution of *Artemisia* and heather	Relatively warm, but colder and wetter than the present
		Larix and birch forests, frigid shrub. Increasing dark-coniferous species, appearance of elm, flourishing of fern (Polypodiaceae)	Warmer and wetter than preceeding stage, but colder and drier than the present

"thermal maximum" took place in the mid-Boreal: climate was warmer than the present-day one, but, at the same time, colder compared to the Atlantic warming. The main type of vegetation was birch (*Betula*) with admixture of elm (*Ulmus*). According to Mikishin et al. (1997), this phenomenon might have been caused by a significant influence of the warm Tsushima Current which could penetrate much northward along the western coast of Sakhalin Island because of an incomplete opening of La Perouse (Soya) Strait (the strait between Hokkaido and Sakhalin) at 9000–8300 ^{14}C yr BP. In southeastern Sakhalin, the Boreal warming was weaker (Mikishin and Gvozdeva, 1996). A similar viewpoint was substantiated by Korotky et al. (1997) who stressed that sea level was about 20 m below its present position at that time. These authors confirmed that the Boreal climate in its optimum phase was warmer than present-day, thus causing a disappearance of permafrost in the coastal zone of western and southern Sakhalin, an enhancement of eolian processes, development of *Betula–Ulmus* forests characterized by maximum participation of the grass pollen during the Holocene history (up to 60%), and inlaid distribution of spruce (*Picea*) forests.

Based on a study of the Holocene deposits on the coast of western and southern Sakhalin and using numerous C 14 dates, Korotky et al. (1997) showed that climatic optimum occurred at the chronological level of 5000–7000 ^{14}C yr BP with a well-expressed within-Atlantic warming. At that time, broad-leaved forests were distributed north up to 49°N (near Shakhtersk Town). *Abies–Picea* taiga flourished in highlands. The birch–spruce (*Betula–Picea*) forests (in lowlands) and *Picea–Abies* taiga (in highlands) formed in the northern Sakhalin (western coast). Broad-leaved trees (oak and elm) also played a limited role in the formation of these forests.

Along the eastern coast of northern Sakhalin, birch–spruce (*Betula–Picea*) forests were common. Forests consisting of birch–alder (*Betula–Alnus*) and larch (*Larix*) formed in the coastal plains; on the southern coast (Aniva Bay), spruce–broad-leaved forests existed. Asymmetry in the distribution of vegetation (more thermophilous in the west, more frigid in the east) can be explained by differences in the hydrothermal regime of the Sea of Japan and the Sea of Okhotsk and by the influence of the branch of the warm Tsushima Current flowing in Tatarsky (Mamiya) Strait (the strait between continent and Sakhalin) on coastal vegetation (Korotky et al., 1997).

According to Igarashi (1997) and Igarashi et al. (2000), about 7500 ^{14}C yr BP *Larix* decreased markedly in north Sakhalin, and climate began to change from cold/dry to warm/moist, and ca. 6000 ^{14}C yr BP almost the same climatic conditions as the present began to exist. Since about 6000 ^{14}C yr BP, *Picea*-dominated taiga has flourished till present.

In southeastern Sakhalin, the Atlantic period was characterized by an uneven temperature regime with three "waves" of warmth (Mikishin and Gvozdeva, 1996; see Tables 10.1 and 10.3). The "First wave" took place about 7200 ^{14}C yr BP when the predominant type of vegetation became broad-leaved elm–birch forests with an admixture of thermophilous *Tilia*, *Juglans*, *Fraxinus*, *Phellodendron*, etc. Less common were decidious alder–birch (*Alnus–Betula*) forests located at the slopes of mountain ridges. Warming was accompanied by the wetting of climate, as evidenced by the predominance of sedge (*Carex*) and fern *Osmunda*. Then, since 6200–6300 ^{14}C yr BP, an insignificant cooling of climate occurred and it continued up to 5800–5900 ^{14}C yr BP. Broad-leaved forests were replaced by small-decidious and dark-coniferous trees. The most optimum climatic conditions occurred in southeastern Sakhalin about 5400 ^{14}C yr BP. At that time, broad-leaved elm–oak (*Ulmus–Quercus*) and oak–walnut (*Quercus–Juglans*) forests were widely distributed. The Korean cedar (*Pinus koraiensis*) appeared on the slopes of the Susunaisky Ridge, a species growing now only in the continental area of the Far East. Chronological boundaries of the main climatic optimum are tentatively considered to be between 5800 and 5100 ^{14}C yr BP (Mikishin and Gvozdeva, 1996). After that a cooling detected at about 5100 ^{14}C yr BP took place, temperatures were similar to present, but humidity of climate markedly increased compared to the previous epoch. Dark-coniferous *Pinus–Picea* forests were widely distributed during the cooling episode. The cooling changed into a substantial warming similar to the thermal maximum of 5400 ^{14}C yr BP. The main type of vegetation was elm–oak forests, especially oak, which reached its maximum distribution since the beginning of the Holocene. Most likely, winters were mild and heavy snowfall occurred. Dark-coniferous (*Picea–Pinus*) forests did not play any significant role in vegetation; grasslands were dominated by sedge (*Carex*), cereals, wormwood, and motley grass. According to Igarashi and Igarashi (1998), mixed forest composed mainly of *Picea jezoensis* and/or *Picea glehnii*, *Abies sachalinensis*, *Betula*, and *Alnus* coexisting with *Larix gmelini*, *Pinus pumila*, *Quercus*, and *Alnus* developed in southern Sakhalin between 4600 ^{14}C yr BP and 300–400 ^{14}C yr BP. At the end of the Atlantic a cooling again occurred which was followed by a more significant cooling in the Subboreal, but climate was still warmer than the present. Mikishin and Gvozdeva (1996) believe that the climate of the Atlantic was similar to the present climate of Hokkaido for most of this period in southeastern Sakhalin; their reconstructions of quantitative parameters are given in Table 10.2 along with those of Burashnikova et al. (1982) and Korotky et al. (1997). In the periods of within-Atlantic coolings, annual average temperatures decreased by 1–2°C, as compared to warm stages, but only by 5100 ^{14}C yr BP did they approached present-day conditions (Mikishin and Gvozdeva, 1996). As one can see from comparison of climatic parameters in Table 10.2, data obtained by analogy (comparison of distribution and dominants of mid-Holocene vegetation with Recent analogous vegetation in Hokkaido and other areas; Mikishin and Gvozdeva, 1996; Korotky et al., 1997) and by statistical methods of treatment of fossil pollen (Burashnikova et al., 1982) may differ substantially, although local climatic differences related to relief, influence of coastal currents, etc. should also be taken into consideration.

Sakaguchi (1989) studied three Holocene pollen diagrams from the western coast of Sakhalin (the samples were collected before World War II) and found that the reconstruction of paleoclimate is difficult due to remarkable fluctuations resulting from wildfires, which strongly influenced the Sakhalin forests. However, his data refer to the Late Holocene.

New data on the vegetational and climatic history were recently obtained from a study of the Holocene deposits of Kunashir(i) Island, southern Kurile Islands. A comprehensive paleogeographical investigation was based on more than 10 sections (Korotky et al., 1995, 1996b,c, 1998, 1999, 2000a).

During the climatic optimum dated at the level of 6400–5000 ^{14}C yr BP, broad-leaved/dark-coniferous and broad-leaved forests (*Quercus, Juglans, Ulmus, Phellodendron, Corylus*) were common. The beginning of the Atlantic (about 7100 ^{14}C yr BP) is characterized by the development of birch; about 6500–6300 ^{14}C yr BP oak expansion took place. According to the above-mentioned authors, not only global warming influenced the course of natural processes in the island, but also a warm current that was displaced northwardly about 4700–5900 ^{14}C yr BP. Two warm stages recognized in the Atlantic were divided by a mild cooling at the level of 5600–5700 ^{14}C yr BP; at that time, alder forests flourished on coastal plains. In general, dry and cool climate in Kunashiri Island changed to warm and moist about 7000–6500 ^{14}C yr BP, later than on Hokkaido, and highest sea level position was about 6500–6300 ^{14}C yr BP (Korotky et al., 1999, 2000a).

A cooling recorded at the boundary of the Atlantic–Subboreal led to the formation of large dune fields (up to 20 m high) and peatlands. This cold stage is about 4500–4800 years in age. The climate and landscapes in the early stage of the Subboreal were similar to those of the Atlantic period, but they significantly changed in the second half of the former period. While the Early Subboreal is characterized by the development of coniferous–broad-leaved forests with a predominance of *Abies* in the central part of the Kunashiri Island and broad-leaved forests in the southern part, the second half of the Subboreal was a time of shifting of coniferous–broad-leaved forests to the southern part of the island and appearance of dark-coniferous forests dominated by spruce (*Picea*) in the north.

On Iturup Island located to north of Kunashiri Island, the climatic optimum coincides with the Atlantic period; at that time, the central part of the island was covered by nemoral oak–broad-leaved forests with participation of thermophilous Amur cork-tree, horn beam, walnut, beech, ash-tree, and lilac which shows evidence of a substantial rise in summer temperatures (Korotky et al., 2000b). The sea level was 3.5 m above the present-day position.

2.2. Japan and Korea

Traditional chronological units of the Holocene based on archaeological studies accepted in Japan include the Jomon, Yayoi and Kofun periods, and the historical epoch. Correlation between European post-glacial chronology and climatic fluctuations and Japanese historical and cultural chronology are given by Sakaguchi (1982, 1983) and are shown in Figure 10.2. According to this scheme, the warmest stage of the Holocene climate occurred in Japan about 7600–6500 ^{14}C yr BP (Early Jomon warmest stage); from the second half of the early stage to the beginning of the late stage of Jomon the climate became warm, from the middle part of the late

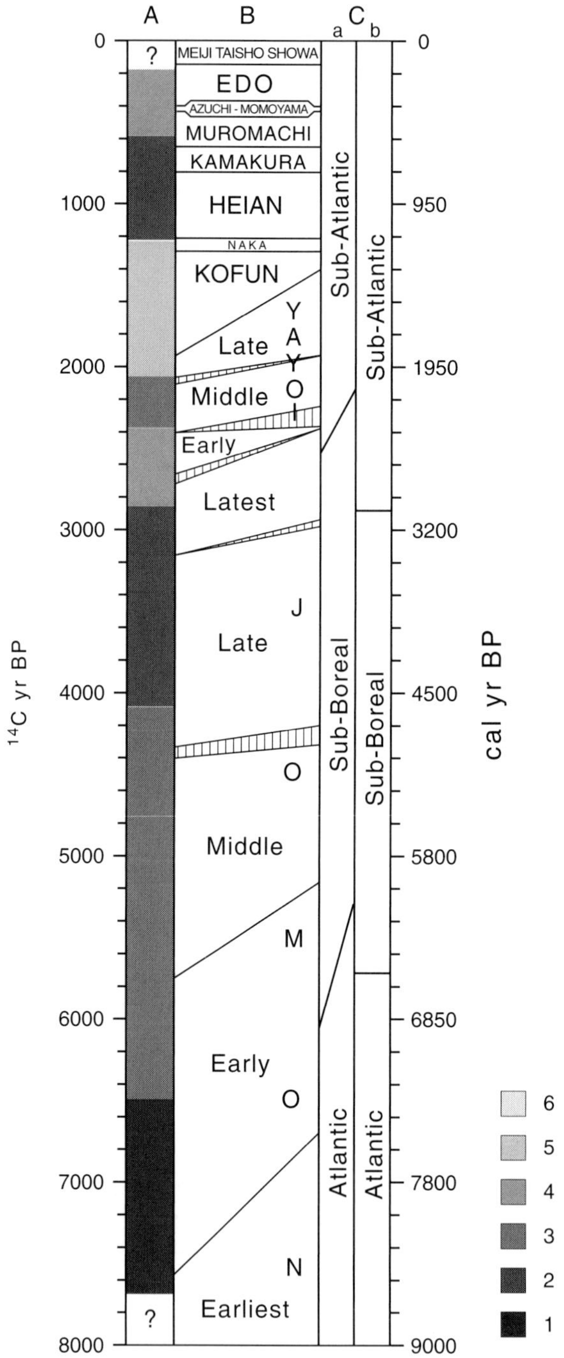

Figure 10.2. Long-term climatic fluctuations during the past 8000 years in Japan (after Sakaguchi, 1983). A – Climatic history; B – historical/cultural changes; C – variants of Blytt–Sernander scheme; 1 – warmer; 2 – warm; 3 – unstable warm; 4 – cold; 5 – unstable cold; and 6 – range of corrected data.

stage to the first half of the latest stage it became warm and unstable warm, and in the second half of the latest stage it became cold. Climatic conditions of the Earliest Jomon were never repeated in the Holocene, and, thus, the climatic optimum was a unique event. However, the cold episodes of the mid-Holocene established by Mikishin and Gvozdeva (1996) for Sakhalin are not evident in the Sakaguchi (1982, 1983) scheme. Moreover, as mentioned above, the most optimum conditions of the Holocene in Sakhalin were found in levels of between 5800 and 5100 or between 5000 and 7000 ^{14}C yr BP, while Sakaguchi (1983) believes that the Early Jomon warmest stage ended about 6500 ^{14}C yr BP. Three minima of the paleotemperature curve of the Ozegahara peat column (150 km north of Tokyo, 36° 56′N and 139° 14′E) mentioned in the text (Sakaguchi, 1983, p. 9) and studied in detail by the pollen method, namely, 6650, 5650, and 4300 ^{14}C yr BP, are not shown in the long-term climatic fluctuations scheme (Fig. 10.2). They partly coincide with those in Sakhalin (Table 10.1).

According to generalized data of Gohara (1976), the climatic optimum is dated at 6000 ^{14}C yr BP with a temperature slightly higher than that of the present. The fossil coral reefs (Numa Coral Bed, about 6000 ^{14}C yr BP [Numa age]) referred to this period were found in the southern part of Boso Peninsula in 1911 by the famous Japanese paleontologist M. Yokoyama (Gohara, 1976). However, the radiocarbon ages of the fossil corals and the associated mollusks near Tokyo (Numa Bed) indicate that the optimum conditions might have lasted for several thousand years, from about 8000 to 3000 ^{14}C yr BP (Hamada, 1977). The Numa coral assemblage formed under warmer conditions than the present oceanological situation of Tokyo Bay, is similar to those of Shikoku and/or Kyushu (Hamada, 1977).

Fujii and Fuji (1967) divided the Holocene climatic cycle into seven phases, and one of them, phase C, corresponds to the Atlantic and early part of the Subboreal in Europe; its age is between 8000 and 4000 or 3000 ^{14}C yr BP (=post-glacial climatic optimum, according to these authors). At that time, the forest zone was 200–300 m higher, and the average temperature was 2°C or 3°C higher than at present. Thus, the authors accept a broad chronological definition of the climatic optimum. Most likely, as in case of northward-extended Holocene distribution of corals in Japan, the warm climatic conditions continued for several thousand years, as exemplified by the development of vegetation, but this climate trend had a more complicated structure, namely, an alternation of short warm and cold epochs.

At about 8000 ^{14}C yr BP, oak abruptly became dominant in many parts of Hokkaido, and it decreased a little at 5000 ^{14}C yr BP (Igarashi et al., 1993). During the interval from 6000 to 1000 ^{14}C yr BP, *Quercus* forests shifted to the present subalpine zone under a warmer and moister climate than present (Igarashi, 2000). The warm, moist climate caused by the inflow of the warm Tsushima Current was the factor that caused the increase in cool-temperate forests at ca. 8000 ^{14}C yr BP; the climatic optimum in Hokkaido, therefore, began at ca. 8000 ^{14}C yr BP and it began to decline after its peak at ca. 6000 ^{14}C yr BP, with cooler climate arriving approximately 5000 ^{14}C yr BP (Igarashi, 1993); climate in Hokkaido has been cool since 2000 ^{14}C yr BP (Igarashi and Takahashi, 1985; Igarashi et al., 2001). However, about 7000 ^{14}C yr BP *Picea* and *Abies* expanded in the island, suggesting climate deterioration. Cool and moist climate reconstructed for this time can be correlated with a cold interval around 7500 ^{14}C yr BP named mesoglaciation and known worldwide (Igarashi, 1994). Sakaguchi (1989) suggested that the cold climate changed abruptly to a warm climate at 8800 ^{14}C yr BP in Hokkaido. Igarashi and Kumano (1974) recognized in Hokkaido at the level of about 7000 ^{14}C yr BP a *Picea–Betula–Myrica* pollen zone which characterizes climatic conditions as cooler compared to the present but, in contrast to the terrestrial conditions, warm-water mollusks lived in the surrounding sea. The overlying *Quercus–Juglans* zone is dominated by the temperate broad-leaved forests; the climate was warmer than the present. This is in contradiction with pollen data on Sakhalin where first warming occurred about 7200 ^{14}C yr BP (Table 10.1). In the northeastern Hokkaido, the oyster settlements were widely developed at the level of 5000–6000 ^{14}C yr BP (Ohshima et al., 1972; Matsushima, 1982a,b). Several warm-water subtropical bivalve mollusks at this time invaded the area of Kushiro Bay (eastern Hokkaido) and Nemuro (northeastern Hokkaido), but they do not live in these areas at present (Matsushima, 1984). The hydroclimatic conditions under which the so-called thermally anomalous molluscan fauna (TAMA) existed on the northern and eastern coasts of Hokkaido in the mid-Holocene can be compared with those of the present-day Mutsu Bay (northern Honshu) (Matsushima and Yamashiro, 1992). According to Matsushima and Ohshima (1974), the minimum temperature of surface waters during climatic optimum (5000–6000 ^{14}C yr BP) is estimated to be about 5°C higher than the present temperature on the Sea of Okhotsk side of Hokkaido.

In numerous shell middens on the coast of Hokkaido, some warm-water mollusks (which disappeared in the Late Holocene) were tentatively referred to the late Earliest to the middle Early Jomon stage (5000–6000 ^{14}C yr BP) (Akamatsu, 1969). It was found later that the species–indicators of the climatic optimum first appeared on the Sea of Japan side of Hokkaido at about 8000 (or 7500) ^{14}C yr BP and contemporaneously they reached Cape Soya; they appeared about 6800 ^{14}C yr BP on the Sea of Okhotsk side of Hokkaido and invaded the Pacific coast (Erimo Cape and Uchiura Bay) of the island at about 6000 ^{14}C yr BP (Akamatsu and Kitagawa, 1983; Takagi et al., 1990; Akamatsu et al., 1995). A similar chronological pattern of the immigration of the thermophilous mollusks in the mid-Holocene in Hokkaido was revealed by Matsushima (1984); warm-water species appeared on the Sea of Japan side of the island 7000 ^{14}C yr BP, and on the Pacific side ca. 6000 ^{14}C yr BP. This reflects peculiar oceanographic changes around Japan during the Holocene, i.e., the formation of the system of currents we will discuss below.

Thus, the substantial characteristic features of the climate and biota of the Middle Holocene in Hokkaido were (a) the discrepancy in migration rates of marine animals along the eastern and western coasts of the island and (b) the difference between the pattern of change of air temperature and the degree of hydroclimatic warming; the latter occurred later.

In Honshu Island, the warmest period (2–3°C higher than today) occurred about 7000–4000 ^{14}C yr BP, and vegetation was in a stage of equilibrium (Tsukada, 1986).

At that time (R II period, according to this author), the upper limit of the *Fagus* zone was 400 km higher than its modern level; southwestern Japan was characterized by warm-temperate evergreen forests (*Cyclobalanopsis, Castanopsis, Camellia*, and *Podocarpus*), and northwestern Japan by cool-temperate forests (*Fagus, Quercus, Juglans, Ulmus, Pterocarya*, and *Tilia*) (Tsukada, 1986). The age of 8000 ^{14}C yr BP corresponds to the expansion of *Fagus crenata* north of 40°N, indicating the development of a maritime climate in Japan, but only ca. 6500 ^{14}C yr BP evergreen broad-leaved trees such as *Quercus* (*Cyclobalanopsis*) and *Castanopsis* increased rapidly and climate became warmer than before (Yasuda, 1984). The development of maritime conditions between 8000 and 8500 ^{14}C yr BP also resulted in heavy snowfall.

The climatic changes by one way or another influenced sea level changes, and the sea level reached maximal high position in the mid-Holocene. The traces of the high stand and related paleoenvironmental changes in the course of global warming are studied in detail in many areas of Japan (see for reviews Pirazzoli, 1978; Ota et al., 1990; Umitsu, 1991). However, local peculiarities of the climate during the mid-Holocene are known insufficiently.

Based on the data of Matsushima (1984), who studied the Holocene molluscan faunal changes and related environmental changes, the tropical mollusks now living in the south of Kyushu and Taiwan appeared in southwestern Japan at about 7000 ^{14}C yr BP and reached southern Kanto (the plain around Tokyo Bay) at 6500–6000 ^{14}C yr BP, although subtropical species appeared as early as 9000 ^{14}C yr BP in southern Kanto and extended their distribution to northern Honshu at about 6000–5500 ^{14}C yr BP. Warm embaymental conditions were documented based on the molluscan fossils from the Taito-zaki Formation (Boso Peninsula) dated 6400–7300 ^{14}C yr BP (Ohara and Taira, 1974). Studies of the Holocene environmental history in areas near Tokyo City show that the highest stage of the sea level was probably younger than about 5950–5540 ^{14}C yr BP (this might be correlated with maximal warming) (Ando, 1986). At the same time, in the central Kanto Plain, oyster reefs were formed over a wide area at 6500–5300 ^{14}C yr BP, indicating favorable hydroclimatic conditions for thermophilous marine animals (Endo et al., 1989). The maximum stage of the Jomon transgression in Tokyo Bay is estimated by Kosugi (1989) at 6500–5300 ^{14}C yr BP, indirectly suggesting maximum warming as well. The highest sea level of the transgression based on radiocarbon dating of molluscan shells was determined to be between 6500 and 5500 ^{14}C yr BP in central Japan (Mikawa Bay) (Matsushima, 1989) and between 6400 and 5500 ^{14}C yr BP in northeastern Japan (Shimokita Peninsula) (Matsushima and Nara, 1987). In general, these estimates correspond to those for the entire Japanese coast: about 6500–6000 ^{14}C yr BP (Ota et al., 1990), 6500–5000 ^{14}C yr BP (Umitsu, 1991), or 6000–5000 ^{14}C yr BP (Fuji, 1982). However, if even maximal warming indicated by marine organisms generally coincided with the time of maximum stand of the sea level, we have to take into account that sea surface warming may not reflect air temperatures because of the influence of warm or cold currents. Thus, the trend of paleotemperature changes for terrestrial environments and their dynamics can be different from that of sea surface temperature changes. Besides a lack of coincidence of hydroclimatic warming with the course of air temperatures mentioned for Hokkaido at 7000 ^{14}C yr BP (Igarashi and Kumano, 1974), an analogous event was also documented for the Sea of Japan coast of Honshu (Wakasa Bay, central Japan). The pollen analysis of the Holocene Takahama Shell Bed dated between 4000 and 3500 ^{14}C yr BP showed that climate in that area was very similar to the present, but the fossil shell composition of the bed has some resemblance to the living fauna along the Kii Peninsula where climate is warmer than that of Takahama Shell Bed (Umeda and Nakagawa, 1993). Four species of gastropods and bivalves dated 4000–2000 ^{14}C yr BP from the shell bed cannot be found in the present Wakasa Bay, being locally extinct (Nakagawa et al., 1993a,b).

Despite the fact that a majority of Japanese paleogeographers agree that the highest sea level during the Holocene occurred around 6000 ^{14}C yr BP, Fujimoto (1990, 1993) obtained three sea level curves from Matsushima and Oku-Tokyo (Paleo-Tokyo) Bays (Pacific side) and Nanao-nishi Bay (Noto Peninsula; Sea of Japan side) showing a similar trend with the highest sea level at about 3500 ^{14}C yr BP. The sea level at that time seems to have stood 1.0–1.5 m above the present. In this case, the highest sea level stand and "the peak of warmth" in Japan and adjacent areas do not coincide. This is inconsistent with geoarchaeological data demonstrating that many ancient shell middens were present along the former coast of the Japanese Islands during the mid-Holocene and they were located near the margin of the uplands where early peoples settled (Sakaguchi, 1983). If the sea level 5000–6000 ^{14}C yr BP was several metres lower than at present (Fujimoto, 1990), it is difficult to explain why ancient people gathered mollusks at a great distance from their settlements and then carried them inland. Sakaguchi (1983) reported that the shell middens increased explosively with the beginning of the late stage of the Jomon period (maximum warming) and suddenly decreased with the closing of this stage. Based on radiocarbon dating of shell middens from environs of Tokyo, Taira (1980) showed that even minor fluctuations of sea level correspond in time to climatic events and changes of ocean surface temperature in Japan: positive sea level oscillations ("transgressions") occur at periods of northward incursion of warm waters, and vice versa, times of southward retreat of the warm waters correspond to "regressions" (Table 10.4).

Similar to Fujimoto's (1990) viewpoint, Shimoyama et al. (1986) documented the highest sea level to be about 3300 ^{14}C yr BP in the Itoshima Lowland (Kyushu). However, in other Kyushu localities, Shimoyama and Shuto (1978) detected a three-meter higher sea level between 5800 and 6000 ^{14}C yr BP, or "at least three high sea levels are suggested in the Jomon transgression ... about 6,000 y.B.P., about 4,700 y.B.P. and about 3,100 y.B.P., respectively" (Shimoyama, 1989). Kawana and Pirazzoli (1990) did not find evidence of higher Holocene sea level stands in the South Ryukyus (Irabu and Shimoji Islands), and they suggested that the sea level reached its present position some 1000–2000 ^{14}C yr BP and has remained almost stable since that time.

Table 10.4. Correlation between the Holocene sea level oscillations in Japan and relative ocean surface temperatures in Eastern Asia (after Taira, 1980).

Sea level stage (^{14}C yr BP)	Age of sea level stage (^{14}C yr BP0	Age of sea surface warming (w)/cooling (c)
Regression	8700–7500	7800–7100 (c)
Transgression	7500–6700	7100–6100 (w)
Regression	6700–6000	6100–5900 (c)
Transgression	6000–5300	5900–4700 (w)
Regression	5300–4300	4700–4300 (c)
Transgression	4300–3900	4300–3900 (w)
Regression	3900–3500	3900–3600 (c)
Transgression	3500–3100	3600–2700 (w)
Regression	3100–3000	2700–2000 (c)
Transgression	3000–2800	–

Although a detailed evaluation of sea level change is beyond the scope of this paper, we would merely like to emphasize once again that a close correlation between the sea level and climatic changes should not be neglected.

Paleoclimatic and paleoenvironmental reconstructions in Korea for the last 18,000 years are based on a limited data set, compared to extensive studies of the Chinese coast, the Japanese Islands, and the Russian Far East, and much more detailed and substantial studies are required as pointed out in a review of Park and Yi (1995). In general, the mid-Holocene time is believed to be a warm and dry period (Kong, 1994). The post-glacial hypsithermal period is defined in a time span between 8000 and 4500 ^{14}C yr BP and is characterized by the dominance of pine (*Pinus* (*Diploxylon*)) and oak (*Quercus*), a gradual increase of decidious broad-leaved trees in central Korea, and the appearance of thermophilous evergreen broad-leaved trees in southern Korea (Kong, 1994). Within the climatic optimum (hypsithermal), the three stages were recognized, but their chronological levels are unknown (Yi et al., 1996a,c). Han et al. (1996) and Yi et al. (1996b) also suggested that sea level was higher than present-day on the western coast of Korea during the mid-Holocene, which is in contradiction with viewpoint of Park (1987). The former authors interpreted gravel deposits with inclusion of shells of the oyster *Crassostrea gigas* found 4 and 5–6 m above the present sea level in some coastal areas as formed during periods of increasing storm activity of the Yellow Sea. The North Korean geologists also suggested a high stand of sea level occurred at 6000 ^{14}C yr BP (Ryu and Sin, 1993). In the neighboring Liaoning Province of China (coast of the Yellow Sea), a warm period occurred at the level of 8000–3000 ^{14}C yr BP and was characterized by a humid temperate climate with an annual mean temperature of about 13°C (3–5°C higher than present) in its early phase (Early Takushan period = Atlantic) (Chen et al., 1978). Rapid warming in China began at 8500–8300 ^{14}C yr BP, and deviation of annual mean temperatures from present ones about 7000–6000 ^{14}C yr BP was roughly estimated at about 2°C in the Yangtze Valley, and 3°C in northern and Northeastern China (Shi et al., 1993). The Holocene highest sea level in eastern China occurred within 6500–5000 ^{14}C yr BP (Shi et al., 1994). In general, climatic optimum in China is placed between 8500 and 3000 ^{14}C yr BP (Zhang, 1995).

2.3. Primorye and Priamurye

The beginning of the mid-Holocene (Boreal period) in Primorye is characterized by the appearance of birch–broad-leaved (*Betula–Quercus*) and birch–elm (*Betula–Ulmus*) forests (the latter occurred in northern part of the province) with an admixture of various thermophilous trees (*Phellodendron, Juglans, Tilia, Carpinus, Fraxinus, Aralia*) and the Korean cedar (*Pinus koraiensis*) in mountain areas (Korotky et al., 1980, 1988; Golubeva and Karaulova, 1983; Shumova, 1995). Non-forest space was occupied by wormwood associations. It is believed that annual average and average July air temperatures were higher by 1–2°C compared to the present, average January temperatures were 0.5–1°C above the present; and the amount of precipitation was close to present or slightly lower (Shumova, 1995). A cooling of climate took place at a chronological level of 8000–8500 ^{14}C yr BP corresponding to the Novosantchugian cold episode of Siberia (Golubeva and Karaulova, 1983; Shumova, 1995). At that time (the second half of the Boreal), the areas occupied by steppes increased along with a decline of forests and impoverishment of their floral composition. Forest-steppe of birch–oak type with an inclusion of linden (*Tilia*), elm (*Ulmus*), and hazel (*Corylus*) was the predominant kind of vegetation. Alder–birch (*Alnus–Betula*) forests and heaths (*Erycales*) with larch (*Larix*) were distributed along the seacoast. According to Shumova (1995), the climate in the Late Boreal phase was colder and drier than the present; annual average as well as average July and average January temperatures were, respectively, 2, 1, and 2–3°C lower compared to the present values.

A majority of paleogeographers agree that the Atlantic period was the time of maximum warming and wetting of the climate in Primorye. The range of estimates of climatic parameters for the mid-Holocene of the Russian Far East is presented in Table 10.5. As one can see, excess of annual average temperatures varies between 2 and 6°C; all these figures are based on the palynological analysis. Vertical natural belts in the southern Sikhote-Alin Mountains shifted 600–1000 m upward. The main type of vegetation were polydominant broad-leaved forests (with the predominance of oak), and with an increased role, compared with the present, of thermophilous trees such as *Carpinus* and *Juglans*. The share of their pollen reached 10% or even 15%. A number of other species whose pollen is rarely found in recent surface samples were characteristic elements of the mid-Holocene vegetation in Primorye – *Acer, Tilia, Fraxinus, Syringa, Aralia* (Korotky and Karaulova, 1976; Golubeva and Karaulova, 1983; Verkhovskaya and Kundyshev, 1993, 1995; Gvozdeva et al., 1997; Korotky, 1998). The Sikhote-Alin' Mountains characterized at present by dark-coniferous taiga were covered with cedar/broad-leaved and

Table 10.5. Climatic estimates of the mid-Holocene in the Russian Far East.

Area	Excess of average annual temperature (as compared to the present) (°C)	Precipitation (mm/year)	Source
Southern Far East	4–6	"More humid"	Korotky et al. (1988)
	3–5	–	Korotky et al. (1980, 1996a)
Latitude of Vladivostok City (43° 07′N)	3–4	More than 1000	Korotky (1994)
Northern Sikhote-Alin' Mountains	3–5	–	Korotky et al. (1981)
Kievka Bay (southern Primorye)	2	"More humid" or above 50–100 as compared to the present	Kuzmina et al. (1987); Shumova and Klimanov (1989); Shumova (1995)

broad-leaved forests with an admixture of oak. Boreal (temperate) dark-coniferous trees (*Picea* and *Abies*) occurred very rarely because their pollen is volatile and it is detected in small quantities in samples of deposits. Analogous vegetation grew in the northern regions of Primorye, but it additionally included the birch *Betula schmidti* (Golubeva and Karaulova, 1983). The forest coverage in Primorye was maximal as compared to the previous and subsequent stages of the Holocene except, probably, for the extreme southwestern part of the territory – a coastal lowland near the Talmi Lagoon (located near the Russian/North Korean border) where motley grass meadows were distributed at 5300–5600 or even since 6000 ^{14}C yr BP as well as at present (Golubeva and Karaulova, 1983). According to Verkhovskaya and Kundyshev (1995), this type of vegetation formed later, at the very end of climatic optimum, when global warming and dry conditions took place.

A cooling in the beginning of the Subboreal (4200–4700 ^{14}C yr BP, according to Korotky (1994)) brought about a decline of thermophilous broad-leaved species and a wide distribution of dark-coniferous trees, e.g., spruce (*Picea*), *Abies* and the Korean cedar (*Pinus*), and the appearance of frigid shrubs. There is no agreement concerning the type of vegetation at that time. According to Shumova (1995), spruce–oak (*Picea–Quercus*) and cedar–oak (*Pinus–Quercus*) dominated and they tended to occur in the mountain areas in southern Primorye; larch (*Larix*) and alder (*Alnus*) forests occupied near-shore areas. The lowering of annual average temperatures reached 2°C compared to the present values; the amount of precipitation was similar to or a little lower than that of the preceding period (Shumova, 1995). Based on other data obtained from archaeological sites of the Zaysanovskaya culture

(Verkhovskaya and Kundyshev, 1993), the climate in southern Primorye at ca. 4000–4600 ^{14}C yr BP was much drier than the recent climate: forest-steppe and steppe occupied extensive areas; birch and hazel (*Corylus*) were the dominant forest vegetation. Eolian processes increased on the seacoast.

A warming of the mid-Subboreal that followed the Early Subboreal cooling was significant, but not as pronounced as that of the Atlantic. This time was also drier than the Atlantic as confirmed by quantitative and qualitative analysis of moisture-loving thermophilous broad-leaved trees (Korotky et al., 1980, 1988; Golubeva and Karaulova, 1983). According to Shumova (1995), excess of annual average temperatures were 1–2°C compared to the present values. These figures are obviously too low. Declines of broad-leaved trees in forests was caused by the dry climate and not by low temperatures; the Subboreal warming in southern Sakhalin 3300–3700 ^{14}C yr BP was similar in its temperature parameters to the Atlantic warming (Mikishin and Gvozdeva, 1996) or even stronger than the latter. Oak forests were dominated in Primorye at that time. It was a period of maximum distribution and flourishing of oak in the Holocene in the Russian Far East: *Quercus* forests were displaced northwardly as far as Sovetskaya Gavan' Town (nearly 50°N) along the coast of the Sea of Japan. In the northern Sikhote-Alin' Mountains, dark-coniferous forests were distributed more widely when compared to the Atlantic (Korotky et al., 1980; Golubeva and Karaulova, 1983).

The history of the development of the vegetation and climate of the Lower Priamurye in the Holocene was reconstructed based on more than 20 sections and outcrops (Sokhina et al., 1978). The authors distinguished five phases of the vegetational development, but the primary shortcoming of this investigation was the small number of radiocarbon determinations, and, thus, the chronological frames of the phases can be re-considered upon appearance of new data.

The second phase of this scheme corresponds to the Boreal period. At that time, cedar (*Pinus*)–broad-leaved forests with the participation of birch in the south, and coniferous and birch forests (with a small inclusion of broad-leaved trees) in the northern areas were common. Lowlands in the valley of the Amur River were occupied by larch, sedge (*Carex*), and bog-moss (*Sphagnum*) swamps. In the northern continental regions, larch–dark-coniferous forests with an admixture of birch and rare broad-leaved trees as well as shrub vegetation dominated. Probably, the Boreal vegetation in the near-mouth area of the Amur River was more "northern" in character than described by the above authors. As evidenced by data of Mikishin et al. (1987), in this area small-decidious and larch forests were predominant about 8400–9300 ^{14}C yr BP; this is confirmed by the findings of massive trunks and stumps buried under more younger peats. Dark-coniferous trees were sparse, and thermophilous broad-leaved species did not grow in this area.

The third phase coincides with the boundary between the Boreal and Atlantic and is characterized by a cooling of climate. This phase is more expressed in the southern regions of Priamurye where *Abies–Picea* (with an admixture of the Korean cedar) increased in the mountains; birch forests and shrub–bog-moss swamps occurred in the northern areas.

The fourth phase is correlated with the Holocene optimum, i.e., the Atlantic, and partly with the Subboreal (6000–3000 ^{14}C yr BP). This phase is characterized by the maximum warming and further advancement of the broad-leaved forests to the north, and by upward migration of forest vegetation on the mountain slopes. The upper limit of the forest zone in the Yam-Alin' Range (53.5°N) shifted 300–400 m upward (Korotky et al., 1988). Birch–broad-leaved forests thrived in the southern regions of Priamurye; dark-coniferous forests advanced to high part of mountain slopes. In the northern coastal areas (the mouth of the Amur River), cedar–broad-leaved and *Abies–Picea* forests were developed, but birch-dominated forests with sparse growth (and with rare inclusion of broad-leaved species) were distributed in the plains. Dark-coniferous (with an admixture of small-decidious and broad-leaved species and larch) forests were distributed in the northern continental areas. Unfortunately, no precise estimates of temperature parameters have been obtained for Priamurye.

2.4. Mid-Holocene paleoceanography of the Sea of Japan

The Sea of Japan (East Sea in Korean terminology) is a marginal, deep basin enclosed by an island arc, which separates it from the Pacific Ocean. The Sea of Japan is directly connected with the Sea of Okhotsk by the Tatarsky (Mamiya) Strait (sometimes called "the Gulf of Tartary"), and with the East China Sea by the Tsushima and Korea Straits (between Korea and Kyushu Island). Tsugaru (between Honshu and Hokkaido) and La Perouse (or Soya; between Sakhalin and Hokkaido) Straits connect the sea with the Pacific Ocean. The sea is rather isolated as the straits mentioned are shallow; the maximum depth of Korea Strait is 150 m, and Tsugaru Strait, 200 m; Nevelskogoko Strait (northernmost Tatarsky Strait) has a sill depth of 5 m, La Perouse Strait, 53 m. This suggests that the deep basin of the Sea of Japan has no exchange with deep Pacific waters (the maximum depth of the sea is 3695 m in its eastern part (Nishimura, 1983), or 3670 m (Yurasov and Yarichin, 1991)). The Sea of Japan is unique among the enclosed seas of the northwestern Pacific in having strong ocean currents in the form of powerful streams, and we will briefly describe the present-day system of currents based on Nishimura (1983) and Yurasov and Yarichin (1991).

The Sea of Japan is divided by a frontal zone (or "polar front") into two distinct areas: the southeastern warm-water area washed by the Tsushima Current with three branches (T-1, coastal branch flowing along western Japan; T-2 and T-3, offshore branches; T-3 is called the East Korean Current in Russian oceanographical literature), and the northwestern area dominated by the cold Liman(ian), Primorskoye, and North Korean Currents. This division is clearly reflected in the biota and biological production of the sea (Nishimura, 1965a,b, 1966, 1968, 1969). A variant of the modern system of currents in the Sea of Japan is shown in Figure 10.3. The Tsushima Current enters the sea particularly through the Korea Strait and flows northeastward in three meanders including the East Korean

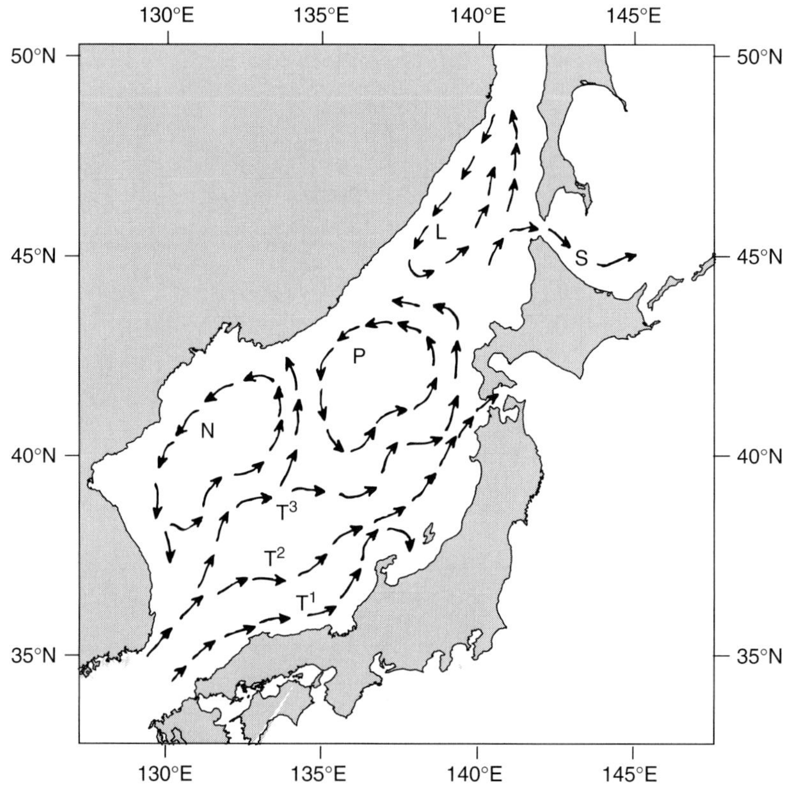

Figure 10.3. A scheme of the present-day system of currents in the Sea of Japan (modified after Nishimura, 1983). N – North Korean Current (cold); T1-T3 – Tsushima Current (warm; T1 – coastal branch; T2 and T3 – offshore branch); P – Primorskoe Current (cold); L – Liman Current (cold); and S – Soya Current (warm).

Current. According to Russian authors (Yurasov and Yarichin, 1991), only two stable branches can be recognized in the warm area of the sea. The Tsushima Current flows out into the Pacific Ocean, mostly through the Tsugaru Strait and partly through the La Perouse (Soya) Strait into the Sea of Okhotsk but its northernmost part reaches middle Sakhalin Island in Tatarsky Strait. The northern cold area of the Sea of Japan is washed by three cold currents collectively known in Japanese literature as the Liman Current, but the latter by itself is a current washing continental coast of the northern Sea of Japan – between Tatarsky Strait and mid-Primorye. Because the Liman Current bears no similarity in its hydrophysical characteristics to waters of the Amur River estuary as well as the cold Primorskoye Current flowing south of the Liman Current, it was proposed that the latter be renamed the Schrenck Current (Yurasov and Yarichin, 1991). It is believed that the cold currents whose volume transport and speed are much smaller than those of the Tsushima Current are counter- or compensation currents of the latter (Nishimura, 1983).

The system of currents influenced significantly the climatic conditions (air temperatures, humidity, etc.), distribution of coastal and open-sea organisms and, probably, cultural dynamics of along-shore settlements of the ancient people. Estimations of time of the appearance of warm currents in different areas of the Sea of Japan are controversial and are based chiefly on diatom analysis, oxygen isotope analysis of benthic and planktonic foraminifers, distributions of the foraminifers alone, and molluscan assemblages considerations. According to Chinzei and Oba (1986), the inflow of the Tsushima Current started after 9500 ^{14}C yr BP, and, at the level of 6300 ^{14}C yr BP, low δ18O value as well as planktonic foraminifers and other microfossils rich in warm-water species indicate a predominant influence of the Tsushima Current in the southeastern Sea of Japan. The influx of the Tsushima Current might have occurred on a full scale since 8000 ^{14}C yr BP and it caused a remarkable wetting in Japan, but the strongest influx was after 6000 ^{14}C yr BP (Koizumi, 1987, 1989). After that time, four peaks of "diatome temperature" were detected in the mid- and late Holocene. Yi et al. (1997) estimated the time of full circulation of water mass between the East Sea (Sea of Japan) and the East China Sea ("the South Sea [of Korea] and Okinawa Sea"), i.e., the full influence of Tsushima Current on the southern part of the Sea of Japan (Ulleung Basin) occurred about 8000 ^{14}C yr BP. Igarashi (1993) also supported the viewpoint that the full-scale onset of the Tsushima Current was approximately 8000 ^{14}C yr BP. Gorbarenko et al. (1995) assumed that oceanographic parameters are similar to those of the present onset in the Sea of Japan some 7000 ^{14}C yr BP. The rising sea surface temperatures are recorded from 6000 ^{14}C yr BP by study of the ratio of planktonic foraminifers (Gorbarenko, 1991). The isotopic values of mollusk shells collected from archaeological sites and natural shell beds in central Japan (the Pacific side) indicate that the water temperature was highest about 7000 ^{14}C yr BP, and then became lower at around 4500–4000 ^{14}C yr BP (Chinzei et al., 1987b). Gorbarenko and Southon (2000) concluded that the warmest conditions in the Sea of Japan occurred at 6500–6000 ^{14}C yr BP. A brief cooling event that occurred about 4000 ^{14}C yr BP is registered also for the Okinawa area (Ujiie, 1997) and the northeastern South China Sea (Wei et al., 1997, 1998; see also Jian et al., 1998). Sakaguchi (1987) also recorded a cold episode at 4500 ^{14}C yr BP in the Ozegahara peat bog. The same trend seems to be shown for the subtropical Kuroshio front along the Pacific coast of Japan based on the oxygen and ecological analyses of microfossil assemblages; an increase of warm-water species continued until about 6500 ^{14}C yr BP when temperature record shows a value much higher than present (Chinzei et al., 1987a). Taira (1975) established that ca. 7000 ^{14}C yr BP, the Kuroshio temperatures were up to 8°C or averaging 6°C above the present values. This estimate seems to be too high. Likewise, Matsushima and Ohshima (1974), based on an analysis of molluscan assemblages, argued that minimum surface temperature of the southern Sea of Okhotsk was 5°C higher than at present. Taira's (1975) estimates were based on the oxygen isotope analysis of molluscan shells and corals. Chinzei et al. (1987b) demonstrated that direct conversion of isotope values into the water temperatures yields higher temperature values than expected, i.e., the minimum summer temperature for 7000 ^{14}C yr BP would be 35°C, which is unnatural for this area (Boso Peninsula).

Sakaguchi et al. (1985) found a tropical–subtropical (in zonal-geographical terminology used hereafter; for explanation see Lutaenko, 1993) species of bivalve mollusk, *Trapezium liratum*, in a core obtained in Tokoro Plain (northern Hokkaido) at the level of 8520±120 ^{14}C yr BP. Sakaguchi (1992) suggested that this finding indicates the birth of the warm La Perouse (Soya) Current, a branch of the Tsushima Current. Thus, the Tsushima Current reached the Sea of Japan side of Hokkaido before 8500 ^{14}C yr BP, which accords with data of M. Akamatsu and his co-workers (Akamatsu and Kitagawa, 1983; Akamatsu et al., 1995) on the appearance of warm-water mollusks in the western Hokkaido at about 7500–8000 ^{14}C yr BP. However, the warm-water mollusks invaded the Hokkaido coast of the Sea of Okhotsk at about 6800 ^{14}C yr BP (Takagi et al., 1990). These differences may be explained by insufficient geochronological evidence and the controversy will be settled with increasing of accelerated mass spectrometry (AMS) datings of molluscan shells.

Taira (1979) suggested an abrupt Holocene oceanic warming started between 9000 and 8000 ^{14}C yr BP. At 9000–7800 ^{14}C yr BP (T-1 transgressive stage, according to his scheme), warm waters were introduced into the Sea of Japan, Yellow Sea, and northwestern Pacific, but it seems that the Tsushima Current did not move northward at that time, at least, the current did not reach Hokkaido yet (Taira, 1992). The warmest interval in relative surface temperatures occurred 5900–4700 ^{14}C yr BP, the Tsushima Current reached north Hokkaido, and warm waters were first imported into the Sea of Okhotsk through the Soya Strait, and at 4300–3900 ^{14}C yr BP the Kuroshio Current retreated southward (Taira, 1992); this nearly corresponds to an observed lowering of temperature on the Pacific side of Japan during 4500–4000 ^{14}C yr BP (Chinzei et al., 1987b) and a cooling event around 4000–2000 ^{14}C yr BP based on planktonic foraminifera from the Okinawa Trough and the South China Sea (Jian et al., 1996). The warming of surface waters in the Sea of Japan seems to occur later than that of the Pacific side of Japan because of cold deep circulation, meltwater discharge from the Siberia, and winter sea ice in the northern part of the sea (Taira, 1992). A pronounced, but relatively short period regional cooling (called Kuromatsunai cold episode 2) was detected at ca. 9000 ^{14}C yr BP; it was caused by cold meltwater spreading over the Seas of Okhotsk and Japan because the maximum solar radiation brought about an enormous volume of permafrost meltwater from the Amur River basin, Primorye, Sakhalin, and Hokkaido (Sakaguchi, 1992). In this case, Taira's (1992) scenario explaining the discrepancy between Holocene current patterns around Japan is possible. For instance, Oba (1997) demonstrated that the Kuroshio front migrated northward passing off Boso Peninsula (Tokyo) at 10,000 ^{14}C yr BP, but the Tsushima was not active at that time.

Another important feature of the mid-Holocene paleoceanography is a difference in the rate of penetration of warm currents not only along the Pacific and Sea of Japan sides of Japan, but also along the island and continental coasts of the sea, as was demonstrated by using molluscan assemblages (Taira and Lutaenko, 1993). In the Early Holocene, the coasts of North Korea and Primorye were washed by intensified cold currents of the Liman (Schrenck), Primorskoye, and North Korean Currents, and thereby their cold waters acted as a barrier to any northward flow of warm waters. This seems to be supported by the lack of subtropical bivalve mollusks in the Early Holocene deposits along the Primorye (Evseev, 1981). We suggested that about 7000–6000 ^{14}C yr BP, the East Korean Current, a branch of the Tsushima, moved northward at about 40°N, and subtropical bivalve mollusks reached Peter the Great Bay (northwestern Sea of Japan) (Taira and Lutaenko, 1993). The meandering stream of the Tsushima Current, T-3 offshore stream (=East Korean Current), is known to be strongest (Nishimura, 1983). The possibility of the East Korean Current axis shifting northeastward in the mid-Holocene is confirmed by the fact that even at present, during some "warm" years, the T-3 can be observed near Peter the Great Bay, as evidenced by the findings of some pelagic subtropical and tropical animals like some fishes, etc. (Ivankov, 1996). The intensification of the East Korean Current in the mid-Holocene led to the appearance in the northwestern Sea of Japan not only of subtropical, but also tropical–subtropical bivalve mollusks (whose geographical ranges are extended southward to the Philippines, Vietnam, and Indonesia) which formed stable populations with annual reproduction, species such as *Anadara inaequivalvis, T. liratum*, and *Dosinia penicillata* (Lutaenko, 1991a, 1993). They settled in bays characterized by intense summer warming, which is necessary for successful reproduction (winter cooling in itself does not prevent warm-water fauna from living in temperate latitudes (Scarlato, 1981)). Thus, a combination of such factors as the considerable indentation of the coast (ria type of bays with shallow water semi-enclosed areas in their tops) and penetration of the warm Tsushima waters to the northwestern Sea of Japan which intensified the effect of local warming had resulted in the formation of subtropical-type molluscan fauna in this area during the Middle Holocene (Lutaenko, 1991b, 1993). The correlation and relative role of warm currents and geomorphological processes in the appearance/disappearance of TAMAs have been discussed in the literature (DeVries and Wells, 1990; Sandweiss et al., 1996, 1997; DeVries et al., 1997; Wells and Noller, 1997), illustrating the importance of correct interpretation of paleomalacological data when reconstructing Holocene oceanographical conditions. The example of Peter the Great Bay mid-Holocene TAMA shows that three species of warm-water bivalves became extinct in the course of the Late Holocene coolings (distributional ranges of two of them are shown in Fig. 10.4), while embaymental environments were still existing. This means that the climatic changes, and not solely changes in coastal morphology, are responsible for local extinctions of warm-water species.

The Kuroshio Current probably extended to the Hokkaido coast earlier than 5900–4700 ^{14}C yr BP, since the first warm-water mollusks appeared here about 7500–8000 ^{14}C yr BP. However, there is still no direct evidence on the chronological ranges of existence of thermophilous species on Hokkaido due to a lack of AMS datings. Fifteen AMS dates on three "extinct" (locally disappeared) species of

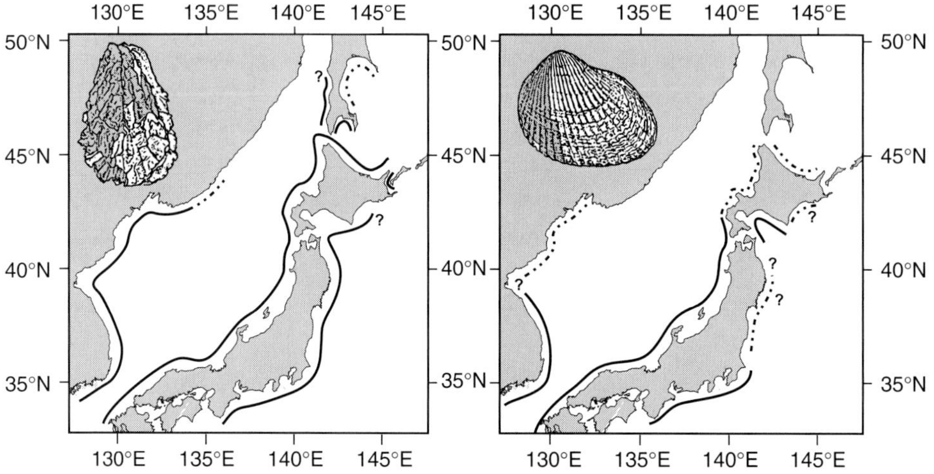

Figure 10.4. The distribution of the TAMA's elements – warm-water bivalve mollusks *Anadara kagoshimensis* (right) and *Crassostrea gigas* (left) during the mid-Holocene in the Sea of Japan and adjacent areas.

bivalves from the coast of Peter the Great Bay coast demonstrate that their Holocene ranges lie between 7140 and 1260 ^{14}C yr BP (taking into account a reservoir effect) (Jones and Kuzmin, 1995; Kuzmin, 1995). As mollusks reflect the effect of the East Korean Current, we can assume that the current penetrated to the northwestern Sea of Japan about 7000 ^{14}C yr BP, which is 500–1000 years later compared to Hokkaido. A comparison of the mid-Holocene TAMAs from the shell middens of Hokkaido and Primorye (Rakov et al., 1996; Rakov and Lutaenko, 1997) revealed a difference in species composition: at least five species found in Hokkaido have never lived in Peter the Great Bay. The oyster, *Crassostrea gigas*, invaded the coast of Terpenye Bay (~50°N) in Sakhalin Island (Sea of Okhotsk side) (Akamatsu and Ushiro, 1992) and also penetrated to the coast of middle Primorye (Vladimir Bay – V.A. Rakov, personal communication) (Fig. 10.4). Another example is provided by the subtropical mollusk, *Anadara broughtoni*, discovered in the Neolithic shell midden in Chertovy Vorota Cave (Khudik, 1991). A refuge for the oyster exists in the northern Tatarsky Strait, in De Kastri Bay (Scarlato, 1981), clearly illustrating the mid-Holocene strong influence of the Tsushima Current flowing along the eastern side of Sakhalin, as the continental coastal area was occupied by the cold Schrenck Current. These new data make it possible to reject an early interpretation postulating an absence of subtropical species of bivalve mollusks on the mid-Primorye coast (Lutaenko, 1993).

Pletnev et al. (1987) suggested that an increase of convective mixing of water masses occurred during the Atlantic in the eastern part of the Tatarsky Strait despite the influence of the Tsushima Current. This resulted in a lack of stable stratification and warming of surface water layer manifested in a high (for the period of climatic optimum) population density of diatoms (due to upwelling of nutrients to photic zone) and an increase of the level of carbonate compensation (strong dissolution of planktonic foraminifers). However, the boundary of winter drift ice shifted northward during the mid-Holocene compared to at present in the area of Tatarsky Strait (Korotky et al., 1997).

Quantitative estimates of hydroclimatic warming of surface waters in the Sea of Japan and adjacent areas the Sea of Okhotsk are given in Table 10.6 and are shown in Figure 10.5. Many authors estimated an excess of annual average SSTs as compared to the present values to be 1–2°C in different areas of the Sea of Japan or the entire basin. Troitskaya (1974) did not define clearly the temperature parameters, and, based on the comparison of the species composition of benthic foraminifers, gave an excessively high figure (Table 10.6). Unfortunately, this estimate was cited by subsequent authors without any substantiation (Pushkar, 1979; Korotky and Vostretsov, 1998). Rakov (1995) estimated the July–August

Table 10.6. Hydroclimatic estimates of surface waters of the Sea of Japan and the Sea of Okhotsk in the period of climatic optimum of the Holocene (based on data of Russian authors).

Area	Water temperature parameters	Excess as compared to the present (°C)	Source
Sea of Japan	Annual average SSTs	1–2	Pletnev (1985); Pletnev and Grebennikova (1991)
	Annual average SSTs	1–2	Korotky et al. (1988)
	SSTs	3–8	Korotky (1994)
Yamato Rise (southern Sea of Japan)	SSTs	1–2	Pletnev (1978)
Tatarsky (Mamiya) Strait	SSTs	2	Pletnev et al. (1987)
Tatarsky Strait and northern Sea of Japan	SSTs	2	Korotky et al. (1997)
Peter the Great Bay (northwestern Sea of Japan)	July–August shallow water temperatures	3–11	Rakov (1995) (based on Fig. 8)
Peter the Great Bay (northwestern Sea of Japan)	Shallow water temperatures	10	Troitskaya (1974)
Southern Sakhalin shelf (southern Sea of Okhotsk)	Annual average SSTs	5–7	Cherepanova et al. (1988)

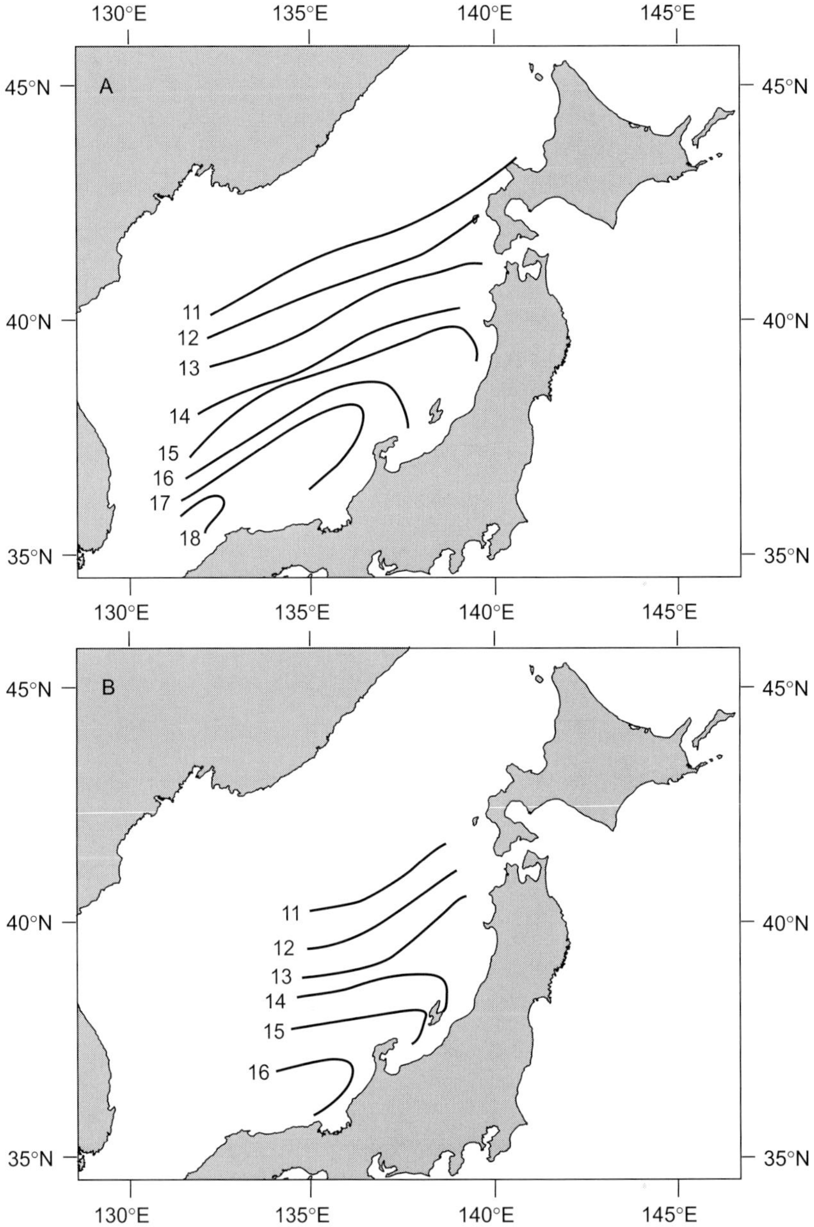

Figure 10.5. Present-day (A) and the mid-Holocene (B) annual average SSTs (C) in the Sea of Japan as evidenced by the planktonic foraminifers (modified after Pletnev, 1982, 1985).

temperatures in shallow inlets of Peter the Great Bay to be 3–11°C higher than the present-day values, and, moreover, he suggested that this area was devoid of shore ice during the winter time in the mid-Holocene (see also Rakov and Vostretsov, 1998) (at present, shallow water bays and lagoons are completely covered by seashore ice in winter). This conclusion is based on the temperature tolerance of a locally extinct element of TAMA, *Meretrix lusoria*. However, this species inhabits the intertidal and upper subtidal zone of the Yellow Sea (Pak, 1985; Bernard et al., 1993) including its upper part, the Bohai Sea, which is covered with ice during the winter at present. Chinzei (1993) showed that paleotemperature analysis based on the existence of the extant molluscan taxa or associations incorporates a large degree of uncertainty because mollusks react to overall environmental conditions (not only temperature) in an intricate way. Some problems related to interpretation of the Holocene assemblages of mollusks were discussed by Maeda (1978) with examples from Osaka and Tokyo Bays (the Pacific coast of Japan); despite the warming in the mid-Holocene, thermophilous species did not inhabit these areas due to the narrow opening of the bays, and, possibly, limited larvae exchange between the open and semi-enclosed, protected environments. Thus, the molluscan data should be used with caution for the precise estimations of the sea surface temperatures.

Yi et al. (1996b) suggested that a series of "old" spits (mid- and late Holocene in age) consisting of gravel mixed with reworked oyster shells (beach driftage) discovered on the western coast of Korea should be interpreted as traces of storm or, at least, storm-influenced deposits. Their origin is believed to be related to a global warming which caused high storm frequency (Yi et al., 1996b; Lutaenko, 2001). We explained in a similar way the origin of a high gravel terrace with abundant TAMA's elements, *Anadara inaequivalvis*, found near the Russian/North Korean border, on the northern coast of Talmi Lagoon (Lutaenko, 1997). The height of this terrace (old beach ridge) is about 4 m above the present sea level. Molluscan fossils from the "old" terrace were dated by both conventional and AMS methods (Alekseev and Golubeva, 1980; Jones and Kuzmin, 1995), and their ages are 5320 $\pm$ 45 (OS-3026), 5360 $\pm$ 35 (OS-3028), 6000 $\pm$ 130 (GIN-759b), and 5630 $\pm$ 110 (GIN-759a) ^{14}C yr BP. This provides an example of the geomorphic imprint of the mid-Holocene storm activity in the Sea of Japan between 5000 and 6000 ^{14}C yr BP. At present there is no evidence of storm accumulation of coarse-grained deposits in the two above areas of Korea and Primorye.

The suggestion about increased storm activity during the mid-Holocene in the Asian marginal seas seems to be confirmed by the analysis of data on the prevalence of different types of coastal accumulation throughout the Holocene in Japan, Primorye, and Sakhalin (Afanasiev, 1992). This author used more than 400 radiocarbon datings of coastal terraces characterized by different sedimentological composition. According to the scheme compiled (Fig. 10.6), there were three synchronous phases of storm and storm-influenced accumulation in the above areas, i.e., at 5300–6000, 3000–4300, and 1400–1700 ^{14}C yr BP. In general, they coincide with the main warmings in the Russian Far East; the first phase of storm accumulation is in close accord with maximum stage of warmth in Sakhalin at the level of 5900–5100 (Table 10.1). It is likely that increased storm frequency was one of the paleoceanographic features in the Sea of Japan along with intensification of the warm currents.

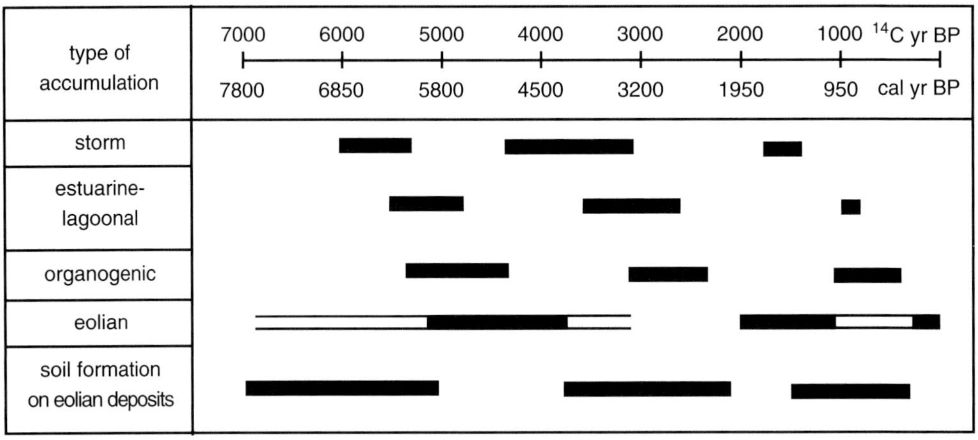

Figure 10.6. Temporal predominance of the coastal sedimentation processes throughout the Holocene in Japan, Sakhalin, and Primorye (after Afanasiev, 1992).

3. Cultural dynamics in the mid-Holocene in the basin of the Sea of Japan

3.1. Japanese Archipelago

Neolithic Jomon culture represents a classic case of continuous cultural development over a long time – roughly ten thousands years – in isolated or almost isolated territory. The abundance and great variability of its archaeological assemblages have ensured a special place for Jomon in the archaeology and prehistory of the Japanese Archipelago (Yamanouchi, 1937, 1964; Sugiyama, 1942; Watanabe, 1966; Kidder and Esaka, 1968; Iofan, 1974; Aikens and Higuchi, 1982; Vasil'evsky et al., 1982; Pearson, 1986, 1992; Nishida, 1989; Kobayashi, 1992a,b, 2004; Ishikawa and Suzuki, 1994; Aikens, 1995; Takahashi et al., 1997). The problems of systematization, periodization, and dating of Jomon culture sites are an area of concentration for many research efforts. Some of the most accepted versions of the dating of Jomon culture's stages are represented in the Table 10.7. Despite of some differences between the versions, chronological frameworks for the Jomon culture overall correspond to the Early and Middle Holocene periods. Within these frameworks the earliest stages of Jomon (Incipient, Initial) correlate with the Early Holocene and the beginnings of Middle Holocene; the early and middle stages (Early and Middle Jomon) closely correspond to the mid of Middle Holocene; and the late stages (Late and Final Jomon) correspond to the end of the Middle Holocene and the beginnings of Late Holocene.

In the context of our topic, the Early, Middle, and Late Jomon stages are most interesting as the ones representing cultural dynamics during the Middle Holocene period. Cultural dynamics are reflected in characteristic traits of every stage.

Table 10.7. Chronological scales of Jomon culture.

Jomon culture stages	Versions of dating according to			
	Watanabe (1966)	Ikawa-Smith (1980)	Pearson (1992)	Hatsushige and Mitsunori (1996)
Incipient		13,000–9500	12,500–10,000	12,000–9000
Initial	10,750–6200	9500–7500	10,000–7000	9000–6000
Early	3400–2800	7500–5600	7000–4500	6000–5000
Middle	2800–2230	5600–4500	4500–3500	5000–4000
Late	1800–1050	4500–3000	3500–3000	4000–3000
Final	1100–300	3000–2300	3000–2400	3000–2400

3.1.1. Early Jomon

The changes in Jomon culture connected with its Early stage are concerned mainly with mode of life, subsistence, and material culture. Early Jomon sites (Fig. 10.7) give sure evidence for the establishment of stabilized or long-lived settlements. The dwelling type was the semi-subterranean pit-house of rectangular shape with two or three rows of wooden pillars along long axes supporting the walls and roof. The appearance of stable long-lived settlements may be considered as an indicator of the establishment of a sedentary mode of life. In the previous Incipient and Initial Jomon periods a mobile or semi-sedentary style of existence dominated (Nishida, 1989).

The subsistence of Early Jomon followed the traditions of previous times and consisted of the hunting and fishing of terrestrial and maritime resources. Geographic variability in the occurrence of food resources in Japan resulted in various local patterns of subsistence. The population of seacoast areas was oriented primarily to marine resources, while the habitants of inland zones intensively exploited terrestrial food sources. This subsistence variability was preserved until the Final Jomon stage (Aikens and Higuchi, 1982, pp. 182–186; Pearson, 1992, pp. 65–66; Takahashi et al., 1997).

A significant innovation appeared during the Early Jomon stage, the cultivation of edible plants like bottle gourds (*Lagenaria siceraria*), beans (*Vigna angularis*), burdock (*Arctium*), and others. Cultivated plants formed a relatively small part of the diet and did not require a lot of effort in cultivation; intensive gathering rather than the agriculture predominated (Pearson, 1992, p. 68; Takahashi et al., 1997).

The Early stage was the time when the technical complex, or set of tools for the hunting, fishing, and gathering way of life that characterizes Jomon culture, was established. Main types of tools that existed from the Early Jomon to the Final Jomon stage included flaked triangular-shaped arrowpoints and spearheads with concave bases often made of obsidian or other volcanic material, handled scrapers,

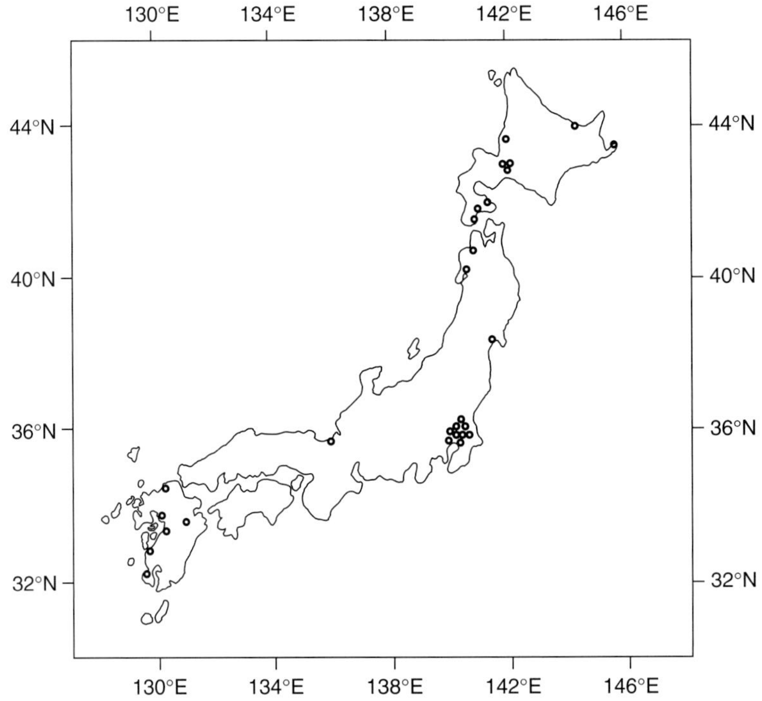

Figure 10.7. Map of location of most known sites of Early Jomon stage, Japan (after Kuzmin et al., 1998).

stone knives with flaked edges, chipped stone adzes, polished stone adzes, fishnet sinkers, fishing spears, fishhooks, toggle harpoons, wooden dugout canoe and paddles, pointed digging sticks, wooden containers, and other items.

Ceramic assemblages of the late Early Jomon stage reflect certain changes in pottery-making standards. After a total dominance of simple unrestricted conical-shaped and roughly designed vessels during the Incipient, Initial, and most of Early Jomon, some new forms of pottery, characterized by elaborate modeling of the orifice and new principles of ornamentation appeared by the end of the Early Jomon stage. Several ceramic style zones located in Honshu, Kyushu and Hokkaido formed during Early Jomon (Yamanouchi, 1964, pp. 40–66; Aikens and Higuchi, 1982, pp. 185–186; Pearson, 1992, pp. 70–71; Aikens, 1995). At the same time, the pottery-making craft of Early Jomon preserved and developed some traditions from previous periods. Cord-impressed designs that began during the Incipient stage continued throughout Jomon, and became one of the most constant and distinctive characteristics of the culture.

Lacquer processing of wood also marks the Early Jomon stage. Lacquered wooden objects were found at Torihama Shell Mound, Yoneizumi (Pearson, 1992, p. 123). Lacquering, which began many thousands of years ago, became one of the most popular of traditional Japanese crafts.

3.1.2. Middle Jomon

The term *Jomon culture* is most typically associated with the impressive images of the Middle Jomon stage – ceramic vessels of splendid design, giant shell mounds, cemeteries, and a variety of ritual objects. Japanese archaeological science recognizes the Middle Jomon as a period when significant changes occurred in Jomon culture (Aikens and Higuchi, 1982, pp. 137–164, 329; Nishida, 1989; Kobayashi, 1992a, 2004; Pearson, 1992, pp. 64–127; Takahashi et al., 1997). On one hand, Middle Jomon's traditions followed those of Early Jomon but, on the other, many new traits separate Middle Jomon from the previous stage. Archaeological evidence indicates changes occurred in all parts of prehistoric life, including in subsistence, social organization, material production, and spiritual culture.

Two parallel processes took place during the Middle Jomon stage – an increase in the sizes and numbers of shell mounds and in the sizes and density of settlements – which reflect population growth (Fig. 10.8). Both trends were most intensive in northeastern and central Japan, on Honshu Island (Tohoku and Kanto Districts). The shell mounds' sizes began to increase rapidly from the end of Early Jomon and reached a maximum after 5000 ^{14}C yr BP. For example, Nakazato shell mound site (Tokyo) has shall deposits about 200 m in diameter and 3–4 m in thickness (Takahashi et al., 1997, p. 116). The number of pit dwellings in the sites of terminal Early Jomon and Middle Jomon age reaches several hundred, as in the case of the San'naimaruyama site (Aomori),

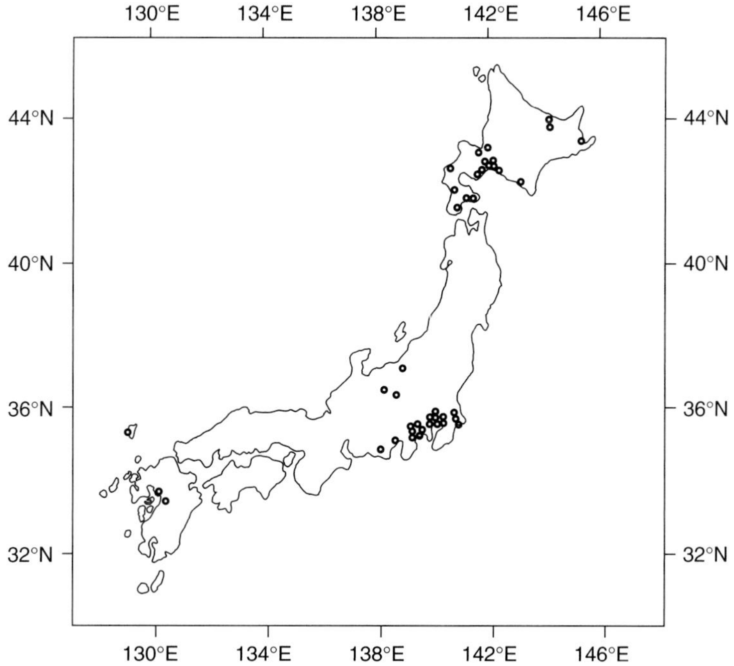

Figure 10.8. Map of location of most known sites of Middle Jomon stage, Japan (after Kuzmin et al., 1998).

where 600 pits were detected (Takahashi et al., 1997, p. 124). The sizes of house structures increased significantly, reaching 200 and more square meters in some cases.

A burial system that was practiced through the Final Jomon stage formed in the Middle Jomon. The temporal sequence of the Jomon burial system is documented most completely in eastern Japanese sites. The pattern of burials was relatively complex and reflected social patterns in Jomon communities (Kobayashi, 1992b, 2004, pp. 99–136; Pearson, 1992, pp. 80–81; Yamada, 1996).

The changes in material production patterns are most obvious in pottery-making activity. First, in the Middle Jomon stage the variety of forms and sizes of ceramic vessels increased greatly in comparison with the previous period (Fig. 10.9). Second, the Middle Jomon's pottery is characterized by a complication of design. Relief decoration on vessel rims and walls required a high level of pottery-making skill. The ceramics of the most sophisticated flame-like style may be called *Jomon's Baroque* – they represent the apex of the pottery-making art during the Japanese Neolithic (Fig. 10.10). Third, new technologies of surface treatment appeared and developed, such as polishing and ocher painting. By the end of the Middle Jomon stage the technologies of polishing and painting came to be used more frequently.

The innovations of Middle Jomon pottery-making indicate increasing labor intensity and improvements in craft skills. The pottery of the Middle Jomon is an outstanding cultural phenomenon (Yamanouchi, 1964; Iofan, 1974, pp. 12–18; Sahara, 1979; Harris, 1997; Kobayashi, 2004, pp. 19–72; etc.).

The last part of the Middle Jomon stage was a time of broad distribution of various types of ritual and prestige objects (Yamanouchi, 1964; Iofan, 1974, pp. 12–20; Pearson, 1992, pp. 82–85; Kobayashi, 2004, pp. 137–168). These included ceramic vessels of the *Katsusaka* type with anthropomorphic and zoomorphic designs, ceramic anthropomorphic figurines or *dogu*, and stone rods and bar-like objects called *sekibo*. Some sites contain the remains of ritual activities, such as anthropomorphic figurines in human burial areas on the site of Tanabatake, cases of broken ceramic figurines scattered over wide areas on the site of Shakado, and stone rods buried under the floor or displaced near the hearth in houses, etc.

3.1.3. Late Jomon

Archaeological records of the Late Jomon stage (Fig. 10.11) demonstrate mostly the continuity and gradual development of the main tendencies of Middle Jomon. The Late Jomon stage was not marked by outstanding events similar to those which occurred in Middle Jomon, such as a rapid increase in the numbers and sizes of shell mounds and settlements, or the extremely ornate style of the ceramics. Archaeological contexts of the Late Jomon stage are not as impressive as those of Middle Jomon; however, they too contain certain information about cultural dynamics in that time.

The subsistence pattern based on hunting, fishing, gathering, and partial plant cultivation followed the experience and traditions of previous periods. A tendency toward a reduction in size of shell mounds began to develop, although in some local

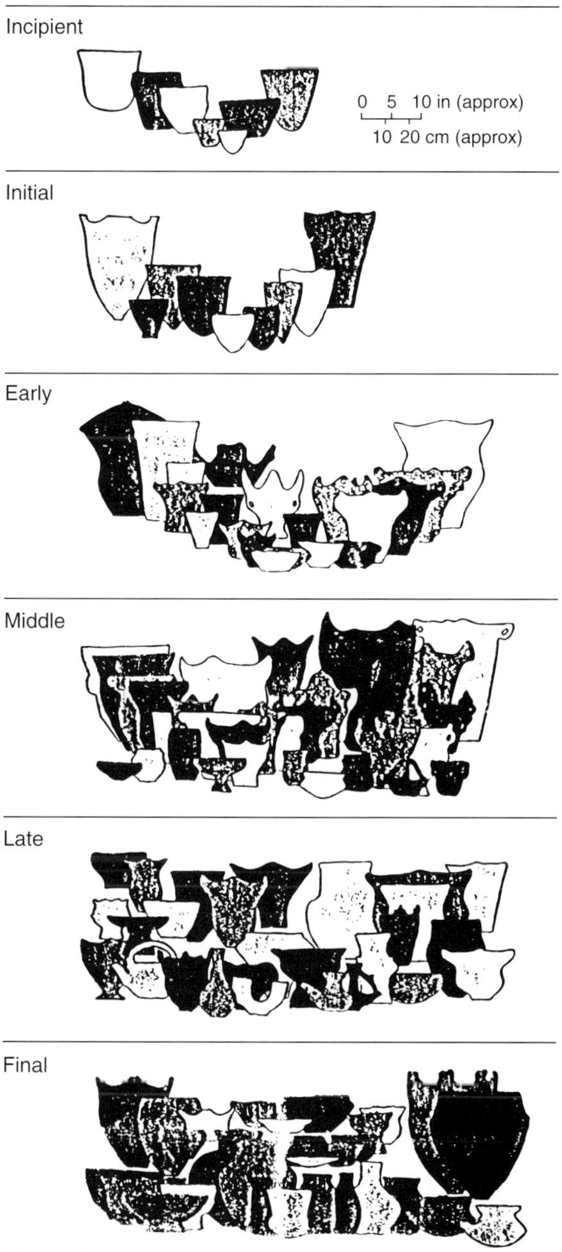

Figure 10.9. Table of development of Jomon pottery forms (after Pearson, 1992).

areas giant shell mounds continued to be accumulated. The Kasori-minami shell mound (Chiba), for example, is estimated to be 9500 m^2 in area and 5500 m^3 in volume (Suzuki, 1989).

The density of Late Jomon settlements was lower than the density during Middle Jomon. The number of sites of Late or Final Jomon in central Japan is less than

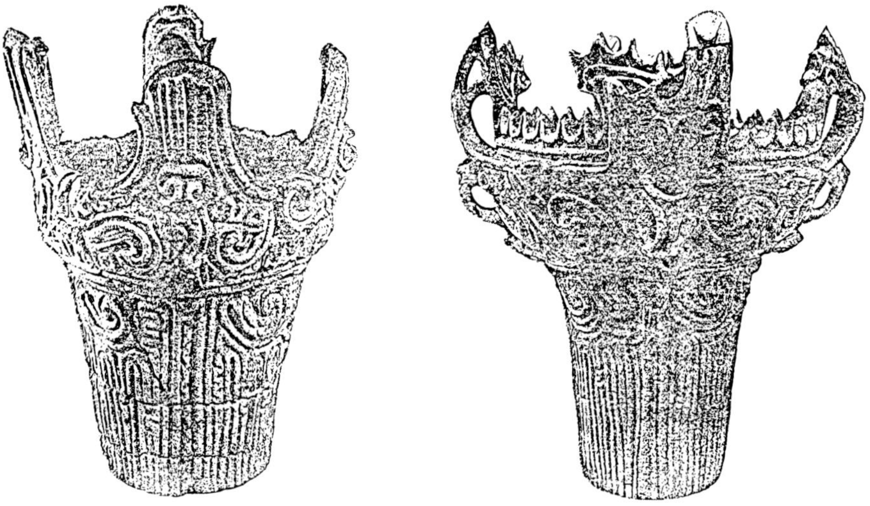

Figure 10.10. Ceramic vessels of flame-like style, Middle Jomon stage.

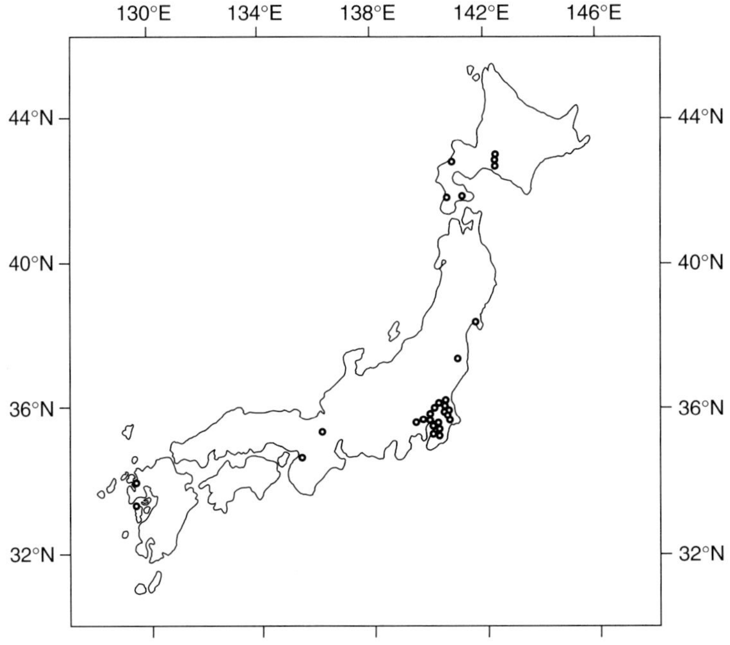

Figure 10.11. Map of location of most known sites of Late Jomon stage (after Kuzmin et al., 1998).

1/3rd that of Middle Jomon sites (Komiyama, 1996). At the same time Late Jomon sites give interesting evidences of inter-village communication and co-operation; communal burials and fish weirs, for example, were used by the habitants of several neighboring villages (Nishida, 1989; Pearson, 1992, pp. 79–80).

In the Late Jomon ritual practice expanded further. New types of ritual anthropomorphic ceramic figurines appeared, and a most remarkable image is a *snow-goggle* figure in ornamented cloth and a crown-like headdress (Aikens and Higuchi, 1982, pp. 165, 170; Pearson, 1992, p. 84).

Temporal changes are most obvious in the pottery-making traditions of the Late Jomon (Aikens and Higuchi, 1982, pp. 164–181; Pearson, 1992, pp. 73–75). The ceramic vessels of that time were less splendid and sophisticated than those of Middle Jomon. However, the invention and distribution of some new types – especially those with complicated morphology (Fig. 10.9), as well as the elaboration of ceramic paste, surface treatment, and firing technology – reflected progressive development of the pottery-making craft. Pottery design became less ornate and more elegant while preserving some essential principles of the previous period (Fig. 10.12). It may be said that Late Jomon's potters added new quality to their ceramics by improving all stages of the production process.

The changes and innovations of the Early, Middle, and Late Jomon stages took place within a cultural context that was under continual change, what we here call a *development.* The tempo and scope of these cultural changes varied appreciably. In the Early Jomon some important changes and innovations began and developed gradually. The period encompassing the end of the Early Jomon and most of the Middle Jomon was one of the intensive cultural change. During the Late Jomon,

Figure 10.12. Ceramic vessel of Late Jomon stage.

cultural dynamics had a progressive character yet lacked the splashes and intensity, which marked the Middle Jomon.

3.1.4. Korean Peninsula

Beginning from 8000 ^{14}C yr BP Neolithic cultures united under the common name *Chulmun* were spread over all parts of the Korean Peninsula (Fig. 10.13). The most distinctive trait of these cultures was pottery decorated predominantly with comb-impressed or comb-incised patterns (Arimitsu, 1962; Nelson, 1975, 1990, 1992, 1993, pp. 58–109; Larichev, 1978; Pak, 1979, pp. 28–52; Choe and Bale, 2002). The Korean term *Chulmun* means "the comb". At present, archaeological remains of Chulmun populations have been discovered and excavated in various parts of the Korean Peninsula. Chronological frameworks for Chulmun cultural complexes developed using radiocarbon dating extend from 8000 to 2000 ^{14}C yr BP with the most of the dated sites belonging to the interval from 6000 to 3500 ^{14}C yr BP (Nelson, 1993, pp. 58–59, 64–65). Based on recent investigations some researchers place the beginning of the Korean Neolithic at around 10,000 ^{14}C yr BP. According to this view, Neolithic assemblages are attributed to the Bissalmuneui cultural period, which has an upper temporal limit of around 3500 ^{14}C yr BP. The term

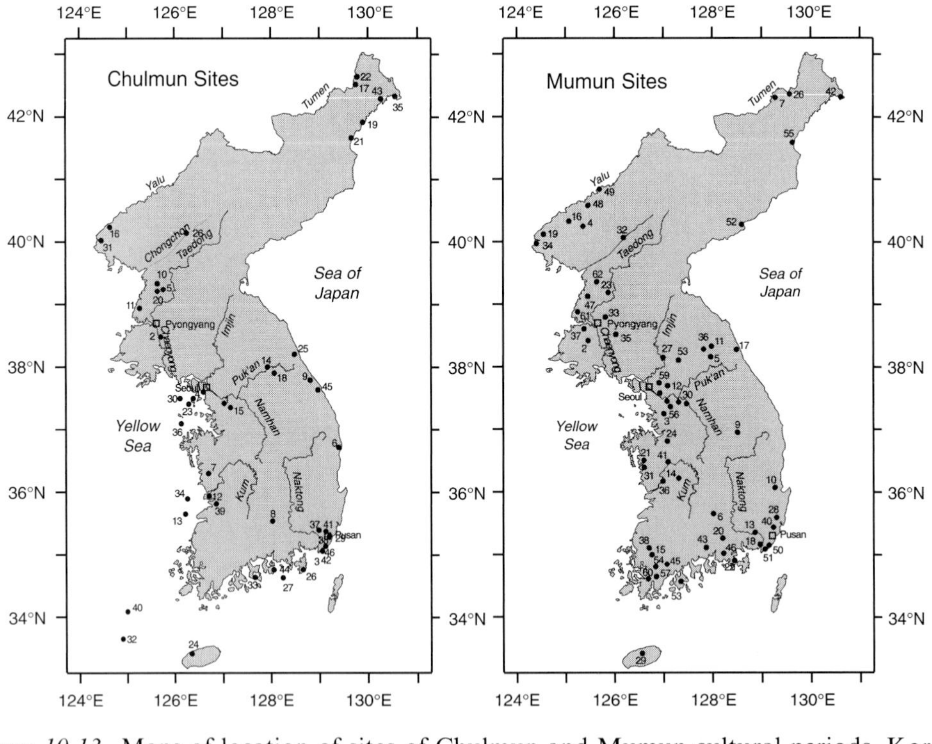

Figure 10.13. Maps of location of sites of Chulmun and Mumun cultural periods, Korean Peninsula (after Nelson, 1993).

"Bissalmuneui" is considered an analogy to "Chulmun." The time from 10,000 to 8000 ^{14}C yr BP is considered the Incipient stage of Bissalmuneui (Choe and Bale, 2002).

Chulmun cultures are characterized by long-lived village settlements with pit-houses of oval-like or round-like shape. The subsistence pattern was complex consisting of land and marine hunting, fishing, gathering, and edible plant cultivation, with the importance of each depending on natural conditions in certain areas. The set of tools included stone-chipped arrowheads, scrapers, gravers, drills, polished axes and adzes, grinding slabs and stones, mortars, pestles, hoes, weights (net-sinkers), and various bone and antler implements.

Chulmun style pottery has a simple morphological structure, consisting mostly of conical-shaped forms with flat or round bottoms. Its technological features are thin fragile walls, rough surface treatment, and low-temperature firing. Typical variants of decoration were vertical zigzag pattern made by comb-incising and horizontal rows of comb impressions (Fig. 10.14). Several regional styles of Chulmun pottery tradition have been distinguished (Nelson, 1993, pp. 61–62; Choe and Bale, 2002).

Around 5500–5000 ^{14}C yr BP settlements with ceramics of a new style known as Mumun appeared in the Korean peninsular (Nelson, 1993, pp. 110–163; Choe and Bale, 2002). Distinctive traits of Mumun pottery assemblages compared with the preceeding Chulmun style include the occurrence of complicated forms with a developed morphological structure, geometric designs or the absence of any design,

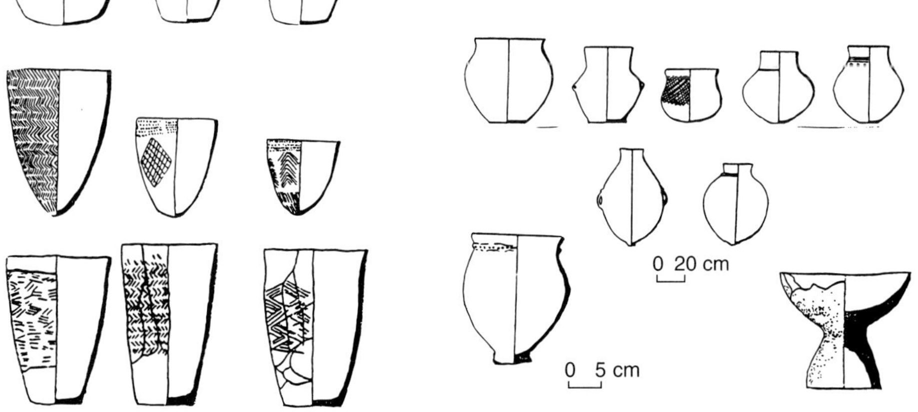

Figure 10.14. Ceramics assemblages of Chulmun cultural tradition (A) and Mumun cultural tradition (B), Korean Peninsula.

and the appearance of new technologies of surface treatment and firing (Larichev, 1978, pp. 54–56, 73–75; Nelson, 1993, pp. 116–123; Choe and Bale, 2002, p. 96) (Fig. 10.14). Typical Mumun vessels forms have well-formed structural parts such as the neck and body, and red painted and polished walls. Obviously, the Mumun pottery style was not connected closely with the preceding Chulmun style, and represented another cultural tradition of pottery-making.

Comparing the locations of Chulmun and Mumun ceramics sites, it is evident that the areas settled were the same in both cases. However, settlement density is markedly higher for the period of Mumun sites (Fig. 10.13). According to C.M. Nelson, Mumun sites are dated from 4800 to 2800 ^{14}C yr BP, with many of the sites falling into the interval from 4100 to 3750 ^{14}C yr BP (Nelson, 1993, pp. 113–116). The period of overlap for the dates of Chulmun and Mumun sites corresponds mainly to the interval 5000–4000 ^{14}C yr BP. C.P. Choe and M. Bale date the period of co-existing of Bissalmuneui (Chulmun) and Mumun cultural traditions from 3500 to 3000 ^{14}C yr BP (Choe and Bale, 2002).

Some important changes and innovations in subsistence, settlement pattern, social structure, and spiritual culture are connected with the period of Mumun sites. The most progressive innovation in subsistence was the beginning of rice cultivation. Recent investigations connect small-scale cultivation of millet, barley, and some other plants with Chulmun, or Bissalmuneui, cultural context while the appearance of rice agriculture seems to be attributed to Mumun and a new population (Choe and Bale, 2002). It was the background for the further development of agricultural economy in the Korean peninsular. The oldest evidence for rice growing was discovered in western Korean sites dated to about the middle to second half of the fifth millennium BP. The rice that was found is not indigenous to the Korean peninsular. This plant began to be cultivated about 8500–7000 ^{14}C yr BP in neighboring China, and it is from there that it came to the Korea region (Nelson, 1993, p. 163; Kong, 1994; Choe and Bale, 2002).

The beginning and spread of rice cultivation caused the invention of new types of tools. The best known and most common types are stone polished harvesting knives that are found in agricultural regions of Eastern Asia and neighboring areas from 5000 to 3000 ^{14}C yr BP (Pearson, 1976; Nakamura, 1993; Nelson, 1993, pp. 123–126). The Mumun period was the time of the blossoming of abrasive stone processing, when polished tools replaced chipped ones almost completely.

Settlement and dwelling patterns changed during the Mumun period. Mumun houses look more developed and formal compared with Chulmun ones. Semi-subterranean pit-houses are clearly rectangular in shape, with lines of postholes along the structure's long axis, as well as the presence of storage pits and other interior constructions. The sizes of dwellings increased, reaching 50 m^2 in some cases. New type of settlements consisting of one or two large house structures appeared. At the same time, very large sites like Soktalli existed, covering an area near 100,000 m^2 and consisting of at least a 100 dwellings (Nelson, 1993, pp. 138–144).

The most impressive phenomenon of the Mumun period is represented by the megalithic construction of dolmens. The oldest dolmen discovered comes from the site of Yangsuri and is dated to 3900 ^{14}C yr BP. Dolmens constructed of massive stone slabs spread very widely from the north to the south in the Korean Peninsula from 5000 to 4000 ^{14}C yr BP, and were accompanied by Mumun style pottery. Three types of dolmens are distinguished: (1) the table type, (2) the go-table type, and (3) the unsupported cap-like type. The dolmens are interpreted as ritual objects associated with burial ceremony (Larichev, 1978, pp. 56–81; Pak, 1979, pp. 58–60; Nelson, 1993, pp. 147–152).

By the end of Mumun period the first bronze items had appeared in the Korean Peninsula. It was a great innovation, the origin of which seems to have been located outside the Korea region – probably in more western and northern centers of bronze metallurgy such as in China and southern Siberia (Pak, 1979, pp. 52–73; Nelson, 1990, 1993, pp. 132–138, 159–161).

Archaeology provides evidence for complex changes in cultural dynamics in the Korean Peninsula beginning about 5000 ^{14}C yr BP. The changes in culture reflected most obviously in the substitution of ceramics styles were accompanied by changes in subsistence, tool-making technology, and social organization and settlement patterning. The latter may be interpreted as a *development* generally. The changes in the field of spiritual culture are represented by such remarkable innovations as megalithic funeral cult constructions, the dolmens.

At present, a commonly accepted model explaining and linking all these changes with each other has not yet been developed in Korean archaeology. The main problem is to distinguish the changes which had an independent character occurring in place and the ones which were influenced from outside. Among the latter, the appearance of rice cultivation accompanied by some new types of stone tools looks the most obvious and China is considered as the source of these innovations (Nelson, 1993, pp. 157–163). It seems likely that the appearance of Mumun style pottery was also connected with cultural impulses from Northeastern China (Nelson, 1993, p. 158), although some researchers propose that independent local transformations of pottery-making traditions occurred about 5000 ^{14}C yr BP (Choe, 1982). The origin of dolmens is mysterious: stone burial constructions were spread over a broad area of Eastern, South, and Southeastern Asia from 5000 to 3000 BP (Nelson, 1993, pp. 147–151, 158–159, 1995, pp. 206–221), but the Korean peninsular was where the greatest concentration of dolmen megaliths occurred. Dolmens are thus not know with certainty to be an outside innovation; at the same time, the local background for the appearance of dolmens is not yet known.

3.2. Primorye

In the Primorye region, Neolithic cultures are fixed within existing chronological frameworks from ca. 7500 to 3500 ^{14}C yr BP (Fig. 10.15). About 400 Neolithic sites are known, although only around 30 have been studied fairly completely. Basic conceptions of Primorye region's Neolithic cultural systematization and periodization were worked out from the 1960s to the 1990s (Andreev, 1960; Okladnikov,

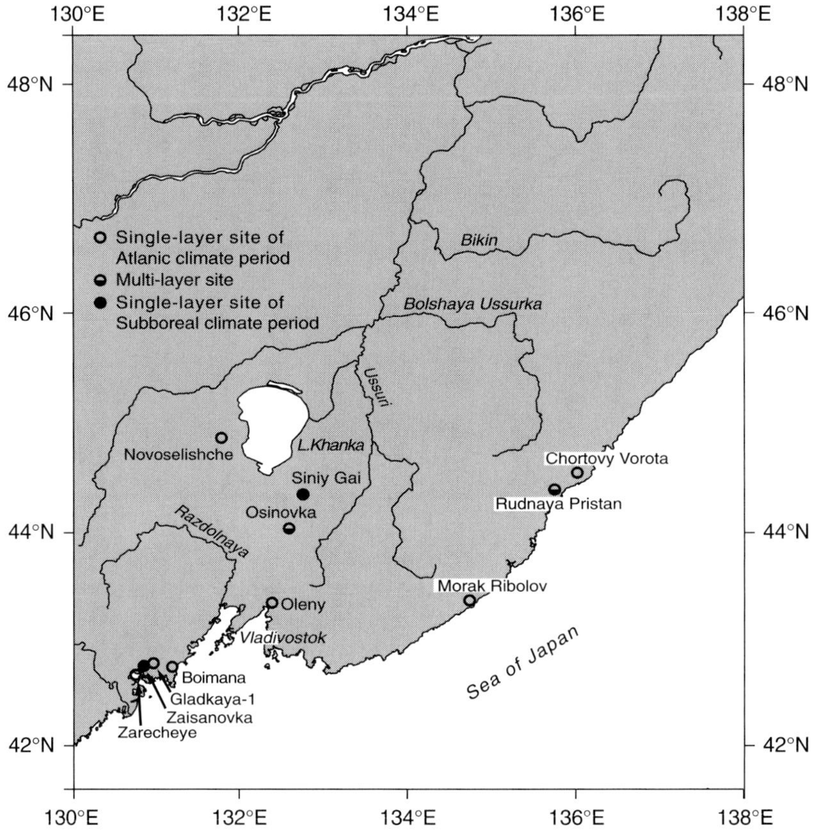

Figure 10.15. Map of location of basic Neolithic (mid-Holocene) sites of Primorye.

1970; Brodyansky, 1979, 1987; Andreeva, 1987, 1991; Dyakov, 1992; Popov et al., 1997; Vostretsov, 1998). At present the process of discovering and explaining new archaeological sites is continuing. In this chapter, we summarize the most widely adopted framework for cultural systematization of the Neolithic.

3.2.1. Rudninskaya culture

The Rudninskaya culture occupies the northeastern portions of the central Primorye region. The culture is dated to 7500–6000 ^{14}C yr BP. The sites are represented by settlements and seasonal camps. Important sites include the lower horizon of the multi-layered settlement Rudnaya Pristan, and a one-dwelling settlement in the Chertovy Vorota cave. Semi-subterranean pit-houses were of rectangular-like shape and no more than ca. 30–35 m^2 in extent, had a central pitted fireplace, and arrangements of postholes.

The lithic industry is characterized by a combination of the Late Paleolithic tradition of blade technology and Neolithic technologies employing bifacial processing and abrasion. The set of tools included arrowheads, spearheads, knives, scrapers, drills, polished axes, and adzes. Abrasive techniques were applied to the making of stone ornaments such as pendants and beads. Bone items compose of a significant part of the archaeological assemblage of Rudninskaya culture, and include toothed harpoons, points, and ornaments. The remains of basketry made from plant fibers, including fragments of fishing nets, were discovered in the Chertovy Vorota cave site.

Distinctive traits of the ceramics are a coiling method of forming, the dominance of vessels of simple form with unmarked orifices and flat bottoms, mineral-tempered ceramic paste, slipping as surface treatment technology, and low-temperature firing. The pottery was decorated mainly by stamping. The most popular designs imitated the pattern of a fishing net (*Amurian weaving*) or fish scales.

The population which carried the traditions of Rudninskaya culture had a complex subsistence system based on the hunting, fishing, and gathering of terrestrial resources (Andreeva, 1991; Dyakov, 1992).

3.2.2. Boismanskaya culture

The sites of the Boismanskaya culture are distributed over the seacoastal area of southwestern Primorye and are represented by settlements with shell mounds. According to current data the chronological limits of this culture are from about 7000 to 5000 ^{14}C yr BP. Basic sites are Boismana-1 and Boismana-2 (Popov et al., 1997; Vostretsov, 1998; Moreva and Popov, 2003). Archaeological assemblages of the Boismanskaya culture are similar to those in northeastern Korea belonging to the Neolithic Chulmun cultural tradition.

Settlements of the Boismanskaya culture are characterized by semi-subterranean pit-houses that are rectangular in shape and small in size, from ca. 9 to 12 m^2 in area. Shell mounds are up to 1.10 m in thickness. A remarkable feature of Boismanskaya culture is the presence of burial grounds in the shell mounds. The burial grounds consist of one central burial surrounded by several in a circular pattern. All individuals were buried in flexed position. Boismanskaya culture's population is attributed to an Arctic race according to anthropological research (Popov et al., 1997).

The lithic industry was based on the production and utilization of blade-like and amorphous flakes in tool-making. Bifacially retouched chipped tools were dominant while abrasive technology use was fairly restricted. Stone tools included arrowheads, spearheads, knives, drills, adzes and axes, net-sinkers, and some other types. Bone tools were preserved perfectly in the shell mounds and included toothed and turned harpoons, fishhooks, fish-sparkles, and points (Fig. 10.16).

Standard traits of the Boismanskaya culture's pottery are an absolute dominance of simple vessel morphology with unmarked orifices, plain wall profiles, an absence of artificial tempering of the ceramic paste excluding occasional cases of shell-tempering, a combination of coiling and molding, an absence of special surface treatment, low-

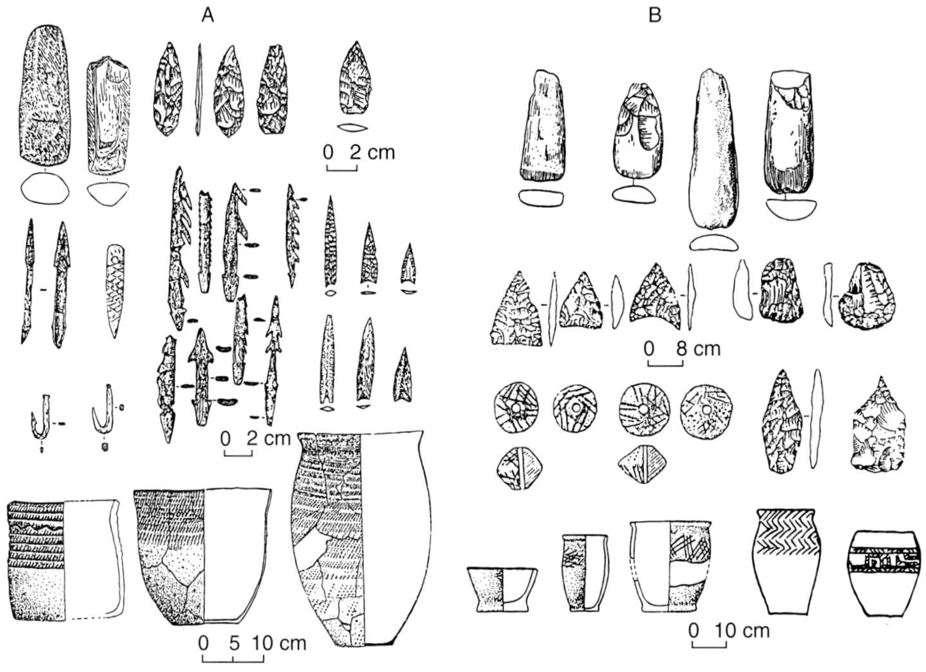

Figure 10.16. Archaeological assemblages of Boismanovskaya culture (A) and Zaisanovskaya culture (B), Primorye.

temperature firing, and comb-impressed designs (Zhushchikhovskaya, 1998). It must be noted that the earliest stage of the pottery tradition found recently at Boismana-2 site is characterized by vessels with pointed bottoms, which over time become flat-bottomed forms (Moreva and Popov, 2003).

A special category of the finds are art creations and ritual objects like bone zoomorphic figurines, bone and shell ornaments like bracelets and pendants, and stone anthropomorphic masks.

The subsistence of the Boismanskaya culture was based on the hunting, fishing, and gathering of marine resources, which formed the main part of the diet (Yoneda et al., 1998; Vostretsov, 2001). Some researchers have suggested that the Boismanovskaya culture's populations invented the practice of aquaculture, specifically the intentional cultivation of marine mollusks (Brodyansky and Rakov, 1992, 1996).

3.2.3. Zaisanovskaya culture

The area of the Zaisanovskaya culture includes southern, eastern, central, and southwestern Primorye. The dates of the Zaisanovskaya culture fall into the interval from 5370 ± 40 to 3090 ± 35, concentrating mainly from 4900 ± 200 to 3635 ± 30 (Popov et al., 1997, p. 39; Kuzmin et. al., 1998, pp. 16–17; Vostretsov et al., 2003b).

The sites of the Zaisanovskaya culture are represented by long- and short-lived settlements. Important sites are Oleny, Zaisanovka-1, Rudnaya Pristan (upper layer), Valentin-Peresheek, Siny Gai (lower layer), and Novoselishche-4 (lower layer) (Andreev, 1957; Brodyansky, 1979, 1987; Andreeva, 1987; Dyakov, 1992; Kluiev and Zhushchikhovskaya, 1996). Some interesting sites were discovered most recently such as Krounovka-1 (lower layer) that contain important records for our understanding of the Zaisanovskaya culture (Vostretsov et al., 2003a,b). Two types of houses are distinguished – semi-subterranean houses and ground or surface structures. The sizes of houses are ca. 60 m^2, markedly exceeding the sizes of houses of the Rudninskaya and Boismanskaya cultures. Cemeteries and burial areas remain undiscovered at present.

The lithic industry demonstrates certain changes in comparison with the previous period. Stone assemblages contain various types of polished tools, instruments made of chipped flakes, and microliths. Obsidian was used widely for making of small chipped tools. Besides the usual Neolithic axes and adzes, arrowheads, spearheads, knives, drills, and scrapers, new implements appeared, including ground stone and hoes, and are widely distributed.

Pottery assemblages of the Zaisanovskaya culture were marked by the dominance of artificial or natural sandy pastes, morphological variability, and the development of special surface technologies like slipping and polishing. New techniques and motifs of design began to be used, including incised vertical zigzags, meander-like and sometimes curve-lined patterns, and dotted impressions (Zhushchikhovskaya, 2005, pp. 120–123). A new category of ceramic artifact, the spindle whorl, also appeared during the period of the Zaisanovskaya culture. These artifacts, together with the textile impressions on ceramic vessels' bottoms, indicate the development of weaving craft (Fig. 10.16). Several local variants of the Zaisanovskaya pottery-making tradition may be distinguished. The most obvious differences are between the eastern site assemblages and those to the west and southwest. In general, western and southwestern pottery-making standards are more elaborate and advanced than on the eastern sites.

The main branches of the Zaisanovskaya culture subsistence were terrestrial hunting and gathering. River fishing played a less important role. It must be emphasized that seacoast sites do not contain evidence for intensive exploitation of marine resources. Based on the finds of stone net-sinkers on some seacoast sites, marine gathering may be assumed in few cases (Vostretsov, 2001). An important recent discovery is the identification of agricultural remains. Paleobotanical evidence indicating plant cultivation was found at the site of Krounovka-1 (lower layer) dated to 4640 ± 40, the earliest evidence for agriculture in the Primorye region (Vostretsov et al., 2003b). Some other sites also contain paleobotanical remains of plant cultivation (Popov et al., 1997, p. 39).

The chronological sequence of Primorye's Neolithic cultures is confirmed by stratigraphical observations on some multi-layered sites like Rudnaya Pristan, Osinovka, Pereval, and Boismana-2. On these sites the horizons of Zaisanovskaya culture overlie those of the Rudninskaya and Boismanskaya cultures. The lithological characteristics of Neolithic cultural horizons differ. The assemblages of the Rudninskaya and Boismanskaya cultures occur in brown loams, which accumulated during the Atlantic period of the Holocene, from ca. 7000 to 5000 ^{14}C yr BP. In most cases, the remains of the Zaisanovskaya culture are associated with the yellow and gray sands and sandy loams which formed in the relatively cool and dry climate of the post-Atlantic period of the Holocene, after ca. 5000 ^{14}C yr BP (Verkhovskaya and Kundyshev, 1993).

The Rudninskaya and Boismanskaya cultures differ one from another in their cultural characteristics, although both are roughly contemporaneous chronologically, and represent the same stage of the Neolithic. The assemblages of the Zaisanovskaya culture reflect the next stage of the development of Neolithic technology in lithic industry and pottery-making. The widening of the area the culture encompassed, and the increasing sizes of settlements and houses probably indicate an intensification of social processes in that period. The evidences for progressive changes are most obvious in the assemblages of the late phase of the Zaisanovskaya culture, and include the probable beginnings of agriculture, and increased technological specialization in the pottery-making craft.

The traditions of the Zaisanovskaya culture did not follow directly from those of the Boismanskaya and Rudninskaya cultures. Some elements of the Boismanskaya culture are present in Zaisanovskaya culture assemblages – generally in ceramic styles – although they do not play a leading role. Many traits of the Zaisanovskaya culture do not appear to have a local origin and may be interpreted as the result of the appearance of a new population in the Primorie region about 5000 ^{14}C yr BP. Continental regions of Eastern Asia located to the west from Primorye, such as southern and eastern Manchuria, are considered the most probable native land or source area for the Zaisanovskaya culture's traditions. In that area the Neolithic Lower Houwa, Fuhe, and Zhaobaogou cultures existed from ca. 7500 to 5000 ^{14}C yr BP. In their assemblages one can detect the traits which are similar to characteristics of the Zaisanovskaya culture: the wide usage of the obsidian in the lithic industry, a high percentage of the tools connected with plant gathering and the initial steps toward plant cultivation, morphological variability in pottery including vessels with marked orifices and the presence of bowl-like vessels, the popularity of incised vertical zigzag and meander-like motifs in pottery decoration, and relief cornice-like thickening along the vessel rim (Li Tzjian, 1994; Nelson, 1995, pp. 21–146).

3.3. Lower Amur region (Priamurye)

The Amur River is one of the biggest water arteries of Eurasia and the largest in the territory of the Russian Far East. Its length is about 3000 km, with a width along its lower course of 3–5 km, and about 15 km in the estuary. It streams from the west to the east and enters into the Okhotsk Sea. Two characteristics of the Amur River are most important in the context of our topic. Firstly, it is a natural water *road* for peoples moving through the region. Secondly, the Amur, especially its lower course, is a river providing highly favorable conditions for the fishery, and for peoples exploiting fish resources.

Researchers connect the origin of early Neolithic traditions in this area with the Osipovskaya culture, which is attributed to the final Pleistocene–early Holocene and dated to around 13,000–10,000 ^{14}C yr BP. Within the chronological interval 9000–3500 ^{14}C yr BP several Neolithic cultures are distinguished (Fig. 10.17). In general, Neolithic horizons were discovered on multi-layered sites which tend to be located in river valleys (Okladnikov, 1984; Derevyanko and Medvedev, 1995a, 1996; Medvedev, 1995, 2003). The following cultures are most interesting in the context of mid-Holocene climate and culture change.

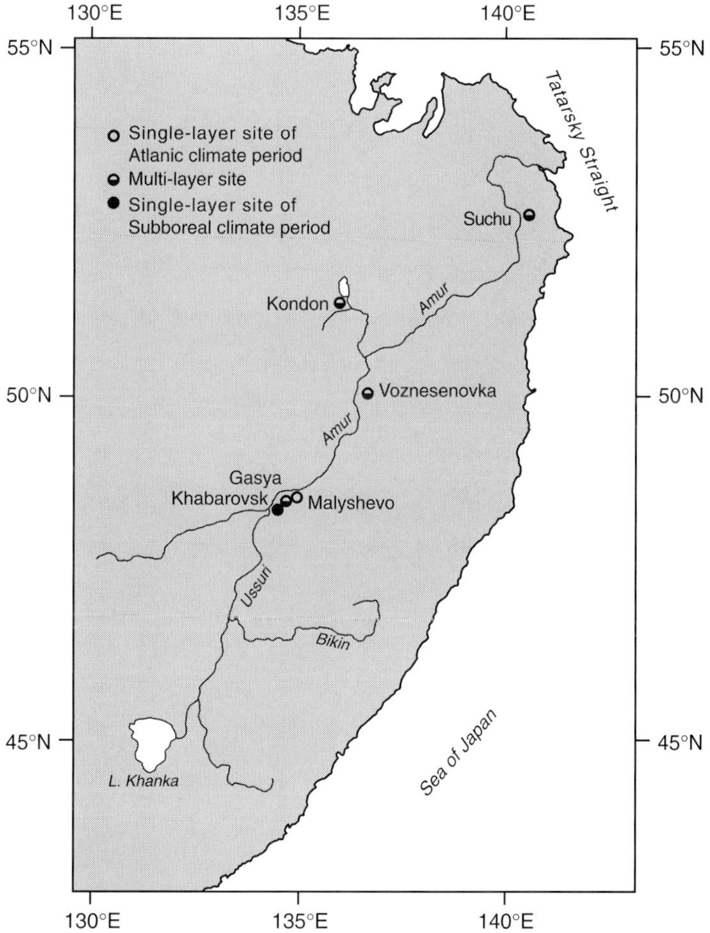

Figure 10.17. Map of location of basic Neolithic (mid-Holocene) sites of Lower Amur River region (Priamurye).

3.3.1. Malyshevskaya culture

This culture, recognized on the site of Malyshevo, and in the lower layer of Voznesenovskoye, and at some other locations, is dated to ca. 7000–5000 ^{14}C yr BP, to the climatic optimum phase of the Holocene. Semi-subterranean pit-houses were present, and were round or oval in shape and ranged in size from 50 to 270 m^2. Characteristic details of the interior of these structures were a hearth in central part of the floor, earthen platforms along the walls, and a clay cover on the floor. The lithic industry combined blade techniques, bifacial processing of chipped tools, and abrasive techniques. The pottery tradition is represented by conical-shaped flat-bottomed vessels made by the coiling method, fired at low temperature, and decorated by comb-stamping. The subsistence of the Malyshevskaya culture's people was based on river fishing and land hunting.

3.3.2. Kondonskaya culture

This culture is probably related to the Malyshevskaya culture and existed in Lower Amur region from around 6000 to 4000 ^{14}C yr BP. Major sites are Kondon and the lower horizon of Suchu (Okladnikov, 1984; Derevyanko and Medvedev, 1996; Medvedev, 2003). Semi-subterranean pit-houses were present and had round- and oval-like shapes with posthole construction, a hearth in the central part of the structure, and storage pits. The lithic industry combined archaic blade technology with the widespread usage of bifacial retouch for the making of arrowheads, spearheads, knives, and scrapers, and abrasive technologies for manufacturing axes and adzes.

The pottery-making tradition was more developed in comparison with the one in the Malyshevskaya culture. Besides conical-shaped flat-bottomed vessels with unrestricted orifices, forms with separated body and orifice parts appeared. Several kinds of paste tempers were used, including sand, grog (crushed sherds), and sometimes organic matter. In the surface treatment technology polishing began to be used. A great variety of stamped patterns were characteristic, with most typical variants net-like impressions called *Amurian weaving*, or scale-like impressions (Fig. 10.18). The pottery-making tradition of the Kondonskaya culture is similar to the ceramics tradition of the Primorie region's Rudninskaya culture, especially in the field of ornamentation. This is considered as an indicator of contacts or connections between the two cultures, which existed in neighboring regions within close time limits.

The main emphasis in subsistence for the Kondonskaya culture peoples was fishing. The tools used in fishing are common on sites, and some storage pits containing fish bones have been discovered.

3.3.3. Proto-Voznesenovskaya culture

This cultural tradition was recognized recently based on typological analyses of the assemblages from some multi-layered sites (Mylnikova, 1992; Zhushchikhovskaya,

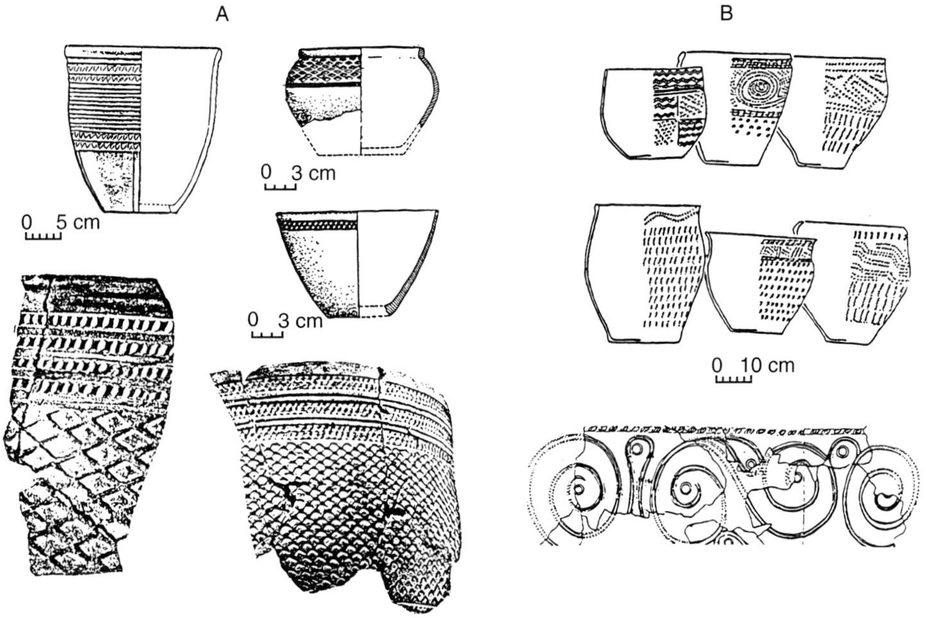

Figure 10.18. (A) Pottery of Kondonskaya culture: ceramic vessels (above) and decorated fragments (below). (B) Pottery of Voznesenovskaya culture: ceramic vessels (above) and pattern of curve-lined design (below).

1992, 2005, pp. 128–133; Shevkomud, 1996, 1999). The culture seems to date to around 5000–4000 ^{14}C yr BP based on several radiocarbon dates and on relative correlation of assemblages. The remains of a single dwelling have been excavated, a semi-subterranean pit-house of oval shape with a niche-like entrance in one of the pit's walls.

The most remarkable feature of the Proto-Voznesenovskaya culture is its ceramic assemblage, whose technology, morphology, and use of design differ from the pattern common in the Lower Amur region. Shell tempered paste technology and fine comb-impressed vertical zigzag ornamentation are original to the Proto-Voznesenovskaya culture tradition and do not have local roots. At the same time these and some other features of Proto-Voznesenovskaya ceramics have analogues in Neolithic sites dating from 6000 to 4000 ^{14}C yr BP located in northern China in a similar overall landscape setting (Nishida, 1987; Nelson, 1995, pp. 132–135; Alkin, 1996). Some specialists in the field of pottery technology link shell-tempering with the ceramics of mobile, non-sedentary communities, based on the behavior of organic-tempered ceramic pastes in the process of pottery-making and use, and data drawn from ethno-archaeological observations (Reid, 1989; Skibo et al., 1989). The lithic assemblages of the Proto-Voznesenovskaya culture are characterized by a leading role of retouched combined tools made on flakes and split fragments.

3.3.4. Voznesenovskaya culture

The major sites defining this culture are Voznesenovskoye, and some dwellings at the Kondon multi-layered settlement. Semi-subterranean pit-houses are of oval- or rectangular-like shape with a central hearth. On some sites dwellings connected with ritual or cult functions were discovered (Derevyanko and Medvedev, 1995b).

The lithic industry is characterized by the total dominance of a bifacial retouching technique. Combined tools, arrowheads, scrapers, and knives were very carefully made. The abrasive technique was also used for the making of axes and adzes, knives, and arrowheads. The archaic blade technology disappeared entirely.

Pottery-making was marked by new ceramic paste and surface treatment technologies, and sophisticated designs. Grog temper (i.e., crushed sherds) was used, vessel walls were treated by fine polishing or painted with red ocher. Refined geometric curve-lined style blossomed in the pottery-making of the Voznesenovskaya culture (Fig. 10.18). It represents the earliest occurrence of the Amur region's famous curve-lined ornamental art, which has survived in various crafts of the region's native peoples (Okladnikov, 1981). Unique samples include curve-lined anthropomorphic images on ceramic vessels that were probably used in non-utilitarian functions. The Voznesenovskaya culture's ceramic tradition was for the prehistoric cultures of the Russian Far East, the same as the Middle Jomon ceramic tradition was for the Japanese Neolithic – it represented the *peak* of development of Neolithic Far Eastern pottery-making.

The subsistence of the Voznesenovskaya culture's bearers was oriented toward exploiting riverine resources. Stone pebble net-sinkers and storage pits containing fish bones indicate the intensive development of the fishery. A remarkable trait of Voznesenovskaya culture's sites is evidences for ritual activity, indicated by the presence of probable special sacral places within the sites, and fine ceramic vessels with anthropomorphic images.

The chronology of the Voznesenovskaya culture is not entirely clear. Based on current data it may be supposed that this cultural tradition existed around 4500–3500 ^{14}C yr BP. As a whole, the Voznesenovskaya culture looks more advanced than other Neolithic cultures of the Lower Amur region. This is reflected in various kinds of archaeological records. Stratigraphical observations at the multi-layered site of Gasya show a sequence deposit of the Malyshevskaya and Voznesenovskaya cultures (Derevyanko and Medvedev, 1995a). The horizons of the Malyshevskaya culture are connected with light-yellow loams accumulated during the Atlantic period of Holocene, while the horizons of Voznesenovskaya culture are layered above in grayish-brown sandy loam formed during the post-Atlantic period.

Complicated cultural processes occurred in the Lower Amur region during the Neolithic, which are not clearly understood at present. The Neolithic cultures of the Lower Amur region are not linked in a continuous succession. One can distinguish the stage of local cultural development represented by the Malyshevskaya and Kondonskaya cultures at 7000–5000 ^{14}C yr BP from the assemblages of the

Proto-Voznesenovskaya culture dated to about 5000 ^{14}C yr BP. Late Voznesenovskaya culture may be estimated as the *zenith* of cultural evolution of the Lower Amur region's population during the Neolithic. Some researches consider Voznesenovskaya culture to be the result of mixing of traditions brought by Proto-Voznesenovskaya migrants with those of local populations. According to this version Proto-Voznesenovskaya assemblages are considered a separate variant of Voznesenovskaya culture (Shevkomud, 1996, 1999). In contrast with other regions in the Japan Sea basin area in the mid-Holocene, the Lower Amur region is not characterized by any serious changes or innovations in subsistence pattern. Fishing was the stable economic base of all cultural groups occupying the banks of the great river during the period in question.

3.4. Sakhalin Island

This is the final geographical area in our consideration of mid-Holocene climate and cultural dynamics in the Japan Sea basin area. The events which took place in the northern part of Sakhalin Island are most relevant to this subject. Sites attributed to the Neolithic Imtchin culture were recognized in the river valleys of northern Sakhalin (Fig. 10.19) (Shubina, 1987; Vasilevsky, 1992, 1995, 2003; Zhushchikhovskaya and Rakov, 1994; Zhushchikhovskaya and Shubina, 2006). Long-lived settlements consisted of round-shaped houses excavated from 0.5 to 1.2 m into the earth, with a niche-like entrance and a hearth in the central part of the floor. The primary sites of the Imtchin culture are Imtchin II and Imtchin XII. According to corrected radiocarbon dates, the temporal limits of Imtchin culture seem to be from ca. 5000 to 4000–3500 ^{14}C yr BP. Most recent opinion considers the Imtchin culture to be a complicated unit including a number of different cultural traditions (Vasilevsky, 2003).

Stone assemblages of the Imtchin culture contain the tools used mainly in hunting and fishing activities, such as arrowheads, spearheads, knives, and scrapers. The technology of tool-making was based on flaking and bifacial retouching. Abrasive technology was used for the processing of axes and adzes. The most characteristic features of Imtchin ceramics are a shell-tempered paste, a simple pot-like shape for the vessels, fine comb-impressed vertical zigzag designs, and modeling of the vessels' rims with a band-like cornice grooved horizontally (Fig. 10.20).

Imtchin culture traditions are close to those of the Proto-Voznesenovskaya culture of the Lower Amur region. The similarity is most obvious for the ceramic assemblages and in-house construction. It is reasonable to consider the Imtchin culture of northern Sakhalin in connection with migration processes which occurred in the Lower Amur region about 5000 ^{14}C yr BP. The mouth of the Amur river is separated from northern Sakhalin by the narrow Tatarsky strait, and this is the area of concentration of Proto-Voznesenovskaya culture sites. The radiocarbon dates from Imtchin culture sites support the assumption about their close relationship to and possible origin from Proto-Voznesenovskaya culture.

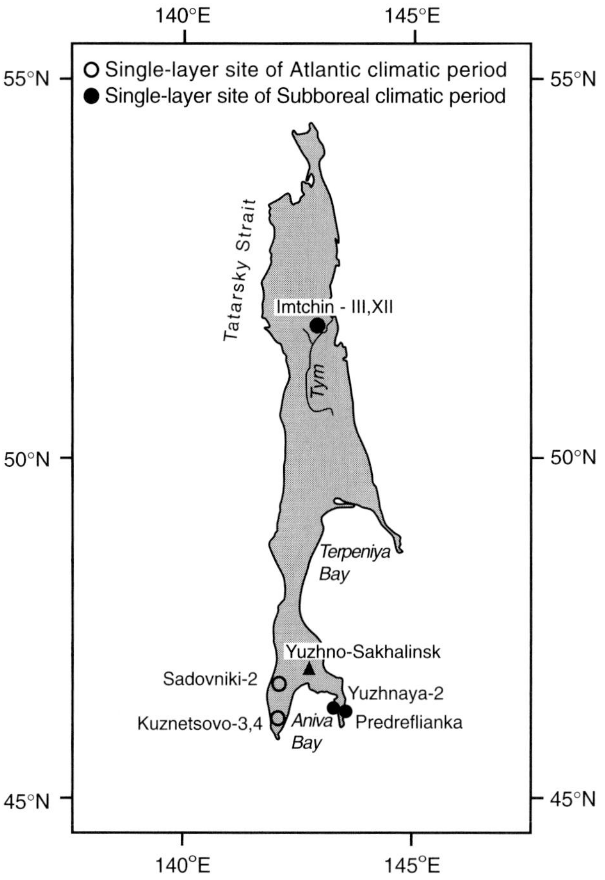

Figure 10.19. Map of location of basic Neolithic (mid-Holocene) sites of Sakhalin Island.

Active settlement of northern Sakhalin began with the coming of the Imtchin culture bearers. Probably, there were several waves of population moving from the Lower Amur region to Sakhalin Island. The early complexes of the Imtchin culture most closely resemble the assemblages of the Proto-Voznesenovskaya culture of the Lower Amur region. The late Imtchin complexes contain features probably related to Voznesenovskaya culture, such as well-made pottery decorated with the in curve-lined style.

Cultural processes which occurred in the southern part of Sakhalin Island were of another character. The first Neolithic culture recognized in this area is Yuzhno-Sakhalinskaya, with sites located mostly in the southwestern seacoast (Fig. 10.19). Major sites include Sadovniki-2 and Kuznetsovo-4 (Shubin et al., 1982; Vasilevsky, 1992, 1995). Calibrated radiocarbon dates indicate that the temporal limits of Yuzhno-Sakhalinskaya culture are mainly from 7500 to 6500 ^{14}C yr BP.

Semi-subterranean pit-houses of rectangular-like shape are present, whose size was up to 100 m^2 in some cases. A hearth was located near the wall opposite the

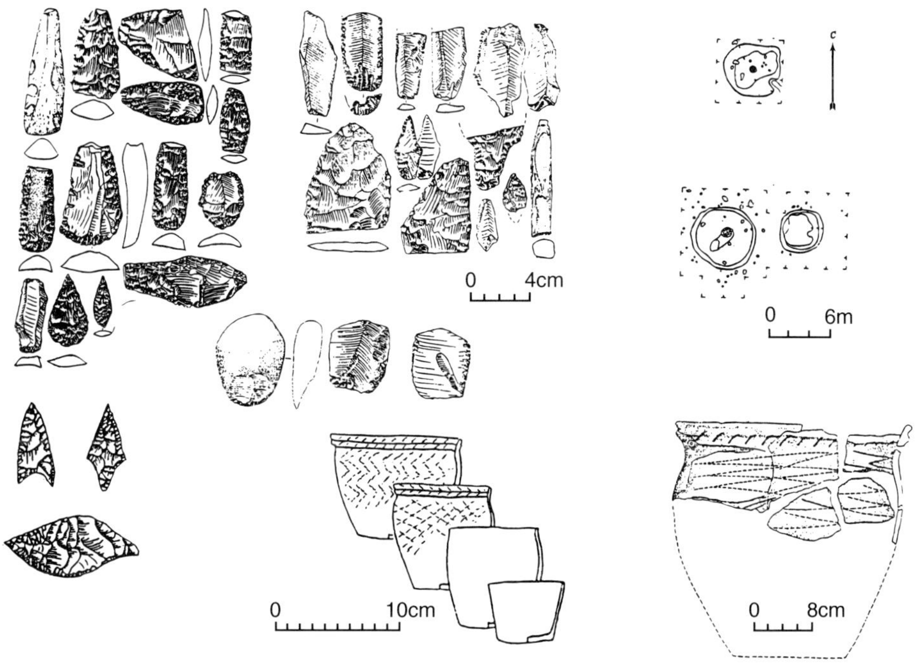

Figure 10.20. Archaeological assemblage of Imtchinskaya culture, Sakhalin Island.

entrance, bench-like earthen projections extended along the walls, and the entrances were corridor-shaped.

The lithic assemblages include retouched tools made on blades and blade-like flakes. Only a few partially polished tools have been found. As a whole, the stone artifacts are characterized by a combination of Paleolithic and Neolithic technical traits.

Ceramics of the Yuzhno-Sakhalinskaya culture represent the earliest pottery-making tradition in southern Sakhalin Island (Zhushchikhovskaya, 1997; Zhushchikhovskaya and Shubina, 2006). Its technological features include plant-organic-tempered paste, a *slab construction* method of forming, an absence of any surface treatment, and very low temperature firing (about 500°C). The most distinctive trait is the box-like shape of the ceramic vessels, a form that is unusual for Far Eastern prehistoric pottery. Box-shaped vessels occur in some early ceramics assemblages of northern Japan dated to 13,000–10,000 ^{14}C yr BP according to some researchers (Suda, 1995). Organic-tempered box-like ceramics are recognized in early assemblages from eastern North America where it has been suggested that the form is related to wooden containers (Griffin, 1965).

The subsistence of Yuzhno-Sakhalinskaya culture was based on hunting, fishing, and gathering. Although the main area of the Yuzhno-Sakhalinskaya culture is located in southern Sakhalin, recently some evidence for these cultural assemblages was discovered in the northern part of the island (Vasilevsky, 2003).

Archaeological records of southern part of Sakhalin Island do not provide evidence for developments following the Yuzhno-Sakhalinskaya cultural tradition in later cultural contexts. By around 4000–3500 ^{14}C yr BP, new population groups appeared in southern Sakhalin and settled mainly in the southeastern part of the island (Fig. 10.19). The Anivskaya Neolithic culture representing these new populations is radiocarbon dated to 3560 ± 140–2320 ± 160 ^{14}C yr BP. Major sites are Yuzhnaya and Predreflianka (Vasilevsky, 1992, 1995). The settlements consist of semi-subterranean pit-houses of rectangular-like shape with pitted fireplaces bordered with stones, and postholes in the floor, which was covered by a clay layer. In some cases, remains of carbonized wooden construction with the traces of a clay cover extending along the walls are preserved.

The set of tools recovered indicate subsistence was oriented to hunting, fishing, and gathering. Pottery assemblages of Anivskaya culture have not any traits related to the traditions of Yuzhno-Sakhalinskaya culture. Characteristics of the ceramics include a sand-tempered paste, a coiling method for forming, a flat-bottomed conical vessel shape with an unrestricted or slightly restricted orifice, slipping used for surface treatment, and low temperature firing. Ceramic design is standard: vertical or inclined cord impressions on the vessel walls. This type of decoration has close analogues in the pottery-making Neolithic Jomon culture of the neighboring Japanese isles.

Archaeological assemblages on Sakhalin Island reflect differing cultural traditions in the northern and southern parts of the island during the Neolithic. The northern Neolithic culture was connected to that of the Lower Amur River region, while the southern Neolithic culture was related to that of the prehistoric habitants of the Japanese Archipelago. Mid-Holocene cultural dynamics differed in the northern and southern parts of Sakhalin Island. The period of the appearance and existence of the Imtchinskaya culture in the north corresponded to the hiatus between the Yuzhno-Sakhalinskaya and Anivskaya cultures in the south. Given the limited archaeological research on Sakhalin Island as a whole, the development of explanatory models that could correlate and link cultural events in this region during mid-Holocene is a major goal for future research.

4. Discussion and conclusions

Despite the fact that there is a chronological disagreement in the interpretation of paleoclimatic events in the basin of the Sea of Japan, the data presented and discussed above can be summarized as follows.

The first warming of the mid-Holocene took place between 7200 and 6200 ^{14}C yr BP in Sakhalin, although the maximum warming was between 5000 and 6000 ^{14}C yr BP. In Japan, taking into account information on the maximum high stand of the sea level, the peak of warmth appears to have been ca. 6500–5500 or 6500–5000 ^{14}C yr BP, although other evidence suggests the warmest stage in Japan was between ca. 8000 and 6000 ^{14}C yr BP. No exact information on the chronology of warm/cold stages in the mid-Holocene is available for Korea, Primorye, and Priamurye.

A significant cooling in Japan was detected between 4000 and 4500 ^{14}C yr BP. This corresponds to the Early Subboreal cooling in Sakhalin Island at about 4400 ^{14}C yr BP, to that in Primorye about 4200–4700 ^{14}C yr BP, and in the Kurile Islands approximately 4500–4800 ^{14}C yr BP.

The Tsushima Current's inflow in the Sea of Japan started after 9000–9500 ^{14}C yr BP, but the full scale influx of the current might occur approximately 8000 ^{14}C yr BP. Maximum *oceanic* warming around Japan seems to have occurred ca. 6500–7000 ^{14}C yr BP or about 6000–4700 ^{14}C yr BP. The warming of the surface waters of the Sea of Japan seems to occur later than on the Pacific Ocean (i.e., eastern) side of Japan. Nevertheless, the first warm-water mollusks appeared in Hokkaido at ca. 8500–7500 ^{14}C yr BP on the Sea of Japan side of the island, earlier than the maximum surface water warming that occurred, and at about 7000 ^{14}C yr BP in the northwestern Sea of Japan. The effective use of mollusks as indicators of hydro-climatic warming and warm/cold currents requires as many direct AMS dates as possible on individual species.

A period of high storm frequency, corresponding to maximum warming in Japan and Sakhalin, may have taken place about 5300–6000 ^{14}C yr BP.

The "climatic optimum" or, at least, the time span between 5000 and 6000 ^{14}C yr BP, is characterized in the Russian Far East by annual average temperatures in excess of 2–6°C and more humid conditions. Annual average SSTs in the Sea of Japan were 1–2°C higher compared to the present.

Cultural dynamics in the basin of the Sea of Japan and adjacent areas during the mid-Holocene period may be summarized as follows. Cultural changes took place in all parts of the research area. However, the character and content of the changes were diverse in each region. Three types of cultural changes are distinguished (Table 10.8). Types 1 and 2 differ one from another in basic cultural context. At the same time, they are similar in the broad spectra of changes, which occurred in the subsistence, demography, social pattern, material, and spiritual culture. Most of the observed changes were progressive in character and may be defined as developmental: the increasing of productivity and appearance of new forms of

Table 10.8. Types of cultural dynamics in the mid-Holocene in the Sea of Japan basin.

Type	Region	Type's content
Type 1	Japanese Archipelago	Stability of basic cultural context. Changes and innovations in subsistence, demography, material culture, spiritual culture
Type 2	Korea Peninsula	Changing of basic cultural context. Changes and innovations in subsistence, demography, material culture, spiritual culture
Type 3	Russian Far East	Changing of basic cultural contexts. Local changes and innovations in subsistence, material culture, spiritual culture

subsistence, the growth of population, an elaboration of social structure, advancements in tool-making and pottery-making technology, and increased sophistication of ritual practice.

Types 2 and 3 differ by the shifting of basic cultural contexts most probably caused by migration processes, based on existing archaeological records.

The chronological correlation of cultural dynamics reveals a synchronity in the changes in the Korean Peninsula and the Russian Far East (Fig. 10.21). Major changes in basic cultural contexts begin about 5000 ^{14}C yr BP for the Korean peninsular, the Primorie region, the Lower Amur River region, and Sakhalin Island.

The temporal occurrence of cultural dynamics in the Japanese Archipelago in the mid-Holocene is a more complicated task because of lack of a commonly accepted chronological scale for Jomon culture (Table 10.7). Only approximate estimations

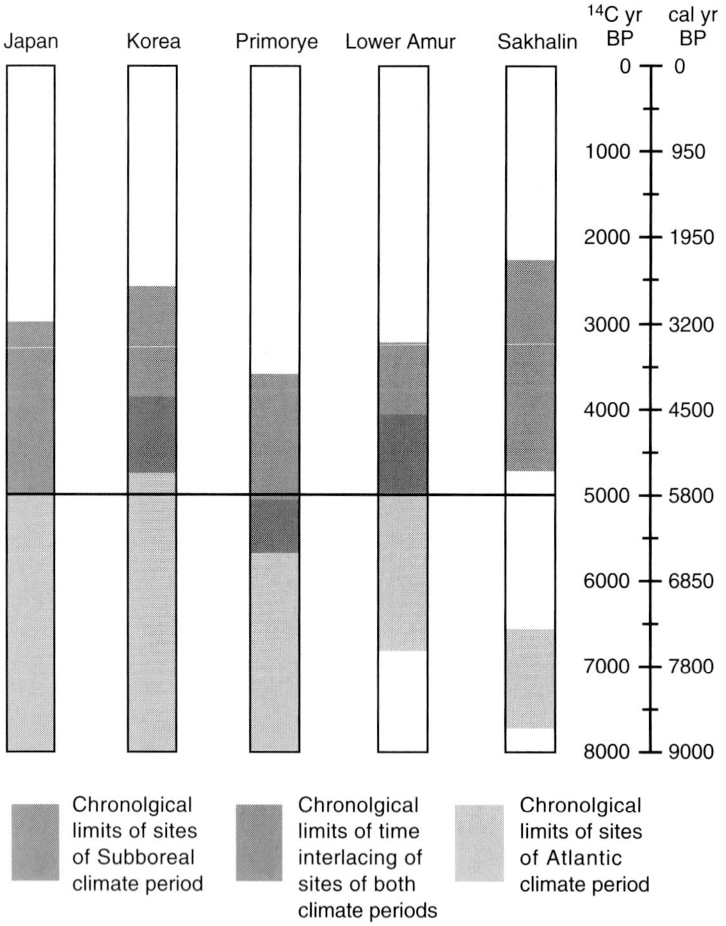

Figure 10.21. Correlation table of chronology of cultural changes in mid-Holocene in the basin of Sea of Japan. Chronological divisions of Japanese Neolithic (Jomon culture) are given according to Hatsushige and Mitsunori (1996).

of the amount of time corresponding to observed cultural changes can be currently recognized. As noted above, the beginnings of the changes were connected with the Early Jomon stage, with the greatest changes during the Middle Jomon stage. Early Jomon is dated to 7500–5600 ^{14}C yr BP in some "early" chronological frameworks and to 5400–4800 ^{14}C yr BP in "later" frameworks; correspondingly, most early frameworks place Middle Jomon to ca. 5600–4500 ^{14}C yr BP and most late ones to ca. 4800–4230 ^{14}C yr BP. A recent chronology of Jomon sites in the Kanto Plain of Central Japan dates the Early stage to 6000–4400 ^{14}C yr BP and the Middle stage to 5300–3200 ^{14}C yr BP (Kuzmin et al., 1998, p. 102).

Taking into account the variability in the various chronological scales of Jomon culture, it may be found that the Early Jomon stage corresponds entirely or partially to the interval from ca. 6000 to 5000 ^{14}C yr BP, and the beginnings of the Middle Jomon stage between the middle of the fifth millennium BP and the beginnings of 4th millennium BP. Thus, the probable cultural changes observed in the Early and Middle Jomon stages fall into the interval from ca. 6000 to 5000 ^{14}C yr BP and continue to some extent after 5000 ^{14}C yr BP (Fig. 10.21).

Examining the chronological correlation between climatic-environment changes and cultural dynamics during the mid-Holocene in the basin of the Sea of Japan leads to the conclusion that the most intensive cultural changes seem to be after the period of maximum climatic warming about 5000 ^{14}C yr BP. This was the final stage of *maximum* warming and the beginnings of a series of significant cultural changes in all parts of the Japan Sea basin. Cultural events in the mainland part of the research area and in northern Sakhalin Island differed from the events which took place in the Japanese Archipelago in the mid-Holocene. In the mainland part of the research area, including in the Korean Peninsula, Primorye and the Lower Amur River regions, and in northern Sakhalin Island, cultural changes were connected with migration processes. The general direction of the migrations was from the continental territories of Eastern Asia (Northeastern China in particular) to the areas bordering the Sea of Japan. The Neolithic cultures of Northeastern China seem to be the sources or origins of population groups which began to move out from their nuclear regions about 5000 ^{14}C yr BP. After 5000 ^{14}C yr BP new populations, bringing new cultural traditions, appeared in the Korean Peninsula, Primorye, the Lower Amur River region, and in northern Sakhalin Island. Basic cultural contexts changed in all of these regions.

What could be the reasons for migrations during certain parts of the mid-Holocene in Eastern Asia? Assuming that climatic change in Northeastern China was similar to what was occurring in the Japan Sea basin, it is probable that favorable environmental and ecological circumstances caused economic and demographic increases during the period of maximum warming in the mid-Holocene. Critical situations could arise when the productive capacity of the landscape in certain areas was pressed by the increasing population. For economies based on food gathering rather than food production (i.e., agriculture) the movement of groups to new territories might be an optimal survival strategy. We are inclined to connect the appearance of the Zaisanovka Neolithic culture of the Primorye region with a movement of population groups and their associated cultural traditions from neighboring Northeastern China.

At present the study of prehistoric migrations in the Far Eastern area and adjacent territories is only in its initial stages. The model outlined here is a first attempt to link and explain cultural events which took place in the Russian Far East and the Korean Peninsula during the mid-Holocene through a combination of factors, including migration and relationships between settlement and environment.

Mid-Holocene cultural dynamics of the Japanese Archipelago were apparently caused by the independent evolution of local populations under the influence of climatic and environmental processes. During the period of *maximum* warming, from ca. 6000 to 5000 ^{14}C yr BP, favorable conditions appeared for an increasing role for marine resources and intensive plant-gathering. Maximum sea levels at 5000 ^{14}C yr BP greatly shaped the orientation to a marine economy. The increasing subsistence productivity resulted in greater stability for this form of subsistence, with a resulting growth of population and an elaboration of social structure that were reflected in changes in settlement and house patterns. Elaboration of social structure resulted in increased time and labor resources applied to activities which were not involved directly in food supplying, such as pottery-making, or ritual practice. The correlation between economic and subsistence blossoming and stabilization, and the development of more complex social structure and the intensification of pottery-making and some other crafts, has been demonstrated in a broad range of prehistoric and ancient cultures (Feinmann et al., 1981; Underhill, 1991, 1992; Nelson, 1992; Zhushchikhovskaya, 1994, 2005; etc.). The outstanding development of Middle Jomon ceramic assemblages, including the production of vessels with highly sophisticated relief designs, took a lot of labor by skilled craftsmen. This labor-intensive activity was facilitated by the economic rise that occurred during the end of Early Jomon and the Middle Jomon periods.

An elaboration of ritual practice at this time is reflected in the great increase in the number and variety of ritual objects that are found; some Middle Jomon burial practices were also labor intensive. It is interesting to note that the impressive appearances of labor- and time-intensive activities connected with ritual practice in the mid-Holocene period are also present in the archaeological records of other parts of the research area. These include the massive stone constructions or dolmens in the Korean Peninsula, and the sacral places and ceramic vessels with unusual anthropomorphic designs of the Voznesenovskaya culture of the Lower Amur River region. It seems likely that the mid-Holocene was characterized by a blossoming of ritual activities in the prehistoric cultures of the Sea of Japan basin.

Acknowledgments

We would like to thank Dr. D. Sandweiss and Dr. K. Maasch (Institute for Quaternary Studies, University of Maine, Orono) for their efforts to organize the

FERCO Conference on Climate and Culture at 3000 yr BC (October 1998). Many colleagues provided us with useful literature: Prof. K. Taira (Hokkaido University of Education, Asahigawa), Prof. Y. Sakaguchi (Tokyo), Dr. M. Akamatsu (Historical Museum of Hokkaido, Sapporo), Dr. K. Amano (Joetsu University of Education, Joetsu), Prof. K. Chinzei (Kyoto University, Kyoto), Dr. Y. Matsushima (Kanagawa Prefectural Museum, Yokohama), Dr. A. Suzuki (Hokkaido University of Education, Iwamizawa), S. Fujii (Fujii Laboratory for Geological Environment, Toyama), Dr. K.-Y. Wei (National Taiwan University, Taipei), Dr. H.-I. Yi and Dr. D.-H. Shin (Korea Ocean Research & Development Institute, Ansan), Dr. S. Shimoyama (Kyushu University, Fukuoka), Prof. Sarah M. Nelson (Denver State University), Mr. Masahiro Fukuda (University of Tsukuba, Japan), Dr. S.A. Gorbarenko (Pacific Oceanological Institute, FEB RAS, Vladivostok) and Dr. N.G. Razjigaeva (Pacific Institute of Geography, FEB RAS, Vladivostok). Part of the English text was improved by Ms. T.N. Koznova (Department of Scientific Information, Institute of Marine Biology, Far East Branch of the Russian Academy of Sciences, Vladivostok), and by Dr. David G. Anderson. This work was supported by a grant from the Russian Science Support Foundation to K. Lutaenko, and by grants from the Asia-Pacific Network for Global Change Research (APN)ARCP2006-FP14-Adrianov and from the Russian Academy of Sciences "Reaktsia morskoy bioty na izmenenie climata i okruzhaushei sredy".

References

Afanasiev, V. V. 1992., Coastal evolution of seas of the Far East in the Holocene. In *Coast Evolution in the Conditions of the Uplift of the Ocean Level*, edited by N. A. Aybulatov and S. A. Lukyanova, pp. 166–174. Moscow State University, Moscow. In Russian.

Aikens, C. M., 1995. First in the World: the Jomon Pottery of early Japan. In *The Emergence of Pottery: Technology and Innovation in Ancient Societies*, edited by W. K. Barnett and J. W. Hoopes, pp. 11–22. Smithsonian Institution Press, Washington and London.

Aikens, C. M., and T. Higuchi, 1982. *Prehistory of Japan*. Academic Press, New York.

Akamatsu, M., 1969. Molluscan assemblages of shell mounds in Hokkaido – with special reference to the so-called Jomon transgression. *Earth Science* 23:107–117. In Japanese with English abstract.

Akamatsu, M., and Y. Kitagawa, 1983. Holocene shell beds in the northern district of the Ishikari lowland, Hokkaido. *The Annual Report of the Historical Museum of Hokkaido* 11:35–45. In Japanese with English abstract.

Akamatsu, M., and H. Ushiro, 1992. A note on the aeo-Atlantic stage in the middle age in Hokkaido and South Sakhalin. In *Preliminary Reports on "Research Project of the Historical and Cultural Exchange of the North" in 1991*, pp. 91–108. Historical Museum of Hokkaido, Sapporo. In Japanese with English abstract.

Akamatsu, M., N. Yamazaki, and T. Arakawa, 1995. Faunal characteristics of Holocene molluscan fossils in Hokkaido – examples of the Ishikari lowland and around the Uchiura Bay. *Bulletin of the Historical Museum of Hokkaido* 23:7–18. In Japanese with English abstract.

Alekseev, M. N., and L. V. Golubeva, 1980. On the stratigraphy and paleogeography of the Upper Pleistocene of South Primorye. *Bulletin of the Comission on Studies of the Quaternary Period* 50:96–107. In Russian, title translated.

Alkin, S. V., 1996. Early Neolithic cultures of northeastern China. In *Late Paleolithic–Early Neolithic, Eastern Asia and Northern America* (International Symposium Materials, March 22–25, 1994, Vladivostok, Russia), edited by N. A. Kononenko, pp. 7–10. Institute of History, Archaeology and Ethnography of the Peoples of the Far East, FEB RAS and Tohoku Fukushi University, Vladivostok. In Russian, title translated.

Ando, K., 1986. Holocene paleoenvironments and the highest sea-level based on the diatom assemblages in Minuma lowland, Saitama prefecture. *The Quaternary Research (Japan)* 25(3):165–176. In Japanese with English abstract.

Andreev, G. I., 1957. The Site Zaisanovka 1 in the Primorye Region. *Soviet Archaeology* 2:121–145. In Russian with Franch abstract.

Andreev, G. I., 1960. Some aspects of the cultures of Southern Primorye of III–I mil. BC. *The Materials and Records of the Archaeology of the USSR* 86:136–161. In Russian, title translated.

Andreeva, Zh. V., editor, 1987. *Valentin-Peresheek–the Settlement of Prehistoric Ore-Miners.* Nauka, Moscow. In Russian, title translated.

Andreeva, Zh. V., editor, 1991. The Neolithic of the Southern Far East: ancient settlement in the Cave Chertovy Vorota. Nauka, Moscow. In Russian, title translated.

Arimitsu, K., 1962. *The Comb Patterned Pottery of Korea.* Department of Archaeology Publication 3. Kyoto University, Kyoto.

Bernard, F. R., Y. Y. Cai, and B. Morton, 1993. *Catalogue of the Living Marine Bivalve Mollusks of China.* Hong Kong University Press, Hong Kong.

Borzenkova, I. I., 1990. On the climatic changes in the Late Wurm–Holocene (Last 20 Thousand Years). *Transactions of the State Hydrological Institute (Leningrad)* 348:70–78. In Russian, title translated.

Brodyansky, D. L., 1979. The problem of periodization and chronology of the Neolithic of Primorye region. In *Ancient Cultures of Siberia and the Pacific Basin*, edited by R. S. Vasil'evsky, pp. 110–115. Nauka, Novosibirsk. In Russian, title translated.

Brodyansky, D. L., 1987. *Introduction to Far Eastern Archaeology.* Far East State University Press, Vladivostok. In Russian, title translated.

Brodyansky (as Brodianski), D. L., and V. A. Rakov, 1992. Prehistoric aquaculture on the Western Coast of the Pacific. In *Pacific Northeast Asia in Prehistory. Hunter-Fisher-Gatherers, Farmers, and Sociopolitical Elites*, edited by C. M. Aikens and S. N. Rhee, pp. 27–31. Washingthon State University Press, Pullman.

Brodyansky (as Brodianski), D. L., and V. A. Rakov, 1996. Marine adaptation of population and producing economy in the Neolithic Age on the Coast of Primorye. *Bulletin of the Far Eastern Branch, Russian Academy of Sciences* 1:124–130. In Russian with English abstract.

Burashnikova, T. A., M. V. Muratova, and I. A. Suetova, 1982. Climatic model of the Soviet Union Territory during the Holocene Optimum. In *Evolution of the Environment at the USSR Territory during Late Pleistocene and Holocene*, edited by A. A. Velichko, I. I. Spasskaya, and N. A. Khotinsky, pp. 245–251. Nauka, Moscow. In Russian.

Chen, C.-H., Y.-C. Lu, and C. T. Shen, 1978. Development of natural environment in the southern part of Liaoning Province during the last 10,000 years. *Scientia Sinica* 21(4):516–532.

Cherepanova, M. V., T. A. Grebennikova, A. S. Astakhov, and V. V. Shapovalov, 1988. Stratigraphy and paleogeography of the shelf and continental slope of Southern Sakhalin. In *Coastal Zone of the Far Eastern Seas in the Pleistocene*, edited by A. M. Korotky and V. S. Pushkar, pp. 93–108. USSR Academy of Sciences, Far East Branch, Vladivostok. In Russian, title translated.

Chinzei, K., 1993. Comments of paleotemperature analysis based on the molluscan species. *Fossils* 55:49–51. In Japanese with English abstract.

Chinzei, K., K. Fujioka, F. Kitazato, I. Koizumi, T. Oba, M. Oda, M. Okada, T. Sakai, and Y. Tanimura, 1987a. Postglacial environmental change of the Pacific Ocean off the coasts of Central Japan. *Marine Micropaleontol* 11:273–291.

Chinzei, K., H. Koike, T. Oba, Y. Matsushima, and H. Kitazato, 1987b. Secular changes in the oxygen isotope ratios of molluscs shells during the Holocene of Central Japan. *Palaeogeography, Palaeoclimatolagy, Palaeoecology* 61:155–166.

Chinzei, K., and T. Oba, 1986. Oxygene isotope studies of the Deep Sea sediments around Japan. *Recent Progress of Natural Sciences in Japan (Quaternary Research)* 11:35–43.

Choe, C. P., 1982. The diffusion route and chronology of Korean plant domestication. *Journal of Asian Studies* 41(3):519–529.

Choe, C. P., and M. T. Bale, 2002. Current perspectives on settlement, subsistence, and cultivation in prehistoric Korea. *Arctic Anthropology* 39:95–121.

Derevyanko, A. P., and V. E. Medvedev, 1995a. *The Study of Gasya Site.* Nauka, Novosibirsk. In Russian with English abstract; title translated.

Derevyanko, A. P., and V. E. Medvedev, 1995b. Sacral places on Suchu Island – new type of Far Eastern archaeological sites. In *Survey of the Results of Field and Analytical Researches of Archaeologists, Ethnologists and Anthropologists of Siberia and Far East in 1993*, edited by A. P. Derevyanko and V. E. Larichev, pp. 228–238. Siberian Branch of the Russian Academy of Sciences, Novosibirsk. In Russian, title translated.

Derevyanko, A. P., and V. E. Medvedev, 1996. Suchu Island – Unique archaeological site of the Far East. In *Archaeology of Northern Pacific,* edited by A. L. Ivliev, N. N. Kradin, and I. S. Zhushchikhovskaya, pp. 214–221. Dalnauka, Vladivostok. In Russian with English abstract.

DeVries, T. J., L. Ortlieb, A. Diaz, L. Well, and C. Hillaire-Marcel, 1997. Determining the Early History of El Nino. *Science* 276:965–966.

DeVries, T. J., and L. E. Well, 1990. Thermally-anomalous Holocene molluscan assemblages from Coastal Peru: evidence for paleogeographic, not climatic change. *Palaeogeography, Palaeoclimatology, Palaeoecology* 81:11–32.

Dyakov, V. I., 1992. *Multi-layered site Rudnaya Pristan' and the Periodization of the Neolithic cultures of Primorye Region.* Dalnauka, Vladivostok. In Russian, title translated.

Endo, K., M. Kosugi, M. Matsushita, N. Miyaji, R. Hishida, and T. Takano, 1989. Holocene environmental history in and around the Paleo-Nagareyama Bay, Central Kanto Plain. *The Quaternary Research (Japan)* 28(2):61–77. In Japanese with English abstract.

Evseev, G. A., 1981. *Communities of Bivalve Mollusks in Post-Glacial Deposits of Shelf of the Sea of Japan.* Nauka, Moscow. In Russian, title translated.

Feinmann, G. M., S. Upham, and G. Lightfoot, 1981. The production step measure: an ordinal index of labor impact in ceramic manufacture. *American Antiquity* 46(1):871–884.

Fuji, N., 1982. Chronostratigraphic subdivision of the postglacial deposits in the Japanese Islands. *Striae* 16:24–35.

Fujii, S., and N. Fuji, 1967. Postglacial sea level in the Japanese Islands. *Journal of Geosciences, Osaka City University* 10:43–51.

Fujimoto, K., 1990. Reexamination of Late Holocene sea-level changes in Japan. *The Science Reports of the Tohoku University, 7th Series (Geography)* 40(2):37–70.

Fujimoto, K., 1993. Late Holocene sea-level changes in Hiyogawa lowland facing Nanao-Nishi Bay, Noto Peninsula, Central Japan. *The Quaternary Research (Japan)* 32(1):1–12. In Japanese with English abstract.

Gohara, Y., 1976. Climatic fluctuations and sea level changes during the latest Pleistocene and Early Holocene. *Pacific Geology* 11:87–97.

Golubeva, L. V., and L. P. Karaulova, 1983. Vegetation and climatostratigraphy of Pleistocene and Holocene of the USSR Far East South. *Transactions of the Geological Institute, USSR Academy of Sciences* 366:1–143. In Russian.
Gorbarenko, S. A., 1991. Stratigraphy of Late Quaternary sediments in Seas of Japan and Okhotsk and their paleoceanological conditions. In *The INQUA International Symposium on Stratigraphy and Correlation of Quaternary Deposits of the Asian and Pacific Regions*, edited by CCOP Technical Secretariat, pp. 11–24. CCOP (Coordinating Committee for Offshore Prospecting).
Gorbarenko, S. A., S. G. Pliss, J. R. Southon, M. Kashgarian, N. B. Verkhovskaya, and A. S. Kundyshev, 1995. Detailed carbonate stratigraphy of the Japan Sea sediments during last Glaciation – Holocene. *Terrestrial, Atmospheric and Oceanic Sciences (Taiwan)* 6(1):103–113.
Gorbarenko, S. A., and J. R. Southon, 2000. Detailed Japan sea paleoceanography during the last 25 kyr: constraints from AMS dating and δ 18O of planktonic foraminifera. *Palaeogeography, Palaeoclimatology, Palaeoecology* 156:177–193.
Griffin, J. B., 1965. Ceramic complexity and cultural development. In *Ceramics and Man*, edited by F. R. Matson, pp. 104–113. University of Chicago Press, Chicago.
Gvozdeva, I. G., S. A. Gorbarenko, V. A. Rakov, K. A. Lutaenko, E. I. Shornikov, and Yu. A. Mikishin, 1997. *Primorye Paleoenvironment in Middle and Late Holocene Based on the Complex Study of the Shkotovo Outcrop*. Pacific Oceanological Institute, Far East Branch of the Russian Academy of Sciences, Vladivostok. In Russian with English abstract.
Hamada, T., 1977. The Holocene corals of raised reefs of Japan. *Memoirs de Bureau Recherche Geologique et Miniers* 89:389–395.
Han, S. J.,D.-H. Shin, H.-I. Yi, and J. H. Chun, 1996. Evidence of high sea-level stands in the Korean Coast during the Holocene Epoch. In *Abstracts of Papers Presented at the 1996 Annual Meeting, International Geological Correlation Program, Project 367 (Late Quaternary Coastal Records of Rapid Change: Application to Present and Future Conditions), Sydney, November 4–14, 1996*, edited by C. V. Murray-Wallace, P. S. Roy, and K. H. Lenham, p. 31.
Harris, V., 1997. Jomon pottery in ancient Japan. In *Pottery in the Making (World Ceramic Traditions)*, edited by I. Freestone and D. Gaimster, pp. 20–25. British Museum, London.
Hatsushige, O., and T. Mitsunori, 1996. *Nihon Kokogaku Yogo Jikon (Japanese Archaeological Dictionary)*. Kashiva Shobo, Tokyo. In Japanese.
Igarashi, Y., 1993. History of environmental change in Hokkaido from the viewpoint of palynological research. In *Biodiversity and Ecology in the Northernmost Japan*, edited by S. Higashi, A. Osawa, and K. Kanagawa, pp. 2–19. Hokkaido University Press, Sapporo.
Igarashi, Y., 1994. Quaternary forest and climate history of Hokkaido, Japan, from marine sediments. *Quaternary Science Reviews* 13:335–344.
Igarashi, Y., 1997. Vegetation and climate history in the circum Okhotsk region since the Last Glacial Maximum. In *Cryospheric Studies in Kamchatka I*, pp. 29–37. Institute of Low Temperature Science, Hokkaido University, Sapporo.
Igarashi, Y., 2000. Vegetation history around the Kyogoku mire, southwestern Hokkaido, during the last 13,000 years. *Japanese Journal of Ecology* 50:99–110. In Japanese with English abstract.
Igarashi, Y., S. G. Hu, M. Okamura, H. Matsuoka, M. Murayama, T. Murata, and N. Kobayashi, 2001. Dynamics of vegetation and climate during these 50,000 tears along the Sea of Okhotsk Region, Hokkaido, based on fossil pollen assemblages from bottom sediments of the Lake Abashiri. In *The 16th International Symposium on Okhotsk Sea and Sea Ice, 4–8 February, Mombetsu, Hokkaido, Japan: Abstracts*, pp. 228–234. Okhotsk Sea & Cold Ocean Research Association, Mombetsu.

Igarashi, Y., and T. Igarashi, 1998. Late Holocene vegetation history in South Sakhalin, northeast Asia. *Japanese Journal of Ecology* 48:231–244. In Japanese with English abstract.

Igarashi, Y., T. Igarashi, H. Daimaru, O. Yamada, T. Miyagi, K. Matsushita, and K. Hiramatsu, 1993. Vegetation history of Kenbuchi Basin and Furano Basin in Hokkaido, North Japan, since 32,000 yrs BP. *The Quaternary Research (Japan)* 32(2):89–105. In Japanese with English abstract.

Igarashi, Y., and S. Kumano, 1974. Holocene climatic change in the lowland area of north of Sapporo, Japan. *The Quaternary Research (Japan)* 13(2):48–53. In Japanese with English abstract.

Igarashi, Y., T. Sagayama, T. Higake, and M. Fukuda, 2000. Late Quaternary environmental change in Central and North Sakhalin, Russia. *Journal of Geography* 109(2):165–173. In Japanese with English abstract.

Igarashi, Y., and N. Takahashi, 1985. Origin and vegetational succession of upland bogs in the Daisetsu Mountains, Central Hokkaido. *The Quaternary Research (Japan)* 24(2):99–109. In Japanese with English abstract.

Ikawa-Smith, F., 1980. Current issues in Japanese archaeology. *American Scientist* 68(2):134–145.

Iofan, N. A., 1974. *The Culture of Ancient Japan*. Nauka, Moscow. In Russian, title translated.

Ishikawa, H., and K. Suzuki, editors, 1994. *Japanese History: New View (Prehistoric Period)*. Shinjin Butsu Oraisha, Tokyo. In Japanese.

Ivankov, V. A., 1996. Change of climate and dynamics of the coast water ichtyocenoses structure. In *Global Changes Studies at the Russian Far East, Abstracts of Workshop, September 21–23, 1996, Vladivostok*, pp. 14–15. Dalnauka, Vladivostok.

Jian, Z., B. Li, U. Pflaumann, and P. Wang, 1996. Late Holocene cooling event in the Western Pacific. *Science in China, Series D* 39(5):543–550.

Jian, Z., Y. Saito, P. Wang, B. Li, and R. Chen, 1998. Shifts of the Kuroshio axis over the last 20000 years. *Chinese Science Bulletin* 43(12):1053–1056.

Jones, G. A., and Ya. V. Kuzmin, 1995. Radiocarbon AMS dating of the thermophilous mollusc shells from Peter the Great Gulf coast. In *Complex Studies of the Holocene Sections on Peter the Great Gulf Coast (Sea of Japan)*, edited by Ya. V. Kuzmin, pp. 34–38. Bagira-Press, Moscow. In Russian with English abstract.

Kawana, T., and P. Pirazzoli, 1990. Re-examination on the Holocene emerged shorelines in Irabu and Shimoji Islands, the South Ryukyus, Japan. *The Quaternary Research (Japan)* 28(5):419–426.

Khotinsky, N. A., 1977. *Holocene of the Northern Eurasia*. Nauka, Moscow. In Russian with English abstract.

Khotinsky, N. A., 1982. Holocene chronological levels: problems of the Holocene paleogeography. In *Evolution of the Environment at the USSR Territory during Late Pleistocene and Holocene,* edited by A. A. Velichko, I. I. Spasskaya, and N. A. Khotinsky, pp. 142–147. Nauka, Moscow. In Russian.

Khotinsky, N. A. and S. S. Savina, 1985. Paleoclimatic schemes (of the USSR territory in the boreal, atlantic and subboreal periods of Holocene. *Proceedings of the USSR Academy of Sciences, Geographical Series* 4:18–34. In Russian.

Khudik, V. D., 1991. Taxonomic composition and results of paleoecological analysis of mollusks from the Neolithic Site Chertovy Vorota. In *The Neolithic of the Southern Far East: Ancient Settlement in the Cave Chertovy Vorota*, edited by Zh. V. Andreeva, pp. 212–214. Nauka, Moscow. In Russian, title translated.

Kidder, J. E., and T. Esaka, 1968. *Prehistoric Japanese Arts: Jomon Pottery*. Kodansha, Tokyo.

Klimanov, V. A., 1989a. Cyclicity and quasiperiodicity of climatic fluctuations in the Holocene. In *Paleoclimates of the Postglacial and Holocene*, edited by A. A. Frolova, pp. 29–33. Nauka, Moscow. In Russian, title translated.

Klimanov, V. A., 1989b. Paleoclimatic reconstructions at the USSR territory in the main thermal maxima of the Holocene. *Transactions of the Institute of Geology and Geophysics, Siberian Branch of the USSR Academy of Sciences* 657:131–136. In Russian, title translated.

Kluiev, N. A., and I. S. Zhushchikhovskaya, 1996. On the main tendencies of modern archaeology of primorye region, Russian Far East. *The Society of North-Eurasian Studies Newsletter* 7:17–21.

Kobayashi, T., 1992a. Regional organization in the Jomon period. *Arctic Anthropology* 29(1): 82–95.

Kobayashi, T., 1992b. Patterns and levels of social complexity in Jomon Japan. In *Pacific Northeast Asia in Prehistory. Hunter-Fisher-Gatherers, Farmers and Sociopolitical Elites*, edited by M. Aikens and S. N. Rhee, pp. 91–98. Washington State University Press, Pullman.

Kobayashi, T., 2004. *Jomon Reflections: Forager Life and Culture in the Prehistoric Japanese Archipelago*, Oxford: Oxbow Books.

Koizumi, I., 1987. Pulses of Tsushima current during the Holocene. *The Quaternary Research (Japan)* 26(1):13–25. In Japanese with English abstract.

Koizumi, I., 1989. Holocene pulses of diatom growths in the warm Tsushima current in the Japan Sea. *Diatom Research* 4(1):55–68.

Komiyama, T., 1996. Site location from the Middle to Final Jomon period in the Chubu highlands, Central Japan. In *Archaeology of Northern Pacific*, edited by A. L. Ivliev, N. N. Kradin, and I. S. Zhushchikhovskaya, pp. 347–351. Dalnauka, Vladivostok.

Kong, W.-S. 1994. The vegetational history of Korea during the Holocene period. *The Korean Journal of Quaternary Research* 8(1):9–22.

Korotky, A. M., 1994. Fluctuation of the Sea of Japan level and landscapes of coastal zone (development stages and tendencies). *Bulletin of the Far Eastern Branch, Russian Academy of Sciences* 3:29–42. In Russian with English abstracts.

Korotky, A. M., 1998. Development of Muravyev-Amurski Peninsula's and its nearest territories' environment in Quaternary period. *Bulletin of the Far Eastern Branch, Russian Academy of Sciences* 1:12–26. In Russian with English abstracts.

Korotky, A. M., T. A. Grebennikova, V. S. Pushkar, N. G. Razjigaeva, V. G. Volkov, L. A. Ganzey, L. M. Mokhova, V. B. Bazarova, and T. R. Makarova, 1996a. *Climatic Changes of South Far East Area in Late Cenozoic (Miocene-Pleistocene)*. Russian Academy of Sciences, Far East Branch, Vladivostok. In Russian with English abstract.

Korotky, A. M., and L. P. Karaulova, 1976. Radiocarbon (C14) control of the Quaternary stratigraphy of Primorye. In *Geomorphology and Quaternary Geology of the Far East*, edited by A. M. Korotky and A. P. Kulakov, pp. 98–111. Far East Science Centre, USSR Academy of Sciences, Vladivostok. In Russian, title translated.

Korotky, A. M., L. P. Karaulova, and T. S. Troitskaya, 1980. Quaternary deposits of Primorye. Stratigraphy and paleogeography. *Transactions of the Institute of Geology and Geophysics, Siberian Branch of the USSR Academy of Sciences* 429:1–234. In Russian, title translated.

Korotky, A. M., S. P. Pletnev, V. S. Pushkar, T. A. Grebennikova, N. G. Razjigaeva, N. D. Sakhebgareeva, and L. M. Mokhova, 1988. *Development of Natural Environment of the Southern Far East (Late Pleistocene–Holocene)*. Nauka, Moscow. In Russian, title translated.

Korotky, A. M., V. S. Pushkar, T. A. Grebennikova, N. G. Razjigaeva, L. P. Karaulova, L. M. Mokhova, L. A. Ganzey, T. A. Cherepanova, V. B. Bazarova, V. G. Volkov, and N. N. Kovalyukh, 1997. *Marine Terraces and Quaternary History of Sakhalin Island Shelf*. Dalnauka, Vladivostok. In Russian with English abstract.

Korotky, A. M., N. G. Razjigaeva, T. A. Grebennikova, L. A. Ganzey, L. M. Mokhova, and V. B. Bazarova, 1996c. Landscape types and coastal environments of Kunashiri Island in Holocene. In *Global Changes Studies at the Russian Far East, Abstracts of Workshop, September 21–23, 1996, Vladivostok*, pp. 19–20. Dalnauka, Vladivostok.

Korotky, A. M., N. G. Razjigaeva, T. A. Grebennikova, L. A. Ganzey, L. M. Mokhova, V. B. Bazarova, and L. D. Sulerzhitsky, 1995. Holocene marine terraces of Kunashiri Island, Kurile Islands. *The Quaternary Research (Japan)* 34(5):359–375.

Korotky, A. M., N. G. Razjigaeva, T. A. Grebennikova, L. A. Ganzey, L. M. Mokhova, V. B. Bazarova, and L. D. Sulerzhitsky, 1998. Holocene marine terraces of Kunashir Island. *Geology of the Pacific Ocean* 17(1):29–45. In Russian with English abstract.

Korotky, A. M., N. G. Razjigaeva, T. A. Grebennikova, L. A. Ganzey, L. M. Mokhova, V. B. Bazarova, L. D. Sulerzhitsky, and K. A. Lutaenko, 1999. Holocene deposits and paleogeography of Kunashir Island (Kurile Islands). *Geology of the Pacific Ocean* 18(1):25–40. In Russian with English abstract.

Korotky, A. M., N. G. Razjigaeva, T. A. Grebennikova, L. A. Ganzey, L. M. Mokhova, V. B. Bazarova, L. D. Sulerzhitsky, and K. A. Lutaenko, 2000a. Middle- and Late-Holocene environments and vegetation history of Kunashir Island, Kurile Islands, Northwestern Pacific. *The Holocene* 10(3):311–331.

Korotky, A. M., N. G. Razjigaeva, T. A. Grebennikova, L. M. Mokhova, V. B. Bazarova, L. D. Sulerzhitskiy, and K. A. Lutaenko, 2000b. Late Pleistocene–Holocene sedimentation and paleolandscapes of the Kurilka River Basin (Iturup Island, Kuriles). *Geology of the Pacific Ocean* 19(5):61–77. In Russian with English abstract.

Korotky, A. M., N. G. Razjigaeva, L. M. Mokhova, L. A. Ganzey, T. A. Grebennikova, and V. B. Bazarova, 1996b. Coastal dunes–indicators of global cooling, Kunashiri Island, Kuril Islands. *Geology of the Pacific Ocean* 15(1):51–59. In Russian with English abstract.

Korotky, A. M., and Yu. E. Vostretsov, 1998. Geographical, environmental and cultural dynamics in the Middle Holocene in Peter the Great Bay. In *The First Fishers in Peter the Great Bay. Environment and Ancient Man in Boisman Bay*, edited by Yu. E. Vostretsov, pp. 9–29. Institute of History, Archaeology and Ethnography of the Peoples of the Far East, Vladivostok. In Russian with English abstract.

Korotky, A. M., V. I. Vysotchin, and I. G. Gvozdeva, 1981. Permafrost of Sikhote-Alin and its dynamycs in the Late Wurm-Holocene. In *Landscape Geophysical Investigations in the Far East*, edited by Sh. Sh. Gasanov and I. I. Krylov, pp. 45–57. USSR Academy of Sciences, Far East Science Centre, Vladivostok. In Russian, title translated.

Kosugi, M., 1989. Coastline and adjacent environments of the former Tokyo Bay in the Holocene. *Geographical Review of Japan* 62-A(5):359–374. In Japanese with English abstract.

Kuzmin, Ya. V., 1995. Paleoenvironment of Peter the Great Gulf in Mid-Holocene. In *Complex Studies of the Holocene Sections on Peter the Great Gulf Coast (Sea of Japan)*, edited by Ya. V. Kuzmin, pp. 44–61. Bagira-Press, Moscow. In Russian with English abstract.

Kuzmin, Ya. V., S. V. Alkin, A. Ono, H. Sato, T. Sakaki,S. Matsumoto, K. Orimo, and S. Ito, 1998. *Radiocarbon chronology of the Stone Age of Northeast Asia*. Pacific Institute of Geography of the Far East Branch, Russian Academy of Sciences, Vladivostok. In Russian with English, Chinese, Korean and Japanese abstracts.

Kuzmina, N. N., G. M. Shumova, E. I. Polyakova, and G. N. Nedesheva, 1987. Paleogeographical reconstructions of the Holocene of North-Western Coast and Shelf of the Sea of Japan. *Proceedings of the USSR Academy of Sciences, Geographical Series* 4:78–89. In Russian.

Larichev, V. E., 1978. The Neolithic and Bronze Age of Korea. In *Siberia, Central and Eastern Asia in the Past: The Neolithic and Epoch of the Metal*, edited by V. E. Larichev, pp. 9–87. Nauka, Nauka. In Russian, title translated.

Liao, X., A. Street-Perrott, and J. F. B. Mitchell, 1994. GCM experiments with different cloud parameterization: comparisons with paleoclimatic reconstructions for 6000 years B.P. *Palaeoclimates, Data and Modelling* 1(1):99–123.

Lutaenko, K. A., 1991a. On Mid-Holocene migrations of warm-water mollusks in the Sea of Japan. *Doklady Akademii Nauk SSSR* 321(3):596–598. In Russian.

Lutaenko, K. A., 1991b. On the origin of warm-water elements of Malacofauna of Peter the Great Bay, Sea of Japan. *Marine Biology (Vladivostok)* 1:12–20. In Russian with English abstract.

Lutaenko, K. A., 1993. Climatic optimum during the Holocene and the distribution of warm-water mollusks in the Sea of Japan. *Palaeogeography, Palaeoclimatology, Palaeoecology* 102:273–281.

Lutaenko, K. A., 1997. The Holocene paleoenvironments in Talmi lagoon, Northwestern part of the East/Japan Sea as revealed by faunal changes. In *International Workshop on Paleoceanography and Paleoclimatology, Abstracts, February 20–28, 1997, Ansan*, edited by S.-J. Han and H.-I. Yi, pp. 36–38. KORDI (Korea Ocean Research and Development Institute), Ansan.

Lutaenko, K. A., 2001. The Holocene mollusks of Garolim Bay (Western Coast of Korea): taphonomic and paleogeographic significance. *The Bulletin of the Russian Far East Malacological Society* 5:39–61. In Russian with English abstract.

Maeda, Y., 1978. Holocene transgression in Osaka Bay. Environmental changes in the Osaka Bay area during the Holocene. Part III. *Journal of Geosciences, Osaka City University* 21(5):53–63.

Matsushima, Y., 1982a. Radiocarbon ages of the Holocene marine deposits along Kucharo Lake, Northern Hokkaido. *Bulletin of the Kanagawa Prefectural Museum (Natural Science)* 13:51–66. In Japanese with English abstract.

Matsushima, Y., 1982b. The Radiocarbon age of the molluscan fossils from the alluvial deposits, along Pashikuru-Numa, the Pacific Coast of Hokkaido. *Bulletin of the Kushiro Museum* 9:1–8. In Japanese.

Matsushima, Y., 1984. Shallow marine molluscan assemblages of postglacial period in the Japanese Islands – its historical and geographical changes induced by the environmental changes. *Bulletin of the Kanagawa Prefectural Museum (Natural Science)* 15:37–109. In Japanese with English and German abstract.

Matsushima, Y., 1989. Holocene marine environmental change in the inner part of Mikawa Bay, Central Japan. *Bulletin of the Kanagawa Prefectural Museum (Natural Science)* 18:23–34. In Japanese with English abstract.

Matsushima, Y., and M. Nara, 1987. Radiocarbon ages of the molluscan shells from the Holocene marine deposits in the Tanabu Plain, Shimokita Peninsula, Northeast Japan, with considerations on their related problems. *Bulletin of the Kanagawa Prefectural Museum (Natural Science)* 17:57–72. In Japanese with English abstract.

Matsushima, Y., and K. Ohshima, 1974. Littoral molluscan fauna of the Holocene climatic optimum (5,000–6,000 y BP) in Japan. *The Quaternary Research (Japan)* 13 (3):135–159. In Japanese with English abstract.

Matsushima, Y., and J. Yamashiro, 1992. Radiocarbon ages of the molluscan fossils from the Holocene deposits in Kushiro Moor, Hokkaido. *Bulletin of the Kanagawa Prefectural Museum (Natural Science)* 21:37–43. In Japanese with English abstract.

Medvedev, V. E., 1995. On the problem of initial and Early Neolithic of lower Amur River region. In *Survey of the Results of Field and Analythical Researches of Archaeologists, Ethnologists and Anthropologists of Siberis and Far East in 1993*, edited by A. P. Derevyanko and V. E. Larichev, pp. 228–238. Siberian Division of Russian Academy of Scences, Novosibirsk. In Russian, title translated.

Medvedev, V. E., 2003. Academician A.P. Okladnikov and the Neolithic of lower Priamurye region: the evolution of ideas. In *The Problems of Archaeology and Paleoecology of Northern, Eastern and Central Asia*, edited by A. P. Derevyanko, pp. 164–171. Institute of Archaeology and Ethnography, Siberian Branch of the Russian Academy of Sciences, Novosibirsk.

Mikishin, Yu. A. 1996. Temperature maxima and natural changes of Sakhalin Island in the Holocene. In *Global Changes Studies at the Russian Far East, Abstracts of Workshop, September 21–23, 1996, Vladivostok*, pp. 24. Dalnauka, Vladiovostok.

Mikishin, Yu. A., and I. G. Gvozdeva, 1996. *The Natural Evolution in the South-Eastern Part of Sakhalin Island in Holocene*. Far East State University Press, Vladivostok. In Russian with extended English abstract.

Mikishin, Yu. A., I. G. Gvozdeva, and M. M. Pevzner, 1997. Boreal thermal maximum of the Holocene in Sakhalin and the new data on the "Uandi" section. In *Global Change Studies at the Far East, Abstracts of Workshop, September 1–4, 1997, Vladivostok*, pp. 23–26. Dalnauka, Vladivostok.

Mikishin, Yu. A., T. I. Petrenko, I. G. Gvozdeva, and G. G. Rasova, 1987. Stratigraphy of the 5 m–high terrace of the Tchlia Lake. In *Palynology of the East of the USSR*, edited by V. A. Markevich, pp. 94–101. Far East Science Centre, USSR Academy of Sciences, Vladivostok. In Russian.

Moreva, O. L., and A. N. Popov, 2003. Cultural attribution of sharp-bottomed pottery from site Boismana-2. In *The Problems of Archaeology and Paleoecology of Northern, Eastern and Central Asia*, edited by A. P. Derevyanko pp. 176–179. Institute of Archaeology and Ethnography, Siberian Branch of the Russian Academy of Sciences, Novosibirsk.

Mylnikova, L. N., 1992. *Pottery-Making of Neolithic Population of Lower Amur Region*. Thesis/ PhD dissertation. Novosibirsk. In Russian, title translated.

Nakagawa, T., O. Fukuoka, S. Fujii, M. Chiji, and T. Nakamura, 1993b. Fossil shell assemblages in the Holocene Takahama shell bed discovered at Takahama-Cho, Western Fukui Prefecture, Central Japan. *Monograph of the Fukui City Museum of Natural History* 1:1–113. In Japanese with English abstract.

Nakagawa, T., O. Fukuoka, S. Fujii, and T. Nakamura, 1993a. Fossil shell assemblages and 14 C–ages on the Holocene shell bed discovered at Takahama-Cho, Western Fukui Prefecture, Central Japan. *The Quaternary Research (Japan)* 32(2):75–87. In Japanese with English abstract.

Nakamura, G., 1993. The origins of rice cultuvation: problems associated with finds of rice and Ishibocho in the Northern Indian subcontinent. *Bulletin of the Ancient Orient Museum* 14:53–117.

Nelson, S. M., 1975. The subsistence base of Middle Han sites of the Chulmun Period. *Asian Perspectives* 18(1):5–14.

Nelson, S. M., 1990. Korean interpretations of Korean archaeology. *Asian Perspectives* 27(2):185–191.

Nelson, S. M., 1992. The question of agricultural impact on sociopolitical development in prehistoric Korea. In *Pacific Northeast Asia in Prehistory. Hunter-Fisher-Gatherers, Farmers and Sociopolitical Elites*, edited by M. Aikens and S. N. Rhee, pp. 179–184. Washington State University Press, Pullman.

Nelson, S. M., 1993. *The Archaeology of Korea*. Cambridge University Press, Cambridge.

Nelson, S. M., 1995. *The Archaeology of Northeast China*. Routledge, London and New York.

Nishida, Y., 1987. Shell tempered pottery from Elasu, Northeast China. *Bulletin of the Department of Archaeology, University of Tokyo* 6:45–56.

Nishida, M., 1989. *Historical View of Jomon Ecology*. Tokyo Daigaku Shuppankai, Tokyo. In Japanese.

Nishimura, S., 1965a. The zoogeographical aspects of the Japan Sea. Part I. *Publications of the Seto Marine Biological Laboratory* 13(1):35–79.
Nishimura, S., 1965b. The zoogeographical aspects of the Japan Sea. Part II. *Publications of the Seto Marine Biological Laboratory* 13(2):81–101.
Nishimura, S., 1966. The zoogeographical aspects of the Japan Sea. Part III. *Publications of the Seto Marine Biological Laboratory* 13(5):365–384.
Nishimura, S., 1968. The zoogeographical aspects of the Japan Sea. Part IV. *Publications of the Seto Marine Biological Laboratory* 15(5):329–352.
Nishimura, S., 1969. The zoogeographical aspects of the Japan Sea. Part V. *Publications of the Seto Marine Biological Laboratory* 17(2):67–142.
Nishimura, S., 1983. Okhotsk Sea, Japan Sea, East China Sea. In *Ecosystems of the World 26. Estuaries and Enclosed Seas*, edited by B. H. Ketchum, pp. 375–401. Elsevier.
Oba, T., 1997. Migration of the Kuroshio current since the Last Glacial Age based on the oxygen isotopic ratios of foraminiferal tests in five cores off Central Japan. In *International Workshop on Paleoceanography and Paleoclimatology, Abstracts, February 20–28, 1997, Ansan*, edited by S.-J. Han and H.-I. Yi, pp. 39–44. KORDI (Korea Ocean Research and Development Institute), Ansan.
Ohara, S., and K. Taira, 1974. Molluscan remains from the Taito-Zaki formation. *Journal of the College of Arts and Sciences, Chiba University* B-7:43–53 + 5.
Ohshima, K., S. Yamaguchi, and H. Satoh, 1972. Alluvial shell beds along Kucharo Lake, Hokkaido. *Journal of the Geological Society of Japan* 78:129–135. In Japanese with English abstract.
Okladnikov, A. P., 1970. The Neolithic of Siberia and the Far East. In *Stone Age in the USSR Territory*, edited by A. A. Formosov, pp. 172–193. Nauka, Moscow and Leningrad. In Russian, title translated.
Okladnikov, A. P., 1981. *Ancient Art of the Amur River*. Aurora Press, Leningrad. In Russian and English.
Okladnikov, A. P., 1984. The pottery of prehistoric site Kondon, Priamurye. Nauka, Novosibirsk. In Russian, title translated.
Ota, Y., M. Umitsu, and Y. Matsushima, 1990. Recent Japanese research on relative sea level changes in the Holocene and related problems – review of studies between 1980 and 1988. *The Quaternary Research (Japan)* 29(1):31–48. In Japanese with English abstract.
Pak, I. Z., 1985. *Mollusks of Korea*. Kim Il Sung University Press, Pyongyang. In Korean, title translated.
Pak, M. N., 1979. *The Essays of Early Korean History*. Moscow State University, Moscow. In Russian, title translated.
Park, Y. A., 1987. Chapter 25. Continental shelf sedimentation. In *Geology of Korea*, edited by D. Lee, pp. 406–426. Kyohak-sa Publishing Company, Seoul.
Park, Y. A., and H.-I. Yi, 1995. Late Quaternary climatic changes and sea-level history along the Korean coast. *Journal of Coastal Research Special Issue* 17:163–168.
Pearson, R. J., 1976. Japan, Korea and China: the problem of defining continuities. *Asian Perspectives* 19(1):176–189.
Pearson, R. J., editor, 1986. *Windows on the Japanese Past: Studies in Archaeology and Prehistory*. Center for Japanese Studies, University of Michigan, Ann Arbor.
Pearson, R. J., editor, 1992. *Ancient Japan*. Smithsonian Institution Press, Washington, DC.
Pirazzoli, P., 1978. High stands of Holocene sea levels in the Northwest Pacific. *Quaternary Research* 10(1):1–29.
Pletnev, S. P., 1978. On the stratigraphy and paleogeography of the Late Pleistocene deposits of underwater Yamato Rise (Sea of Japan). In *Paleontology and Stratigraphy of the Cenozoic*

Deposits of Japan and Philippine Seas, edited by N. P. Vassilkovsky, pp. 94–102. USSR Academy of Sciences, Far East Science Centre, Vladivostok. In Russian, title translated.

Pletnev, S. P., 1982. Distribution of planktonic foraminifers in the recent deposits of the Sea of Japan. In *Recent Deposition of Sediments and Quaternary Morpholitogenesis of the Far East*, edited by S. P. Pletnev and V. S. Pushkar, pp. 80–89. USSR Academy of Sciences, Far East Science Centre, Vladivostok. In Russian, title translated.

Pletnev, S. P. 1985. *Stratigraphy of Bottom Deposits and Paleogeography of the Sea of Japan in Late Quaternary Period (According to Planctonic Foraminifera)*. USSR Academy of Sciences, Far East Science Centre, Vladivostok. In Russian, title translated.

Pletnev, S. P., and T. A. Grebennikova, 1991. Paleoceanographic Situation of the Sea of Japan in Late Quaternary time as evidenced by biostratigraphic data. In *Stratigraphy and Correlation of the Quaternary Deposits of Asia and the Pacific Region*, edited by G. I. Khudyakov, pp. 202–204. Nauka, Moscow. In Russian, title translated.

Pletnev, S. P., T. A. Grebennikova, A. I. Botsul, and V. I. Babanakova, 1987. An estimation of hydroclimatic conditions of Tatarsky Strait in postglacial time. In *Paleogeographical Investigations in the Far East*, edited by A. M. Korotky and V. S. Pushkar, pp. 75–83. USSR Academy of Sciences, Far East Branch, Vladivostok. In Russian, title translated.

Popov, A. N., T. I. Chikisheva, and E. G. Shpakova, 1997. *The Boisman Archaeological Culture of the Southern Primorye: The Multilayered Site Boisman-2*. Institute of Archaeology and Ethnography, Siberian Branch of the Russian Academy of Sciences Press, Novosibirsk. In Russian with English abstract.

Punning, J.-M. K. and A. V. Raukas, 1985. Paleogeography of the Late Quaternary time of Northern Europe. Itogi Nauki I Tekhniki. *Seriya Paleogeographiya* 2:1–215. In Russian, title translated.

Pushkar, V. S., 1979. Biostratigraphy of deposits of the Late Anthropogene of the Southern Far East (based on diatom analysis). Nauka, Moscow. In Russian, title translated.

Rakov, V. A., 1995. Malacofauna of the Early Neolithic shell middens from Peter the Great Gulf Coast. In *Complex Studies of the Holocene Sections on Peter the Great Gulf Coast (Sea of Japan)*, edited by Ya. V. Kuzmin, pp. 38–44. Bagira-Press, Moscow. In Russian with English abstract.

Rakov, V. A., and K. A. Lutaenko, 1997. The Holocene molluscan fauna from shell middens on the coast of Peter the Great Bay (Sea of Japan): paleoenvironmental and biogeographical Significance. *The Western Society of Malacologists, Annual Report* 29:18–23.

Rakov, V. A., A. N. Popov, and K. A. Lutaenko, 1996. The comparative analysis of molluscan assemblages from the Neolithic sites of the Primorye region and Hokkaido (preliminary onformation). In *Late Paleolithic–Early Neolithic, Eastern Asia and Northern America (International Symposium Materials, March 22–25, 1994, Vladivostok, Russia)*, edited by N. A. Kononenko, pp. 200–203. Institute of History, Archaeology and Ethnography of the Peoples of the Far East, FEB RAS and Tohoku Fukushi University, Vladivostok. In Russian with English abstract.

Rakov, V. A., and Yu. E. Vostretsov, 1998. Mollusks Gathering. In *The First Fishers In Peter the Great Bay. Environment and Ancient Man in Boisman Bay*, edited by Yu. E. Vostretsov, pp. 241–275. Institute of History, Archaeology and Ethnography of the Peoples of the Far East, Vladivostok. In Russian with English abstract.

Reid, K. C., 1989. The material science perspective on hunter-gatherer pottery. In *Pottery Technology: Ideas and Approaches*, edited by G. Bronitsky, pp. 167–180. Boulder Press, London and San Francisco.

Ryu, J. M., and Y. U. Sin, 1993 Chapter 5. Marine geology. Section 5. Fracture structures and fluctuation of sea-level surface in Quaternary. In *Geology of Korea*, edited by R. J. Paek,

H. G. Kan, G. P. Jon, Y. M. Kim, and Y. H. Kim, pp. 441–450. Foreign Languages Books Publishing House, Pyongyang.
Sahara, M., 1979. Jomon doki 2 (Jomon pottery 2). In *Nihon no genshi bijutsu 1979*. Vol. 2.
Sakaguchi, Y., 1982. Climatic variability during the Holocene Epoch in Japan and its causes. *Bulletin of the Department of Geography, University of Tokyo* 14:1–27.
Sakaguchi, Y., 1983. Warm and cold stages in the past 7600 years in Japan and their global correlation–especially on climatic impacts to the global sea level changes and the ancient Japanese history. *Bulletin of the Department of Geography, University of Tokyo* 15:1–31.
Sakaguchi, Y., 1987. Japanese prehistoric culture flourished in forest-grassland mixed areas. *Bulletin of the Department of Geography, University of Tokyo* 19:1–19.
Sakaguchi, Y., 1989. Some pollen records from Hokkaido and Sakhalin. *Bulletin of the Department of Geography, University of Tokyo* 21:1–17.
Sakaguchi, Y., 1992. Cooling of Hokkaido around 9000 BP caused by permafrost meltwater burst. *Bulletin of the Department of Geography, University of Tokyo* 24:1–6.
Sakaguchi, Y., K. Kashima, and A. Matsubara, 1985. Holocene marine deposits in Hokkaido and their sedimentary environments. *Bulletin of the Department of Geography, University of Tokyo* 17:1–17.
Sandweiss, D. H., J. B. Richardson, E. J. Reitz, H. B. Rollins, and K. Maasch, 1996. Geoarchaeological evidence from Peru for a 5000 years B.P. Onset of El Nino. *Science* 273:1531–1533.
Sandweiss, D. H., J. B. Richardson, E. J. Reitz, H. B. Rollins, and K. Maasch, 1997. Determining the early history of El Nino: a response. *Science* 276:966–967.
Scarlato, O. A., 1981. *Bivalve Mollusks of Temperate Latitudes of Western Portion of the Pacific Ocean*. Guide-Books on the Fauna of the USSR Published by the Zoological Institute, *USSR Academy of Sciences* 126:1–479. In Russian, title translated.
Shevkomud, I. Y., 1996. On relative chronology of Neolithic complexes with curve-lined ornament of the lower Amur. In *Archaeology of Northern Pacific*, edited by A. L. Ivliev, N. N. Kradin, and I. S. Zhushchikhovskaya, pp. 369–72. Dalnauka, Vladivostok.
Shevkomud, I. Y., 1999. Late Neolithic of north-eastern lower Amur area: the assemblages with the pottery of comb design and curve-lined design. Thesis/PhD dissertation, Novosibirsk.
Shi, Y., Z. Kong, S. Wang, L. Tang, F. Wang, T. Yao, X. Zhao, P. Zhang, and S. Shi, 1993. Mid-Holocene climates and environments in China. *Global and Planetary Changes* 7:219–233.
Shi, Y., Z. Kong, S. Wang, L. Tang, F. Wang, T. Yao, X. Zhao, P. Zhang, and S. Shi, 1994. Climates and environments of the Holocene megathermal maximum in China. *Science in China, Series B* 37(4):481–493.
Shimoyama, S., 1989. The Quaternary beds in the Fukuoka plain with special reference to the magnitude of the Jomon transgression. *Science Reports of the Department of Geology, Kyushu University* 16(1):37–58. In Japanese with English abstract.
Shimoyama, S., and T. Shuto, 1978. Quaternary molluscan fossil assemblages from Arato, Fukuoka city. *Science Reports of the Department of Geology, Kyushu University* 13(1):47–59. In Japanese with English abstract.
Shimoyama, S., Y. Sato, and H. Noi, 1986. Holocene deposits and fossil molluscan assemblages from Itoshima lowland in the Itoshima coastal plain, Kyushu. *Science Reports of the Department of Geology, Kyushu University* 14(4):143–162. In Japanese with English abstract.
Shubin, V. O., O. A. Shubina, and S. V Gorbunov, 1982. *The Neolithic Culture of Southern Sakhalin*. Regional Museum Press, Yuzhno-Sakhalinsk. In Russian, title translated.
Shubina, O. A., 1987. *The Houses of Imtchin Neolithic Culture's Settlements*. Regional Museum Press, Yuzhno-Sakhalinsk. In Russian, title translated.

Shumova, G. M., 1995. The natural environment of Southeasterm Primorye during the Holocene climatic optimum (based on cores from the Kievka Bight). In *Complex Studies of the Holocene Sections on Peter the Great Gulf Coast (Sea of Japan)*, edited by Ya. V. Kuzmin, pp. 25–34. Bagira-Press, Moscow. In Russian with English abstract.

Shumova, G. M., and V. A. Klimanov, 1989. Vegetation and climate of the postglacial and Holocene of the Coastal Zone of Middle Primorye. In *Paleoclimates of the Postglacial and Holocene*, edited by N. A. Khotinsky, pp. 154–160. Nauka, Moscow. In Russian, title translated.

Skibo, J. M., M. B. Schiffer, and K. C. Reid, 1989. Organic tempered pottery: an experimental study. *American Antiquty* 54:122–146.

Sokhina, E. N., T. D. Boyarskaya, A. P. Okladnikov, V. I. Roslikova, and A. I. Chernyuk, 1978. *Section of Latest Deposits of the Lower Priamurye.* Nauka, Moscow. In Russian, title translated.

Suda, R., 1995. Some aspects of recent research during the transitional period in Tohoku and Hokkaido, Northern Japan. In *The Origin of Ceramics in Eastern Asia*, edited by H. Kajiwara, pp. 105–116. Tohoku Fukushi University, Sendai. In Japanese and English.

Sugiyama, S., 1942. *The History of Fiber Technology in Prehistoric Japan: The Origin.* Yuzanaku Press, Tokyo. In Japanese.

Suzuki, K., 1989. *Archaeology of Shell Middens.* Tokyo Daigaku Shuppankai, Tokyo. In Japanese.

Svitoch, A. A., T. N. Voskresenskaya, Z. V. Aleshinskaya, N. S. Bolikhovskaya, N. A. Bolshakov, N. I. Glushankova, V. S. Gunova, O. V. Denisenko, V. P. Stepanov, and G. M. Shumova, 1988. Latest Deposits and Paleogeography of the Pleistocene of Sakhalin. VINITI (All-Union Institute of Scientific and Technical Information), Moscow. Unpublished thesis, available from VINITI as document N 9068-1988. In Russian, title translated.

Taira, K., 1975. Temperature variation of the "Kuroshio" and crustal movements in eastern and southwestern Asia 7,000 years B.P. *Palaeogeography, Palaeoclimatology, Palaeoecology* 17:333–338.

Taira, K., 1979. Holocene migrations of the warm-water front and sea-level fluctuations in the Northwestern Pacific. *Palaeogeography, Palaeoclimatology, Palaeoecology* 28:197–204.

Taira, K., 1980. Radiocarbon dating of shell middens and Holocene sea-level fluctuations in Japan. *Paleogeography, Palaeoclimatology, Palaeoecology* 32:79–87.

Taira, K., 1992. Holocene paleoceanographic changes in Japan. *Reports of the Taisetsuzan Institute of Science, Hokkaido University of Education* 27:1–7. In Japanese with English abstract.

Taira, K., and K. A. Lutaenko, 1993. Holocene palaeoceanographic changes in the Sea of Japan. *Reports of the Taisetsuzan Institute of Science, Hokkaido University of Education* 28:65–70. In Japanese with English abstract.

Takagi, T., M. Akamatsu, and T. Takahashi, 1990. Holocene molluscan assemblages from the Northern Ishikari lowland, Hokkaido, Japan. *The Annual Report of the Historical Museum of Hokkaido* 18:1–17. In Japanese with English abstract.

Takahashi, R., T. Toizumi, and K. Yasushi, 1997. Current studies of the Jomon archaeology. In *Archaeological Studies of Japan*, pp. 105–137. The Japanese Archaeological Association, Tokyo.

Troitskaya, T. S., 1974. Migration sequence of benthos foraminifera complexes in Holocene sediments of the Amur Bay (the Sea of Japan). *Transactions of the Institute of Geology and Geophysics, Siberian Branch of the USSR Academy of Sciences* 84:30–40. In Russian.

Tsukada, M., 1986. Vegetation in prehistoric Japan: the last 20,000 years. In *Windows on the Japanese Past: Studies in Archeology and Prehistory*, edited by R. J. Pearson, pp. 11–56. Center for Japanese Studies, University of Michigan, Ann Arbor.

Tzjian, L., 1994. Preliminary studying of the Incised pottery from Eastern Area of Dunbei. In *Ancient Cultures of Southern Siberia and Northeastern China*, edited by V. E. Larichev, pp. 67–82. Nauka, Novosibirsk. In Russian, title translated.

Ujiie, H., 1997. Prominent course changes of the Kuroshio current during the latest Quaternary: mainly based on the Okinawa Trough Piston Core Analysis. In *International Symposium on Quaternary Environmental Change in the Asia and Western Pacific Region, Program and Abstracts, October 14–17, 1997, The University of Tokyo, Tokyo*, p. 128. The University of Tokyo, Tokyo.

Umeda, M., and T. Nakagawa, 1993. Fossil-pollen assemblage of the Holocene Takahama shell bed at Takahama-Cho, western part of Fukui Prefecture, Central Japan. *Monograph of the Fukui City Museum of Natural History* 1:141–149. In Japanese with English abstract.

Umitsu, M., 1991. Holocene sea-level changes and coastal evolution in Japan. *The Quaternary Research (Japan)* 30(2):187–196.

Underhill, A. P., 1991. Pottery production in Cheifdoms: the Longshan period in Northern China. *World Archaeology* 23(1):12–27.

Underhill, A. P., 1992. Regional growth of cultural complexity during the Longshan period of northeast Asia in *Prehistory*. In *Hunter-Fisher-Gatherers, Farmers and Sociopolitical Elites*, edited by M. Aikens and S. N. Rhee, pp. 173–178. Washington State University Press, Pullman.

Vasilevsky, A. A., 1992. Neolithic cultures of Sakhalin Island (brief survey). *The Archaeology of Hokkaido (Hokkaido-Koukogaku)* 28:115–136.

Vasilevsky, A. A., 1995. The correction of carbon datings and absolute chronology of archaeological cultures of Sakhalin Island. *Regional Bulletin, Yuzhno-Sakhalinsk (Kraevedchesky Bulletin)* 2:93–110. In Russian, title translated.

Vasilevsky, A. A., 2003. *Stone Age of Sakhalin Island.* Thesis/PhD dissertation, Novosibirsk.

Vasil'evsky, R. S., E. L. Lavrov, and C. S. Bu 1982. *The Cultures of Stone Age of Northern Japan.* Nauka, Novosibirsk. In Russian, title translated.

Velichko, A. A., and V. A. Klimanov, 1990. Climatic zonality of the Northern Hemisphere 5 or 6 thousand years ago. *Proceedings of the USSR Academy of Sciences, Geographical Series* 5:38–52. In Russian.

Verkhovskaya, N. B., and A. S. Kundyshev, 1993. Southern Primorie environment during Neolithic and Early Iron Age. *Bulletin of the Far Eastern Branch, Russian Academy of Sciences* 1:18–26. In Russian with English abstracts.

Verkhovskaya, N. B., and A. S. Kundyshev, 1995. Vegetation of Peter the Great [Bay] Coast during the optimal phase of the Holocene. In *Complex Studies of the Holocene Sections on Peter the Great Gulf Coast (Sea of Japan)*, edited by Ya. V. Kuzmin, pp. 8–17. Bagira-Press, Moscow. In Russian with English abstract.

Vostretsov, Yu. E., editor, 1998. *The First Fishermen in Peter the Great Gulf (Man and Nature in Boisman Bay).* Vladivostok: Dal'nauka. In Russian, title translated.

Vostretsov, Yu. E., 2001. The nature and man in Southern Primorye region in Middle Holocene. *Bulletin of the Far Eastern Branch, Russian Academy of Sciences* 4:92–112.

Vostretsov, Yu. E., E. I. Gelman, M. Komoto, K. Miyamoto, H. Obata, 2003a. New ceramics assembalge of Neolithic Site Krounovka 1 in Primorye. In *The Problems of Archaeology and Paleoecology of Northern, Eastern and Central Asia*, edited by A. P. Derevyanko, pp. 86–93. Institute of Archaeology and Ethnography, Siberian Branch of the Russian Academy of Sciences, Novosibirsk.

Vostretsov, Yu. E., E. A. Sergusheva, M. Komoto, K. Miyamoto, and H. Obata, 2003b. Current data on early agriculture in Primorye: Neolithic assemblage of Site Krounovka 1. In *The Problems of Archaeology and Paleoecology of Northern, Eastern and Central Asia*, edited by

A. P. Derevyanko, pp. 373–378. Institute of Archaeology and Ethnography, Siberian Branch of the Russian Academy of Sciences, Novosibirsk.

Watanabe, N., 1966. Radiocarbon dates of the Jomon and Yayoi periods in Japan. *The Quaternary Research (Japan)* 5:158–168.

Wei, K.-Y., M.-Y. Lee, W. Duan, C. Chen, and C. H. Wang, 1998. Palaeoceanographic change in the northeastern South China Sea during the last 15000 years. *Journal of Quaternary Science* 13(1):55–64.

Wei, K.-Y., T.-N. Yang, and C.-Y. Huang, 1997. Glacial–Holocene calcareous nannofossils and paleoceanography in the Northern South China Sea. *Marine Micropaleontology* 32: 95–114.

Wells, L. E., and J. S. Noller, 1997. Determining the early history of El Nino. *Science* 276:966.

Yamada, Y., 1996. The social structure of Jomon period (from mortuary practice of Eastern Japan). In *Archaeology of Northern Pacific*, edited by A. L. Ivliev, N. N. Kradin, and I. S. Zhushchikhovskaya, pp. 342–347. Dalnauka, Vladivostok.

Yamanouchi, S., 1937. The main divisions and subdivisions of Jomon pottery. *Prehistoric Archaeology (Shenshi Kokogaku)* 1(1):28–32.

Yamanouchi, S., editor, 1964. *Primitive Arts of Japan. Vol. 1. Jomon Pottery*. Kodansha, Tokyo.

Yasuda, Y., 1984. Oscillations of climatic and oceanographic conditions since the Last Glacial Age in Japan. In *The Evolution of the East Asian Environment*, edited by R. O. Whyte, pp. 397–413. Hong Kong University Press, Hong Kong.

Yi, H.-I., S.-J. Han, J.-J. Bahk, J.-G. Lee, and J.-H. Chun, 1997. Late Quaternary paleoceanographic and paleoclimatic history from the Ulleung, East Sea of Korea (Japan Sea). *Ocean Research* 19(3):297–310.

Yi, H.-I., S.-J. Han, and D.-H. Shin, 1996b. Holocene high sea-level stands (climatic optimum)? Storm-surge deposits? Or are both relared? In *The 20th Annual Meeting and International Conference of the Korean Quaternary Association, Proceedings, Ansan, December 14, 1996*, pp. 27–28. Hanyang University, Ansan.

Yi, H.-I., S.-J. Han, and D.-H. Shin, 1996c. Late Quaternary monsoons and water circulation of the Northwest Pacific marginal seas (i.e., Yellow Sea, East China Sea, South Sea and East Sea of Korea). In *The 20th Annual Meeting and International Conference of the Korean Quaternary Association, Proceedings, Ansan, December 14, 1996*, pp. 29–32. Hanyang University, Ansan.

Yi, H.-I., S.-J. Han, D.-H. Shin, and J.-H. Chun, 1996a. Late Quaternary evolution and sea-level changes of the Yellow Sea and East China Sea Basin, Northwest Pacific marginal sea. In *Korea-China International Seminar on Holocene and Late Pleistocene Environments in the Yellow Sea Basin, Abstracts, November 20–22, 1996, Seoul*, pp. 64–66. Seoul National University, Seoul.

Yoneda, M., Ya. V. Kuzmin, M. Morita, A. N. Popov, T. I. Chikisheva, Y. Shibata, and E. G. Shpakova, 1998. The reconstruction of prehistoric diet after constant isotopes of carbon and nitrogene in the collagen of the bones from Neolithic cemetery Boismana-2, Primorye region. *The Humanities in Siberia* 3:9–13. In Russian, title translated.

Yurasov, G. I., and V. G. Yarichin, 1991. *Currents of the Sea of Japan*. USSR Academy of Sciences, Far East Branch, Vladivostok. In Russian, title translated.

Zhang, P., 1995. Climates and environments in mid-Holocene and during the last 2000 years in China. In *China Contribution to Global Change Studies, Report No. 2*, edited by D. Ye and H. Lin, pp. 133–136. Science Press, Beijing.

Zhushchikhovskaya, I. S., 1992. The Neolithic shell-tempered pottery in the sites of the Southern Far East. *Regional Bulletin, Yuzhno-Sakhalinsk (Kraevedchesky Bulletin)* 2:150–155. In Russian, title translated.

Zhushchikhovskaya, I. S., 1994. Pottery-making of prehistoric cultures of Southern Far East: paleo-economical characteristics. In *The Essays of Archaeology of Prehistoric Epoch of the Far East*, edited by Zh. V. Andreeva, pp. 148–204. Nauka, Moscow. In Russian, title translated.

Zhushchikhovskaya, I. S. 1997. On early pottery-making of Russian Far East. *Asian Perspectives* 36(2):159–174.

Zhushchikhovskaya, I. S., 1998. The pottery of Boismana 1 site. In *The First Fishers in Peter the Great Bay. Environment and Ancient Man in Boisman Bay*, edited by Yu. E. Vostretsov, pp. 123–196. Institute of History, Archaeology and Ethnography of the Peoples of the Far East, Vladivostok. In Russian with English abstract.

Zhushchikhovskaya, I. S., 2005. *Prehistoric pottery-making of the Russian Far East*, In BAR International Series 1434. Archaepress, Oxford.

Zhushchikhovskaya, I. S., and V. A. Rakov, 1994. On results of identification of molluscan shells in the paste of prehistoric Far Eastern pottery. *Regional Bulletin, Yuzhno-Sakhalinsk (Kraevedchesky Bulletin)* 4:112–120. In Russian, title translated.

Zhushchikhovskaya, I. S., and O. A. Shubina, 2006. Pottery-making and culture history of Neolithic Sakhalin. University of Oregon Anthropological Papers 65, pp. 91–128.

Climate Change and Cultural Dynamics: A Global Perspective on Mid-Holocene Transitions
David G. Anderson, Kirk A. Maasch and Daniel H. Sandweiss (Editors)

Chapter 11

Mid-Holocene climatic and cultural dynamics in Northern Europe

Wibjörn Karlén[1] and Lars Larsson[2]
[1]*Institutionen för Kulturgeografi, Box 513, Uppsala, 751 20 Sweden*
[2]*University of Lund, Department of Archaeology and Ancient History, Sandgatan 1, Lund, S-223 50 Sweden*

Abstract

Detailed studies of changes in the altitude of the pine tree limit, dendrochronology, glacial advances (revealed by an increased influx of glacial rock flour), and frontal moraines show that the mid-Holocene climate of Scandinavia fluctuated markedly. Major warm periods are dated to 7200–6500 cal yr BP and occasionally between 6100 and 5000 cal yr BP. Even the relatively cool period between 6500 and 6100 cal yr BP may have had short, slightly warmer periods; during the warm interval between 6100 and 5000 cal yr BP, relatively cool events occurred. The warmest periods appear to have occurred around 5800, 5400, and 5000 cal yr BP. Even after 5000 cal yr BP, the climate was relatively warm on several occasions but never reached the level of the previous warm events. The view presented here differs distinctly from the commonly accepted climate model which claims that no major changes in the climate took place during the mid-Holocene.

In Scandinavian archaeology there is a general agreement today that the social changes in the mid-Holocene showed considerable dynamism. The transition from the Atlantic chronozone to the Subboreal is contemporaneous with the shift from the Mesolithic to the Neolithic at 6000 cal yr BP. Changes in forest composition due to elm disease facilitated the introduction of agriculture and correlate with a major change in the material culture. Another natural explanation that has played a major role in the discussion concerns the increasingly dramatic rises in sea level. Large Late Mesolithic sites along the coast were submerged and replaced by small hamlets at the coast as well as further inland. Earthen long barrows and later megalithic tombs replaced the large cemeteries. The wetlands were used as sacrificial sites with deposits mainly of flint axes. Despite the great changes, we can trace the development of archaeological cultures in most regions within southern Scandinavia from the Mesolithic to the Neolithic.

1. Introduction

Recently, many scientists and politicians have discussed possible global warming resulting from the anthropogenic release of greenhouse gases. Numerous papers on this topic consider a warming by a few degrees centigrade as a disaster. However, changes of this magnitude are nothing new: there is good evidence that rapid changes in temperature of the same magnitude have taken place on several occasions during the Holocene. Did these changes affect the preferred sites for human habitation during the mid-Holocene? We briefly review a commonly used model for the Holocene climate, summarize results indicating frequent fluctuations in temperature with an amplitude of a few degrees centigrade, and discuss possible implications.

The accepted view of the Holocene climate in Scandinavia, as well as elsewhere, is largely based on a model that the biologist Blytt created and Sernander later modified (Blytt, 1882). This model introduced the frequently used terms Preboreal, Boreal, Atlantic, Subboreal, and Subatlantic (Berglund, 1968; Mangerud et al., 1974; Robertsson, 1994). Each period was thought to exhibit a climate typical for it and much effort was focused on the dating of the boundaries (Nilsson, 1964).

Observations of changes in the decomposition of peat were interpreted as an indication of changes in climate (Granlund, 1932; Lundqvist, 1957). However, much important climatic information that the peat stratigraphy could have yielded was not generally accepted and was not entered into the common description of the climate. One reason for this disregard of the information indicating changes in climate was that the dating techniques then available did not prove the contemporaneity of surfaces recognized at several sites.

Much effort has long focused on an elaboration of the Blytt–Sernander model (Mangerud et al., 1974; Robertsson, 1994). With the improvement of the pollen technique and the application of transfer functions, calculations of the actual changes in temperature have become possible. These calculations have shown a decrease of approximately 2°C in temperature between the mid-Holocene and the present. This estimated temperature change is similar to that suggested by Andersson (1902), who studied the distribution of hazel during the Holocene. The estimate is also similar to one based on the maximum distribution of fossil pine in the Swedish mountains (Karlén and Kuylenstierna, 1996a; Karlén, 1998).

The Blytt–Sernander model, in its later versions, did not indicate any marked changes in the climate – the temperature was inferred to have increased slowly to a maximum around 6000–5000 cal yr BP (4000–3000 BC) and thereafter decreased without major fluctuations before about 2000 cal yr BP (0 BC/AD). After this date the inferred temperature fluctuated slightly (Berglund, 1968; Nesje and Kvamme, 1991; Robertsson, 1994). This reconstruction led a group of scientists to suggest that these periods, which from the beginning were based on inferred changes in climate, should be used as chronostratigraphic terms with boundaries defined by ^{14}C dates (Mangerud et al., 1974).

Many scientists accept the view that the Holocene climate of Scandinavia was stable and consider this perspective confirmed by the Greenland ice core data. Changes in the climate during the Holocene are without doubt small compared with the fluctuations during the Pleistocene, but that does not mean that Holocene climate change lacks significance.

In addition to the previously mentioned recurrence surfaces in peat bogs, several Scandinavian studies show that climate fluctuations occurred frequently and with amplitudes probably larger than shown by the ice core data from Greenland. Fluctuations in the Scandinavian climate are known from studies of lake level changes in southern Scandinavia (Gaillard, 1985), and, in the north, from dated glacial expansions, changes in pine tree limit, and dendrochronology (Karlén, 1976, 1982; Nesje and Dahl, 1991; Nesje and Kvamme, 1991; Briffa and Schweingruber, 1992; Briffa et al., 1992; Karlén and Kuylenstierna, 1996a; Grudd et al., 2002; Hormes et al., 2004). These changes in climate would have affected the length of the growing season by perhaps as much as two months, if the annual temperature amplitude during the mid-Holocene was similar to that of today (Fig. 11.1). These fluctuations were possibly large enough to affect the survival of communities or at least affect the preference for a certain type of site even if marine transgressions with an attendant flooding of fishing grounds were important.

In this chapter, we review information about changes in the Holocene and then compare these data with information about cultural changes in Scandinavia and surrounding areas. The reconstruction of temperature extremes is only semi-quantitative, but the timing of changes in climate can be estimated from dates on the maximum pine tree limit, glacial advances, and variations in tree ring width. One reason why it has been difficult to calculate precisely the changes in the temperature is that the magnitude of the land uplift after the last deglaciation is not well known for the mountain area, which is the area from which most information has been obtained (Karlén, 1976). Without the combined data from the high tree limit, which indicates warm summer temperature, and from glacial advances, which indicate cold summers or high winter precipitation, it would have been impossible to conclude that major fluctuations in climate occurred.

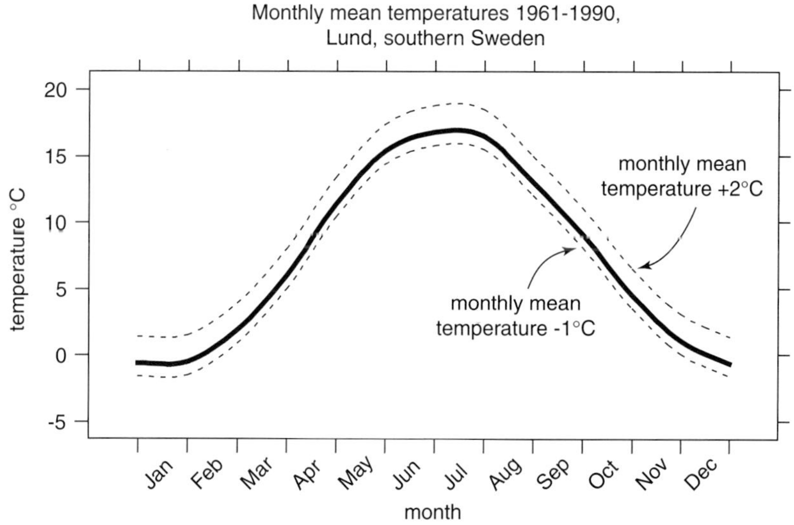

Figure 11.1. The mean annual monthly temperature for Lund, southern Sweden, 1961–1990 (bold line) (Alexandersson et al., 1991). In addition a temperature 2°C higher and 1°C lower than today is indicated (dotted lines). During warm events the mean monthly temperature may never have been below the freezing point.

The maximum mid-Holocene summer temperature was close to 2°C warmer than at present. During periods of glacial advances the temperature was probably about as cool as it was during the centuries preceding the last one hundred years of warming climate, e.g. close to 1°C cooler than at present. The amplitude between the warm and cold events is therefore between 2 and 3°C.

2. Fluctuations in the Scandinavian climate

The changes in climate presented below have been inferred mainly from studies of tree limit and glacial advances published during the last several decades (Denton and Karlén, 1973; Karlén, 1976, 1982; Karlén and Kuylenstierna, 1996a,b; Davis and Bohling, 2001; Grudd, 2006; Hormes et al., 2004; Matthews et al., 2005; and references in these chapters). The opinions expressed here do not conform entirely to those expressed in some of the original chapters (e.g. Nesje and Dahl, 1991; Nesje and Kvamme, 1991; Kullman, 1995; Dahl and Nesje, 1994).

The period around AD 1750, when historical records show glacial advances in Norway, was once considered the coldest part of the Holocene. With new dating techniques, particularly the ^{14}C dating techniques used on buried organic debris and pro-glacial lacustrine sediments, we now know that glacial advances took place even before AD 1750 in Scandinavia (e.g. Karlén, 1976; Karlén and Matthews, 1992; Matthews and Karlén, 1992; Nesje and Dahl, 1991; Dahl and Nesje 1994).

The dates for the climate changes presented here are based mainly on ^{14}C dates of pinewood found above the present pine tree limit (altitude in metre above the present local, pine tree limit). Dates on glacial advances obtained by the lichenometry technique, as well dates on lacustrine sediments and organic debris beneath moraines, are important because they show that glacial advances occurred during a number of periods throughout the entire Holocene. However, the precision of the lichenometric technique is limited because of the small number of calibration points for the growth rate beyond AD 1700 (Karlén, 1976, 1981; Karlén and Black, 2002). Information about glacial fluctuations, which is both detailed and continuous, is based on the facts that the erosion of wet-based glaciers depends largely on glacier size and that the rock-flour released by erosion is trapped in lakes downstream of the glacier. Variations in rock-flour sedimentation leave a continuous record that reveals periods of enhanced glacial erosion and glacial withdrawal. Although glacial rock flour content in lacustrine sediments yields data about glacial fluctuations, the dating of sediments is sometimes problematic.

Fragments of pine found above the present alpine tree limit have been discussed in the literature since the beginning of this century (Smith, 1911; Lundqvist, 1959, 1969; Selsing and Wishman, 1984; Kullman, 1995). The fossil pine remnants are sometimes found on the ground, but more commonly the wood is preserved in small lakes and in peat. Before the introduction of the ^{14}C technique, the wood fragments could not be dated and it was assumed that the greater pine distribution occurred during a climatic optimum in the mid-Holocene.

Variations in the pine tree limit yield information about summer temperature (Hustisch, 1958). Approximately 250 pine logs or log fragments found at a number of sites in the Scandinavian mountain range have now been ^{14}C dated. When possible, the time for the establishment of the pine above the present tree limit has been dated. Because the present pine tree limit has south–north and west–east gradients, arranging the dates by age and altitude in masl does not show any pattern. If, instead, the altitudes are calculated in reference to the present pine tree limit it becomes obvious that pine became established during a number of periods that were limited in time, and that glacial advances took place between these warm events. An error of up to a few hundred years is always possible when using the ^{14}C dating technique. The present pine tree limit is difficult to determine in some areas because of factors such as variations in climate governed by local radiation conditions, for instance "cold air lake" formation causing extreme low winter temperature or the presence of large bodies of water, which reduce summer temperature locally. In addition, it is possible that near the tree limit in the southern Central Swedish mountains, burning of the forest to ensure better hunting may have affected the altitude of the tree limit (Kullman, 1976). This process may also complicate estimates of the present tree limit altitude in some areas. However, in spite of these shortcomings, it is obvious that the dates on fossil pine from above the present pine tree limit occur in clusters.

Glacial advances follow periods of positive mass balance; e.g. periods when the amount of solid precipitation is greater than the amount of snow and ice which melts away during the summer. An increase in mass balance can be a result of either low summer temperature or increased winter precipitation. Observations of glaciers in Sweden show that low summer temperature is the usual reason for positive mass balance at least in continental areas (Holmlund et al., 1996). However, during occasional years a strongly increased winter precipitation may determine the annual mass balance. Winter precipitation is a factor of importance for glaciers along the west coast of Norway.

Dendrochronology is based on the measurement of tree ring width or wood density (Briffa and Schweingruber, 1992; Briffa et al., 1992). These factors vary sensitively with changes in summer climate in areas such as northern Sweden. Because pinewood can be preserved for a long period, it has been possible to construct a chronology covering the last 7400 years (Grudd et al., 2002). Dendrochronology can reveal details about changes in climate, but because of non-climatic effects, such as growth rate changes associated with aging, the technique does not reveal long-term trends.

2.1. Changes in the Scandinavian summer temperature

The inferred Holocene climate presented below is mainly based on information summarized in Figure 11.2. The upper panel of the diagram shows the periods during which pine germinated above the present pine tree limit. For this diagram an

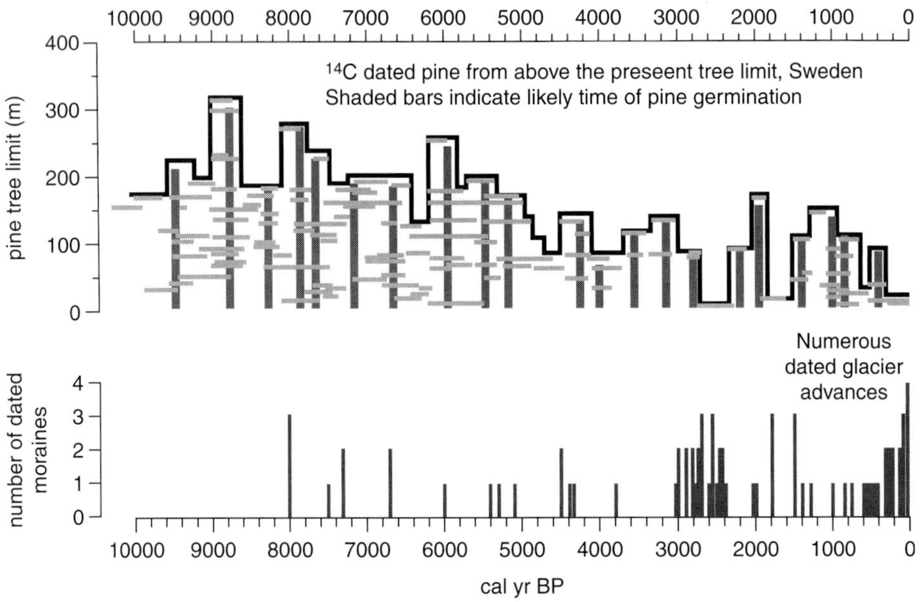

Figure 11.2. Inferred variation in the Scandinavian climate during the Holocene. Dated changes in pine tree limit during mid-Holocene is enlarged. The lower panels show age of glacier moraines dated by the lichenometric method (Karlén, 1976) and periods for which there are information about glacier advances obtained from pro-glacier lacustrine sediments and dates on soils buried beneath glacier moraines.

envelope based on the midpoint of calibrated dates (Calib 3.0 program) is used. The lower panels show the dates on glacial expansions obtained by lichenometry in northern Sweden as well as the periods for which pro-glacial sediments, ^{14}C dates obtained on glacial silt in peat, and dates obtained on soils buried beneath glacial moraines indicate the occurrence of glacial expansions. Note that even during long periods of relatively cold climate, pine tree establishment occurred infrequently at altitudes above the present.

The events around 9000 cal yr BP (7050 BC) are the first ones for which a cluster of dates indicates a high pine tree limit. Tree limit reached around 300 m above the present one. The land uplift for the mountain area is not well known, but after 8700 cal yr BP it is believed to be of the order of 100 m (Karlén, 1976). Assuming a lapse rate of 0.6°C/100 m of altitude change, the climate must have been distinctly warmer than at present (1–1.5°C). This warm period was followed by the coldest period during the entire Holocene known in Scandinavia as well as at several localities abroad, e.g. Greenland, where ice cores show that the temperature may have dropped 4.5°C below the present temperature (Alley et al., 1997; O' Brian et al., 1995). Even if the cold climate only lasted for a few hundred years around 8200 cal yr BP (6250 BC), several glaciers studied reached their Holocene maximum extent at this time. The event is very distinct in pro-glacial, lacustrine sediment studies.

The climate warmed shortly after this cold event but the tree limit did not reach as high as it had before. Around 7900 and again around 6000 cal yr BP (5950 and 4050 BC) the tree limit was again as high as 200 m above the present limit. A few glaciers advanced between these events (around 7400 cal yr BP, 5450 BC) but this was not a major advance. Between 6700 and 6000 cal yr BP (4750 and 4050 BC) tree limit was mostly low. Lacustrine sediments show that glacial advances took place during this period. Although this was probably not a particularly cold period, the cool climate lasted for about 700 years, a period long enough to cause glacial advances even if the excess of accumulation was limited.

Between 6100 and 5000 cal yr BP (4150 and 3050 BC) other optima in the climate occurred. The tree limit reached close to 200 m above the present one and this high limit was only to a minor extent a result of later land uplift. A tree limit expansion of up to 200 m only occurred during a few short intervals and the climate may have been relatively cool between these short events. Glacial advances are dated to around 5400 cal yr BP (3450 BC). This seems to be the beginning of a major change in the climate (cf. the boundary between Atlantic and Subboreal time in the Blytt–Sernander model; Mangerud et al., 1974). After a short warmer period around 5000 cal yr BP (3050 BC), the climate became distinctly cooler and a glacial advance is dated to 4900 cal yr BP (2950 BC). Around 4400 cal yr BP (2450 BC) the climate was briefly warmer, but there is only evidence for the tree limit reaching an altitude of between 80 and 130 m above the present during short periods. Major glacial expansions around 4100 cal yr BC (2150 BC) indicate that climate became distinctly cooler between at least some of these relatively warm events.

A relatively warm event is dated to around 3500 cal yr BP (1550 BC). Single dates indicate minor expansions of the tree limit around 3100 and 2800 cal yr BP (1150 and 850 BC). During the period between 3000 and 2000 cal yr BP (1050 and 50 BC) the climate was cold on repeated occasions, and glaciers expanded frequently and became larger than they were even during the Little Ice Age.

By 2000 cal yr BP (50 BC) a brief warming caused a short cessation in the glacial advances, but by 1800 cal yr BP (AD 150) new advances occurred. Around 1500 cal yr BP (AD 450) the tree limit reached 200 m above the present and a short but distinct period of glacial advances took place around 1200 cal yr BP (AD 750) just before the so-called Medieval Warm period around 1000 cal yr BP (AD 950). This warm period was followed by the well-known Little Ice Age, which lasted up to the end of the 19th century AD.

Dendrochronology data reveal that the climate during the last several hundred years, which is often called "The Little Ice Age", fluctuated and includes both several cold periods as well as warm periods (Briffa et al., 1992; Briffa, 2000).

2.2. Mid-Holocene climatic dynamics in northern Europe

Archaeological studies particularly from southern Sweden reveal distinct cultural changes during a relatively short period within one or a couple of hundred years coinciding with the mid-Holocene at 6000 cal yr BP. This change is a clear combination of regional innovations and continental European contacts. The change affected the economy, the material culture and the worldview of the societies. The change from hunter-gatherer to farming societies involved several complex processes with different regional implications within Scandinavia and the Baltic area. However, these transitions were mainly accomplished within the existing societies. Immigration seems to have been small but of major importance. The economic change might to some extent be a result of changes in the climate. A quick change to a warmer climate made it easier to cultivate the plants imported into southern Scandinavia from the continent.

The differences between hunter-gatherer societies in southern and northern Scandinavia were not so great just before the mid-Holocene. Unlike more southerly regions, however, there were no significant changes in the economy in northern Scandinavia after the mid-Holocene. Along much of the eastern Baltic coast, it was not until much later that agriculture spread to the north and these areas were incorporated into a larger Scandinavian community.

3. Mid-Holocene cultural dynamics in northern Europe

In northern Europe, especially southern Scandinavia, the transition from the Atlantic chronozone to the Subboreal – or the cultural shift from the Mesolithic to the Neolithic at 6000 cal yr BP – has long been perceived as a particularly interesting period in prehistory. Here, archaeology and palaeoecology have had an intimate association that is almost unparalleled in other parts of Europe. In the mid-19th century, Danish scholars divided the Stone Age into a hunter-gatherer phase and an agricultural phase. At the turn of the century, the former was given the name Ertebølle culture after a shell midden site in western Denmark. The agricultural phase was called the Funnel Beaker culture after the shape of the most frequently used pot.

3.1. From hunter-gathering to farming

Scandinavian archaeologists today generally agree that the social changes in the mid-Holocene showed considerable dynamism, but there is still discussion about whether the two cultures really developed diachronically or were partly synchronic (Fig. 11.3). This debate is due in part to a misunderstanding, since scholars have had different views of what characterizes a hunter-gatherer society and a farming society. The main differences of opinion, however, have concerned the interpretation of different excavation results (Rowley-Conwy, 1984, 2004; Jennbert, 1986; Zvelebil, 1986; Andersen and Johansen, 1987; Larsson, 1987b, 1992b 2007; Madsen, 1987; Nielsen, 1987; Andersen, 1993; Price et al., 1995; Tilley, 1996; Kristiansen and Fischer, 2002). In many surveys of European prehistory one can

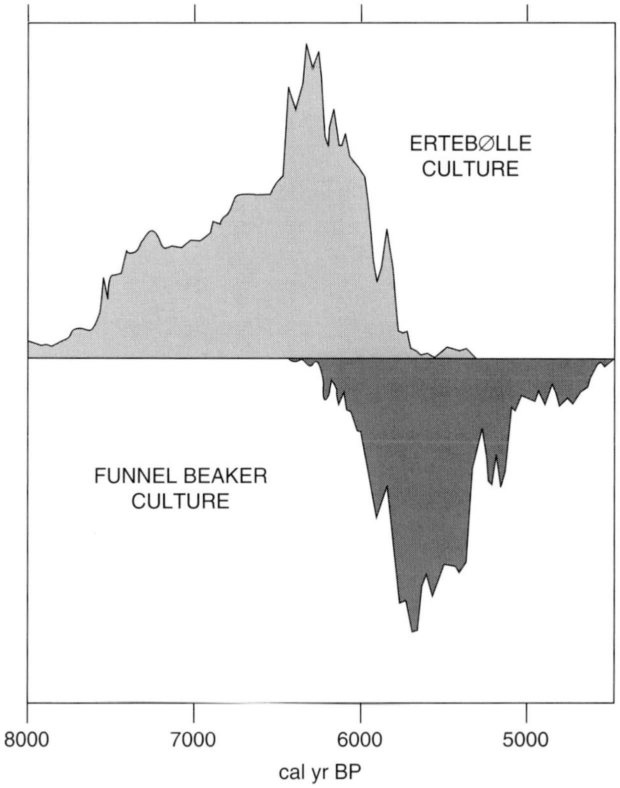

Figure 11.3. The chronological relations between the Ertebølle culture and the Funnel Beaker culture of southern Scandinavia based upon the radiocarbon datings (from Persson 1998).

glimpse surprise on the part of scholars that tillage and animal husbandry did not spread to Scandinavia earlier than the mid-Holocene (Hodder, 1990; Whittle, 1996). Agricultural societies in the form of the Linear Band Pottery culture existed as early as 7500 cal yr BP in the northern parts of continental Europe. In addition, the Linear Band Pottery culture made a thrust along the River Oder, almost reaching the Baltic Sea (Bogucki, 1998). In spite of this, there were no successors, whether in this area or further north, for roughly a thousand years. The suggested explanation is that the agricultural communities were confronted with hunter-gatherer communities with a large population and very stable conditions based on a good supply of food, which was the basis for a well-functioning social system. The introduction of farming required much more labour, for which there was no incentive for many generations.

Today's researchers play down the significance of social changes caused by changes in natural conditions. There is no doubt that scholars ascribed too much importance to these factors in the past. In all societies there are not only economic conditions but also a world-view with direct and indirect application to the social order. The world-view can withstand a great deal of change caused by internal conflicts and external influence. One or usually more interacting factors can lead to increased pressure on the preconditions – both actual and mental – for the existing social structure so that a society is forced to accept such radical changes that they cannot be accommodated within this social framework. These changes lead to the formation of new societies based on new conditions and a revised world-view. Such a process appears to have happened during the mid-Holocene.

3.2. The elm decline

What, then, are the factors that have been perceived as significant for changes during the mid-Holocene? Among those that have been adduced, several are directly or indirectly related to a change in climate. The explanatory model put forward in the 1960s was based on interpretations of pollen analyses (Iversen, 1973). In the mid-Holocene the pollen composition in bog layers shows a large decline in elm. Shortly afterwards the proportion of oak and other broad-leaved trees also decreases. Quick-growing trees such as birch and hazel replace them. This shift in forest composition is interpreted as showing that the first farmers kept their animals in enclosures and used foliage as fodder; elm was the most attractive foliage for this purpose. At a later stage, the increase of hazel and birch is a clear indication of forest regeneration in areas cleared by means of slash-and-burn agriculture.

Today, however, the elm decline is interpreted as a result of elm disease. This phenomenon can be detected all over northern Europe without any demonstrable association with the introduction of farming (Friman, 1997). Moreover, analyses of cow dung from an early agricultural site in Switzerland have shown that elm was not the primary tree for fodder: ash, lime, and willow were totally predominant (Rasmussen, 1991). The fact that humans did not cause the elm decline, however, does not mean that this change had no significance for the spread of agriculture. Elm grows in nutrient-rich soil, casts a broad shadow, and was one of the most important trees in the forests of southern Scandinavia (Fig. 11.4). A forest afflicted by elm disease is transformed into an area of dead tree trunks with rapidly flourishing bushes and other undergrowth (Fig. 11.5C). Burning the dead trees was a labour-saving way to acquire large areas for cultivation and pasture. The ravages of elm disease may have seemed like a gift from heaven to people at the time, perhaps a sign from a higher power that they could intensify the change to a new economy and a new social order (Larsson, 2003b).

3.3. Sea level changes

Another natural explanation that has played a major role in the discussion concerns the increasingly dramatic rises in sea level (Fig. 11.5B). At the end of the Mesolithic there were several almost cyclically recurring transgressions (Christensen, 1995).

Figure 11.4. The results of a recent elm decline in southernmost Sweden.

The highest, which in southern Scandinavia reached about 5 m above today's sea level, occurred in the mid-Holocene. Late Mesolithic settlements can be found in sheltered lagoons and at estuaries where there was a nearby supply of fresh water, brackish water, and salt water with plenty of fish, the staple diet (Larsson, 1987a).

One factor that has steered our perception of the stable social organization in the Late Mesolithic 8000–6000 cal yr BP is the occurrence of large cemeteries right beside settlement sites of significant size (Larsson, 1989, 2004). There are several sites in southern Scandinavia where these cemeteries occur. They give an important insight into highly varied mortuary practices with distinct regional differences. Cemeteries with many burials and long-term settlement sites indicate prolonged use of resources in a geographically defined area with a permanent or semi-permanent settlement structure (Rowley-Conwy, 1998).

A large rise in sea level meant that the areas beside the settlements were transformed into open bays and the number of fish species fell, as did the number of islands

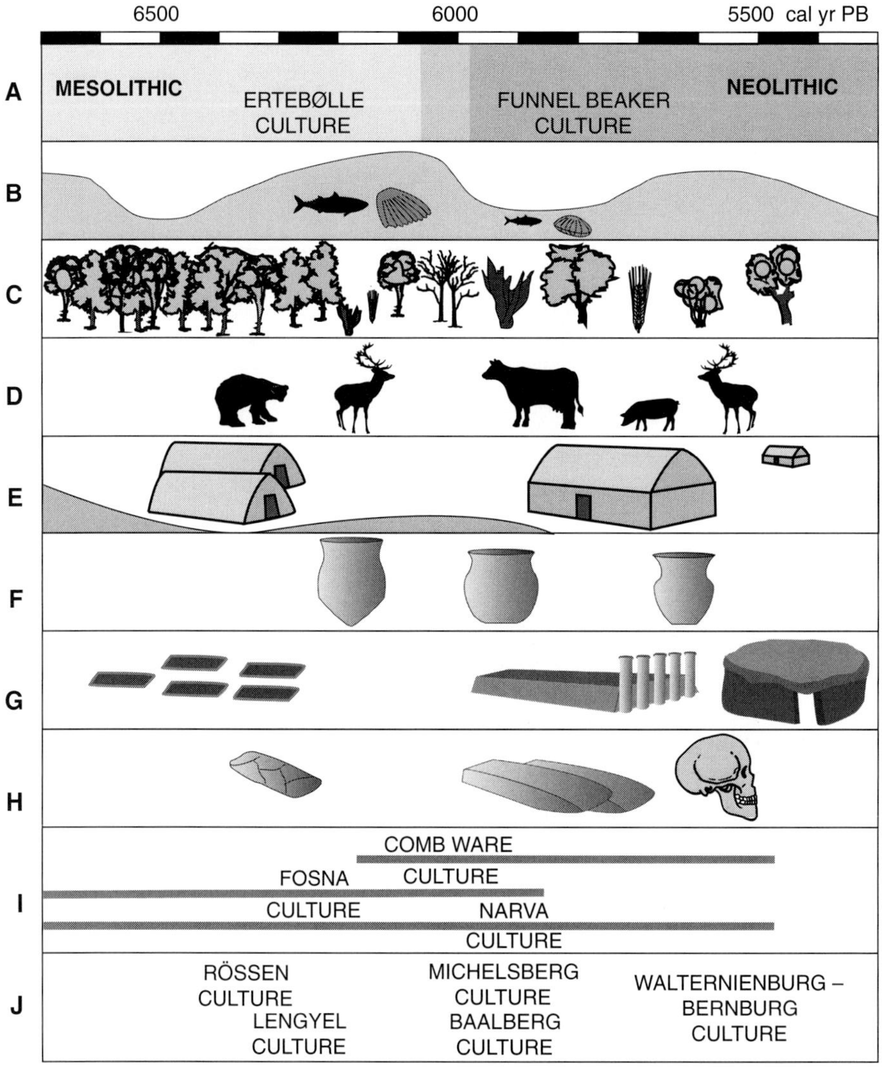

Figure 11.5. Cultural changes during the mid-Holocene in southern Scandinavia. A. Division of Late Mesolithic and Early Neolithic in southern Scandinavia; B. Sea level changes; C. Changes from dense forests to an open forest by elm decease and intentional fires; D. Changes of mammals; E. Changes in settlement structure; F. Changes of vessel types; G. Changes from large cemeteries to long barrows and megalithic tombs; H. Deposition of axes and human sacrifices; I. The cultural situation in the Eastern Baltic region; and J. Cultural situation on continental Europe.

providing resting places for seals and the shoals where mollusks grew. At the coastal sites in western Denmark one can see how the large assemblages of oyster shells in kitchen middens from the Late Mesolithic were replaced by thinner layers of common sea mussels in the early Neolithic 6000–5400 cal yr BP (Andersen, 1991, 1995, 2000).

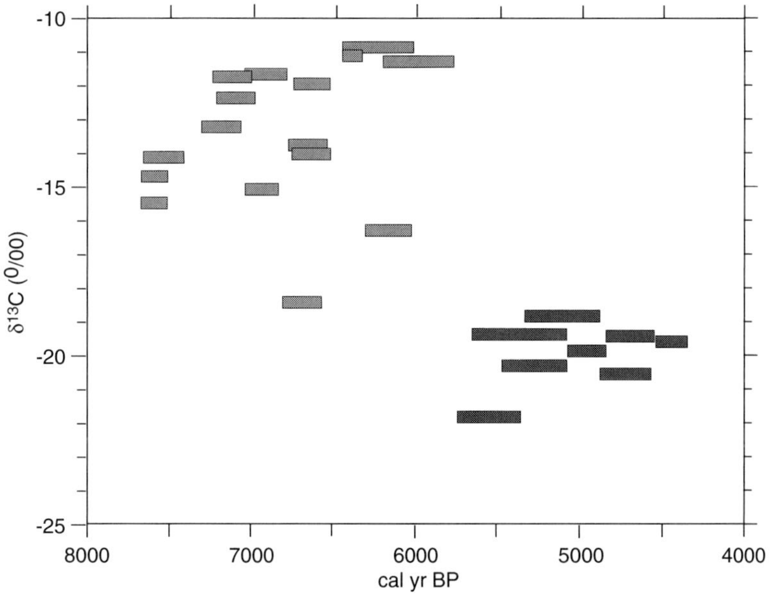

Figure 11.6. The value of ^{13}C in humans and dogs from coastal sites dated to the Ertebølle culture (light grey) and Funnel Beaker culture (dark grey) respectively (from Persson, 1998).

At settlement sites in southern Scandinavia there is also a significant decline in the proportion of fish bones; in certain areas people seem to have stopped fishing altogether. This change in the choice of food is clearest from measurements of the content of ^{13}C in human bones (Fig. 11.6). A low value was proof of a significant proportion of marine food. At a cemetery in Denmark the deceased from the Late Mesolithic had a ^{13}C value of −12 ‰, which shows that the food was of marine origin (Noe-Nygaard, 1988). Another body buried a few metres away but dated to the Early Neolithic shows a value of –22 ‰, indicating that this person's food was found on land (Larsson, 1991; Richards et al., 2003). Throughout the period of use, the cemetery was beside the coast. Both animals and plants that live on and use land made up a greater proportion of the human diet from the Early Neolithic onwards. The importance of the sea as a source of food clearly declined in favour of what could be produced on land.

In the mid-Holocene one can also detect a spread of settlement from the coast towards areas further inland (Fig. 11.5E). In the Late Mesolithic the rich coast provided conditions for large, lasting settlements for several families. The rise in sea level meant that the coast could no longer feed as dense a settlement. Some people could stay on, while others moved inland (Fig. 11.7). Farming meant that new resources could be used. The result was increased but more dispersed settlement (Larsson, 1987a). A farm with one or a couple of families became the smallest unit for several centuries to come (Larsson, 1998a). Graves are found in isolation or in a small number (Andersson, 2004) and in long barrows (Fig. 11.5G). Long barrows, a form of grave found throughout continental Northern Europe from

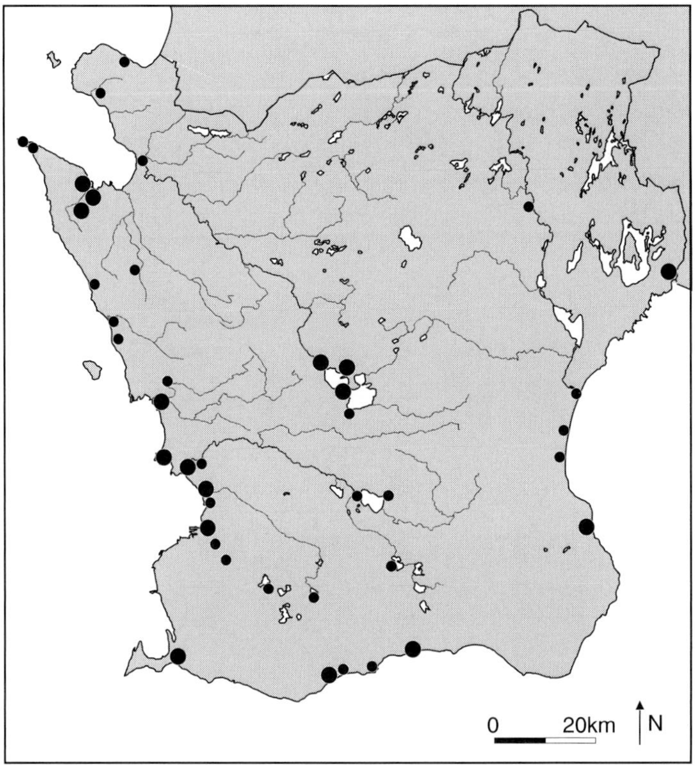
0
20km
N

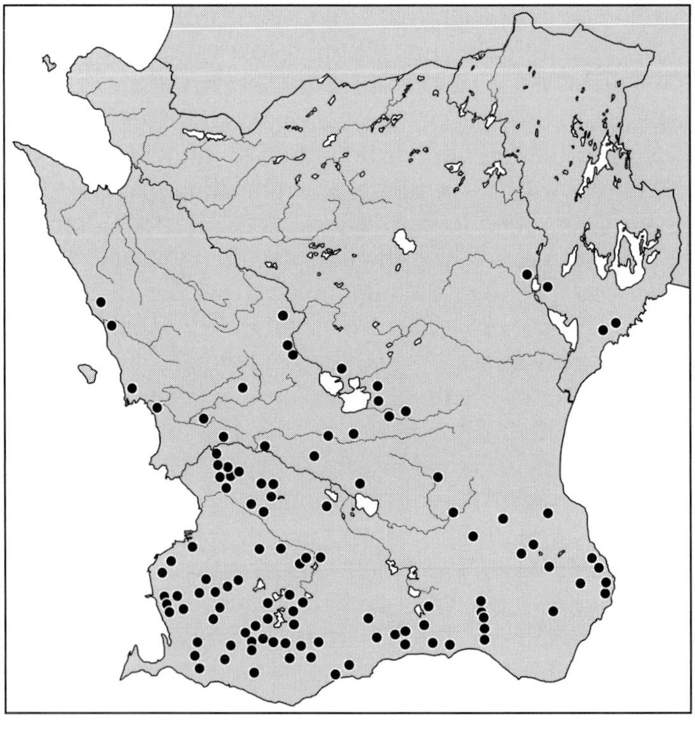

England to Poland, begin to occur immediately after the introduction of the Funnel Beaker culture. Barrow tombs consist of long but low, often trapezoidal banks of earth that can reach lengths of up to a few hundred metres but are relatively narrow. The graves are usually found in the broader, eastern part of the barrow, ending in a facade marked by large poles (Madsen, 1979; Midgley, 1985; Larsson, 1992a). This type of grave – the first monumental form – occurs in both Denmark and southernmost Sweden (Kristensen, 1991; Liversage, 1992; Larsson, 2003a).

3.4. Migration or internal changes

For many years, change in the mid-Holocene was thought to be associated with a massive immigration of farmers from continental Europe. In recent decades, however, the prevailing view has been that there was little migration.

It is not sufficient to import domesticated animals or grain if one wants to become a good stockbreeder or cultivator. First-hand knowledge of how to look after animals or to sow and harvest crops must be conveyed through demonstration by people familiar with these skills. The knowledge could have been passed on from one community to another through intermarriage (Jennbert, 1984). The large cemeteries in the Late Mesolithic give us a great amount of skeletal material. The graves in the Early Neolithic are few. We are therefore forced to compare a large number of skeletons from the time before the mid-Holocene with a small number from the time immediately after the mid-Holocene. It is clear that there was considerable variation in the physiognomy of the Mesolithic people. In general, however, they had a coarser skeletal structure than Early Neolithic people (Bennike and Alexandersen, 1997). Significant differences can also be traced in the size of teeth, with Early Neolithic people generally having finer teeth. This trend could be seen as support for the view that there was considerable immigration, but it need not be the case. The people who came to southern Scandinavia may have been few in number. Through their knowledge they could have acquired a dominant position in society and have had a higher value as marriage partners. They could also give their children better conditions in which to grow up and thus reduce infant mortality. Genetic characteristics which dominated among the few but influential newcomers could thus have spread in a short time and thus made a significant impact on the population.

Figure 11.7. The difference in the location of settlement between the cultures can be exemplified by the distribution of polished stone axes belonging to the Late Ertebølle culture (above) and polished flint axes from the Early Funnel Beaker culture (below). Legend: small dots mark single finds and large dots mark sites with more than five axes (from Jennbert, 1984).

3.5. Material culture changes

A very distinct change in the mid-Holocene affected material culture. This change is a clear combination of regional innovations and continental European contacts. Pottery was introduced from the south as early as 6700 cal yr BP. The decoration on the pots and the manufacturing technique indicate close contacts with the Rössen culture in present-day central Germany. New pot shapes and other modes of ceramic decoration arose around 6000 cal yr BP (Fig. 11.5F). The construction technique also changed. These forms were probably developed in northern Germany through influence from the south. Polished flint axes replaced polished rock axes or unpolished flint axes. The Funnel Beaker culture replaced the Ertebølle culture. The change appears to have been quick and pronounced at least in some parts of south-western Scandinavia. In western Danish shell middens and at eastern Danish bog sites, dating shows that this process of change may have been as short as a few decades (Andersen, 1991; Fischer, 2002).

People in southern Scandinavia were integrated into a larger European community. There were, however, earlier contacts between the hunter-gatherer societies of southern Scandinavia and the farmers on the continent (Fig. 11.8). Around 7000 cal yr BP distinct signs of contacts have already appeared in the form of axes in southern Scandinavia, which resemble forms on the continent (Klassen, 2004). Direct imports of a special shaft-hole axe occurred in the latter part of the Late Mesolithic (Fischer, 1982). The pottery that occurs about 700 years before the transition from Mesolithic to Neolithic is likewise direct evidence of contacts. The presence of grain impressions in Ertebølle pottery suggests that there may also have been limited cultivation (Welinder, 1998). The change from hunter-gatherer to farming societies probably involved several complex processes dependent on time and space. A division of the process into three stages is commonly accepted (Zvelebil, 1998a). During the first stage, the availability phase, a small part of the subsistence, less than 5%, can be related to domesticates and cultigens. In the second stage, the substitution phase, the domesticates and cultigens make up between 5 and 50%, while in the third stage, the consolidation phase, they are more than 50%. The substitution phase is less common in archaeological as well as ethnological sources (Zvelebil, 1998b). The change from the availability phase to the consolidation phase might have taken place at different speeds in different parts of southern Scandinavia. With the start of the Neolithic, certain tool forms begin to show specialization. Large, four-sided polished flint axes, probably a southern Scandinavian innovation, were made in tens of thousands. This was the first case of mass manufacture. Most of the axes are about 20 cm long, but some exceed 40 cm in length (Olausson, 1983). Making these tools requires great knapping skills and a supply of large lumps of flint. At some places in southern Scandinavia there are finds of mines where big pieces of flint were extracted from the chalk (Olausson et al., 1980; Rudebeck, 1998). Remains of axe manufacture were found at the mineshafts. Knapping specialists must have operated at the mines, knapping the

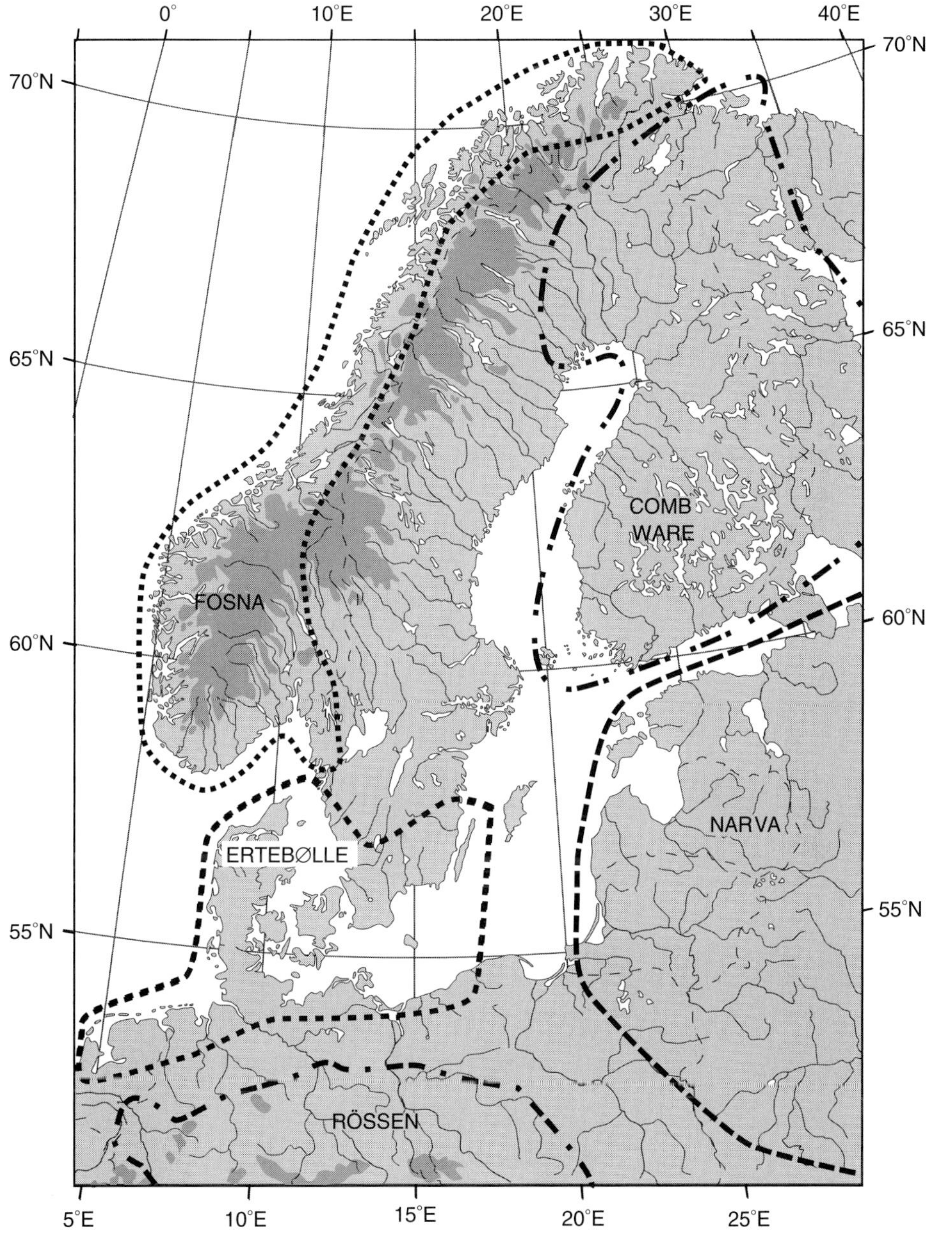

Figure 11.8. Cultural relations in northern Europe during the Late Atlantic period.

axes and then probably letting other people do the simpler but time-consuming work of polishing them. However, people living scattered on their farms must have known quite a lot about flint working, as is evident from the many repairs to axes damaged by use.

3.6. Wetland depots

If one were to characterize southern Scandinavian archaeology in some way, a special feature would be the interest devoted to finds from wetlands, whether large or small. Most of the originally nutrient-rich lakes were somewhere in the process of filling up during the mid-Holocene because of the accumulation of dead plant parts from the flourishing vegetation on the bottoms and the shores. Some still had open water and were therefore attractive to humans for fishing and gathering. Excavations have uncovered small reed huts with bark floors, dated to the Late Mesolithic, located in marshy zones beside open water (Larsson, 1990, 1998b). These huts were mainly used during the summer as temporary camps for collecting hazelnuts for the winter, and they have a tradition lasting several thousand years.

A feature of special importance in the mid-Holocene is that the wetlands were used as sacrificial sites (Fig. 11.5H). Even during the Late Mesolithic there are a few finds suggesting votive rituals (Larsson, 1990). Shaft-hole axes of continental origin, which must have had a high value, have been found by modern peat cutters. Really rich votive deposits, however, begin to appear at the very start of the Neolithic (Karsten, 1994; Larsson, 2007). Axes deposited in pairs appear to have been a common sacrifice, but there are also sacrifices of individual objects and mass offerings of several tools or several different kinds of object. Pots are also common votive gifts (Koch, 1998).

Human sacrifices in Northern Europe are associated with the well-preserved and hence well-known bodies from the Early Iron Age, sacrificed during the centuries around the birth of Christ (Glob, 1969). There are in addition a large number of skeletons or parts of skeletons of people found in bogs. Most of them were women or children. ^{14}C analyses show that a significant proportion of these human remains are from the start of the Neolithic (Bennike and Ebbesen, 1987; Koch, 1998; Larsson, 1998b). Humans could therefore be sacrificed as early as the start of the Neolithic. This observation means that the conceptual world involved demands not just to give up pieces of material culture but also human lives.

3.7. Regions within south-western Scandinavia

In the Late Mesolithic one can distinguish different local, regional, and interregional groupings in southern Scandinavia. Certain artifact forms have a special regional distribution, such as harpoons and stone axes (Vang Petersen, 1984). The west and the east of southern Scandinavia are clearly distinguished. It is of particular interest that the same division largely continues during the earliest phase of the Neolithic, as reflected in pot shapes and ceramic decoration, as well as in axe types (Larsson, 1984). Here we may be dealing with tribal areas with a fixed structure surviving over a long time, despite extensive cultural changes. Distinctive local features occur during the Mesolithic in certain forms of flint axe. In the Neolithic it is possible to distinguish similar local groups by the shape and decoration of pots.

3.8. Middle and northern Scandinavia

Only a few centuries after the introduction of farming in southern Scandinavia, the new material culture reached central Sweden at the latitude of Stockholm (Fig. 11.9) (Berglund, 1985; Sundström, 2003). The situation on the west coast, both in the Late Mesolithic and in the Early Neolithic, is extremely unclear (Kihlstedt et al., 1997). Agriculture was probably introduced here as early as in east central Sweden. This is also the case for the coastal area in southern Norway (Berglund, 1985; Nygaard, 1989).

In northernmost Scandinavia, extensive archaeological work in the last few decades has given us a greatly expanded knowledge base. Along the coast of the North Atlantic there are large settlement sites with several sunken house foundations (Olsen, 1994). These may have been village-like communities in the period before the mid-Holocene. At that time resource use was also extended to include the inland as well. Rock carvings begin to occur in concentrations at several river estuaries (Helskog, 1988), suggesting that certain sites along the coast were places of assembly for the people of the coast and the inland alike. Joint rituals were held here, partly to avoid confrontations between different groups.

In connection with the mid-Holocene, changes in material culture began to appear. Pottery was introduced, along with new stone tools, suggesting changed contact routes: instead of communication along the coast to the southwest, there were now relations with the inland to the southeast. The houses with sunken floors became slightly bigger but still remained relatively small, 15–20 m^2 (Olsen, 1994). Specialization in hunting and gathering became more common. It is possible that the same population used both the coast and the interior through seasonally based migrations.

3.9. North-eastern Europe

During the few centuries before the mid-Holocene, pottery was introduced in Finland (Edgren, 1993). The special decoration has given its name to the culture: Comb Ware (Fig. 11.8). As in southern Scandinavia, Finland offers a case of outside influence: the combed ware came from southeast Europe. In very recent years evidence of this culture has also been found in northernmost Sweden (Halén, 1994). Also associated with the Comb Ware culture are graves, sometimes assembled in cemeteries as in southern Scandinavia (Edgren, 1993, 2006). The differences between hunter-gatherer societies in the south and the north were not so great just before the mid-Holocene. Unlike more southerly regions, there were no significant changes in the economy in northern Scandinavia (Halén, 1994). Along much of the eastern Baltic coast, it was not until about 1500 years later that agriculture spread to the north and these areas were incorporated into a larger Scandinavian community, if not completely then at least through networks.

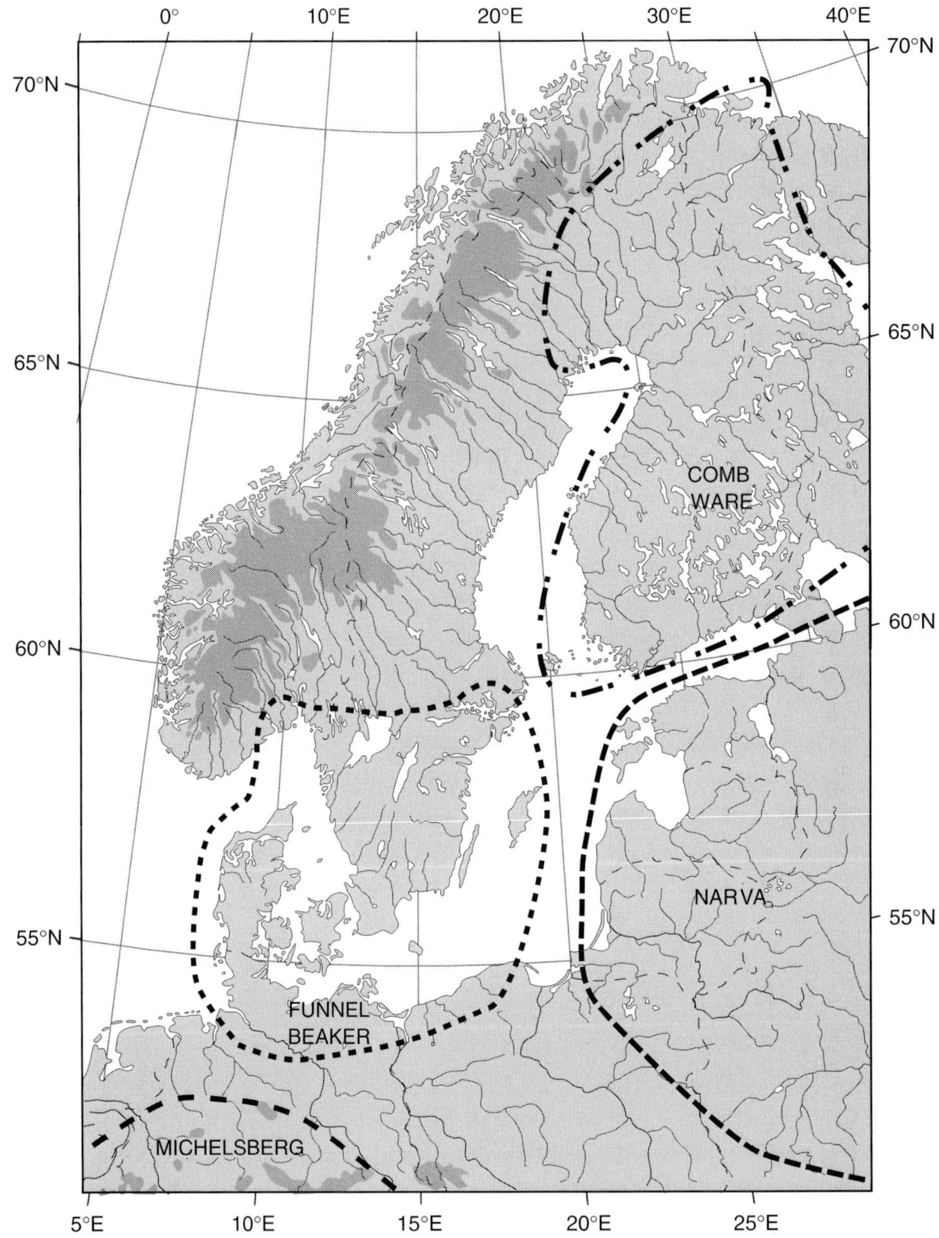

Figure 11.9. Cultural relations in northern Europe during the earliest part of the Subboreal period.

As regards the south-eastern part of the Baltic coast, there is obvious influence from the Ertebølle culture. If one also includes northern Poland as a whole, one can see how agricultural communities such as the successors of the Linear Band culture, in the form of the Lengyel culture, adopted greater variation in resource use, with elements of hunting and gathering (Bogucki, 1998). The succeeding Funnel Beaker culture settled on till and sandy soils where agriculture had not previously been practiced.

The terminological confusion concerning the Baltic is particularly obvious. Pottery arises around 7500 cal yr BP, roughly a thousand years earlier than in southern Scandinavia. This is probably due to contacts with farming communities in the Ukraine (Timofeev, 1998). Occasional bones of domesticated animals are perceived as examples of contacts with farming societies further to the south. However, there was no real agriculture here until later in the Neolithic.

4. Summing up

Because the cultural dynamics of southern Scandinavia were so striking during a relatively short period coinciding with the mid-Holocene, so far we have focused entirely on this period. To find explanatory models for the change from societies wholly dominated by hunting-gathering to those with varying elements of agriculture, it is necessary to apply a longer historical perspective.

The Mesolithic climate, to the extent that it has been considered, has been regarded as relatively stable and hence of little or no significance for cultural change. As shown above, the ecological variables that have been viewed as significant during the mid-Holocene are changes in the composition of the forest and shifts in sea level. The results obtained by determining changes in the tree line in the northern Swedish Mountains suggest that more climatic variables should be taken into consideration.

There were considerable variations in climate during the Mesolithic, and the last part of the Mesolithic (the Late Atlantic) seems to have been characterized by a relatively long period of colder climate. Precisely in the mid-Holocene there was a quick change to a warmer climate, a change that raised the temperature by about 2°C in a short time. The change in the growing season in particular may have made it easier to cultivate the plants imported into southern Scandinavia from the continent. The climatic change meant that sowing could be started several weeks earlier than today. This extended growing season would have significantly facilitated cultivation. Although the internal and external social relations are of great importance for understanding the change from hunting-gathering to farming, a striking change in climate such as that marked by the shift in the tree line may be an important factor in our understanding of why the change took place in the mid-Holocene and not in some other period.

5. Conclusions

Evidence obtained from dating of changes in the alpine tree limit and glacier size variations indicates that major changes in the climate have taken place during the entire Holocene, including the mid-Holocene. These changes were of the magnitude 1–2°C, which is likely to have changed the length of the growing season by several weeks. During warm intervals, the monthly mean temperature is likely to have remained above the freezing point even during the cold months. A striking cultural change took place during a relatively short period coinciding with the mid-Holocene: the transition from societies wholly dominated by hunting-gathering to those with varying elements of agriculture. In present research the significance of social changes caused by changes in natural conditions is played down. However, a quick change to a warmer climate during the mid-Holocene made it easier to cultivate the plants imported into southern Scandinavia from the continent.

References

Alexandersson, H., C. Karlström, and S. Larsson-McCann, 1991. Temperature and Precipitation in Sweden 1961-1990, Reference Normals. Swedish Meteorological and Hydrological Institute Report no. 81: p. 87.

Alley, R. B., P. A. Mayewski, T. Sowers, M. Stuiver, K. C. Taylor, and P. U. Clark, 1997. Holocene climatic instability: a prominent, widespread event 8200 yr ago. *Geology* 25: 483–486.

Andersen, S. H., 1991. Norsminde. A "Kokkenmodding" with Late Mesolithic and Early Neolithic occupation. *Journal of Danish Archaeology* 8:13–40.

Andersen, S. H., 1993. Bjørnsholm, a stratified Kokkenmodding on the Central Limfjord, North Jutland. *Journal of Danish Archaeology* 10:59–96.

Andersen, S. H., 1995. Coastal adaptation and marine exploitation in Late Mesolithic Denmark – with special emphasis on the Limfjord region. In *Man and Sea in the Mesolithic: Coastal Settlement above and below Present Sea Level*, edited by A. Fischer, pp. 41–66. Oxbow Monograph 53. Oxbow, Oxford.

Andersen, S. H., 2000 'Køkkenmøddinger' (Shell Middens) in Denmark; a urvey. *Proceedings of the Prehistoric Society* 66:361–384.

Andersen, S. H., and E. Johansen, 1987. Ertebølle revisited. *Journal of Danish Archaeology* 5:31–61.

Andersson, G., 1902. Hasseln i Sverige fordom och nu. *Sveriges Geologiska Undersökning* Serie Ca., No. 3.

Andersson, M., 2004. *Making Place in the Landscape.* Early and Middle Neolithic societies in two west Scanian valleys. National Heritage Board. Skånska spår– arkeologi längs Västkustbanan. Riksantikvarieämetet, Lund.

Bennike, P., and V. Alexandersen, 1997. Danmarks urbefolkning. *Nationalmuseets Arbejdsmark* 1997:143–156.

Bennike, P., and K. Ebbesen, 1987. The bog find from Sigersdal. Human sacrifice in the Early Neolithic. *Journal of Danish Archaeology* 5:85–115.

Berglund, B. E., 1968. Vegetationsutvecklingen i Norden efter istiden. *Sveriges Natur* 1968: 31–52.

Berglund, B. E., 1985. Early agriculture in Scandinavia: research problems related to pollen-analytical studies. *Norwegian Archaeological Review* 18(1–2):77–90.

Blytt, A., 1882. Die Theorie der wechselnden kontinentalen und insulaten Klimate. *Botanisches Jahrbuch* 2:1–50. Leipzig.

Bogucki, P., 1998. Holocene climatic variability and early agriculture in temperate Europe: the case of Northern Poland. In *Harvesting the Sea, Farming the Forest: The Emergence of Neolithic Societies in the Baltic Region*, edited by M. Zvelebil, R. Denell, and L. Domanska, pp. 77–85. Sheffield Archaeological Monographs 10. Sheffield Academic Press, Sheffield.

Briffa, K. R., 2000. Annual climate variability in the Holocene: interpreting the message of ancient trees. *Quaternary Science Reviews* 19:87–105.

Briffa, K. R., and F. H. Schweingruber, 1992. Recent Dendro-climatic evidence of Northern and Central European summer temperatures. In *Climate Since A.D. 1500*, edited by R. S. Bradley and P. D. Jones, pp. 366–392. Routledge, London and New York.

Briffa, K. R., P. D. Jones, T. S. Bartholin, D. Eckstein, F. H. Schweingruber, W. Karlén, P. Zetterberg, and M. Eronen, 1992. Fennoscandian summers from A.D. 500: temperature changes on short and long timescales. *Climatic Dynamics* 7:111–119.

Christensen, C., 1995. The Littorina trangression in Denmark. In *Man and Sea in the Mesolithic: Coastal Settlement above and below Present Sea Level*, edited by A. Fischer, pp. 15–22. Oxbow Monograph 53. Oxbow, Oxford.

Dahl, S. O., and A. Nesje, 1994. Holocene glacier fluctuations at Hardangerjökulen, Central Southern Norway: a high resolution composite chronology from lacustrine and terrestrial deposits. *The Holocene* 4:269–277.

Davis, C., and G. C. Bohling, 2001. The search for patterns in ice-core temperature curves. In *Geological Perspectives of Global Change*, edited by L. C. Gerhard, W. E. Harrison, and B. M. Hanson, pp. 213–229. American Association of Petroleum Geologists.

Denton, G. H., and W. Karlén, 1973. Holocene climatic variations – their pattern and possible cause. *Quaternary Research* 3:155–205.

Edgren, T., 1993. Den förhistoriska tiden. *Finlands historia*, del 1:11–270. Ekenäs.

Edgren, T., 2006. Kolmhaara. Reconsidered. Some new observations concerning the Neolithic Burial practice in Finland. In *Back to the Origin. New research in the Mesolithic-Neolithic Zvejnieki Cemetery and Environment, Northern Latvia*, edited by L. Larsson and I. Zagorska, pp. 327–336. Act Archaeologica Lundensia, Series in 8°, N°. 52. Almqvist & Wiksell International, Lund.

Esper, J., R. J. S. Wilson, D. C. Frank, A. Moberg, H. Wanner and J. Luterbacher, 2005. Climate: past ranges and future changes. *Quaternary Science Reviews* 24(20–21): 2164–2166.

Fischer, A., 1982. Trade in Danubian Shaft-Hole Axes and the introduction of Neolithic economy in Denmark. *Journal of Danish Archaeology* 1:7–12.

Fischer, A., 2002. Food for feasting? An evaluation of explanations of the Neolithisation of Denmark and Southern Sweden. In *The Neolithisation of Denmark. 150 years of Debate*, edited by A. Fischer and K. Kristiansen, pp. 343–393. J. R. Collis Publications, Sheffield.

Friman, B., 1997. Neolithization and "Classical" Elm decline: a synthesis of two debates. *Lund Archaeological Review* 2(1996):5–16.

Gaillard, M.-J., 1985. Postglacial paleoclimatic changes in Scandinavia and Central Europe. A tentative correlation based on studies of lake-level fluctuations. *Ecologia Mediterranea* 11:159–175.

Glob, P. V., 1969. *The Bog People: Iron-Age Man Preserved.* Thames and Hudson, London.

Granlund, E., 1932. De svenska högmossarnas geologi. Deras bildningsbetingelser, utvecklingshistoriaoch utbredning jämte sambandet mellan högmossbildning och försumpning. *Sveriges Geologiska Undersökningar* No. 373.

Grudd, H., 2006. Tree rings as sensitive proxies of past climate change. Thesis, Department of Physical Geography and Quaternary Geology, Stockholm University.

Grudd, H., K. R. Briffa, W. Karlén, T. S. Bartholin, P. D. Jones, and B. A. Kromer, 2002. 7400-year tree-ring chronology in Northern Swedish Lapland: natural climatic variability expressed on annual to millennial timescales. *The Holocene* 12:657–665.

Halén, O., 1994. Sedentariness during the Stone Age of Northern Sweden: in the light of the Alträsket Site, c. 5000 B.C., and the Comb Ware Site Lillberget, c. 3900 B.C. source critical problems of representativity in archaeology. *Acta Archaeologica Lundensia* 4:20. Almquist & Wiksell International, Lund.

Helskog, K., 1988. Helleristningene i Alta. Spor etter ritualer og daligliv i Finnmarks forhistorie. Alta Museum, Alta.

Hodder, I., 1990. *The Domestication of Europe*. Cambridge University Press, Cambridge.

Holmlund, P., W. Karlén, and H. Grudd, 1996. Fifty years of mass balance and glacier front observations at the Tarfala Research Station. *Geografiska Annaler* 78A:105–114.

Hormes, A., W. Karlén, and G. Possnert, 2004. Radiocarbon dating of Palaeosol components in Moraines in Lapland, Northern Sweden. *Quaternary Science Reviews* 23:2031–2043.

Hustich, I., 1958. On the recent expansion of the Scotch Pine in Northern Europe. *Fennia* 82: 1–25. Helsinki.

Iversen, J., 1973. The development of Denmark's nature since the Last Glacial. *Geological Survey of Denmark.* V(7-C):11–126.

Jennbert, K., 1984. *Den produktiva gåvan.* Tradition och innovation i Sydskandinavien för omkring 5 300 år sedan. *Acta Archaeologica Lundensia* 4:16. Almqvist & Wiksell International, Lund.

Jennbert, K., 1986. Neolithisation – a Scanian Perspektive. *Journal of Danish Archaeology* 4: 196–197.

Karlén, W., 1976. Lacustrine sediments and thee-limit variations as indicators of Holocene climatic fluctuations in Lappland: Northern Sweden. *Geografiska Annaler* 58A:1–34.

Karlén, W., 1981. A comment on John A. Matthews's article regarding ^{14}C dates of glacial variations. *Geografiska Annaler* 63A:19–21. Stockholm.

Karlén, W., 1982. Holocene glacier fluctuations in Scandinavia. In *Holocene Glaciers*, edited by W. Karlén, pp. 26–34. *Strie* 18. Uppsala.

Karlén, W., 1998. Climate variations and the enhanced greenhouse effect. *AMBIO* 27(4):270–274.

Karlén, W., and J. Black, 2002. Estimates of Lichen growth-rate in Northern Sweden. *Geografiska Annaler* 84A:225–232.

Karlén, W., and J. Kuylenstierna, 1996a. On solar forcing of Holocene climate: evidence from Scandinavia. *The Holocene* 6:359–365.

Karlén, W., and J. Kuylenstierna, 1996b. Evidence from the Scandinavian tree line since the Last Ice Age. In *The Global Warming Debate. The European Science and Environment Forum*, edited by J. Emsley, p. 288.

Karlén, W., and J. A. Matthews, 1992. Reconstructing Holocene glacier variations from glacial lake sediments: studies from Östra Peninsula and Jostedalsbreen-Jotunheimen, Southern Norway. *Geografiska Annaler* 74A:327–348.

Karsten, P., 1994. Att kasta yxan i sjön. En studie över rituell tradition och förändring utifrån skånska neolitiska offerfynd. *Acta Archaeologica Lundensia* 8:23. Lund: Almqvist & Wiksell International, Lund.

Kihlstedt, B, M. Larsson, and B. Nordqvist, 1997. Neolitisering en i Syd-Väst- och Mellansveige – social och ideologosk förändring. In *Regionalt och interregionalt. Stenåldersundersökningar i Syd- och Mellansverige*, edited by M. Larsson and E. Olsson, pp. 85–133. Riksantikvarieämbetets arkeologiska undersökningar, Skrifter 23. Riksantikvarieämbetet, Stockholm.

Klassen, L., 2004. *Jade und Kupfer.* Undersuchungen zum Neolithisierungsprozess im westlichen Ostseeraum unter besonderer Berücksichtigung der Kulturentwicklung Europas 5500-3500 BC. *Jutland Archaeological Society Vol. 47.* Jutland Archaeological Society, Aarhus.

Koch, E., 1998. Neolithic bog pots from Zealand, Mon, Lolland and Falster. *Nordiske Fortidsminder*, Serie B, Vol. 16. Det Konglige Nordiske Oldtidselskab, Kobenhavn.

Kristensen, I. K., 1991. Storgården IV: an Early Neolithic long barrow near Fjelso, North Jutland. *Journal of Danish Archaeology* 8:72–87.

Kristansen, K., and A. Fischer, editors, 2002. *The Neolithisation of Denmark. 150 Years of Debate*. JR Collis Publications, Sheffield.

Kullman, L., 1976. Recent trädgränsdynamik i V Härjedalen. *Svensk Botanisk Tidskrift* 70: 107–137. Stockholm.

Kullman, L., 1995. Holocene tree-limit and climate history from the Scandes Mountains, Sweden. *Ecology* 76:2490–2502.

Larsson, L., 1987a. Some aspects of cultural relationship and ecological conditions during the Late Mesolithic and Early Neolithic. In *Theoretical Approaches to Artefacts, Settlement and Society*, edited by G. Burenhult, A. Carlsson, Å. Hyenstrand, and T. Sjovold, pp. 165–176. BAR International Series 366. BAR, Oxford.

Larsson, L., 1989. Late Mesolithic settlements and cemeteries at Skateholm, Southern Sweden. In *The Mesolithic in Europe. Papers Presented at the Third International Symposium Edinburgh 1985*, edited by C. Bonsall, pp. 367–378. John Donald Publishers, Edinburgh.

Larsson, L., 1990. The Mesolithic of Southern Scandinavia. *Journal of World Prehistory* 4: 257–309.

Larsson, L., 1991. Coastal adaptation in the early and middle Holocene of Southern Scandinavia. *Journal of Korean Ancient Historical Society Hanguk Sanggosa Hakbo* 8:93–118.

Larsson, L., 1992a. Facade for the dead. A preliminary report on the excavation of a long barrow in Southern Scania. *Papers of the Archaeological Institute University of Lund* 1991–1992:45–56.

Larsson, L., 1998a. Neolithic societies and their environment in Southern Sweden: a case study. In *Understanding the Neolithic of North-Western Europe*, edited by M. Edmonds and C. Richards, pp. 428–455. Cruithne Press, Glasgow.

Larsson, L., 1998b. Prehistoric Wetland Sites in Sweden. In *Hidden Dimensions: The Cultural Significance of Wetland Archaeology*, edited by K. Bernick, pp. 64–82. UBC Press, Vancover.

Larsson, L., 2003a. Monumentala gravformer i det äldsta bondesamhället. University of Lund, Department of Archaeology and Ancient History Report Series no. 83. Department of Archaeology and Ancient History, Lund.

Larsson, L., 2003b. The Mesolithic of Sweden in Retrospective and Progressive Perspective. In *Mesolithic on the Move. Papers Presented at the Sixth International Conference on the Mesolithic in Europe, Stockholm 2000*, edited by L. Larsson, H. Kindgren, K. Knutsson, D. Loeffler, and A. Åkerlund, pp. XXII–XXXII. Oxbow, Oxford.

Larsson, L., 2004. The Mesolithic period in southern Scandinavia: with special reference to burials and cemeteries. In *Mesolithic Scotland and its neighbours. The early Holocene prehistory of Scotland and its British and Irish context and some northern European Perspectives*, edited by A. Saville, pp. 371–392. Society of Antiquaries of Scotland, Edinburgh.

Larsson, L., 2007. The ritual use of wetlands during the Neolithic: a local study in Southernmost Sweden. In *Wetland Archaeology and Environments: Regional Issues, Global Perspectives*, edited by M. Lillie and S. Ellis, pp. 79–90. Oxbow, Oxford.

Larsson, M., 1984. Tidigneolitikum i Sydvästskåne: Kronologi och bosättningsmönster. *Acta Archaeologica Lundensia* 4:17. Almqvist & Wiksell International, Lund.

Larsson, M., 1987b. Neolithization in Scania – A funnel beaker perspective. *Journal of Danish Archaeology* 5:244–247.

Larsson, M., 1992b. The Early and Middle Neolithic funnel beaker culture in the Ystad area (Southern Scania). In *The Archaeology of the Cultural Landscape: Field Work and Research in a South Swedish Rural Region*, edited by L. Larsson, J. Callmer, and B. Stjernquist, pp. 91–159. *Acta Archaeologica Lundensia* 4(19). Almqvist & Wiksell International, Lund.

Liversage, D., 1992. Barkar: long barrows and settlements. *Arkaologiske studier IX*. Akademisk Forlag, Kobenhavn.

Lundqvist, G., 1959. C14-Daterade Tallstubbar från fjällen. *Sveriges Geologiska Undesökning*, Ser. Ba, 18:147. Stockholm.

Lundqvist, J., 1957. C^{14}-Dateringar av rekurensytor i Värmland. *Sveriges Geologiska Undersökning* Serie C, 554:22. Stockholm.

Lundqvist, J., 1969. Beskrivning till jordartskarta över Jämtlands län. *Sveriges Geologiska Undersökning*, Ser. Ca, 45:418. Stockholm.

Madsen, T., 1979. Earthen long barrows and timber structures: aspects of the Early Neolithic mortuary practice in Denmark. *Proceedings of the Prehistoric Society* 45:301–320.

Madsen, T., 1987. Where did all the hunters go? An assessment of an epoch-making episode in Danish prehistory. *Journal of Danish Archaeology* 5:229–239.

Mangerud, J., S. T.Andersen, B. E. Berglund, and J. J. Donner, 1974. Quaternary stratigraphy of Norden, a proposal for terminology and classification. *Boreas* 3:109–128.

Matthews, J. A., and W. Karlén, 1992. Asynchronous neo-glaciation and Holocene climatic change reconstructed from Norwegian glaciolacustrine sedimentary sequences. *Geology* 20:991–994.

Matthews, J.A., M. S. Berrisford, P. Q. Dresser, A. Nesjr, S. O. Dahl, A. E. Bjune, J. Bakke, H. John, B. Birks, Ö Lie, L. Dumayne-Peaty, and C. Barnett, 2005. Holöcene glacier history of Björnbreen and climatic reconstruction in Central Jotunheimen, Norway, based on proximal glaciofluvial stream-bank mires. *Quaternary Science Reviews* 24:67–90.

Midgley, M. S., 1985. *The Origin and Function of the Earthen Long Barrows in Northern Europe*. BAR International Series 259. BAR, Oxford.

Moberg, A., D. M. Sonechkin, K. Holmgren, N. M. Datsenko, and W. Karlén, 2005, Highly variable Northern Hemisphere temperatures reconstructed from low- and high-resolution proxy data. *Nature* 433:613–617.

Nielsen, P. O., 1987. The beginning of the Neolithic – assimilation or complex change? *Journal of Danish Archaeology* 5:240–243.

Nesje, A., and S. O. Dahl, 1991. Holocene glacier variations of Blåisen, Hardangerjökulen, Central South Norway. *Quaternary Research* 35:25–40.

Nesje, A., and M. Kvamme, 1991 Holocene glacier and climate variations in Western Norway: evidence for early Holocene glacier demise and multiple neoglacial events. *Geology* 19: 610–612.

Nilsson, T., 1964. Standardpollendiagramme und C^{14}-datirungen aus dem Ageröds Mosse im mittleren Schonen. *Lunds Universitets Årsskrifter* N.F. Avd. 2, Bd. 59(7):52. Lund.

Noe-Nygaard, N., 1988. ^{13}C values of dog bones reveal the nature of changes in Manés food resources at the Mesolithic-Neolithic transition, Denmark. *Isotope Geoscience* 73: 87–96.

Nygaard, S. E., 1989. The Stone Age of Northern Scandinavia: a review. *Journal of World Prehistory* 3:71–116.

O'Brien, S. R., P. A. Mayewski, L. D. Meeker, D. A. Meese, M. S. Twickler, and S. I. Whitlow, 1995. Complexity of Holocene climate as reconstructed from a Greenland ice core. *Science* 270:1962–1964.

Olausson, D., 1983. Lithic technological analysis of the thin-butted flint axe. *Acta Archaeologica* 53:1–87.

Olausson, D., E. Rudebeck, and U. Säfvestad, 1980. Die südschwedischen Feuersteingruben-Ergebnisse und Probleme. In *5000 Jahre Feuersteinbergbau. Die Suche nach dem Stahl der*

Steinzeit, edited by G. Weisgeber, pp. 183–204. Veröffentlichungen aus dem Deutschen Bergbau-Museum Bochum, nr. 22. Bochum: Deutschen Bergbau-Museum Bochum.

Olsen, B., 1994. *Bosetning og Samfunn i Finnmarks forhistorie*. Universitetsforlaget, Oslo.

Persson, P., 1998. *Neolitikums början: Undersökningar kring jordbrukets introduktion i Nordeuropa*. Doctoral dissertation. Department of Archaeology, Göteborg University, Göteborg.

Price, T. D., A. B. Gebauer, and L. H. Keeley, 1995. The spread of farming into Europe North of the Alps. In *Last Hunters First Farmers*, edited by D. T. Price and A. B. Gebauer, pp. 95–126. School of American Research Press, Santa Fe.

Rasmussen, P., 1991. Leaf-foddering of livestock in the Neolithic: archaeobotanical evidence from Weier, Switzerland. *Journal of Danish Archaeology* 8:51–71.

Richards, M. P., D. T. Price, and E. Koch, 2003. Mesolithic and Neolithic subsistence in Denmark: new stable isotope data. *Current Anthropology* 44:288–295.

Robertsson, A. M., 1994. Vegetationens utveckling. In *Berg och jord, Sveriges Nationalatlas*, edited by C. Fréden, pp. 136–137.

Rowley-Conwy, P., 1984. The laziness of the short-distance hunter: the origins of agriculture in Western Denmark. *Journal of Anthropological Archaeology* 3:300–324.

Rowley-Conwy, P., 1998. Cemeteries, seasonality and complexity in the Ertebølle of Southern Scandinavia. In *Harvesting the Sea, Farming the Forest: The Emergence of Neolithic Societies in the Baltic Region*, edited by M. R. Zvelebil, R. Denell, and L. Domanska, pp. 193–202. Sheffield Archaeological Monographs 10. Sheffield Academic Press, Sheffield.

Rowley-Conwy, P., 2004. How the West was lost. A reconsideration of agricultural origins in Britain, Ireland, and Southern Scandinavia. *Current Anthropology* 45(Supplement):S83–S113.

Rudebeck, E., 1998. Flint extraction, axe offerings, and the value of cortex. In *Understanding the Neolithic of North-Western Europe*, edited by M. Edmonds and C. Richards, pp. 312–327. Cruithne Press, Glasgow.

Selsing, L., and E. Wishman, 1984. Mean summer temperatures and circulation in a South-West Norwegian mountain area during the Atlantic period, based upon changes of the Alpine pine-forest limit. *Annals of Glaciology* 5:127–132.

Smith, H., 1911. Postglaciala regionförskjutningar i norra Härjedalensoch södra Jämtlands fjälltrakter. *Geologidka Föreningen Förhandlingar i Stockholm* 33:503–530.

Sundström, L., 2003. Det *hotade kollektivet.* Neolitiseringsprocessen ur ett östsvenskt perspektiv. Kust till kust-bok 6. Coast to Coast project, Uppsala.

Tilley, C., 1996. *An Ethnography of the Neolithic: Early Prehistoric Societies in Southern Scandinavia.* Cambridge University Press, Cambridge.

Timofeev, V. I., 1998. The beginning of the Neolithic in the Eastern Baltic. In *Harvesting the Sea, Farming the Forest: The Emergence of Neolithic Societies in the Baltic Region*, edited by M. R. Zvelebil, R. Denell, and L. Domanska, pp. 225–236. Sheffield Archaeological Monographs 10. Sheffield Academic Press, Sheffield.

Vang Petersen, P., 1984. Chronological and regional variation in the Late Mesolithic of Eastern Denmark. *Journal of Danish Archaeology* 3:7–18.

Welinder, S., 1998. Pre-Neolithic farming in the Scandinavian Peninsula. In *Harvesting the Sea, Farming the Forest: The Emergence of Neolithic Societies in the Baltic Region*, edited by M. R. Zvelebil, R. Denell, and L. Domanska, pp. 165–173. Sheffield Archaeological Monographs 10. Sheffield Academic Press, Sheffield.

Whittle, A., 1996. *Europe in the Neolithic: The Creation of New Worlds.* Cambridge University Press, Cambridge.

Zvelebil, M., 1986. Mesolithic societies and the transition to farming: problems of time, scale and organisation. In *Hunters in Transition. Mesolithic Societies of Temperate Eurasia and their Transition to Farming*, edited by M. Zvelebil, pp. 5–15. Cambridge University Press, Cambridge.

Zvelebil, M., 1998a. Agricultural frontiers, Neolithic origins, and the transition to farming in the Baltic Basin. In *Harvesting the Sea, Farming the Forest: The Emergence of Neolithic Societies in the Baltic Region*, edited by M. R. Zvelebil, R. Denell, and L. Domanska, pp. 9–27. Sheffield Archaeological Monographs 10. Sheffield Academic Press, Sheffield.

Zvelebil, M., 1998b. What's in a name: the Mesolithic, the Neolithic, and social change of the Mesolithic-Neolithic Transition. In *Understanding the Neolithic of North-Western Europe*, edited by M. Edmonds and C. Richards, pp. 1–36. Cruithne Press, Glasgow.

Climate Change and Cultural Dynamics: A Global Perspective on Mid-Holocene Transitions
David G. Anderson, Kirk A. Maasch and Daniel H. Sandweiss (Editors)

Chapter 12

Mid-Holocene cultural adaptations to central Maine

David Sanger[1,]*, Heather Almquist[2] and Ann Dieffenbacher-Krall[3]
[1]*Department of Anthropology, University of Maine, Orono, ME 04469, USA*
[2]*Department of Geography, University of Montana, Missoula, MT 59812, USA*
[3]*Climate Change Institute, University of Maine, Orono, ME 04469, USA*

Abstract

Like much of northeastern North America, central Maine experienced a lengthy, warm and dry mid-Holocene period. Cultures changed throughout the period; however, it is not clear to what extent climate change induced shifts in culture. Maine's aboriginal populations practiced hunting, fishing, and gathering in mid-Holocene times and like other people with this subsistence pattern probably adopted cultural mechanisms, such as prey switching and small-scale population movements, to compensate for local shortages. A decade-long effort has produced a Holocene record of cultural events, vegetation, wetland evolution, lake levels, and moisture balance at comparable temporal and spatial scales. Following deglaciation, wetlands of various types dominated central Maine. In the Orono area, we have a detailed record of wetland evolution from open water, through Typha*-dominated marshes, to* Sphagnum *peatlands. A major vegetation change occurred around 5800 cal yr BP with the crash of* Tsuga canadensis *(eastern hemlock) and a replacement by northern hardwoods, which provided more mast foods and potentially more game. At the same time, lower lake and river levels probably resulted in decreased fish habitat and more difficult canoe travel. After 4000 cal yr BP, water levels rose quickly and more archaeological sites are recorded, although we do not know if human populations actually increased. Traditionally in Maine, fluctuations in artifact forms have suggested culture change. At present, the role of climate change in affecting these measures of cultural variability is unclear.*

1. Introduction

At the time of first contact by Europeans around AD 1600, the well-watered, mixed softwood and hardwood forests of northeastern North America supported aboriginal peoples with varied life styles (Trigger, 1978). Some, such as the Iroquoian speakers arrayed around the eastern Great Lakes and the St. Lawrence River valley, depended heavily on horticulture, especially *Zea mays* (maize), *Phaseolus* sp. (beans.), and *Cucurbita* sp. (squash), supplemented by hunting, fishing, and gathering. East of the Iroquois dwelt Algonquian-speaking hunters and gatherers, who

*Corresponding author. *E-mail address:* sanger@maine.edu (David Sanger).

also fished. However, there can be little doubt that in the mid-Holocene, the entire area supported hunters and gatherers dependent on naturally occurring resources. In keeping with our theoretical position involving culture and local environment interaction, we restrict our analysis to the central Penobscot River valley, Maine, where we have the most detailed archaeological and paleo-environmental data. Some of the conclusions may have broader implications, of course.

Central Maine is an ecological tension zone and mid-Holocene environments were unlike modern ones. In our assessment of the relationships between climate and culture in mid-Holocene times, we examine those aspects of climate and the environment that are most likely to impact humans at a hunting and gathering stage of subsistence. This includes vegetation critical to supporting prey animals, and aquatic environments, in which humans found plant and animal resources.

Anthropologists have long recognized that hunters and gatherers the world over have a capacity to adapt to changing environmental conditions (e.g., Kelly, 1995). For the mid-Holocene Northeast hunter-gatherers, these characteristics might have involved small, band-organized social units, with an ability to switch prey rapidly, and to organize people on the landscape to avoid over-exploitation during periods of reduced game or other resources, and with a propensity to modify social groupings with relative ease in order to mitigate localized, resource shortfalls (Leacock, 1954; Tanner, 1979).

Although these attributes are best documented for the Boreal Forest inhabitants of southern Quebec (Leacock, 1954), there are hints of comparable behavior among the historic tribes of the Maine. Frank Speck (1940, p. 207), the major ethnographer for the Penobscot people of central Maine noted, "When a family [hunting and trapping] territory became overcrowded, or when the game in it became too scarce to support its occupants, a division was made among members of the band." This broad-spectrum approach to subsistence coupled with sociological flexibility served to reduce risk. It also means that correlations between culture change and climate may be difficult to document, especially if the climatic events are non-catastrophic, are localized rather than widespread, and do not require major technological changes. The use of social organization to help mitigate adverse impacts of environmental change may leave few artifact traces in hunter-gatherer archaeological records. Additions and deletions to the artifact inventory may be reflective of widespread horizon styles that operate within strictly social domains that have minimum impact on actual procurement efficiency. In other words, changes in the archaeological record reflected only in tool categories may be totally disconnected from shifts in environment.

Without a doubt, the best evidence for human adaptations to mid-Holocene environments would rest with the reconstruction of subsistence patterns based on analysis of food remains in the sites. Unfortunately, in the acidic forest soils (spodosols) of the Northeast, little to no bone preservation occurs unless the bone is burned (calcined). Even then, some bone elements preserve better than others, as demonstrated by controlled burning and other experiments (Knight, 1985). Lists of faunal elements from a number of New England mid-Holocene sites include taxa readily available from the wetlands (Spiess, 1992; Spiess and Mosher, 2006). However, the vagaries of preservation preclude firm statements regarding prey choice that may be due to changes in the local environment alone. In some instances, however, variations in the environment do appear to be reflected in site location and contents.

The pre-European period in the Northeast is divided into sub-periods, based originally on cultural stages defined by the appearance of distinctive index artifacts, such as pottery, or a projectile point style. With the advent of radiocarbon dating, these stages evolved into periods, but still retained some of the artifact-driven stage concept. As a result, the terminology and temporal boundaries vary slightly within the broader region.

People entered the area sometime around 13,000 cal yr BP, a millennia or two following deglaciation of the Northeast. This colonizing culture is called Paleoindian. The succeeding period, known as the Archaic, extends from about 11,000 cal yr BP to the advent of ceramics in the region, sometime around 3000 cal yr BP, which ushers in the Woodland or Ceramic Period. This chapter is concerned with the middle, or Archaic period.

2. Previous environment/culture interpretations

Interest in the relationships between humans and their environment has a long history in northeastern North America. William Ritchie (1965, 1971) noted that with the increase of northern hardwood trees in upstate New York, new cultures developed and the region appeared more heavily populated. Ritchie attributed the growth in human population to an increase in carrying capacity for mast–food–dependent animals, such as *Odocoileus virginianus* (whitetail deer) adapted to the hardwood forests. Similarly, James Fitting (1968) argued for an early Holocene "boreal forest" element in the landscapes south of the Great Lakes. He then used that ill-advised reconstruction to explain the apparent scarcity of humans during the early and middle Holocene. In Maine, Sanger (1977) also questioned the lack of evidence for people and developed a series of hypotheses that included the traditional vegetation explanation. The list also included concerns over the amount and kind of archaeological research, archaeologists' abilities to recognize mid-Holocene cultural assemblages, and changing river gradients in an area of important anadromous fish resources.

These early attempts reflected the ecological paradigm that drove much archaeological research at the time. As naïve and even deterministic as the models may appear today, the problems lay more with the kinds of data available, than with the basic assumption that hunters, fishers, and gatherers are always dependent on the local environment for their livelihood. In the last 30–40 years, environmental scientists have added greatly to the amount of data available, and have filled in large geographical gaps in our knowledge. In addition to vegetation history, environmental reconstructions now include discussions of lake and river levels and climate.

Further, in the intervening years, the archaeological record has expanded enormously, and filled in once large hiatuses in the cultural record.

3. The cultural record

Archaeologists operating in northeastern North America (Fig. 12.1) traditionally have relied on a few key artifacts in good contexts to typify archaeological culture types. Good context, unfortunately, has been elusive generally; many sites in the area reflect multiple occupations in shallow and highly compressed cultural horizons (Wright, 1995).

However, Maine archaeologists have excavated stratified sites that permit better separation of culture types and associated radiocarbon dates (e.g., Petersen, 1991; Sanger, 1996; Sanger et al., 2001; Sanger, 2006). In Maine, the perception of an apparent scarcity of people in mid-Holocene times changed radically. Petersen and Putnam (1992) listed no fewer than 50 radiocarbon dates pre-dating 6700 cal yr BP. Research since has revealed many more sites in this period. The issue then became one of population numbers relative to later periods, not presence or absence. Although Petersen and Putnam (1992) discussed various population density models based on modern analogues, none can overcome the problem of differential site preservation in a geologically dynamic environment (Kelley, 2006).

In central Maine, known pre-6700 cal yr BP sites survive only on a limited suite of landforms, mostly riverbank locations where sedimentation protected the sites.

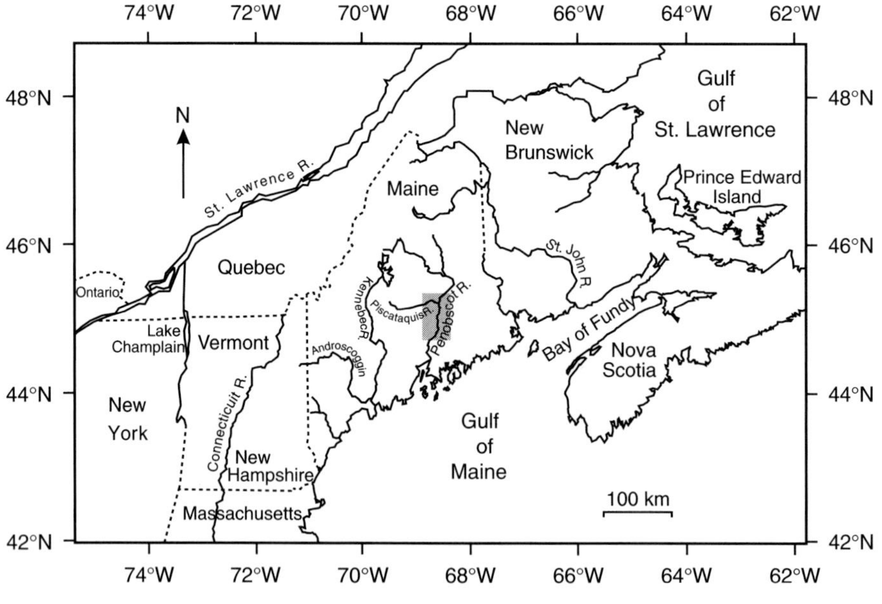

Figure 12.1. Northeastern United States and adjacent Canada. Central Maine, subject of this article, is shaded.

These settings are closely correlated with natural ponding events linked to bedrock-sill dams often in association with tributaries (Putnam, 1994; Sanger et al., 2001; Kelley and Sanger, 2003; Kelley, 2006). Site surveys in meandering river valleys featuring "cut and fill" riverbanks may reveal only recent sites (Sanger and Newsom, 2000). Upland sites are rare to non-existent, and no rock-shelter sites have been reported. A strong case can be made, however, for more sites, artifacts, and presumably more people, after 5700 cal yr BP (Anonymous, 1999).

The archaeological record for coastal Maine also depends on regional geological history. After a low stand of minus 50–60 m at approximately 13,000 cal yr BP, sea levels rose rapidly, generally slowing during the Holocene (Barnhardt et al., 1995). Although artifacts have been recovered from scallop fishing nets (Sanger, 1988; Crock et al., 1993), and occasional specimens potentially dating to about 6700 cal yr BP occur on beaches, the oldest intact coastal sites date no earlier than about 5700–4500 cal yr BP (Bourque, 1995). The very few sites known in the latter period display a full maritime adaptation in addition to dependence on terrestrial resources.

Beyond Maine, the situation varies. A recent review of potential pre-6700 cal yr BP sites in the state of New York reveals a surprising lack of sites and artifacts considered to be characteristic of the time period (Funk, 1996). Archaeological surveys, some of which involved detailed overland transects required by cultural resource management, indicate very few old sites. From one survey that located 630 sites, only 3% can be attributed to the period 11,000–6700 cal yr BP. In another study, an examination of over 25,000 bifaces, considered to be the best index fossils, identified less than 5% from the same early period.

A few pre-6700 cal yr BP sites occur in Vermont, most notably John's Bridge (Thomas, 1992) in the Lake Champlain drainage. The incidence of early sites increases in southern New Hampshire and Massachusetts. Dincauze (1976) reported on the Neville site near Manchester, NH, a stratified riverbank site with radiocarbon dates to about 8500 cal yr BP. Others also occur (e.g., Robinson et al., 1992).

In Canada, pre-6700 cal yr BP Archaic sites have been identified in southern Ontario (Ellis et al., 1991; Wright, 1995), whereas in the Province of Quebec, some are found on the north side of the Gulf of St. Lawrence (Wright, 1995; Pintal, 1998; Plourde, 2000; Pintal, 2006; Plourde, 2006). In the Maritime Provinces, Deal et al. (2006) described a few artifacts from a collection made at Gaspereau Lake, Nova Scotia, that may, on typological grounds, date to the target period. To conclude this brief review of the archaeological evidence, we stress the highly variable nature of the site inventory and the vagaries of site preservation. The question relative to carrying capacity potential needs to be re-phrased from the former presence or absence of people issue. In this chapter, we focus on known sites, rather than gaps in the record, and then attempt to assess the impact of environment change, concentrating our efforts on the situation in central Maine, the region for which we have the greatest amount of data, on both culture and local environment.

4. Wetland adaptations

Archaeological sites in central Maine tend to be clustered around wetland environments, and only rarely in the uplands (Anonymous, 1999; Sanger, 2006). The interior of Maine features many wetland settings, including rivers, lakes, and other habitually wet landforms such as marshes and bogs. In a series of papers, Nicholas (1991, 1992, 1998) called attention to the importance of marshes and other wetlands in the Northeast for their food resources and abilities to support humans.

Many of Maine's wetlands are of the *Sphagnum* peat varieties that hold little attraction for hunters and gatherers. Other wetlands with more open water, such as marshes and fens, provided mammals such as *Castor canadensis* (beaver), *Ondatra zibethicus* (muskrat), *Alces alces* (moose), and *Odocoileus virginianus*, together with the potential for fish, birds, and emergent aquatic plants (Nicholas, 1991). Because of the perceived importance of wetlands to aboriginal peoples of the area, we undertook a detailed reconstruction of wetland evolution around Orono (Fig. 12.2)

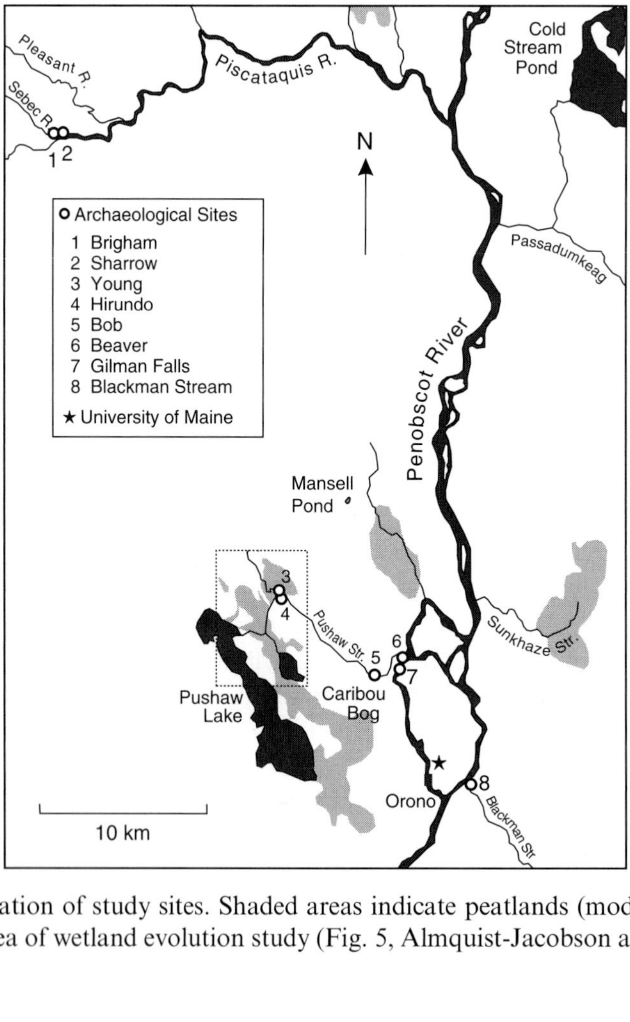

Figure 12.2. Location of study sites. Shaded areas indicate peatlands (modern extent). Dotted box shows area of wetland evolution study (Fig. 5, Almquist-Jacobson and Sanger, 1999).

and a number of sites in the Milford Reservoir region (Almquist-Jacobson and Sanger, 1999). This, combined with a history of upland vegetation (Almquist-Jacobson and Sanger, 1995) and a study of lake levels in Mansell Pond (Almquist et al., 2001), provided the basis for an evaluation of wetland and adjacent upland areas and their ability to sustain hunters and gatherers. Ongoing research into past river flood events tends to confirm water level history predicated by the lake level studies (Kelley and Sanger, 2003; Kelley, 2006).

The many rivers of Maine provided food resources in addition to major highways for aboriginal people. In this well-watered landscape, travel by canoe was much preferable to overland treks through dense forests. The well-known *Betula papyrifera* (white or paper birch) bark canoe of the Northeast was ideally suited to travel the rivers and lakes (Adney and Chapelle, 1964). Light in weight, capable of carrying heavy loads, and sturdy and easily repaired in the field, these craft formed the material culture core of the historically documented pattern of adaptation for the Penobscot people of central Maine (Speck, 1940). For us, it is axiomatic that birch bark canoes played a key role in adaptation to the interior waterways of central Maine.

The waterways provided access to upriver spawning grounds for many species of anadromous fish, including *Pomolobus pseudoharengus* (alewife), *Alosa sapidissima* (shad), *Acipenser* sp. (sturgeon), *Roccus saxatilis* (striped bass), *Osmerus mordax* (smelt), and *Salmo salar* (Atlantic salmon). *Anguilla rostrata* (American eels) proceeded down the rivers in the autumn to reach their spawning beds in the Atlantic Ocean. Non-migratory fish and amphibians also occurred.

5. Paleoecology

Using evidence from pollen, charcoal, and altitudinal limits, several researchers have estimated that the northeastern United States and adjacent Canada experienced a thermal maximum from about 6800–4400 cal yr BP (Davis et al., 1980; Prentice et al., 1991; Webb et al., 1993). Mean annual temperatures during this period may have been about 2°C higher, and mean annual precipitation may have been about 400 mm (about 30%) less than today (Davis et al., 1980; Prentice et al., 1991). Evidence for mid-Holocene declines in lake levels has been noted for northern Maine (Nurse, 2003; Dieffenbacher-Krall and Nurse, 2005), Massachusetts (Newby et al., 2000; Shuman et al., 2001), and southern Quebec (Lavoie and Richard, 2000). Pollen evidence suggests the regional climate gradually became cooler and moister after about 4400 cal yr BP, though several lake-level studies indicate that maximum dryness persisted until almost 3000 cal yr BP (e.g., Davis et al., 1980; Jackson and Whitehead, 1991; Prentice et al., 1991; Spear et al., 1993; Jacobson and Dieffenbacher-Krall, 1995; Shuman et al., 2001; Muller et al., 2003; Nurse, 2003; Dieffenbacher-Krall and Nurse, 2005).

Important shifts in water resources occurred in the region during the mid-Holocene Period. Mansell (Hatch) Pond (45.0417°N, 68.7333°W) provides a record of lake level changes in central Maine since about 9000 cal yr BP (Fig. 12.3)

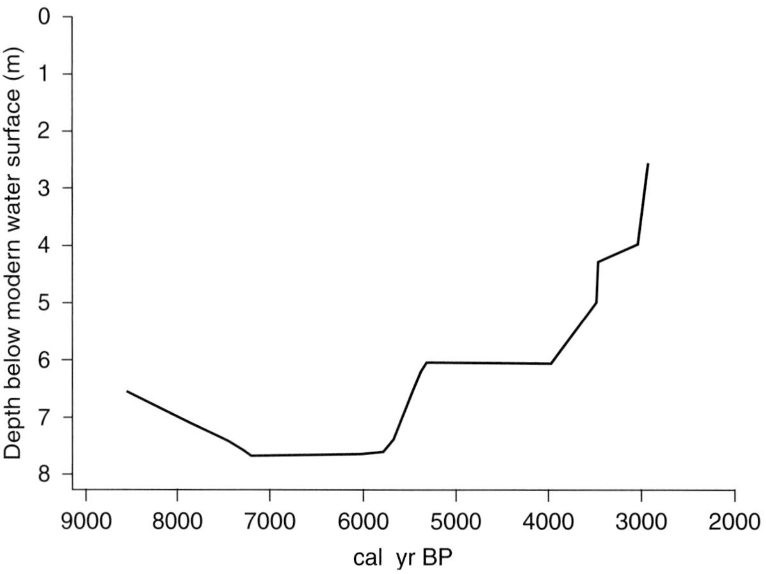

Figure 12.3. Lake level of Mansell Pond through mid-Holocene (redrawn from Almquist et al., 2001).

(Almquist et al., 2001). Sedimentary changes in eight cores from a littoral to deep water transect determined Holocene changes in the lake's water level. *Sphagnum* peat and *gyttja* (lacustrine algal sediment) units were identified in the cores based on texture, macrofossils, and organic content, with transitions between peat and *gyttja* units radiocarbon dated. On the basis of the premise that *Sphagnum* peat accumulated above the lake's water surface, and *gyttja* was deposited below the water surface, Almquist et al. (2001) determined that by 8750 cal yr BP the water level was between 6.5 and 7.5 m below the modern level (Fig. 12.3). The water level continued to drop until about 7000 cal yr BP, reaching a low of between 7 and 9 m below modern, and it remained low until after 6000 cal yr BP. By 5000 cal yr BP, the water level had risen to between 5.5 and 6.5 m below modern. Between 4000 and 3000 cal yr BP, the water level rose by 2–3 m, probably reaching a depth of 4 m below modern by 3000 cal yr BP, and rising to 2.5 m below modern shortly thereafter.

Increased precipitation and runoff would have increased flow in streams and rivers of the region. As water levels rose, many lakes in the region found new outlets or increased flow from existing outlets. Connectivity among lakes and navigability of rivers and streams would have increased, affecting the potential for movement of people, fish, and other animals.

Studies of peatland development in the Milford Drainage Basin at central Maine showed that peatland initiation began, on previously inundated sites, between 11,200 and 8800 cal yr BP (Fig. 12.4; Sanger and MacKay, 1973; Hu and Davis, 1993; Almquist-Jacobson and Sanger, 1999). From about 6800 cal yr BP until 1000 cal yr BP, the extent of wetland areas remained fairly stable. The development

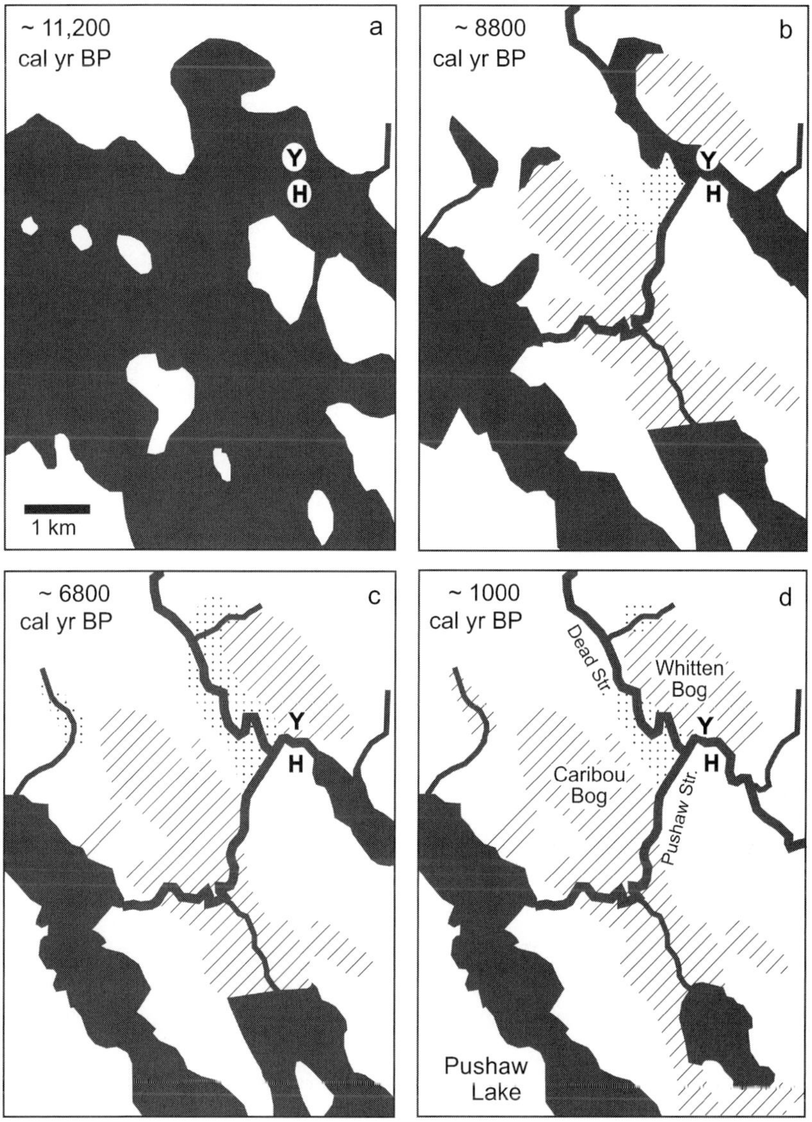

Figure 12.4. Wetland type and distribution in Milford Drainage Basin at (a) circa 11,200 cal yr BP, (b) circa 8800 cal yr BP, (c) circa 6800 cal yr BP, and (d) circa 1000 cal yr BP. Black indicates open water, dotted areas are marsh, and hatched areas are peatlands. Archaeological sites (Y = Young, H = Hirundo) are shown (redrawn from Almquist-Jacobson and Sanger, 1999).

of the peatlands and marshes in this area appears to have been controlled primarily by autogenic sedimentary processes rather than by climate.

During the period from 6000 to 4000 cal yr BP, the forests of much of northeastern North America responded both to these changing climate conditions and to a decline in *Tsuga canadensis* (eastern hemlock) populations. Around 5500–5400 cal yr BP, *Tsuga* as a percentage of the total terrestrial pollen decreased dramatically throughout the northeast and Midwest, possibly because of a pathogen or insect infestation (Davis, 1981; Allison et al., 1986), or a decline in effective precipitation to the region (Shuman et al., 2001; Nurse, 2003; Shuman et al., 2004). In less than a century, *Tsuga* pollen values dropped from 20–30 to 5–10% in Maine (Anderson et al., 1992; Almquist-Jacobson and Sanger, 1995; Schauffler, 1998; Nurse, 2003), from over 30 to less than 10% in southwestern New Brunswick (Mott, 1975), and from 30 to 5% in New Hampshire (Davis, 1981). Forming a dense canopy, *Tsuga* prevents the development of a diverse understory or the establishment of ruderal and hardwood trees, e.g., *Betula* (birch), *Populus* (aspen), *Fagus* (beech), *Quercus* (oak), and *Acer* (maple). In Maine and New Brunswick, the decline of *T. canadensis* stands allowed hardwood species to expand (Mott, 1975; Anderson et al., 1986; Anderson et al., 1992). Previously closed *T. canadensis* forests in central Maine were replaced by hardwood stands consisting chiefly of *Betula* and *Fagus* after 5400 cal yr BP (Fig. 12.5, Almquist-Jacobson and Sanger, 1995). *Acer rubrum* (red maple) stands developed, probably on wet sites, and *Pinus strobus* (eastern white pine) and *Quercus* stands most likely occupied the drier sites.

As the climate became moister, the abundance of *Pinus strobus* began to decrease in Maine and southern Ontario, presumably because fire became too infrequent to permit widespread establishment of the seedlings, which require mineral soil (Jacobson and Dieffenbacher-Krall, 1995). Cooling of the climate over the past 4000 years has allowed boreal taxa, such as *Picea* (spruce) and *Abies* (fir) to move southward. *Picea* appears to have remained sparse in the region until around 1500 years ago when *Picea* pollen percentages increased sharply (Tolonen, 1983; Anderson et al., 1986; Gajewski, 1987; Kellogg, 1991; Anderson et al., 1992; Almquist-Jacobson and Sanger, 1995; Schauffler and Jacobson, 2002). Thus, the tree species composition of the forests experienced by the people of mid-Holocene Maine varied considerably from that found shortly before European contact.

6. Archaeology

6.1. The Gulf of Maine Archaic tradition

In common with much of the Northeast, Maine has evidence for late Pleistocene Paleoindian sites, a technology that featured fluted bifaces, and a distinctive suite of chipped stone tools made on scarce cherts and other silicates. Spiess et al. (1998) listed site locations, major artifact classes, and the few animal remains recovered. No transition to the succeeding Archaic period technology has been documented; indeed, the technologies could scarcely be more distinct (Sanger et al., 2003). Robinson (1992) characterized the early Holocene lithic technology as the Gulf of Maine Archaic [lithic] tradition (GMAt), which currently has no named cultural

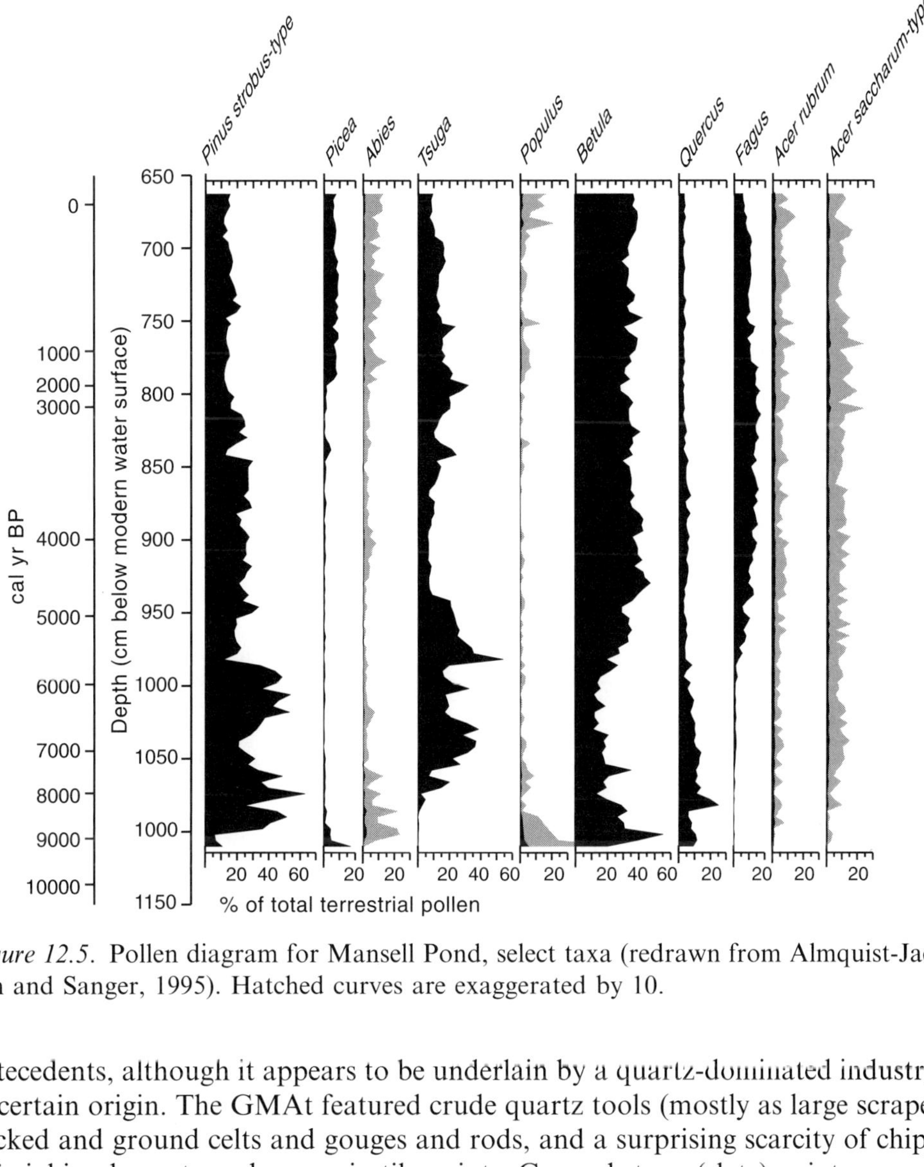

Figure 12.5. Pollen diagram for Mansell Pond, select taxa (redrawn from Almquist-Jacobson and Sanger, 1995). Hatched curves are exaggerated by 10.

antecedents, although it appears to be underlain by a quartz-dominated industry of uncertain origin. The GMAt featured crude quartz tools (mostly as large scrapers), pecked and ground celts and gouges and rods, and a surprising scarcity of chipped bifacial implements such as projectile points. Ground stone (slate) points occurred. In place of the high-quality silicates and cherts of the Paleoindian Period, during the GMAt, the emphasis lay on the utilization of quartz and low-grade metamorphic rocks readily available in much of central Maine (Sanger et al., 2001; Sanger, 2006). A red ocher burial practice was associated (Robinson, 1996, 2001, 2006). In southern New England, the GMAt bumps up against, and was replaced by, a biface tradition described as an extension of the Atlantic slope biface macro-tradition that stretched south into the mid-Atlantic region and into New England (Dincauze, 1976; Sanger, 2006). According to Plourde (2006), GMAt influences extended into the St. Lawrence Valley of Quebec.

GMAt site locations in Maine are inland – defined as above head of tide – as anticipated given the sea-level rise history of the Gulf (Barnhardt et al., 1995). Major published sites in the central Penobscot River drainage (Fig. 12.2) include the Brigham and Sharrow sites on the Piscataquis River (Petersen et al., 1986; Petersen, 1991); the Blackman Stream site on the Penobscot (Sanger et al., 1992); and the Beaver and Gilman Falls sites at the confluence of Pushaw Stream and the Stillwater River (Sanger, 1996; Sanger et al., 2001). Occupation began as early as 10,600 cal yr BP at Sharrow. At Gilman Falls, a major occupation (zone 3) dated between 8200 and 7200 cal yr BP included abundant evidence for quarrying and manufacture of tools from local, low-grade metamorphic rocks, such as phyllite and granofels (Sanger et al., 2001). The assemblage also includes use of quartz for scrapers. Faunal remains from these sites indicate the importance of wetland resources.

6.2. The Laurentian tradition

About 6700 cal yr BP, a major technological addition occurred in central Maine. For several previous millennia, the local cultures rarely manufactured chipped bifaces from cryptocrystalline rocks, despite the local availability of a porphyritic rhyolite, known locally as "felsite." Added to the basic GMAt lithic tool kit by 6700 cal yr BP were semi-lunar ground slate knives (or ulus), plummets, and more elaborated, better-finished ground slate points. By about 6400 cal yr BP, and possibly earlier, large, side-notched bifaces made from chipped stone appeared at sites such as Brigham and Sharrow (Petersen et al., 1986; Petersen, 1991), and in sites near Orono, such as Hirundo (Sanger et al., 1977), Bob (Mack et al., 2002), and Gilman Falls (Sanger, 1996, Sanger et al., 2001). In the St. Croix watershed, near the Maine-New Brunswick International border, they occur in surface collections and in a good archaeological context at the Narrows site radiocarbon dated in features to around 5700 cal yr BP (Cox, 1991). These large, side-notched bifaces show great similarities with Otter Creek points, the premier hallmark of the Laurentian tradition (Funk, 1988), especially the Vergennes phase, named after sites such as KI near Vergennes on the Otter Creek, a tributary of Lake Champlain, Vermont.

At a time when archaeological research in much of the Northeast was in its infancy, William Ritchie (1965) had developed the basic culture sequence for New York State. One of his earliest cultural definitions was the Laurentian tradition, which he recognized from sites in upstate New York, adjacent Vermont, Maine, and Canada. He defined it on the basis of the distinctive bifaces as well as pecked and ground stone implements. Locally, sites of this type also contained many copper artifacts (e.g., at Morrison Island and Allumettes Island in the Ottawa River [Clermont and Chapdelaine, 1998; Clermont et al., 2003; Chapdelaine and Clermont, 2006]). Ritchie (1965, p. 79) referred to the Laurentian culture type as a basic adaptation, "an extensive Archaic cultural continuum," to the Northeastern northern hardwood forests with cultural contacts to the south and west. This view is shared in more recent overviews (e.g., Funk, 1988; Wright, 1995; Sanger and Newsom, 2000; Sanger, 2006). In time, Ritchie recognized several variations of the Laurentian tradition, all sharing broad-bladed bifaces, as opposed to some narrow-bladed bifaces of the near contemporaneous Lamoka tradition. Of the four Laurentian tradition phases, only the first, the Vergennes phase, makes any sense for central and eastern Maine, and even then its utility is equivocal (Sanger, 2006).

If defined by the presence of side-notched bifaces alone, the Vergennes phase is poorly represented west of the Kennebec River. Significantly, these bifaces rarely occur on the coast (Cox, 1991). Two possibilities exist. First, sites of a suitable age, 5700 cal yr ago or earlier, have not survived the erosive events of rising sea levels. Second, and more likely in our estimate, is a view that regards the Laurentian tradition as an interior, wetland-adapted culture with most sites occurring east of the Kennebec River. The interior adaptation model accords better with Bourque's (1995) assertion that the narrow-stemmed points of occupation 1 and 2 at Turner Farm, Penobscot Bay, represent a long period of coastal adaptation and lithic technological relationships with Atlantic littoral cultures (Dincauze, 1976). Such points are rare in inland sites located much above head of tide.

The presence of Otter Creek points in stratified sites and, in particular, the assemblage at the Narrows site, suggested to Cox (1991) that an *in situ* hypothesis for Maine variant of the Vergennes phase has merit. That which archaeologists in Maine have called the Laurentian tradition probably represents the addition of the Otter Creek bifaces to a lithic system that has roots extending back several millennia: in other words, the Gulf of Maine Archaic tradition, or something similar (Cox, 1991; Robinson, 1992; Petersen, 1995; Sanger, 1996; Sanger and Newsom, 2000; Sanger, 2006). Otter Creek point variants persisted in central Maine until about 5300 cal yr BP, after which time they disappeared from the local sequence (Petersen, 1991; Mack et al., 2002).

Maine archaeologists recognize the long-standing cultural boundary at the Kennebec River. Perhaps not coincidentally, this region currently represents a significant transition zone between temperate and northern vegetation types in Maine. Indeed, more than 60 plant taxa attain either their northern or southern limit in a zone bounded on the east by the Penobscot and west by the Kennebec (McMahon, 1990). Further, the incidence of wetlands drops off sharply in the rolling uplands and sandy sediments of western Maine (Thompson and Borns, 1985). Thus, there may be a causal link between kinds of adaptation, in part reflected in material culture, between those sites in the Penobscot and those in the Kennebec drainage (Sanger, 2006).

To conclude, this brief overview of early central Maine Archaic period history traces the development of hunting, fishing and gathering cultures adapted to the interior waterways and wetlands of Maine. In its near-final form, known as the Laurentian tradition, it adopted broad-bladed, side-notched bifaces (Otter Creek points) from cultures in the St. Lawrence Valley. Throughout the five millennia, there are no strong indications of an interruption in the way of life, although various artifact classes and forms changed through time. The physical environment was not static. After a review, we assess some of the potential impacts on the cultural environment.

7. Culture and environment linkages

In this section, we assess the potential impacts of climatic and environment changes on humans. Obviously, the reverse effect may occur, and probably has occurred in southern New England where there is abundant historical evidence for deliberate forest burning by aboriginal horticulturists (Cronon, 1983). To date, there is no compelling evidence for any deliberate burning in most of Maine prior to the European presence. Any assessment of change in the natural systems relative to human populations must include a rationale for one having an influence on the other. In other words, why should a climatic perturbation or environmental change impact hunters and gathers living in an ecosystem with considerable biological diversity, especially if, as we suspect, cultural coping devices had evolved to buffer negative aspects of environmental variability? Scale also enters the equation. Hunters and gatherers depend on localized resources. This suggests that climatic and environmental reconstructions must be at the same scale, that is, at the level of the catchment area that a band might reasonably exploit (Dincauze, 1981).

7.1. Vegetation and climate changes

The *T. canadensis* (hemlock) forest "crash" of around 5400 cal yr BP opened up the forests to a diverse understory and the development of the northern hardwoods, prime habitat for deer. The proposed warmer temperatures and less precipitation of the thermal maximum period may have permitted deer to cope better with the Northeast winters, as snow depths control partially their ability to survive winters (Kelsall and Prescott, 1971). There may have been other impacts brought about by the hemlock decline, such as increased land erosion and runoff leading to higher flood events until the forests reestablished (Putnam, 1994). The geoarchaeological record in the Penobscot Valley does suggest an increase in alluvium at this time (Kelley and Sanger, 2003; Kelley, 2006). Although more floods may have represented little more than a nuisance to people, these periods of sedimentation have proven critical for our ability to reconstruct past cultures by providing stratigraphic separation between artifact assemblages.

7.2. Wetland evolution

Many wetlands in central Maine evolved toward *Sphagnum* peatlands throughout the Holocene (Fig. 12.4). As noted, the mid-Holocene cultural adaptation pattern involved exploitation of wetlands for their varied resources. Hirundo site, one of the most intensively utilized site localities (Sanger et al., 1977), was occupied almost as soon as Pushaw Lake waters receded, circa 10,000 cal yr BP, and continued in use into the Contact period. Although the once extensive *Typha* marshes shrank in size and became *Sphagnum* peatlands, parts of Whitten Bog remained marshes until quite recently (Fig. 12.4; Almquist-Jacobson and Sanger, 1999). A short canoe trip up Dead Stream takes one from the Hirundo site to the Whitten Bog marsh areas in a matter of minutes. Reconstruction at this scale is pertinent to human activities.

Two aquatic mammals, muskrat and beaver, dominate the Hirundo site faunal assemblage (Knight, 1985). The post-hemlock decline vegetation changes recorded in the Mansell Pond pollen diagram (Almquist-Jacobson and Sanger, 1995) would have enhanced preferred beaver diet – river-edge hardwoods such as *Populus* and *Acer rubrum*. More beaver means more beaver ponds, which, in turn, provide muskrat habitat. From a canoeist's perspective, beaver dams provide another bonus: dams create a series of stepped ponds that inundate rapids and maintain high water levels for summer travel, a period when many smaller waterways become non-navigable (Cook, 1985).

The correlation between the long-standing cultural boundary between the Kennebec and Penobscot drainages and the occurrence of the vegetation transition zone, when combined with the much lower frequency of wetlands west of the Kennebec River, suggests that there may well be an important causal link (Sanger, 2006).

7.3. Water levels

Changes in water levels previously noted at Mansell Pond and in other Northeast lakes have impacted the nature of the archaeological record directly. Although the chronology varies somewhat (Almquist et al., 2001; Dieffenbacher-Krall and Nurse, 2005), most Northeastern lakes experienced low levels in mid-Holocene times, a reflection perhaps of lower precipitation – by as much as 30% (Davis et al., 1980; Prentice et al., 1991) – combined with higher summer evaporation rates. Mid-Holocene archaeological sites situated on the margins of lakes would have been inundated by rising water levels after the low stands, which achieved their highest levels at European arrival. Such sites are invisible in a traditional land-based site survey unless mid-Holocene levels are restored. Underwater archaeology, an expensive enterprise not yet undertaken systematically, is a possibility.

Given the interconnected system of waterways in central Maine and the importance of these for local cultures (Speck, 1940), lowered water levels during the mid-Holocene must have had an impact. As the late Pleistocene drainage systems cut through the glacial deposits to achieve equilibrium with falling sea levels, they encountered erosion-resistant bedrock and other obstacles such as boulder lags from various ice-contact features (Putnam, 1994; Sanger et al., 2001; Kelley and Sanger, 2003; Kelley, 2006). These formed the many rapids that characterize central

Maine rivers. High water levels expedite canoe travel through these rapids; low levels mean carries (portages) and prolonged drags through boulder-strewn riverbeds (Cook, 1985). If the low mid-Holocene water levels demonstrated at Mansell Pond and other lakes accurately reflect moisture availability generally, travel by water must have been more difficult, especially during the summer.

Geoarchaeological studies of sediment history in the Orono region demonstrate periods of flood events, recognized by rapid development of alluvium, followed by periods of infrequent high floods during which time forest soils developed (Sanger et al., 2001; Kelley and Sanger, 2003; Kelley, 2006). Radiocarbon chronologies for several sedimentary sequences coincide closely with the reconstructed water-level curve from nearby Mansell Pond (Fig. 12.3).

Water levels, connections to lakes, temperatures, and flow rates affect fish migrations, access to spawning grounds, and habitat for immature fish. As outlined earlier, fish played an important dietary role in Maine. Various species of anadromous fish require correct water depths and volumes (flow) and temperature. One widely studied species is the Atlantic salmon (*Salmo salar*), which is currently undergoing intensive international scrutiny as part of an effort to restore natural breeding populations. Low water-flow regimes impede upriver migration of adults ready to spawn, whereas prolonged river water temperatures above 27°C – achieved in several Maine summers recently – prove lethal to salmon (Shepard, 1995). Although we cannot reconstruct precise water flows and temperatures for mid-Holocene times, the evidence for less water, combined with suggestions of higher ambient temperatures, probably had an adverse impact on salmon populations. Other key species, such as shad and alewives that tend to "run" in the spring during high water, may not have been affected (Sanger and Newsom, 2000). As Carlson (1988, 1992) noted, salmon bones rarely occur in archaeological sites of pre-European times. However, they became more common, predictably, during the colder Little Ice Age when water levels were higher (e.g., Robinson and Cowie, 1992).

8. Conclusions

As our ability to resolve the past in parts of the Northeast increases, broad-scale correlations between cultural and environmental change become less convincing. This is not to imply that hunters and gatherers were immune to environmental constraints. Our conclusion is that though changes did occur, aboriginal people could mitigate, or at least minimize, the impacts of environmental perturbations, through prey switching and moving people from affected areas. That said, any short-term population declines and territorial abandonment, had either occurred, are currently beyond our abilities to resolve archaeologically. Archaeologists have no tool such as pollen analysis with which to record absolute differences in human populations, whereas the current chronologies cannot provide reliable decadal resolution of the kind we feel would be required to track human responses to climate shifts in this area.

Explanation for demise of the Paleoindian culture at the end of the Pleistocene remains contentious (Spiess et al., 1998). However, after the onset of the Gulf of Maine Archaic culture type early in the Holocene, and for several millennia thereafter, there is no evidence for an abrupt shift in material culture, or in the way in which people adapted to the region. In central Maine, the basic wetland adaptation pattern, which probably emerged with the post-Paleoindian Period, remained almost unchanged. The lengthy persistence of the associated red ocher burial tradition reinforces notions of stability in the symbolic nature of culture (Sanger, 2006). Even the appearance of the Laurentian tradition, defined almost entirely on the basis of a new projectile point technology, had no apparent impact unless it was an impression of an increased number of sites, which occur on modern floodplains. The wetland adaptation pattern was probably the key to the maintenance of hunting and gathering lifestyles highly tuned to the Northeast. As long as the wetlands proved productive, humans could flourish in their presence. As the wetlands evolved into less useful peat bogs, a modification to the traditional pattern may have involved the greater use of riverine resources, which in turn became possible with increased water levels that affected fish resources. Higher water levels during the Ceramic period (roughly 3000–450 years ago) may account for the much greater incidence of sites in the headwaters of rivers systems in northern Maine in this period (Anonymous, 1999), a combination of better fish habitat and territory more readily accessed by canoes.

Finally, we are very cognizant of the fact that our interdisciplinary research involving vegetation, wetland, riverbank sediment regimes, and archaeology has been undertaken in a relatively restricted area – within a few kilometers of the University of Maine in the central Penobscot Valley. That is fully in keeping with our philosophy regarding scale and the need to match environmental change at a level commensurate with the exploitation, or catchment zone, of hunters and gatherers. New lake level studies in northern Maine indicate that though the precise history varies, water levels have generally been on the rise from mid-Holocene times onward. Future research will judge if our observations and conclusions can be extended beyond the limited central Penobscot Valley region.

Acknowledgments

We wish to acknowledge the support of Bangor Hydro-Electric Company for much of the paleo-environmental and archaeological research reported in this chapter. To our colleagues at the University of Maine and many, many students, we owe a great debt. And finally, we wish to thank the conference organizers for the kind invitation to participate in this project, and for their editorial assistance.

References

Adney, E. T., and H. I. Chapelle, 1964. *The Bark Canoes and Skin Boats of North America.* Smithsonian Institution, Washington, DC.

Allison, T. D., R. E. Moeller, and M. B. Davis, 1986. Pollen in laminated sediments provides evidence for a mid-Holocene forest pathogen outbreak. *Ecology* 67:1101–1105.

Almquist-Jacobson, H., and D. Sanger, 1995. Holocene climate and vegetation in the Milford drainage basin, Maine, U.S.A., and their implications for human history. *Vegetation History and Archaeobotany* 4:211–222.

Almquist-Jacobson, H., and D. Sanger, 1999. Paleogeographic changes in wetland and upland environments in the Milford drainage basin of central Maine in relation to Holocene human settlement history. In *Current Northeast Paleoethnobotany*, edited by J. P. Hart, pp. 177–190. New York State Museum Bulletin 494, Albany.

Almquist H., A. C. Dieffenbacher-Krall, R. Flanagan-Brown, and D. Sanger, 2001. Holocene record of lake levels of Mansell Pond, central Maine, U.S.A. *The Holocene* 11(2): 189–201.

Anderson, R. S., R. B. Davis, N. G. Miller, and R. Stuckenrath, 1986. History of late- and post-glacial vegetation and disturbance around Upper South Branch Pond, northern Maine. *Canadian Journal of Botany* 64:1977–1986.

Anderson, R. S., G. L. Jacobson Jr., R. B. Davis, and R. Stuckenrath, 1992. Gould Pond, Maine: late-glacial transitions from Marine to upland environments. *Boreas* 21:359–371.

Anonymous, 1999. *Maine Prehistoric Survey Site Forms*. Maine Historic Preservation Commission, Augusta.

Barnhardt, W. A., W. R. Gehrels, D. F. Belknap, and J. T. Kelley, 1995. Late Quaternary relative sea-level change in the western Gulf of Maine: evidence for a migrating glacial forebulge. *Geology* 23:317–320.

Bourque, B. J., 1995. *Diversity and Complexity in Prehistoric Maritime Societies*. Plenum Press, New York.

Carlson, C. C., 1988. Where's the salmon? A reevaluation of the role of anadromous fishes to aboriginal New England. In *Holocene Human Ecology in Northeastern North America*, edited by G. Nicholas, pp. 47–80. Plenum Press, New York.

Carlson, C. C., 1992. *The Atlantic Salmon in New England Prehistory and History: Social and Environmental Implications*. Ph.D. thesis. University of Massachusetts, Amherst, MA.

Chapdelaine, C., and N. Clermont, 2006. Adaptation, continuity, and change in the middle Ottawa Valley: a view from the Morrison and Allumettes Island late Archaic sites. In *The Archaic of the Far Northeast*, edited by D. Sanger, and M. Renouf, pp. 191–219. University of Maine Press, Orono.

Clermont, N., and C. Chapdelaine, 1998. *Île Morrison: Lieu Sacré et Atelier de l'Archaïque dans l'Outaouais'*. Recherches Amérindiennes au Québec, Montréal.

Clermont, N, C. Chapdelaine, and J. Cinq-Mars, 2003. Île aux Allumettes. *Collection Paléo-Québec*. Recherches Amérindiennes au Québec, Montréal.

Cook, D. S., 1985. *Above the Gravel Bar*. Milo Printing Company, Milo, ME.

Cox, S. L., 1991. Site 95.20 and the Vergennes phase in Maine. *Archaeology of Eastern North America* 19:135–161.

Crock, J. G., J. B. Petersen, and R. M. Anderson, 1993. Scalloping for artifacts: a biface and plummet from eastern Blue Hill Bay, Maine. *Archaeology of Eastern North America* 20:179–192.

Cronon, W., 1983. *Changes in the Land: Indians, Colonists, and the Ecology of New England*. Hill and Wang, New York.

Davis, M. B., 1981. Outbreaks of forest pathogens in Quaternary history. In *Proceedings of the Fourth International Palynological Conference, 1976–1977*, edited by H. Birks, Vol. 3, pp. 216–227. Birbal Sahni Institute of Paleobotany, Lucknow, India.

Davis, M. B., R. W. Spear, and L. C. K. Shane, 1980. Holocene climate of New England. *Quaternary Research* 14:240–250.

Deal, M., D. Rutherford, B. Murphy, and S. Buchanan, 2006. Rethinking the Archaic sequence for the Maritime provinces. In *The Archaic of the Far Northeast*, edited by D. Sanger, and M. Renouf, pp. 253–283. University of Maine Press, Orono.

Dieffenbacher-Krall, A. C., and A. M. Nurse, 2005. Late-glacial and Holocene record of lake levels of Mathews Pond and Whitehead Lake, Northern Maine, USA. *Journal of Paleolimnology* 34:283–310.

Dincauze, D. F., 1976. *The Neville Site.* Harvard University Press, Cambridge.

Dincauze, D. F., 1981. Paleoenvironmental reconstruction in the Northeast: the art of multidisciplinary science. In *Foundations of Northeast Archaeology*, edited by D. Snow, pp. 51–96. Academic Press, New York.

Ellis, C. J., S. Wortner, and W. Fox, 1991. Nettling: an overview of an Early Archaic "Kirk corner-notched cluster" site in southwestern Ontario. *Canadian Journal of Archaeology* 15:1–34.

Fitting, J., 1968. Environmental potential and the postglacial readaptation in eastern North America. *American Antiquity* 33:441–445.

Funk, R. E., 1988. The Laurentian concept: a review. *Archaeology of Eastern North America* 16:1–41.

Funk, R. E., 1996. Holocene or hollow scene? The search for the earliest Archaic cultures in New York State. *The Review of Archaeology* 17:11–25.

Gajewski, K., 1987. Climatic impacts on the vegetation of eastern North America during the past 2000 years. *Vegetation* 68:179–180.

Hu, F.-S., and R. B. Davis, 1993. Postglacial development of a Maine bog and environmental implications. *Canadian Journal of Botany* 73:628–649.

Jackson, S. T., and D. R. Whitehead, 1991. Holocene vegetation patterns in the Adirondack Mountains. *Ecology* 72:641–653.

Jacobson, G. L., Jr., and A. Dieffenbacher-Krall, 1995. White pine and climate change: Insights from the Past. *Journal of Forestry* 93:39–42.

Kelley, A. R., 2006. *Archaeological Geology in the Central Penobscot Valley, Maine.* Ph.D. thesis. University of Maine, Orono.

Kelley, A. R., and D. Sanger, 2003. Post-glacial development of the Penobscot River Valley: implications for geoarchaeology. In *Geoarchaeology of Landscapes in the Glaciated Northeast*, edited by D. Cremeens, and J. Hart, pp. 119–133. York State Museum Bulletin 497, Albany.

Kellogg, D. C., 1991. *Prehistoric Landscapes, Paleoenvironments, and Archaeology of Western Muscongus Bay, Maine.* Ph.D. thesis. University of Maine, Orono.

Kelly, R. L., 1995. *The Foraging Spectrum: Diversity in Hunter-Gatherer Lifeways.* Smithsonian Institution Press, Washington.

Kelsall, J. P., and W. Prescott, 1971. *Moose and Deer Behaviour in Snow in Fundy National Park, New Brunswick.* Report 15. Canadian Wildlife Service, Ottawa.

Knight, J. A., 1985. *Differential Preservation of Calcined Bone at the Hirundo Site, Alton.* M.S. thesis. University of Maine, Orono.

Lavoie, M., and P. J. H. Richard, 2000. Postglacial water-level changes of a small lake in southern Quebec, Canada. *The Holocene.* 10:621–634.

Leacock, E., 1954. The Montagnais "Hunting Territory" and the fur trade. *Memoirs of the American Anthropological Association*, 78. Menasha, WI.

Mack, K. E., D. Sanger, and A. R. Kelley, 2002. Archaeological investigations at the Bob Site, Pushaw Stream, Maine. *Occasional Papers in Maine Archaeology*, 12. Maine Historic Preservation Commission, Augusta.

McMahon, J. S., 1990. *The Biophysical Regions of Maine: Patterns in the Landscape and Vegetation.* M.S. thesis. University of Maine, Orono.

Mott, R. J., 1975. Palynological studies of lake sediment profiles from southwestern New Brunswick. *Canadian Journal of Earth Sciences* 12:273–288.
Muller, S. D., P. J. H. Richard, J. Guiot, J. L. de Beaulieu, and D. Fortin, 2003. Postglacial climate in the St. Lawrence lowlands, southern Quebec: pollen and lake-level evidence. *Palaeogeography, Palaeoclimatology, Palaeoecology* 193:51–72.
Newby, P. E., P. Killoran, M. R. Waldorf, B. N. Shuman, R. S. Webb, and T. Webb III, 2000. 14,000 years of sediment, vegetation and water-level changes at the Makepeace Cedar Swamp, southeastern Massachusetts. *Quaternary Research* 53:352–368.
Nicholas, G. P., 1991. Putting wetlands into perspective. *Man in the Northeast* 42:29–38.
Nicholas, G. P., 1992. Directions in wetlands research. *Man in the Northeast* 43:1–9.
Nicholas, G. P., 1998. Wetlands and hunter-gatherers: a global perspective. *Current Anthropology* 39:720–731.
Nurse, A. M., 2003. *12,000-Year Record of Lake-Level and Vegetative Change at Mathews Pond, Piscataquis County, Maine, USA*. M.S. thesis. University of Maine, Orono.
Petersen, J. B., 1991. Archaeological testing at the Sharrow site: a deeply stratified early to late Holocene cultural sequence in central Maine. *Occasional Papers in Maine Archaeology*, 8, Maine Historic Preservation Commission, Augusta.
Petersen, J. B., 1995. Preceramic archaeological manifestations in the Far Northeast: a review of current research. *Archaeology of Eastern North America* 23:207–230.
Petersen, J. B., and D. E. Putnam, 1992. Early Holocene occupation in the central Gulf of Maine region. In *Early Holocene Occupation in Northern New England. Occasional Papers in Maine Archaeology*, 9, edited by B. S. Robinson, J. B. Petersen, and A. K. Robinson, pp. 13–61. Maine Historic Preservation Commission, Augusta.
Petersen, J. B., N. D. Hamilton, D. Putnam, A. E. Spiess, R. Stuckenrath, C. A. Thayer, and J. A. Wolford, 1986. The Piscataquis archaeological project: a late Pleistocene occupational sequence in northern New England. *Archaeology of Eastern North America* 14:1–18.
Pintal, J.-Y., 1998. Aux Frontiéres de la Mer: La Préhistoire de Blanc-Sablon. *Collection Patrimoines Dossiers 102*, Municipalité de Blanc-Sablon, Blanc-Sablon.
Pintal, J.-Y., 2006. The archaic sequence of the St. Lawrence lower north shore, Quebec. In *The Archaic of the Far Northeast*, edited by D. Sanger, and M. Renouf, pp. 105–138. University of Maine Press, Orono.
Plourde, M., 2000. *Cap-de-Bon-Désir: Is the Early Archaic in the Eastern St. Lawrence Valley Related to the Gulf of Maine Archaic Tradition?* Paper presented at the Canadian Archaeological Association, Ottawa.
Plourde, M., 2006. The Cap-de-Bon-Désir site: a new regional variation of the Gulf of Maine Archaic tradition. In *The Archaic of the Far Northeast*, edited by D. Sanger, and M. Renouf, pp. 139–160. University of Maine Press, Orono.
Prentice, I. C., P. J. Bartlein, and T. Webb, III, 1991. Vegetation and climate change in eastern North America since the last glacial maximum. *Ecology* 72:2038–2056.
Putnam, D. E., 1994. Vertical accretion of flood deposits and deeply stratified archaeological site formation in central Maine, U.S.A. *Geoarchaeology* 9:467–502.
Ritchie, W. A., 1965. *The Archaeology of New York State*. New York Natural History Press, Garden City, NY.
Ritchie, W. A., 1971. The Archaic in New York. *New York State Archaeological Association Bulletin* 52:2–12.
Robinson, B. S., 1992. Early and middle Archaic period occupation in the Gulf of Maine region: mortuary and technological patterning. In *Early Holocene Occupation in Northern New England. Occasional Papers in Maine Archaeology*, 9, edited by B. S. Robinson,

J. B. Petersen, and A. K. Robinson, pp. 63–116. Maine Historic Preservation Commission, Augusta.

Robinson, B. S., 1996. A regional analysis of the Moorehead burial tradition: 8500–3700 B.P. *Archaeology of Eastern North America* 24:95–148.

Robinson, B. S., 2001. *Burial Ritual, Groups, and Boundaries on the Gulf of Maine: 8600–3800 B.P.* Ph.D. thesis. Brown University, Providence.

Robinson, B. S., 2006. Burial ritual, technology, and cultural landscape in the Far Northeast: 8600–3700 B.P. *The Archaic of the Far Northeast*, edited by D. Sanger, and M. Renouf, pp. 341–381. University of Maine Press, Orono.

Robinson, B. S., and E. P. Cowie, 1992. Analysis of faunal remains recovered during Phase II testing of the weston project. In *Archaeological Phase II Testing of the Weston Project, (FERC No. 2325), Somerset County. Maine*, edited by E. P. Cowie, and J. B. Petersen, Report on file. Maine Historic Preservation Commission, Augusta.

Robinson, B. S., J. B. Petersen, and A. K. Robinson, editors, 1992. Early Holocene Occupation in Northern New England. *Occasional Papers in Maine Archaeology*, 9, Maine Historic Preservation Commission, Augusta.

Sanger, D., 1977. Some thoughts on the scarcity of archaeological sites in Maine between 10,000 and 5,000 years ago. *Maine Archaeological Society Bulletin* 17:18–25.

Sanger, D., 1988. Maritime adaptations in the Gulf of Maine. *Archaeology of Eastern North America* 16:81–99.

Sanger, D., 1996. Gilman Falls site: implications for the early and middle Archaic of the Maritime Peninsula. *Canadian Journal of Archaeology* 20:7–28.

Sanger, D., 2006. An introduction to the Archaic of the Maritime Peninsula. In *The Archaic of the Far Northeast*, edited by D. Sanger, and M. Renouf, pp. 221–252. University of Maine Press, Orono.

Sanger, D., W. R. Belcher, and D. C. Kellogg, 1992. Early Holocene occupation at the Blackman Stream site, central Maine. In *Early Holocene Occupation in Northern New England. Occasional Papers in Maine Archaeology*, 9, edited by B. S. Robinson, J. B. Petersen, and A. K. Robinson, pp. 149–161. Maine Historic Preservation Commission, Augusta.

Sanger, D., R. B. Davis, R. G. MacKay, and H. W. Borns Jr., 1977. The Hirundo archaeological project–an interdisciplinary approach to central Maine prehistory. In: *Amerinds and their Paleoenvironments in Northeastern North America Annals of the New York Academy of Sciences*, edited by W. B. Newman, and B. Salwen, Vol. 288, pp. 457–471. New York Academy of Sciences, New York.

Sanger, D., A. R. Kelley, and H. Almquist, 2003. Geoarchaeology and cultural interpretation in the lower Penobscot Valley, Maine. In *Geoarchaeology of Landscapes in the Glaciated Northeast*, edited by D. Cremeens, and J. Hart, pp. 135–150. *New York State Museum* Bulletin 497, New York State Museum, Albany.

Sanger, D., A. R. Kelley, and H. Berry, IV, 2001. Geoarchaeology at Gilman Falls: a middle Archaic quarry and habitation in central Maine, U.S.A. *Geoarchaeology: An International Journal* 16(6):633–665.

Sanger, D., and R. G. MacKay, 1973. The Hirundo archaeological project preliminary report. *Man in the Northeast* 6:21–29.

Sanger, D., and B. Newsom, 2000. Middle Archaic in the lower Piscataquis river, and its relationship to the Laurentian tradition in central Maine. *The Maine Archaeological Society Bulletin* 40:1–22.

Schauffler, M., 1998. *Paleoecology of Coastal and Interior Picea (Spruce) Stands in Maine*. Ph.D. dissertation. University of Maine, Orono.

Schauffler, M., and G. L. Jacobson Jr., 2002. Persistence of coastal spruce refugia during the Holocene in northern New England, USA, detected by stand-scale stratigraphies. *Journal of Ecology* 90:235–250.

Shepard, S. L., 1995. *Atlantic Salmon Spawning Migrations in the Penobscot River, Maine: Fishing, Flows and High Temperatures*. M.S. thesis. University of Maine, Orono.

Shuman, B., J. Bravo, J. Kaye, J. A. Lynch, P. Newby, T. Webb III, 2001. Late Quaternary water-level variations and vegetation history at Crooked Pond, southeastern Massachusetts. *Quaternary Research* 56:401–414.

Shuman, B., P. Newby, Y. Huang, and T. Webb III, 2004. Evidence for the close climate control of New England vegetation history. *Ecology* 85:1297–1310.

Spear, R. W., M. B. Davis, and L. C. K. Shane, 1993. Late Quaternary history of low- and mid-elevation vegetation in the White Mountains of New Hampshire. *Ecological Monographs* 64:85–107.

Speck, F. G., 1940. *Penobscot Man*. University of Pennsylvania Press, Philadelphia.

Spiess, A. E., 1992. Archaic period subsistence in New England and the Atlantic Provinces. In *Early Holocene Occupation in Northern New England. Occasional Papers in Maine Archaeology*, 9, edited by B. S. Robinson, J. B. Petersen, and A. K. Robinson, pp. 163–185. Maine Historic Preservation Commission, Augusta.

Spiess, A. E., J. W. Bradley, and D. Wilson, 1998. Paleoindian occupation in the New England-Maritimes region: beyond cultural ecology. *Archaeology of Eastern North America* 26: 201–264.

Spiess, A. E., and J. Mosher, 2006. Archaic period hunting and fishing around the Gulf of Maine. In *The Archaic of the Far Northeast*, edited by D. Sanger, and M. Renouf, pp. 385–410. University of Maine Press, Orono.

Tanner, A., 1979. *Bringing Home Animals*. C. Hurst, London.

Thomas, P. A., 1992. The early and middle Archaic periods as represented in western Vermont. In *Early Holocene Occupation in Northern New England*, edited by B. S. Robinson, J. B. Petersen, and A. K. Robinson, pp. 187–203. *Occasional Papers in Maine Archaeology*, 9, Maine Historic Preservation Commission, Augusta.

Thompson, W. B., and H. W. Borns Jr., 1985. *Surficial Geologic Map of Maine*. Maine Geological Survey, Augusta, Maine.

Tolonen, M., 1983. Pollen evidence of vegetational change following early European settlement of Monhegan Island, Maine, northeastern U.S.A. *Boreas* 12:201–215.

Trigger, B. G., 1978. *Handbook of North American Indians—Northeast*, Vol. 15. Smithsonian Institution, Washington.

Webb, T. III, P. J. Bartlein, S. P. Harrison, and. K. H. Anderson, 1993. Vegetation, lake levels, and climate in eastern North America for the past 18,000 Years. In *Global Climates Since the Last Glacial Maximum*, edited by H. E. Wright Jr., J. E. Kutzbach, T. Webb III, W. F. Ruddiman, F. A. Street-Perrott, and P. J. Bartlein, pp. 415–467. University of Minnesota Press, Minneapolis.

Wright, J. V., 1995. *A History of the Native Peoples of Canada, vol. 1 (10,000–1,000 B.C.)*. Canadian Museum of Civilization, Ottawa.

Climate Change and Cultural Dynamics: A Global Perspective on Mid-Holocene Transitions
David G. Anderson, Kirk A. Maasch and Daniel H. Sandweiss (Editors)

Chapter 13

Mid-Holocene cultural dynamics in southeastern North America

David G. Anderson[1], Michael Russo[2] and Kenneth E. Sassaman[3]

[1]*Department of Anthropology, University of Tennessee, 250 South Stadium Hall, Knoxville, TN 37996-0720, USA*
[2]*National Park Service, Southeastern Archeological Center, Tallahassee, FL 32310, USA*
[3]*Department of Anthropology, University of Florida, 1112 Turlington Hall, P.O. Box 117305, Gainesville, FL 32611-7305, USA*

Abstract

The Middle Archaic period in the Southeast, corresponding to the Mid-Holocene era, from 8000 to 5000 ^{14}C yr BP (ca. 8900–5750 cal yr BP) is a time of appreciable culture change. During this interval monumental construction began in a number of areas, long-distance exchange networks emerged, evidence for warfare appeared, and experimentation with agriculture was initiated. These trends continued and accelerated during the ensuing Late Archaic, and it was at the very start of this period, soon after 5000 ^{14}C yr BP (ca. 5750 cal yr BP), that pottery appeared. Variability is evident in the size and complexity of southeastern Middle Archaic societies, something that appears linked to changes in population interaction, climate, and resource structure. During the Mid-Holocene, use of the southeastern Coastal Plain decreased dramatically, and extensive use of shellfish resources appeared for the first time along the major rivers of the interior and in coastal areas. With the onset of essentially modern climate and resource structure after 5000 ^{14}C yr BP (ca. 5750 cal yr BP), a dramatic increase in regional population levels is indicated.

1. Introduction

To archaeologists in southeastern North America, the Mid-Holocene is considered interchangeable with the Middle Archaic period, from 8000 to 5000 radiocarbon years before present (^{14}C yr BP). It is widely viewed as the time of human adaptation to the climate interval variously known as the Hypsithermal, Altithermal, Atlantic, or Climatic Optimum. In calibrated time, 8000–5000 ^{14}C yr BP is roughly 8900–5750 cal yr BP or 6900–3750 B.C. In this paper both uncalibrated radiocarbon years and calibrated or calendar years are employed; calibrations were done using the Calib 4.0 program (Stuiver et al., 1998a, 1998b). A brief review of events in the Southeast during the Archaic, and particularly the Middle Archaic, follows (Fig. 13.1).

The Early Archaic is traditionally dated in the region from 10,000 to 8000 ^{14}C yr BP (ca. 11,450–8900 cal yr BP) (Smith, 1986; Anderson and Sassaman, 1996, 2004).

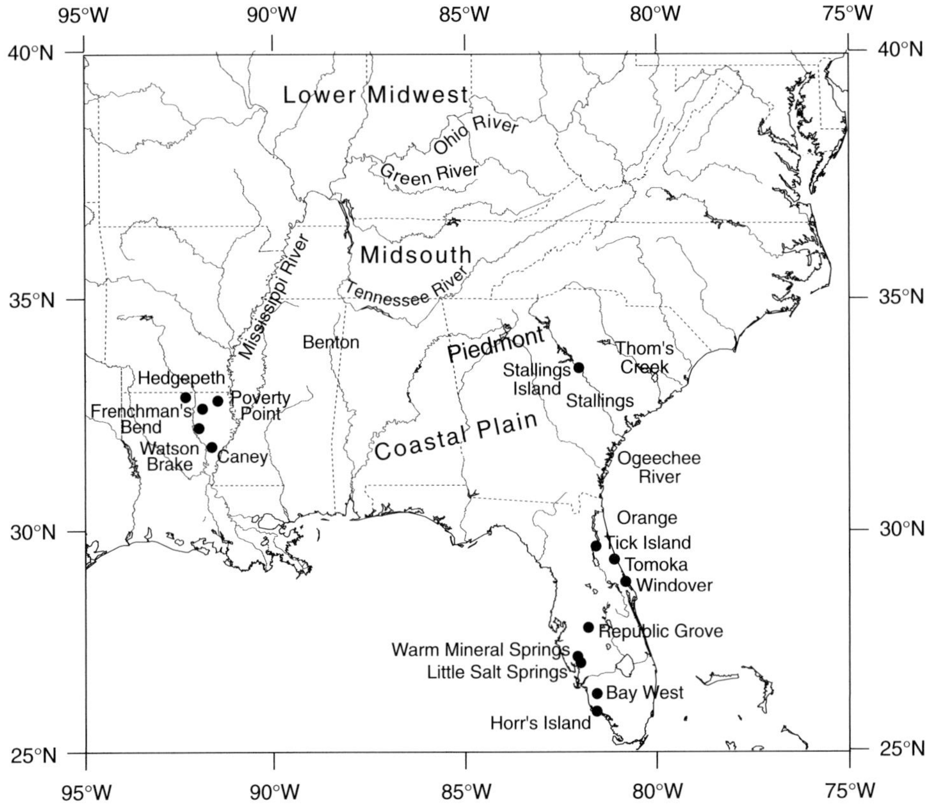

Figure 13.1. Location of archaeological sites and cultures mentioned in the text.

Sites are recognized by the occurrence of successive side- and corner-notched and bifurcate-based points. Early Archaic lithic assemblages were characterized by formal chipped tools fashioned from high-quality stone. Over time carefully crafted formal tools were replaced by more casual tools made on locally available materials. Mixed hardwoods forest were present across much of the region, creating favorable environments for hunting and gathering populations in both riverine and interriverine settings. Widespread use of wild plant foods is indicated by carbonized remains as well as ground and pecked stone plant processing tools. People organized themselves in small bands. Periodically, meetings between widely ranging bands occurred in favored settings, tying together people over large areas, and facilitating the spread of information, mates, and materials. Over time, annual ranges are thought to have grown progressively smaller, so that by the end of the era groups in some areas were likely restricted to within portions of river systems. Large numbers of sites are observed over the landscape.

The Middle Archaic in the Southeast is dated from 8000 to 5000 ^{14}C yr BP (ca. 8900–5750 cal yr BP) (Bense, 1994; Sassaman, 1995, 2005a; Sassaman and Anderson, 1995, 1996, 2004; Anderson and Sassaman, 2004). Interior sites are identified by an array of point forms, reflecting an increasingly diversified cultural landscape. The replacement of mixed hardwood forests by pine forests and cypress swamps across the southeastern Coastal Plain consolidated peoples in river valleys where hardwood forests remained. In these restricted environments, regional population levels appear to have stabilized or decreased somewhat. Large earth and shell midden sites with dense occupational debris and numerous burials appear along some of the major drainages of the Midsouth and lower Midwest, encompassing northern Alabama and Mississippi, all of Tennessee and Kentucky, and the areas just to the north of the Ohio River in southern Illinois, Indiana, and Ohio (Smith, 1986; Steponaitis, 1986; Bense, 1994; Russo, 1996a, 1996b; Sassaman and Anderson, 1996; Milner, 2004a, 2004b). Occupied for extended parts of the annual cycle, these sites served as aggregation loci and possibly as special burial areas (cf. Claassen, 1996; Milner and Jefferies, 1998). While widely ranging foraging groups were still present, they are found in regions outside the lower Midwest and Midsouth, such as in portions of the South Atlantic Slope in Georgia and the Carolinas, or in the Trans-Mississippi uplands of Louisiana, eastern Texas, and southwestern Arkansas (Anderson, 1996; Sassaman, 1995).

Long-distance exchange networks spanning large portions of the eastern United States, including the southeast, emerged by ca. 7500 cal yr BP (Jefferies, 1996, 2004). Goods in circulation included shell from the southern coasts and copper from the Great Lakes. Some items like bone pins, bannerstones (presumed atlatl weights), and elaborate bifaces were circulating, often over smaller areas, indicating more localized exchange networks were also operating. Interaction and exchange enhanced the status of network participants and may have helped reduce conflict and subsistence uncertainty by creating ties between groups.

The construction of earthen mound complexes in the Lower Mississippi Valley and earthen and shell mound complexes in Florida arose between ca. 6000 and 5000 ^{14}C yr BP (ca. 6800–5750 cal yr BP) (Russo, 1994a, 1994b, 1996a, 1996b; Saunders et al., 1997, 2005). These centers are thought to represent the communal action of large numbers of people, the development of more complex organizational forms, and an increased need to define territories and/or alliance networks. Territorial circumscription is suggested by the appearance of evidence for conflict – burials with embedded projectile points, scalping marks, and parry fractures – in some parts of the midcontinent, in the Midsouth and lower Midwest. Variability in mortuary treatment indicates appreciable differences in status were emerging, although evidence of heritable ranking has not been identified. That is, achieved rather than ascribed or hereditary positions are all that are thought to have been present at this time.

All the trends initiated in the Middle Archaic continued to grow in scale over the course of the Late Archaic, from 5000 to 3000 ^{14}C yr BP (ca. 5750–3200 cal yr BP) (Smith, 1986; Sassaman and Anderson, 2004). During this interval essentially modern climate, sea level, and vegetation communities emerged. Mound construction, long-distance prestige-goods exchange, and warfare expanded, culminating in dramatic cultural expressions like Poverty Point, Stallings Island, Green

River/Indian Knoll, Orange, and Horr's Island (Fig. 13.1). Shellfish use in the Midsouth and lower Midwest continued, and use of coastal resources becomes widespread. A major increase in regional population levels is indicated, with sites found in all parts of the landscape.

Wild plant foods were collected in increasing quantity, and by the end of the Late Archaic, between 4000 and 3000 ^{14}C yr BP (ca. 4450–3200 cal yr BP), morphological changes indicative of domestication are evident in a number of local species such as goosefoot, sumpweed, sunflower, and gourds (Smith, 1992). Pottery appeared in the Stallings culture of Georgia and South Carolina about 4500 ^{14}C yr BP (ca. 5150 cal yr BP) and soon thereafter in the Thom's Creek and Orange cultures of South Carolina and Florida, respectively (Sassaman, 1993, 2005a). These cultures, as well as unnamed Ten Thousand Island, Bonita, Cottage, and Reed cultures in South Florida, built extensive large-scale shell and earthworks including mounds and rings (Russo, 1996b, 2004, 2006a). By 3400 ^{14}C yr BP (ca. 3650 cal yr BP) the mounds and earthworks of the Poverty Point culture, a vast exchange network centered on the Lower Mississippi Valley, were among the largest ever built in the East (Gibson, 1996, 2000; Sassaman, 2005b) (Fig. 13.2).

Any summary treatment of the Mid-Holocene Southeast must sacrifice detail for generalization. The long-held notion that Archaic populations were part of a gradual evolutionary trend toward ever-more efficient adaptation to the southeastern environment (e.g., Caldwell, 1958) ignores many of the nonlinear environmental and cultural trends apparent at subregional scales of observation. Likewise, the notion that Mid-Holocene prehistory reflects, in part, cultural responses to a pan-regional warm, dry climate (e.g., Smith, 1986, p. 24; Bense, 1994, p. 74) runs counter to paleoenvironmental research which documents increased moisture in portions of the lower Southeast (e.g., Sassaman, 1995, pp. 141–143; Dye, 1996, pp. 141–142; Watts et al., 1996). Effects of sea level rise were likewise very different on the Atlantic and Gulf coasts. Modern perspectives on the Mid-Holocene must therefore be sensitive to the tremendous natural and cultural diversity of the era. In attempting to explain this diversity, we must be careful not to regard observed differences simply as expressions of subregionalization or cultural isolation, as if divergent trends mimicked processes of speciation. Histories of population interaction at scales sometimes spanning most of the Southeast must be integrated with data on local conditions and adaptations to fully appreciate the conditions resulting in the Southeast's marked cultural variation during this era.

2. Emergent riverine economies and cultural complexity in the midcontinent

The Mid-Holocene was a period of dramatic cultural change in the Midsouth and in the immediately adjacent lower Midwest. During this period ceremonial shell/earthen mound construction was initiated, long-distance exchange networks spanning much of the region appeared, new tool forms, such as bannerstones and grooved axes were adopted, and there was increased evidence for interpersonal

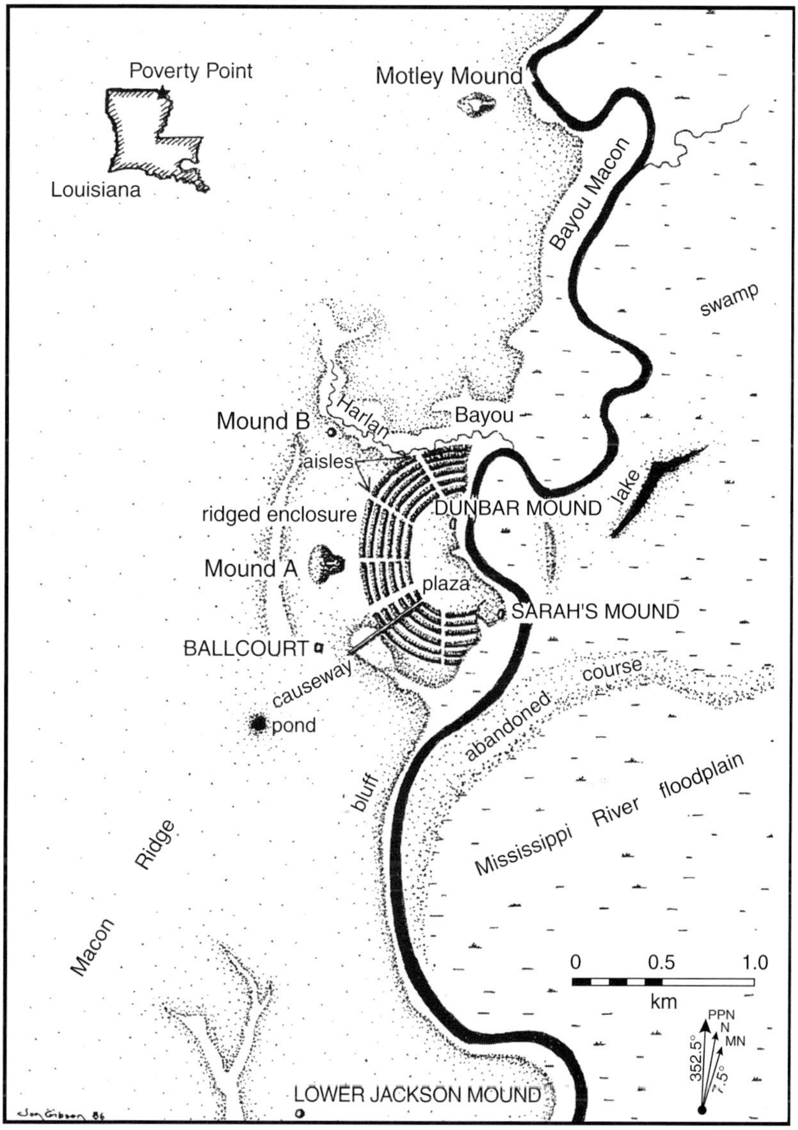

Figure 13.2. The Poverty Point site, Louisiana (drawn by Jon Gibson, from Gibson, 2000, p. 82, courtesy University Press of Florida).

violence or warfare (Griffin, 1967; Smith, 1986, 1996; Steponaitis, 1986; Jefferies, 1995, 1996, 2004; Sassaman, 1996, 2000, 2004, 2005a, 2005b; Sassaman and Anderson, 1996, 2004; Anderson and Smith, 2003a, 2003b; Anderson and Sassaman, 2004; Gibson and Carr, 2004; Milner, 2004a, 2004b). All these factors indicate local cultures were growing in scale and organizational complexity.

Climate appears to have helped shape some of these developments. In the lower Midwest and Midsouth, Mid-Holocene climate appears to have been hotter and dryer than at present, leading to a reduction in upland vegetation, increased surface erosion, and aggrading floodplains (Knox, 1983; Delcourt and Delcourt, 1987, 2004; Jacobson et al., 1987; Webb, 1987, 1988; Webb et al., 1993; Schuldenrein, 1996; Stein, 2005). Formation of backwater slough habitat enhanced floodplain productivity, while shoal environments favored freshwater shellfish. These warming and drying trends may have rendered riverine areas more favorable and upland areas less favorable to human populations.

As early as 7300 ^{14}C yr BP (ca. 8100 cal yr BP), some populations in the Midsouth began to occupy riverine sites for extended stays, or otherwise returned regularly enough to create midden deposits. Shellfish began to be collected in significant numbers (Marquardt and Watson, 1983, 2005; Claassen, 1996; Dye, 1996). Some of the earliest shell midden sites (which are actually typically accumulations of both shell and earth, or are sometimes middens with occasional scattered shell) were occupied by Morrow Mountain phase populations of the Midsouth, beginning about 7000 ^{14}C yr BP (ca. 7800 cal yr BP) (Dowd, 1989). Over the ensuing millennium, evidence for substantial architecture and intensive occupation appears at terrace-edge and lakeside locations in the Lower Midwest (e.g., Jefferies and Butler, 1982; Brown and Vierra, 1983), as well as an increasing number of riverine sites across the midcontinent.

The large shell middens of the Midsouth, such as those along the Green River in Kentucky and the Tennessee River of northern Alabama, are assumed to reflect aggregation loci where group ceremony and ritual occurred (Claassen, 1991a, 1991b, 1996; Sassaman, 1993, 2005a; but see Milner and Jefferies 1998; Crothers 1999, 2004; Milner, 2004a, 2004b; Marquardt and Watson, 2005, p. 636, for different or "minimalist" opinions about the extent of ceremony that may have occurred). Burial in shell mounds also likely served to demarcate territories and bind people together in an era of increasing population and competition, as reflected in the widespread evidence for prestige goods exchange (Brown, 1985; Johnson and Brookes, 1989; Jefferies, 1995, 1996, 2004; Meeks, 1999) and the somewhat more restricted evidence for conflict (Smith, 1996). Much of the long-distance exchange appears to have been driven by the efforts of individuals to enhance their personal status, although there is no evidence for hereditary leadership positions. Achieving success in warfare may have been another means of enhancing personal status, just as it was in later prehistory and during the early historic era in the region (e.g., DePratter, 1983, pp. 44–67; Dye, 1990; Anderson, 1994, pp. 132–135). The occurrence of burials and cemeteries at many sites in the Midsouth, particularly when coupled with evidence for long-distance exchange, is seen as evidence for an increasing complex and sedentary lifestyle (Marquardt and Watson, 1983, 2005; Marquardt, 1985). Archaeologists debate whether overall conditions "pushed" or "pulled" Mid-Holocene populations into riverine zones, that is, whether peoples were forced to adopt floodplain resources due to desiccation of the uplands, or were attracted to them as part of a general strategy of increasing sedentism (cf. Brown and Vierra, 1983, pp. 167, 190; Carmichael, 1977).

Just as earthen mounds are present in some parts of the region and not in others (see below), shell middens are also unevenly distributed across the interior

Southeast, and are in fact absent in some areas where mollusks were prevalent (Claassen, 1996; see also Peacock, 1998, 2002). One reason shell middens are unevenly distributed along the coastal portions of the region is likely because they are drowned or destroyed. But in the interior Southeast it appears shell middens simply were not accumulated in some areas. Shell midden/mound construction, if indeed a symbolic strategy to link people together in regional alliances or to demarcate territories, may have been necessary only where population levels were high and competition for resources was intense. The presence of large shell middens at some sites may indicate particularly intense competition. The mounding of earth and shell in the Midsouth at this time parallels comparable developments in coastal areas and the Lower Mississippi Valley. As we shall see, collective ceremonial behavior was sufficiently sophisticated to construct enduring large-scale monuments in many parts of the region (Russo, 1996a, 1996b, 2004, 2006a; Gibson and Carr, 2004; Sassaman, 2004; J. Saunders, 2004).

In the Midsouth, Morrow Mountain phases were replaced after about 6500 ^{14}C yr BP (ca. 7400 cal yr BP) by a series of local cultural traditions, one of the most distinctive of which is the Benton phase of the upper Tombigbee, middle Tennessee River, and middle Cumberland river area, dating from ca. 6000 to 5000 ^{14}C yr BP (ca. 6850–5750 cal yr BP), with most of the dates falling between ca. 5700 and 5200 ^{14}C yr BP (ca. 6500–6000 cal yr BP) (Johnson and Brookes, 1989; Jefferies, 1996, pp. 228–230; Meeks, 1999). Exchange connections in this "Benton Interaction Sphere" are subregional in scale, extending over no more than a few hundred kilometers. Exotic artifacts from much greater distances, such as Great Lakes copper or marine shell from the Gulf or Atlantic coasts, are rare. Benton is characterized by elaborate burial ceremonialism, however, involving the placement of caches of large and elaborate bifaces with the dead. It has been suggested that interaction networks helped tie people together, and emerged in part to alleviate subsistence uncertainty (Braun and Plog, 1982). Subsistence uncertainty in the Mid-Holocene Southeast may have been brought about by rising regional population levels and broad scale climatic conditions that combined to place pressure on available food resources, although whether this was actually the case remains to be documented empirically. Throughout the Midsouth, riverine adaptations continued at varying scales through the Late Archaic period. The overall trend was for relatively intensive riverine economies, possibly territorially and ethnically circumscribed, with intergroup interactions ranging from violent to peaceful.

The intensive collection of a wide range of wild plants is documented during the Mid-Holocene in the Midsouth and lower Midwest, and morphological changes indicative of domestication, such as an increase in seed size or a decrease in seed coat thickness, are evident in several local species by the end of the Late Archaic period, from ca. 4000 to 3000 ^{14}C yr BP (ca. 4450–3200 cal yr BP) (Smith, 1992). Local plant species of this "Eastern Agricultural Complex" include sunflower (*Helianthus annus*), sumpweed (*Iva annua*), goosefoot (*Chenopodium berlandieri*), maygrass (*Phalaris caroliniana*), knotweed (*Polygonum erectum*), little barley (*Hordeum pusillum*), and local cucurbits or gourds. Intensive collection of these plants, a process that would have eventually led to domestication, probably dates back to the Mid-Holocene.

Eastern North America is one of the few places in the world where plant domestication can be explored in detail, along with the changes in culture that occurred with a shift from hunting and gathering to agriculture. Exactly how the process of domestication occurred is unknown. There is no doubt that plant resources were intensively exploited by Archaic populations in Eastern North America. Domestication is thought to have occurred in areas of greatest population density and pressure, as part of an effort to maximize subsistence resources. There is some evidence to suggest that as domesticated plant foods became increasingly important, the use of shellfish declined in the interior (Claassen, 1991a, 1991b; but see Peacock, 1998, 2002). Exactly how the process of domestication occurred, and how it effected the collection of other types of subsistence resources, however, is not well understood.

One theory posits that plants of the Eastern Agricultural Complex were first gathered, and then encouraged to grow, in the disturbed habitats found in and near the major earth and shell mound sites of the Midsouth and lower Midwest (Smith, 1992). These plants thrive in such habitats, and their close proximity to burgeoning human populations would have prompted intensifying exploitation over time, leading to domestication (Smith, 1986, 1987, 1989, 1992). An alternative to this "weedy floodplain" hypothesis suggests that these plant foods were instead collected in adjoining upland areas, where the peoples living at lowland shell middens would have likely ranged from time to time, perhaps seasonally (Fritz, 1990; Gremillion, 1996, 2002; Marquardt and Watson, 2005, p. 630). This counter argument has arisen because there is comparatively little evidence for domesticates at early shell midden sites, and a great deal of evidence for them in upland cave and rockshelter settings. These distributions warrant explanation, since they do not appear to reflect factors of differential preservation, and since appreciable effort has been directed to the recovery of plant remains in recent research at Shell Mound Archaic sites in the Midsouth (Marquardt and Watson, 1983; Gremillion, 1996; Crothers, 1999; Marquardt and Watson, 2005, pp. 630–631); the only common domesticates found in floodplain Archaic shell midden settings, gourds, may have been used for containers and fishing floats rather than as a food source (Fritz, 1999; Marquardt and Watson, 2005, pp. 630–631).

The increased attention given to the local starchy and oily seeds during the middle and later Holocene may have been due, in part, to increased subsistence pressures created by growing, and somewhat circumscribed, human riverine populations. Interestingly, while domesticates were assuming an increasing importance in the diet in some parts of the region, in other areas hunting and gathering continued. In particular, there is little evidence for cultivation until much later in time in the Lower Mississippi Valley, across thinly populated areas of the Atlantic and Gulf Coastal Plains, and in a number areas on the Gulf and Atlantic shores occupied by peoples who never became reliant on agriculture (e.g., Yarnell and Black, 1985; Fritz and Kidder, 1993; Gremillion, 2002). The need for domesticates may have been unnecessary in areas rich in natural resources, or where population density was low or only minimally circumscribed.

3. The initiation of mound building and other monumental construction

As Mid-Holocene population levels grew, competition and interaction between groups appears to have likewise increased, as indicted by the appearance of exchange networks, warfare, as well as the construction of complex mound centers. One of the most exciting archaeological discoveries in recent years is the recognition that earthen mound construction in the southeastern United States extends well into the Mid-Holocene, to as far back as 6000 ^{14}C yr BP (ca. 6800 cal yr BP). With a few enigmatic exceptions, initial mound building in the Southeast had long been thought to date primarily to the Woodland period, after about 3000 ^{14}C yr BP (ca. 3200 cal yr BP).

An exceptional and until recently poorly understood precursor to the Woodland earthen mound building tradition was the Poverty Point culture of northeast Louisiana and vicinity, dating to about 3400–2900 ^{14}C yr BP (ca. 3650–3050 cal yr BP) (Gibson, 2000; Kidder, 2001). So unusual was this complex of pre-Woodland sites with elaborate mounds that for many years, researchers openly acknowledged that it did not "fit" with archaeological understanding of southeastern prehistory (Ford and Webb, 1956, p. 14; Gibson, 1996, p. 288). The Poverty Point type site is one of the largest earthen mound complexes ever built in North America, and includes a massive bird effigy mound some 21 meters high, a number of smaller mounds, and six semi-circular rings some 1200 meters across (Gibson, 2000; Kidder, 2001). A number of contemporaneous mound centers are present in this part of the Lower Mississippi Valley, making up an extensive interaction network that made use of raw materials from across much of the lower Southeast and up the Mississippi River into the Midwest. How such large and complex mound centers could arise among presumably preagricultural, egalitarian hunting-gathering populations was a puzzle for many years.

Massive earthen mound constructions predating Poverty Point by up to three millennia have now been documented at a number of locations in the Lower Mississippi Valley (Russo, 1994a, 1996a; Saunders et al., 1997, 2005). Early mound complexes dating between ca. 5500 and 5000 ^{14}C yr BP (ca. 6300–5750 cal yr BP) include Caney, Frenchman's Bend, Hedgepeth, and Watson Brake (Saunders et al., 1994, 1997, 2005; Russo, 1996a; J. Saunders, 2004) (Figs. 13.3 and 13.4). These are not isolated small mounds, but complexes whose construction and maintenance required the cooperative interaction of large numbers of people. Watson Brake, for example, which was built between ca. 5400 and 5000 ^{14}C yr BP (ca. 6250–5750 cal yr BP), includes 11 mounds up to 7 m in height connected by a circular earthen embankment some 280 m across. As at Poverty Point, appreciable occupational debris is evident. Paleosubsistence analyses undertaken to date suggest Watson Brake was used seasonally rather than year-round (Saunders et al., 1997), although

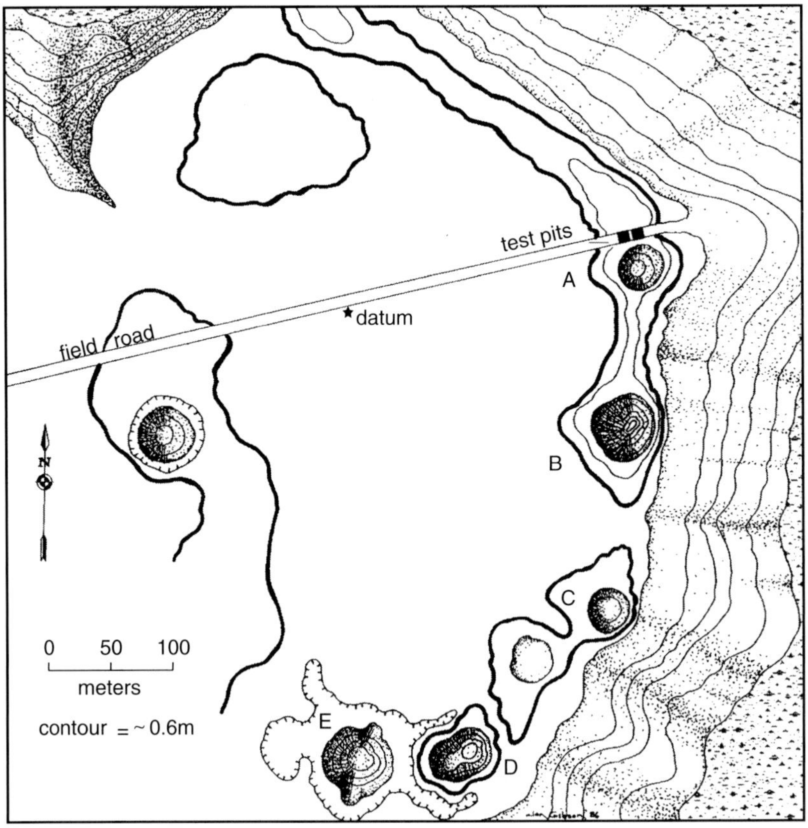

Figure 13.3. The Caney Mounds, Louisiana (drawn by Jon Gibson, adopted from Gibson, 1994, p. 173, courtesy *Southeastern Archaeology*).

it must be cautioned that the area that has been examined at the site is extremely small. No evidence for long-distance exchange or elite (or any) burials has been found at the site (J. Saunders, 2004).

To some, Poverty Point is now no longer an enigma, but rather a high point in a long and quite probably sacred and symbolically charged tradition of mound building in the region (Gibson, 1996, 2000; Clark, 2004; Sassaman, 2005a, 2005b; Sassaman and Heckenberger, 2004). Little is currently known about the Mid-Holocene cultures that initiated this tradition, but it appears they made extensive use of rich local wetland habitats. Competition for status between individuals and groups may have driven some of this construction activity (cf., Anderson, 2002, Gibson, 2004, J. Saunders, 2004). Minimally, the mound complexes suggest the involvement of peoples over a large area, and may demarcate the emergence of more complex forms of social organization, such as tribal-level entities, rather than the band-level groupings of 25–50 people that were long assumed by scholars to have been the primary social grouping of Early and Middle Archaic populations (Bender, 1985; Anderson, 2002, 2004). That is, the complex mound centers of the Middle Archaic may reflect the actions of many band-sized groups, or tribal

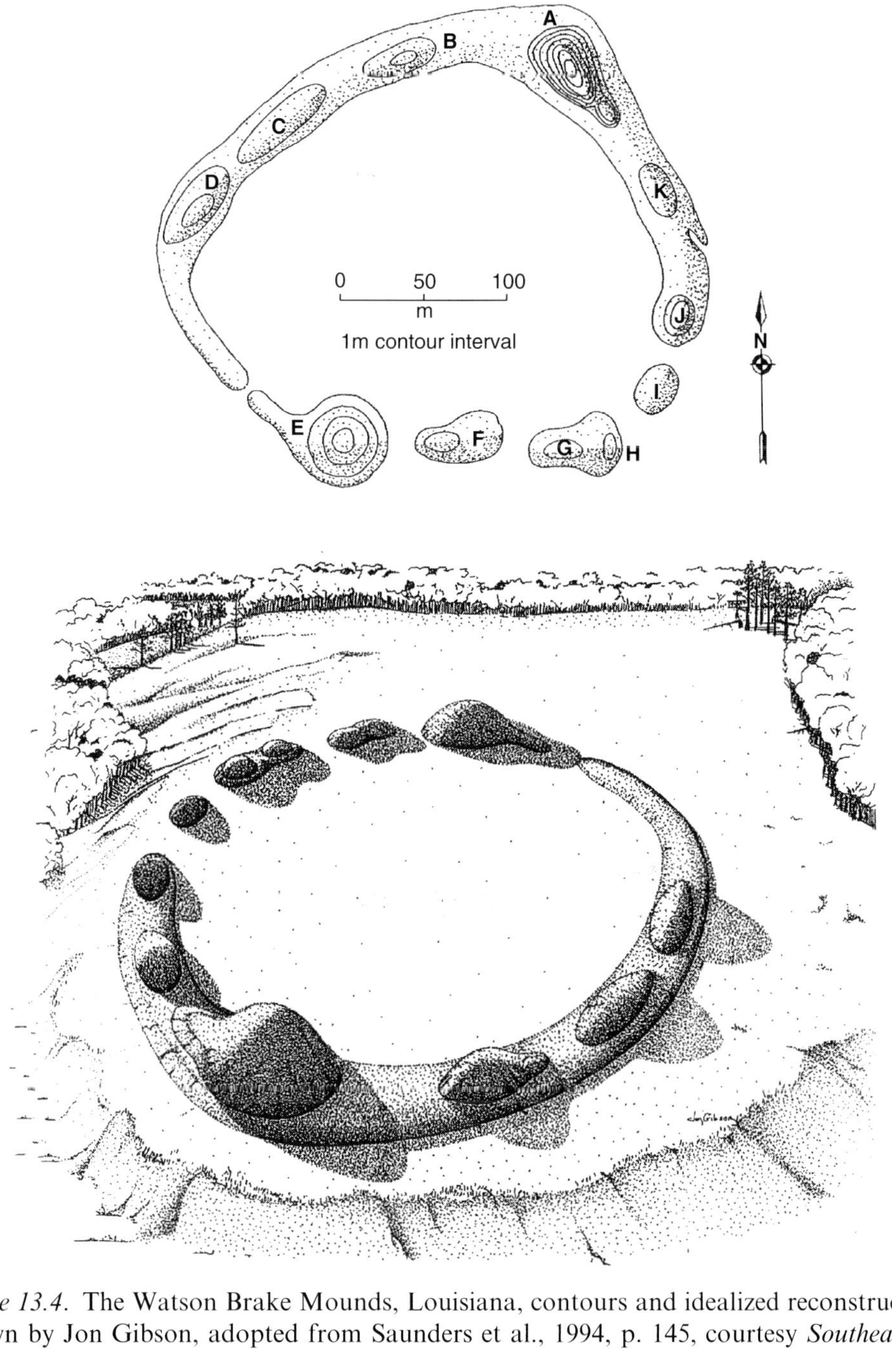

Figure 13.4. The Watson Brake Mounds, Louisiana, contours and idealized reconstruction (drawn by Jon Gibson, adopted from Saunders et al., 1994, p. 145, courtesy *Southeastern Archaeology*).

segments, bound together into a new social formation by the collective ceremonial activity represented by the construction of these earthen mound complexes, and the communal feasting, ritual and other behavior that likely took place at them. Mound construction and associated ritual, in this view, helped create pan-tribal social institutions linking peoples from across large areas together (Anderson, 2002, 2004; Widmer, 2004). Besides enhancing the status of tribal leaders, interaction and exchange over large areas likely helped reduce the possibility of warfare and alleviate subsistence stress, by creating ties between different groups (Braun and Plog, 1982; Jefferies, 2004). When resources in one area grew scarce, the existence of alliances would facilitate temporary group relocation into more favored areas, until the shortfall passed.

Early mound construction has been identified in several other parts of the Eastern North America, notably in coastal areas where, as in the Midsouth, both shell and earth were sometimes used to create monumental accumulations in or near villages or aggregation sites. In terms of architectural complexity, area, and volume, the largest of the coastal shell ring sites compare favorably with the Middle Archaic mound complexes of the lower Mississippi valley (Russo, 2006a, 2006b). The largest coastal ring sites contain up to 14 individual rings of heights up to 6 m and are associated with other earthen/shell constructions including mounds, causeways, ridges, ramps, paths, plazas, and extensive sheet and mounded middens (Russo, 2006a). In total, over 50 ring sites have been identified in South Carolina, Georgia, and Florida (Russo, 2006a) (Figs. 13.5 and 13.6). Horr's Island in southwest Florida is one of the largest, consisting of a complex arrangement of mounded

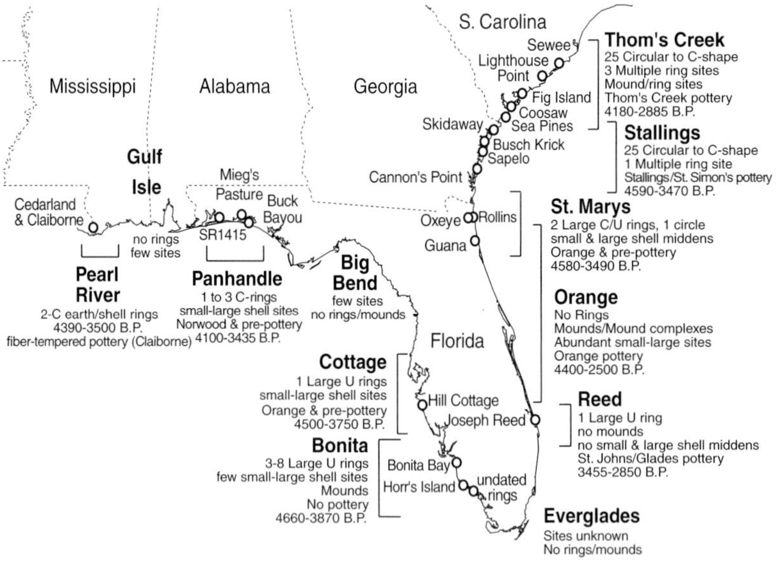

Figure 13.5. Southeastern United States coastal shell ring and other culture areas. Radiocarbon age ranges are conventional dates and do not include standard deviations (courtesy National Park Service, from Russo, 2006a).

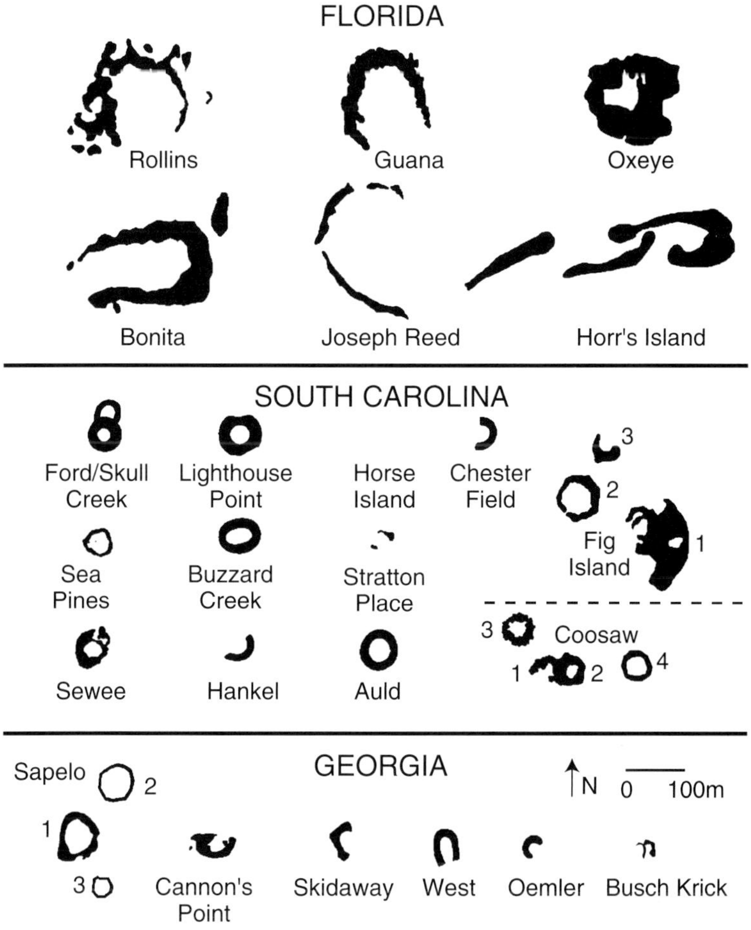

Figure 13.6. Footprints of Middle and Late Archaic shell ring and midden sites from Florida, Georgia, and South Carolina (courtesy National Park Service, from Russo, 2006a).

shell rings and conical shell/earth mounds built between 4600 and 5000 cal yr B.P. (Russo, 1994b, 1996b). Paleosubsistence analyses indicate occupation of the site was year-round – the earliest evidence for sedentism found to date in eastern North America. Even earlier shell and earthen mounds are known from northeast Florida, at Tomoka (Piatek, 1994), Tick Island (Russo, 1994b, pp. 106–108; Aten, 1999), Thornhill Lake (Endonino, 2003), Bluffton (Sears, 1960), and at Hontoon Island and vicinity (Sassaman, 2003; Randall and Sassaman, 2005), among other locales. Ceremonial functions have been suggested for these sites, and for many of the Archaic shell rings of the South Atlantic coast that date from ca. 5000 to 3000 B.P. (Russo, 1996b, 1999, 2004, 2006a; Russo and Heide, 2001, 2002a, 2002b; Saunders and Russo, 2002).

At mound sites in Florida, burials represent the obvious ceremony at Tick Island, Bluffton, and Thornhill Lake. The ceremonial use of other Florida and coastal early mound sites remains to be definitively established. Although burials or isolated human remains have been found in a number of mounds, they often appear intrusive, while other mounds lack burials altogether (Russo, 1991). Unfortunately, most mounds have not been intensively investigated to draw any conclusions as to their use (Dickel, 1992; Piatek, 1994; Randall and Sassaman, 2005). At Horr's Island Mounds A and B, burials were identified, but post-dated the construction of the mounds. This suggests that if the mounds associated with the ring site were not originally built for burial, they were certainly used for burial within a few generations of their construction. That is, they seem to have become ancestral architectural memorabilia whose original purpose may have changed through time to become a place of ceremonial burial (e.g., Dillehay, 1990, p. 233; Russo, 1991, 2006a).

As for the ceremonial use of shell rings, it is unclear whether burial was one of the ceremonies practiced. Certainly, complete burials have never been found in the rings themselves and it remains something of a mystery as to what peoples of the Orange, Stallings, Thom's Creek, and other ring-producing cultures were doing with their dead. Isolated remains are common, but consist mostly of teeth, which do not in and of themselves argue strongly for a mortuary function for shell rings. Of the 30 citations of human remains recovered from 15 shell ring sites, Russo (2006a) found that possibly only five occurrences of isolated remains came from secure ring contexts.

One ceremony with more substantial evidence is feasting. While some portions of shell rings evidence construction using the refuse of quotidian meals (e.g., Trinkley, 1985; Thompson, 2006), most rings are made from massive pilings of shell resulting from large-scale, short-term feasts (Russo, 2004; R. Saunders, 2002, 2004). Ethnographically, large-scale feasts among tribal societies are associated with ceremony and ritual, and, hence, the case for shell rings as ceremonial sites has been forwarded (Russo, 2004; R. Saunders, 2004). The presence of both large- and small-scale food deposits suggests that rings served multiple functions, including both habitation and feasting, which may have co-occurred or changed through time (Russo, 2006a; Thompson, 2006).

Whereas increased population and tribal warring were common elsewhere in the Southeast, the competitive feasting activities of transegalitarian groups on local scales likely served to mitigate conflicts, such as along the South Carolina coast where shell rings and ring complexes are densely packed. Mounds associated with ring sites, and constructional asymmetries found within most of the ring sites suggest, minimally, situational status differences among the ring builders, perhaps tolerated by an otherwise egalitarian ethnic only in times of ceremony and feasting. At some ring sites, complexity and asymmetries in mound and ring architecture suggest hierarchical social asymmetries that would not subsequently again be seen until the Woodland period (Russo, 1999, 2002, 2004, 2006a; R. Saunders, 2002). That none of the rings and mounds have revealed evidence of caching, storage, or burial with exotic goods obtained from involvement in geographically distant trade networks suggests that any incipient ranking that may have occurred at shell ring sites was dissimilar from the aggrandizing chiefdoms that would come to typify the later Southeastern socio-political landscape.

4. The persistence of interriverine forager adaptations

Concurrent with the more complex societies evident in several parts of the Southeast during the later Mid-Holocene, less complex social groups were also present, usually in more geographically marginal areas on the region's peripheries (Sassaman, 1983, 1991, 1995, 2001a, 2001b; Amick and Carr, 1996; Anderson, 1996). These tended to be highly mobile, small-scale foraging groups that appear to have made wide-ranging and generalized use of upland habitats. Some occupational expressions may be seasonal encampments of river-based populations. On both the eastern and western margins of the region, however, they appear to represent distinctive adaptations, local societies perhaps at no more than a band level of social organization (Sassaman, 1991, 1995; Anderson, 1996, p. 165). The large numbers of Middle Archaic sites in the uplands of western Louisiana, for example, appear to reflect a continuation of relatively uncomplicated foraging adaptations from the Early Archaic period (Anderson and Smith, 2003b, pp. 369–377), although closer to the Mississippi River such sites may be the encampments of the same populations that built mound complexes like Watson Brake.

The persistence of organizationally uncomplicated foraging populations is best documented along the South Atlantic Slope of Georgia and the Carolinas (Sassaman, 1983, 1985, 1991, 1995; Sassaman and Anderson, 1995). Throughout the Piedmont and into the Blue Ridge physiographic provinces, for example, sites with Morrow Mountain points are distributed across interriverine landforms. These sites are highly redundant, consisting of few formal tools, a high level of expedient tools and minimal evidence for midden. Riverine sites in this region rarely contain midden deposits, shellfish, or evidence for intensive or continuous occupation. These sites are thought to reflect a relatively uncomplicated foraging adaptation to Ridge and Valley and Piedmont habitats; uncomplicated, at least, when compared with cultural developments in the interior midsouth (Sassaman, 1995, pp. 182–183, 191; see also Anderson, 1996, p. 164–165). Vegetational history across this expanse is not well documented, although some have suggested that oak-hickory forests remained dominant throughout the Mid-Holocene (Delcourt and Delcourt, 1981, 1987; Jacobson et al., 1987).

Differences in respective environments may account for the pronounced differences in cultural expressions, with Midsouth Shell Mound Archaic groups adapted to patchy, high-yield riverine habitats, and Piedmont/Ridge and Valley groups adapted to dispersed, homogenous, lower-yield oak-hickory forest habitats. However, the presence of Morrow Mountain points in both areas also encourages consideration of possible sociocultural or historical factors that might have been operating (Sassaman, 1995, 2001a, 2001b, 2005a). The intensive economies and presumably stressful circumstances Midsouth groups faced (i.e., warfare, circumscription) may have resulted in some population fissioning and movement. Morrow Mountain groups on the South Atlantic Slope may well have opted out of the complex sociopolitical environment of the Midsouth or, given their marginal geographic location, may not have been compelled to participate in this way of life. Thus, by the Mid-Holocene, historical conditions at the regional scale appear to have had a significant effect on adaptations on the local scale. To understand cultural developments in one part of the region, we need to be aware of conditions over a much larger area (Sassaman, 1995).

While Piedmont habitats teemed with activity during the Morrow Mountain phase on the South Atlantic Slope, even if by groups less complexly organized than those in the interior of the continent, the adjacent Coastal Plain was unquestionably underutilized. Large numbers of sites are known in this area during the preceding Early Archaic period (Anderson and Hanson, 1988; Sassaman et al., 1990; Anderson, 1996). Beginning no later than 8000 ^{14}C yr BP (ca. 8900 cal yr BP), pine began to expand in the interriverine areas at the expense of mast-producing species such as oak and hickory, and cypress in riverine zones of the Coastal Plain, something attributed to the global warming (resulting in both increased aridity and increased storm frequency) occurring at this time and to an increase in fires accidentally or intentionally set by Indian populations (Delcourt and Delcourt, 1987, 2004; Webb et al., 1993; Watts et al., 1996; Goman and Leigh, 2004). The expansion of pine forests, which produced lower quantities of food resources of use to human populations, appears to have led to a general depopulation of the interior Coastal Plain and an increased use of Piedmont and other interior areas, and probably made coastal areas more attractive as well (Anderson, 1991; Kowalewski, 1995, pp. 162–165; Sassaman 1995, 2001a). This population relocation is perhaps the most obvious impact climate change had on regional cultures during the Mid-Holocene.

Although drier conditions may have prevailed during the Mid-Holocene in the Midwest and parts of the Southeast, sediment records from the Atlantic Coastal Plain of North Carolina, South Carolina, and northwest Georgia suggest that moist conditions locally were uninterrupted (Seielstad, 1994; Watts et al., 1996; Goman and Leigh, 2004). Even the numerous interriverine wetlands known as Carolina Bays appear to have been well watered throughout this interval (Brooks et al., 1996). Other support is found in analysis of paleochannel morphology. For example, large meandering paleochannels on the floodplain of the Ogeechee River in southeast Georgia show wet paleoclimate during the early to Mid-Holocene from roughly 8500 to 4500 ^{14}C yr BP (ca. 9500–5150 cal yr BP) (Leigh and Feeney, 1995). All these data corroborate paleoclimate simulations and reconstructions derived from global circulation models and pollen studies. In particular, simulations indicate that the lower Southeast U.S. was much wetter from 9000 to 3000 ^{14}C yr BP (ca. 10,200–3200 cal yr BP) than at present because of increased solar radiation and intensified summer monsoon conditions (Kutzbach, 1987).

Thus, despite the diminished regional availability of mast resources, ponds and rivers throughout the Coastal Plain were potentially rich, albeit localized, resource patches of aquatic species, as well as mesic and hydric vegetation. Given this, one would be hard pressed to argue that Coastal Plain habitats were underutilized during the Mid-Holocene because it was an impoverished "pine barren" (e.g., Larson, 1980). Instead, limited use of the province, coupled with increased use of adjacent Piedmont habitat, may be traced to differences in resource structure, not composition. The increased range of soil moisture and surface water conditions evident in the Coastal Plain encouraged more heterogeneous landscapes compared to the relatively homogenous Piedmont (Hoover and Parker, 1991). Human use of the patchier Coastal Plain would have required levels of settlement permanence and logistical organization (sensu Binford, 1980) well beyond that necessary in the Piedmont. That productive Coastal Plain habitat was underutilized suggests that cultural traditions of land use emphasized freedom from constraints on mobility and economic demands, not limited local resource potential (Sassaman, 2001a, 2001b).

5. Pond, river, and coastal settlement in Florida and adjoining areas

The general lack of surface water in peninsular Florida during the early Holocene encouraged settlement around sinks and ponds. By shortly after 8000 ^{14}C yr BP (ca. 8900 cal yr BP), these occupations involved the interment of humans in pond/bog settings at places such as Windover (Doran and Dickel, 1988; Doran et al., 1988; Doran, 2002). Archaic basketry, canoes, carved wooden sculpture, tools, utensils, and clothing, as well as superbly preserved human tissue have all been found in wet sites, of which cemeteries like Windover and Little Salt Springs in Florida are the best known examples (Clausen et al., 1979; Doran et al., 1988; Doran, 2002; Purdy, 1988, 1991). As more well preserved human remains are found, DNA analyses may help delimit genetic relationships between populations. Recent wet-site work has also shown that the use of dugout canoes in Florida has appreciable antiquity. The earliest specimen currently known dates to roughly 6000 ^{14}C yr BP (ca. 6800 cal yr BP), and numerous canoes have been found dating between 5000 and 2300 cal yr BP (Newsom and Purdy, 1990; Milanich, 1994, p. 70; Wheeler et al., 2003a, 2003b, pp. 543–544).

The overall moister conditions of lower southeastern environments in the Mid-Holocene were not apparently present or of much significance in early Mid-Holocene Florida, where karst substrate and sea level were critical factors determining the availability of surface water. Some of peninsular Florida's cenotes and sinkholes have produced evidence for fluctuating water levels, and their impact on human populations, back to the late Pleistocene. At the 60-m deep Little Salt Springs (Clausen et al., 1979), water rose to present levels by 8500 ^{14}C yr BP (ca. 9500 cal yr BP), but fell some 8 m by 5000 ^{14}C yr BP (ca. 5750 cal yr BP). A drying trend recorded in peat communities marginal to the cenote peaked at about

7000 ^{14}C yr BP (ca. 7800 cal yr BP). Drops in water levels in the Everglades are further evidence of later Holocene fluctuations.

The interment of humans in subaqueous graves at places such as Little Salt Springs, Windover, Bay West, and Republic Groves, in what appear to have been marked cemeteries/burial areas, suggests territorial marking perhaps associated with either permanent or repeated, intensive occupation of the surrounding area (Russo, 1996b; Doran, 2002). The latter two sites have produced radiocarbon samples and diagnostic bifaces dating from about 7000 to 5700 ^{14}C yr BP (ca. 7800–6475 cal yr BP). The Newnan projectile point type from these sites is also among the more widespread types in Florida, found not only at pond sites, but also across much of the interior "uplands" of north-central Florida. Unlike the Coastal Plain of eastern Georgia and the Carolinas, the interior uplands forests of Florida would not be dominated by southern pine communities until the very end of the Mid-Holocene about 5000–4500 ^{14}C yr BP (ca. 5750–5150 cal yr BP) (Watts et al., 1996). Thus, despite potentially drier conditions, interior Florida appears to have supported a richer mast canopy than points farther north, and hence encouraged widespread, albeit low-intensity use of the interior uplands when compared to the riverine and coastal zones at the end of the Mid-Holocene.

The St. Johns River of east Florida is the earliest riverine environment in the lower Southeast with the massive midden deposits that characterize the extensive shellfishing that first occurred in the Mid-Holocene. The river is an extremely low gradient stream flowing south to north paralleling the Atlantic coast for 200 miles. While nearly dry during the Early Archaic period, the river valley flooded under rising sea levels and increased artesian flow during the Mid-Holocene allowing productive estuaries to expand at its northern mouth, vast lakes to develop in its center, and extensive marshes to spread across its southern-most reaches. By 6000 cal yr BP (McGee and Wheeler, 1994, p. 344), the central region of the river valley was being extensively exploited for fish and banded mystery snail (*Viviparus georgianus*); the northern tidewaters were exploited for oyster, clam, coquina, and fish by 5600 cal yr BP (Russo, 1992; Russo and Saunders, 1999); and by 4300 cal yr BP the southern marshes were home to numerous communities living off of mussel and other freshwater fisheries (Russo, 1986). Archaic shell midden sites along the river are far more abundant than along the coast where studies indicate that the large sites with monumental architecture in this region were occupied year-round (Russo, 1992; Russo and Ste. Claire, 1992). Unfortunately, no successful seasonality measure of the mystery snail and mussels that dominate the St. Johns middens has been developed, and determining permanent occupation along the river has been difficult. Nonetheless, the limited data have indicated permanent occupation of the river by Archaic groups (Russo, 1986; Russo and Ste. Claire, 1992, p. 344; Russo et al., 1992; Newsom, 1994, pp. 414–416; Wheeler et al., 2003a, pp. 154–155).

While the earliest Florida Atlantic coast site dates to 5600 cal yr BP (Russo and Saunders, 1999), even earlier settlement occurred on the Gulf coast of Florida. Shell middens dating as early as 6700 cal yr BP have been identified on high dune formations (Milanich et al., 1984, p. 270; Russo, 1996a, p. 263). By 5000 cal yr BP mound and shell ring complexes began to appear and were widespread from South Carolina to Mississippi by 4000 cal yr BP (Trinkley, 1980a, 1980b; Webb, 1982, p. 34; Bruseth, 1991, p. 15; Thomas and Campbell, 1991; Russo, 1996a, 1996b, 2006a; Russo and Heide, 2001, 2002a, 2002b; Saunders and Russo, 2002). Compared to normative assumptions about the development of the modern coastlines and their initial settlement (i.e., that extended settlement along the littoral was unlikely until seas levels stabilized near their present level), the oldest of these dates are extremely early for the coastal zone of the Southeast (cf. Rouse, 1951; Widmer, 1988). It is thus unclear if shell midden sites dating to 6700 or 5600 cal yr BP represent the earliest coastal occupations of the Mid-Holocene or the earliest extant remnants of coastal settlement whose initial occupations remain unknown.

Microtidal conditions (less than 2 meter tidal range) characterize the entire coast of the Florida peninsula today, and almost certainly in the Mid-Holocene as well. Karstified carbonate strata and a general lack of relief characterizes the peninsula, and relatively little sediment reaches the coast. These conditions appear to have enabled some of the earlier coastal sites in Florida to survive alluvial burial (i.e., flooding by rising sea levels) better than sites in the barrier island chains of Georgia and the Carolinas and along the central Gulf coast, where the earliest sites are more recent, between 4000 and 4500 cal yr BP. Also helping to account for Florida's durable early coastal sites are their locations on unusually high topographic formations adjacent to, but largely unassailed by rising seas (Milanich et al., 1984; Russo, 1991, 1996b).

Unfortunately, except for these few sites in Florida, evidence for Mid-Holocene use of marine resources has likely been lost by sea level rise in other areas of the Southeast (Blanton, 1996; Russo, 1996b). The Mid-Holocene was generally a time of moderate sea level rise. During the early Holocene, sea levels in the Southeast rose at a rate of as much as 1 cm a year, dropping to as little as 25 mm per year by about 8000 ^{14}C yr BP (ca. 8900 to cal yr BP). From 6000 to about 3000 ^{14}C yr BP (ca. 6800–3200 cal yr BP), sea level rose at a rate of no more than 3 mm a year (Colquhoun and Brooks, 1986). On the Atlantic coast, the slow sea level rise combined with a relatively steep offshore gradients precluded significant ocean transgression over the shoreline, and estuaries and barrier islands approximated the configurations they currently hold. On the Gulf shore, a lower gradient under a transgressing sea likely resulted in the submergence and erosion of all but those archeological sites situated on the rare topographic high. While some have suggested that prior to 4500 ^{14}C yr BP (ca. 5150 cal yr BP), the rate of sea level rise was too rapid to promote productive habitat for shellfish and other coastal food resources (DePratter and Howard, 1980; Widmer, 1988), it now appears productive estuaries dating to the Mid-Holocene were present across the Southeast coasts as evidenced by dense midden sites on both dry and submerged lands, submerged paleo-oyster bars, and the presence estuarine shellfish in paleo-shorelines (Stapor et al., 1988, pp. 192–202; Faught, 1996; Russo, 1996b). Certainly, particular configurations of coastal islands and mainland shores have changed to varying degrees since 4500 ^{14}C yr BP (ca. 5150 cal yr BP). But during the Middle and Late Archaic productive estuaries, if not ubiquitous, did gain footholds along the Southeast coasts.

Pottery first appears about 4500 ^{14}C yr BP (ca. 5150 cal yr BP) in the Stallings culture of the middle Savannah River, and soon thereafter in the Orange culture along the St. Johns River and the Thom's Creek culture of South Carolina (Trinkley, 1980a, 1980b; Sassaman, 1993; Russo, 2006a). Some archeologists have suggested that early appearance of pottery in parts of the Southeast was linked causally to or resulted from the surpluses provided by the early shellfishing economies (Goodyear, 1988; Sassaman, 1993, pp. 216–217; Russo, 2006a, 2006b). However, intensive shellfishing arose hundreds of years before the earliest pottery appeared in many other areas of the Southeast (Russo, 1992, 1996a, 1996b; Russo and Saunders, 1999; Morey et al., 2002) indicating, perhaps, a sufficient, but not necessary linkage between pottery and intensified shellfish production.

6. Population distributions

The incidence of Archaic sites across the Southeast also provides clues about Mid-Holocene settlement (Anderson, 1996). As of the mid-1990s, ca. 180,000 sites had been recorded across the region, of which over 32,000 have Archaic components. An increase in the total number of sites occurs over the course of the Archaic, from roughly 7000 in the Early Archaic to 10,000 in the Middle Archaic to 15,000 in the Late Archaic. Standardizing for time, the proportional occurrence of sites in the Early, Middle, and Late Archaic subperiods is actually 1.00 to 0.98 to 2.11. That is, a slight decrease in site incidence occurred from the Early to Middle Archaic, followed by a fairly dramatic increase from the Middle to Late Archaic. Although care must be used when equating site numbers with population levels, the data suggest regional population levels may have stabilized fairly early on and then remained uniform or even dropped slightly through the Mid-Holocene, with marked growth only coming in the Late Archaic (see also Milner, 2004a, 2004b, p. 28–29). The data also indicate that Middle Archaic sites are not distributed evenly over the region.

Large numbers of Middle Archaic sites are reported in the Piedmont of North Carolina, South Carolina, and Georgia (Kowalewski, 1995, pp. 160–167; Anderson, 1996, pp. 163–165). As noted, this area appears to have been occupied by highly mobile foraging groups using an expedient lithic technology. These site concentrations thus probably do not represent dense populations but, instead, the remains of small, organizationally uncomplicated groups ranging widely over the landscape (Kowalewski, 1995; Sassaman, 1995). Large numbers of Middle Archaic sites are also present in Louisiana in the interior uplands to the west of the river, that appear to represent the remains of a highly mobile foraging adaptation similar to that noted in the South Appalachian area (Anderson, 1996; Saunders and Allen, 1997; Anderson and Smith, 2003a, 2003b, p. 367–377).

In the interior Midsouth, in contrast, Middle Archaic sites are somewhat more restricted, with concentrations along the major drainages, some in areas where extensive shell and earthen midden deposits have been found, such as along the Green and Tennessee Rivers. The major site concentrations in the Midsouth are thought to represent the territorial or integrative centers of individual groups. Site incidence over the remainder of the Midsouth may reflect the seasonally exploited foraging ranges of these groups, with areas of low site density possibly indicative of buffer zones between distinct societies. Other major concentrations of Middle Archaic sites associated with massive mound/midden complexes and evidence for extended occupations occur in Louisiana near the Mississippi River and its tributaries, and along the Gulf coast and St. Johns River of Florida.

The Gulf and Atlantic Coastal Plain across much of the region, from southeastern Mississippi to northern North Carolina, has a very low incidence of Middle Archaic sites, a reversal of the pattern noted in the Early Archaic period. As noted, the expansion of pine forests in the Coastal Plain beginning about 9000 cal yr BP is thought to have made this area less attractive for settlement, prompting a relocation of populations into the interior. Finally, Late Archaic sites occur in large numbers over almost every part of the Southeast, suggesting appreciable population increase following the close of the Mid-Holocene, and considerable landscape infilling.

7. Conclusions

As noted in recent syntheses (Anderson, 1995; Sassaman and Anderson, 1996, 2004; Anderson and Sassaman, 2004; Gibson and Carr, 2004), our understanding of Archaic occupations in the Southeast is increasing rapidly, with much of the information gained from cultural resource management projects. Monumental architecture dating more than 5000 cal yr BP is now well documented in Louisiana and Florida (Russo, 1996a), and shell rings of the Atlantic and Gulf coasts (Russo and Heide, 2001), once widely regarded as merely refuse heaps, are now accepted by many as ritual deposits (Russo, 2004). Floor plans of a number of Archaic structures have been documented, showing them to be, not surprisingly, more substantial than one might expect of small-scale, mobile societies (Sassaman and Ledbetter, 1996). Middle and Late Archaic exchange and interaction is attracting increasing attention, and there is evidence that the scale at which regular interaction occurred between groups was hundreds of kilometers in extent, particularly in the Midsouth (Jefferies, 1995, 1996, 2004). Local participation in long-distance exchange, however, appears to have varied appreciably over the region (Claassen, 1996). The incidence of extralocal materials varied appreciably, something that probably reflects the actions and abilities of individual network participants, who likely shaped the extent of local involvement.

We now have the capability to develop paleotemperature reconstructions spanning most of the Holocene with a resolution of a decade or less, using high precision uranium–thorium dating of banded speleothem calcite, coupled with carbon and oxygen isotope composition analyses (Dorale et al., 1992). Such analyses, when conducted in various parts of the karst-rich southeastern landscape, will provide measures of mean annual temperature for most or all of the period of human occupation. Such data can be tied directly into analyses linking effective temperature and technological organization (Binford, 1980; Cable, 1982, 1996). Dendrochronology also appears to offer the potential to delimit annual variation in rainfall well back into the Holocene which, with paleotemperature information, should provide precise information about Mid-Holocene climate in parts of the region (Stahle et al., 1988; Miller et al., 2006).

As we have seen, differing patterns of land use and different levels of organizational complexity are evident during the Middle Archaic period in the Southeast. These include (1) foragers of low social complexity and high mobility in the South Appalachian Piedmont and the interior uplands of western Louisiana; (2) a pattern of infrequent use of the Coastal Plain from North Carolina to Mississippi; (3) possible logistically based year-round coastal and riverine adaptations in Florida, coupled with seasonal foraging in interior areas; (4) complex, riverine occupations along major drainages of the Midsouth and in the lower Mississippi Valley, possibly with territorial buffers as well as less complex systems in intervening areas; and (5) the construction of monumental architecture and the emergence of tribal-level forms of social organization in some areas. The Mid-Holocene cultural landscape was a varied one, and also changed appreciably over time. At the beginning of the period band level societies were ubiquitous, while at the end societies of varying levels of size and complexity were present in different parts of the landscape.

The varying levels of social complexity observed over the Mid-Holocene Southeast were variously related to differences in regional physiography, resource structure, climate, intensity of intergroup interaction, and the historical traditions prevalent in each area. Areas rich in resources where interaction potential was greatest, such as in the major river drainages of the Midsouth, or areas rich in subsistence resources, such as in the lower Mississippi Valley or the Atlantic and Gulf coasts of Florida, tended to have the highest levels of sociopolitical complexity, and complex social formations developed early in these settings. In areas where interaction potential might have been lower, or subsistence resources more uncommon, cultural systems of less complexity are evident. Of course, what is meant by cultural "complexity" during the Mid-Holocene has itself been variously defined, with a diverse array of approaches used to explore the subject (cf., Anderson, 2002, 2004, 2005; Clark, 2004; Gibson, 2004; Russo, 2004; Sassaman, 2004, 2005a, 2005b; Sassaman and Heckenberger, 2004; J. Saunders, 2004). Appreciable variability also characterizes the societies of the region. The earliest mound sites in the lower Mississippi Valley and peninsular Florida, as well as the early shell ring sites in Georgia, South Carolina, and Florida, for example, lack much evidence for artifacts obtained through long-distance exchange, a pattern quite unlike that in the Middle and Late Archaic shell middens of the Midsouth, where artifacts of copper and marine shell are more frequently found in burials. But the extent of monumental construction was, in some cases, far greater than that found in contemporary Midsouth and even many subsequent regional societies. None of these Middle Holocene cultures, however, have the markedly elaborated burials and other evidence for self-aggrandizing behavior or hereditary leadership groups common to many subsequent Woodland or Mississippian era cultures. The Mid-Holocene Southeast was a varied cultural landscape, within which a number of distinct historical trajectories were operating. To understand occupations in any one area, we must have an appreciation for climatic and cultural developments at a range of geographic and temporal scales. Research of this kind is just emerging, which is fitting. Two decades ago we had no idea monumental architecture was being built during the Mid-Holocene in the Southeast. It will be interesting to see what we know about life during this period given another two decades of research.

Acknowledgments

The authors wish to thank George L. Jacobson, Jr. for his thoughts on the initial version of this manuscript, as well as Dan Sandweiss and Kirk Maasch for their encouragement and patience during the production effort. All radiocarbon ages were calibrated using the Calib 4.4 program (Stuiver et al., 1998a, 1998b; McCormac et al., 2002).

References

Amick, D. S., and P. J. Carr, 1996. Changing strategies of lithic technological organization. In *Archaeology of the Mid-Holocene Southeast*, edited by K. E. Sassaman and D. G. Anderson pp. 41–56. University Press of Florida, Gainesville.

Anderson, D. G., 1991. The bifurcate tradition in the South Atlantic Region. *Journal of Middle Atlantic Archaeology* 7:91–106.

Anderson, D. G., 1994. *The Savannah River Chiefdoms: Political Change in the Late Prehistoric Southeast.* The University of Alabama Press, Tuscaloosa, AL.

Anderson, D. G., 1995. Recent advances in Paleoindian and Archaic period research in the southeastern United States. *Archaeology of Eastern North America* 23:145–76.

Anderson, D. G., 1996. Modeling regional settlement in the Archaic period southeast. In *Archaeology of the Mid-Holocene Southeast*, edited by K. E. Sassaman and D. G. Anderson, pp. 157–176. University Press of Florida, Gainesville.

Anderson, D. G., 2002. Evolution of tribal social organization in the southeast. In *The Archaeology of Tribal Societies*, edited by W. A. Parkinson, pp. 246–277. International Monographs in Prehistory, Ann Arbor.

Anderson, D. G., 2004. Archaic mounds and the archaeology of southeastern tribal societies. In *Signs of Power: The Rise of Cultural Complexity in the Southeast*, edited by J. L. Gibson and P. J. Carr, pp. 270–299. The University of Alabama Press, Tuscaloosa.

Anderson, D. G., 2005. Why California? The relevance of California ethnography to the archaeology of the Eastern Woodlands. In *Engaged Anthropology Research Essays on North American Archaeology, Ethnobotany, and Museology*, edited by M. Hegmon and B. S. Eiselt, pp. 200–225. Anthropological Papers, Museum of Anthropology, University of Michigan, 94. Ann Arbor, Michigan.

Anderson, D. G., and G. T. Hanson, 1988. Early Archaic settlement in the southeastern United States: a case study from the Savannah River valley. *American Antiquity* 53:262–286.

Anderson, D. G., and K. E. Sassaman, editors. 1996. *The Paleoindian and Early Archaic Southeast*. The University of Alabama Press, Tuscaloosa.

Anderson, D. G., and K. E. Sassaman, 2004. Early and Middle Holocene Periods, 9500–3750 B.C. *Southeast Volume, Smithsonian Handbook of North American Indians*, edited by R. D. Fogelson, pp. 87–100. Smithsonian Institution, Washington, DC.

Anderson, D. G., and M. T. Smith, 2003a. Problems of the past: perspectives on eastern North American archaeology from prehistory to the seventeenth-century In *Blackwell Companion to Colonial American History*, edited by V. Daniel, pp. 1–24. Blackwell Publishers, Ltd., Oxford.

Anderson, D. G., and S. D. Smith, 2003b. *Archaeology, History, and Predictive Modeling Research at Fort Polk, 1972–2002*. The University of Alabama Press, Tuscaloosa.

Aten, L. E., 1999. Middle Archaic ceremonialism at Tick Island, Florida: Ripley P. Bullen's 1961 excavations at the Harris Creek site. *The Florida Anthropologist* 53:131–200.

Bender, B., 1985. Prehistoric developments in the American midcontinent and in Brittany, northwest France. In *Prehistoric Hunter-Gatherers: The Emergence of Cultural Complexity*, edited by T. D. Price and J. A. Brown, pp. 21–57. Academic Press, Orlando, FL.

Bense, J. A., 1994. *Archaeology of the Southeastern United States Paleoindian to World War I*. Academic Press, New York.

Binford, L. R., 1980. Willow smoke and dogs' tails: hunter-gatherer settlement systems and archaeological site formation. *American Antiquity* 45:4–20.

Blanton, D. B., 1996. Accounting for submerged Mid-Holocene archaeological sites in the southeast: a case study from the Chesapeake Bay estuary, Virginia. In *Archaeology of the Mid-Holocene Southeast*, edited by K. E. Sassaman and D. G. Anderson, pp. 200–221. University Press of Florida, Gainesville.

Braun, D. P., and S. Plog, 1982. Evolution of "tribal" social networks: theory and prehistoric North American evidence. *American Antiquity* 47:504–525.

Brooks, M. J., B. E. Taylor, and J. A. Grant, 1996. Carolina Bays and Holocene landscape evolution on the upper Coastal Plain of South Carolina. *Geoarchaeology* 11:481–504.

Brown, J. A., 1985. Long term trends to sedentism and the emergence of complexity in the American midwest. In *Prehistoric Hunter-Gatherers: The Emergence of Cultural Complexity*, edited by T. D. Price and J. A. Brown, pp. 201–231. Academic Press, Orlando, FL.

Brown, J. A., and R. K. Vierra, 1983. What happened in the Middle Archaic? introduction to an ecological approach to Koster site archaeology. In *Archaic Hunters and Gatherers in the American Midwest*, edited by J. Phillips and J. A. Brown, pp. 165–195. Academic Press, New York.

Bruseth, J. E., 1991. Poverty Point development as seen at the Cedarland and Claiborne Sites, southern Mississippi. In *The Poverty Point Culture: Local Manifestations, Subsistence Practices, and Trade Networks*, edited by K. M. Byrd, pp. 7–25. *Geoscience and Man* 29:7–25.

Cable, J. S., 1982. Organizational variability in Piedmont hunter-gatherer lithic assemblages. In *The Haw River Sites: Archaeological Investigation at Two Stratified Sites in the North Carolina Piedmont*, edited by S. R. Claggett and J. S. Cable, pp. 637–688. Commonwealth Associates, Inc. Report R-2386, Jackson, Michigan.

Cable, J. S., 1996. The Haw River Sites revisited: implications for modeling terminal late glacial and early Holocene hunter-gatherer settlement systems in South Carolina. In *The Paleoindian and Early Archaic Southeast*, edited by D. G. Anderson and K. E. Sassaman, pp. 107–148. University of Alabama Press, Tuscaloosa.

Caldwell, J. R., 1958. *Trend and Tradition in the Prehistory of the Eastern United States*. American Anthropological Association Memoir 88. Menasha, Wisconsin.

Carmichael, D. L., 1977. Preliminary archaeological survey of Illinois uplands and some behavioral considerations. *Midcontinental Journal of Archaeology* 2:219–251.
Claassen, C. P., 1991a. Gender, shellfishing, and the Shell Mound Archaic. In *Engendering Archaeology*, edited by J. Gero and M. Conkey, pp. 276–300. Basil Blackwells, Oxford.
Claassen, C. P., 1991b. New hypotheses for the demise of the Shell Mound Archaic. In *The Archaic Period in the Mid-South*, edited by C. McNutt, pp. 66–72. Mississippi Department of Archives and History Archaeological Report 24. Jackson, Mississippi.
Claassen, C. P., 1996. A consideration of the social organization of the Shell Mound Archaic. In *Archaeology of the Mid-Holocene Southeast*, edited by K. E. Sassaman and D. G. Anderson, pp. 235–258. University Press of Florida, Gainesville.
Clark, J. E., 2004. Surrounding the sacred: geometry and design of early mound groups as meaning and function. In *Signs of Power: The Rise of Cultural Complexity in the Southeast*, edited by J. L. Gibson and P. J. Carr, pp. 162–213. The University of Alabama Press, Tuscaloosa.
Clausen, C. J., A. D. Cohen, C. Emeliani, J. A. Holman, and J. J. Stipp, 1979. Little Salt Spring, Florida: a unique underwater site. *Science* 203:609–614.
Colquhoun, D. J., and M. J. Brooks, 1986. New evidence from the southeastern U.S. for eustatic components in the Late Holocene sea levels. *Geoarchaeology* 1:275–291.
Crothers, G. M., 1999. Archaic period subsistence and economy: the Green River shell midden sites of Kentucky. Unpublished Ph.D. dissertation, Department of Anthropology, Washington University, St. Louis.
Crothers, G. M., 2004. The Green River in comparison to the lower Mississippi valley during the Archaic: to build mounds or not to build mounds? In *Signs of Power: The Rise of Cultural Complexity in the Southeast*, edited by J. L. Gibson and P. J. Carr, pp. 86–96. The University of Alabama Press, Tuscaloosa.
Delcourt, P. A., and H. R. Delcourt, 1981. Vegetation maps for eastern North America: 40,000 years B.P. to present. In *Geobotany*, edited by R. Romans, pp. 123–166. Plenum Press, New York.
Delcourt, P. A., and H. R. Delcourt, 1987. *Long Term Forest Dynamics of the Temperate Zone: A Case Study of Late-Quaternary Forests in Eastern North America.* Springer, New York.
Delcourt, P. A., and H. R. Delcourt, 2004. *Prehistoric Native Americans and Ecological Change: Human Ecosystems in Eastern North America since the Pleistocene*. Cambridge University Press, Cambridge.
DePratter, C. B., 1983. Late Prehistoric and early historic chiefdoms in the Southeastern United States. Unpublished Ph.D. dissertation, Department of Anthropology, University of Georgia, Athens, Georgia.
DePratter, C. B., and J. Howard, 1980. Indian occupation and geologic history of the Georgia coast: a 5,000 year summary. In *Excursions in Southeastern Geology: The Archaeology of the Georgia Coast*, edited by J. D. Howard, C. B. DePratter, and R. W. Frey, pp. 1–65. Geological Society of America, Guidebook 20. Atlanta.
Dickel, D., 1992. An archaeological and historical survey of Bonita Springs, Parcel Three, Lee County, Florida. AHC Technical Report 43. Miami, FL.
Dillehay, T. D., 1990. Mapuche ceremonial landscape, social recruitment and resource rights. *World Archaeology* 22:223–241.
Dorale, J. A., I. A. Gonzalez, M. K. Reagan, D. A. Pickett, M. T. Murrell, and R. G. Baker, 1992. A high-resolution record of Holocene climate change in speleothem calcite from Cold Water Cave, northeast Iowa. *Science* 258:1626–1630.
Doran, G. H., editor. 2002. *Windover: Multidisciplinary Investigations of an Early Archaic Florida Cemetery*. University Press of Florida, Gainesville.

Doran, G. H., and D. N. Dickel, 1988. Multidisciplinary investigations at the Windover site. In *Wet Site Archaeology*, edited by B. A. Purdy, pp. 263–289. Telford Press, Caldwell, NJ.

Doran, G. H., D. N. Dickel, W. E. Ballinger Jr., O. F. Agee, P. J. Laipis, and W. W. Hauswirth, 1988. Anatomical, cellular and molecular analysis of 8,000-yr-old human brain tissue from the Windover archaeological site. *Nature* 323:803–806.

Dowd, J. T., 1989. *The Anderson Site: Middle Archaic Adaptation in Tennessee's Central Basin.* Tennessee Anthropological Association Miscellaneous Paper 13.

Dye, D. H., 1990, Warfare in the interior southeast: the De Soto expedition in the interior. In *Columbian Consequences, Vol. 2: Archaeological and Historical Perspectives on the Spanish Borderlands East*, edited by D. H. Thomas, pp. 211–222. Smithsonian Institution Press, Washington, DC.

Dye, D. H., 1996. Initial riverine adaptation in the midsouth: an examination of three Middle Holocene shell middens. In *Of Caves and Shell Mounds*, edited by K. C. Carstens and P. J. Watson, pp. 140–158. University of Alabama Press, Tuscaloosa.

Endonino, J., 2003. Pre-ceramic Archaic burial mounds along the St. Johns River, Florida. Paper presented at the 59th Annual Meeting of the Southeastern Archaeological Conference, Charlotte, North Carolina.

Faught, M. K., 1996. Clovis origins and underwater prehistoric archaeology in Northwestern Florida. Unpublished Ph.D. dissertation, Department of Anthropology, University of Arizona, Tucson.

Ford, J. A., and C. H. Webb 1956 *Poverty Point: A Late Archaic Site in Louisiana.* American Museum of Natural History, Anthropological Papers 46, part 1. New York.

Fritz, G. J., 1990. Multiple pathways to farming in precontact eastern North America. *Journal of World Prehistory* 4:387–435.

Fritz, G. J., 1999. Gender and the early cultivation of gourds in Eastern North America. *American Antiquity* 64:417–429.

Fritz, G. J., and T. R. Kidder, 1993. Recent investigations into prehistoric agriculture in the lower Mississippi valley. *Southeastern Archaeology* 12:1–14.

Gibson, J. L., 1994. Before their Time? Early Mounds in the Lower Mississippi Valley. *Southeastern Archaeology* 13:162–181.

Gibson, J. L., 1996. Poverty Point and greater southeastern prehistory: the culture that did not fit. In *Archaeology of the Mid-Holocene Southeast*, edited by K. E. Sassaman and D. G. Anderson, pp. 288–305. University Press of Florida, Gainesville.

Gibson, J. L., 2000. *The Ancient Mounds of Poverty Point Place of Rings.* The University Press of Florida, Gainesville, FL.

Gibson, J. L., 2004. The power of beneficent obligation in first mound-building societies. In *Signs of Power: The Rise of Cultural Complexity in the Southeast*, edited by J. L. Gibson and P. J. Carr, pp. 234–253. The University of Alabama Press, Tuscaloosa.

Gibson, J. L., and P. J. Carr, editors. 2004. *Signs of Power: The Rise of Cultural Complexity in the Southeast.* The University of Alabama Press, Tuscaloosa.

Goman, M., and D. S. Leigh, 2004. Wet early to Middle Holocene conditions on the upper Coastal Plain of North Carolina, USA. *Quaternary Research* 61:256–264.

Goodyear, A. C., 1988. On the Study of Technological Change. *Current Anthropology* 29: 320–323.

Gremillion, K. J., 1996. The paleoethnobotanical record for the Mid-Holocene southeast. In *Archaeology of the Mid-Holocene Southeast*, edited by K. E. Sassaman and D. G. Anderson, pp. 99–114. University Press of Florida, Gainesville.

Gremillion, K. J., 2002. The development and dispersal of agricultural systems in the Woodland period southeast. In *The Woodland Southeast*, edited by D. G. Anderson and R. C. Mainfort Jr., pp. 483–501. The University of Alabama Press, Tuscaloosa.
Griffin, J. B., 1967. Eastern North American archaeology: a summary. *Science* 156:175–191.
Hoover, S. R., and A. J. Parker, 1991. Spatial components of biological diversity in landscapes of Georgia, USA. *Landscape Ecology* 5:125–136.
Jacobson, G. L., Jr., T. Webb III, and E. C. Grimm, 1987. Patterns and rates of vegetation change in eastern North America from full glacial to Mid-Holocene time. In *The Geology of North America, Vol. K-3, North America and Adjacent Oceans During the Last Deglaciation*, edited by W. F. Ruddimen and H. E. Wright Jr., pp. 277–288. Geological Society of America, Boulder, Colorado.
Jefferies, R. W., 1995. Late Middle Archaic exchange and interaction in the North American midcontinent. In *Native American Interaction: Multiscalar Analyses and Interpretations in the Eastern Woodlands*, edited by M. S. Nassaney and K. E. Sassaman, pp. 73–95. University of Tennessee Press, Knoxville.
Jefferies, R. W., 1996. The emergence of long-distance exchange networks in the Southeastern United States. In *Archaeology of the Mid-Holocene Southeast*, edited by K. E. Sassaman and D. G. Anderson, pp. 222–234. University Press of Florida, Gainesville.
Jefferies, R. W., 2004. Regional scale interaction networks and the emergence of cultural complexity along the northern margins of the southeast. In *Signs of Power: The Rise of Cultural Complexity in the Southeast*, edited by J. L. Gibson and P. J. Carr, pp. 71–85. The University of Alabama Press, Tuscaloosa.
Jefferies, R. W., and B. M. Butler, 1982. *The Carrier Mills Archaeological Project: Human Adaptation in the Saline Valley, Illinois*. Center for Archaeological Investigations, Southern Illinois University, Research Paper 33. Carbondale, Illinois.
Johnson, J. K., and S. O. Brookes, 1989. Benton points, turkey tails, and cache blades: Middle Archaic exchange in the midsouth. *Southeastern Archaeology* 8:134–145.
Kidder, T. R., 2001. Mapping Poverty Point. *American Antiquity* 67:89–101.
Knox, J. C., 1983. Responses of river systems to Holocene climate. In *Late Quaternary Environments of the United States, part 2, The Holocene*, edited by H. E. Wright Jr., pp. 26–41. University of Minnesota Press, Minneapolis.
Kowalewski, S. A., 1995. Large-scale ecology in aboriginal Eastern North America. In *Native American Interactions: Multi-Scalar Analyses and Interpretations in the Eastern Woodlands*, edited by M. S. Nassaney and K. E. Sassaman, pp. 147–173. The University of Tennessee Press, Knoxville.
Kutzbach, J. E., 1987. Model simulations of the climatic patterns during the deglaciation of North America. In *The Geology of North America, Vol. K-3, North America and Adjacent Oceans During the Last Deglaciation*, edited by W. F. Ruddimen and H. E. Wright Jr., pp. 425–446. Geological Society of America, Boulder, Colorado.
Larson, L. H., 1980. *Aboriginal Subsistence Technology on the Southeastern Coastal Plain During the Late Prehistoric Period.* Ripley P. Bullen Monographs in Anthropology and History 2. University Press of Florida, Gainesville.
Leigh, D. S., and T. P. Feeney, 1995. Paleochannels indicating wet climate and lack of response to lower sea level, Southeast Georgia. *Geology* 23:687–690.
Marquardt, W. H., 1985. Complexity and scale in the study of fisher-gatherers: an example from the eastern United States. In *Prehistoric Hunter-Gatherers: The Emergence of Cultural Complexity*, edited by T. D. Price and J. A. Brown, pp. 5–98. Academic Press, Orlando, FL.

Marquardt, W. H., and P. J. Watson, 1983. The Shell Mound Archaic of western Kentucky. In *Archaic Hunters and Gatherers in the American Midwest*, edited by J. Phillips and J. Brown, pp. 323–339. Academic Press, Orlando, FL.

Marquardt, W. H., and P. J. Watson, editors. 2005. *Archaeology of the Middle Green River Region, Kentucky*. Institute of Archaeology and Paleoenviromental Studies Monograph 5. University Press of Florida, Gainesville, FL.

McCormac, F. G., P. J. Reimer, A. G. Hogg, T. F. G. Higham, M. G. L. Baillie, J. Palmer, M. Stuiver, 2002. Calibration of the radiocarbon time scale for the southern hemisphere: AD 1850-950. *Radiocarbon* 44:641–651.

McGee, R. M., and R. J. Wheeler, 1994. Stratigraphic excavations at Grove's Orange Midden, Lake Monroe, Volusia County, Florida: methodology and results. *The Florida Anthropologist* 47:333–349.

Meeks, S. C., 1999. The "Function" of stone tools in prehistoric exchange systems: a look at Benton interaction in the Midsouth. In *Raw Materials and Exchange in the Mid-South*, edited by E. Peacock and S. O. Brookes, pp. 29–43. Mississippi Department of Archives and History, Archaeological Report No. 29, Jackson.

Milanich, J. T., 1994. *Archaeology of Precolumbian Florida*. University Press of Florida, Gainesville.

Milanich, J. T, J. Chapman, A. S. Cordell, S. A. Hale, and R. A. Marrinan, 1984. Prehistoric development of Calusa society in southwest Florida: excavations on Useppa Island. In *Perspectives on Gulf Coast Prehistory*, edited by D. D. Davis, pp. 258–314. University Press of Florida, Gainesville.

Miller, D. L., C. I. Mora, H. D. Grissino-Mayer, C. J. Mock, M. E. Uhle, and Z. D. Sharp, 2006. Tree-ring oxygen isotope records of tropical cyclone activity. *Proceedings of the National Academy of Sciences of the United States of America* 103(39): 14294–14297.

Milner, G. R., 2004a. Old mounds, ancient hunter-gatherers, and modern archaeologists. In *Signs of Power: The Rise of Cultural Complexity in the Southeast*, edited by J. L. Gibson and P. J. Carr, pp. 300–316. The University of Alabama Press, Tuscaloosa.

Milner, G. R., 2004b. *The Mound Builders: Ancient Peoples of Eastern North America*. Thames & Hudson, London.

Milner, G. R., and R. W. Jefferies, 1998. The Read Archaic Shell Midden in Kentucky. *Southeastern Archaeology* 17:119–132.

Morey, D. F., G. M. Crothers, J. K. Stein, J. P. Fenton, and N. P. Herrmann, 2002. The Fluvial and Geomorphic Context of Indian Knoll, an Archaic Shell Midden in West-Central Kentucky. *Geoarchaeology* 17:521–553.

Newsom, L. A., 1994. Archaeobotanical data from Groves' Orange Midden (8VO2601), Volusia County Florida. *The Florida Anthropologist* 47:404–417.

Newsom, L. A., and B. A. Purdy, 1990. Florida canoes: a maritime heritage from the past. *The Florida Anthropologist* 43:164–180.

Peacock, E., 1998. Fresh-water Mussels as indicators of prehistoric human environmental impact in the Southeastern United States. Unpublished Doctoral Thesis, Department of Archaeology and Prehistory, The University of Sheffield, England.

Peacock, E., 2002. Shellfish use during the Woodland period in the middle south. In *The Woodland Southeast*, edited by D. G. Anderson and R. C. Mainfort Jr., pp. 444–460. The University of Alabama Press, Tuscaloosa.

Piatek, B. J., 1994. The Tomoka mound complex in northeast Florida. *Southeastern Archaeology* 13:109–118.

Purdy, B. A., editor. 1988. *Wet Site Archaeology*. Telford Press, Caldwell, NJ.

Purdy, B. A., editor. 1991. *The Art and Archaeology of Florida's Wetlands*. CRC Press, Boca Raton, FL.
Randall, A. R., and K. E. Sassaman, 2005. *St. Johns Archaeological Field School 2003-2004: Hontoon Island State Park*. Technical Report 6. Laboratory of Southeastern Archaeology, Department of Anthropology, University of Florida, Gainesville.
Rouse, I., 1951. *A Survey of Indian River Archeology, Florida*. Yale University Press, New Haven.
Russo, M., 1986. The coevolution of environment and human exploitation of faunal resources in the Upper St. Johns River Basin. Unpublished MA Project Report, Department of Anthropology, University of Florida, Gainesville.
Russo, M., 1991. Archaic sedentism on the Florida Coast: a case study from Horr's Island. Unpublished Ph.D. dissertation, Department of Anthropology, University of Florida. University Microfilms, Ann Arbor, Michigan.
Russo, M., 1992. Chronologies and cultures of the St. Marys region of northeast Florida and southeast Georgia. *The Florida Anthropologist* 45:107–126.
Russo, M., 1994a. A brief introduction to the study of Archaic mounds in the southeast. *Southeastern Archaeology* 13:89–93.
Russo, M., 1994b. Why we don't believe in Archaic ceremonial mounds and why we should: the case from Florida. *Southeastern Archaeology* 13:93–109.
Russo, M., 1996a. Southeastern preceramic Archaic ceremonial mounds. In *Archaeology of the Mid-Holocene Southeast*, edited by K. E. Sassaman and D. G. Anderson, pp. 259–287. University Press of Florida, Gainesville.
Russo, M., 1996b. Southeastern Mid-Holocene coastal settlements. In *Archaeology of the Mid-Holocene Southeast*, edited by K. E. Sassaman and D. G. Anderson, pp. 177–199. University Press of Florida, Gainesville.
Russo, M., 1999 Non-symmetrical traits in shell rings: towards an understanding of circular community dynamics. Paper presented at the 56th Annual Meeting of the Southeastern Archaeological Conference, Pensacola, FL.
Russo, M., 2002. Architectural features at Fig Island. In *The Fig Island Ring Complex (38CH42): Coastal Adaptations and the Question of Ring Function in the Late Archaic*, edited by R. Saunders and M. Russo, pp. 85–97. Submitted to the South Carolina Department of Archives and History (Grant 45-01-16441), Columbia.
Russo, M., 2004. Measuring shell rings for social inequality. In *Signs of Power: The Rise of Cultural Complexity in the Southeast*, edited by J. L. Gibson and P. J. Carr, pp. 26–70. The University of Alabama Press, Tuscaloosa.
Russo, M., 2006a. Archaic Shell Rings of the Southeast. U.S. National Historic Landmark Theme Study submitted to the National Park Service, Washington, DC.
Russo, M., 2006b. Fig Island Shell Rings. National Historic Landmark Nomination submitted to the National Park Service, Washington, DC.
Russo, M., and D. Ste. Claire, 1992. Tomoka Stone: Archaic period coastal settlement in east Florida. *The Florida Anthropologist* 45:336–346.
Russo, M., and G. Heide, 2001. Shell rings of the Southeast US. *Antiquity* 75:491–492.
Russo, M., and G. Heide, 2002a. The Joseph Reed shell ring. *The Florida Anthropologist* 55:67–87.
Russo, M., and G. Heide, 2002b. Mapping the Sewee Shell Ring. Submitted to the Francis Marion and Sumter National Forests, South Carolina by the Southeast Archeological Center, Tallahassee.
Russo, M., and R. Saunders, 1999. Identifying the early use of coastal fisheries and the rise of social complexity in shell rings and Arcuate Middens on Florida's Northeast Coast. Project report submitted to the National Geographic Society, Washington, D.C.

Russo, M., B. A. Purdy, L. A. Newsom, and R. M. McGee, 1992. A reinterpretation of Late Archaic adaptations in central-east Florida: Groves' Orange Midden (8-Vo-2601). *Southeastern Archaeology* 11:95–108.

Russo, M., A. S. Cordell, and D. L. Ruhl, 1993. *The Timucuan Ecological and Historical Preserve Phase III Final Report*. Florida Museum of Natural History, Gainesville, Florida. SEAC Accession Number 899, Tallahassee, FL.

Sassaman, K. E., 1983. Middle and Late Archaic settlement in the South Carolina Piedmont. Unpublished M.A. Thesis, Department of Anthropology, University of South Carolina, Columbia.

Sassaman, K. E., 1985. A preliminary typological assessment of MALA hafted bifaces from the Pen Point site, Barnwell County, South Carolina. *South Carolina Antiquities* 17:1–17.

Sassaman, K. E., 1991. Adaptive flexibility in the Morrow Mountain phase of the Middle Archaic period. *South Carolina Antiquities* 23:31–41.

Sassaman, K. E., 1993. Early pottery in the Southeast: tradition and innovation in cooking technology. University of Alabama Press, Tuscaloosa.

Sassaman, K. E., 1995. The cultural diversity of interactions among Mid-Holocene societies of the American Southeast. In *Native American Interaction: Multiscalar Analyses and Interpretations in the Eastern Woodlands*, edited by M. S. Nassaney and K. E. Sassaman, pp. 174–204. University of Tennessee Press, Knoxville.

Sassaman, K. E., 1996. Technological innovations in economic and social contexts. In *Archaeology of the Mid-Holocene Southeast*, edited by K. E. Sassaman and D. G. Anderson, pp. 57–74. University Press of Florida, Gainesville.

Sassaman, K. E., 2001a. Articulating hidden histories of the Mid-Holocene in the southern Appalachians. In *Archaeology of the Appalachian Highlands*, edited by L. P. Sullivan and S. C. Prezzano, pp. 103–121. University of Tennessee Press, Knoxville.

Sassaman, K. E., 2001b. Hunter-gatherers and traditions of resistance. In *The Archaeology of Traditions: Agency and History Before and After Columbus*, edited by T. R. Pauketat, pp. 218–236. University Press of Florida, Gainesville.

Sassaman, K. E., 2003. St. Johns Archaeological Field School 2000–2001: Blue Spring and Hontoon Island State Parks. Technical Report 4. Laboratory of Southeastern Archaeology, Department of Anthropology, University of Florida, Gainesville.

Sassaman, K. E., 2004. Complex hunter–gatherers in evolution and history: a North American perspective. *Journal of Archaeological Research* 12:227–280.

Sassaman, K. E., 2005a. Structure and practice in the Archaic Southeast. In *North American Archaeology*, edited by T. R. Pauketat and D. D. Loren, pp. 79–107. Blackwell Publishing, Ltd., Malden, MA.

Sassaman, K. E., 2005b. Poverty Point as structure, event, process. *Journal of Archaeological Method and Theory* 12:335–364.

Sassaman, K. E., and D. G. Anderson, 1995. *Middle and Late Archaic Archaeological Records of South Carolina: A Synthesis for Research and Resource Management,* revised 2nd edition. Savannah River Archaeological Research Papers 6. Savannah River Archaeological Research Program, South Carolina Institute of Archaeology and Anthropology, University of South Carolina, Columbia.

Sassaman, K. E., and D. G. Anderson, editors. 1996. *The Archaeology of the Mid-Holocene Southeast*. University Press of Florida, Gainesville.

Sassaman, K. E., and D. G. Anderson, 2004. Late Holocene period, 3750 to 650 B.C. *Southeast volume, Smithsonian Handbook of North American Indians*, edited by R. D. Fogelson, pp. 101–114. Smithsonian Institution, Washington, DC.

Sassaman, K. E., and M. J. Heckenberger, 2004. Crossing the symbolic Rubicon in the southeast. In *Signs of Power: The Rise of Cultural Complexity in the Southeast*, edited by J. L. Gibson and P. J. Carr, pp. 214–233. The University of Alabama Press, Tuscaloosa.

Sassaman, K. E., and R. J. Ledbetter, 1996. Middle and Late Archaic architecture. In *Archaeology of the Mid-Holocene Southeast*, edited by K. E. Sassaman and D. G. Anderson, pp. 75–95. University Press of Florida, Gainesville.

Sassaman, K. E., M. J. Brooks, G. T. Hanson, and D. G. Anderson, 1990. Native American prehistory of the Middle Savannah River Valley: a synthesis of archaeological investigations on the Savannah River site, Aiken and Barnwell counties, South Carolina. South Carolina Institute of Archaeology and Anthropology, University of South Carolina, Savannah River Archaeological Research Papers 1.

Saunders, R., 2002. Summary and conclusions. In *The Fig Island Ring Complex (38CH42): Coastal Adaptations and the Question of Ring Function in the Late Archaic*, edited by R. Saunders and M. Russo, pp. 154–159. Submitted to the South Carolina Department of Archives and History (Grant 45-01-16441), Columbia.

Saunders, J. W., 2004. Are we fixing to make the same mistake again? In *Signs of Power: The Rise of Cultural Complexity in the Southeast*, edited by J. L. Gibson and P. J. Carr, pp. 146–161. The University of Alabama Press, Tuscaloosa.

Saunders, R., 2004. The stratigraphic sequence at Rollins shell ring: implications for ring function. *The Florida Anthropologist* 57:249–268.

Saunders, J. W., and T. Allen, 1997. The Archaic period. *Louisiana Archaeology* 22:1–30.

Saunders, R., and M. Russo, editors. 2002. *The Fig Island Ring Complex (38CH42): Coastal Adaptations and the Question of Ring Function in the Late Archaic*. Submitted to the South Carolina Department of Archives and History (Grant 45-01-16441), Columbia.

Saunders, J. W., T. Allen, and R. T. Saucier, 1994. Four Archaic? Mound complexes in Northeast Louisiana. *Southeastern Archaeology* 13:134–153.

Saunders, J. W., R. D. Mandel, R. T. Saucier, E. T. Allen, C. T. Hallmark, J. K. Johnson, E. H. Jackson, C. M. Allen, G. L. Stringer, D. S. Frink, J. K. Feathers, S. Williams, K. J. Gremillion, M. F. Vidrine, and R. Jones, 1997. A mound complex in Louisiana at 5400-5000 years before the present. *Science* 277:1796–1799.

Saunders, J. W., R. D. Mandel, C. G. Sampson, C. M. Allen, E. T. Allen, D. A. Bush, J. K. Feathers, K. J. Gremillion, C. T. Hallmark, H. E. Jackson, J. K. Johnson, R. Jones, R. T. Saucier, G. L. Stringer, and M. F. Vidrine, 2005. Watson Brake: a Middle Archaic mound complex in northeast Louisiana. *American Antiquity* 70:631–668.

Schuldenrein, J., 1996. Geoarchaeology and Mid-Holocene landscape history. In *Archaeology of the Mid-Holocene Southeast*, edited by K. E. Sassaman and D. G. Anderson, pp. 3–27. University Press of Florida, Gainesville.

Sears, W. H., 1960. The Bluffton burial mound. *The Florida Anthropologist* 13:55–60.

Seielstad, C. A., 1994. Holocene environmental history at Chadderton Springs on the Southern Coastal Plain of Georgia. Unpublished M.A. Thesis, Department of Geography, University of Georgia, Athens.

Smith, B. D., 1986. The archaeology of the southeastern United States: from Dalton to deSoto, 10,500-500 B.P. *Advances in World Archaeology* 5:1–92.

Smith, B. D., 1987. The independent domestication of indigenous seed-bearing plants in Eastern North America. In *Emergent Horticultural Economies of the Eastern Woodlands*, edited by W. Keegan, pp. 3–48. Center for Archaeological Investigations, Southern Illinois University, Occasional Paper 7. Carbondale, IL.

Smith, B. D., 1989. Origins of Agriculture in eastern North America. *Science* 246:1566–1571.

Smith, B. D., 1992. *Rivers of Change: Essays on Early Agriculture in Eastern North America*. Smithsonian Institution Press, Washington, DC.

Smith, M., 1996. Bioarchaeological inquiry into Archaic period populations of the Southeast: trauma and occupational stress. In *Archaeology of the Mid-Holocene Southeast*, edited by K. E. Sassaman and D. G. Anderson, pp. 134–154. University Press of Florida, Gainesville.

Stahle, D. W., M. K. Cleaveland, and J. G. Hehr, 1988. North Carolina climate changes reconstructed from tree rings: A.D. 372 to 1985. *Science* 240:1517–20.

Stapor, F. W., Jr., T. D. Mathews, and F. E. Lindfors-Kearns, 1988. Episodic barrier island growth in Southwest Florida: a response to fluctuating Holocene sea level? *Miami Geological Society Memoir* 3:149–202.

Stein, J. A. K., 2005. Environment of the Green River Sites. In *Archaeology of the Middle Green River Region, Kentucky*, edited by W. H. Marquardt and P. J. Watson, pp. 19–39. Institute of Archaeology and Paleoenviromental Studies Monograph 5. University Press of Florida, Gainesville, Florida.

Steponaitis, V. P., 1986. Prehistoric archaeology in the southeastern United States, 1979–1985. *Annual Reviews in Anthropology* 15:363–404.

Stuiver, M., P. J. Reimer, and T. F. Braziunas, 1998a. High precision radiocarbon age calibrations for terrestrial and marine samples. *Radiocarbon* 40:1127–1151.

Stuiver, M., P. J. Reimer, E. Bard, J. W. Beck, G. S. Burr, K. A. Hughen, B. Kromer, F. G. McCormac, J. van der Plicht, and M. Spurk, 1998a. INTCAL98 radiocarbon age calibrations, 24,000-0 cal BP. *Radiocarbon* 40:1041–1083.

Thomas, P. M. Jr., and L. J. Campbell, 1991. The Elliott's point Complex: new data regarding the localized Poverty Point expression on the northwest Florida Gulf coast, 2000 B.C–500 B.C. In *The Poverty Point Culture Local manifestations, Subsistence Practices, and Trade Networks*, edited by Kathleen M. Byrd, pp 103–199. Geoscience and Man, Vol. 29, Louisiana State University, Baton Rouge, Louisiana.

Thompson, V. D., 2006. Questioning complexity: the prehistoric hunter-gatherers of Sapelo Island, Georgia. Ph.D. dissertation, University of Kentucky, Lexington.

Trinkley, M. B., 1980a. A typology of Thom's Creek pottery for the South Carolina coast. *South Carolina Antiquities* 12:1–135.

Trinkley, M. B., 1980b. Investigations of the Woodland period along the South Carolina Coast. Unpublished Ph.D. dissertation, Department of Anthropology, University of North Carolina, Chapel Hill.

Trinkley, M. B., 1985. The form and function of South Carolina's early Woodland shell rings. In *Structure and Process in Southeastern Archaeology*, edited by R. S. Dickens Jr. and H. T. Ward, pp. 102–118. University of Alabama Press, Tuscaloosa.

Watts, W. A., E. C. Grimm, and T. C. Hussey, 1996. Mid-Holocene forest history of Florida and the Coastal Plain of Georgia and South Carolina. In *Archaeology of the Mid-Holocene Southeast*, edited by K. E. Sassaman and D. G. Anderson, pp. 28–38. University Press of Florida, Gainesville.

Webb, C. H., 1982. The Poverty Point culture. *Geoscience and Man* 17. Baton Rouge, Louisiana.

Webb, T., III, 1987. The appearance and disappearance of major vegetational assemblages: long-term vegetational dynamics in eastern North America. *Vegetatio* 69:177–187.

Webb, T., III, 1988. Eastern North America. In *Vegetation History*, edited by B. Huntley and T. Webb, pp. 383–414. Kluwer Academic Publishers, Dordrecht, Netherlands.

Webb, T., III, P. J. Bartlein, S. P. Harrison, and K. H. Anderson, 1993. Vegetation, lake levels, and climate in eastern North America for the past 18,000 years. In *Global Climate Since the Last Glacial Maximum*, edited by H. E. Wright Jr., J. E. Kutzbach, T. Webb III,

W. F. Ruddiman, F. A. Street-Perrott, and P. J. Bartlein, pp. 415–467. University of Minnesota Press, Minneapolis.

Wheeler, R. J., C. L. Newman, and R. M. McGee, 2003a. A new look at the Mount Taylor and Bluffton sites, Volusia County, with an outline of the Mount Taylor culture. *The Florida Anthropologist* 56:132–157.

Wheeler, R. J., J. J. Miller, R. M. McGee, D. Ruhl, B. Swann, and M. Memory, 2003b. Archaic period canoes from Newnans Lake, Florida. *American Antiquity* 68:533–551.

Widmer, R. J., 1988. *The Evolution of the Calusa: A Nonagricultural Chiefdom on the Southwest Florida Coast*. The University of Alabama Press, Tuscaloosa.

Widmer, R. J., 2004. Explaining sociopolitical complexity in the foraging adaptations of the southeastern United States: the roles of demography, kinship, and ecology in cultural evolution. In *Signs of Power: The Rise of Cultural Complexity in the Southeast*, edited by J. L. Gibson and P. J. Carr, pp. 234–253. The University of Alabama Press, Tuscaloosa.

Yarnell, R.A. and M.J. Black, 1985. Temporal trends indicated by a survey of Archaic and Woodland plant food remains from Southeastern North America. *Southeastern Archaeology* 4:93–106.

Climate Change and Cultural Dynamics: A Global Perspective on Mid-Holocene Transitions
David G. Anderson, Kirk A. Maasch and Daniel H. Sandweiss (Editors)

Chapter 14

Mid-Holocene culture and climate on the Northwest Coast of North America

Madonna L. Moss[1], Dorothy M. Peteet[2] and Cathy Whitlock[3]
[1]*Department of Anthropology, 1218 University of Oregon, Eugene, OR 97403-1218*
[2]*NASA/Goddard Institute for Space Studies, 2880 Broadway, New York, NY 10025; Lamont Doherty Earth Observatory, Room 204 New Core Lab, Palisades, NY 10964*
[3]*Department of Earth Sciences, Montana State University, Bozeman, MT 59717*

Abstract

On the Northwest Coast of North America, the middle Holocene was a time of changing climate and culture. In this chapter, we review the paleoclimatic and archaeological records of this region with the intent of approaching possible causal relationships between them. Early Holocene archaeological sites are relatively few, containing artifact assemblages predominantly of chipped stone, and only rarely, faunal remains. The mid-Holocene climate of the Northwest Coast was cooler and wetter than the early Holocene, but warmer and somewhat drier than today. Archaeologists have observed that compared to the early Holocene, during the mid-Holocene the number of archaeological sites increased, their average size was larger, and shell middens became common, preserving bone and antler technologies as well as abundant faunal remains. Archaeologists have traditionally viewed 5800 cal yr BP (5000 ^{14}C yr BP) as a major turning point in the prehistory of the Northwest Coast. Many archaeologists have perceived the supposed dramatic cultural changes as related to environmental changes including stabilization of sea levels, shellfish beds, and salmon runs. This review demonstrates that archaeological sites with mid-Holocene components are not common, nor are they well known. Nevertheless, a time of cultural transition does appear to be indicated at about 4850 cal yr BP (4300 ^{14}C yr BP). Mid-Holocene climate change undoubtedly affected Northwest Coast societies, and we suggest a few ways in which changing climate may have affected some of the key resources upon which people relied. Limitations of both the paleoclimatic and archaeological records, however, preclude all but a preliminary treatment of these issues.

1. Introduction

For our purposes, the Northwest Coast of North America encompasses the region extending from northern California to southern Alaska (Figs. 14.1–14.3). Across this region, the middle Holocene was a time of changing climate and culture. Our review and evaluation of the archaeological and paleoclimatic records for this time period was initially stimulated by the invitation to attend the FERCO International Conference on Climate and Culture at 3000 B.C. held at the University of Maine, in October 1998.

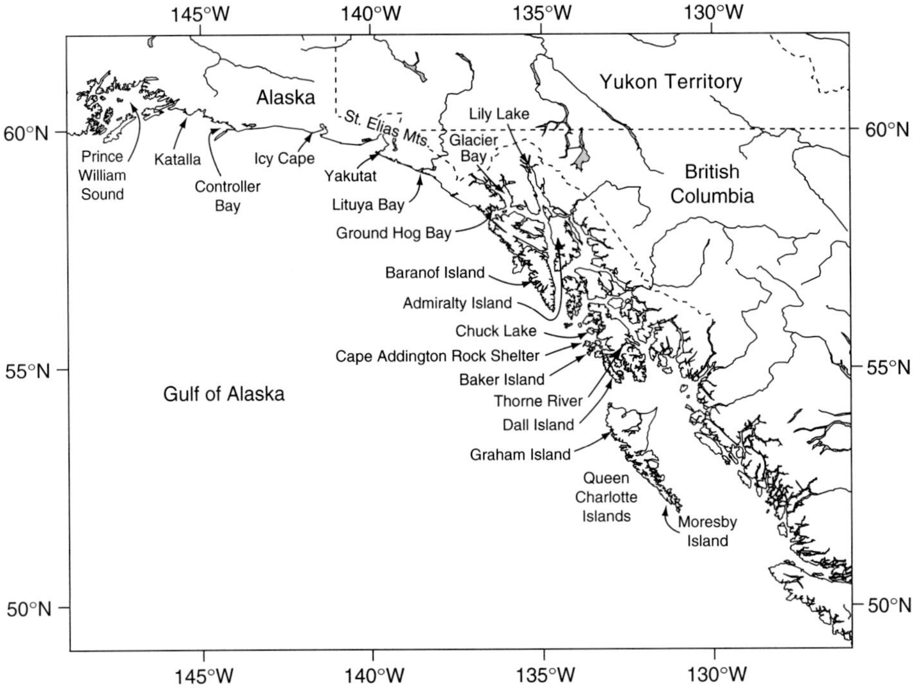

Figure 14.1. Place names and sites mentioned in the text, Alaska to the Queen Charlotte Islands.

In their 1996 study, Sandweiss et al. (1996) identified geoarchaeological evidence for El Niño-Southern Oscillation (ENSO) off the coast of Peru. At the FERCO conference and elsewhere in this book, strong evidence with relatively tight chronological control supporting the mid-Holocene onset of ENSO was presented for other localities in the Pacific basin, including Ecuador, Chile, the Galapagos Islands, and Australia. Beyond this southern Pacific basin region, however, ENSO is not easily identified, and in some world regions, it does not adequately explain the pattern of climate variability observed on interannual time scales. In their report of the conference results, Sandweiss et al. (1999) revised the chronology of events to employ calibrated radiocarbon ages in an attempt to promote global comparisons. The date for the onset of ENSO in the southern Pacific basin was revised to 5800 cal yr BP. Sandweiss et al. (1999) also re-framed the discussion to consider the possible causal connections among changes in mid-Holocene climate and culture more broadly.

In this chapter, we present a synthesis of mid-Holocene environmental history based on records of pollen, plant macrofossils, charcoal, limnologic, glacial, and marine sediments. The available data indicate that the environmental changes during the transition from early Holocene to mid-Holocene and then late Holocene were registered throughout the Northwest Coast. Superimposed on these long-term shifts were climate variations that took place more locally on annual-to-centennial time scales. As will be shown, the temporal resolution of available paleoclimatic evidence

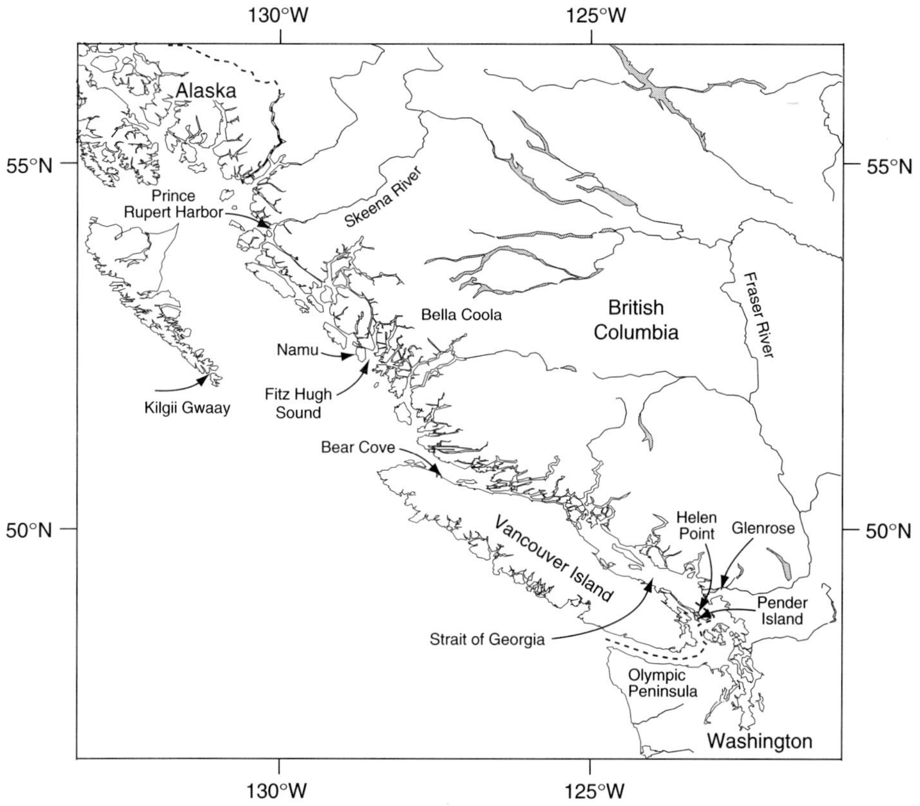

Figure 14.2. Place names and sites mentioned in the text, British Columbia to Washington.

precludes identification of ENSO events in the mid-Holocene, although a few records suggest a change in climate variability in the late Holocene that may represent the onset of ENSO. We also review the archaeological record and find evidence for change sometime after 4850 cal yr BP. The mid-Holocene on the Northwest Coast was one of those times of transition that Northwest Coast archaeologists invoke rather frequently, but always in terms of broad generalizations. 5800 cal yr BP (5000 ^{14}C yr BP) has been a key turning point in the "storyline" of Northwest Coast prehistory, and archaeologists often write about cultural developments before and after this juncture. A critical review of the limited archaeological record of this period, however, indicates that the Mid-Holocene period is not well understood. A time of cultural transition, or a change in the preservation of the archaeological record itself, appears to be indicated sometime after 4850 cal yr BP (4300 ^{14}C yr BP). An even more widespread transition, characterized by direct ties to ethnographic groups in some subregions, occurred after 3800 cal yr BP (3500 ^{14}C yr BP).

Mid-Holocene climate change undoubtedly affected Northwest Coast societies, but at this time we cannot demonstrate direct causal linkages. For example, some archaeologists have claimed that sea level stabilization and newly productive

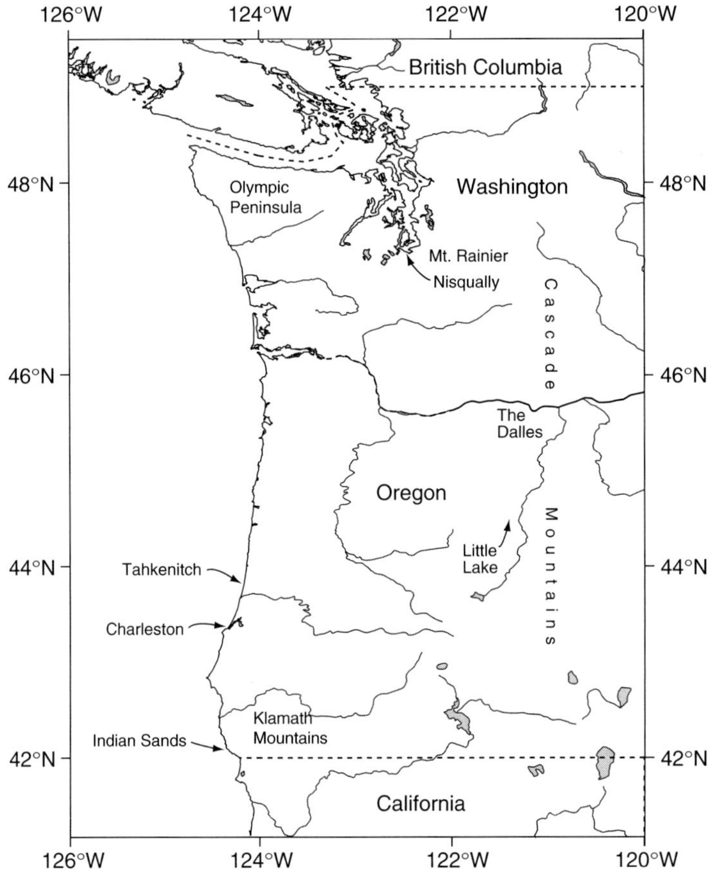

Figure 14.3. Place names and sites mentioned in the text, Washington and Oregon.

salmon populations during the mid-Holocene strongly influenced culture. Unambiguous evidence for these environmental changes is scarce, and how these may have derived from climate change remains unknown. To the extent feasible, we suggest how changing climate may have affected some of the key resources upon which people depended. Finally, we briefly discuss some avenues for future research that could yield data capable of approaching these unanswered questions.

2. Mid-Holocene paleoclimate

2.1. Overview

Over the last half-century, inferences about the environments of the northwest Pacific Coast during the mid-Holocene (ca. 7000–3500 cal yr BP) have been drawn from many sources, including pollen and plant macrofossil records that have been used to infer vegetation changes; charcoal records that provide information of past fire activity; chironomid and other proxies for paleolimnological conditions; records of glacial activity and treeline fluctuations; and stable isotope, pollen, and geochemical records from coastal marine cores. The timing of mid-Holocene climate changes is limited by the resolution of the existing studies and the accompanying radiocarbon stratigraphy. Unfortunately, much of the published data relies upon bulk radiocarbon chronology, and thus more precise chronology awaits accelerator mass spectrometry (AMS) ^{14}C ages. Here we summarize existing paleoenvironmental records from northern California to Alaska that provide information on the climate of the mid-Holocene period. Radiocarbon dates in the published studies have been converted to calendar years, using Calib 4.4 (Stuiver et al., 1998a).

Climate change occurs on multiple temporal and spatial scales, ranging from those that result from variations in the Earth's orbital configuration on 10^3–10^4 years to seasonal variations that occur within a single year and are highly spatially specific. The challenge in paleoenvironmental research is to disentangle the causes of and responses to climate variations on these different scales, recognizing that (1) such variations are superimposed upon each other, and (2) biotic and physical systems respond with different sensitivities and rates to various climate forcings (Bartlein, 1988; Whitlock and Grigg, 1999; Overpeck et al., 2003). Mid-Holocene climate variations occurring on interannual-to-centennial time scales were superimposed on millennial-scale changes in the seasonal cycle of insolation, which affected the entire hemisphere. To understand the impact of the short-term climate variability, it is necessary to recognize the influence of the long-term changes in shaping mid-Holocene environmental conditions.

In the early Holocene, between 11,000 and 7000 cal yr BP, summer insolation in the North America was $>6\%$ higher in summer and $>7\%$ lower in winter than at present as a result of the increased tilt of the Earth and the occurrence of perihelion in summer (Berger, 1978; Kutzbach and Guetter, 1986). Paleoclimate model simulations suggest that this amplification of the seasonal cycle of insolation increased temperature and decreased effective moisture in northwestern North America (e.g., COHMAP Members, 1988; Kutzbach et al., 1998; Bartlein et al., 1998). In addition, summer drought was intensified in the early Holocene because of the expansion of the northeastern Pacific subtropical high-pressure system (Heusser et al., 1985; Whitlock, 1992; Thompson et al., 1993; Bartlein et al., 1998). At most sites along the North Pacific coast, the early Holocene vegetation was characterized by more xerophytic species and higher fire frequencies than at present. Upper treeline in coastal mountains lay above its present elevation (Pellatt and Mathewes, 1997; Pellatt et al., 2000), and areas of prairie and steppe were expanded at low elevations (Barnosky et al., 1987). A general trend towards cooler and wetter conditions in the mid- and late Holocene is attributed to the attenuation of the amplification of the seasonal cycle and its effect on regional climate. As summer insolation decreased from a maximum at 11,000 to 9000 cal yr BP, summer temperature decreased, effective precipitation increased, and the northeastern Pacific subtropical high pressure gradually weakened. At 5800 cal yr BP, temperatures were warmer than today but cooler than before and the intensity of summer drought was greater than at present but less than in the early Holocene. Modern climatic conditions were established in the last 3000–2000 cal yr BP in most regions.

The paleovegetation response to this large-scale climate forcing is clearly expressed in pollen records throughout the Northwest Coast (see reviews by Heusser, 1977, 1983; Whitlock, 1992; Thompson et al., 1993; Whitlock and Brunelle, 2007). Charcoal records, on the other hand, suggest considerable geographic variation in fire activity during the Holocene that reflects subregional differences in climate as well as the local influences of topography, substrate, and fuel characteristics (Marlon et al., 2006; Whitlock et al., in press). Most sites show increases in charcoal and woody pollen abundance from low levels in the late-glacial period to highest levels in the late Holocene, suggesting that charcoal levels are governed by fuel characteristics and ultimately by climate and vegetation (Marlon et al., 2006), but, on finer time scales, there is little synchrony in fire occurrence.

Neoglaciation, a period of heightened glacial activity, began as early as 5800 cal yr BP (5000 ^{14}C yr BP) in the Canadian Cordillera and Cascade Range (Luckman et al., 1993; Clague and Mathewes, 1996). These initial advances in the mid-Holocene were followed by widespread glacial activity beginning at about 3800 cal yr BP (3500 ^{14}C yr BP) (Mann and Hamilton, 1995) in the St. Elias Mountains (Denton and Karlen, 1977), the Lituya Bay area (Mann and Ugolini, 1985), British Columbia (Ryder and Thompson, 1986; Osborn and Luckman, 1988), and Washington (Burbank, 1981). This glacial evidence is consistent with a cooler, wetter late Holocene compared with the rest of the Holocene.

Specific regions are discussed below, with a focus on the transition from the early to the middle Holocene.

2.2. Northern California

Analysis of diatoms, alkenones, pollen, and geochemistry in cores off the northern California coast suggests changes from the early to mid-Holocene and late Holocene that imply shifts in the strength of the California current (Barron et al., 2003). The period from 8200–3200 cal yr BP was marked by sea-surface temperatures that were 1–2°C cooler than the early Holocene and a relatively strong, broad California current. At 5150 cal yr BP, coastal redwood and alder pollen appear in the marine sediments and suggest increased effective moisture and the development of the southern coastal temperate rain forest. At 3200 cal yr BP, an increase of 1°C in sea-surface temperatures, based on the alkenone record, is attributed to a warming of fall and winter conditions and a weaker California Current. Intervals of pine pollen alternating with alder and redwood pollen suggest rapid changes in effective moisture and seasonal temperature that are attributed to an enhanced ENSO cycle in the last 3500 cal yr (Barron et al., 2003).

Pollen records from the Klamath Mountains of northern California and southern Oregon show vegetation histories that are strongly modulated by local characteristics (aspect, elevation, and geologic substrate) (Mohr et al., 2000; Briles et al., 2005;

Daniels et al., 2005), but in general they indicate the same trends as the marine records. High-elevation forests were composed primarily of pine, huckleberry, oak, and sagebrush in the early Holocene. After ca. 5600 cal yr BP, fir and mountain hemlock increased in abundance, suggesting a cooling trend or increased effective moisture. At wet sites in the Klamath region, a shift from xerophytic taxa, including oak, occurred after 4000 cal yr BP when more mesophytic species, including pine, fir, and Douglas-fir, were better represented. Charcoal records from northern California and southern Oregon show a consistent long-term trend in fire activity (Marlon et al., 2006). Pollen and charcoal records from the Sierra Nevada also show a decline in xerophytic and fire-adapted species as well as fire frequency after the early Holocene (Smith and Anderson, 1992; Brunelle and Anderson, 2003). Higher-than-present fire activity is noted throughout northern California at ca. 1000 cal yr BP.

2.3. Coastal Oregon

Pollen and charcoal data from the Oregon Coast Range suggest a similar shift in climate from the early Holocene to the mid-Holocene (Worona and Whitlock, 1995; Long et al., 1998; Long and Whitlock, 2002; Long, 2003). Pollen data from Little Lake indicate a forest dominated by Douglas-fir prior to 6850 cal yr BP. Red alder grew in areas of frequent disturbance, oak occupied the driest sites, and bracken ferns were present in forest openings. The vegetation was probably similar to the driest forests of the eastern Coast Range and western Cascade Range today. The charcoal data suggest that fires were frequent in the early Holocene and probably helped to maintain the xerophytic elements within the forest. In the mid-Holocene (6850–2750 cal yr BP at Little Lake), forest composition shifted towards more mesophytic and disturbance-sensitive taxa, including western red cedar, western hemlock, red alder, and sword ferns. The forest became more closed and fires were less frequent than before. This general decrease in fire activity is also noted along the Oregon Coast in association with an expansion of Sitka spruce and western hemlock in the last 4000 cal yr (Long and Whitlock, 2002). Similarly, a record from the western Cascade Range suggests that warmer-than-present conditions in the early Holocene allowed low-elevation forest species to move upslope of their present ranges (Sea and Whitlock, 1995); modern forests of western hemlock, Douglas-fir, and fir were established at about 4500 cal yr BP, when the conditions became cooler and effectively wetter than before.

2.4. Western Washington

Several pollen and macrofossil records are available from the Olympic Peninsula and Puget Trough region that provide evidence for greater-than-present summer drought in the early Holocene and a shift to cool, humid conditions in the mid- and late Holocene. Most sites show this shift to cool humid conditions as an increase in mesophytic taxa between 8000–7000 cal yr BP. In the southern Puget Trough, western red cedar became more abundant in the mid-Holocene and spread throughout humid lowland settings. This species was present in the northern Puget Trough between 6850 and 5700 cal yr BP, at 6000 cal yr BP in southern British Columbia (Wainman and Mathewes, 1987), and on the north-central coast of British Columbia at 4000 cal yr BP (Hebda and Mathewes, 1984). On the Olympic Peninsula and Puget Trough, the spread of western red cedar was followed by an increase in western hemlock and Sitka spruce (Heusser, 1977; McLachlan and Brubaker, 1995). In some locations in the Puget Lowland, western white pine became common (Barnosky, 1981; Cwynar, 1987), and in dry settings, areas of oak savanna shrank in size and were replaced by Douglas-fir and western hemlock forest (Hibbert, 1979). The available charcoal records suggest a decrease in fire activity in the mid-Holocene compared with the early Holocene (Cwynar, 1987), and fewer fires still in the late Holocene.

A plant macrofossil record from middle-elevation montane forests on Mount Rainier indicates the presence of subalpine fir, Douglas-fir, western white pine, noble fir, and lodgepole pine between 6850 and 3650 cal yr BP (Dunwiddie, 1986). This assemblage suggests warmer conditions and more frequent fires in the middle Holocene than at present. After 3650 cal yr BP, late-successional species, such as western hemlock, mountain hemlock, and Alaska-cedar, became important. The transition corresponds in time with the onset of Neoglaciation (Burbank, 1981), and cooler conditions and reduced fire frequency probably enabled the spread of late-successional species.

2.5. British Columbia

Hebda (1995) and Brown and Hebda (2003) summarize the vegetation and climate history of British Columbia from several pollen and charcoal records. The mid-Holocene is characterized as a warm and moist "mesothermic" period (7700–5150 cal yr BP) which followed an earlier warm and dry interval and was replaced by cool, moist conditions in the late Holocene. The mid-Holocene increase in moisture led to an expansion of western red cedar and western hemlock in the lowlands, and western hemlock and mountain hemlock at higher elevations (Hallett et al., 2003). Cooler water temperatures decreased fire activity, renewed glacial activity, and shifted upper treeline downslope in the mid- and late Holocene (Clague and Mathewes, 1989; Pellatt and Mathewes, 1994; Hebda, 1995; Pellatt et al., 2000; Brown and Hebda, 2003). Abrupt increases in moisture in the mid-Holocene are inferred from the expansion of muskeg on the Queen Charlotte Islands (Heusser, 1995) and bog expansion on Vancouver Island (Brown and Hebda, 2003). In southern British Columbia, fires became less frequent in wet western regions but apparently more common in dry eastern regions during the middle and late Holocene. Farther north along the coast, the early Holocene featured relatively large fires, but as conditions became wetter in the late Holocene, fires were confined to south-facing slopes (Gavin et al., 2003). In the drier Okanogan region of south-central British Columbia, the influence of climate as opposed to local controls was more evident in the late Holocene than in the middle Holocene (Gavin et al., 2006).

2.6. Southeastern Coastal Alaska

Heusser's (1960) pollen records from southeastern Alaska show a similar climatic pattern to that seen in British Columbia, with warmer, wetter conditions than previously (but drier than today) characterizing the mid-Holocene. Increases in mid-Holocene moisture are evident from the sediment stratigraphy (alder woody peat to sedge peat) and from increases in pollen of western hemlock. Lily Lake (Chilkat Peninsula) shows a similar pollen stratigraphy, but Cwynar (1990) does not discuss the stratigraphy with reference to climate.

At Pleasant Island in Glacier Bay, Alaska, Hansen and Engstrom (1996) interpret the mid-Holocene as a time of accelerated muskeg development, with the reappearance of lodgepole pine and a decline in hemlock and spruce about 7700–3650 cal yr BP (6900–3400 ^{14}C yr BP). However, they interpret the early Holocene as characterized by warmth and increasing precipitation, in contrast to the evidence of a warm, dry climate from areas both north and south (Heusser, 1960, 1995; Mathewes, 1985; Peteet, 1986, 1991; Hebda, 1995).

At Muskeg Cirque, Lituya Bay (Mann, 1983), western hemlock increases about 7950 cal yr BP (7100 ^{14}C yr BP) and pine expansion takes place after an earlier disappearance. Sedimentation rates also increase in the core, suggesting muskeg increase and increased moisture. Slate Mesa and Pike Lakes core analyses from the Yakutat area (Peteet, 1986, 1991), exhibited a mid-Holocene shift towards a cooler, wetter climate at 6850 cal yr BP (6000 ^{14}C yr BP), followed by cooler and even wetter conditions in the late Holocene beginning at 3800 cal yr BP (3500 ^{14}C yr BP). The mid-Holocene increase in western hemlock along with changes from woody peat to sedge suggests muskeg expansion regionally at that time.

2.7. Southcentral Coastal Alaska

At Munday Creek, Icy Cape, Heusser (1960) interpreted the pollen record as indicating mid-Holocene warmth and increased moisture compared with the early Holocene. Peteet (1986) re-cored the site and produced a radiocarbon-dated pollen and macrofossil stratigraphy that demonstrated changes in sedimentation rate and macrofossils that supported this climate inference. A temperature and precipitation reconstruction from the site suggests climate variations that are broadly consistent with the patterns of Holocene climate change in western Washington and coastal British Columbia (Heusser et al., 1985; Mathewes and Heusser, 1981). Both regions show early Holocene aridity followed by mid-Holocene cooling and increased moisture.

In both Controller Bay and Katalla (Heusser, 1960; Sirkin and Tuthill, 1987), and at Prince William Sound (Heusser, 1960), the mid-Holocene was dominated by alder, because the conifer species apparently were limited by migration distances southeastward (Heusser, 1960, 1995). The development of muskeg sometime after 8950 cal yr BP (8000 ^{14}C yr BP) during the mid-Holocene is implied by increases in sedge and sphagnum in the pollen record, suggesting wetter conditions.

2.8. Paleoclimate summary

The mid-Holocene of the Northwest Coast appears to have been cooler and wetter than the early Holocene, but warmer and somewhat drier than today. Paleoenvironmental data from the region suggest a shift first in precipitation ca. 7450–6000 cal yr BP, followed by cooler temperatures ca. 5000–3500 cal yr BP. These shifts resulted in substantial, although apparently gradual changes in the region's vegetation, some of which must have had direct impacts on Northwest Coast inhabitants. For example, the distribution and abundance of western red cedar, a key industrial resource used by Northwest Coast peoples prehistorically, changed as a result of increased temperature and effective moisture during the Holocene. Hebda and Mathewes (1984) suggest a direct linkage between climate change, the expansion of this conifer, and cultural adaptations.

Although large-scale shifts in Holocene climate are strongly expressed throughout the region, decadal-to-centennial climate variations are hard to discern and often show more irregular spatial patterns. For example, fire reconstructions suggest effectively drier conditions, at ca. 4000 and 1000 cal yr BP in the Klamath region, at 5500, 4000, 2200, and 1000 cal yr BP in the Sierra Nevada region, and at 850–350 cal yr BP in the Coast Range of British Columbia (Mohr et al., 2000; Brunelle and Anderson, 2003; Gavin et al., 2003). Likewise, tree-ring and glacial records indicate cooler conditions in recent centuries, but the onset of these events in Alaska, the Coast Range, and the Canadian Rockies is not synchronous (Luckman, 2002). Few records have the sampling resolution required to compare interannual-to-centennial patterns of environmental and biotic variability in different regions, and the paucity of data limits an assessment of ENSO's influence in the early and middle Holocene. Nonetheless, there is tantalizing evidence of increased climate variability in some records that may be caused by the onset of ENSO activity in the late Holocene (Barron et al., 2003; Markgraf and Diaz, 2000).

3. Archaeological models

Before turning to the archaeological evidence itself, we consider Knut Fladmark's paleoecological model of Northwest Coast prehistory that has profoundly influenced the way in which archaeologists have interpreted the cultural evidence for the middle Holocene time period. We briefly re-evaluate Fladmark's model in light of more recent discussions on mid-Holocene culture change.

3.1. Fladmark's (1975) model

Fladmark's (1975) paleoecological model of Northwest Coast prehistory is probably the most influential work related to our understanding of mid-Holocene culture change. In this ambitious study, Fladmark provided an environmental and ecological explanation for the entire course of Northwest Coast prehistory as it was known at that time. He characterized the 11,400–5800 cal yr BP (10,000–5000 ^{14}C yr BP) period as one of environmental instability, changing sea levels, unstable river gradients, and fluctuating climate. Archaeological sites dating from this time were thought to contain only chipped stone artifacts and little in the way of features. Fladmark considered this evidence of a "sub-climax" generalized adaptation. After 5800 cal yr BP (5000 ^{14}C yr BP), sea levels were thought to have stabilized, initiating both environmental and cultural change. Fladmark (1975, p. 262) wrote,

> After 5,000 [^{14}C yr] B.P., with the stabilization of the land-sea interface and the attainment of climax salmon productivity, shell-midden sites begin to appear along the entire coast. Shell-midden accumulations are the result of dense winter population aggregates existing on the expenditure of stored energy derived from the fall salmon fisheries, and reflect a settlement type characteristic of the ethnographic Northwest Coast cultural pattern. Stone grinding, ornaments, and art work occur more or less simultaneously along the coast with the advent of shell-middens.

Fladmark, and perhaps most Northwest Coast archaeologists, consider salmon to have been the single most important food resource on the Northwest Coast. Fladmark (1975, p. 195) claimed that "most salmon populations did not achieve climax productivity until at least 5,000 [^{14}C] B.P." Fladmark (1975, pp. 198–199) also mentioned that:

> peak sockeye and coho (salmon) productivity could not have been achieved until major drainage systems developed stable run-off and sedimentation patterns. Optimum spawning and maturation conditions depend upon regular seasonal discharge and sedimentation rates; any fluctuations away from normal stream behavior are detrimental to climax salmon populations.

Fladmark (1975, pp. 203–207) did concede that pink, chum, and chinook salmon may have been present in the region during the Wisconsin glaciation and that they would not have been particularly susceptible to fluctuations in run-off and changing stream gradients. However, he wrote, "it is difficult to conceive of any salmon, particularly sockeye and coho, attaining full productivity prior to the complete stabilization of stream gradients about 5,000 [^{14}C] B.P." (Fladmark, 1975, p. 207).

Fladmark originally did not emphasize changes in shellfish abundance as being causally related to shell midden accumulation. He wrote (Fladmark, 1975, p. 253), "the advent of shell-midden build-up at 5,000 [^{14}C] B.P. can be seen as not a function of preservation or increasing abundance of shell-fish, but instead the result of a shift to the winter village settlement pattern following the development of peak salmon productivity" (underlining in the original). A decade later, however, Fladmark (1986, p. 54) suggested that "all intertidal and anadromous resources were kept below climax productivity as a result of pronounced sea level instability which prevailed on the coast before about 4500 [^{14}C] years ago."

3.2. Re-evaluating Fladmark's model

Other archaeologists have proposed models to explain the development of Northwest Coast cultural complexity (see review in Matson and Coupland, 1995, pp. 145–154; Moss and Erlandson, 1995). Most models are not, however, anchored so explicitly to cultural developments at 5800 cal yr BP (5000 ^{14}C yr BP). Using the archaeological evidence that has accumulated over the last 25 years, we re-evaluate the key ingredients of Fladmark's model (1975, 1982) outlined above.

3.2.1. The appearance of shell middens

In 1975, no shell-midden components dated older than 5800 cal yr BP (5000 ^{14}C yr BP) were known from the Northwest Coast. We now know a number of shell-bearing components dating to the early Holocene: the 9750 cal yr BP (9500 ^{14}C yr BP) site of Kilgii Gwaay in southern Haida Gwaii (Fedje et al., 2001), the 9150 cal yr BP (8200 ^{14}C yr BP) shell midden at Chuck Lake (Locality 1) in southeast Alaska (Ackerman, 1992), and the 8600 cal yr BP (8440 ^{14}C yr BP) shell scatter at Indian Sands in Oregon (Moss and Erlandson, 1998a). As shown in Table 14.1, at least 14 other Northwest Coast shell components have yielded dates older than 5800 cal yr BP (5000 ^{14}C yr BP); the best known of these are: the 6850 cal yr BP (6000 ^{14}C yr BP) Cohoe Creek site (Ham, 1990; Fedje and Christensen, 1999; Christensen and Stafford, 2005), the 6850 cal yr BP (6000 ^{14}C yr BP) component at Namu in central British Columbia (Cannon, 1991; Carlson, 1998), and the 5900 cal yr BP (5100 ^{14}C yr BP) shell layer at Tahkenitch Landing in Oregon (Minor and Toepel, 1986). Although it is not well-known, the lowest levels of shell midden at the Boardwalk site in Prince Rupert Harbor have not been excavated; Ken Ames (2004) has estimated that based on accumulation rates, these may be as much as 8000 ^{14}C years old. Fladmark now acknowledges that shell middens by themselves are insufficient evidence of winter village occupation (Matson and Coupland, 1995, p. 148). It remains true that most shell middens are younger than 5800 cal yr BP (5000 ^{14}C yr BP), but how much younger? Although a number of sites with shell components date to 5800 cal yr BP, most shell middens are younger than 4850 cal yr BP (4300 ^{14}C yr BP). In addition, as we shall see, a significant proportion of the 5800 cal yr BP old sites are not shell middens.

Table 14.1. Archaeological sites on the Northwest Coast of North America with components dated to 6000–4400 ^{14}C yr BP.

Site Name	Location	Uncorrected ^{14}C age[a]	Dated material	Calendar age range[b]	Description	Reference
Ground Hog Bay	Icy Straits, AK	5360 ± 90	charcoal	6280–4530	microblades, bifaces, no fauna	Ackerman et al. (1979)
		4155 ± 95	charcoal		oldest date on component is early Holocene	
Lake Eva	Baranof Island, AK	5780 ± 90	charcoal	6720–6200	lithic site, no diagnostic artifacts	Swanson and Davis (1983)
		5500 ± 70	charcoal			
		5520 ± 100	charcoal			
Campen Midden	Killisnoo/ Admiralty	6050 ± 90	shell	6280–5560	small shell midden, dates only	Moss (unpublished)
	Island, AK	5520 ± 60	shell			
Thorne River Basket	Prince of Wales Island, AK	5360 ± 60	spruce root	6270–6000	isolated spruce root basket found in intertidal mud	Fifield (1995, pers. comm.)
Chuck Lake	Heceta Island, AK				small scale testing only	Ackerman et al. (1985)
Locality 2		5140 ± 90	shell	5300–4980	chipped stone, including microblades	
Locality 3		5240 ± 90	shell	5440–5210	"may not be directly ass'd with midden"	
Wolf's Lair	Baker Island, AK	4440 ± 60	wood club	5280–4880	wood club in sea cave	Moss and Erlandson (2000)
Kit'n'Kaboodle Cave	Dall Island, AK	5070 ± 70	shell	5660–4410	north rockshelter	Erlandson and Moss (2004)
		4910 ± 100	shell		north cave entrance	
		4660 ± 70	shell		north rockshelter, microblade core	
Naden Harbour (GaUd3)	Graham Island, B.C.	5860 ± 60	not specified		shell midden	Fedje et al. (2005a)

Continued

Table 14.1. continued

Site Name	Location	Uncorrected ^{14}C age[a]	Dated material	Calendar age range[b]	Description	Reference
Blue Jackets Creek	Graham Island, B.C.	5260 ± 440	charcoal	6400–5600	shell midden	Sutherland, CAA ^{14}C database
Cohoe Creek	Graham Island, B.C.	6150 ± 70	charcoal	7200–5600	shell midden, lithic quarry microblades, bifaces, bone tools	Ham (1990)
		5715 ± 90	shell			
		4990 ± 110	charcoal			
Cohoe Creek A Midden	Graham Island, B.C.	5680 ± 100	charcoal	6200–5600	shell midden, Late Moresby	Christensen and Stafford (2005)
		5590 ± 50	charcoal			Christensen and Stafford (2005)
		5380 ± 40	charcoal			Christensen and Stafford (2005)
		5320 ± 60	charcoal			Fedje and Christensen (1999)
		5290 ± 40	charcoal			Christensen and Stafford (2005)
		5260 ± 40	charcoal			Christensen and Stafford (2005)
		5230 ± 40	charcoal			Christensen and Stafford (2005)
		5090 ± 50	charcoal			Christensen and Stafford (2005)
		5000 ± 70	charcoal			Fedje and Christensen (1999)
		4970 ± 60	charcoal			Fedje and Christensen (1999)

		4930 ± 50	charcoal			Christensen and Stafford (2005)
		4900 ± 80	charcoal			Fedje and Christensen (1999)
		5990 ± 60	shell			Fedje and Christensen (1999)
		5650 ± 60	shell			Fedje and Christensen (1999)
		5570 ± 50	shell			Fedje and Christensen (1999)
		5550 ± 60	shell			Fedje and Christensen (1999)
Cohoe Creek A2 Midden	Graham Island, B.C.	5370 ± 70	charcoal	6300–5890	shell midden, Late Moresby	Fedje and Christensen (1999)
		5790 ± 60	shell			Fedje and Christensen (1999)
Cohoe Creek B Midden	Graham Island, B.C.	4420 ± 60	charcoal	5100–4820	shell midden, Late Moresby	Fedje and Christensen (1999)
		4390 ± 70	charcoal			Christensen and Stafford (2005)
		5020 ± 60	shell			Fedje and Christensen (1999)
		4890 ± 70	shell			Fedje and Christensen (1999)

Continued

Table 14.1. continued

Site Name	Location	Uncorrected ^{14}C age[a]	Dated material	Calendar age range[b]	Description	Reference
Strathdang kwun	Graham Island, B.C.	5740 ± 60	charcoal	6600–5100	shell midden, Late Moresby	Fedje and Christensen (1999)
		5330 ± 60	charcoal			
		4520 ± 60	charcoal			
		6000 ± 50	shell			
		5990 ± 70	shell			
		5810 ± 60	shell			
		5240 ± 60	shell			
Lawn Point	Graham Island, B.C.	5750 ± 110	charcoal	6720–6410	Component 4, microblade industry Late Moresby	Fladmark (1986)
Kasta	Moresby Island, B.C.	6010 ± 95	charcoal	6980–6000	Level 4, microblade industry	Fladmark (1986)
		5420 ± 100	charcoal		Level 3, Late Moresby	
Lyell Bay East	Lyell Island, B.C.	5350 ± 60	charcoal	6200–5750	microblades, Late Moresby	Fedje and Christensen (1999)
		5030 ± 40	charcoal		lithic site	
Arrow Creek 1	Moresby Island, B.C.	5650 ± 90	charcoal	6530–6310	no shell, chipped stone	Fedje et al. (1996)
		5650 ± 70	charcoal		Late Moresby	
Dodge Point	Lyell Island, B.C.	5490 ± 80	charcoal	6390–6200	cobble tools only	Fedje et al. (1996)
Kitandach	Prince Rupert Harbor, B.C.	4965 ± 95	charcoal	5890–4870	Period III: "shallow midden accumulations"	MacDonald and Inglis (1981)
		4460 ± 120	charcoal		and "restricted site areas." Cobble tools, bone, antler shell artifacts, ornaments	MacDonald (1983)

Dodge Island	Prince Rupert Harbor, B.C.	4790±100	charcoal	6500–5330	Period III (see above)	MacDonald and Inglis (1981)
		5555±140	charcoal			MacDonald (1983)
		4875±125	charcoal			MacDonald, CAA ^{14}C database
Lachane	Prince Rupert Harbor, B.C.	4630±105	charcoal	5570–4870	Period III (see above)	MacDonald and Inglis (1981)
		4455±80	charcoal			MacDonald (1983)
Ridley Island	Prince Rupert Harbor, B.C.	4390+80	charcoal	5750–5300	shell midden, dates considered too old	Inglis, CAA ^{14}C database
		4510±60	charcoal			
Paul Mason	Skeena River, B.C.	5050±140	charcoal	5930–4450	Bornite phase, obsidian microblades cobble tools, chipped stone	Coupland (1988)
		4745±195	charcoal			
		4655±130	charcoal			
		4395±130	charcoal			
		4350±320	charcoal			
Namu[c]	Fitz Hugh Sound, B.C.	6060±100	charcoal	7150–5610	Period 2-isolated shell lenses, pebble tools, bifaces microliths antler, bone artifacts	Carlson (1996), Cannon (1991)
		5740±100	charcoal			
		5700±360	charcoal			
		5590±100	bone			
		5590±90	charcoal			
		5240±90	charcoal			
		5170±90	charcoal			
		4775±130	bone	5450–4410	Period 3-massive shell deposits bone/antler tools dominate	
		4700±125	bone			
		4680±160	bone			
		4540±140	charcoal			

Continued

Table 14.1. continued

Site Name	Location	Uncorrected ^{14}C age[a]	Dated material	Calendar age range[b]	Description	Reference
King Island	Fitz Hugh Sound, B.C.	6140 ± 50	shell	6370–6200	shell midden	Cannon, CAA ^{14}C database
Fougner Bay	Fitz Hugh Sound, B.C.	5290 ± 60	charcoal	6100–6000	shell midden	Cannon, CAA ^{14}C database
Joashila	Bella Coola, B.C.	5340 ± 100	charcoal	6215–6015	lithic component	Hobler, CAA ^{14}C database
Bear Cove	Vancouver Island, B.C.	4576 ± 39	bone	ca. 5300	pebble and bifacial tools	Carlson (2003)
		4470 ± 60	bone	5310–4870	no shell, but rich marine fauna	
Beaver Harbor	Vancouver Island, B.C.	5275 ± 110	charcoal	6300–6100	shell midden, date considered too old	Carlson, CAA ^{14}C database
Deep Bay	Vancouver Island, B.C.	4860 ± 180	charcoal	5840–5330	date on scattered charcoal in possible clay living floor; one associated artifact; other dates are substantially younger	Pratt (1992)
Ts'ishaa	Benson Island, B.C.	5050 ± 60	charcoal	5920–4830	shell midden associated with higher sea level	McMillan and St. Claire (2005)
		4470 ± 70	charcoal			
		4430 ± 80	charcoal			
		4415 ± 35	charcoal			
Glenrose Cannery	Fraser River, B.C.	8150 ± 250	charcoal	9470–6360	Old Cordilleran, pebble tools, bifaces, bone/antler, ornaments, little shell, faunal remains predate 5000 BP.	Matson (1976)
		5730 ± 125	charcoal			

Glenrose Wet	Fraser River, B.C.	4590 ± 50	wood stake	5440–4300	wood stakes preserved in mud	Eldridge and Acheson (1992)
		4370 ± 60	wood stake			Moss and Erlandson (1998b)
		4260 ± 70	wood stake			
		3950 ± 60	wood stake			
St. Mungo	Fraser River, B.C.	4480 ± 90	charcoal	5365–4950	St. Mungo-11 dates from the component are younger; some shell; expedient chipped stone, decorated ground stone, some bone/antler tools	Pratt (1992)
		4440 ± 80	charcoal			
Pender Canal[c]	Gulf of Georgia, B.C.	ca. 5000–4500		5570–4070	Mayne Phase, "early midden" chipped stone, midden burial, labrets assigned to "main midden" assigned to "early midden"	Carlson (1987)
		5170 ± 220	bone			Carlson and Hobler (1993)
		4580 ± 550	bone			
		4430 ± 170	bone			
Crescent Beach	Surrey, B.C.	4440 ± 80	charcoal	5290–4870	Charles/St. Mungo; 11 other dates are substantially younger; some shell	Pratt (1992)

Continued

Table 14.1. continued

Site Name	Location	Uncorrected ^{14}C age[a]	Dated material	Calendar age range[b]	Description	Reference
Helen Point	Mayne Island, B.C.	5420 ± 230	charcoal	6440–5930	Mayne Phase, very little shell; bifaces, quartz and obsidian microblades, ground and chipped slate, labrets, antler, rock slab features	Carlson (1970) Carlson (1998, pers. comm.)
45-PI-72	Nisqually, WA	6130 ± 70	shell	6390–5930	shell midden	Wessen (1998, pers. comm.)
		5260 ± 70	charcoal			
Tahkenitch I	Douglas County, OR	5100 ± 70	charcoal	5920–5750	estuarine shellfish & fish chipped stone, bone/ antler tools	Minor and Toepel (1986)

[a] Sites listed here have components that date between 6000 and 4400 ^{14}C yr BP. In some cases, components encompass materials older and/or younger than the mid-Holocene. This list does not include all radiocarbon dates for individual sites. The focus here is on the mid-Holocene record, but if a single component is of long duration, bracketing dates are included.

[b] Age ranges are expressed in cal yr BP. For most sites, they were established by calibrating individual dates using Calib (Stuiver et al., 1998a). For Alaska and British Columbia shell dates, ΔR value of 278 ± 50 was used (Moss et al., 1989). For Washington, ΔR value of 240 ± 50 was used (Erlandson and Moss, 1999). For six sites, calibrated age ranges were estimated by using a table derived from the InCal98 calibration curve (Stuiver et al., 1998b), available at http://www.rlaha.ox.ac.uk/orau/intcal98.14c.

[c] The dates on human bone from Namu and Pender Canal were treated as mixed marine and atmospheric samples. The ΔR value of 278 ± 50 was used, and based on the $\delta\ ^{13}C$ values on bone samples from coastal British Columbia (Chisholm et al., 1983), these were estimated to be 90% marine carbon.

3.2.2. Sea level change

As evidence has accumulated since Fladmark's study, we know that sea level histories must be reconstructed at local scales. Since the time Fladmark proposed his original model, the magnitude of Holocene sea level change has also been reduced. For example Fladmark estimated that sea level at Prince Rupert Harbor at 8350 cal yr BP (7500 ^{14}C yr BP) was 40–50 m above modern, while Clague (1983, p. 337) showed that by 9500 cal yr BP (8500 ^{14}C yr BP), it was within 2–3 m of modern (Matson and Coupland, 1995, p. 148). Recent work on the Queen Charlotte Islands ("Gwaii Haanas" or Moresby Island in particular) demonstrates that sea level was the same as it is now about 10,500 cal yr BP (9300 ^{14}C yr BP), 15 m higher at 10,100 cal yr BP (8900 ^{14}C yr BP), relatively stable between 10,100–6300 cal yr BP (8900 and 5500 ^{14}C yr BP), and falling gradually since 6300 cal yr BP (5500 ^{14}C yr BP) (Fedje et al., 1996, p. 133). On Gwaii Haanas, then, sea levels were close to modern in the early Holocene, with gradual sea level change – not stabilization – after 6300 cal yr BP (5500 ^{14}C yr BP). For example, sites on Haida Gwaii dating to the Late Moresby tradition (Table 14.1) occur on raised beaches 12–15 m above current sea level, while sites dating to the subsequent Early Graham tradition occur on raised beaches approximately 7 m above current sea level (Fedje and Mackie, 2005, pp. 160–161; Fedje et al., 2005c).

Along the Cascadia subduction zone, from Vancouver Island south to northern California, sea levels have been affected by earthquakes over the last 2000 or more years. While we can generalize to state that marine transgression brought the sea in most areas to near modern levels by 4500 cal yr BP (4000 ^{14}C yr BP), the magnitude and timing of sea level change varies considerably from place to place along the Northwest Coast, depending on many variables (Mobley, 1988). For example, the Cape Addington Rockshelter did not become available for human occupation until sometime after 1950 cal yr BP (2000 ^{14}C yr BP) (Moss, 2004). The generalization that eustatic sea levels stabilized at 5800 cal yr BP seems of less value viewed in this context, particularly with regard to predicting salmon abundance and distribution.

3.2.3. Salmon productivity

Salmon remains have been found in late Pleistocene and early Holocene paleontological localities in southeast Alaska and on the Queen Charlotte Islands (Heaton and Grady, 2003; Fedje and Mathewes, 2005; Wigen, 2005, p. 108). Early Holocene salmon bones have been found at the Kilgii Gwaay and Richardson Island archaeological sites on the Queen Charlottes (Fedje et al., 2005b; Mackie et al., 2004; Mackie and Sumpter, 2005). In addition, Early Holocene settlement sites on the Charlottes are located closer to salmon streams than expected, suggesting that salmon were a key resource during this time (Mackie and Sumpter, 2005). Columbia River salmon were abundant enough to attract heavy use by humans at The Dalles during the early Holocene (Cressman, 1960; Butler, 1990; Butler and

O'Connor, 2004). Circumstantial evidence at the Milliken site indicates that Fraser River salmon were also the focus of early Holocene occupation (Carlson, 1998). Both The Dalles and Milliken sites are located in the interior along large rivers, where sea level change would not be expected to have much of an impact. The location of the Thorne River site, however, strongly suggests salmon procurement at ca. 8350 cal yr BP (7500 ^{14}C yr BP) (Holmes et al., 1989). We would expect that salmon using this drainage on Prince of Wales Island in southeast Alaska would be affected by changing stream gradients and sea levels, yet the fish apparently were available to humans during the period of site occupation. The strongest evidence for human reliance on pre-5800 cal yr BP old salmon runs comes from Namu on the central coast of British Columbia, for which Cannon (1991) and Carlson (1998) assert a salmon-based economy dating between 7450 –and 5800 cal yr BP (6500 and 5000 ^{14}C yr BP). Although regional evidence is limited, the prehistoric use of salmon prior to 5800 cal yr BP appears to contradict Fladmark's claim that stable sea levels after that time were a prerequisite for salmon use. Granted, the prehistoric geographic distribution and abundance of the various salmon species are not yet documented. Interestingly, weir or trap fishing may have begun as early 5300 cal yr BP (4600 ^{14}C yr BP) at the Glenrose wet site (Eldridge and Acheson, 1992). Although at least 10 Northwest Coast wood stake fishing weir sites are older than 3200 cal yr BP (3000 ^{14}C yr BP), the vast majority of weir sites date after 2600 cal yr BP (2500 ^{14}C yr BP) (Moss and Erlandson, 1998b; Byram, 2002). How strongly these cultural developments relate to technological, social, environmental, or taphonomic variables is not yet known.[1]

Despite these issues, Fladmark's model retains great appeal for Northwest Coast archaeologists. It is the foundation for recent statements, such as that by Coupland (1998, p. 50) that "maritime adaptations – evolved on the central Northwest Coast between 5000 and 4500 [^{14}C yr] BP." Fladmark's model helps tell a coherent story about the course of Northwest Coast prehistory (Moss and Erlandson, 1995). Now, we turn more directly to the mid-Holocene sites themselves.

4. The archaeological record

4.1. 5000 year old archaeological sites

Since this review was initially prompted in response to Sandweiss and Maasch's directive to consider culture and climate at "3000 B.C." (5800 cal yr BP), Moss

[1] For the interior Columbia Basin, Chatters et al. (1995) estimated the effects of climatic warming on salmon fisheries. They found conditions for salmon production prior to 4350 cal yr BP (3900 ^{14}C yr BP) to be poor, between 3650 and 2350 cal yr BP (3400 and 2300 ^{14}C yr BP) to be optimal, and after 1100 cal yr BP (1200 ^{14}C yr BP) to be good, equivalent to current climatic conditions. These shifts, however, are not easily extrapolated to coastal environments, as they derive from changing fluvial environments in the interior Northwest.

focused on sites ranging from 6850 to 5000 cal yr BP (6000 to 4400 ^{14}C yr BP). Other archaeologists involved in this FERCO project have assessed a wider time frame, due to limitations in chronological resolution of local records (e.g., Chapter by Kennett et al. in this book).

Identifying mid-Holocene sites on the Northwest Coast is easier said than done, because of several methodological problems. Because we rely on radiocarbon ages derived from charcoal, shell, and bone samples in this section of the chapter, calibration is essential. Only the ranges of calibrated dates are presented in Table 14.1 due to differences in reporting and changes in calibration curves over the years. Further, on the Northwest Coast, ancient trees can grow to be more than 1000 years old, so charcoal dates can be affected by Schiffer's (1986) old wood problem. The shell dates are affected by both geographic and temporal variations in the marine reservoir effect. Several of the dates used in this review were on human bone (from the Namu and Pender Canal sites), an additional source of imprecision. Dates on human bones from Bluejackets Creek were not included because glue was used to stabilize the bones (Sutherland, as cited in Canadian Archaeological Association C14 Database, http://www.canadianarchaeology.ca/). The notes to Table 14.1 specify the ways in which we addressed these and other problems.

Over the approximately 1500 mile (2400 km) length of the Northwest Coast considered in the archaeological review, extending from Yakutat Bay in the north to the California border in the south, we could find only 38 archaeological components dated between 6850 and 5000 cal yr BP (6000 and 4400 ^{14}C yr BP). 5000 cal yr BP (4400 ^{14}C yr BP) was chosen as a cut-off point, because after that time, the number of dated archaeological components appears to increase. Of the 38 components, 11 contain lithic artifacts only and 20 have shell midden components. Three sites (two wet sites and one sea cave) contain wood or basketry remains. Four sites do not fall unambiguously in one of these categories. Nine sites include evidence of microblade technologies, sometimes considered diagnostic of the early Holocene (but see Moss et al., 1996). Other lithic sites contain cobble tools, bifaces, and other attributes of Carlson's (1996) Pebble Tool tradition also known as the Old Cordilleran (Matson and Coupland, 1995). The lithic assemblages, in both shell-bearing and lithics-only sites, demonstrate cultural continuity with the early Holocene. Of the shell-bearing sites, the amounts of shell vary considerably, due to combinations of cultural and non-cultural factors. The range of shellfish use, however, does not appear to represent a "sea change" in economic adaptation. At multi-component sites in Prince Rupert Harbor and at Glenrose, the mid-Holocene shell deposits contain mostly mussels, while subsequent components at these same sites contain proportionately more clams. This is not a regional phenomenon, however, with the early Holocene Chuck Lake shell midden containing predominantly butter and littleneck clams, the same species abundant in the mid-Holocene Locality 2 and 3 middens (Ackerman et al., 1985).

Regarding vertebrate fauna available since the early Holocene, the Kilgii Gwaay assemblage shows that rockfish, lingcod, dogfish, sea otter, harbor seal, river otter, sea lion, alcids, albatross, and other seabirds and waterfowl were available on

Haida Gwaii by 9750 cal yr BP (9500 ^{14}C yr BP) (Fedje et al., 2001). The early Holocene Chuck Lake Locality 1 assemblage shows an equally diverse set of resources available in southeast Alaska at 9150 cal yr BP (8200^{14}C yr BP). At Bear Cove – despite the virtual absence of shell – porpoise/dolphin, northern fur seal, Steller sea lion, sea otter, rockfish, Pacific cod, herring, ratfish, dogfish, salmon, sculpin, greenling, lingcod, ducks, loon, gull, cormorant, and common murre are represented in Component I, dating to 9000–5000 cal yr BP (8020–4470 ^{14}C yr BP; Carlson, 1979, 2003). At Glenrose between 6850 and 5800 cal yr BP (6000 and 5000 ^{14}C yr BP), marine resources included harbor seal, loon, western grebe, diving ducks, flatfish, eulachon, and salmon, in addition to abundant elk and deer. At Namu, 7450–5800 cal yr BP (6500–5000 ^{14}C yr BP), we find a similarly broad faunal inventory with salmon making up over 80% of the identified fish. In their 1990 review, Hebda and Frederick (1990, p. 330) state that as early as 7450 cal yr BP (6500 ^{14}C yr BP), vertebrate and invertebrate marine resources upon which Northwest Coast societies depended were well established. Yet they go on to reinforce Fladmark's model by stating that after 5800 cal yr BP (5000 ^{14}C yr BP),

> there is an apparent dramatic increase in the number of marine remains in archaeological deposits: this is the time when vast quantities of shellfish remains are deposited in massive shell middens (Fladmark, 1982, p. 110). At previously occupied sites shell layers occur on top of primarily shell-free layers. There is often a temporal hiatus of as much as 1,000 to 1,500 years between these deposits. The transition from shell-free to shell-rich matrices appears abrupt. In sites where these deposits mark the initial occupation, shellfish remains are heavily concentrated in the earliest layers (Hebda and Frederick, 1990, p. 332).

To support this statement, Hebda and Frederick (1990) cite evidence from the following sites: St. Mungo, Glenrose, Pender Canal, Helen Point, Yuquot, Bear Cove, Namu, Prince Rupert Harbor, and Rosie's Rockshelter. Yet the relevant occupation levels at St. Mungo, Glenrose, Yuquot, and Rosie's Rockshelter date after 5150 cal yr BP (4500 ^{14}C yr BP). Massive and dense shell middens in Prince Rupert Harbor post-date 3200 cal yr BP (3000 ^{14}C yr BP). Age estimates for the Bear Cove III shell midden range from 5300 to 950 cal yr BP (4400 to 1000 ^{14}C yr BP) (Carlson, 2003, p. 72). The earliest component at Helen Point contained "only minute amounts of highly fragmented shell" (Carlson, 1970, p. 114). The age of the "main midden" at Pender Canal is 5150 cal yr BP (4500 ^{14}C yr BP) (Carlson and Hobler, 1993, p. 38), although the standard deviation on the ^{14}C date is ± 550. Not mentioned above is the Tahkenitch Landing site, but the 5900 cal yr BP (5100 ^{14}C yr BP) shell facies here cannot be considered a "massive" shell midden. Namu does have a substantial shell midden containing "vast quantities of shellfish remains," dated to 5600–5150 cal yr BP (4800–4500 ^{14}C yr BP). The Chuck Lake Locality 3 midden is rich in shell, but is not a large deposit. It is true that some large shell middens accumulated after 5800 cal yr BP (5000 ^{14}C yr BP), but it would be more accurate to say that most such sites post-date 4850 cal yr BP (4300 ^{14}C yr BP).

This brief review leads us to suggest that there was no dramatic cultural change on the Northwest Coast at 5800 cal yr BP (5000 ^{14}C yr BP). The available evidence indicates continuity with the early Holocene. The stone tool technologies represented in the North Coast microblade complex, the Pebble Tool tradition, and the Old Cordilleran continue. The basic faunal inventory shows no dramatic change, although we have insufficient data to address fine-grained subsistence changes within subregions. Further, the number of Northwest Coast sites does not increase substantially until sometime after 4850 cal yr BP (4300 ^{14}C yr BP). More complete analyses of the post-4850 cal yr BP record are necessary to establish the chronological threshold when bone and antler technologies and ground slate and polished stone technologies become significant proportions of artifact assemblages. More substantial evidence of culture change appears to have occurred after 3800 cal yr BP (3500 ^{14}C yr BP), but that is beyond the scope of this chapter. By that time, the evidence for increased populations and greater sedentism on the Northwest Coast is clear and widespread.

5. Mid-Holocene environmental change

In the passage quoted above, Hebda and Frederick (1990, p. 332) mention the phenomenon of shell layers occurring on top of primarily shell-free layers. They also allude to a temporal hiatus of 1000–1500 years between these deposits at some sites. Certainly, there are taphonomic reasons for initial shell layers to deteriorate in the naturally acidic Northwest Coast forest soils more quickly than subsequent layers built into a partially neutralized soil matrix (see Stein (1992) for another explanation of this process in a late Holocene site). Interestingly, the older shell midden deposits at Boardwalk were left unexcavated because of their waterlogged condition and great depth (Ames, 2004). Could what appears to be a temporal hiatus evident in some sites be related to sea level change or climate-driven erosional events?

We have already presented the evidence for mid-Holocene climate change. The intensity or duration of summer drought decreased in the mid-Holocene, compared with that of the early Holocene, and declined further in the late Holocene. The expansion of western red cedar is consistent with these trends as this species requires abundant precipitation and moist conditions that make drought stress and frequent forest fires unlikely. Hebda and Mathewes (1984) suggest that western red cedar survived south of the glacial margin at the end of the Pleistocene, and then spread north along the coast of British Columbia. During the early Holocene, western red cedar was rare, but between 6850–5700 cal yr BP and 4000 cal yr BP, it increased in abundance from the Puget Trough to the north-central coast of British Columbia in a time transgressive pattern. On Anthony Island (southern Haida Gwaii), cedar-type pollen "was essentially absent" until after 4300 ^{14}C yr BP (Hebda et al., 2005, p. 67). Hebda and Mathewes (1984, p. 712) state that Northwest Coast cultural developments were "environmentally constrained by limited occurrence and abundance" of red cedar before 5800 cal yr BP. Based on the ethnographic importance of cedar – used to build houses, canoes, furniture, all sorts of bowls, boxes, baskets, and other implements and tools – they infer that the establishment of abundant western red cedar was one of the primary stimuli to the development of the Northwest Coast material culture known ethnographically. This hypothesized relationship between western red cedar and culture change has received surprisingly little attention by archaeologists, nor have the responses of other resource populations to mid-Holocene climate changes been studied.

6. Recent fluctuations in climate and resources

Sandweiss and Maasch (1998, pers. comm.) have characterized the onset of ENSO as a "global environmental transformation." ENSO is an oscillation in which the atmosphere and ocean interact to produce a slow, irregular variation between two extremes, a warm "El Niño" phase and a cool "La Niña" phase (Pacific Northwestern Climate Impacts Group, 1999). During an El Niño event, sea surface temperatures are unusually high in the central and eastern Pacific, coastal sea surface temperatures are high along the tropical coast of South America and usually along the coasts of both Americas, and rainfall patterns shift, bringing drought to usually rainy areas and rain to usually dry areas. During a La Niña event, sea surface temperatures are unusually low in the central and eastern Pacific, coastal sea surface temperatures are low along the tropical coast of South America and usually along the temperate coasts of both Americas, and normal rainfall patterns intensify (Pacific Northwestern Climate Impacts Group, 1999).

As discussed in earlier sections, at the current time, we have no paleoclimatic evidence that records ENSO events on the Northwest Coast during the mid-Holocene, but there is some suggestion that interannual-decadal variability was more significant in the late Holocene. The expression of climate variability on this time scale would show an irregular spatial pattern, judging by 20th century instrumental observations. Consideration of the recent impacts of ENSO in the 20th century provides a source of analogies that might be useful in understanding how climate change may have affected key resource populations that were of central importance to Northwest Coast societies in the past. In this section, we briefly look at recent ENSO impacts on marine communities in order to speculate on some of the possible effects of prehistoric climate change.

McGowan et al. (1998) recently surveyed the effects of 20th century El Niño episodes on marine organisms in the North Pacific. Although El Niño events vary in intensity and are not always environmentally disastrous, the severest episodes have directly impacted entire ecosystems. McGowan et al. (1998, p. 210) identified 12 warm episodes in the California Current since 1916, with "remarkable" warm events between 1957 and 1958, and between 1982 and 1984. The authors also documented a clear increase in sea surface temperatures from California to the Gulf of Alaska since 1977. Warm surface waters have blocked the upwelling of nutrient-rich cold waters, resulting in an overall decline of primary marine productivity. Levels of plankton have decreased by 70% in some areas. In California, kelp beds have become greatly diminished, and abalone, squid, and anchovy populations have drastically declined. El Niño events have been disastrous for commercial fishing off British Columbia, Washington, and Oregon, with coho and chinook salmon undergoing significant decline. Oregon's three most important seabird species suffered poor breeding years, and pup counts for California sea lions and northern fur seals dropped substantially. The large warming episodes in the California Current system are clearly linked to El Niños, and the geographic position of the current itself has shifted over time.

Climatic fluctuations in the Gulf of Alaska do not appear to be as closely related to El Niño. However, a surge of warm weather in the 1980s is linked to the wholesale death of young among northern fur seals and Steller sea lions in parts of Alaska. Sea lions, fur seals, common murres, kittiwakes, ocean perch, herring, rockfish, shrimp, king crab, and other crustaceans declined. But some Gulf of Alaska fish stocks actually increased, including "spectacular shifts upward" in sockeye and pink salmon (McGowan et al., 1998, pp. 214–215). Pollock, hake, and Pacific cod populations appear to have increased as well. McGowan et al. also suggest that the last 20 years of warmer weather may signal a more significant climate change than a temporary El Niño effect (see also Chavez et al., 2003).

El Niño events also have caused changes in the geographic distribution of marine organisms. Pacific mackerel and market squid migrating north from California are found in greater numbers off the Oregon, Washington, and British Columbia coasts during El Niño years. Northern elephant seals have recently established a new breeding colony near Charleston, Oregon, thought to be a result of the 1997 El Niño event (Hodder et al., 1998; Jan Hodder, 1997, pers. comm.). Northern fur seals and California sea lions have also shifted their distributions to the north (Ono, 1995, p. 114). El Niño affects pinnipeds through a number of mechanisms: (1) increased sea level and heightened storm activity can increase pup mortality at breeding sites, (2) sea temperature changes can alter prey distribution causing changes in pinniped foraging patterns and also emigration, and (3) changes in mortality and fertility cause longer term changes in pinniped population structure (Trillmich et al., 1991, pp. 258–260).

In addition to interannual variability, climate has fluctuated on decadal time scales in the north Pacific. The Pacific (inter)Decadal Oscillation or PDO (Mantua et al., 1997) differs from ENSO in that 20th century PDO "events" persisted for 20–30 years, while typical ENSO events persisted for 6–18 months. The effects of PDO are most visible in the North Pacific, while the opposite is true for ENSO. Neither the relation between ENSO and the PDO nor the mechanism to explain the decadal climate variability is known. Nonetheless, ocean–atmospheric interactions in the north Pacific region on decadal time scales seem to affect marine biological and hydrological responses, as evidenced by variations in Alaska salmon production and streamflow on Pacific coastal rivers (Hare and Francis, 1995; Mantua et al., 1997; Gedalov and Smith, 2001; Chavez et al., 2003).

During the recent El Niño (1997–1998), catastrophic coastal erosion attracted great media attention as luxury condominiums and mobile homes in Oregon and California fell into the sea. The 1982–1983 El Niño was also severe. When warm water moves along the equator eastward towards Peru, there is a wave-like bulge in sea level (Komar, 1997, p. 118). The water level changes associated with these sea level waves can be large; for example, in February, 1983, sea level averaged almost 20 cm above the previous maximum for that month (Komar, 1997, p. 120). Obviously, such high sea levels, especially during the winter storm season, can result in catastrophic coastal erosion. During the 1997–1998 El Niño, four months of sustained sea levels 10 cm above normal resulted in 12 m of coastal retreat on northeastern Haida Gwaii (Barrie et al., 2005, pp. 18–19). Interestingly, Carlson (1997) has suggested that sea level rise about 5800 cal yr BP (5000 ^{14}C yr BP) on the central coast may have wiped out much of the shoreline archaeological record at this time, leaving only landward edges of sites – not shell dumps – preserved. The erasure of certain portions or components of archaeological sites may give the erroneous impression of cultural abandonment or of an abrupt cultural change. Instead of cultural change, in some cases we may have shoreline changes affecting site preservation. Of course, not all winter storm surges on the Northwest Coast are associated with ENSO (Schwing et al., 1999).

Could the onset of ENSO variations at 5800 cal yr BP have initiated cooler and wetter conditions off the Northwest Coast? Did the modern pattern of cyclic storms with heavy precipitation begin at this time? Could these climatic conditions directly affect the environmental suitability of the region for the culturally important western red cedar? Did higher wave energies during such periods impede maritime subsistence or remove some archaeological components from the record? Could El Niño have affected the distribution of economically important, fish, pinnipeds, and seabirds? Could "climatic chaos" (Sandweiss and Maasch, 1998, pers. comm.) after 5800 cal yr BP have caused a delay in Northwest Coast cultural developments that were underway earlier? Or were human populations still small, mobile, and flexible enough not to have suffered much of an impact? Could ENSO-driven marine erosion have caused the wholesale destruction of many archaeological sites dated to the mid-Holocene along some parts of the coast? At present, we have little data to address these questions, and high-resolution records that preserve interannual and interdecadal climate variability are clearly needed to compare the effects on the southern Northwest Coast adjacent to the California Current with those on the northern Northwest Coast, which is affected by the Aleutian Low. Tentatively, it would seem that the environmental impacts in the south would have somewhat negative consequences for resource populations and the people who depended on them, but not necessarily at the same time in different regions.

Even though ENSO has attracted a great deal of scientific (and public) attention in recent years, it is by no means completely understood. The recent expression of ENSO and its relation to the Pacific Decadal Oscillation pattern on decadal time scales are confounded by global warming due to the build-up of greenhouse gases. These human-induced impacts on climate reduce the utility of projecting current climate shifts – and their impacts on resource populations – onto the past (but see Sandweiss et al., 2004). Climate change during the mid-Holocene was more closely tied to orbital variations, although the role of anthropogenic impacts on climate – as might be associated with the shift to agriculture in many parts of the world – has recently been suggested. Clearly, the causes and effects of changing climate today could be vastly different from those of the past.

6.1. Tentative conclusion

Undoubtedly, mid-Holocene climate change on the Northwest Coast affected the course of cultural development. By ca. 7000 cal yr BP for most of the coast, the climate was cooler and wetter than the early Holocene, although warmer and somewhat drier than today. Between 4500 and 3500 cal yr BP, the cool wet conditions of the present day were established. Interannual and interdecadal climate variability is registered in the 20th century, and the onset of such variations is not known with any certainty. The mid-Holocene expansion of western red cedar in the southern region probably had significant implications for culture. However, western red cedar was present on the landscape before this time, and the pollen and archaeological records are presently inadequate to demonstrate a causal relationship between red cedar expansion and culture change. Significant culture change on the Northwest Coast does not occur until after ca. 4900 cal yr BP during a time of gradually changing climate. While we appreciate the important role that climate may have played, the lack of fine chronological resolution precludes establishing a clear relationship between climate and culture on the Northwest Coast.

Northwest Coast archaeologists have traditionally viewed 5800 cal yr BP (5000 ^{14}C yr BP) as the threshold of environmental stabilization and a major turning point in the region's prehistory. This review tentatively suggests that mid-Holocene climate change was initiated earlier, and that on millennial time scales, Holocene changes were gradual. We presently do not have much high-resolution paleoclimatic evidence on shorter time scales to assess the registration or impact of ENSO or decadal variability for the Northwest Coast. In addition, the variable effects of isostasy along the northern Northwest Coast and tectonism along southern Northwest Coast led to highly dynamic sea levels during much of the Holocene, suggesting that this key littoral ecotone was never really stable.

The archaeological evidence for the mid-Holocene time period is quite scarce. The data available for ca. 5800 cal yr BP demonstrate cultural continuity with both earlier and later periods. Only after ca. 4900 cal yr BP do we see an increase in the number and size of sites, increasing sedentism, significant technological change, and evidence for emerging cultural complexity on a regional scale. These changes intensify and become more widespread after 3800 cal yr BP. Northwest Coast culture change clearly was not synchronous with climate change.

For the southern hemisphere, data suggest an onset of ENSO at 5800 cal yr BP. Available data appear insufficient to address whether or not ENSO drove or was part of the mid-Holocene climate change experienced on the Northwest Coast of North America. Based on the record elsewhere in the world, however, it appears to be a possibility. Our knowledge of contemporary ENSO is limited; some scientists have recently asserted that the traditional climatological definition of El Niño is equator-centric and that a more complete definition for regions outside the tropics is needed (Schwing et al., 1999). Even today, it is risky to attribute changes in marine resource populations directly to El Niño, and both short-term and long-term effects differ in tropical and temperate latitudes (Trillmich et al., 1991). If mid-Holocene ENSO variability on the Northwest Coast is identified at some point in the future, we would expect this time period to be characterized by substantial short-term climatic fluctuations, not "stability." We should also expect that such ENSO events had different consequences on the northern and southern Northwest Coast, depending on the proximity to the dynamic California Current system. The climatic signals should differ for these two subregions.

To address the numerous questions raised in this chapter, we would need better chronological resolution of both paleoenvironmental and archaeological records. Study of ocean or glacial cores might be useful in more precisely identifying climate change. Study of fish scale accumulations representing pelagic species sensitive to climate change (sardines and anchovies) are currently providing relatively high-resolution records for the California and Humboldt current areas (Baumgartner et al., 1998), and stable isotope records in lakes are used to infer salmon recruitment in lakes (Finney et al., 2000). Van Geen and Takesue (1999) have documented the intensity of coastal upwelling in response to variation in alongshore winds in these same regions, by studying foraminifera and mollusks in estuarine sediments and archaeological sites.

Such studies would ideally be paired with regional analyses of the local archaeological records. Again, fine chronological control would be necessary in well-stratified sites where discrete temporal components could be isolated and precisely dated. The linkages between climate and culture are not direct, so more attention must be paid to classes of mediating data. In some cases, it may be possible to identify the remains of food animals in archaeological sites that are outside their modern ranges as a result of climate change (see Lutaenko, 1993, and this volume; Sandweiss et al., 1996, and this volume). On the Northwest Coast, Crockford (1997) compiled identifications of northern bluefin tuna in archaeological assemblages from along southern British Columbia and Washington. The presence of tuna in these northerly waters suggests that at the time these archaeological sites were occupied, sea temperatures were somewhat higher than their 20th century average. Stable isotopic analyses of archaeofaunal pinnipeds have potential for detailing how prehistoric distributions of sea mammals differed from that of today (e.g., Burton et al., 2001; Gifford-Gonzalez et al., 2005; Moss et al., 2006).

Finally, more frequent data-sharing and collaboration between paleoecologists and archaeologists, such as that initiated here, is an obvious first step in approaching the effects of climate change on culture. For the Northwest Coast, the contributions of other specialists, particularly invertebrate, fisheries, and marine mammal biologists, will also be essential.

Acknowledgements

Moss and Peteet would like to thank Dan Sandweiss and Kirk Maasch for inviting us to the FERCO conference in Orono in October, 1998. Although we missed the occasion to meet each other in person, we appreciate the opportunity to work together on this chapter. This circumstance has also prompted Moss and Whitlock to work together. Despite the fact that our offices once occupied the same building on the University of Oregon campus, this effort represents our first collaboration. Moss is grateful to Quentin Mackie for permission to cite some of his unpublished work, and to Jon Erlandson for making her participation in the FERCO conference possible. We thank Kirk Maasch for preparing the three maps included in this chapter. We also appreciate the hard work and persistence of editor, David Anderson, who has brought this publication project to fruition.

References

Ackerman, R. E., 1992. Earliest stone industries on the North Pacific Coast of North America. *Arctic Anthropology* 29(2):18–27.

Ackerman, R. E., T. D. Hamilton, and R. Stuckenrath, 1979. Early culture complexes of the northern Northwest Coast. *Canadian Journal of Archaeology* 3:195–209.

Ackerman, R. E., K. C. Reid, J. D. Gallison, and M. E. Roe, 1985. Archaeology of Heceta Island: A survey of 16 timber units in the Tongass National Forest, southeastern Alaska. *Center for Northwest Anthropology Project Report 3,* Washington State University, Pullman.

Ames, K. M., 2004. The Prince Rupert Harbor sequence revisited. Paper presented at the 57th annual Northwest Anthropological Conference, Eugene, OR, March 24–27.

Barnosky, C. W., 1981. A record of late quaternary vegetation from Davis Lake, southern Puget Lowland, Washington. *Quaternary Research* 16:221–239.

Barnosky, C. W., P. M. Anderson, and P. J. Bartlein, 1987. The northwestern U.S. during deglaciation: Vegetational history and paleoclimatic implications. In *North America and Adjacent Oceans during the Last Deglaciation*, edited by W. F. Ruddiman and H. E. Wright Jr., pp. 289–321. Geological Society of America, Boulder.

Barrie, J. V., K. W. Conway, H. Josenhans, J. J. Clague, R. W. Mathewes, and D. W. Fedje, 2005. Late quaternary geology of Haida Gwaii and surrounding marine areas. In *Haida Gwaii: Human History and Environment from the Time of Loon to the Time of the Iron People*, edited by D. W. Fedje and R. W. Mathewes, pp. 7–20. UBC Press, Vancouver.

Barron, J. A., L. Heusser, T. Herbert, and M. Lyle, 2003. High-resolution climatic evolution of coastal northern California. *Paleoceanography* 18:1020–1034.

Bartlein, P. J., 1988. Late-tertiary and quaternary paleoenvironments. In *Vegetation History*, edited by B. Huntley and T. Webb III, pp. 113–152. Kluwer Academic Publishers, Dordrecht.

Bartlein, P. J., P. M. Anderson, K. H. Anderson, M. E. Edwards, R. S. Thompson, R. S. Webb, T. Webb III, and C. Whitlock, 1998. Paleoclimate simulations for North America over the past 21,000 years: Features of simulated climate and comparisons with paleoenvironmental data. *Quaternary Science Reviews* 17:549–585.

Baumgartner, T., V. Ferreira, D. Holmgren, A. Soutar, M. Salamanca, L. Ortlieb, D. Gutierrez, R. Escribano, and S. Mayor, 1998. Developing a paleoclimate transect for interhemispheric comparison over the past 2000 years from high-resolution marine sites along the Eastern Pacific Margin. Meeting abstract from Pole-Equator-Pole: Paleoclimate of the Americas, March 16–20, Merida, Venezuela.

Berger, A., 1978. Long-term variations of daily insolation and Quaternary climate changes. *Journal of Atmospheric Sciences* 35:2362.

Briles, C.E., C. Whitlock, and P. J. Bartlein, 2005. Postglacial vegetation, fire, and climate history of the Siskiyou Mountains, Oregon. *Quaternary Research* 64:44–56.

Brown, K. J., and R. J. Hebda, 2002. Ancient fires on southern Vancouver Island, British Columbia, Canada: A change in causal mechanisms at about 2,000 ybp. *Environmental Archeology* 7:1–12.

Brown, K. J., and R. J. Hebda, 2003. Coastal rainforest connections disclosed through a Late quaternary vegetation, climate, and fire history investigation from the Mountain Hemlock Zone on southern Vancouver Island, British Columbia, Canada. *Review of Palaeobotany and Palynology* 123:247–269.

Brunelle, A., and R. S. Anderson, 2003. Sedimentary charcoal as an indicator of late-Holocene drought in the Sierra Nevada and its relevance to the future. *The Holocene* 13:21–28.

Burbank, D. W., 1981. A chronology of late Holocene glacier fluctuations on Mount Rainier, Washington. *Arctic and Alpine Research* 13:369–386.

Burton, R. K., J. J. Snodgrass, D. Gifford-Gonzalez, T. Guilderson, T. Brown, and P. L. Koch, 2001. Holocene changes in the ecology of northern furs seals: Insights from stable isotopes and archaeofauna. *Oecologia* 128:107–115.

Butler, V. L., 1990. Distinguishing natural from cultural salmonid deposits in Pacific northwest North America. Unpublished PhD dissertation, Department of Anthropology, University of Washington, Seattle.

Butler, V. L. and J. E. O'Connor, 2004. 9000 years of salmon fishing on the Columbia River, North America. *Quaternary Research* 62:1–8.

Byram, R. S., 2002. Brush fences and basket traps: The archaeology and ethnohistory of tide-water weir fishing on the Oregon Coast. Unpublished PhD dissertation, Department of Anthropology, University of Oregon, Eugene.

Cannon, A., 1991. *The Economic Prehistory of Namu.* Department of Archaeology, Simon Fraser University, Publication 19, Burnaby, B.C.

Carlson, C., 1979. The early component at Bear Cove. *Canadian Journal of Archaeology* 3: 177–194.

Carlson, C. C., 2003. The Bear Cove fauna and the subsistence history of the Northwest Coast maritime culture. In *Archaeology of Coastal British Columbia, Essays in Honour of Professor Philip M. Hobler*, edited by R. L. Carlson, pp. 65–86. Archaeology Press, Simon Fraser University, Burnaby, B.C.

Carlson, R. L., 1970. Excavations at Helen Point on Mayne Island. *BC Studies* 6–7:113–125.

Carlson, R. L., 1987. The 1986 excavations at the canal Site (DeRt 1 and DeRt 2). Ms. on file, Heritage Conservation Branch, Victoria.

Carlson, R. L., 1996. Early Namu. In *Early Human Occupation in British Columbia*, edited by R. L. Carlson and L. Dalla Bona, pp. 83–102. University of British Columbia Press, Vancouver.

Carlson, R. L., 1997. Early maritime adaptations on the Northwest Coast. *Paper presented at the annual meeting of the American Association for the Advancement of Science*, February 15, Seattle.

Carlson, R. L., 1998. Coastal British Columbia in the light of North Pacific maritime adaptations. *Arctic Anthropology* 35(1):23–35.
Carlson, R. L., and P. M. Hobler, 1993. The Pender Canal excavations and the development of Coast Salish culture. *BC Studies* 99:25–52.
Chatters, J. C., V. L. Butler, M. J. Scott, D. M. Anderson, and D. A. Neitzel, 1995. A paleoscience approach to estimating the effect of climatic warming on salmonid fisheries of the Columbia River Basin. *Canadian Special Publications of Fisheries and Aquatic Sciences* 121:489–496.
Chavez, F. P., J. Ryan, S. E. Lluch-Cota, and M. C. Niquen, 2003. From Anchovies to Sardines and back: Multidecadal change in the Pacific Ocean. *Science* 229:217–220.
Chisholm, B. S., D. E. Nelson, and H. P. Schwarcz, 1983. Marine and terrestrial protein in prehistoric diets on the British Columbia coast. *Current Anthropology* 24:396–398.
Christensen, T. and J. Stafford, 2005. Raised beach archaeology in Northern Haida Gwaii: Preliminary results from the Cohoe Creek site. In *Haida Gwaii: Human History and Environment from the Time of Loon to the Time of the Iron People*, edited by D. W. Fedje and R. W. Mathewes, pp. 245–273. UBC Press, Vancouver.
Clague, J. J., 1983. Glacio-isostatic effects of the Cordilleran Ice Sheet, British Columbia, Canada. In *Shorelines and Isostasy*, edited by D. E. Smith and A. G. Dawson, pp. 321–346. Academic Press, San Diego.
Clague, J. J. and R. W. Mathewes, 1989. Early Holocene thermal maximum in western North America: New evidence from Castle Peak, BC. *Geology* 17:277–280.
Clague, J. J. and R. W. Mathewes, 1996. Neoglaciation, glacier-dammed lakes, and vegetation change in northwestern British Columbia. *Arctic and Alpine Research* 28:10–24.
COHMAP Members, 1988. Climatic changes of the last 18,000 years: Observations and model simulations. *Science* 241:1043–1052.
Coupland, G., 1988. *Prehistoric Cultural Change at Kitselas Canyon.* Canadian Museum of Civilization, Mercury Series, Archaeological Survey of Canada Paper No. 138, Ottawa.
Coupland, G., 1998. Maritime adaptation and evolution of the developed Northwest Coast pattern on the central Northwest Coast. *Arctic Anthropology* 35(1):36–56.
Cressman, L. S., 1960. Cultural sequence at The Dalles, Oregon: A contribution to Pacific Northwest prehistory. *Transactions of the American Philosophical Society* 50(10):1–108.
Crockford, S. J., 1997. Archeological evidence of large Northern Bluefin Tuna, *Thunnus thynnus*, in coastal waters of British Columbia and Northern Washington. *Fishery Bulletin* 95:11–24.
Cwynar, L. C., 1987. Fire and the forest history of the North Cascade Range. *Ecology* 68: 791–802.
Cwynar, L. C., 1990. A late quaternary vegetation history from Lily Lake, Chilkat Peninsula, southeast Alaska. *Canadian Journal of Botany* 68:1106–1112.
Daniels, M.L., R. S. Anderson, and C. Whitlock, 2005. Vegetation and fire history since the Late Pleistocene from the Trinity Mountains, northwestern California, USA. *The Holocene* 15(7):1062–1071.
Denton, G. H., and W. Karlen, 1977. Holocene glacial and treeline variations in the White River Valley and Skolai Pass, Alaska and Yukon territory. *Quaternary Research* 7:63–111.
Dunwiddie, P. W., 1986. A 6000-year record of forest history on Mount Rainier, Washington. *Ecology* 67:58–68.
Eldridge, M., and S. Acheson, 1992. The antiquity of fish weirs on the Southern Coast: A response to Moss, Erlandson, and Stuckenrath. *Canadian Journal of Archaeology* 16: 112–116.
Erlandson, J. M., and M. L. Moss, 1999. The systematic use of radiocarbon dating in archaeological surveys in coastal and other erosional environments. *American Antiquity* 64:431–443.

Erlandson, J. M., and M. L. Moss, 2004. Archaeological Investigations at Kit'n'kaboodle Cave (49-CRG-46), Dall Island, Southeast Alaska. Paper presented at the 57th annual Northwest *Anthropological Conference*, Eugene, OR, March 24–27.

Fedje, D. W., and T. Christensen, 1999. Modeling paleoshorelines and locating early Holocene coastal sites in Haida Gwaii. *American Antiquity* 64:635–652.

Fedje, D. W., and Q. Mackie, 2005. Overview of cultural history. In *Haida Gwaii: Human History and Environment from the Time of Loon to the Time of the Iron People*, edited by D. W. Fedje and R. W. Mathewes, pp. 154–162. UBC Press, Vancouver.

Fedje, D. W., and R. W. Mathewes (editors), 2005. *Haida Gwaii: Human History and Environment from the Time of Loon to the Time of the Iron People*. UBC Press, Vancouver.

Fedje, D. W., C. Christensen, H. Josenhans, J. B. McSporran, and J. Strang, 2005a. Millennial tides and shifting shores: Archaeology on a dynamic landscape. In *Haida Gwaii: Human History and Environment from the Time of Loon to the Time of the Iron People*, edited by D. W. Fedje and R. W. Mathewes, pp. 163–186. UBC Press, Vancouver.

Fedje, D. W., A. P. Mackie, J. B. McSporran, and B. Wilson, 1996. Early period archaeology in Gwaii Haanas: Results of the 1993 field programme. In *Early Human Occupation in British Columbia*, edited by R. L. Carlson and L. Dalla Bona, pp. 133–150. University of British Columbia Press, Vancouver.

Fedje, D. W., A. P. Mackie, R. J. Wigen, Q. Mackie, and C. Lake, 2005b. Kilgi Gwaay: An early maritime site in the south of Haida Gwaii. In *Haida Gwaii: Human History and Environment from the Time of Loon to the Time of the Iron People*, edited by D. W. Fedje and R. W. Mathewes, pp. 187–203. UBC Press, Vancouver.

Fedje, D. W., M. P. R. Magne, and T. Christensen, 2005c. Test excavations at raised beach sites in southern Haida Gwaii and their significance to Northwest Coast archaeology. In *Haida Gwaii: Human History and Environment from the Time of Loon to the Time of the Iron People*, edited by D. W. Fedje and R. W. Mathewes, pp. 204–244. UBC Press, Vancouver.

Fedje, D. W., R. Wigen, Q. Mackie, C. R. Lake, and I. D. Sumpter, 2001. Preliminary results from investigations at Kilgii Gwaay: An early Holocene archaeological site on Ellen Island, Haida Gwaii, British Columbia. *Canadian Journal of Archaeology* 25:98–120.

Finney, B. P., I. Gregory-Eaves, J. Sweetman, M. S. V. Douglas, and J. Smol, 2000. Impacts of climatic change and fishing on Pacific salmon abundance over the past 300 years. *Science* 290:795–799.

Fladmark, K. R., 1975. *A Paleoecological Model for Northwest Coast Prehistory*. Archaeological Survey of Canada Mercury Series 43, Ottawa.

Fladmark, K. R., 1982. An introduction to the prehistory of British Columbia. *Canadian Journal of Archaeology* 6:95–156.

Fladmark, K. R., 1986. Lawn Point and Kasta: Microblade sites on the Queen Charlotte Islands, British Columbia. *Canadian Journal of Archaeology* 10:37–58.

Gavin, D. G., L. B. Brubaker, and K. P. Lertzman, 2003. An 1800-year record of the spatial and temporal distribution of fire from the west coast of Vancouver Island, Canada. *Canadian Journal of Forest Research* 33:573–586.

Gavin, D.G., F. S. Hu, K. P. Lertzman, and P. Corbett, 2006. Weak climatic control of forest fire history during the late Holocene. *Ecology* 87:1722–1732.

Gedalov, Z., and D. J. Smith, 2001. Interdecadal climate variability and regime-scale shifts in Pacific North America. *Geophysical Research Letters* 28:1515–1518.

Gifford-Gonzalez, D., S. D. Newsome, P. L. Koch, T. P. Guilderson, J. J. Snodgrass, and R. K. Burton, 2005. Archaeofaunal insights on Pinniped–Human interactions in the Northeastern Pacific. In *The Exploitation and Cultural Importance of Sea Mammals*, edited by G. G.

Monks, pp. 19–38. *Proceedings of the 9th Conference of the International Council of Archaeozoology*, Durham, August, 2002. Oxbow Books, Oxford, UK.
Hallett, D. J., D. S. Lepofsky, R. W. Mathewes, and K. P. Lertzman, 2003. 11,000 years of fire history and climate in the mountain hemlock rain forests of southwestern British Columbia based on sedimentary charcoal. *Canadian Journal of Forest Research* 33:292–312.
Ham, L. C., 1990. The Cohoe Creek site: A Late Moresby Tradition shell midden. *Canadian Journal of Archaeology* 14:199–221.
Hansen, B. C. S., and D. R. Engstrom, 1996. Vegetation history of Pleasant Island, southeastern Alaska, since 13,000 yrs BP. *Quaternary Research* 46:161–175.
Hare, S. R., and R. C. Francis, 1995. Climate change and salmon production in the Northeast Pacific Ocean. *Climate Change and Northern Fish Populations*, edited by R. J. Beamish, pp. 357–372. *Canadian Special Publications of Fisheries and Aquatic Sciences* 121. National Research Council of Canada, Ottawa.
Heaton, T. H., and F. Grady, 2003. The Late Wisconsin vertebrate history of Prince of Wales Island, Alaska. In *Ice Age Faunas of North America*, edited by B. W. Schubert, J. I. Mead, and R. W. Graham, pp. 17–53. Indiana University Press, Bloomington.
Hebda, R., 1995. British Columbia vegetation and climate history with focus on 7 ka BP. *Geographie physique et Quaternaire* 49:55–79.
Hebda, R. J., and S. G. Frederick, 1990. History of marine resources of the Northeast Pacific since the last glaciation. *Transactions of the Royal Society of Canada*, series I, I:319–342.
Hebda, R. J. and R. W. Mathewes, 1984. Holocene history of cedar and native Indian cultures of the North American Pacific coast. *Science* 225:711–713.
Hebda, R. J., M. G. Pallatt, R. W. Mathewes, D. W. Fedje, and S. Acheson, 2005. Vegetation history of Anthony Island, Haida Gwaii, and its relationship to climate change and human settlement. In *Haida Gwaii: Human History and Environment from the Time of Loon to the Time of the Iron People*, edited by D. W. Fedje and R. W. Mathewes, pp. 59–76. UBC Press, Vancouver.
Heusser, C. J., 1960. *Late Pleistocene Environments of North Pacific North America*. American Geographical Society Special Publication 35, New York.
Heusser, C. J., 1977. Quaternary paleoecology of the Pacific slope of Washington. *Quaternary Research* 8:282–306.
Heusser, C. J., 1983. Vegetation history of the northwestern United States, including Alaska. In *Late Quaternary Environments of the United States*, edited by S. C. Porter, pp. 239–258. University of Minnesota Press, Minneapolis.
Heusser, C. J., 1995. Late quaternary vegetation response to climatic-glacial forcing in North Pacific America. *Physical Geography* 16:118–149.
Heusser, C. J., L. Heusser, and D. M. Peteet, 1985. Late quaternary climate changes on the American North Pacific coast. *Nature* 315:485–487.
Hibbert, D. M., 1979. Pollen analysis of late-quaternary sediments from two lakes in the Southern Puget Lowland, Washington. M.S. thesis, Department of Geological Sciences, University of Washington, Seattle.
Hodder, J., R. Brown, and C. Cziesla, 1998. The Northern Elephant Seal in Oregon: A pupping range extension and onshore occurrence. *Marine Mammal Science* 14:873–881.
Holmes, C. E., R. J. Dale, and J. D. McMahan, 1989. *Archaeological Mitigation of the Thorne River Site (CRG-177), Prince of Wales Island, Alaska*. Office of History and Archaeology Report No.15, Division of Parks and Outdoor Recreation, Alaska Department of Natural Resources, Anchorage.
Komar, P. D., 1997. *The Pacific Northwest Coast: Living with the Shores of Oregon and Washington*. Duke University Press, Durham.

Kutzbach, J. E., and P. J. Guetter, 1986. The influence of changing orbital patterns and surface boundary conditions on climate simulations for the past 18,000 years. *Journal of Atmospheric Sciences* 43:1726–1759.

Kutzbach, J., R. Gallimore, S. Harrison, P. Behling, R. Selin, and F. Laarif, 1998. Climate and biome simulations for the past 21,000 years. *Quaternary Science Reviews* 17:473–506.

Long, C. J., 2003. Holocene fire and vegetation history of the Oregon Coast Range, USA. Unpublished PhD dissertation, Department of Geography, University of Oregon, Eugene.

Long, C. J., and C. Whitlock, 2002. Fire and vegetation history from the coastal rain forest of the Western Oregon Coast Range. *Quaternary Research*. 58:215–225.

Long, C. J., C. Whitlock, P. J. Bartlein, and S. H. Millspaugh, 1998. A 9000-year fire history from the Oregon Coast Range, based on a high-resolution charcoal study. *Canadian Journal of Forestry* 28:774–787.

Luckman, B. H., 2000. The little ice age in the Canadian Rockies. *Geomorphology* 32:357–384.

Luckman, B. H., G. Holdsworth, and G. D. Osborn, 1993. Neoglacial glacier fluctuations in the Canadian Rockies. *Quaternary Research* 39:144–153.

Luckman, B. H., and R. Villalba, 2001. Assessing the synchroneity of glacier fluctuations in the western cordillera of the Americas during the last millennium. In *Interhemispheric Climate Linkages*, edted by V. Markgraf, pp. 119–140. Academic Press, San Diego.

Lutaenko, K. A., 1993. Climatic optimum during the Holocene and the distribution of warm-water mollusks in the Sea of Japan. *Palaeogeography, Palaeoclimatology, Palaeoecology* 102:273–281.

MacDonald, G. F., 1983. Prehistoric art of the northern Northwest Coast. In *Indian Art Traditions of the Northwest Coast*, edited by R. L. Carlson, pp. 99–120. Archaeology Press, Simon Fraser University, Burnaby.

MacDonald, G. F., and R. I. Inglis, 1981. An overview of the North Coast Prehistory Project (1966–1980). *BC Studies* 48:37–63.

Mackie, A., and I. D. Sumpter, 2005. Shoreline settlement patterns in Gwaii Haanas during the early and late Holocene. In *Haida Gwaii: Human History and Environment from the Time of Loon to the Time of the Iron People*, edited by D. W. Fedje and R. W. Mathewes, pp. 337–371. UBC Press, Vancouver.

Mackie, Q., N. Smith, M. Steffen, D. Fedje, R. Wigen, D. McLaren, and A. Mackie, 2004. Early Holocene maritime adaptations on the Northwest Coast: The evidence from Haida Gwaii. *Paper presented at the 57th annual Northwest Anthropological Conference*, Eugene, OR, March 24–27.

Mann, D. H., 1983. The quaternary history of the Lituya Glacial Refugium, Alaska. Unpublished PhD thesis, University of Washington, Seattle.

Mann, D. H., and T. D. Hamilton, 1995. Late Pleistocene and Holocene paleoenvironments of the North Pacific coast. *Quaternary Science Reviews* 14:449–471.

Mann, D. H., and F. C. Ugolini, 1985. Holocene glacial history of the Lituya District, southeast Alaska. *Canadian Journal of Earth Sciences* 22:913–928.

Mantua, N. J., S. R. Hare, Y. Zhang, J. M. Wallace, and R. C. Francis, 1997. A Pacific interdecadal climate oscillation with impacts on salmon production. *Bulletin of the American Meteorological Society* 78:1069–1079.

Markgraf, V., and H. F. Diaz, 2000. The past ENSO record: A synthesis. In *El Nino and the Southern Oscillation: Multiscale Variability and Global and Regional Impacts*, edited by H. F. Diaz and V. Markgraf, pp. 465–488. Cambridge University Press, Cambridge, UK.

Marlon, J., P. J. Bartlein, and C. Whitlock, 2006. Fire-fuel-climate linkages in the northwestern USA during the Holocene. *The Holocene* 16(8):1059–1071.

Mathewes, R. W., 1985. Paleobotanical evidence for climatic change in southern British Columbia during late-glacial and Holocene time. In *Climatic Change in Canada 5: Critical Periods in the Quaternary Climatic History of Northwestern North America*, edited by C. R. Harington, pp. 397–422; *Syllogeus* 55. Canada National Museum of Natural Sciences, Ottawa.
Mathewes, R. W., and L. E. Heusser, 1981. A 12,000 year palynological record of temperature and precipitation trends in southwestern British Columbia. *Canadian Journal of Botany* 59:707–710.
Matson, R. G., 1976. *The Glenrose Cannery Site.* National Museum of Man, Mercury Series, Archaeological Survey of Canada 52, Ottawa.
Matson, R. G., and G. Coupland, 1995. *The Prehistory of the Northwest Coast.* Academic Press, San Diego.
McGowan, J. A., D. R. Cayan, and L. M. Dorman, 1998. Climate-ocean variability and ecosystem response in the Northeast Pacific. *Science* 281:210–217.
McLachlan, J. S., and L. B. Brubaker, 1995. Local and regional vegetation change on the northeastern Olympic Peninsula during the Holocene. *Canadian Journal of Botany* 73: 1618–1627.
McMillan, A. D. and St. D. E. Claire, 2005. *Ts'ishaa: Archaeology and Ethnography of a Nuu-chah-nulth Origin Site in Barkley Sound.* Archaeology Press, Department of Archaeology, Simon Fraser University, Burnaby, B.C.
Minor, R., and K. A. Toepel, 1986. The archaeology of the Tahkenitch Landing Site: Early prehistoric occupation on the Oregon Coast. Heritage Research Associates Report 46, Eugene, OR.
Mobley, C. M., 1988. Holocene sea levels in Southeast Alaska: Preliminary results. *Arctic* 41:261–266.
Mohr, J. A., C. Whitlock, and C. N. Skinner, 2000. Postglacial vegetation and fire history, eastern Klamath Mountains, California. *The Holocene* 10:587–601.
Moss, M. L., 2004. Archaeological investigation of Cape Addington Rockshelter: Human occupation of the rugged seacoast on the outer Prince of Wales Archipelago, Alaska. University of Oregon Anthropological Paper No. 63, Eugene, OR.
Moss, M. L., J. M. Erlandson, and R. Stuckenrath 1989. The Antiquity of Tlingit Settlement on Admiralty Island, Southeast Alaska. *American Antiquity* 54(3):534–543.
Moss, M. L., and J. M. Erlandson, 1995. Reflections on North American Pacific coast prehistory. *Journal of World Prehistory* 9:1–45.
Moss, M. L., and J. M. Erlandson, 1998a. Early Holocene adaptations on the southern Northwest Coast. *Journal of California and Great Basin Anthropology* 20:13–25.
Moss, M. L., and J. M. Erlandson, 1998b. A comparative chronology of Northwest Coast fishing features. In *Hidden Dimensions, the Cultural Significance of Wetland Archaeology*, edited by K. Bernick, pp. 180–198. University of British Columbia Press, Occasional Paper 11, Vancouver.
Moss, M. L., and J. M. Erlandson, 2000. Wolf's Lair: Middle and late Holocene wooden artifacts from a sea cave on Baker Island, Southeast Alaska. *Canadian Journal of Archaeology* 24:107–128.
Moss, M. L., J. M. Erlandson, R. S. Byram, and R. E. Hughes, 1996. The Irish Creek site: Evidence for a mid-Holocene microblade component on the northern Northwest Coast. *Canadian Journal of Archaeology* 20:75–92.
Moss, M. L., D. Y. Yang, S. D. Newsome, C. F. Speller, I. McKechnie, A. D. McMillan, R. J. Losey, and P. L. Koch, 2006. Historical ecology and biogeography of North Pacific

Pinnipeds: Isotopes and ancient DNA from three archaeological assemblages. *Journal of Island and Coastal Archaeology* 1(2):165–190.

Ono, K. A., 1995. Effects of climate change on marine mammals in the Far North. In *Human Ecology and Climate Change: People and Resources of the Far North*, edited by D. L. Peterson and D. R. Johnson, pp. 105–121. Taylor & Francis, Washington, D.C.

Osborn, G., and B. H. Luckman, 1988. Holocene glacier fluctuations in the Canadian cordillera (Alberta and British Columbia). *Quaternary Science Reviews* 7:115–128.

Overpeck, J. T., C. Whitlock, and B. Huntley, 2003. Terrestrial biosphere dynamics in the climate system: Past and future. In *Paleoclimate, Global Change and the Future*, edited by K. Alverson, R. Bradley, and T. Pedersen, pp. 81–111. Springer, Berlin.

Pacific Northwestern Climate Impacts Group, 1999. El Niño Southern Oscillation (ENSO). Joint Institute for the Study of the Atmosphere and the Oceans University of Washington, Seattle. http://www.earthscape.org/r1/jis01/jis01g.html; accessed August 10, 2006.

Pellatt, M. G., and R. W. Mathewes, 1994. Paleoecology of postglacial treeline fluctuations on the Queen Charlotte Islands, Canada. *Ecoscience* 1:71–81.

Pellatt, M. G., and R. W. Mathewes, 1997, Holocene tree line and climate change on the Queen Charlotte Islands, Canada. *Quaternary Research* 48:88–99.

Pellatt, M. G., M. J. Smith, R. W. Mathewes, I. R. Walker, and S. L. Palmer, 2000. Holocene treeline and climatic change in the subalpine zone near Stoyoma Mountain, Cascade Mountains, southwestern British Columbia, Canada. *Arctic, Antarctic, and Alpine Research* 32: 73–83.

Peteet, D. M., 1986. Vegetational history of the Malaspina Glacier District, Alaska. *Quaternary Research* 25:100–120.

Peteet, D. M., 1991. Postglacial history of lodgepole pine near Yakutat, Alaska. *Canadian Journal of Botany* 69:786–796.

Pratt, H. L., 1992. The Charles culture of the Gulf of Georgia: A re-evaluation of the culture and its three sub-phases. M.A. thesis, University of British Columbia, Vancouver.

Ryder, J. M., and B. Thompson, 1986. Neoglaciation in the southern coast mountains of British Columbia: chronology prior to the late Neoglacial maximum. *Canadian Journal of Earth Sciences* 23:273–287.

Sandweiss, D. H., K. A. Maasch, and D. G. Anderson, 1999. Transitions in the Mid-Holocene. *Science* 283:499–500.

Sandweiss, D. H., K. A. Maasch, F. Chai, C. F. T. Andrus, and E. J. Reitz, 2004. Geoarchaeological evidence for multidecadal natural climatic variability and ancient Peruvian fisheries. *Quaternary Research* 61:330–334.

Sandweiss, D. H., J. B. Richardson III, E. J. Reitz, H. B. Rollins, and K. A. Maasch, 1996. Geoarchaeological evidence from Peru for a 5000 years BP onset of El Niño. *Science* 273:1531–1533.

Schiffer, M. B., 1986. Radiocarbon dating and the "Old Wood" problem: The case of the Hohokam chronology. *Journal of Archaeological Science* 13:13–30.

Schwing, F. B., P. Orton, D. A. Jay, H. Batchelder, and L. K. Rosenfeld, 1999. Conference explores El Niño's relationship to the Northeast Pacific. *EOS, Transactions, American Geophysical Union* 80(11):122–127.

Sea, D. S., and C. Whitlock, 1995. Postglacial vegetation history of the high cascade range, central Oregon. *Quaternary Research* 43:370–381.

Sirkin, L., and S. J. Tuthill, 1987. Late Pleistocene and Holocene deglaciation and environments of the southern Chugach Mountains, Alaska. *Geological Society of America Bulletin* 99: 376–384.

Smith, S. J., and R. S. Anderson, 1992. Late Wisconsin paleoecologic record from Swamp Lake, Yosemite National Park, California. *Quaternary Research* 38:91–102.

Stein, J. K., 1992. *Deciphering a Shell Midden*. Academic Press, San Diego.

Stuiver, M., P. J. Reimer, E. Bard, J. W. Beck, G. S. Burr, K. A. Hughen, B. Kromer, F. G. McCormac, J. v. d. Plicht, and M. Spurk, 1998b. INTCAL98 Radiocarbon Age Calibration, 24,000-0 cal BP. *Radiocarbon* 40:1041–1083.

Stuiver, M., P. J. Reimer, and T. F. Braziunas, 1998a. High-precision radiocarbon age calibration for terrestrial and marine samples. *Radiocarbon* 40:1127–1151.

Swanson, K., and S. D. Davis, 1983. Archaeological investigations at Lake Eva, Baranof Island, Alaska. Ms. on file, Tongass National Forest, Chatham Area, Sitka, AK.

Thompson, R. S., C. Whitlock, P. J. Bartlein, S. P. Harrison, and W. G. Spaulding, 1993. Climatic changes in the western United States since 18,000 yr.BP. In *Global Climates since the Last Glacial Maximum*, edited by H. E. Wright, Jr., J. E. Kutzbach, T. Webb, III, W. F. Ruddiman, F. A. Street-Perrott, and P. J. Bartlein, pp. 468–513. University of Minnesota Press, Minneapolis.

Trillmich, F., K. A. Ono, D. P. Costa, R. L. DeLong, S. D. Feldkamp, J. M. Francis, R. L. Gentry, C. B. Heath, B. J. LeBoeuf, P. Majluf, and A. E. York, 1991. The effects of El Niño on pinniped populations in the Eastern Pacific. In *Pinnipeds and El Niño: Responses to Environmental Stress*, edited by F. Trillmich and K. A. Ono, pp. 247–270. *Ecological Studies* 88. Springer, Berlin.

Van Geen, A. and R. Takesue, 1999. Past and present coastal upwelling along the western Americas. In *Reconstructing Ocean History: A Window into the Future*, edited by F. Abrantes and A. Mix, pp. 399–418. Kluwer Academic/Plenum Publishers, New York.

Wainman, N., and R. W. Mathewes, 1987. Forest history of the last 12,000 years based on plant macrofossil analysis of sediment from Marion Lake, southwestern British Columbia. *Canadian Journal of Botany* 65:2179–2187.

Whitlock, C., 1992. Vegetation and climatic history of the Pacific Northwest during the last 20,000 years: Implications for understanding present-day biodiversity. *The Northwest Environmental Journal* 8:5–28.

Whitlock, C., and L. D. Grigg, 1999. Paleoecological evidence of Milankovitch and Sub-Milankovitch climate variations in the western U.S. during the late Quaternary. In *Mechanisms of Millennial-Scale Global Climate Change*, edited by R. S. Webb, P. U. Clark, and L. D. Keigwin, pp. 227–241. *American Geophysical Union Geophysical Monograph* 112. American Geophysical Union, Washington, D. C.

Whitlock, C. and A. Brunelle, 2007. Pollen records from northwest North America. In *Encyclopedia of Quaternary Science* edited by S. Elias. Elsevier, Amsterdam. (in press).

Whitlock, C., P. J. Bartlein, C. Briles, A. Brunelle, C. Long, and J. Marlon. Long-term relations among fire, fuels, and climate in the northwestern U.S. based on lake-sediment studies. *Journal of International Wildfire Research* (in press).

Wigen, R. J., 2005. History of the vertebrate fauna in Haida Gwaii. In *Haida Gwaii: Human History and Environment from the Time of Loon to the Time of the Iron People*, edited by D. W. Fedje and R. W. Mathewes, pp. 96–115. UBC Press, Vancouver.

Worona, M. A., and C. Whitlock, 1995. Late quaternary vegetation and climate history near Little Lake, central Coast Range, Oregon. *Geological Society of America Bulletin* 107: 867–876.

Climate Change and Cultural Dynamics: A Global Perspective on Mid-Holocene Transitions
David G. Anderson, Kirk A. Maasch and Daniel H. Sandweiss (Editors)

Chapter 15

Middle Holocene climate change and human population dispersal in western North America

Douglas J. Kennett[a,*], Brendan J. Culleton[a], James P. Kennett[b], Jon M. Erlandson[a,c] and Kevin G. Cannariato[d]

[a]*Department of Anthropology, University of Oregon, Eugene, OR 97403-1218, USA*
[b]*Department of Earth Science and Marine Science Institute, University of California, Santa Barbara, CA 93106, USA*
[c]*Museum of Natural and Cultural History, University of Oregon, Eugene, OR 97403-1224, USA*
[d]*Department of Earth Sciences, University of Southern California, Los Angeles, CA 90089, USA*

Abstract

Available climate records in western North America (7000–3800 cal yr BP) indicate a severe dry interval between 6300 and 4800 cal yr BP embedded within a generally warm and dry Middle Holocene. Dry conditions in western North America between 6300 and 4800 cal yr BP correlate with cold to moderate sea-surface temperatures (SST) and relatively high-marine productivity along the Southern California Coast evident in Ocean Drilling Program (ODP) Core 893A/B (Santa Barbara Basin). Based on archeological, linguistic, and genetic data, we argue for a movement of Uto-Aztecan people from western desert environments to the Southern California Coast, including the southern Channel Islands, and into portions of the Central Valley by at least 5500–4500 cal yr BP. We hypothesize that population dispersal from the desert interior was primarily in response to severe and prolonged drought and that people moved selectively to coastal and aquatic habitats because of the ameliorated effects of drought and their overall productivity.

1. Introduction

Multidisciplinary studies employing genetic, linguistic, and archeological data have revolutionized the study of past human migration. Mitochondrial DNA (mtDNA) and Y-chromosome DNA work on extant populations has redefined and focused our view of the original dispersal of anatomically modern humans from Africa between 200 and 150 thousand years (Stringer, 2002; Jobling et al., 2004) and the study of ancient mtDNA extracted from Neanderthal fossils from Europe and the

*Corresponding author. *E-mail address:* dkennett@uoregon.edu (D.J. Kennett).

Caucasus suggest that these Archaic *Homo sapiens* were a genetically distinctive population (Krings et al., 1997; Ovchinnikov et al., 2000). Recent genetic and archeological studies also indicate a single dispersal of anatomically modern humans along the coasts of southern and southeast Asia and into Australia (Mellars, 2006), followed by movement into East Asia (Jin et al., 2002), Europe (Sykes, 1999; Richards et al., 2000; Mellars, 2006), and the Americas (Merriwether and Ferrell, 1996; Bianchi et al., 1997), perhaps in multiple waves (Karafet et al., 1997, 1999), and ultimately to some of the most remote portions of our planet in Oceania (Deka et al., 2000). Creative analysis of multiple lines of linguistic, genetic, and archeological data have mapped subsequent movement, displacement, and re-organization of populations within these geographic areas, often associated with the transition to agriculture (Cavalli-Sforza, 1996; Bellwood, 2001; Diamond and Bellwood, 2003), and the extraction of ancient DNA from well-dated skeletal material has provided specific information about the timing of these migrations (Haak et al., 2005). This work has revitalized the study of human dispersal and its historical importance for understanding broad-scale cultural developments, moving beyond the extreme and reactionary anti-diffusionist paradigms of the late 20th century (Trigger, 1989).

Linguistic diversity and the patchwork distribution of language groups in western North America reflect a complex history of early settlement, *in situ* development, and periodic population movement. A large number of geographically limited language families were recorded along the Pacific Coast, a product of an early migratory history (Golla, 2000a,b) and great ecological diversity (Nichols, 1992). Larger, more linguistically homogeneous regions such as the desert interior reflect more recent population migration (Kaestle, 1995, 1997, 1998). Recent mtDNA work confirms that one coastal group – the Chumash of Southern California – exhibits a distinctive founding haplogroup D sequence that is rare and primarily found in populations that lived along the coasts of North and South America, supporting the idea of an early coastal dispersal during the colonization of the Americas (Johnson and Lorenz, 2006; Kemp et al., 2007). The Uto-Aztecan language family is widely distributed in western North America from in Southern California and adjacent southern Channel Islands, south of Chumash territory, and extend across Southern California desert areas through much of the Great Basin. This wedge-shaped distribution is interpreted, along with the broader distribution of related language groups down into Mexico, as representing a movement of people from the interior to the coast (Kroeber, 1925; Bright and Bright, 1976), a scenario supported by a recent study of modern mtDNA lineages in California (Johnson and Lorenz, 2006). Estimates for the spread of Uto-Aztecan people derived from glottochronology and archeology range from about 2000 to 7000 cal yr BP (Moratto, 1984; Vellanoweth, 2001; Raab and Howard, 2002).

In this chapter, we synthesize the available genetic, linguistic, and archeological data for the spread of Uto-Aztecan peoples from the desert western interior to the Southern California Coast and argue that this expansion occurred during the

Middle Holocene between about 5500 and 4500 cal yr BP. We also argue, based on genetic and archeological evidence, for an associated spread of Uto-Aztecan people through the Central Valley of California, possibly as far north as the wetland environments within the vicinity of San Francisco Bay. Speakers of Penutian languages later colonized this region and were the ancestral populations to people living in the valley at the time of European contact. We evaluate these data within the context of newly available climatic data for western North America (Fig. 15.1), and argue that the movement of Uto-Aztecan people was stimulated by severe drought conditions across western North America between 6300 and 4800 cal yr BP that reduced terrestrial productivity and drinking water availability. These dry conditions correlate with cold to moderate sea-surface temperatures (SST) and relatively high marine productivity along the Southern California Coast evident in Ocean Drilling Program (ODP) Core 893A/B (Santa Barbara Basin), conditions that would have been particularly attractive to people living in interior areas at this time. To build these arguments we first turn to the available paleoclimatic records and then to the genetic, linguistic, and archeological data.

2. Climate records

2.1. Santa Barbara Basin paleoenvironmental record

Changes in SST and marine productivity during the Holocene have been inferred using various marine sediment records from coastal California (Pisias, 1978, 1979; Heusser et al., 1985; van Geen et al., 1992) including an especially high-resolution Holocene record (Kennett and Kennett, 2000; Cannariato et al., 2003). This Holocene (11,500 cal yr BP to present) record represents the upper 17 m of a 200 m core, a late Quaternary sequence spanning the last 160 thousand years (Site 893A/B), drilled in Santa Barbara Basin as part of the ODP (Ingram and Kennett, 1995; Kennett and Ingram, 1995a,b; Behl and Kennett, 1996; Cannariato et al., 1999; Hendy and Kennett, 1999, 2000) (Fig. 15.2). The sequence consists of laminated sediments deposited at an average rate of ~155 cm/1000 yr. Climatic change through the Holocene is inferred from oxygen isotopic ($\delta^{18}O$) analysis of two planktonic foraminiferal taxa: *Globigerina bulloides*, a surface dweller, and *Neogloboquadrina pachyderma*, a species that lives near the base of the thermocline (~60 m below surface). Our Holocene age model is based on 20 accelerator mass spectrometry (AMS) ^{14}C dates converted to calendar years using a reservoir age of 230 ± 35 years (Ingram and Southon, 1996; Kennett et al., 1997; see Roark et al., 2003 for chronological details). This has provided one of the highest resolution marine Holocene climate sequences in the world: 25 yr intervals from 0 to 3000 cal yr BP and 9000 to 11,000 cal yr BP and 50 yr intervals from 3000–9000 cal yr BP. The high quality of this climate record results from a combination of rapid sedimentation rates, lack of bioturbation, a continuous abundance of foraminifera for geochemical and faunal analyses, and high environmental sensitivity in this region (Kennett and Kennett, 2000).

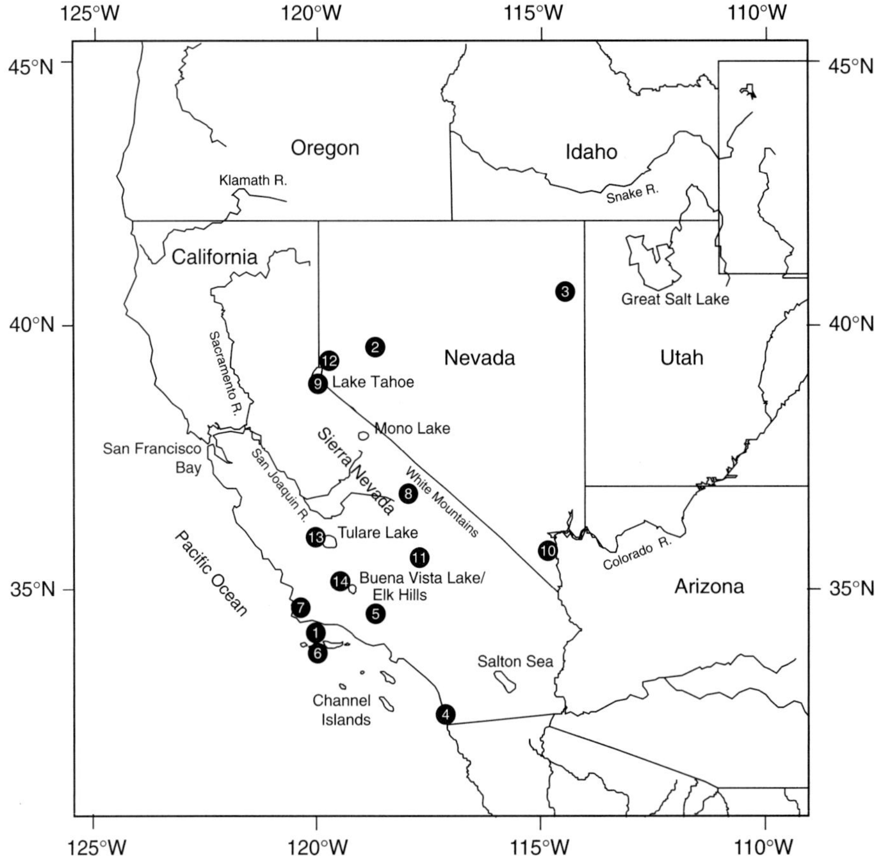

Figure 15.1. Map of western North America showing the locations of the main paleoclimatic records discussed in this contribution. (1) Ocean Drilling Program (ODP), Site 893A/B, Santa Barbara Basin (Kennett and Ingram, 1995b; Huesser and Sirocko, 1997; Kennett and Kennett, 2000); (2) Leonard Rockshelter pollen sequence (Byrne et al., 1979); (3) Ruby Valley pollen record (Thompson, 1992); (4) Archeological pollen sequences (Masters and Gallegos, 1997); (5) Late Holocene tree ring record, coastal Southern California (Larson and Michaelson, 1989); (6) Santa Rosa Island pollen sequence (Cole and Liu, 1994); (7) Union pollen spectra (Morgan et al., 1991); (8) Bristlecone pine tree ring record (LaMarche, 1973, 1974; Hughes and Graumlich, 1996, 2000); (9) Lake Tahoe submerged tree stump record (Lindström, 1990); (10) Southern Great Basin black mat records (Quade et al., 1998); (11) Owens Lake (Benson et al., 2002); (12) Pyramid Lake (Benson et al., 2002); (13) Tulare Lake geomorphology and pollen records (Negrini et al., 2006); (14) Buena Vista Lake and Elk Hills geomorphology (Culleton et al., 2005).

This record reveals millennial-scale oscillations in SST during the Holocene (Fig. 15.2a). Compared with the previous glacial episode (Kennett and Ingram, 1995a), Holocene SSTs were warm (average of ~12.5°C). Three distinct cycles are present in the Middle Holocene with warming between 8200–6300 and 5800–3800 cal yr BP, punctuated by a cool interval from 6300 to 5800 cal yr BP. The coldest SSTs during

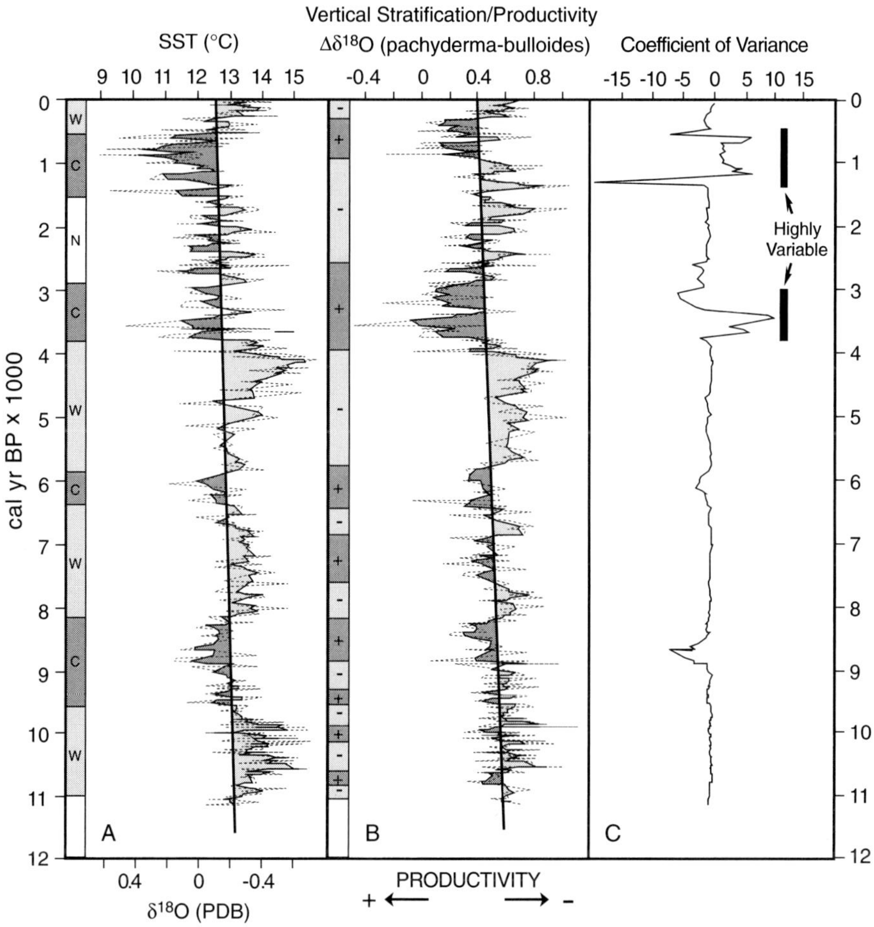

Figure 15.2. Holocene climate record for Santa Barbara Basin. (A) Estimated sea-surface temperature (SST) curve is based on the oxygen isotopic composition of *Globigerina bulloides* (surface-dwelling species of foraminifera) from varved sediments in Santa Barbara Basin. SST estimates are based on Bemis et al. (1998). The SST curve has been normalized for the Early Holocene by removing the oxygen isotopic component resulting from ice volume changes. Bar at left represents warm (w) and cold (c) cycles through the Holocene. (B) Vertical stratification/productivity record inferred from oxygen isotopic differences between *G. bulloides* and *N. pachyderma* (deeper-dwelling planktonic foraminiferal species). Bar at left shows intervals inferred as high (+) or low (–) productivity during the Holocene. (C) Variation in marine climate during the Holocene ($\delta^{18}O$ of *G. bulloides*, 50 year resolution). Coefficient of variance (standard deviation/average) was used to compare each oxygen isotopic measurement to the four surrounding it.

the Middle Holocene are centered on 6000 cal yr BP (~12°C). The warmest Middle Holocene interval occurred between 4500 and 4000 cal yr BP (~15°C), in agreement with Friddell et al. (2003). SSTs between 5800 and 5200 cal yr BP were relatively moderate compared to these warm and cold episodes.

Inferred surface ocean productivity fluctuations occurred during the Holocene (Fig. 15.2b), often synchronously with changes in SST. Changes in marine productivity have been inferred using a marine productivity index. This index is based on temperature differences between surface waters (as measured by the oxygen isotopic composition of surface-dwelling *G. bulloides*) and waters at the base of the thermocline (as measured by the oxygen isotopic composition of *N. pachyderma,* which inhabits the thermocline). Sediment trap studies within Santa Barbara Basin indicate that the isotopic difference between *G. bulloides* and *N. pachyderma* reflects the degree of surface ocean stratification, providing measures of upper water column stability, upwelling intensity, and magnitude of surface ocean productivity (Pak and Kennett, 2002). During the Holocene, inferred warming of surface waters was often associated with cooling at the thermocline, and vice versa, suggesting episodic variations in the intensity of upwelling. During cool episodes, little or no vertical temperature gradient existed between surface and thermoclinal species suggesting that upwelling of deeper, nutrient-rich waters was then especially intense during these intervals. Vertical mixing and inferred high productivity were greatest during the Middle Holocene from 7500 to 6800 cal yr BP and 6500 to 5900 cal yr BP. Reduced vertical mixing and lower marine productivity occurred between 6800 and 6500 cal yr BP, and again between 5900 and 3900 cal yr BP.

2.2. Associated terrestrial climate changes

High-resolution $\delta^{18}O$ and total inorganic carbon (TIC) records from Pyramid and Owens Lake basins reveal at least five distinctive climatic episodes in western North America during the Holocene (Benson et al., 2002; Fig. 3; see Fig. 15.1 for locations). Younger Dryas cooling was followed during the earliest Holocene by drying (11,600–10,000 cal yr BP) except for a brief wet period between 10,400 and 10,200 cal yr BP. Relatively wet conditions occurred during the remaining Early Holocene (10,000–8000 cal yr BP). Under these conditions, Lake Tahoe fed Pyramid Lake via the Truckee River and a substantial body of water existed in the Owens Lake Basin (Benson et al., 2002). A drying trend between 8000–6500 cal yr BP is suggested as lake sizes declined in both basins. $\delta^{18}O$ and TIC records from Pyramid Lake suggest periodic wet intervals between 8000 and 6500 cal yr BP with the most pronounced isotopic excursion between 7000 and 6400 cal yr BP interpreted as a major influx of water from Lake Tahoe. Persistently warm and dry conditions occurred throughout the remainder of the Middle Holocene (6400–3800 cal yr BP). At this time Owens Lake dried completely and water flow from Lake Tahoe to Pyramid Lake was substantially reduced. Wetter conditions generally mark the Late Holocene after ~3000 cal yr BP, but several multidecadal major droughts are known to have occurred between 1500 and 600 cal yr BP (Stine, 1994). These new data are generally consistent with the early work of Antevs (1948, 1952, 1955) who argued that the Middle Holocene (~7000–4500 cal yr BP) was warm and dry across much of western North America, the so-called altithermal or climatic optimum.

This was preceded by the anathermal (10,000–7000 cal yr BP) and followed by the medithermal (4500 cal yr BP to present) intervals marked by generally cool and wet climatic conditions. Dry Middle Holocene conditions in the Great Basin are also suggested by decreased sedimentation rates in the Ruby Valley marshlands of western Nevada between 7700 and 5500 cal yr BP (Thompson, 1992), decreases in spring discharge indicated by the absence of black mats in the southern Great Basin between ~7300 and 2500 cal yr BP (Quade et al., 1998), and changes in the distribution of xeric flora (Hansen, 1947; Bright, 1966; Byrne et al., 1979; Mehringer, 1985; Madsen and Rhode, 1990) and associated fauna (Grayson, 2000).

Axelrod (1981) argued that xeric (dry) vegetation expanded into the San Francisco Bay area in the Early and Middle Holocene and Moratto et al. (1978) identified several dry episodes in the Middle Holocene based on pollen records from California (Birman, 1964; Adam, 1967; Curry, 1969; Sercelj and Adam, 1975; Wood, 1975; Casteel et al., 1977) and correlated these with the bristlecone pine precipitation record from the White Mountains (LaMarche, 1973, 1974; Hughes and Graumlich, 1996, 2000) indicating a significant dry episode between 6000 and 4800 cal yr BP. Recent work in the San Joaquin Valley fills out the interior California climate picture. Geomorphic evidence shows Tulare Lake level fluctuations in the Middle Holocene, with two lowstands between ca. 7800–7000 cal yr BP and 5500–3500 cal yr BP, and desiccation indicated by mudcracks at 5500 cal yr BP (Negrini et al., 2006). The stratigraphic data is corroborated by pollen (sedge/cattail) and algae (*Pediastrum*/*Botryococcus*) spectra indicating that the most brackish conditions in Tulare Lake occurred in the Middle Holocene, with especially poor conditions between 5500–4500 cal yr BP. Through the Holocene, inferred lake levels correlate well between Tulare, Pyramid, and Owens Lakes (Benson et al., 2002; see Negrini et al., 2006, Fig. 12).

Further south in the Buena Vista Basin, Culleton et al. (2005) compiled geomorphic and archeological evidence from Buena Vista Lake and the Elk Hills that suggest gradual, low-energy deposition in the lakes and sloughs from 8000–6000 cal yr BP, followed by general desiccation ca. 6000–5000 cal yr BP indicated by buried calcic slough deposits displaying deep cracks and vegetation established on a subaerial surface. These Middle Holocene muds are overlain by 2–3 m of bedded sands and gravels derived from the surrounding uplands, deposited relatively abruptly at high energy judging from the lack of soil development (originally noted by D.W. Fuqua in Buena Vista Lake sediments in 1961; Hubbs et al., 1962, pp. 231–232). This massive erosion event was the culmination of drought-induced devegetation in the uplands, where today the buried Pleistocene soil is stripped of its A horizon, and Late Holocene sediments overlie the scoured Pleistocene B horizon (Culleton et al., 2005). Radiocarbon dates on freshwater mussel shells and archeological assemblages at Buena Vista Lake and Elk Hills place the event toward the end of the Middle Holocene, which would correlate with wetter conditions that caused a Mono Lake highstand (Stine, 1990, 1994) and the reformation of Owens Lake after 3800 cal yr BP (Benson et al., 2002). Overall, the San Joaquin Valley data correlate well with the driest interval in the Middle Holocene as indicated by the

Pyramid Lake $\delta^{18}O$ record (Benson et al., 2002) and are consistent with some of the most compelling evidence for severe Middle Holocene aridity based on submerged tree stumps in Lake Tahoe. Lindström (1990, also see Harding, 1965; Benson et al., 2002) documented ~20 tree stumps submerged up to 4 m below the current lake level. These trees have been radiocarbon dated to between ~6300 and 4800 cal yr BP and represent a low lake-level stand at that time.

Drought conditions appear to have been less severe in coastal California during the Middle Holocene compared with the interior. Relatively dry conditions in the Santa Barbara region are suggested by high percentages of *Chenopodium* and *Ambrosia* pollen in estuarine deposits on Santa Rosa Island between 5200 and 3250 cal yr BP (Cole and Liu, 1994) and dune building became more widespread on San Miguel Island between 7000–3500 cal yr BP (Erlandson et al., 2005). Pollen evidence from sediment records north of Point Conception also suggests dry conditions peaking in the Middle Holocene (7600–4800 cal yr BP) on the Santa Barbara Coast (Morgan et al., 1991). However, frequency changes in pine and oak pollen (Heusser and Sirocko, 1997) in Santa Barbara Basin (ODP Hole 893a) exhibit no distinct trends during the Middle and Late Holocene and thus climatic interpretations are inconclusive. Also, pollen spectra from estuarine and archeological deposits in coastal San Diego County indicate relatively stable environmental conditions during the Holocene (Masters and Gallegos, 1997).

Relationships between marine and terrestrial climatic conditions on the California Coast are complex, but historical data suggest that these two climate systems are currently closely interrelated (Jones and Kennett, 1999). Late Holocene records indicate that intervals marked by cooler SSTs in Santa Barbara Basin were contemporaneous with low precipitation over parts of western North America (Kennett and Kennett, 2000; Graham et al., 2007). A comparison of Santa Barbara Basin core data with the bristlecone pine record from the White Mountains of eastern California suggests correlation between cool SST and drier conditions during the last 4000 cal yr BP (Fig. 15.3). During this interval, cool SSTs and low precipitation dominate between 4000 and 2300 cal yr BP and again between 1500 and 500 cal yr BP. Warm SSTs and higher precipitation are evident between 2300 and 1500 cal yr BP and again following 500 cal yr BP. Cool SSTs between ~1500 and 500 cal yr BP also correlate with lower precipitation evident in a shorter tree ring record from the coastal ranges of Southern California (Larson and Michaelson, 1989; Kennett and Kennett, 2000). Several other lines of evidence also indicate dry conditions during this interval (Graumlich, 1993; Stine, 1994; Raab and Larson, 1997; Jones et al., 1999).

Middle Holocene relationships between inferred precipitation and SST are much less apparent. Prior to 4000 cal yr BP correlations between SST in the Santa Barbara Basin and the bristlecone pine record, so evident in the Late Holocene, are largely absent, possibly reflecting a general shift in climate sensitivity in southern California at the end of the Middle Holocene. During the Middle Holocene dry conditions throughout western North America (Antevs, 1948; Benson et al., 2002) correspond to warm SSTs in Santa Barbara Basin. Similarly warm SSTs through

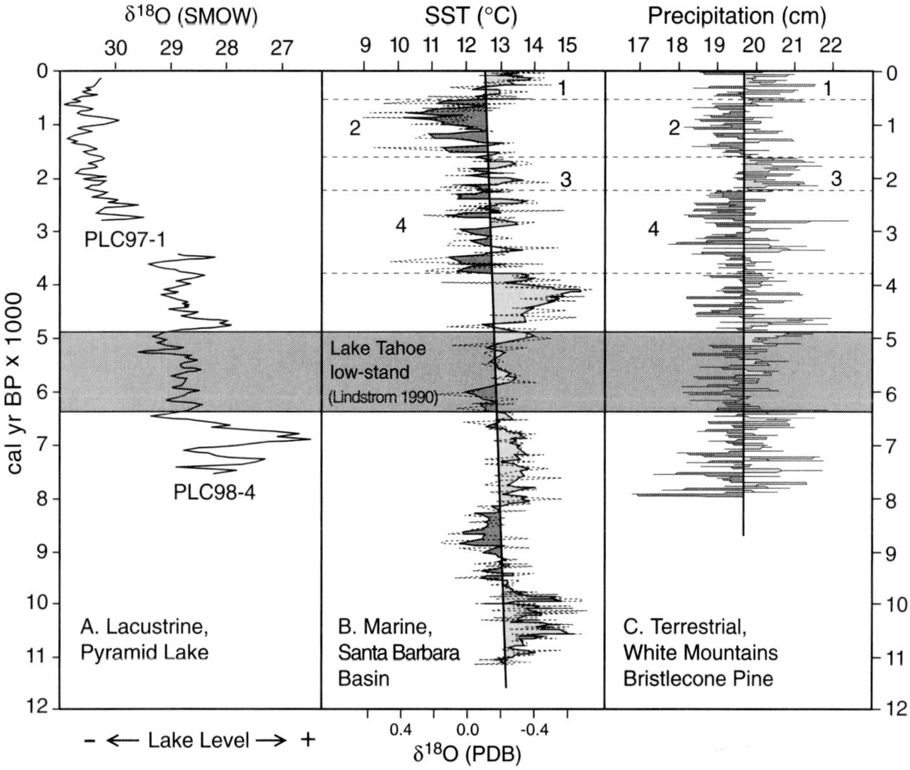

Figure 15.3. Bristlecone pine inferred precipitation (8000 cal yr BP to present) and Pyramid Lake $\delta^{18}O$ record compared with Holocene inferred SST from Santa Barbara Basin. (A) Smoothed (40 yr) $\delta^{18}O$ record for lake carbonates from Pyramid Lake (Cores PLC97-1 and PLC98-4), Western Nevada (Benson et al., 2002). Oscillations in $\delta^{18}O$ are interpreted to largely represent changes in freshwater input into the lakes which correlate to changes in lake size. Late Holocene high $\delta^{18}O$ values (PLC97-1) interpreted by Benson et al. (2002) as representing a phase of cooler, wetter climate; (B) Inferred SST record from Santa Barbara Basin, from Fig. 15.2; (C) Bristlecone pine record of inferred precipitation based on ring width measurements from trees in the White Mountains, California (data from LaMarche, 1973, 1974; Hughes and Graumlich, 2000; see http://www.ncdc.noaa.gov/paleo/drought/drght_graumlich.html). Zones 1 through 4 denote cool/dry (2 and 4) and warm/wet cycles (1 and 4) exhibited by these SST and precipitation records.

the Middle Holocene recorded in another Santa Barbara Basin sequence have been interpreted as evidence for stronger El Niño-Southern Oscillation (ENSO) activity in the Pacific and implied generally wetter conditions in western North America (Friddell et al., 2003). This, however, is inconsistent with indications of widespread Middle Holocene aridity in western North America. Instead, evidence for relatively stable decadal-scale climate variability in Site 893A/B appears to be more consistent with weaker ENSO activity during the Middle Holocene (Sandweiss et al., 1996, 1997, 2001; Overpeck and Webb, 2000; Tudhope et al., 2001; Koutavas et al., 2006).

If so, it follows that generally warmer SSTs at the millennial-scale are not necessarily accompanied by more intense or frequent ENSO activity.

Although inferred relations between SST and precipitation are clearly more complex in western North America, the coolest SST interval (6300–5000 cal yr BP) corresponds with the onset of the driest interval during the Middle Holocene reflected in the bristlecone pine sequence (LaMarche, 1973, 1974), Pyramid Lake $\delta^{18}O$ record (Benson et al., 2002), San Joaquin Valley geomorphology (Culleton et al., 2005; Negrini et al., 2006) and the submerged-stumps from Lake Tahoe (Lindström, 1990; Benson et al., 2002). This suggests that the climate during this interval operated similarly to that of the Late Holocene.

3. Linguistic, genetic, and archeological records

3.1. Linguistic data

Native California's linguistic diversity has long provided the basis for speculation regarding population movements, replacements, and interactions during the last several millennia (Kroeber, 1925). The distribution of major language families (e.g., Hokan, Penutian, and Uto-Aztecan; Fig. 15.4) and their relative linguistic and dialectical differentiation imply a series of population movements in California since the terminal Pleistocene (see Moratto, 1984, pp. 530–574). Hokan languages were thought by Kroeber (1925) to represent the earliest stock, as attested by their relatively disjunct distribution on the north and south coasts (e.g., Pomoan, Salinan, Chumashan), the northern Sierra Nevada (Washo), and the Colorado River and Baja California (Yuman). Speakers of Penutian languages are thought to have entered California's Central Valley from the northeast ca. 4500 cal yr BP, which is signaled archeologically by the Windmiller Pattern in the lower Sacramento Valley (Ragir, 1972). These early Penutians subsequently spread through the Central Valley, the Sierra Nevada foothills, and the central coast, and differentiated into existing language groups of Yokutsan, Miwokan, and Costanoan between ca. 3000–2000 cal yr BP (Moratto, 1984). Later expansions and replacements of Penutians by other Penutian tribes after 1000 cal yr BP are also hypothesized by Moratto (1984, p. 571).

Uto-Aztecan language groups are primarily located on the southern California Coast (Takic), desert interior of the Great Basin (Numic), with an additional pocket in the southern Sierra Nevada (Tubatulabalic, Fig. 15.4). The initial movement from the interior deserts to the Mohave Desert and southern coast is placed at ca. 5000–3000 cal yr BP on the basis of linguistic differentiation between Takic and Numic-Tubatulabalic, where it may be signaled by projectile point traditions common to the Great Basin and Southwest such as Humboldt, Gypsum, and Elko (Moratto, 1984, p. 559; Jennings, 1986; Koerper et al., 1994). Similarities in Yokuts (Penutian) and Uto-Aztecan languages led Nichols (1981) to posit a Uto-Aztecan presence in the southern San Joaquin Valley before the Penutian expansion

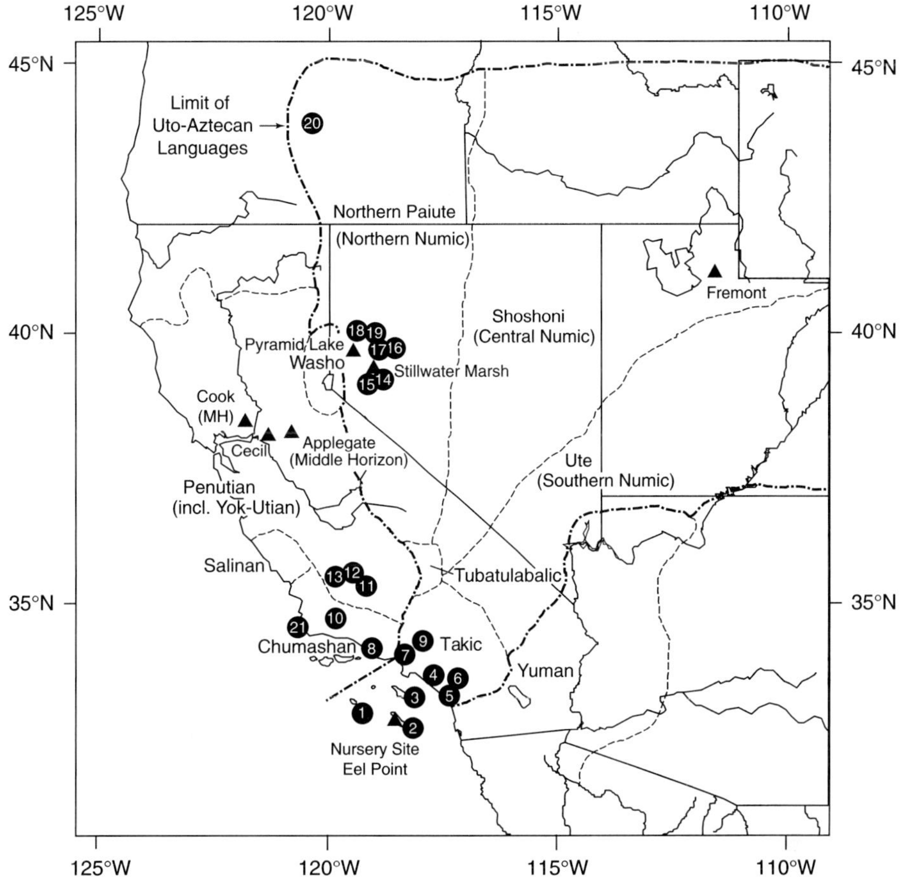

Figure 15.4. Map of western North America showing historic language distributions, prehistoric mtDNA populations (black triangles) and the known distribution of OGR beads that date to between ~5900 and 4700 cal yr BP (numbered). 1 = Celery Site, San Nicolas Island (CA-SNI-351); 2 = Nursery Site, San Clemente Island (CA-SCLI-1215); 3 = Little Harbor, Santa Catalina Island (CA-SCAI-17); 4 = CA-ORA-368; 5 = CA-ORA-667; 6 = CA-ORA-665; 7 = Encino Village (CA-LAN-43); 8 = CA SBA-119; 9 = Vasquez Rocks (CA-LAN-361); 10 = CA-SBA-3404; 11 = Elk Hills (CA-KER-3079/H; KER-3166/H; KER-5404); 12 = Buttonwillow (CA-KER-2720); 13 = McKittrick (CA-KER-824); 14 = Stillwater Marsh; 15 = Hidden Cave; 16 = Silverwater Marsh; 17 = Lovelock Cave; 18 = Shinners Site F; 19 = Kramer Cave; 20 = DJ Ranch (35LK2758); 21 = Hondo Beach, CA-SBA-530. (Data points from Bennyhoff and Hughes, 1987; King, 1990; Howard and Raab, 1993; Vellanoweth, 1995, 2001; Jenkins and Erlandson, 1996; Raab and Howard, 2002; Culleton et al., 2005).

that ultimately reached Buena Vista Lake as late as 1000 cal yr BP (Moratto, 1984, p. 559). Coeval with a (hypothetical) late Penutian spread south, the dispersal of Numic peoples from southeast California into the Great Basin has been argued to have occurred after ca. 1000 cal yr BP as an intensive, low mobility, seed-processing adaptation was adopted by tribes in Owens Valley (Lamb, 1958; Hopkins, 1965; Goss, 1968; Bettinger and Baumhoff, 1982, 1983).

3.2. mtDNA and Uto-Aztecan peoples

A limited number of ancient and modern mtDNA studies in western North America provide some preliminary insights into hypothetical population movements of Uto-Aztecan populations through the Holocene. Linguistic arguments for close genetic relationships between speakers of Uto-Aztecan languages have been challenged by several studies that indicate that the equation of language to genes or culture is not exact (Kemp, 2006). The hypothesized introduction of maize agriculture into the southwest by migrating Uto-Aztecans from northern Mexico ca. 3000 cal yr BP (Hill, 2002) is contradicted by several lines of evidence that show few similarities between modern Uto-Aztecans of the Southwest and Mesoamerica, though they share some linguistic and cultural traditions (Smith et al., 2000; Malhi et al., 2003; Kemp, 2006). In the northern Great Basin, Kaestle and Smith (2001) found genetic discontinuity between prehistoric (primarily 6000–1000 cal yr BP) populations from the Stillwater Marsh and Pyramid Lake sites and extant Northern Paiute people in western Nevada (Fig. 15.4), consistent with the relatively late expansion of Numic peoples into the northern Great Basin hypothesized by Bettinger and Baumhoff (1982, 1983) on the basis of archeological evidence. Archeological populations in California's Central Valley (Fig. 15.4) dating from ca. 3600 cal yr BP (Early Horizon, Windmiller Phase) at the Cecil Site (CA-SJO-112) and ca. 2100–1800 cal yr BP (Middle Horizon) at the Cook (CA-SOL-270) and Applegate (CA-AMA-56) sites are argued to be most similar to extant southern coastal Takic peoples, rather than the modern Yok-Utian groups (Yokuts, Miwok, and Ohlone) that inhabited the region at European contact (Eshleman, 2002). This accords with Nichols's (1981) linguistically-derived hypothesis that Uto-Aztecan peoples (not necessarily *Takic* Uto-Aztecans) inhabited the Central Valley during part of the Middle Holocene, and that the hypothesized Penutian expansion from the northwest Great Basin at 4500 cal yr BP, thought to be manifested in the Windmiller Culture (e.g., Moratto, 1984, p. 553–555), actually occurred much later. Taken together, these studies suggest that the ethnographic distribution of Uto-Aztecan and other peoples in western North America does not reflect past situations, and that several significant population movements have occurred since the Early Holocene.

From the genetic data, what can we infer about the movement of Uto-Aztecan peoples in western North America? Two prehistoric mtDNA populations are known from the ethnographically Takic San Clemente Island: 7 individuals from Eel Point (CA-SCLI-43 Locus C) and 13 from the Nursery Site (CA-SCLI-1215) (Potter, 2004). The burial components at both sites date primarily to the Late Holocene (Eel Point, ca. 3000 cal yr BP; Nursery Site, 1500 cal yr BP), though Potter (2004, p. 51) suggests that some Eel Point burials may date to the Middle Holocene based on the archeological assemblage and ^{14}C dates as early as 4500 cal yr BP. Haplogroup frequencies for the prehistoric San Clemente Island populations and other prehistoric and extant western populations are compiled from the available literature in Table 15.1 (Note: the Middle Horizon group is the aggregate of the Cook and Applegate populations). Pairwise comparisons of genetic similarity for all groups are calculated in Table 15.2 with Fisher's exact test using the population differentiation option of GENEPOP v.3.4 (Raymond and Rousset, 2002) treating the four haplogroups (i.e., A, B, C, and D) as alleles of a single locus. The Fisher's *P* tests the null hypothesis that the two groups are drawn from the same larger population, rejecting it when *P* values are below a critical level (e.g., $P<0.05$ or <0.10). Note that higher *P* values do not indicate greater genetic similarity between groups.

The analysis replicates that of previous studies, which allows us to see the basis for the interpretations of other researchers. Ancient populations in each region cannot be differentiated from each other (e.g., Stillwater Marsh and Pyramid Lake in western Nevada, $P = 0.79$; Cecil Site and Middle Horizon in the Central Valley, $P = 0.88$), as is the case with the San Clemente Island samples (Eel Point and Nursery Site, $P = 0.26$). This suggests some degree of temporal continuity within each region during the first part of the Late Holocene (Kaestle and Smith, 2001; Eshleman, 2002). The prehistoric western Nevada populations are differentiated from most groups except Yok-Utian, Northern Paiute (compared to Stillwater Marsh), and the Nursery Site at the $P < 0.05$ level, which indicated to Kaestle and Smith (2001) that these groups were pre-Numic, and probably Penutian rather than Uto-Aztecan. As Eshleman (2002) found, the prehistoric Central Valley populations are differentiated from all groups except for Takic at the $P < 0.05$ level. Interestingly, they are also not differentiated from the Eel Point population (vs. Cecil Site, $P = 0.058$; vs. Middle Horizon, $P = 0.13$), but are distinct from the roughly contemporaneous western Nevada groups. Comparing Eel Point and Nursery Site groups to other extant and prehistoric populations is less clear-cut. They are each clearly dissimilar to Chumash, Northern Paiute, Fremont, and Yuman (at the $P < 0.10$ level), but neither can be distinguished from Washo and Takic, and the Nursery Site group is also not distinct from Yok-Utian or the western Nevada prehistoric populations. The relative lack of discernment for the San Clemente Island data is probably owed to the small sample sizes involved. That notwithstanding, the results are consistent with San Clemente's occupation by non-Chumash peoples of Uto-Aztecan stock by the beginning of the Late Holocene, probably having settled the island earlier in the Middle Holocene. The dissimilarity to Chumash argues for relatively little genetic communication between these groups in the Middle Holocene, *contra* Potter (2004). The Middle Holocene settlers would have been part of a broader expansion of Uto-Aztecan peoples into coastal Southern California and the Central Valley.

3.3. **Olivella** *grooved rectangle beads and the Uto-Aztecan interaction sphere*

Clear evidence exists for developing cultural interaction extending from the southern Channel Islands to the Los Angeles and Orange County coastal areas and the

Table 15.1. Haplogroup frequencies in extant and prehistoric population pairs in California and the Great Basin.

			A	B	C	D	n	Reference
		Haplogroup:	A	B	C	D	n	Reference
Extant Groups	Language Stock							
	Hokan	Chumash	11	2	3	8	24	Lorenz and Smith (1996); Lorenz et al. (2002)
		Washo	0	15	10	3	28	Lorenz and Smith (1996); Lorenz et al. (2002)
		Yuman	3	59	38	0	100	Malhi et al. (2002)
	Penutian	Yok-Utian	2	5	2	8	17	Lorenz and Smith (1996); Lorenz et al. (2002)
	Uto-Aztecan	Northern Paiute	0	40	9	45	94	Kaestle and Smith (2001)
		Takic	1	6	9	3	19	Lorenz and Smith (1996); Lorenz et al. (2002)
Prehistoric Groups	Region							
	Great Basin	Fremont	0	25	0	0	25	Kaestle and Smith (2001)
		Stillwater Marsh	1	8	0	12	21	Kaestle and Smith (2001)
		Pyramid Lake	2	6	0	10	18	Kaestle and Smith (2001)
	Central Valley	Cecil Site	0	1	9	6	16	Eshleman (2002)
		Middle Horizon	1	4	14	10	29	Eshleman (2002)
	San Clemente Island	Eel Point	1	2	4	0	7	Potter (2004)
		Nursery Site	2	6	2	3	13	Potter (2004)
						Total	411	

Table 15.2. Fisher's exact test *P* (and standard deviation) for extant and prehistoric population pairs. Fisher's exact *P* is the average and standard deviation of five runs of 1000 iterations for each pair using the population differentiation option of Genepop v.3.4 (Raymond and Rousset, 2002), treating the four haplogroups as alleles of a single locus. Significant values (e.g., $P < 0.05$, or $P < 0.10$) reject the null hypothes that the two groups are drawn from the same population.

Extant: Hokan			Penutian	Uto-Aztecan		Prehistoric: Great Basin			Central Valley		San Clemente Island		(n)
Chumash	Washo	Yuman	Yok-Utian	Northern Paiute	Takic	Fremont	Stillwater Marsh	Pyramid Lake	Cecil Site	Middle Horizon	Eel Point	Nursery Site	
x	0	0	0.0760 (0.0020)	0	0.0013 (0.0003)	0	0.0007 (0.0002)	0.0105 (0.0007)	0.0011 (0.0001)	0.0005 (0.0001)	0.0113 (0.00083)	0.0411 (0.0041)	Chumash (24)[a]
	x	0.0319 (0.0012)	0.0050 (0.0008)	0	0.3080 (0.0045)	0.0002 (0.0002)	0.0001 (0.0001)	0.0001 (0.0001)	0.0022 (0.0001)	0.0037 (0.0008)	0.1467 (0.0015)	0.0970 (0.0018)	Washo (28)[a]
		x	0	0	0.0026 (0.0004)	0	0	0	0	0	0.0999 (0.0034)	0.0002 (0.0001)	Yuman (100)[b]
			x	0.0340 (0.0020)	0.0696 (0.0034)	0	0.4084 (0.0042)	0.6546 (0.0022)	0.0229 (0.0022)	0.0506 (0.0023)	0.0374 (0.0016)	0.6441 (0.0025)	Yok-Utian (17)[a]
				x	0	0	0.1154 (0.0058)	0.0288 (0.0018)	0.0001 (0.0001)	0	0	0.0105 (0.0009)	N. Paiute (94)[c]
					x	0	0.0006 (0.0001)	0.0018 (0.0002)	0.1328 (0.0057)	0.3339 (0.0036)	0.7634 (0.0025)	0.2702 (0.0042)	Takic (19)[a]
						x	0	0	0	0	0.0002 (0.00002)	0.0002 (0.0001)	Fremont (25)[c]
							x	0.7906 (0.0017)	0.0001 (0.0001)	0.0004 (0.0002)	0.0001 (0.0001)	0.0729 (0.0045)	Stillwater Marsh (21)[c]
								x	0.0003 (0.0002)	0.0009 (0.0001)	0.0010 (0.0002)	0.1868 (0.0047)	Pyramid Lake (18)[c]
									x	0.8837 (0.0015)	0.0581 (0.0016)	0.0110 (0.0008)	Cecil Site (16)[d]
										x	0.1274 (0.0026)	0.0305 (0.0011)	Middle[d] Horizon (29)
											x	0.2624 (0.0046)	Eel Point (7)[e]
												x	Nursery Site (13)[e]

Haplogroup frequency data are from:
[a] Lorenz and Smith (1996) & Lorenz et al. (2002).
[b] Malhi et al. (2002).
[c] Kaestle and Smith (2001).
[d] Eshleman (2002).
[e] Potter (2004).

Great Basin between about 5500 and 4500 cal yr BP (Vellanoweth, 2001). The best indicator of increased interaction between these spatially disparate areas is the distribution of *Olivella* grooved rectangle (OGR) beads produced on the southern Channel Islands or adjacent mainland coast. King (1990) pointed out that the known spatial distribution of OGR beads generally overlaps with the historic distribution of Uto-Aztecan peoples in western North America, as defined by Kroeber (1925). Howard and Raab (1993) and Raab (1997) were among the first to point out the presence of this rare bead form on the southern Channel Islands. Vellanoweth (1995) reported the presence of OGR beads on San Nicolas Island (CA-SNI-161), along with *Olivella* bead manufacturing debris. Outside of Southern California, Jenkins and Erlandson (1996) documented OGR beads in the northern Great Basin (south-central Oregon). Beads of this kind have also been found at other sites in the western Great Basin (Bennyhoff and Hughes, 1987; Vellanoweth, 1995, 2001; Raab and Howard, 2002). This distribution is wedge-shaped with its terminus on the southern California Coast (south of Malibu) and offshore islands (see Fig. 15.4; Koerper, 1979; Raab, 1997; Raab and Howard, 2002).

More recent work has expanded the known range of OGR beads beyond the ethnographic distribution of Uto-Aztecan languages (Fig. 15.4). A few have been documented in historic Chumash territory (Rincon Point, CA-SBA-119, Bennyhoff and Hughes, 1987; Honda Beach, CA-SBA-530, Lebow et al., 2002; Xonxon'ata, CA-SBA-3404, Hildebrandt, 2004, p. 64). Seven examples are known from the west side of Kern County in the southern San Joaquin Valley, which was inhabited by Yokuts tribes of the Penutian language stock when the Spanish arrived. Culleton et al. (2005) noted that these beads were found on older landforms flanking the former wetlands of Buena Vista Slough, the outlet for Buena Vista and Kern lake that flows north to Tulare Lake and ultimately into the San Joaquin Delta. These OGR bead sites are mainly open-air surface deposits on the Elk Hills, some with clear Late Holocene components, but their antiquity is corroborated by a direct AMS date on one OGR (of three recovered) from CA-KER-5404 of 5300–5000 cal yr BP (2 sigma; Beta-118254), and association with freshwater mussel shell at KER-3166/H dated to 5300–4800 cal yr BP and 5500–4900 (2 sigma; Beta-108267 and Beta-116693). These two sites and other OGR bead sites in the Elk Hills vicinity are also on Pleistocene – to Early Holocene – age landforms (KER-824, Bramlette et al., 1982; and, KER-3079/H Locus C, Culleton et al., 2005) or are partly buried by presumed terminal Middle Holocene sediment (KER-2720, Sutton, 1996; collection viewed by BJC at CSU Bakersfield). This suggests that OGR beads were circulating in larger numbers in the Middle Holocene southern San Joaquin Valley than in the other ethnographic non-Uto-Aztecan areas, such as Chumash territory. Vellanoweth (2001) noted the presence of three OGR beads near the Elk Hills at the time he wrote, and suggested that they represented part of a trade network that ran through the valley from Tejon Pass, and along trails up the west and east sides of the Sierra Nevada into the Uto-Aztecan Great Basin. An alternative explanation is that the Middle Holocene inhabitants of the Buena Vista Basin were speakers of Uto-Aztecan languages, and the anomaly of the OGRs in ethnographic Penutian lands reflects an earlier distribution of Uto-Aztecan peoples (cf. Nichols, 1981; Moratto, 1984).

4. Discussion

Marine climate data from the Santa Barbara Basin indicates that SSTs oscillated during the Middle Holocene between warm and cold states. In general, SSTs were relatively warm during the Middle Holocene, supporting interpretations of Friddell et al. (2003), except for one distinct cold interval between about 6300 and 5800 cal yr BP. More moderate SSTs are evident in this record from 5800 to 5000 cal yr BP. Inferred high marine productivity between 6300 and 5800 cal yr BP corresponds with the coldest SSTs during the Middle Holocene. Climatically influenced changes in terrestrial environments along the coast during the Middle Holocene appear to have been less drastic than in the interior, particularly in central and northern California (Jones and Waugh, 1997). Dry climatic conditions persisted throughout much of the Middle Holocene in eastern California and the Great Basin (Benson et al., 2002) with the driest interval occurring between 6300 and 5000 cal yr BP (LaMarche, 1973, 1974; Lindström, 1990).

The distribution of OGR beads from the southern Channel Islands across southern California and into the western and northern Great Basin suggests increased interaction among these peoples between about 5900 and 4700 cal yr BP. This exchange may have reduced the risk of resource shortfalls associated with dry environmental conditions during the Middle Holocene (see Larson et al., 1994; Kennett and Kennett, 2000 for Late Holocene examples of this phenomenon). The distribution of these beads seems to reflect the establishment of a new trade conduit, perhaps related to the inferred migrations of people from southern California desert environs to the southern Channel Islands (Grenda and Altschul, 1995; Potter, 2004; also see Warren, 1968 and Mikkelsen et al., 2000 for more generalized ideas of Middle Holocene movement from the interior to the coast). Persistently dry conditions in western North America between 6300 and 5000 cal yr BP would have expanded desert environments of southern California, displacing some groups to more humid and productive coastal and interior wetland regions.

We suggest that some Uto-Aztecan groups were displaced as conditions in the southern California desert became dryer and less productive. Some groups may have moved into the southwestern Great Basin where conditions were dry but less severe than those in the southern California desert areas. The western Great Basin (specifically the Carson and Humboldt sinks in western Nevada) appear on genetic grounds to have been occupied by Penutian peoples (Kaestle and Smith, 2001), whose presence may have forced migrating Uto-Aztecans to the southern coast and into the Central Valley from the south. Dry conditions were ameliorated on the coast by maritime influences and the environment provided a range of additional resources not available in the desert interior. The wetlands and lakes of the San Joaquin Valley were clearly affected by Middle Holocene aridity, yet still would have provided more favorable conditions than the interior desert of the southeast. Archeological populations from the Sacramento/San Joaquin Delta and Sierra Nevada foothills that bear closest genetic affinity with modern Takic groups (Eshleman, 2002), and the distribution of OGR beads in the southern San Joaquin Valley, offer evidence of pre-Penutian Uto-Aztecan populations during the Middle Holocene. Such migrations by early Uto-Aztecan groups may have promoted the flow of trade goods, such as OGR beads, over vast areas of western North America.

This interpretation implies a modification of the Uto-Aztecan interaction sphere as developed by Howard and Raab (1993), Jenkins and Erlandson (1996), Vellanoweth (1995, 2001) and Raab and Howard (2002). Marshaling the genetic, linguistic, and archeological data from the Central Valley, the presence of OGR beads outside of the historic distribution of Uto-Aztecan languages is readily explained by an early Uto-Aztecan presence in that region (Nichols, 1981; Moratto, 1984, p. 559). If a Penutian expansion into central California did not occur ca. 4500–4000 cal yr BP, as suggested by most linguistic reconstructions, this places the Penutians in western Nevada at the time OGR beads were circulating there (Fig. 15.4). This accords well with the mtDNA from Pyramid Lake and Stillwater Marsh, which shows ancient populations most similar to California Penutian groups, and unlike the later Numic peoples who occupied the Great Basin historically (Kaestle and Smith, 2001). So, the conspicuous cluster of OGR beads in Middle Holocene western Nevada may not comprise part of an ethnolinguistically-defined cultural interaction sphere among Uto-Aztecans, but perhaps reflect trade interaction between groups at the northwestern frontier of Uto-Aztecan territory. Raab and Howard (2002, p. 595) describe just this possibility:

> "It is also unrealistic on logical grounds to expect a linguistic boundary to be "impermeable" to the movement of various kinds of materials, including beads. For these reasons, we would expect to find OGR beads on both sides of any linguistic frontier. The model presented here does not predict an absence of OGR beads outside of the Uto-Aztecan area; rather it predicts significantly higher frequencies within this area".

Thus, the movements of people from the interior to the coast left a legacy of intergroup networks that allowed the communication of coastal trade items deep into Uto-Aztecan lands and beyond, perhaps into early Penutian territory in the Great Basin. Both increased exchange and migration may have been behavioral responses to unstable, dry conditions in western North America. Increased interaction and migrations of Uto-Aztecan speaking people to California's Central Valley, southern coast, and offshore islands had a profound effect on the evolutionary trajectory of coastal peoples in these regions and are fundamental in explaining the differences observed at historic contact between peoples of Southern California.

5. Conclusions

Articulating multiple lines of archeological, paleoclimatic, linguistic, and genetic evidence from western North America demonstrates the linkages between environmental change and human adaptive response during the arid Middle Holocene.

We argue that declining terrestrial resource abundance in the desert interior between 6300–4800 cal yr BP stimulated the movement of Uto-Aztecan populations toward more productive aquatic habitats on the southern California Coast and the Central Valley. The options for migration and settlement were likely limited by the presence of early Penutian groups aggregated around the marshes and lakes of western Nevada (e.g., Pyramid Lake and Stillwater Marsh), and the predecessors of the Chumash on the Santa Barbara mainland and northern Channel Islands, groups whose local aquatic environments were sufficiently productive to buffer the effects of terrestrial resource decline. In addition, and concomitant with population dispersal, extensive trade networks among Uto-Aztecan and other groups are signaled by the distribution of *Olivella* grooved rectangle beads dated to 5500–4500 cal yr BP from the southern coast, the southern San Joaquin Valley, and the western Great Basin. Inter-regional exchange of goods helped mitigate localized resource shortfalls, and may have developed within systems of kin relations and intermarriage between Uto-Aztecan groups, as were other exchange systems between California and the Great Basin (see Jackson and Ericson, 1994). Taken as a whole, the Uto-Aztecan response to Middle Holocene climate change comprised several inter-related elements: dispersal toward more stable and productive habitats, risk-minimization through resource exchange, both of which relied upon maintaining cooperative social networks that acted over great distances in western North America.

Acknowledgments

Our archeological research was supported by the National Science Foundation (SBR-9521974, Kennett; SBR-9731434, Erlandson) and the National Park Service (Grant#1443CA8120-96-003, Kennett). Channel Islands National Park provided transportation and logistical support necessary to conduct field research on the northern Channel Islands. The research by J. Kennett was supported by the National Science Foundation (Marine Geology and Geophysics) and the Western Regional Center, National Institute for Global Environmental Change, Department of Energy. Culleton was supported during the preparation of this manuscript by a National Science Foundation Graduate Research Fellowship. Brian M. Kemp provided guidance on genetic analyses and statistical treatment of genetic data. We thank S. McClure for her help in compiling the reference list and K. Thompson and H. Berg for their technical assistance. We thank Bill Hildebrandt and an anonymous reviewer for useful comments on a previous version of this paper.

References

Adam, D. P. 1967. Late-Pleistocene and recent palynology in the Central Sierra Nevada, California. In *Quaternary Paleoecology*, edited by E. J. Cushing and H. E. Wright, pp. 275–301. Yale University Press, New Haven.

Antevs, E., 1948. Climatic changes and pre-White Man in the Great Basin, with emphasis on glacial times and post-glacial times. *University of Utah Bulletin* 38:168–191.

Antevs, E., 1952. Climatic history and the antiquity of man in California. *University of California Archaeological Survey Reports* 16:23–31.

Antevs, E., 1955. Geologic-climatic dating in the West. *American Antiquity* 20:317–335.

Axelrod, D. I., 1981. Holocene climatic changes in relation to vegetation disjunction and speciation. *American Naturalist* 117:847–870.

Behl, R. J., and J. P. Kennett, 1996. Brief interstadial events in the Santa Barbara Basin, NE Pacific, during the past 60 kyr. *Nature* 379:243–246.

Bellwood, P., 2001. Early agriculturalist population diasporas? Farming, languages, and genes. *Annual Review of Anthropology* 30:181–207.

Bemis, B. E., H. J. Spero, J. Bijima, and D. W. Lea, 1998. Reevaluation of the oxygen isotopic composition of planktonic foraminifera: Experimental results and revised paleotemperature equations. *Paleoceanography* 13(2):150–160.

Bennyhoff, J. A., and R. E. Hughes, 1987. Shell bead and ornament exchange networks between California and the Western Great Basin. *Anthropological Papers of the American Museum of Natural History* 64(2). New York.

Benson, L., M. Kashgarian, R. Rye, S. Lund, F. Paillet, J. Smoot, C. Kester, S. Mensing, D. Meko, and S. Lindström, 2002. Holocene multidecadal and multicentennial droughts affecting Northern California and Nevada. *Quaternary Science Reviews* 21:659–682.

Bernal, P. A., and J. A. McGowan, 1981. Advection and upwelling in the California Current. In *Coastal Upwelling*, edited by F. A. Richards, pp. 381–399. American Geophysical Union, Washington, D.C.

Bettinger, R. L., and M. A. Baumhoff, 1982. The Numic spread: Great Basin cultures in competition. *American Antiquity* 47:485–503.

Bettinger, R. L., and M. A. Baumhoff, 1983. Return rates and intensity of resource use in Numic and prenumic adaptive strategies. *American Antiquity* 48:830–884.

Bianchi, N., G. Bailliet, C. Bravi, R. Carnese, F. Rothhammer, and V. Martinez-Marignac, 1997. Origin of Ameridian Y-chromosome as inferred by the analysis of six polymorphic markers. *American Journal of Physical Anthropology* 102:79–89.

Birman, J. H., 1964. Glacial geology across the crest of the Sierra Nevada, California. *Geological Society of America Special Paper* 75.

Bramlette, A. G., J. F. Hayes, and S. Stewart, 1982. Archaeological investigations at CA-KER-824 near McKittrick, California. On file, Southern San Joaquin Valley Information Center of the California Historical Resources Information System, California State University, Bakersfield.

Bright, R. C., 1966. Pollen and seed stratigraphy of Swan Lake, Southeastern Idaho: Its relation to regional vegetational history and to Lake Bonneville history. *Tebiwa* 9:1–47.

Bright, W., and M. Bright, 1976. Archaeology and linguistic in prehistoric Southern California. In *Variation and Change in Language: Essays by William Bright*, edited by A. S. Dil, pp. 189–205. Stanford University Press, Stanford.

Byrne, R., C. Busby, and R. F. Heizer, 1979. The Altithermal revisited: Pollen evidence from the Leonard Rockshelter. *Journal of California and Great Basin Anthropology* 1:280–294.

Cannariato, K. G., J. P. Kennett, and R. J. Behl, 1999. Biotic response to late Quaternary rapid climate switches in Santa Barbara Basin: Ecological and evolutionary implications. *Geology* 27:63–66.

Cannariato, K. G., J. P. Kennett, D. J. Kennett, J. S. Revenaugh, and B. A. Malmgren, 2003. *Millennial and Centennial Variability of the California Current System during the Holocene*

Recorded in Santa Barbara Basin. EOS Trans. AGU, 84(46), Fall Meeting Supplement, Abstract OS22C-03.

Casteel, R. W., D. P Adam, and J. D. Sims, 1977. Late-Pleistocene and Holocene remains of *Hysterocarpus* (Tule Perch) from Clear Lake, California, and inferred Holocene temperature fluctuations. *Quaternary Research* 7:33–143.

Cavalli-Sforza, L. L., 1996. The spread of agriculture and nomadic pastoralism: insights from genetics, linguisitics and archaeology. In *The Origins and Spread of Agriculture and Pastoralism in Eurasia*, edited by D. R. Harris, pp. 51–69. Smithsonian Institution Press, Washington, D.C.

Cole, K. L., and G. W. Liu, 1994. Holocene paleoecology of an estuary on Santa Rosa Island, California. *Quaternary Research* 41:326–335.

Culleton, B. J., T. L. Jackson, and S. C. Brewer, 2005. Cultural Response to Environmental Change in the Buena Vista Basin: Archaeological Data Recovery at Eight Sites on the Former Naval Petroleum Reserve No. 1 (Elk Hills), Kern County, California. Prepared by Pacific Legacy, Inc., Central Coast Division, Santa Cruz, California. Report prepared for ICF Consulting, Fairfax, Virginia. On file, Pacific Legacy, Inc., Central Coast Division, Santa Cruz, California.

Curry, R. R., 1969. Holocene climate and glacial history the Central Sierra Nevada, California. *Geological Society of America Special Paper* 123:1–47.

Deka, R., B. Su, L. Jin, R. Chakraborty, L. Martinson, P. Underhill, P. Oefner, S. T. McGarvey, N. Saha, M.D. Shriver, and J. Chu, 2000. Origins of the Polynesian people: New insights from the Y chromosome. *American Journal of Physical Anthropology* S30:136.

Diamond, J., and P. Bellwood, 2003. Farmers and their languages: The first expansions. *Science* 300:597–603.

Erlandson, J. M., T. C. Rick, and C. Peterson, 2005. A geoarchaeological chronology of Holocene dune building on San Miguel Island, California. *The Holocene* 15:1227–1235.

Eshleman, J. A., 2002. *Mitochondrial DNA and Prehistoric Population Movements in Western North America*. Unpublished Ph.D. dissertation, Department of Anthropology, University of California, Davis.

Friddell, J. E., R. C. Thunell, T. P. Guilderson, and M. Kashgarian, 2003. Increased Northeast Pacific climatic variability during the warm Middle Holocene. *Geophysical Research Letters* 30:141–144.

van Geen, A., S. N. Luoma, C. C. Fuller, R. Anima, H. E. Clifton, and S. Trumbore, 1992. Evidence from Cd/Ca ratios in foraminifera for greater upwelling off California 4,000 years ago. *Nature* 358:54–56.

Golla, V., 2000a. Language families of North America. In *American Past, American Present: Genes and Languages in the Americas and Beyond*, edited by C. Renfrew, pp. 59–72. McDonald Institute for Archaeological Research, Cambridge.

Golla, V., 2000b. Language history and communication strategies in aboriginal California and Oregon. In *Languages of the North Pacific Rim*, edited by O. Miyaoka, pp. 43–64. Faculty of Informatics, Osaka Gakuin University, Suita, Japan.

Goss, J. A., 1968. Culture historical inference from Utaztecan linguistic evidence. *Idaho State Museum Occasional Papers* 22:1–42.

Graham, N. E., M. K. Hughes, C. M. Ammann, K. M. Cobb, M. P. Hoerling, D. J. Kennett, J. P. Kennett, B. Rein, L. Stott, P. E. Wigand, and T. Xu, 2007. Tropical Pacific-mid-latitude teleconnections in medieval times, *Climate Change*, 83(1–2):241–285.

Grayson, D. K., 2000. Mammalian responses to Middle Holocene climatic change in the Great Basin of the Western United States. *Journal of Biogeography* 27:181–192.

Graumlich, L. J., 1993. A 1000-year record of temperature and precipitation in the Sierra Nevada. *Quaternary Research* 39:249–255.

Grenda, D. R., and J. H. Altschul, 1995. Islanders and mainlanders: A regional approach to Channel Island prehistory. *Pacific Coast Archaeological Society Quarterly* 31(3):28–38.

Haak, W., P. Forster, B. Bramanti, S. Matsumura, G. Brandt, M. Tänzer, R. Villems, C. Renfrew, D. Gronenborn, K. Werner Alt, and J. Burger, 2005. First European farmers in 7500-year-old Neolithic sites. *Science* 310:1016–1018.

Hansen, H. P., 1947. Postglacial forest succession, climate and chronology in the Pacific Northwest. *Transactions of the American Philosophical Society* 37:1–130.

Harding, S. T., 1965. Recent variations in the water supply of the Great Basin. *Archives Series Report* 16, University of California, Berkeley.

Hendy, I. L., and J. P. Kennett, 1999. Latest Quaternary North Pacific surface-water responses imply atmosphere-driven climate instability. *Geology* 27(4):291–294.

Hendy, I. L., and J. P. Kennett, 2000. Dansgaard-Oeschger cycles and the California current system: Planktonic foraminiferal response to rapid climate change in Santa Barbara Basin, Ocean Drilling Program hole 893A. *Paleoceanography* 15:30–42.

Heusser, L. E., and F. Sirocko, 1997. Millennial pulsing of environmental change in Southern California from the past 24 k.y.: A record of Indo-Pacific ENSO events? *Geology* 25:243–246.

Heusser, C. J., L. E. Heusser, and D. M. Peteet, 1985. Late Quaternary climatic change on the American North Pacific Coast. *Nature* 315:485–487.

Hildebrandt, W. R., 2004. *Xonxon'ata, in the Tall Oaks: Archaeological and Ethnohistory of a Chumash Village in the Santa Ynez Valley*. Contributions in Anthropology, Number 2, Santa Barbara Museum of Natural History.

Hill, J. H., 2002. Proto-Uto-Aztecan cultivation and the northern devolution. In *Examining the Farming/Language Dispersal Hypothesis*, edited by P. Bellwood and C. Renfrew, pp. 331–340. McDonald Institute for Archaeological Research, University of Cambridge.

Hopkins, N. A., 1965. Great Basin prehistory and Uto-Aztecan. *American Antiquity* 31(1):48–60.

Howard, W. J., and L. M. Raab, 1993. Olivella grooved rectangle beads as evidence of an early period Southern Channel Islands interaction sphere. *Pacific Coast Archaeological Society Quarterly* 29(3):1–11.

Hubbs, C. L., G. S. Bien, and H. E. Suess, 1962. La Jolla natural radiocarbon measurements II. *Radiocarbon* 4:204–238.

Hughes, M. K., and L. J. Graumlich, 1996. Climatic variations and forcing mechanisms of the last 2000 years. In *Multi-Millenial Dendroclimatic Studies from the Western United States. NATO ASI Series,* Vol. 141, pp. 109–124.

Hughes, M. K, and L. J. Graumlich, 2000. *Multi-Millennial Nevada Precipitation Reconstruction*. International Tree-Ring Data Bank. IGBP PAGES/World Data Center-A for Paleoclimatology Data Contribution Series #2000-049. NOAA/NGDC Paleoclimatology Program, Boulder, CO, USA.

Ingram, B. L., and J. P. Kennett, 1995. Radiocarbon chronology and planktonic-benthic foraminiferal 14C age differences in Santa Barbara Basin sediments, Hole 893A. In *Proceedings of the Ocean Drilling Program, Scientific Results*, edited by J. P. Kennett, J. G. Baldauf, and M. Lyle, Vol. 146, pp. 19–27. Ocean Drilling Program, College Station, TX.

Ingram, B. L., and J. R. Southon, 1996. Reservoir ages in Eastern Pacific coastal and estuarine waters. *Radiocarbon* 38:573–582.

Jackson, T. L., and J. E. Ericson, 1994. Prehistoric exchange systems in California. In *Prehistoric Exchange Systems in North America*, edited by T. G. Baugh and J. E. Ericson, pp. 385–415. Plenum Press, New York.

Jenkins, D. L., and J. M. Erlandson, 1996. *Olivella* grooved rectangle beads from a Middle Holocene site in the Fort Rock Valley, Northern Great Basin. *Journal of California and Great Basin Anthropology* 18:296–302.

Jennings, J. D., 1986. Prehistory: Introduction. In *Great Basin*, edited by W. L. D'Azevedo, pp. 113–119. *Handbook of North American Indians*, Vol. 11, W. C. Sturtevant, General Editor, Smithsonian Institution, Washington, D.C.

Jin, L., M. Seielstad, and C. Xiao 2002. *Genetic, Linguistic and Archaeological Perspectives on Human Diversity in Southeast Asia.* World Scientific Publishing Company, River Edge, NJ.

Jobling, M. A., M. Hurles, and C. Tyler-Smith, 2004. *Human Evolutionary Genetics: Origins, Peoples and Disease.* Garland Press, New York.

Johnson, J. R. and J. G. Lorenz, 2006. Genetics, linguistics, and prehistoric migrations: An analysis of California Indian mitochondrial DNA lineages. *Journal of California and Great Basin Anthropology* 26:31–62.

Jones, T. L., and D. J. Kennett, 1999. Late Holocene sea temperatures along the Central California Coast. *Quaternary Research* 51:74–82.

Jones, T. L., and G. Waugh, 1997. Climatic consequences or population pragmatism?: A Middle Holocene prehistory of the Central California Coast. In *Archaeology of the California Coast during the Middle Holocene*, edited by J. M. Erlandson and M. A. Glassow, pp. 111–128. Perspectives in California Archaeology 4. Institute of Archaeology, University of California, Los Angeles.

Jones, T. L., G. M. Brown, L. M. Raab, J. L. McVickar, W. G. Spaulding, D. J. Kennett, A. York, and P. L. Walker, 1999. Environmental imperatives reconsidered: Demographic crises in Western North America during the medieval climatic anomaly. *Current Anthropology* 40:137–169.

Kaestle, F. A., 1995. Mitochondrial DNA evidence for the identity of the descendants of the prehistoric stillwater marsh populations. In *Bioarchaeology of the Stillwater Marsh: Prehistoric Human Adaptation in the Western Great Basin*, edited by C. S. Larsen and R. L. Kelly, pp. 73–80. American Museum of Natural History, Washington, D.C.

Kaestle, F. A., 1997. Molecular analysis of ancient native American DNA from Western Nevada. *Nevada Historical Society Quarterly* 40:85–96.

Kaestle, F. A., 1998. *Molecular Evidence for Prehistoric Native American Population Movement: The Numic Expansion.* Unpublished dissertation, University of California, Davis, CA.

Kaestle, F. A., and D. G. Smith, 2001. Ancient mitochondrial DNA evidence for prehistoric population movement: The numic expansion. *American Journal of Physical Anthropology* 115:1–12.

Karafet, T. M., S. L. Zegura, J. Vuturo-Brady, O. Posukh, L. Osipova, and V. Wiebe, 1997. Y Chromosome markers and trans-Bering Strait Dispersals. *American Journal of Physical Anthropology* 102: 301–314.

Karafet, T. M., S. L. Zegura, O. Posukh, L. Osipova, A. Bergen, and J. Long, 1999. Ancestral Asian source(s) of New World Y-chromosomes from Melanesia and Australia. *American Journal of Human Genetics* 68:173–190.

Kemp, B. M., 2006. *Mesoamerica and Southwest Prehistory, and the Entrance of Humans into the Americas: Mitochondrial DNA Evidence.* Unpublished Ph.D. dissertation, Department of Anthropology, University of California, Davis.

Kemp, B. M., R. S. Malhi, J. McDonough, D. A. Bolnick, J. A. Eshleman, O. Rickards, C. Martinez-Labarga, J. R. Johnson, J. G. Lorenz, E. J. Dixon, T. E. Fifield, T. H. Heaton, R. Worl, and D. G. Smith, 2007. Genetic analysis of early Holocene skeletal remains from Alaska and its implications for the settlement of the Americas. *American Journal of Physical Anthropology* 132:605–621.

Kennett, D. J., B. L. Ingram, J. M. Erlandson, and P. Walker, 1997. Evidence for temporal fluctuations in marine radiocarbon reservoir ages in the Santa Barbara Channel, Southern California. *Journal of Archaeological Science* 24:1051–1059.

Kennett, D. J., and J. P. Kennett, 2000. Competitive and cooperative responses to climatic instability in Coastal Southern California. *American Antiquity* 65:379–395.

Kennett, J. P., and B. L. Ingram, 1995a. Paleoclimatic evolution of Santa Barbara Basin during the last 20 k.y.: Marine evidence from Hole 893A. In *Proceedings of the Ocean Drilling Program, Scientific Results*, edited by J. P. Kennett, J. G. Baldauf, and M. Lyle, Vol. 146, pp. 309–325. Ocean Drilling Program, College Station, TX.

Kennett, J. P., and B. L. Ingram, 1995b. A 20,000 year record of ocean circulation and climate change from the Santa Barbara Basin. *Nature* 377:510–514.

King, C. D., 1990. Evolution of Chumash Society: A comparative study of artifacts used for social system maintenance in the Santa Barbara Channel Region before A.D. 1804. Garland Publishing, Inc., New York.

Koerper, H. C., 1979. On the question of the chronological placement of Shoshonean presence in Orange County, California. *Pacific Coast Archaeological Society Quarterly* 15(3):69–84.

Koerper, H. C., A. B. Schroth and R. D. Mason, 1994. Morphology and temporal projectile point types: Evidence from Orange County, California. *Journal of California and Great Basin Anthropology* 16(1):81–105.

Koutavas, A., P. B. deMenocal, G. C. Olive, and J. Lynch-Steiglitz, 2006. Mid-Holocene El Niño–Southern Oscillation (ENSO) attenuation revealed by individual foraminifera in Eastern Tropical Pacific sediments. *Geology* 34(12):993–996.

Krings, M., A. Stone, R. W. Schmitz, H. Krainitzki, M. Stoneking, and S. Pääbo, 1997. Neanderthal DNA sequences and the origin of modern humans. *Cell* 90:19–30.

Kroeber, A. L., 1925. Handbook of the Indians of California. *Bureau of American Ethnology Bulletin* 78. Washington, D.C.

LaMarche, V. C., 1973. Holocene climatic variations inferred from treeline fluctuations in the White Mountains, California. *Quaternary Research* 3:632–660.

LaMarche, V. C., 1974. Paleoclimatic inferences from long tree-ring records. *Science* 183:1043–1048.

Lamb, S. M., 1958. Linguistic prehistory of the Great Basin. *International Journal of American Linguistics* 24:95–100.

Larson, D., J. R. Johnson, and J. C. Michaelson, 1994. Missionizing among the Coastal Chumash of Central California: A study of risk minimization strategies. *American Anthropologist* 96:263–299.

Larson, D., and J. C. Michaelson, 1989. Climatic variability: A compounding factor causing culture change and prehistoric populations of Central California. Manuscript on file, Department of Anthropology, California State University, Long Beach.

Lebow, C. G., D. R. Harro, R. L. McKim, and C. Denardo, 2002. *Final Archaeological Excavations at the Hondo Beach Site (CA-SBA-530) Vandenberg Air Force Base, Santa Barbara County, California*. Final Report submitted to the Department of Air Force, Brooks Airforce Base, Texas.

Lindström, S., 1990. Submerged tree stumps as indicators of mid-Holocene aridity in the Lake Tahoe Basin. *Journal of California and Great Basin Anthropology* 12:146–157.

Madsen, D. B., and D. Rhode, 1990. Early Holocene pinyon (*Pinus monophylla*) in the Northeastern Great Basin. *Quaternary Research* 33:94–101.

Malhi, R. S., H. M. Mortensen, J. A. Eshleman, B. M. Kemp, J. G. Lorenz, F. A. Kaestle, J. R. Johnson, C. Gorodezky, and D. G. Smith, 2003. Native American mtDNA prehistory in the American Southwest. *American Journal of Physical Anthropology* 120:108–124.

Masters, P. M., and D. R. Gallegos, 1997. Environmental change and coastal adaptations in San Diego County during the Middle Holocene. In *Archaeology of the California Coast during the Middle Holocene*, edited by J. M. Erlandson and M. A. Glassow, pp. 11–22. Perspectives in California Archaeology 4. Institute of Archaeology, University of California, Los Angeles.

Mehringer, P. J., Jr., 1985. Late Quaternary pollen records from the interior Pacific Northwest and Northern Great Basin of the United States. In *Pollen Records of Late-Quaternary North America Sediments*, edited by V. M. Bryant, Jr., and R. G. Holloway, pp. 167–189. American Association of Stratigraphic Palynologists, Dallas, TX.

Mellars, P., 2006. Going east: New genetic and archaeological perspectives on the modern human colonization of Eurasia. *Science* 313:796–800.

Merriwether, A., and R. E. Ferrell, 1996. The four founding lineage hypothesis for the New World: A critical reevaluation. *Molecular Phylogenetics and Evolution* 5:241–246.

Mikkelsen, P., W. Hildebrandt, and D. Jones, 2000. Prehistoric adaptations on the shores of Morro Bay Estuary, excavations at Site CA-SLO-165, Morro Bay, California. *Occasional Paper* No. 14. San Luis Obispo County Archaeological Society.

Moratto, M. J., 1984. *California Archaeology*. Academic Press, New York.

Moratto, M. J., T. F. King, and W. B. Woolfenden, 1978. Archaeology and California's climate. *Journal of California Anthropology* 5(2):147–161.

Morgan, A., L. S. Cummings, and J. L. Rudolph, 1991. Paleoenvironmental change. In *Resource Use and Settlement in the Santa Ynez River Valley*, edited by C. F. Woodman, J. L. Rudolph, and T. P. Rudolph, pp. 65–101. Science Applications International Corporation, Santa Barbara, CA.

Negrini, R. M., P. E. Wigand, S. Draucker, K. Gobalet, J. K. Gardner, M. Q. Sutton, and R. M. Yohe II, 2006. The Rambla highstand shoreline and the Holocene lake-level history of Tulare Lake, California, USA. *Quaternary Science Reviews* 25:1599–1618.

Nichols, J., 1992. *Linguistic Diversity in Space and Time*. University of Chicago Press, Chicago.

Nichols, M. J. P., 1981. Old California Uto-Aztecan. *Survey of California and Other Indian Languages, Reports* 1:5–41.

Ovchinnikov, Igor V., A. Götherström, G. P. Romanoval, V. M. Kharitonov, K. Lidén, and W. Goodwin, 2000. Molecular analysis of Neanderthal DNA from the northern Caucasus. *Nature* 404:490–493.

Overpeck, J. T., and R. S. Webb, 2000. Non-glacial rapid climate events: Past and future. *Proceedings of the National Academy of Sciences* 97:1335–1338.

Pak, D. K., and J. P. Kennett, 2002. A foraminiferal isotopic proxy for upper water mass stratification. *Journal of Foraminiferal Research* 32:319–327.

Pisias, N., 1978. Paleoceanography of the Santa Barbara Basin during the last 8,000 years. *Quaternary Research* 10:366–384.

Pisias, N., 1979. Model for paleoceanographic reconstructions of the California Current during the last 8,000 years. *Quaternary Research* 11:373–386.

Potter, A., 2004. *The Genetic Affinities of the Prehistoric People of San Clemente Island, California: An Analysis of Ancient DNA*. Unpublished Ph.D. dissertation, Department of Anthropology, University of Oregon.

Quade, J., R. M. Forester, W. L. Pratt, and C. Carter, 1998. Black mats, spring-fed streams, and late-glacial-age recharge in the Southern Great Basin. *Quaternary Research* 49(2):129–148.

Raab, L. M., 1992. An optimal foraging analysis of prehistoric shellfish collecting on San Clemente Island, California. *Journal of Ethnobiology* 12:63–80.

Raab, L. M., 1997. The Southern Channel Islands during the Middle Holocene: trends in maritime cultural evolution. In *Archaeology of the California Coast during the Middle Holocene*, edited by J. M. Erlandson and M. A. Glassow, pp. 23–34. Perspectives in California Archaeology 4. Institute of Archaeology, University of California, Los Angeles.

Raab, L. M., and W. J. Howard, 2002. Modeling cultural connections between the Southern Channel Islands and Western United States: The Middle Holocene distribution of Olivella grooved rectangle beads. In *The Fifth California Islands Symposium*, edited by D. R. Brown, K. C. Mitchell and H. W. Chaney, pp. 590–597. U.S. Department of the Interior Minerals Management Service, Pacific OCS Region.

Raab, L. M., and D. O. Larson, 1997. Medieval climatic anomaly and punctuated cultural evolution in coastal Southern California. *American Antiquity* 62:319–336.

Ragir, S. R., 1972. The Early Horizon in Central California Prehistory. *Contributions of the University of California Archaeological Research Facility* 15.

Raymond, M., and F. Rousset GENEPOP on the Web. Population Differentiation Option, http://wbiomed.curtin.edu.au/genepop/.

Richards, M., V. Macaulay, E. Hickey, E. Vega, B. Sykes, V. Guida, C. Rengo, D. Sellitto, F. Cruciani, T. Kivisild, R. Villems, M. Thomas, S. Rychkov, O. Rychkov, Y. Rychkov, M. Golge, D. Dimitrov, E. Hill, D. Bradley, V. Romano, F. Cali, G. Vona, A. Demaine, S. Papiha, C. Triantaphyllidis, G. Stefanescu, J. Hatina, M. Belledi, A. Di Rienzo, A. Novelletto, A. Oppenheim, S. Norby, N. Al-Zaheri, S. Santachiara-Benerecetti, R. Scozzari, A. Torroni, and H. J. Bandelt, 2000. Tracing European founder lineages in the near Eastern mtDNA pool. *American Journal of Human Genetics* 67(5):1251–1276.

Roark E. B, B. L. Ingram, J. Southon, and J. P. Kennett, 2003. Holocene foraminiferal radiocarbon record of paleocirculation in the Santa Barbara Basin. *Geology* 31(4):379–382.

Sandweiss, D. H., J. B. Richardson, III, E. J. Reitz, H. B. Rollins, and K. A. Maasch, 1996. Geoarchaeological evidence from Peru for a 5,000 BP onset of El Niño. *Science* 273: 1531–1533.

Sandweiss, D. H., J. B. Richardson, III, E. J. Reitz, H. B. Rollins, and K. A. Maasch, 1997. Determining the beginning of Niño: Response. *Science* 276:966–967.

Sandweiss, D. H., K. A. Maasch, R. L. Burger, J. B. Richardson, III, H. B. Rollins, A. Clement, 2001. Variation in Holocene El Niño frequencies: Climate records and cultural consequences in ancient Peru. *Geology* 29(7):603–606.

Sercelj, A., and D. P. Adam, 1975. A late Holocene pollen diagram from Near Lake Tahoe, El Dorado County, California. *U. S. Geological Survey Journal of Research* 3:737–745.

Smith, D. G., J. Lorenz, B. K. Rolfs, R. L. Bettinger, B. Green, J. Eshleman, B. Schultz, and R. Malhi, 2000. Implications of the distribution of Albumin Naskapi and Albumin Mexico for New World prehistory. *American Journal of Physical Anthropology* 111:557–572.

Stine, S., 1990. Late Holocene fluctuations of Mono Lake, Eastern California. *Palaeogeography, Palaeoclimatology, Palaeoecology* 78:333–381.

Stine, S, 1994. Extreme and persistent drought in California and Patagonia during medieval time. *Nature* 369:546–549.

Stringer, C. B., 2002. Modern humans origins: Progress and prospects. *Philosophical Transactions of the Royal Society of London B* 357:563–579.

Sutton, M. Q., 1996. A charmstone cache from the Southern San Joaquin Valley. *Pacific Coast Archaeological Society Quarterly* 32(4):41–54.

Sykes, B., 1999. The molecular genetics of European ancestry. *Philosophical Transactions of the Royal Society of London* 354:131–139.

Thompson, R. S., 1992. Late Quaternary environments in Ruby Valley, Nevada. *Quaternary Research* 37:1–15.

Trigger, B. G., 1989. *A History of Archaeological Thought*. Cambridge University Press, New York.

Tudhope, A. W., C. P. Chilcott, M. T. McCulloch, E. R. Cook, J. Chappell, R. M. Ellam, D. W. Lea, J. M. Lough, and G. B. Shimmield, 2001. Variability in the El Niño-Southern oscillation through a glacial-interglacial cycle. *Science* 291:1511–1517.

Vellanoweth, R. L., 1995. New evidence from San Nicolas Island concerning the distribution and manufacture of *Olivella* grooved rectangle beads. *Pacific Coast Archaeological Society Quarterly* 32(1):13–22.

Vellanoweth, R. L., 2001. AMS radiocarbon dating and shell bead chronologies: Middle Holocene trade and interaction in Western North America. *Journal of Archaeological Science* 28: 941–950.

Warren, C. N., 1968. Cultural tradition and ecological adaptation on the Southern California Coast. In *Archaic Prehistory in the Western United States*, edited by C. Irwin-Williams, pp. 1–14. *Eastern New Mexico University Contributions in Anthropology* 1(3).

Wood, S. H., 1975. *Holocene Stratigraphy and Chronology of Mountain Meadows, Sierra Nevada, California*. Unpublished Ph.D. dissertation, Department of Geology, California Institute of Technology.

Subject Index